# Energy and Water: Sustainable Development

PROCEEDINGS OF THEME D

## Water for A Changing Global Community
The 27th Congress of the
International Association for
Hydraulic Research

XXVII IAHR Congress

Hosted by the
Water Resources Engineering Division of the
American Society of Civil Engineers

Series Editors: Forrest M. Holly Jr. and Adnan Alsaffar
Theme Editor: John S. Gulliver
Theme Co-Editor: Pierre-Louis Viollet

San Francisco, California
August 10-15, 1997

Published by the
**ASCE** American Society of Civil Engineers
345 East 47th Street
New York, New York 10017-2398

Abstract:

This proceedings, *Energy and Water: Sustainable Development*, contains papers presented at the 27th Congress of the International Association for Hydraulic Research (IAHR) entitled "Water for a Changing Global Community." These papers explore applications of hydraulic research to situations where natural waters are involved in the production or use of energy. Due to the continuously increasing demands on water use, new hydraulic analysis techniques that have proven successful within the current constraints associated with energy production and natural water usage are discussed. These techniques are explored within these topics: 1) Advances in computerized hydraulics; 2) tidal and wave energy production; 3) hydropower operations and optimization; 4) pumping station operations and optimization; 5) dam safety; 6) management of reservoir releases; 7) management of reservoir sedimentation; and 8) fish passage.

Library of Congress Cataloging-in-Publication Data

International Association for Hydraulic Research. Congress (27th : 1997 : San Francisco, Calif.)
Energy and water : sustainable development : proceedings of theme D : the 27th Congress of the International Association for Hydraulic Research : San Francisco, California, August 10-15, 1997 / volume editor, John S. Gulliver, volume co-editor, Pierre-Louis Viollet.
   p.    cm. -- (Water for a changing global community)
"Hosted by the Water Resources Engineering Division of the American Society of Civil Engineers."
Includes indexes.
ISBN 0-7844-0274-4
1. Water-power--Congresses.  I. Gulliver, John S.  II. Viollet, Pierre-Louis.  III. Title.  IV. Series: International Association for Hydraulic Research. Congress (27th : 1997 : San Francisco, Calif.). Water for a changing global community.
TC147.I57  1997   97-17971
627'.8--dc21                CIP

Any statements expressed in these materials are those of the individual authors and do not necessarily represent the views of ASCE, which takes no responsibility for any statement made herein. No reference made in this publication to any specific method, product, process or service constitutes or implies an endorsement, recommendation, or warranty thereof by ASCE. The materials are for general information only and do not represent a standard of ASCE, nor are they intended as a reference in purchase specifications, contracts, regulations, statutes, or any other legal document.
ASCE makes no representation or warranty of any kind, whether express or implied, concerning the accuracy, completeness, suitability, or utility of any information, apparatus, product, or process discussed in this publication, and assumes no liability therefore. This information should not be used without first securing competent advice with respect to its suitability for any general or specific application. Anyone utilizing this information assumes all liability arising from such use, including but not limited to infringement of any patent or patents.

Photocopies. Authorization to photocopy material for internal or personal use under circumstances not falling within the fair use provisions of the Copyright Act is granted by ASCE to libraries and other users registered with the Copyright Clearance Center (CCC) Transactional Reporting Service, provided that the base fee of $4.00 per article plus $.25 per page is paid directly to CCC, 222 Rosewood Drive, Danvers, MA 01923. The identification for ASCE Books is 0-7844-0274-4/97/$4.00 + $.25 per page. Requests for special permission or bulk copying should be addressed to Permissions & Copyright Dept., ASCE.

Copyright © 1997 by the American Society of Civil Engineers,
All Rights Reserved.
Library of Congress Catalog Card No: 97-17971
ISBN 0-7844-0274-4
Manufactured in the United States of America.

# GENERAL PREFACE

The XXVIIth Congress of the International Association of Hydraulic Research (IAHR) was held at the Hyatt Regency Embarcadero Hotel in San Francisco, California August 10-15, 1997. IAHR is a worldwide independent organization of engineers and scientists interested in or working in fields related to hydraulics and its practical application. The Congress was hosted by ASCE Water Resources Engineering Division. The overall theme of the Congress, "Water for a Changing Global Community", was developed in four subthemes encompassing broad aspects of hydraulics and water- resources research and engineering applications:

Theme A : Managing Water: Coping with Scarcity and Abundance.

Theme B : Environmental and Coastal Hydraulics: Protecting the Aquatic Habitat.

Theme C : Groundwater : An Endangered Resource.

Theme D : Energy and Water : Sustainable Development.

These Themes were complemented by the following Specialty Seminars: Multidirectional Waves and Their Interaction with Structures; Modeling of Turbulent flows; Wind Energy and Windflow Around Structures; Continuing Education and Training; Management of Hydraulic Research; and Hydroinformatics for Control, Management and Risk Assessment. The Congress also included the John F. Kennedy Student Paper Competition.

Over 600 papers were presented in 130 sessions during the five days of the Congress. The papers of each Theme and its keynote lecture are grouped in individual volumes of these Proceedings, with Theme B occupying two volumes. The sixth volume is devoted to the papers of the John F. Kennedy Student Paper Competition, as well as brief summaries of the six Specialty Seminars which occupied 18 sessions.

Papers solicited by topic convenors and those received from the general call for papers underwent an external review process organized by the Theme Editors using several designated reviewers for each paper. Revisions were requested for many papers, and final selections were based on technical quality, relevance to the themes of the Congress, and available sessions for presentation. All papers included in this Proceedings were printed directly from camera-ready manuscripts prepared by the authors, who retain responsibility for the presentation and technical content.

In the interest of international cooperation and broad dissemination of technical developments, IAHR has granted permission for all the papers included in these Proceedings to be eligible for discussion in the ASCE Journals of the Water Resources Engineering Division: *Journal of Hydraulic Engineering, Journal of Hydrologic Engineering and Journal of Irrigation and Drainage Engineering*. The papers are also eligible to be nominated for ASCE Awards. More detailed and substantive manuscripts arising from the work reported in Proceedings papers may be submitted to the above ASCE Journals, as well as to IAHR's *Journal of Hydraulic Research*, for possible publication..

The General Editors would like to acknowledge the financial sponsors of the Congress, including at the time of this writing: U.S. Bureau of Reclamation, Iowa Institute of Hydraulic Research, Bechtel Corporation, WEST Consultants Inc., Parsons Brinckerhoff Quade & Douglas Inc., Golder Associates, Ayres Associates, and Harza Corporation. Grateful appreciation is also extended to The Department of Bioresource Engineering of Oregon State University, The Center for Computational and Hydroscience and Engineering of The University of Mississippi, Bechtel Environmental Inc., St. Anthony Falls Hydraulic Laboratory of the University of Minnesota, The College of Engineering of San Francisco State University, Bechtel Corporation, and particularly the Iowa Institute of Hydraulic Research and Dept. of Civil and Environmental Engineering of the University of Iowa for extensive logistic, travel, and office support of Congress planning and Proceedings preparation. Members of the Local Organizing Committee, including the Theme Editors Marshall English, Sam S.Y. Wang, Angelos Findikakis, and John Gulliver, as well as Phil Burgi, Richard Denton, George Hecker, Young Kim, Tatsuaki Nakato, Clifford Pugh, Patrick Ryan and Hsieh Wen Shen are all deeply thanked for their unselfish efforts and camaraderie during the three-year planning effort. Mary Anne Vigander is thanked for her help in organizing the Accompanying Persons Program. Hollie Boyle and Shiela Menaker of ASCE provided invaluable support for the Congress and Proceedings, as did the staff of the IAHR Secretariat in Delft. Twila Meder of the Iowa Institute of Hydraulic Research cannot be thanked enough for her unselfish devotion to the organization of this Congress. Finally, Gerhard Jirka must be acknowledged for his leadership in successfully attracting the XXVIIth Congress to the United States and initiating Congress planning.

Forrest M. Holly Jr., Chair of the Local Organizing Committee
Adnan Alsaffar, ASCE/WED Program Chair
Iowa City, Iowa, April 1997

# Theme D Editor's Preface

The Energy and Water theme of the XXVIIth IAHR Congress explores applications of hydraulic research to situations where natural waters are involved in the production or use of energy. The use of water for energy production is not new, but the demands of competing water users are continuously increasing. In order to improve the allocation of the water resource under ever-changing conditions, new hydraulic analysis techniques are needed. The goal of the Energy and Water theme is to disseminate new hydraulic analysis techniques that have proven successful within the current constraints associated with energy production and natural water usage. The topics can be grouped into the following categories:

- Advances in computational hydraulics
- Tidal and wave energy production
- Hydropower operations and optimization
- Pumping station operations and optimization
- Dam safety
- Management of reservoir releases
- Management of reservoir sedimentation
- Fish passage

The Energy and Water theme addresses the design and operation of energy generating facilities, the ecological impacts of this operation, and the safety of our energy generating facilities during extreme events. The continuing need to improve our stewardship and meet the requirements of multiple resource users demands research in these areas. The Energy and Water theme of the XXVIIth IAHR Congress is designed to respond to these needs.

The topics were convened by Michael Rogers, John Laboon, Rollin Hotchkiss, Steven Daly, Robert Frederking, Michael Abbot, R. Schilling, Larry Weber, Rodney Wittler, Vahid Alavian, Mamood Nagash, Young Kim, and Rex Elder. These topic convenors were primarily responsible for the organization of the Energy and Water theme. The editors would also like to thank the many reviewers of the papers submitted to Theme D. Finally, the Chairman of the Local Organizing Committee, Forrest Holly, and the ASCE/WED Program Chair, Adnan Alsaffar, deserve a special acknowledgment for their excellent leadership and planning of the Congress and Proceedings.

John S. Gulliver, Theme D Editor
Pierre-Louis Viollet, Co-editor
Minneapolis, Minnesota, April 1997

# CONTENTS

## THEME D: ENERGY AND WATER: SUSTAINABLE DEVELOPMENT

*Theme Keynote Address:*    Charles D.D. Howard, Canada
"*Imagineering Hydropower and Water Resources*"    *1*

*Session D.1*
**REHABILITATION OF HYDRAULIC STRUCTURES**
Sponsor: ASCE Committee on Hydraulic Structures
Convenor: Mike Rogers, Harza Corporation, USA

**Hydraulic Study of Secondary Stilling Basins**
TABAN SOWLATI, *University of Windsor, Ontario, Canada,* M.A. MCCORQUODALE, *University of New Orleans, LA,* A.F. SMITH, R. MOULTON, *Grand River Conservation Authority, Cambridge, ON, and* PETER RAE, *ACRES International, Niagara Falls, ON* .................................................. *9*

**French Paper Dam Spillway Improvements**
W. JAMES MAROLD, *Harza Engineers and Scientists, Chicago, IL* .......... *15*

**Hubbart Dam Outlet Works Rehabilitation**
STEVE BAREIS, *U. S. Bureau of Reclamation, Denver, CO, and* LEE GERBIG, *Woodward-Clyde Consultants, Denver, CO* .............................. *21*

**Friant Dam Spillway Rehabilitation**
BRUCE C. MULLER, JR., *U. S. Bureau of Reclamation, Denver, CO* ......... *30*

**Ochoco Dam Spillway Modification**
DOUG STANTON, *U. S. Bureau of Reclamation, Denver, CO* ............... *36*

*Session D.2*
**HYDRAULIC STRUCTURE FAILURES: FAILURE MODES AND LESSONS LEARNED**
Sponsor: ASCE Committee on Hydraulic Structures
Convenor: John Laboon, U. S. Bureau of Reclamation

**Overtopping Breaching of Noncohesive Embankment Dams**
S.E. COLEMAN, R.C. JACK, *and* B.W. MELVILLE, *University of Auckland, Auckland, New Zealand* ............................................... *42*

**Predicting Embankment Dam Breach Parameters-A Needs Assessment**
TONY L. WAHL, *U. S. Bureau of Reclamation, Denver, CO* ............... *48*

**Performance Parameter Evaluation of Potential Hydraulic Structure Failure Modes**
  JAY N. STATELER, U. S. Bureau of Reclamation, Denver, CO .............. 54

**Abrasion/Erosion in Stilling Basins**
  LESLIE HANNA and ELISABETH COHEN, U. S. Bureau of Reclamation,
  Denver, CO ................................................................ 60

**Folsom Spillway Gate Failure and Lessons Learned**
  ROBERT V. TODD, U. S. Bureau of Reclamation, Denver, CO .............. 66


*Session D.7*
**MANAGEMENT OF RESERVOIR SEDIMENTATION**
  Sponsors: IAHR Section III.1 Fluvial Hydraulics and ASCE Committee on Research and Education
  Convenors: Rollin Hotchkiss, University of Nebraska, USA and Giampaolo Di Silvio, Universitá degli Studi di Padova, Italy

**Reservoir Sedimentation in China—Experiences and Lessons**
  ZHIDE ZHOU and XIAOQING YANG, International Research and Training Center on Erosion and Sedimentation, Beijing, China .......................... 72

**Socioeconomic Analysis of Reservoir Sedimentation**
  ROLLIN H. HOTCHKISS, University of Nebraska, Lincoln, NE, and FRANK H. BOLLMAN, Agricultural Industries, Inc., West Sacramento, CA .............. 78

**Sedimentation in Sierra Nevada Reservoirs**
  RICHARD KATTELMANN, University of California, Mammoth Lakes, CA ..... 84

**About Delivery Ratio: How Does it Change in Time and Space?**
  GIAMPAOLO DI SILVIO and ANDREA MARION, University of Padua, Padova, Italy ..................................................................... 90

**Reservoir Sediment Management Related to Sediment Quality and Contamination**
  BERNHARD J. WESTRICH, University of Stuttgart, Stuttgart, Germany ....... 96

**Double-Mass Curve Use on Sedimentology-A Case Study**
  NEWTON DE OLIVEIRA CARVALHO, ELETROBRAS, Rio de Janeiro, Brazil ..102

**The Development of Equilibrium Profiles for Flushing Channels**
  ROBERT H.A. JANSSEN and HSIEH WEN SHEN, University of California, Berkeley, CA .............................................................. 108

**Modeling the Impacts of Reservoir Emptying**
  A. PETITJEAN, F. MAUREL, J.P. BOUCHARD, and J.C. GALLAND, Electricité de France, Chatou, France ..................................................... 114

**Desilting Reservoir Sediment Deposits by Drawdown Flushing**
*JIHN-SUNG LAI, National Taiwan University, Taiwan, R.O.C., and HSIEH WEN SHEN, University of California, Berkeley, CA* ........................... 120

**Non-Equilibrium Sediment Transport Modeling of Sanmenxia Reservoir**
*ALBERT MOLINAS and BAOSHENG WU, Colorado State University, Fort Collins, CO* .................................................. 126

**Desilting of Water Reservoirs in Algeria by Dredging. Study Case: The Fergug Reservoir**
*MOHAMED ERRIH and HABIB BENDAHOU, University of Sciences & Technology of Oran, Oran, Algeria* .................................... 132

**Sediment Management in Buffer Basins at Hydropower Plants**
*H. SCHEUERLEIN and F. EIBL, University of Innsbruck, Innsbruck, Austria* ... 138

**Sediment Management at an Alpine Reservoir**
*H. KNOBLAUCH, G. HEIGERTH, and T. DUM, Technical University of Graz, Graz, Austria* ................................................... 144

**Solution of Sediment Problems for Hydro Cascade on Mountain River**
*ALEXANDER M. PROUDOVSKY, VICTOR B. RODIONOV, VLADIMIR O. SARANCHEV, and MIKHAIL JA. GILDENBLAT, Center for Hydraulic Reserches, NIIES, Moscow, Russia* ........................................... 150

*Session D.9*
**ICE PROBLEMS AT POWER PLANTS AND FIXED STRUCTURES**
Sponsor: IAHR Section III.4 Ice Research and Engineering
Convenors: Steve Daly, Cold Regions Research and Engineering Laboratory, USA and Robert Frederking, National Research Council, Canada

**Effects of Reservoir Regulation on Ice Jam Thickness**
*JON E. ZUFELT, USACRREL, Hanover, NH* ........................... 156

**Ice Effects on Riprap: Small-Scale Tests**
*D.S. SODHI, S. BORLAND, J.M. STANLEY, and C.J. DONNELLY, U. S. Army Cold Regions Research and Engineering Laboratory, Hanover, NH* ............... 162

**Ice Jam Mitigation for Small Streams**
*JAMES H. LEVER, U. S. Army Cold Regions Research and Engineering Laboratory, Hanover, NH* ................................................... 168

**Simulation of Dynamic River Ice Transport and Jamming**
*SHUNAN LU, HUNG TAO SHEN, Clarkson University, Potsdam, NY, and RANDY D. CRISSMAN, New York Power, Authority, Niagara Falls, NY* .............. 174

**Ice Retention with Artificial Islands on the St. Marys River**
*ANDREW M. TUTHILL and KEVIN L. CAREY, Cold Regions Research and Engineering Laboratory, Hanover, NH* .................................*180*

**Ice Control at Locks and Dams**
*F. DONALD HAYNES, USA Cold Regions Research and Engineering Laboratory, Hanover, NH* .................................*186*

**Ice Problems on Hydropower Project Wloclawek (Lower Vistula)**
*WOJCIECH MAJEWSKI, Technical University of Gdansk, Gdansk, Poland* ....*192*

**Model Scale Measurements of Ice Loads on a Submerged Turret Loading (STL) Tanker**
*S. LØSET and Ø. KANESTRØM, Norwegian University of Science and Technology, Trondheim, Norway* .................................*198*

*Session D.15*
**ADVANCES IN COMPUTATIONAL HYDRAULICS**
*Sponsor: IAHR Section I.2 Hydroinformatics*
*Convenor: Michael B. Abbott, International Institute for Hydraulic and Environmental Engineering, The Netherlands*

**Solution of the Saint Venant Equations Through the Use of Riemann Based Methods**
*A.J. CROSSLEY and N.G. WRIGHT, University of Nottingham, Nottingham, UK* .................................*204*

**Time Accurate Simulation of Free Surface Flow Problem**
*WEIXING YUAN, NORBERT RIEDEL, and RUDOLF SCHILLING, University of Technology, Munich, Germany* .................................*210*

**Study on Highly-Accurate Numerical Method for Advection Term**
*KOJI ASAI, Chugoku National Industrial Research, Kure, Japan, TOSHIMITSU KOMATSU, Kyushu University, Fukuoka, Japan, KOICHIRO OHGUSHI, Saga University, Saga, Japan, and KESA YOSHI HADANO, Yamaguchi University, Ube, Japan* .................................*216*

**Implicit TVD Methods for Modelling Discontinuous Channel Flows**
*A.I. DELIS and C.P. SKEELS, University of the West of England, Bristol, UK* ..*222*

**Numerical Simulation of Large Eddy Structure in Hydraulic Jump**
*YOSUKE YAMASHIKI, International Lake Environment Committee, Kusatsu, Japan, TAMOTSU TAKAHASHI, Kyoto University, Uji, Japan, PODALYRO A. DE SOUZA, and KIKUO TAMADA, University of Sao Paulo, Sao Paulo, Brazil* ..........*228*

**Allowance for Secondary Flows in the 3D Numerical Flow Models of Piping Systems**
R. KLASINC, T. DUM, H. KNOBLAUCH, and R. REITBAUER, Technical University of Graz, Graz, Austria .................................234

**Numerical Simulation of Flow in a Channel With a Wavy Sidewall**
THOMAS MOLLS and GANG ZHAO, Southern Illinois University, Carbondale, IL .................................240

**Large Eddy Simulations of the Flow Around a Single Bar and an Array of Bars**
F. HERMANN, R. HOLLENSTEIN, and P. BILLETER, ETH Zürich, Zürich, Switzerland .................................*

**3-D Numerical Solution of Flow in Sine-Generated Meandering Compound Channel**
H.S. JIN, S. EGASHIRA, Ritsumeikan University, Shiga, Japan, and B.Y. LIU, The NEWJEC, Inc., Osaka, Japan .................................246

**3-D Turbulent Flow Simulation for the Tail Water Channel**
JIANYONG GUAN and BIHONG CHEN, Dalian University of Technology, Dalian, China .................................252

**Ritter's Dambreak Waves Revisited**
GUIDO LAUBER and WILLI H. HAGER, VAW, ETH-Zentrum, Zurich, Switzerland .................................258

**Simulation of Waves Generated by Landslides in Vaiont Dam**
MASANORI MICHIUE and OSAMU HINOKIDANI, Tottori University, Tottori, Japan .................................263

**An Extended Depth-Averaged Turbulence Model for Flow Constricted by Cofferdams**
JIAN LIU, NEWJEC Inc., Osaka, Japan, and AKIHIRO TOMINAGA, Nagoya Institute of Technology, Nagoya, Japan .................................269

**2-D Flow and Sediment Simulation for the Flood Regulation of a Reservoir With Water Intake**
FAYI ZHOU, University of Alberta, Alberta, Canada, BIHONG CHEN, Dalian University of Technology, Dalian, China, and CHARLES C.S. SONG, University of Minnesota, Minneapolis, MN .................................275

**3D Numerical Modelling of Sediment Deposition and Bed Changes in a Tunnel-Type Sand Trap**
NILS REIDAR, B. OLSEN, and HILDE MARIE KJELLESVIG, SINTEF Civil and Environmental Engineering, Trondheim, Norway .................................281

---

* Manuscript not available at time of printing

*Session D.16*
HYDROPOWER OPERATIONS AND OPTIMIZATION
Sponsor: IAHR Section II.1 Hydraulic Machinery and Cavitation
Convenors: P. Schilling, Technische Universität München, Germany and Sam Martin, Georgia Institute of Technology, USA

**Increased Output of Old Power Plants: Limitation and Problems to be Solved**
HERMOD BREKKE, The Norwegian University of Science and Technology, Trondheim, Norway .............................................287

**Flow-Induced Multiple-Mode Vibrations of Lightly Damped Rectangular Trashrack Prisms**
K. KERENYI, H. DROBIR, T. STAUBLI, G. DORRER, and N. HEJL, Technische Universität Wien, Wien, Austria.....................................293

**Numerical Analysis of Unsteady Flow on Rotating Buckets for Optimal Operation of Pelton Turbines**
T. KUBOTA and Y. NAKANISHI, Kanagawa University, Yokohama, Japan .....637

**The Analysis of Rotor-Stator Interaction Phenomena in a Pump-Turbine**
Y. QUIAN, Fuji Electric Co., Ltd., Kawasaki City, Japan, and C. ARAKAWA, The University of Tokyo, Tokyo, Japan ................................302

**Head Recovery in the Tailwater of Low Head Hydro Power Stations**
CH. SCHNEIDER, Y. MOCHKAAI, W. KNAPP, and R. SCHILLING, University of Technology, Munich, Germany ....................................308

**Experimental Evaluation of the Compressible Portion of a Liquid Flow Crossing a Cavitating Control Valve**
U. FRATINO and A. F. PICCINNI, Politecnico di Bari, Bari, Italy ...........314

**Design of Throttled Surge Tanks for High-Head Plants. Pressure Wave Transmission and Reflection at a T-Junction with an Orifice in the Lateral Pipe**
R. PRENNER and H. DROBIR, Technische Universität Wien, Wien, Austria ....320

**Surging Phenomena of Twin Surge-Tanks of the Okukiyotsu No. 2 Power Station**
KOICHI FUJINO and KAORU HAGA, Electric Power Development Co. Ltd., Tokyo, Japan ................................................328

**Optimal Unit Commitment Scheduling in Hydropower Systems**
J. YI, J. W. LABADIE, Colorado State University, Ft. Collins, CO, and S. STITT, U. S. Bureau of Reclamation, Denver, CO ............................334

**Real Time Control of River Water Quality for Hydroelectric Operation**
F. WELT, HMS Enérgie Inc., Montréal, Québec, Canada, and R. KAHAWITA,

xii

Ecóle Polytechnique de Montréal, Montréal, Québec, Canada .............340

**The Problem of Small-Scale Hydropower Stations on the Rivers of Central Russia**
S. LACHTCHENOV and V. SEMENKOV, RAO "EES ROSSII", Moscow, Russia 346

**The Development of the Simulated Study on Structural Vibration Induced by Discharge Flow Energy**
GUIFEN LI, XIMIN YUAN, SHUKUN LIU, China Institute of Water Conservancy and Hydroelectric Scientific Research, Beijing City, China, GUANGTAO CUI, Tianjin University, Tianjin City, China ...............................352

*Session D.21*
FISH PASSAGE
Convenor: Larry J. Weber, University of Iowa, USA

**Structural Modifications at Hydro Dams: An Opportunity for Fish Enhancement**
DILIP MATHUR, PAUL G. HEISEY, Normandeau Associates, Drumore, PA, JOHN R. SKALSKI, University of Washington, Seattle, WA, STEVEN G. HAYS, Mid Columbia Consulting, Inc., East Wenatchee, WA, and MARK R. SMITH, U. S. Army Corps of Engineers, Portland, OR .....................................358

**Methodology of Ecological Fishway Design**
TETSURO TSUJIMOTO and NORIKO HORIKAWA, Kanazawa, University, Kanazawa, Japan ...................................................364

**The Virtual Fish Concept: Numerical Prediction of Fish Passage Through Hydraulic Power Plants**
FOTIS SOTIROPOULOS and YIANNIS VENTIKOS, Georgia Institute of Technology, Atlanta, GA .............................................370

**The Regulated River and Sturgeon Spawning Migration**
E. N. DOLGOPOLOVA, Russian Academy of Sciences, Moscow, Russia .......376

**Use of Volga River Flow for Power Generation and Fishery Purposes: Mitigation of Conflicts of Interest**
ALEXANDER ASSARIN, Hydroproject Institute, Moscow, Russia ............382

**Evaluating the Hydraulic Performance of Fish Passage Structures**
BRENT W. MEFFORD and JOE P. KUBITSCHEK, U. S. Bureau of Reclamation, Denver, CO ......................................................387

**Linking Kinematic and Physical Hydraulic Models**
DUNCAN HAY, JAMES STRONACH, Hay & Company Consultants Inc., Vancouver, B.C., Canada, LARRY WEBER, Iowa Institute of Hydraulic Research, Iowa City, IA, and CHARLES SWEENEY, ENSR Consultants, Redmond, WA ...393

Juvenile Fish Passage System at McNary Dam
CINDY PHILBROOK, U. S. Army Corps of Engineers, Walla Walla, WA ...... 399

Hydraulic Investigations Associated with Fish Passage at Red Bluff Diversion Dam, California
JOSEPH P. KUBITSCHEK and BRENT W. MEFFORD, U. S. Bureau of
Reclamation, Denver, CO ......................................... 405

Discharge Equation of Ice Harbor Type Fishway
SHIRO MAENO, Okayama University, Okayama, Japan, G. SAMPATH KUMAR, University of Oulu, Oulu, Finland, HIROSHI NAGO, Okayama University, Okayama, Japan, and HIROMICHI SUETSUGU, The Chugoku Electric Power Co. Inc., Hiroshima, Japan .............................................. 411


*Session D.23*
DAM FOUNDATION EROSION
Convenor: Rodney Wittler, U.S. Bureau of Reclamation

Mechanism of Energy Dissipation and Hydraulic Design for Plunge Pools Downstream of Large Dams
PEIQING LIU, JIZHANG GAO, ZHONGYI LI, and YONGMEI LI, China Institute of Water Resources and Hydropower Research, Beijing, China .............. 417

Local Scouring Downstream of Positive Step Stilling Basins
CORRADO GISONNI, II, Universita di Napoli, Aversa, Italy, and GIACOMO RASULO, Universita di Napoli Federico II, Napoli, Italy ................. 423

Dam Foundation Erosion: Numerical Modeling
GEORGE W. ANNANDALE, TODD LEWIS, Golder Associates Inc., Lakewood, CO, ROD WITTLER, Bureau of Reclamation, Denver, CO, STEVE ABT, and JIM RUFF, Colorado State University, Fort Collins, CO ......................... 429

Dam Foundation Erosion: Behavior of a Free-Trajectory Jet in a Plunge Basin
JEFFREY G. BOHRER and STEVEN R. ABT, Colorado State University, Ft. Collins, CO ................................................... 435

Dam Foundation Erosion: Pit 4 Dam Scale & Prototype Model Test Results and Comparison
R.J. WITTLER, US Bureau of Reclamation, Denver, CO, S.R. ABT, J.F. RUFF, Colorado State University, Fort Collins, CO, and G.S. ANNANDALE, Golder Associates, Lakewood, CO ....................................... 441


*Session D.24*
ADVANCES IN MANAGING RESERVOIR RELEASES
Sponsor: ASCE Committee on Research and Education

Convenor: Vahid Alavian, RANKIN International, USA

**Overview and Planning Considerations for Improving Reservoir Releases**
W. GARY BROCK and J. STEPHENS ADAMS, Tennessee Valley Authority,
Norris, TN ................................................. 447

**EDF Experience in Improving Reservoir Releases for Ecological Purposes**
PH. GOSSE, C. SABATON, F. TRAVADE, Electricité de France, Chatou, France,
and J. EON, La Défense, Paris, France ............................... 453

**Status and Vision of Turbine Aeration**
PATRICK A. MARCH, PAUL N. HOPPING, Tennessee Valley Authority, Norris, TN,
and RICHARD K. FISHER, Jr., Voith Hydro, Inc., York, PA ............... 459

**Effects of the Spillway Operation on the Fishes Habitat: Study of Solutions**
C.M. ANGELACCIO, J.D. BACCHIEGA, C.A. FATTOR, and H. D. BARRIONUEVO,
INCyTH, Ezeiza, Argentina ........................................... 465

**Modeling Approaches for Tailwater Enhancement**
GARY E. HAUSER, MING C. SHIAO, JAMES A. PARSLY, and BRUCE L. YEAGER,
TVA Resource Group, Norris, TN ..................................... 471

**Monitoring Dissolved Oxygen and Total Dissolved Gas in Tailraces**
H. RUCKER and D. HARSHBARGER, Tennessee Valley Authority, Muscle
Shoals, AL ......................................................... 477

**Modifying Reservoir Release Temperatures Using Temperature Control Curtains**
TRACY B. VERMEYEN, US Bureau of Reclamation, Denver, CO ............ 483

**Oxygen Uptake at Barrages of the Elbe Cascade**
P. NOVAK, University of Newcastle upon Tyne, UK, and P. GABRIEL, Czech
Technical University, Prague, Czech Republic ........................ 489

**Influence of Pumped-Storage Power Plant Zarnowiec on Natural Lake
(Lower Reservoir)**
WOJCIECH MAJEWSKI, Technical University of Gdansk, Gdansk, Poland .... 495

**Turbulence and Stability at Interfaces in Stratified Reservoir for Power Generation**
SOTOAKI ONISHI, MITIHIRO SUGII, Science University of Tokyo, Chiba, Japan,
and YUICHI KITAMURA, Electric Power Development Co., Tokyo, Japan .... 501

**A Network Operation of Reservoirs for Enhancement of the Ecological Flushing
Discharge**
NOBUYUKI TAMAI, YOSHI EMURA, and HIRONORI MATSUZAKI, University of
Tokyo, Tokyo, Japan ................................................. 507

**Total Dissolved Gas in the Near-Field Tailwater of Ice Harbor Dam**

STEVEN C. WILHELMS and MICHAEL L. SCHNEIDER, US Army Engineer
Waterway Experiment Station, Vicksburg, MS ........................513

A Unique Approach for Physical Model Studies of Nitrogen Gas Supersaturation
LARRY J. WEBER and CARL MANNHEIM, University of Iowa, Iowa City, IA ..518

Prediction of Dissolved Gas Concentration Downstream of a Spillway
JOSEPH J. ORLINS and JOHN S. GULLIVER, University of Minnesota,
Minneapolis, MN ................................................524


*Session D.25*
PUMP INTAKE STRUCTURES
Convenor: Mahmood Naghash, Bechtel Corporation, USA

Physical Model Study of Several Jet-Breaker Baffle Concepts in Pump Intake Structures with Pipe Inflows
HARTMUT ROSENBERGER, DIETER-HEINZ HELLMANN, University of Kaiserslautern, Kaiserslautern, Germany, and MAHMOOD NAGHASH, Bechtel Corporation, Gaithersburg, MD .....................................530

Influence of Approach Flow Non Uniformity on Vortices at a Pump Intake
G. CONSTANTINESCU and V. C. PATEL, University of Iowa, Iowa City, IA ...537

Effect of Dual Flow Screens on the Hydraulic Performance of Pump
M. PADMANABHAN, D. K. WHITE, and J. LARSEN, Alden Research Laboratory, Inc., Holden, MA ...............................................543

Vortex Suppression in Multiple-Pump Sumps
DEBORAH I. BAUER, TATSUAKI NAKATO, and MATAHEL ANSAR, University of Iowa, Iowa City, IA ............................................549

Measurement of Subsurface Vortices in a Model Pump Sump
V.P. RAJENDRAN and V.C. PATEL, University of Iowa, Iowa City, IA ........555


*Session D.26*
TIDAL AND WAVE ENERGY PRODUCTION
Convenor: Young C. Kim, California State University, USA

A Hybrid System of Wave Power Extraction and Shore Protection
HIDEO KONDO, Muroran Institute of Technology, Muroran, Japan ..........561

Study on Wave Pumping-Up Power Station
SEIYO SHIGEMITSU and AKIRA HIRATSUKA, Osaka Sangyo University,
Osaka, Japan ..................................................566

**An Experimental Study on Effective Conversion of Wave Energy into Potential Energy**
T. KOMATSU, T. OKADA, N. MATSUNAGA, Kyushu University, Kasuga, Japan,
M. HASHIDA, Nippon Bunri University, Oita, Japan, and K. FUJITA, Kyushu
University, Fukuoka, Japan ........................................... 571

**Ocean Wave-Powered Desalination**
MICHAEL E. MCCORMICK, The Johns Hopkins University, Baltimore, MD, and
YOUNG C. KIM, California State University, Los Angeles, CA .............. 577


*Session D.28*
**SPILLWAYS AND OUTLET WORKS**

**Characteristics of Flow Conditions on Stepped Channels**
IWAO OHTSU and YOUICHI YASUDA, Nihon University, Tokyo, Japan ....... 583

**Initiation of Aeration in Stepped Spillways**
CHRISTOBAL MATEOS IGUACEL and VICTOR ELVIRO GARCIA, CEDEX,
Madrid, Spain ..................................................... 589

**Energy Dissipation in Stepped Waterway**
HUBERT CHANSON and L. TOOMBES, The University of Queensland, Brisbane,
Australia ......................................................... 595

**The Effect of Nappe Impact Angle on Aerator Performance**
JALAL ATTARI, Power & Water Institute of Technology, Tehran, Iran, and AMIR
R. ZARRATI, Amir Kabir University of Technology, Tehran, Iran ........... 601

**Design of Spillway Deflectors for Ice Harbor Dam to Reduce Supersaturated Dissolved Gas Levels Downstream**
JAMES D. CAIN, U. S. Army Corps of Engineers, Walla Walla, WA .......... 607

**Air Entrainment in Bottom Outlet Tailrace Tunnels**
JUERG SPEERLI and PETER U.VOLKART, Swiss Federal Institute of Technology,
Zurich, Switzerland ................................................ 613

**The Submerged Hydraulic Jump Downstream Sluice Weir**
RUIWEN PAN and YIMING XU, Yunnan Polytechnic University, Kunming,
China ............................................................ 619

**Control of Hydraulic Jump by Abrupt Drop**
J.E. RICHARDSON, Flow Science, Inc., Los Alamos, NM ................. 625

**Application of Telemac 2D Software to Dimensioning Hydraulic Structures: Case of the Mesce Dam**
E. LAPERROUSAZ, Electricité de France, Bourget-du-Lac, France, C. MOULIN,

*Electricité de France, Chatou, France, J.P. BLAIS, and E. CHIESA, Electricité de France, Bourget-du-Lac, France* .................................. *631*

**Subject Index** ................................................. *643*

**Author Index** ................................................. *649*

# Imagineering Hydropower and Water Resources.
Opening address to Plenary Session D: Water and Energy

By
Charles D. D. Howard, M. ASCE[1]

## Introduction

An organizer of this meeting invited me to speak here after reading "The Control Room of the Future", a series of articles that Jery Stedinger and I wrote for the trade magazine, Hydro Review. These were fanciful pieces that grew out of a challenge from the sponsor of a Hydro Research forum held in 1992. The challenge stretched the limits of our academically trained imagination. One of the organizers of today's conference thought that I might stimulate research into areas you otherwise might not think to explore. He might know that the Hydro Review articles are not all that fanciful. Some of the ideas that Jery and I thought were far out science fiction became established fact, even before the articles were published.

For trained minds like ours, it is a difficult step to invent science fiction - to make the leap to a new paradigm. It's much less taxing to define a research project that is an increment away from the present. It takes effort to research new developments in physics before going to the hydraulics lab or to the computer. It is easier on the brain to follow a well trodden path.

The objective here today is to encourage you to think about research as an open invitation to take discoveries and ideas from other fields, and from your free ranging imagination, and to move those seemingly far fetched ideas towards being the engineering achievements of the future.

## Current Research Priorities for Hydropower

The 1992 Hydro Forum, ancient history from five years ago, had some research ideas that offered incremental improvements. These were itemized in six areas:

1. **Operations.** Advanced monitoring and control tools should be used to improve how projects are operated, to reduce maintenance, and to improve efficiency in the use of water and equipment.

2. **Planning.** A consistent method is need for quantifying all of the benefits from hydropower, including flexibility for power dispatch and electrical system regulation, environmental protection from other nefarious energy sources, and river management.

---

Consulting Engineer, Charles Howard & Associates, Ltd.,Victoria, BC, Canada

3. **Environmental.**      The prime priority is for fish passage research.

4. **Hydromechanical.**    Research focus on variable speed operation of hydro equipment.

5. **Hydrologic Forecasting.**   Research is needed to improve accuracy and timeliness.

6. **Hydraulic Structures.**    Guidelines for installing power at non-power dams.

The 1992 Hydro Forum focused on areas that could produce an immediate payoff to society, with little risk. Today the goal is to discuss "research" that may have a long term payoff, with considerable risk.

## Interdependence of Energy and Water Resources

Water and energy are linked on the supply side and on the demand side. On the supply side, many water projects are paid for by the production of energy, even if the prime purpose is say, flood control, urban water supply, or irrigation. On the demand side, the economic viability of an energy project depends on the market (the load and the price) it will serve, and the rate at which that market might grow. The water demands of energy production, and the electricity demands of water uses, are linked through the costs of water, the costs of electricity, and the affects on the environment.

The way in which people interact with water and energy is changing. Energy projects that depend on water usually are operated by people trained to understand electricity. The people who manage water are trained in water, not energy. The skills and focus of senior people in the energy business will broaden to deal with the responsibility as "Steward" of the river system. The water people will become fully knowledgable of energy production. Responsibilities are broadening as the public and private sectors of society become more fully integrated through regulations and pricing.

There is a parable about the development of water resources in ancient Egypt. This allegedly took place before the dawn of recorded history, so it might be true. An engineer, doing business development on hands and knees in front of the pharaoh, described a grand water resources project that would make the desert green and enrich the nation. A network of canals would carry water to the fields during the flood season, storing it in reservoirs, and doubling the production of agriculture. This would provide wealth, the country and the pharaoh would become rich and powerful. And this would all be done just by developing the water resources that flow freely down to the sea.

The engineer had previously built roads and temples for the pharaoh, and had done a good job. So naturally the pharaoh trusted the engineer and had faith in his ability to carry out this water project. The necessary arrangements were made for funding, permits were granted, and in a few years the project was completed. Soon after that the country became rich with agriculture, and the rest is history. But what happened to the engineer? Why is his name not enshrined on some monument?

According to the parable, one day the high priest brought the news that the son of the pharaoh was seriously ill. The snails in the canal system carried a deadly incurable disease. The engineer could be heard mumbling something about his "terms of reference" as they led him away.

All of that, if it happened at all, is in the distant past. Since then, water projects have displayed sensitivity for the surrounding environment, and for the allocation of benefits within the whole of society. Sometimes this was achieved after extensive public debate and considerable anguish for the project's proponents. The public hearings for the great aqueduct of Rome lasted for one hundred years - it took that long to reach agreement on the allocation of water delivered to the public and to the emperor.

## Where We Are Now

Technology Changes
    Telecommunications by Internet, fax
    Informal global networking
        Non-government organisations
        Ordinary academics
        Individuals
    Inexpensive powerful computers
        Fast CPU, large memory, mass data storage, convenient printing
    Hordes of trained computer users, globally available
    Cheap powerful data acquisition
        Ultrasonics, in situ velocity field measurements
        In situ biological monitoring, fish and wildlife sensors
        Remote sensing: precip, temp, wind, snow, cloud, soil moisture, crops, NEXRAD
        Multichannel Programmable Logical Controllers
        Data Collection Platform (satellite uplinks)
        Inexpensive satellite down links for data
        Inexpensive two way global digital data and device control via Internet

    "Scientific" advances that affect water resources
        Weather forecasting longer range in some places
        Laboratory equipment for turbulence, water chemistry, and biology

Societal changes
    Global integrated markets
    Rise of women in engineering and science
    Maturing of the environmental movements
    Shrinking North American natural resource frontier
    Global immigration, the energy of the new wave
    Reorganization of government priorities
    Steward of the river - a broader scope for water management

## ENERGY AND WATER

### Where We Are Going Right Now

Expanding scope in water management for the private sector
Stronger laws, weaker agencies
Direct public input on decisions and on enforcement
General acceptance of the "Public Trust"
More opportunities for high caliber technical input on water decisions
More scope for technical achievements
    Impact of water's *value* to urban water supply and distribution
    Bottled water, point of use treatment, and costs of urban water quality
    Irrigation's true costs and benefits, the end of subsidy
    Planning for flood and drought management
    Real-time optimization of hydro/thermal electric and water distribution systems
    Scientific management of aquatic biology

### Where We Could End Up

Global hydrology - economic value of weather forecasts
Sensing subsurface hydrology - the value of subsurface storage
Physics of fluid flow - fluid machinery design, and pollution transport
Manufacturing water, desalination, precipitation - developing countries
Aquatic ecology monitoring and management - sustaining existing reservoirs
Civilian Engineers = physics + electronics + computer science + communications + biology

The future will see new institutions that can take advantage of changing technologies and consumer preferences. Society is evolving, sometimes pushing the implementation of new technology, sometimes resisting it. Engineering research should imagine what can be done, then foster debate that will develop public opinion - global climate research is an example of this process. The public needs information for guidance and as a base for setting priorities. Engineering research needs guidance to develop what is wanted, not just what can be done.

### Machinery

Why did it take such a long time for the screw propeller to be introduced, or the propeller turbine? These ideas evolved out of the Archimedean Screw, which is thousands of years old. The Romans had the materials and the technology to build pressurized wood stave pipelines yet they constructed massive aqueducts. Incentives for non-military inventions were dampened by the availability of inexpensive muscle power. Water wheels existed for centuries before someone thought to raise the power output by putting a propeller, a degenerate Archimedean screw, into a pressure pipe.

Today there is a global social, economic, scientific, and technical setting that encourages and rewards new technology. And there have been some remarkable inventions - some based on new knowledge, like superconductivity, and some based on well established knowledge, like the hydraulic combustion engine.

The Economist in May, 1997, describes a hydraulic combustion engine developed at the University of Houston. It uses water to replace the cylinders and pistons of a conventional combustion engine. The goal is to completely eliminate the contact between metal moving parts by eliminating the moving parts. The concept is to introduce a high velocity jet tangentially into a cylinder. The rapidly rotating flow clings to the wall, creating a central cavity which can be filled with fuel and ignited. Expanding gases force the water through turbines in a piping system that leads to another similar cylinder. Ignition in the two cylinders is timed, just like a conventional combustion engine, to create a steady torrent of flow through the closed circulating water system.

The same issue of the Economist reports that on March 12th in Switzerland the Geneva electric utility switched on the first commercial transformer to use superconducting wires. The device was built by a Swiss - Swedish conglomerate using wires made by American Superconductor, partly owned by Electricité de France. It is 20 percent more efficient than a comparable conventional transformer, much more compact and easy to transport, and the coolant, liquid nitrogen, is harmless to the environment.

For several years there has been active research into developing fuel cell technology that would use any hydrocarbon fuel to produce electricity. Hydro Quebec, for one, initially supported this type of research - their interest may have focused on transmitting energy from remote hydro sites to load centres. It is well known that energy can be transmitted more economically, with fewer complications, as hydrogen in a pipeline than as electricity in a wire. They may not have thought through the larger implications, which are very serious for the electric utility industry, and for hydroelectric projects at the upstream end of the long wire to the consumer.

With efficient fuel cells providing electricity in the home and office, and satellite telephone and TV communications, we are only a small move away from a wireless society. The gas main that fuels the kitchen stove, the furnace, and the heat exchanger can also produce electricity right at the point where it is needed. Forward thinking gas and electric utilities are taking this concept seriously. Forward thinking hydraulic engineers could be developing useful futures for cooling towers, hydroelectric dams, and power stations.

## Energy management

Changes in the production and management of energy will affect the aquatic environmental impacts, the operational requirements of hydraulic facilities, and the way in which water uses can be matched to the whims of hydrology.

The North American Electric Reliability Council sets the requirements for the flow of electricity across the interties between the service areas of electric utilities. To meet the reliability criteria, each utility must adjust its generation so that the cumulative flux, positive or negative within a ten minute period, does not exceed the reliability requirement.

In the past this tolerance has been maintained by adjusting generation, usually by using the hydroelectric units in the power system. This affects the design of the generating units, the environmental impacts, and the losses in the transmission system. It is possible to achieve some control by adjusting the load. Instead of starting, stopping, or adjusting, large generating units, the utility could control thousands of individual small loads to minimize the instantaneous Area Control Error (ACE). This is done now to control electric water heaters. But the idea can go much further.

Modern power system control could favour changes in load over changes in generation. With on-line load management, the generation and transmission facilities would tend to be operated continuously at steady outputs. Computers and communications can provide control systems that facilitate distributed energy generation and distributed load regulation.

Local control of loads can be combined with at site generation from supplementary sources. Ambient electric fields generated by the lighting fixtures in office buildings can power office equipment. Helioelectric glass and building cladding can provide solar energy. The possibility exists to develop Virtual Energy sources by optimizing the coordination of generation and loads at many locations.

Monitoring sensors could be built into appliances supplied by the electric utility - customers could purchase discounted appliances with their subscription for electricity. The customer would subscribe for a discount on the cost of electricity by accepting Automatic Load Control (ALC) of the appliances.

At customer sites the operation of electrical appliances could be controlled by a Subscriber Control Module (SCM). This could be programmed by the customer to operate appliances according to time sets or from set points based on monitoring of the customer's premises. The SCM would manage the customer's load on commands from the utility's control computers. Another approach would provide an increment of automatic load control on the customer's premises by monitoring changes in the frequency of the incoming power from the utility. For example, freezer, air conditioning, fans, and refrigerator motor starts could be delayed, or initiated at the most advantage times for the power system.

Load changes could be projected from monitoring of customer premises, appliances, and process computers in industrial facilities. Energy suppliers need the ability to accurately forecast and co-ordinate demands and generation. Generation and transmission facilities could tend to operate continuously at a stable desirable level. The goal would be to achieve overall system efficiency and reliability, reduced maintenance, reduced environmental impact, and lower cost electricity supplies.

## Information and control systems

This past year saw the introduction of the ORBCOM commercial satellite system for two way global communications over the Internet. This new system provides the capability to send and receive digital information to or from any spot on earth. A data collection platform, or any other device, can be located anywhere, can transmit as frequently as necessary, and can be controlled from a personal computer. An e-mail message can command the DCP, to send more information, or to save battery power, or to turn around and face the other way. The general public now has this capability on the desk, with the morning e-mail. This is a global gateway to opportunity for inventions that exploit this new communications technology.

During the past few years some utilities have developed or purchased Decision Support Systems. These include computer models for hydrologic forecasting, probabilistic reservoir storage operations, and deterministic unit commitment and scheduling. Hydrologic data for this software is continually updated by a satellite down link, and operating information is automatically provided from the utility's energy management system.

The existing Decision Support Software provides instructions for maximizing revenue and meeting all of the environmental and contractual constraints. Human intervention is minimized. From here it is a short step to complete automation of water management for other purposes as well. In the time frame of about 40 years the attitude of engineers to computers has moved from widespread opposition, to widespread support, to widespread impatience with the slow rate of progress. Research can push this frontier faster and further by recognizing the multi disciplinary skills that are required to make it move. Water resources management is in about the same state that bioengineering was in when the first electrical engineer went out to get a medical degree.

## Science Fiction - Or Possibilities in Waiting

Now, consider some far out, risky research ideas. Quantum physics eventually will enter the everyday world of hydraulic engineering. This will require a paradigm shift for the deterministic engineer. In the quantum world it is possible to learn what happens by observation, but it may never be possible to understand why. That is the nature of quantum mechanics. As engineers we can live with that, just like we live with the Moody diagram. To paraphrase noble prize winner Richard Feynman - we can know how to calculate the flow of water through a pipe, but we may never know why the calculation usually works.

Last year an article appeared in Physics Review reporting that the phenomena involved with the collapse of a soap bubble involves effects on a molecular scale and at high speed - the probabilistic world of quantum mechanics comes into play. Consider a hydro plant that operates on an entirely new principle, based on quantum mechanics. Generation efficiency would be irrelevant since hydraulic energy would simply initiate a process which takes place at the sub-atomic level.

The device could be a probability warping machine that channels the distribution of photons to a preferential area of each water molecule. This biases the motion that would otherwise be random

- making it more Newtonian-like. It leads to a directional preference for the motion of small particles. The result could be something like an electric charge in the water itself - a charge to be stripped off as a photon bundle. The bundle of light would be gathered up and intensified as the water passes through an optical grid. The light would then flow through optical cable, to be distributes for use directly by customers - electricity would not be involved.

## Conclusion

Think about the difference in technology that took place between, say 1910 and 1950. From horses to aeroplanes. From telegraph to telephone. From DC to AC. Why has there not been a comparable change in the technology of water and energy during the past forty years? Many papers on the program for this conference have titles that are similar to the titles that were common at similar conferences in the 1960's - almost forty years ago. No doubt there have been advances in these topics. But what are the new technologies in the water resources field? What are the new energy production technologies?

Two world wars opened up the floodgates of invention. The military needs of the nations put a premium on inventiveness, and on taking risks. Perhaps there are less destructive ways to move our civilization ahead. We are learning to live in a competitive global economy. The world is full of tourists who carry home ideas, as well as memories. Today the words to watch are deregulation and privatisation. The economic system is struggling to work loose from the stifling cocoon of government, and from the status quo monopolistic control of industry. Perhaps deregulation will provide more stimulus for invention, and initiative, in science and engineering.

## Acknowledgements

Alan Livingstone of Central Maine Power Company contributed the idea about the wireless society, and suggested other improvements. Jery Stedinger of Cornell University pointed out where facts should be clarified more clearly from fiction. Tracey Kenward and Jennifer Ballester of Charles Howard & Associates, Ltd., corrected errors in the manuscript.

## Bibliography

1997 The Economist, *An Engine Made of Water*, p74; *Cool Coils*, p75; April 5th-11th.
1996 Stedinger, J.R; Howard, C.D.D., *Control Room III: Finding Work for Humans*, Hydro Review, August.
1995 Stedinger, J.R; and Howard, C.D.D., *The Control Room of the Future II*, Hydro Review, July.
1993 Stedinger, J.R; Howard, C.D.D., *A Look at the Control Room of the Not-Too-Distant Future*, Hydro Review, August.
1992 *Repowering Hydro: The Renewable Energy Technology for the 21st Century, A North American Hydroelectric Research & Development Forum,* HCI Publications/ Hydro Review, Kansas City, Missouri, September, 1992

# HYDRAULIC STUDY OF SECONDARY STILLING BASINS

TABAN SOWLATI
University of Windsor, ON, Canada, N9B 3P4
J.A. McCORQUODALE
University of New Orleans, LA, 70148
A.F. SMITH and R. MOULTON
Grand River Conservation Authority, Cambridge, ON, N1R 5W6
PETER RAE
ACRES International, Niagara Falls, ON

## ABSTRACT

In spillways with low tailwater levels, the hydraulic jump may not occur on the protected river bed which could result in severe local scour and could endanger the safety of the dam. A traditional solution is to place crushed rock in the scour hole in an attempt to prevent further scour; however, this method is not a permanent solution.

In this research the effect, on the scour profile, of installing a secondary stilling basin downstream of an inadequate primary basin is investigated. The primary stilling basin is retained while the secondary stilling basin (S.S.B) provides for the additional energy dissipation required to prevent excessive scouring in the river bed. This study considered the sensitivity of the scour profile to the tailwater level and end sill slope in the presence of the S.S.B. It was found that a flatter end sill reduced the plunging effect of the flow leaving the S.S.B and reduced the scour depth.

## INTRODUCTION

Mohamed and McCorquodale (1992) found that the short-term scour depth is related to flow regime or type of hydraulic jump that dominates the flow in the scour hole. The experimental studies revealed that the flow during the filling of an empty channel had at least seven distinct regimes or phases, namely: 1) Jet attached to the bed; 2) Breaking wave and adverse hydraulic jump; 3) Unstable moving hydraulic jump on positive slope with a plunging jet; 4) Wave jump and diving jet; 5) Surface jet with entrainment from below (inverted jump); 6) Plunging hydraulic jump (B-jump); 7) Normal free hydraulic jump. The plunging jumps produced the most hazardous scour profiles. Breusers (1966), Dietz (1969) and Zanke (1978) based on experiments using a scale model with small Froude numbers, distinguished four phases in the development of a scour hole: an initial phase, a development phase, a stabilization phase and an equilibrium phase. Veronese (1937) found for plunging jets $y_s + y_t = 3.68 H^{0.225} q^{0.54} \overline{D}^{-0.42}$ for 9 mm $\leq \overline{D} \leq$ 36 mm and q = 10 to 70 L/s. Li (1955) investigated scour below a submerged sluice gate and Ghetti and Zanovello (1954) did the same for a spillway. Both found the scour hole shape to be independent of flow velocity and bed material.

The purpose of this model study was to seek an optimum end sill slope that minimizes the bed scour while preventing undercutting of the end sill. The Shand Dam model was used as a study case.

## EXPERIMENTAL SETUP

A 1:49 scale model of the Shand Dam was constructed in a 4.9 m long by 3.0 m wide by 0.8 m deep basin in the Hydraulic Engineering Laboratory at the University of Windsor. The model consisted of the following parts: a) forebay, b) spillway, c) primary stilling basin, d) secondary stilling basin, and e) a portion of the river downstream of the structure (Figure 1). With the gates fully open, water was delivered to the spillway and through the secondary stilling basin. Experimental work was carried out for various flows (39, 49, 59 and 68 L/s which are 1:100 year, 1:500 year, Regional and 1:10000 year flood, respectively) and end sill slopes (1V:2H, 1V:5H and 1V:6H) in the secondary stilling basin. The model was also run at 141 L/s which represents the PMF.

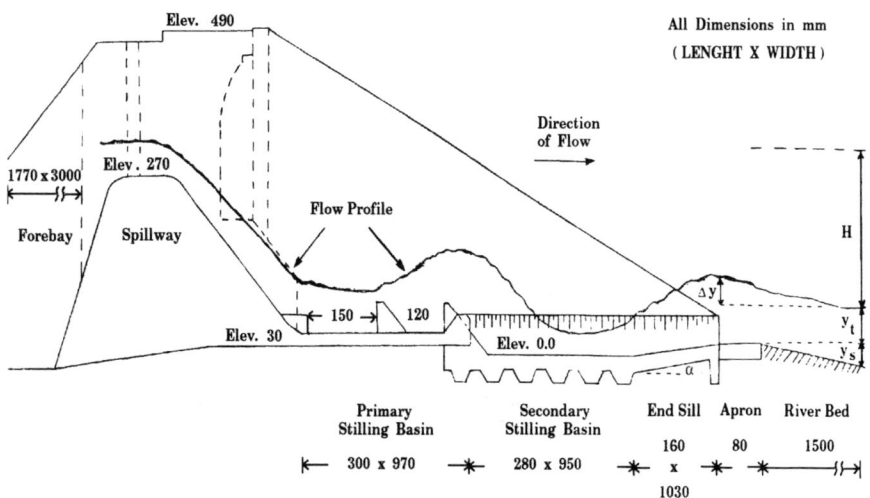

**Figure 1** Cross Section View of Scale Model
( model scale = 1:49 )

All the scour tests were conducted for a standard duration of 30 minutes to study short-term and the initial stages of the long-term scour. The bed material, 3 mm marble, was geometrically scaled so that a rough boundary fully turbulent flow existed. During the test run, scour depths were measured manually at three points. A potentiometer gauge was used, after completion of each test, to measure scour depths for 56 points downstream of the secondary basin.

## EXPERIMENTAL RESULTS AND ANALYSIS

### I. EFFECT OF END SILL SLOPE

MATLAB® program PC version 4.0 was used to process raw data and produce three-dimensional (3-D) plots of the scour profiles. Since input data were discrete measurements on a two dimensional grid, bi-cubic interpolation was applied to fit the scour contours.

Figures 2(a) and (b) show respectively 3-D scour patterns for different end sill slopes of 1V:2H and 1V:6H. The potential scour due to the 1V:2H (Figure 2(a)) is very high and puts the structure at risk; this risk increased with the flow rate. At the 1:10000 year flood, the scour depth was so high that the training walls were endangered and the secondary stilling basin acted like a flip bucket causing the exiting jet to follow a rising and then a plunging trajectory. In addition, the secondary basin with the 1V:2H slope was ineffective in reducing the kinetic energy. As a result, a hydraulic jump occurred on the unprotected bed and caused a large scour hole. The model with an end sill of 1V:6H produced a very low scour depth with practically no undercutting of the end sill (Figure 2(b)). The maximum scour depth was reduced by 85% compared to scour produced in end sill 1V:2H for the Regional flood. In this arrangement, the secondary hydraulic jump was completely contained by the secondary stilling basin. There was no plunging jet and no in-river hydraulic jump for flows up to the Regional flood.

### II. EFFECT OF TAILWATER LEVEL (TWL)

In order to investigate the effect of high TWL, the TWL was raised by 20 mm above the natural TWL (119 mm above the apron at the Regional flood). The lowest TWL was set to produce critical water depth on the river bed (90 mm above apron for Regional flood). The highest TWL resulted in the scour depths being reduced by 28% and 39% for end sill slopes 1V:5H and 1V:2H (Figures 3(a) and (b)), respectively.

Decreasing the TWL caused the maximum scour depth to increase by 45% and 15% for end sill slopes of 1V:5H and 1V:2H (Figures 3(a) and (b)), respectively. The studies on the effect of TWL showed that the difference between water level above the end sill and the TWL, $\Delta y$, is an important parameter (Figure 1). As the TWL increases, $\Delta y$ decreases; as a result, the plunging jet has lower potential energy and causes lower scour at high TWL. However, at low TWL, because of higher $\Delta y$ the jet has more energy to dissipate which increases the scour depths.

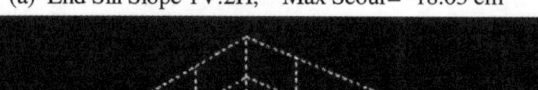
(a) End Sill Slope 1V:2H,   Max Scour= -18.03 cm

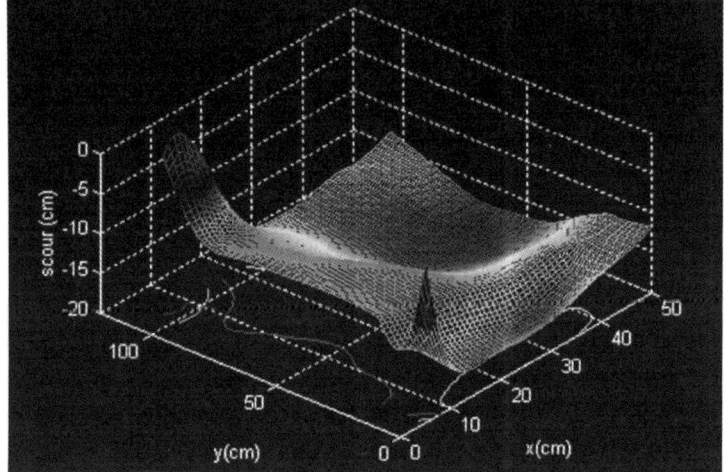

(b) End Sill Slope 1V:6H,   Max Scour= -5.68 cm

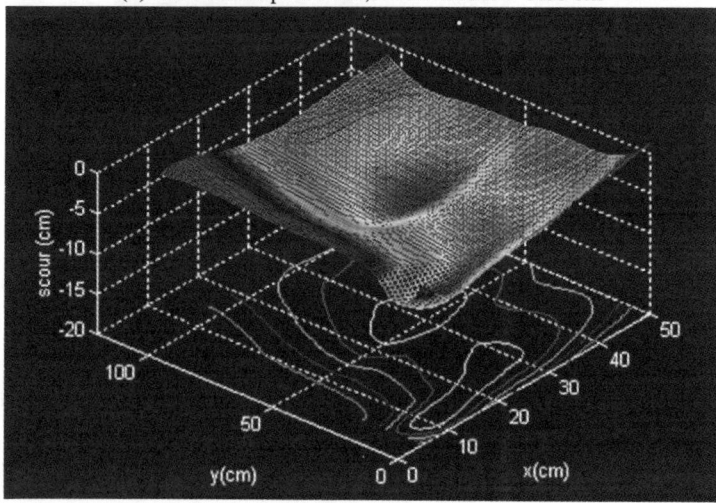

**Figure 2**  Scour Pattern for Regional Flood

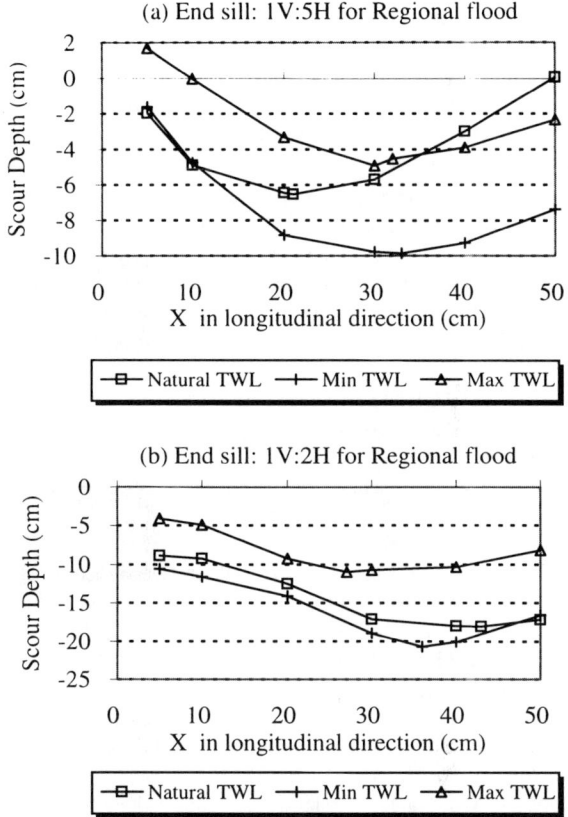

Figure 3 TWL Effect on the Scour Profiles for Regional Flood

GENERALIZATIONS

A dimensional analysis of the variables that control the local scour in this study yielded:

$$\frac{y_s}{H} = f_5(\alpha, F_r, \frac{y_t}{H}, \frac{y_*}{H}) \qquad (1)$$

where $y_* =$ (TWL - Floor Elev. of S.S.B). Multiple non-linear-regression analysis was applied to obtain:

$$\frac{y_s + y_t}{H} = 5.53 F_{r*}^{0.24} \alpha^{0.54} (\frac{y_*}{H})^{0.25} \qquad (2)$$

where $y_t =$ tailwater depth (m); H = [forebay water level - tailwater level (m)] as shown in

Figure 1; $F_r = \dfrac{q}{\sqrt{gHH}}$ and q= [(total discharge) / mean width of S.S.B]; α = end sill angle (Radian) for 0.165 rad < α < 0.463 rad.

## CONCLUSION
The experimental studies revealed that a secondary stilling basin with a low transitional end sill slope produced very small scour and very little undercutting. This improvement was due to a secondary hydraulic jump within the S.S.B, the associated dissipation of energy, a reduced angle of attack at the bed and a favourable reverse roller under the exiting jet. The scour depth was found to be inversely related to the tailwater depth and generally increased with the flow rate for all end sill slopes and TWLs. The best performance was obsereved for an end sill slope of 1V:6H.

## ACKNOWLEDGEMENTS
The model used in this study was constructed as a part of a model study funded by the Grand River Conservation Authority, Cambridge, ON, Canada. Ms. Sowlati was supported by a University of Windsor scholarship and funding from NSERC.

## REFERENCES
Breusers, H. N. C. (1966) - "Conformity and time scale in two dimensional local scour," *Proc. Symp. on model and prototype conformity*, Hydr. Res. Lab., Poona, India, 1-8.
Breusers H. N. C., Raudkivi, A. J. (1991) - "Scouring," *IAHR, Hydraulic structures design manual*, A. A. Balkema/ Rotterdam/ Brookfield.
Dietz, J. W. (1969) - "Kolkbildung im feinem oder leichter Sohlmaterialen bei stromenden Abfluss," *Mitt. Theodor Rehbock Flussbaulab.*, Karlsruhe, Germany, Heft 155, 1-122.
Ghetti, A. (1954) - "The study of bed erosion at weirs by means of small-scale models,"(in Italian) *Univ. of Padua, Inst. Of Hydraulics, Res.* Paper no. 167.
Li, Wen-Hsiung (1955) - "Criteria for similitude of scour below hydraulic structures," *Proc. 6$^{th}$ IAHR Congress*, The Hague, Paper C4.
Mohamed, M. S. and McCorquodale, J. A. (1992) - "Short-term local scour," *Journal of Hydraulic Research*, 30(5), 685-699.
Veronese, A. (1937) - "Erosion de fond en aval d'une de'charge," Univ. de Padova.
Zanke, U. (1978) - "Zusammenhange zwischen stormung und sedimenttransport," Teil 2: Berechnung des sedimenttransportes hinter sohlenstrecken, Sonderfall zweidimensionaler kolk, Heft 48, GermanyBreusers, H. N. C. (1966) - "Conformity and time scale in two dimensional local scour," *Proc. Symp. on model and prototype conformity*, Hydr. Res. Lab., Poona, India, 1-8.

# French Paper Dam Spillway Improvements

### W. JAMES MAROLD, P.E.
### Harza Engineers and Scientists, Chicago, IL, USA

## INTRODUCTION

French Paper Company constructed a Hydroelectric Power Project consisting of a 1.3 MW powerhouse, headrace canal, earth embankment, and wooden timber crib spillway in the St. Joseph river in Niles, Michigan in 1887. The project supplies electric power for operation of the paper plant on the left side of the river. A major repair was made to the wooden timber crib spillway in 1914 after a flood washed out a large portion of the wooden structure. A new concrete ogee shaped structure was placed over the timber crib structure after removal of the wood overflow and wood surfacing. This paper describes the design of the original structure, performance of the improved concrete overflow structure and the design and construction of improvements made in 1995 and 1996 to improve hydraulic operation of the structure.

## ORIGINAL SPILLWAY DESIGN

The design of the original spillway was a timber crib structure with a wooden overflow and wood plank apron. No plans or design calculations were available to determine the details of the original design. However, photographs of the major improvements made in 1914 and drilling performed in 1995 provide sufficient information to interpret most of the design details.

The 1914 photographs indicate that the wooden overflow structure was a vertical wooden timber structure about 5 to 7 feet high, with about a 3-feet thick crest overflow and a 320 to 350 feet wide crest spanning the width of the river banks. The apron was constructed of wood planking about 20 to 30 feet long over the timber crib rockfill. The planking sloped from the bottom of the wooden overflow at a slope of about 6 to 10 H:1V to the downstream edge of the spillway. A lot of riprap size stone existed on the river bed upstream of the vertical structure at the level of the crest of the overflow. This material appeared to extend across the entire width of the riverbed about 100-feet upstream of the spillway. No energy dissipation structure was apparent downstream of the timber planked spillway structure.

The 1995 drilling program consisted of two holes drilled through the crest of the concrete ogee and two holes drilled through the concrete apron. The results of this program indicated that the concrete structure was founded on a timber crib structure which contained large rocks and gravels. The bottom of the timber crib structure is 13 to 18 feet below the bottom of the concrete structure. An interpreted plan and section of the original structure is included as Exhibit 1.

## IMPROVED 1914 SPILLWAY DESIGN

The new concrete ogee spillway and apron constructed in 1914 provided a structure more resistant to erosion and damage during high river flows. Exhibit 1 provides an outline of the concrete structure after removal of the old wooden structure.

The 1995 drilling and testing program provided information on the improved concrete spillway structure. Compressive strength tests on the concrete from samples retrieved from core samples indicated a range of compressive strength from 2580 to 5640 psi with unit weights from 139 to 152 lbs./cu.ft. The timber crib members drilled through were dense strong wood, appeared to be in good condition and were not rotten or decayed. The 1914 photographs of the completed concrete spillway structure indicate that the new concrete structure was constructed over the timber crib after removal of the timber overflow and wooden planking. The large riprap size rock was left in place on the upstream side of the structure.

A blanket of gravel material was also placed over the river bottom on the downstream of the concrete apron. This blanket was constructed at the same level as the downstream edge of the new concrete apron and extended across the entire river bottom for a distance of about 30-feet downstream of the spillway. This material appeared to have a maximum grain size of 3 to 6-inches. The purpose of this blanket may have been to reduce undercutting of the downstream spillway edge during high flows.

# FRENCH PAPER DAM SPILLWAY IMPROVEMENTS

Flash boards were added to the crest of the new concrete structure to raise the river reservoir level about 2-feet higher than previously. The flash boards may have been added to raise the reservoir level for higher net water head on the hydroelectric plant or to raise the crest of the spillway to produce a higher discharge coefficient to pass higher flows.

## PERFORMANCE OF THE IMPROVED 1914 STRUCTURE

The improved concrete timber crib spillway performed well with apparently minor repairs until 1974. Repairs were made to the spillway structure in the fall of 1974 to fill undermined voids below the spillway and to replace some broken out concrete in the surface of the spillway apron. The gravel blanket on the downstream side of the concrete apron had apparently been washed away by this time and undermining of the apron had occurred. The undermining had progressed to a point very near the upstream face of the concrete ogee. Large rocks were required to reduce underseepage flow to allow placement of grout bags to fill the voids below the bottom of the apron. These repairs were made and the structure performed well until 1994. In October 1994, a diving and visual inspection of the downstream edge of the spillway revealed that further undermining had occurred and that some underseepage was occurring at the location of the largest undermined section repaired in 1974. A description of the design studies and construction performed to improve the concrete spillway structure to prevent future undermining and reduce underseepage is presented below.

## DESIGN STUDIES TO REDUCE UNDERMINING

Hydraulic information presented in the 1979 National Dam Safety Program (NDSP) Inspection Report provided a headwater-tailwater rating curve as well as information on historic peak discharges over the spillway. Based on this information, a slightly revised rating curve was developed to reflect a higher discharge coefficient (3.1 versus 2.3) of the ogee spillway geometry than was apparently used to calculate the NDSP rating curve. The new curve assumes that the 2.25 foot high Flash boards trip at about 2784 cfs which is slightly higher than the estimated normal spillway flow of about 2115 cfs. According to plant personnel, the Flash boards usually trip in the spring and fall as a result of log debris moving down the river during flows above normal flows.

The estimated 100-year flood discharge in the NDSP report was 22,500 cfs and the maximum spillway capacity was estimated to be 38,500 cfs. Both these estimates assumed that the Flash boards were completely tripped and the flow discharged over the spillway ogee surface. The Probable Maximum Flood (PMF) was estimated in the NDSP report at 77,100 cfs. The record maximum flow occurred in April 1950 and was estimated at 20,200 cfs.

Computations were then prepared to determine if flows ranging from the estimated normal flow to the spillway capacity would form a hydraulic jump on the surface of the existing apron. These computations indicated that the flow would "sweep- out" and not form a hydraulic jump on the existing apron for all flows. Flow velocities were calculated to vary from 28 to 37 feet per second with depth of flow at the apron varying from 0.2 to 2.6 ft.

Computations were then made to determine the riprap sizes required to resist the sweep-out velocities calculated above to prevent future undermining. These computations indicated that a 50 percent stone diameter between 6 to 10 feet would be required to resist velocities generated by the 100-year and spillway capacity flows. These sizes were judged to be impractical and other methods were evaluated to reduce the high sweepout velocities. Use of energy dissipation blocks and an endsill were then considered. The combined use of an endsill and energy dissipation blocks would reduce the velocities to about 7 to 8 feet per second ,while the use of an endsill only would reduce velocities to about 10 to 12 feet per second for flows up to 22,500 cfs (100-year flood). The use of an endsill combined with riprap was then tentatively selected as the best alternative design. The fifty percent riprap size required to resist the 12 feet per second velocity was calculated to be 24 inches.

Subsequent structural stability calculations indicated that the addition of an endsill would reduce the factor of safety against overturning and sliding because of the higher uplift and impact loading produced by the endsill. As a result of this, the endsill was eliminated and the riprap was designed to resist the velocities of the existing sweepout flow. Instead of providing riprap as large as the calculated sizes indicated above, the individual stones were designed to be grouted with 4000 psi tremie concrete to produce a solid mass of concreted riprap. The solid mass will be able to resist movement during the spillway capacity flow. In order to relieve possible uplift pressures behind the concreted riprap, 6-inch diameter drain pipes installed at 20-foot intervals through the concreted riprap into the supporting bedding material were provided. The integrity of the concreted riprap will require inspection annually

FRENCH PAPER DAM SPILLWAY IMPROVEMENTS 19

for signs of deterioration to verify that the concreted riprap remains intact so that further downstream scour is minimized. This design provided the flexibility to construct the work in the river, without a cofferdam.

## CONSTRUCTION OF THE DESIGNED IMPROVEMENTS

The designed improvements indicated above were constructed in two construction seasons(1995 and 1996) to allow sufficient time to complete the work in the limited low river flow period which typically extends from August to October.

The improvements to prevent further scour below the bottom of the spillway were constructed in the 1995 to reduce the potential for further undermining and possible complete failure of the structure. These improvements consisted of placement of grout bags within the undermined voids below the bottom of the structure, grouting between the bags and placement of the bedding and riprap materials and grouting the riprap together to form the large grouted riprap mass to resist the high sweepout scour velocities. One longitudinal construction joint and one significant and several minor transverse cracks were grouted with non-shrink grout.

Approximately 95 percent of this work was completed prior to resumption of higher flows in late October 1995. About 100-feet of riprap was left ungrouted until the next construction season. It was recognized that spring runoff may move or wash away this section of the ungrouted riprap. The ungrouted riprap performed well throughout the spring runoff until a high intensity 6-inch rain produced about 7,500 cfs flow over the spillway and the ungrouted riprap was washed out. Flow velocities at this discharge were calculated to be 31 feet per second which is much higher than the 12 feet per second velocity required to move the 24-inch sized riprap. The depth of flow over the apron was calculated at about 1 foot which is deep enough to move the riprap.

The 1996 construction season concentrated on placement of an HDPE lining (30 mils thick) over the river bottom upstream of the concrete spillway crest for a distance of 150 to 170-feet. The HDPE was placed over a thin 2-inch thick bedding of pea gravel and covered by a 2-foot thick layer of quarried natural river gravel (6-inch to 1-inch size) to weight the HDPE down and resist erosion. After completion of the HDPE lining, the remaining section of riprap and bedding material lost in the spring flood was replaced and grouted.

20 ENERGY AND WATER

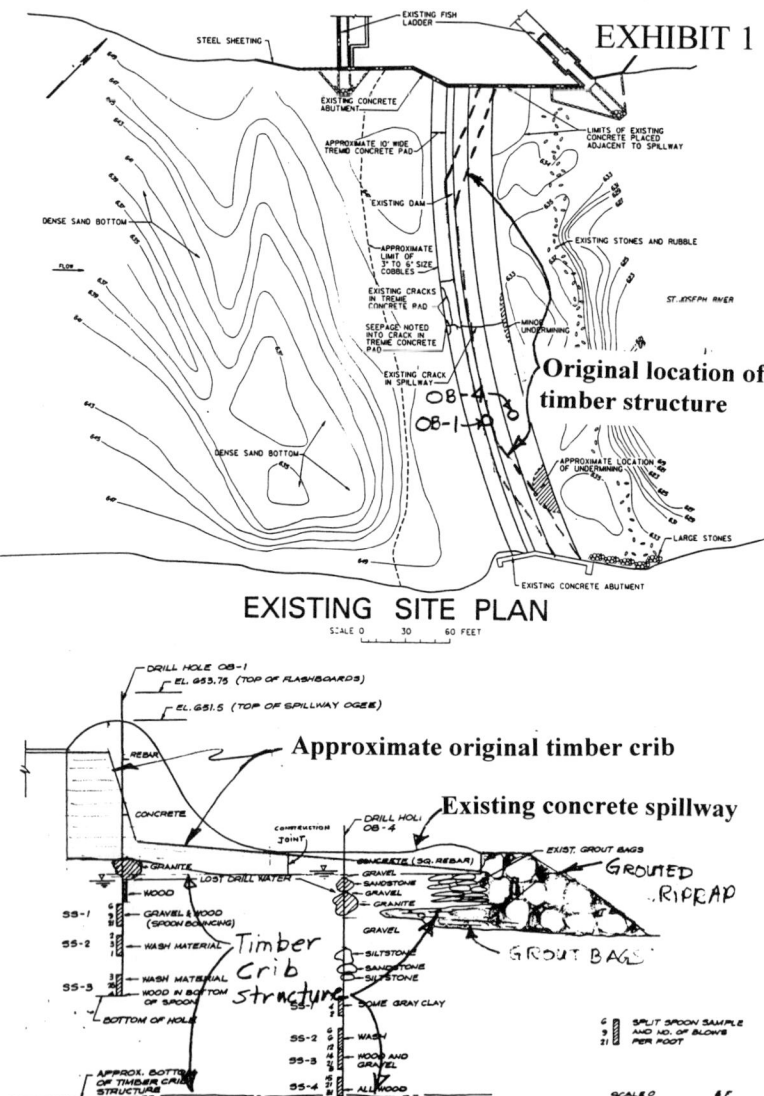

EXHIBIT 1

EXISTING SITE PLAN

Profile at location of OB-1 and OB-4

# HUBBART DAM OUTLET WORKS REHABILITATION

STEVE BAREIS, LEE GERBIG
BUREAU OF RECLAMATION, WOODWARD-CLYDE CONSULTANTS
DENVER, COLORADO, USA

A. BACKGROUND

Hubbart Dam is located in northwestern Montana, about 25 miles southwest of Kalispell. The dam is a variable-radius concrete arch dam with a structural height of 89 feet. The dam is owned and operated by the U.S. Bureau of Indian Affairs, Flathead Agency Irrigation Division (FAID). The dam was constructed in 1923 across the Little Bitterroot River, and has a storage capacity of approximately 12,000 acre-feet. The primary purpose for the dam is to store water for irrigation of portions of the Flathead Indian Reservation.

An ungated free-overflow spillway, is located in the center of the crest. The outlet works extend through the dam near the right (west) abutment. The outlet works consisted of a trashracked intake structure, two 30- by 42-inch emergency slide gates, two 30-inch diameter cast iron conduits through the dam, and two 24-inch balanced needle valves. The needle valves, bypass piping, and controls were enclosed in a reinforced concrete gatehouse attached to the downstream face of the dam, and founded on bedrock.

In 1986, an investigations team from USBR had examined the needle valves and recommended their replacement because of operating difficulties and safety concerns. The investigation followed fatal incidents in 1984 with water-operated needle valves at Bartlett Dam, near Phoenix, Arizona and at a private powerplant in Utah. Subsequently, all USBR designed needle valves were investigated, including the valves at Hubbart Dam.

The Bureau of Indian Affairs made funding available to the Confederated Salish and Kootenai Tribes of the Flathead Nation (CSKT) in 1995, for use in correcting O&M deficiencies on dams included under CSKT's Safety of Dams Program. In April

1995, the Tribes contracted with Woodward-Clyde Consultants of Denver, Colorado to provide designs and specifications for furnishing jet-flow gates to replace the existing needle valves and to rehabilitate the gatehouse. Woodward-Clyde Consultants was selected as the engineering firm because of their experience with replacement of needle valves at other facilities.

B. MECHANICAL DESIGN

A site visit was conducted in March 1995 by Woodward-Clyde engineers, to examine the facilities and discuss various options with the CSKT Safety of Dams Office. The needle valves and gatehouse were in very deteriorated condition and our first discussions centered on the choices of gatehouse valves. There are three choices available for an outlet works operating at 85 feet of head; the fixed-cone valve (with or without a hood), the jet-flow gate (designed by USBR and now available from commercial suppliers), and the clamshell gate (a recent USBR design). Several items taken into consideration were the existing stilling basin, access to the gatehouse, inclement weather, and the remote location and vandalism.

1. The fixed-cone valve, also known as the Howell-Bunger valve, has been used on outlet works of this type since the early 1940's. The fixed-cone valve without a hood has a discharge coefficient of 0.85, and a very wide discharge pattern, requiring a large stilling basin. There is a lot of spray associated with the wide discharge pattern, making winter releases very undesirable. The fixed-cone valve with a hood attached to the moveable sleeve, has a reduced coefficient of discharge between 0.68 and 0.79.

2. The jet-flow gate is essentially a bonnetted slide gate with an orifice and an enlarged discharge pipe, and is an excellent regulating gate. The jet-flow gate has a discharge coefficient of 0.85, and a confined, well aerated, discharge pattern.

3. The clamshell gate is a two-piece gate that rotates over the curved end of the pipe, with a discharge coefficient of nearly 1.00, and does not dissipate energy very well. Low-flow characteristics are a problem, due to the seal design and the wide opening when nearly closed.

It was decided to replace the needle valves with jet-flow gates for the following reasons: (1) The clamshell gate would require a specific design, which would be quite costly, and the stilling basin would possibly require enlargement. (2) The fixed-cone valve without a hood produced too much spray and the discharge cone would carry over to the gatehouse access road alongside the stilling basin. (3) The

# DAM OUTLET WORKS REHABILITATION 23

fixed-cone valve with a hood would be more difficult to install inside the valve house for protection from vandalism, and it would be larger than the jet-flow gate. The jet-flow gate met all the design criteria, with a confined discharge pattern, good low-flow control, good energy dissipation, cavitation free at all openings, excellent discharge capacity, and the ability to be easily installed inside the gatehouse for vandalism and weather protection.

The sizing of the jet-flow gate was very important to meet the existing discharge capacity required for the size of the reservoir and irrigation needs. For the existing needle valves, the flow coefficient C, ranges from O.55 to 0.60. Therefore, since C = 0.84 for the jet-flow gate, the cross sectional area required for matching the flow calculates to a 20.28 inch diameter, measured at the orifice. Twenty-inch jet-flow gates should closely match the existing discharge capacity, and as a matter of interest, the fixed-cone valve with a hood would be 22 inches in diameter.

The existing needle valves were installed at an angle of 15 degrees above the horizontal. The only benefit to this arrangement is to perhaps shorten the stilling basin, by allowing the discharged water to plunge more. It was decided to install the jet-flow gates horizontally, which reduced the flow losses, and made installation easier. Since the discharge pattern is similar, the stilling basin was not modified.

The reuse of original piping required the installation of 30-inch by 24-inch concentric reducers upstream of the jet-flow gates, since the gates are made with a slightly larger inlet diameter for the orifice design to function properly. The jet-flow gates were designed with 30-inch discharge pipes to admit air into the valve and discharge jet, to eliminate cavitation. The discharge pipes were fabricated of stainless steel to eliminate any painting requirements and prevent corrosion. A dielectric gasket set was used between the jet-flow gate and the discharge pipes to prevent galvanic corrosion. With the jet-flow gates located inside the gatehouse, the discharge pipes pass through the downstream wall. The openings in the wall were provided with oversized stainless steel sleeves, and rubber "Link-Seal" wall penetration seals were installed in the space between the sleeves and the discharge pipes. This eliminated any need for expansion couplings in the conduits, and makes future removal of the gates possible.

It was decided to operate the jet-flow gates with a manually-operated hydraulic system, since there was no electrical power available at the dam. The hydraulic operation provides the capability for future remote control. The control panel was mounted on the downstream wall of the gatehouse, between the gates, which provides a convenient location for the operating personnel to observe the flow and the gate position from one place.

## C. GATEHOUSE DESIGN

The existing gatehouse was in very deteriorated condition, and it provided very cramped working areas for the needle valves. The layout of the house was not conducive to the installation of the new gates, and access was troublesome. It was decided to demolish the existing gatehouse and build a new one. The new house was designed for better access and easier future maintenance of the mechanical equipment. The gatehouse was designed as a two story, cast-in-place, reinforced concrete structure, with the entrance door located on the upper level, nearly level with the access road and parking area. The jet-flow gates were to be installed in the lower level, with the hydraulic cylinders rising slightly above the upper level grating. An access hatch would be provided in the roof, so that mechanical equipment could be handled with a mobile crane in the parking area. The parking area would be enlarged by building a retaining wall along the stilling basin and placing fill behind the wall. Another retaining wall would protect the walkway to the gatehouse entrance, and concrete walls around the entrance would protect the door from weather and sliding material on the abutment.

The gatehouse was designed to provide maximum vandalism protection. An oil skimmer, located at the base of the downstream wall will contain any oil spills inside the house. The new gatehouse would be doweled to the dam face, and the concrete roof and walls would be keyed into the dam face. A new concrete floor would provide a better surface for installing mechanical equipment, and provide better drainage of any leakage that may occur.

## D. GATEHOUSE CONSTRUCTION

The CSKT contract for sitework, demolition, and the construction of the gatehouse and appurtenances contained provisions to allow CSKT to install the jet-flow gates while exterior work, such as the retaining walls and earthwork, was being completed by the site Contractor.

The old gatehouse was removed by making full-depth sawcuts through the roof and walls next to the dam face, and at other locations to reduce the concrete into manageable sections. Around the needle valves, concrete was removed with jackhammers, and the valves were lifted out through the downstream walls. Gate valves, elbows, and fittings in the filling and air vent lines were salvaged for re-use with the new gates.

DAM OUTLET WORKS REHABILITATION 25

Following removal of the old concrete, crews began to sawcut and excavate keyways into the dam face, and to drill and grout double rows of no. 6 dowel bars. As the Contractor removed the floor concrete, it was evident that there was not sufficient grade to carry flows from the house to the stilling basin. The river channel was deepened, which allowed positive flow away from the structure.

The Contractor placed the gatehouse walls in one lift, 22-feet high. This required design of the forms by a professional engineer, and special constraints on the speed of placement. Also, the forming for the walls was complicated because of the number of block-outs required.

The specifications recognized the remoteness of the site, and allowed for the use of stabilizing admixtures to slow the set time of the concrete. Concrete was typically four hours old when placed in the forms, with no loss of workability. Twenty-eight day compressive strengths averaged 4,900 psi, with no concrete wasted on this project. By the end of April, the main gatehouse placements were completed, and the retaining walls were under construction. Installation of the gates began on May 4, about three weeks later than originally scheduled. Installation of metalwork, backfill of retaining walls, and reclaiming the site were done concurrently with the gate installations. The sitework was substantially complete on May 17. The Contractor was Noble Excavating Inc., of Libby, Montana.

## E. GATE INSTALLATION

Hilton Valve, Inc., of Redmond, WA fabricated the two, 20-inch diameter jet-flow gates, reducer sections, discharge pipes, and a hydraulic control panel, which were delivered on January 2, 1996.

In early May, a CSKT crew, supervised by a consultant who specializes in hydraulic equipment, began installing valves and piping for the filling line assembly. The filling line is a gated 8-inch pipe which is used to fill either of the 30-inch conduits prior to opening of guard gates. New fittings and air-vacuum valves were assembled on the air vent above each conduit. The air-vacuum valves will allow air to bleed off during filling of the conduits, but will seal when the system is full.

After 30-inch by 24-inch reducer sections had been bolted onto the conduits, the gate bodies were set so the slab could be marked for anchor bolts. Anchor bolts were epoxied into the floor through the gate supports, and later the voids beneath the gate supports were filled with non-shrink grout. The stainless steel discharge pipes were assembled to the gate bodies using dielectric gasket sets. Once the discharge pipes were in place, the Contractor began forming the block-outs in the wall of the gatehouse for second-stage concrete. Stainless steel sleeves were cast into the concrete around each pipe, with a wall penetration seal placed in the annulus between the pipe and sleeve. This will allow for easier removal of the gates in the future.

The gate control panel, which consisted of an hydraulic oil reservoir, two hand-pumps, and a series of control valves, was mounted on the downstream gatehouse wall, on the upper level. Hydraulic hose was routed to each gate, and the gates were exercised in the dry. Both gates were successfully tested by May 23, 1996, and put into operation in June. During the remainder of the year, the CSKT routed the air vent piping outside the gatehouse, installed counterbalancing valves and bypass lines in the hydraulic piping, and put covers over the ends of the discharge pipes.

## F. CONCLUSIONS

The outlet works has performed very well through the first irrigation season. Minor modifications to the stilling basin will be required, because of a slightly different discharge pattern. Some loose rock in the basin formed a small dam, which will be removed and used for riprap along the bank of the basin.

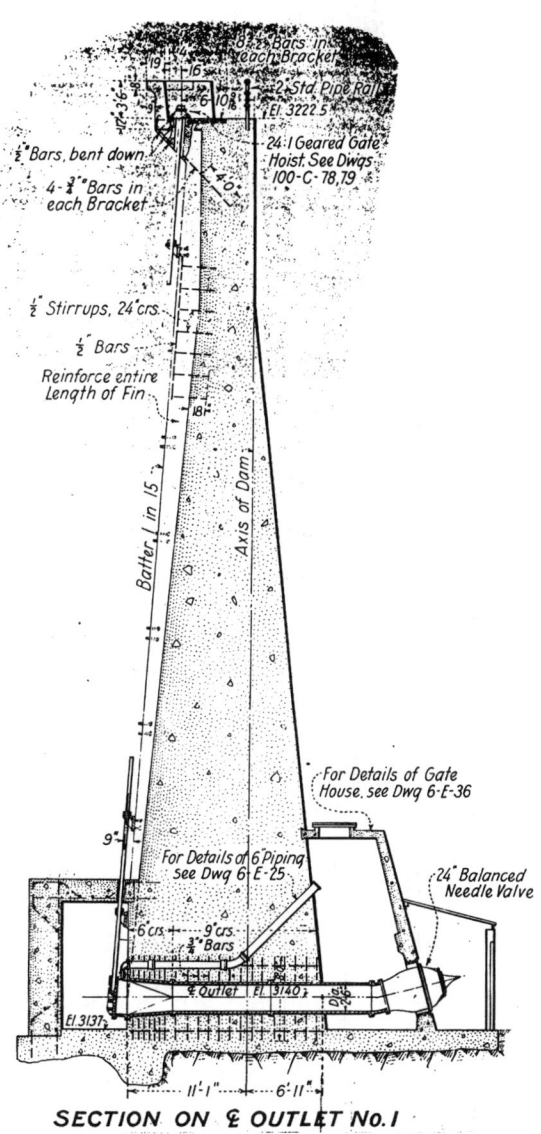

SECTION ON ℄ OUTLET No. I

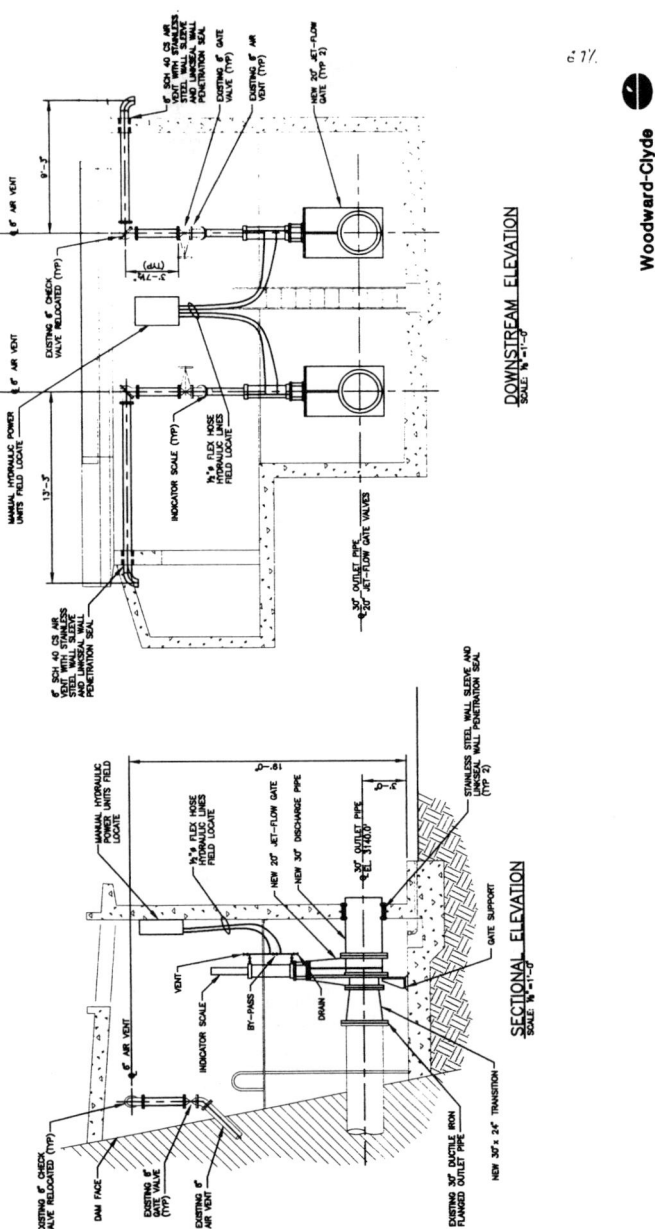

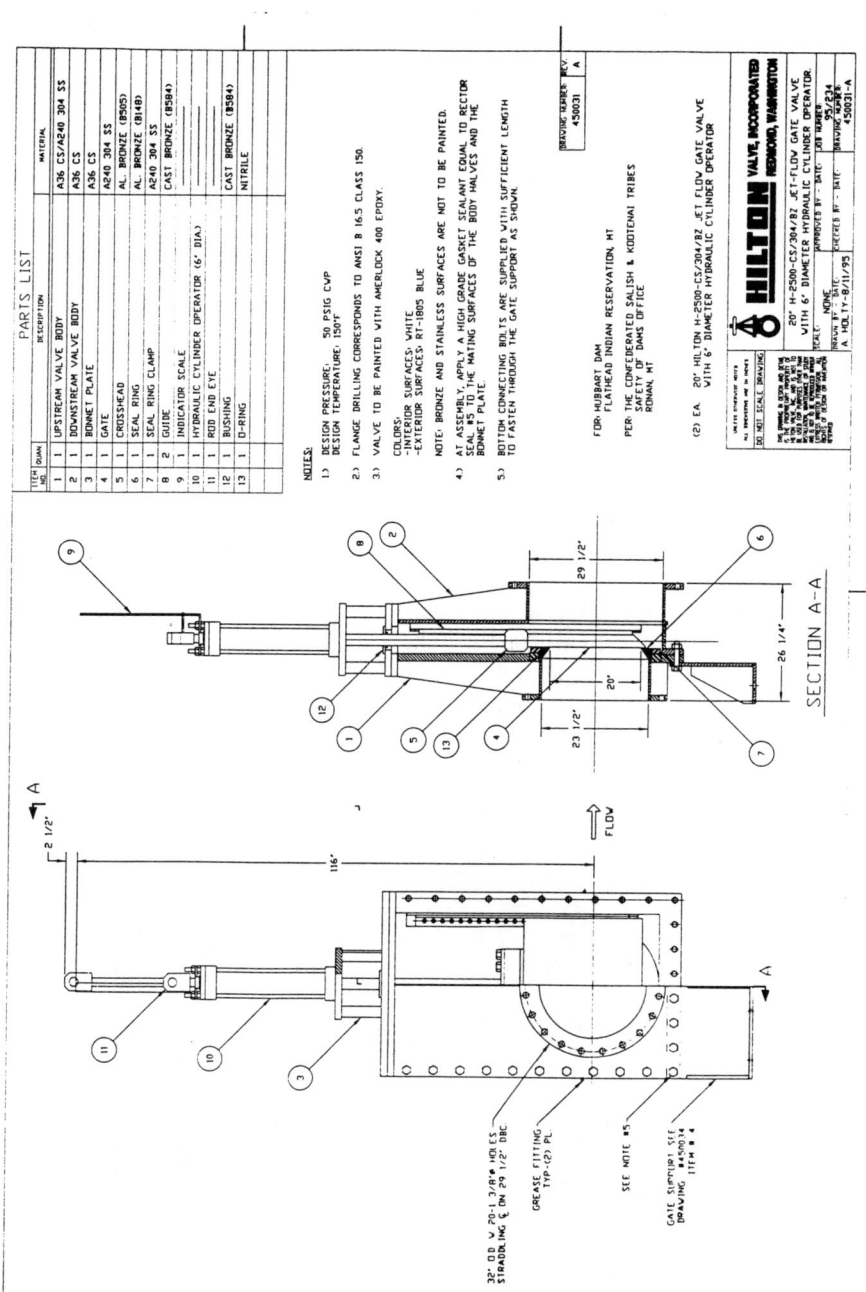

# FRIANT DAM SPILLWAY REHABILITATION

BRUCE C. MULLER, JR.
United States Bureau of Reclamation, Denver, Colorado, USA

Friant Dam is a principal storage feature of the Central Valley Project and is located on the San Joaquin River approximately 20 miles north of Fresno, California. The dam is a concrete gravity structure with a structural height of 319 feet and a crest length of 3,488 feet. The dam was completed in 1942, and reservoir filling commenced in February 1944. The dam serves many purposes including storage and diversion of water for irrigation, flood control, power generation, and recreation. The spillway consists of an ogee-shaped overflow section, chute and stilling basin at the center of the dam. Discharges are controlled by three 100-foot-wide by 18-foot-high drum gates. The crest is at elevation 560 feet and the top of the gates, when raised (closed), is at elevation 578 feet. The capacity of the spillway at normal reservoir elevation 578 feet is 83,000 ft$^3$/s.

PERFORMANCE MONITORING

Friant Dam has been monitored on a regular basis since its completion to ensure that the dam continues to function properly. From the time of completion through the operations and maintenance review inspection on May 1, 1965, the concrete was considered to be in good to excellent condition. However, the inspection on November 1, 1968 identified substantial cracking in the crest of the dam. Near the spillway, the contraction joint between blocks 35 and 36 had opened 1/4 inch and had both horizontal and vertical differential movement. By 1974, the expansion was recognized as an ongoing phenomena and plans for measuring the contraction joint movements between the gravity section blocks and spillway blocks were developed. At the same time, plans were also made for measuring the clearances between spillway piers and the drum gates. In reviewing the construction records of cement and aggregate sources, the movements and cracking of the concrete in the spillway were concluded to be the result of alkali aggregate reaction (AAR). It was also concluded that the volumetric expansion of the concrete due to the reaction were only exhibited in the crest of the dam and appurtenant structures because the mass concrete of the dam contained a pumicite pozzolan which tends to suppress the reaction. The pozzolan was not used in the dam crest or the appurtenant structures because the concrete would not have been able to meet the 28-day strength requirements.

The movement of the concrete in block 42 has been monitored by a plumb line. The data from this plumb line confirms the initiation of movement in approximately 1966. The chemical byproducts of AAR in the existing concrete result in the build-up of internal pressure, volumetric expansion and cracking. Since the lower portions of the spillway concrete are constrained by the mass concrete of the dam, the expansion tends to be limited to the upper portions of blocks 36 through 42. The differential expansion of the blocks and piers leads to the pier plates, which provide the sealing surfaces for the sides of the drum gates, being out of plumb. The 1996 measurements of the pier plates show them to be out of plumb by more than 2 inches on the outside piers (blocks 36 and 42) and by ½ to 3/4 inches on the inner piers.

## OPERATIONS HISTORY

The existing drum gates are normally operated in fully raised (closed) or fully lowered (open) positions. The gates are normally raised in the winter or spring when the reservoir reaches an elevation sufficient to fill the drum gate chamber and float the gates into the raised position. The gates remain in this position until the end of the irrigation season when the reservoir falls below 560 feet and the gates are lowered.

The only significant events in the operation history have been the removal of 1 inch of each of the drum gate seal guards in the late 1970's to allow for concrete expansion, and more extensive modification of the seals in 1986.

## ALTERNATIVES CONSIDERED

As it became apparent that a permanent modification would be necessary to allow continued operation of the gates, several alternatives were considered ranging from a redesign of the seals on the existing drum gates to reconstruction of the spillway. The alternatives considered included:

- **Drum gate rehabilitation** - This alternative would have modified the existing drum gate to better accommodate the movements of the concrete by replacing the existing seals with adjustable seals. While this would be a relatively low cost alternative, continued movement of the structure would have impacted the ability of the gate to function as designed. Regular adjustments of the seals would be necessary, and the deflections could reach a point that the seals would not be able to flex sufficiently as the gates were raised and lowered.

- **Cutting expansion slots in the dam** - One method of accommodating the expansion of alkali reactive concrete has been to cut expansion slots in the existing concrete. The slots would relieve stresses and allow continued expansion of the concrete without all of the movement in blocks 36 and 42 being toward the drum gates.

- **Pier plate reconstruction** - The existing pier plates would be removed and new pier plates would be installed in the original locations.

- **Gate replacement** - The existing gates would be replaced with new gates capable of adapting to the current and future movements of the concrete.

- **Reconstruction of the spillway** - All potentially reactive concrete would be removed and replaced. The drum gates would be removed and replaced with radial gates since they would have to be dismantled to facilitate construction.

After evaluating each alternative based on several factors including cost, expected life of the modifications, confidence in the alternative, improvement in gate operation, resistance to vandalism, ease of gate operation, and resistance to seismic loads, it was decided to rehabilitate the center gate and replace the outer two gates. This hybrid alternative was selected after it was determined that gate rehabilitation was the most cost effective alternative, however the potential deflections at blocks 36 and 42 were considered to be too great to develop an effective sealing system for the outer drum gates.

## DESIGN CONSIDERATIONS

The two types of flow control features considered for replacing the outer drum gates included an inflatable rubber dam and a crest gate. Both types of features require modifications to the existing concrete crest structure in order to accommodate anchorage to the existing dam.

Hydraulic modeling of the rubber dam was performed by Reclamation's Water Resources Research Laboratory to determine the ability of the rubber dam to control discharges at partial gate openings and to determine discharge coefficients for a revised crest shape which would be required to accommodate the rubber dam. The crest gate was not modeled, because the gate leaf would be hinged at the top of the existing concrete cantilever similar to the existing drum gate and could be fabricated to match the existing shape of the drum gate surface which would yield the same discharge coefficients as the existing gates.

A key consideration in the design of the gate replacement options is the ability to accommodate future movement of the spillway piers. The rubber dam option naturally accommodates such movement due to the use of a flexible membrane bladder. The crest gate required that the leaf and seals be designed to prevent binding against the existing pier plates and to provide a positive seal in the raised (closed) position. The gate leaf/seal design is required to accommodate the current deflected shape of the pier plates plus an additional 6 inches of movement at blocks 36 and 42 which would be the additional deflection over the next 100 years at the current rate. In order to ensure that the seals would not require frequent adjustment by Reclamation personnel, the seals are designed to accommodate up to 2 inches of movement.

Anchorage of the replacement gates to the spillway crest was another critical design consideration. The existing drum gates are anchored every 2'-7" to the concrete crest cantilever. The design of the crest gate option includes an anchorage system which connects directly to the existing anchors in the cantilever. The rubber dam option requires an anchor upstream of the existing cantilever to provide a relatively flat area for the rubber bladder. This would require additional concrete to be added and anchored to the upstream side of the cantilever.

The rehabilitation requirements for the center drum gate include redesigned side and downstream seals along with realigning the hinges which connect the drum gate to the crest cantilever. The redesigned seals will allow the gate to function over the range of anticipated future concrete movements. The hinge realignment is necessary to correct a bow in the current hinge alignment. The movement of the hinges has led to several broken bushing shoulders and a few cracks in the hinge castings. The realignment will be accomplished by removing the hinge pins and lifting the gate to separate the mating portions of the hinges. The hinges will be loosened from the concrete cantilever anchors and adjusted to a straight alignment. The gate will then be lowered back into place and connected with new hinge pins and bushings. Several hinges will be salvaged from the demolition of the outer gates to serve as replacements in the case of hinges being damaged while being loosened from the cantilever anchors.

## OPERATIONAL CONSIDERATIONS

Several operational constraints have been considered in developing the requirements for the gate replacement and rehabilitation. These constraints address the need to maintain the basic functions of the spillway and gates both during and after construction.

A critical requirement during construction is to perform the modifications without impacting the storage or delivery of water. After examining the normal fluctuations of the reservoir and flood control constraints, a window of time for performing the modifications was identified which would minimize the risk of impacting water operations at the dam. The reservoir is generally below elevation 560 feet from late summer until February 1 when the gates are needed for flood control. Data from the past 21 years indicated that the contractor would have between four and seven months to perform the modifications. Although the work would have to be well planned, an initial schedule prepared by Reclamation indicated that it would be feasible to complete the work during this season.

A requirement to adjust discharges during large flood events for flood control requires that Reclamation be able to determine the discharges from the modified gates. The discharges from the center drum gate will continue to be determined from water surface elevation and gate position information based on discharge coefficients developed in Reclamation's original hydraulic model study of the spillway. If the outer gates are replaced with crest gates, the gate panels will be shaped to match the existing spillway crest profile which will result in essentially the same discharge rating curve as the existing drum gates. If the outer gates are replaced with rubber dams, additional features must be incorporated into the design to ensure that discharges can be determined accurately. A 1:36 scale model of the rubber dam alternative showed that discharge could be accurately determined based on the rubber dam crest elevation and discharge coefficients for crest heights of 10 feet through 18 feet. For crest heights below 10 feet, the model showed a v-notch forming in the rubber dam which yielded non-uniform discharge across the spillway bay. Since the v-notch results in an unacceptable flow condition, it would be operated such that it could be adjusted for specific discharges in the 10- to 18-foot height range and then opened completely if additional discharge were required. One of the key findings of the model study was that the height of the rubber dam crest must be determined from some form of direct measurement rather than an empirical relationship between internal bladder pressure and crest height. The hydraulic model experienced a hysteresis effect in which there could be different crest heights for the same bladder pressure depending on the operation history leading up to the desired bladder pressure.

In its role as a flood control structure, Friant Dam must be able to release water at any rate ranging from minimum flows to the full capacity of the spillway and river outlet works. While a crest gate would function similar to the existing drum gates, an operating scheme was prepared to address the need for discharges in the ranges where the rubber dam would not be able to function satisfactorily. The operating scheme incorporated the outlet works with a capacity of 16,000 ft$^3$/s and a differential opening of the center drum gate. The difference in discharge between the center drum gate and an individual rubber dam was limited to 25,000 ft$^3$/s to prevent adverse flow conditions in the stilling basin. Use of the outlet works and a differential discharge through the center bay of the spillway provides the ability to regulate the outflows from the dam at any level between no discharge and full capacity.

## CONCLUSIONS

Reclamation's programs which provide for periodic observation of the conditions of its facilities were valuable for identifying changes in the condition of the dam and initiating the process of identifying potential solutions to the alkali aggregate reaction in the spillway concrete and the resulting gate binding. By considering a number of non-traditional alternatives, Reclamation has been able to select an alternative action which allows the spillway to continue to function as designed with minimal cost to the public. Consideration of the operations history of dam and reservoir during the final design process allowed Reclamation to develop specifications for the modifications which are expected to have no significant impacts on the project beneficiaries during the course of the modifications.

Note: On January 13, 1997, a contract was awarded to Gracon Corporation in the amount of $6,962,540 for rehabilitation of the existing spillway. The contractor selected crest gates for replacement of the outer drum gates.

# OCHOCO DAM SPILLWAY MODIFICATION

DOUG STANTON
United States Bureau of Reclamation, Denver, Colorado, USA

BACKGROUND

Ochoco Dam, a feature of the Crooked River Project, is located on Ochoco Creek in central Oregon about 5 miles upstream of the city of Prineville which has a population of approximately 5,000. The dam was originally constructed around 1920 and has undergone several modifications. The purpose of the dam is to provide irrigation water for the Ochoco Irrigation District and flood control in the winter months. Major modifications to the embankment dam were completed in 1994 as well as minor modifications to the outlet works intake which will not be addressed by this paper. This paper will focus on recent modifications to the spillway structure to address dam safety deficiencies.

The spillway is a concrete uncontrolled overflow structure, located on the left abutment of the dam with a 627-foot long, trapezoidal-shaped chute that tapers from 64- to 50-feet-wide. Flow deflector vanes are located at the downstream end which directs discharges to the right away from the adjacent steep hillside. The end section of the spillway chute is supported on piers which suspends this section approximately 30 feet above the natural, unprotected exit channel. The crescent-shaped spillway inlet has a design crest length of 275 feet. The original design capacity of the spillway was 11,200 ft$^3$/s.

DAM SAFETY DEFICIENCIES

The dam safety deficiencies associated with the spillway were the lack of an energy dissipating structure at the downstream end of the chute, and inadequate capacity to pass the probable maximum flood (PMF) without overtopping the spillway walls and the dam embankment. Because of the absence of an energy dissipating structure, spillway flows of any significant magnitude would result in erosion which would expose of an artesian aquifer located downstream of the dam and initiate piping of the foundation and subsequent failure of the dam.

Overtopping of the embankment dam and the spillway walls during flood conditions could also lead to dam failure.

## SPILLWAY MODIFICATIONS CONSIDERED

The three basic options were considered for modifying the spillway to address the dam safety deficiencies, including:
1. A new replacement spillway could be built located on the middle of the dam, and the old spillway would be abandoned.
2. Modify the existing spillway by extending the chute and adding a new conventional hydraulic jump stilling basin.
3. Modify the existing spillway by adding a plunge pool type stilling basin which would be constructed with roller compacted concrete (RCC).

Option 1 was estimated to be the most expensive. It also involved potential problems with the required depths needed for a conventional hydraulic jump type stilling basin relative to the underlying aquifer. Artificial tailwater would need to be created using a downstream structure. It would also raise concerns regarding unobserved piping under the spillway structure.

Option 2 had some significant problems also due to required depths needed for a conventional hydraulic jump stilling basin.

Option 3 was judged the most likely to be successful and appeared to have significant cost savings over both options 1 and 2. However, this type of basin design would need to be verified using a scale hydraulic model due to the anticipated unique shape, geology, topography, and the desired hydraulic characteristics that would be required. The physical model would utilize a trial and error approach for finalizing a suitable configuration.

Based on costs, Option 3 was pursued as the preferred alternative.

## GEOLOGY AND TOPOGRAPHY

Geology and topography downstream of the spillway structure dictated some limiting parameters for the design. Downstream of the existing spillway on the left hillside the slope is approximately a 1 ½ :1 and is composed of alluvial fan and slopewash material overlaying a John Day Tuff which dips sharply to the right. On the right side of the basin area, 15-20 feet of alluvial fan and slopewash material overlays a 15-20 foot layer of fluviolacustrine sediments, which overlays a lower alluvium layer. Far below the alluvium is the John Day Tuff which continues to dip sharply to the right. Geologic investigations revealed that an artesian aquifer is present in the lower alluvium located downstream of the dam and immediately adjacent to the right of the stilling basin area. The aquifer is created by the reservoir and has approximately 70 feet of head above the confining layer of fluviolacustrine sediments. It was concluded that action of high-velocity spillway flows without an energy dissipating structure, for any significant duration could erode the overburden material above the aquifer initiating a foundation piping failure of the embankment dam.

## DESIGN CRITERIA

Various factors were taken into consideration in formulating design criteria to economically address the dam safety deficiencies for the spillway modifications. The critical issues that needed to be addressed included: (1) determining the inflow design flood (IDF) based on downstream inundation considerations and loss of life potential for dam failure scenarios; (2) protecting the aquifer from exposure by erosion; (3) foundation conditions to support the RCC stilling basin and; (4) spillway expansion to prevent dam overtopping. Following is a summary of how these issues were addressed.

Routing of the December general storm PMF hydrograph through the existing spillway resulted in a peak outflow of approximately 52,000 $ft^3/s$. The IDF to be used for sizing the spillway modifications would be determined in conjunction with the studies for downstream inundation, dam overtopping, and spillway chute wall overtopping utilizing the physical hydraulic model.

Inundation studies were first conducted for the area downstream of the dam which included the city of Prineville. The downstream safe channel capacity of Ochoco Creek through Prineville is estimated to be only 1,000 $ft^3/s$. The results of the study indicated that the major portion of the downstream population would be inundated for flows up to 15,000 $ft^3/s$. Additional population were at risk above the 15,000 $ft^3/s$ level, so a warning system was evaluated to determine if adequate warning times could be provided for reducing the loss of life potential in the event of dam failure. The warning system study used the results of the flood routings, potential failure modes, and the hydraulic model to verify that adequate warning time was available, provided early detection monitoring is installed, including rainfall gauges and reservoir inflow monitoring. Modifications required to the existing structure for the 15,000 $ft^3/s$ level were considered relatively minor in comparison with what would need to be done for significantly larger spillway flows up to the PMF.

Flood recurrence intervals were evaluated utilizing both statistical flood hydrology and results of a reconnaissance level paleoflood hydrology study. The recurrence interval for a flood that would produce a 15,000 $ft^3/s$ discharge using statistical hydrology was around a 1,000-yr thunderstorm flood or a 6,500-yr spring or winter flood. The paleoflood studies indicated that this level would be closer to a 15,000-yr flood. After considering all the data, it was judged that 15,000 ft$^3$/s was an acceptable level to consider as the IDF.

A channel degradation study was also conducted for the area downstream of the stilling basin to analyze the potential erosion depths that would result from various flows and durations up to the IDF. A computer model to study sediment erosion was initiated using the outlet channel parameters which included channel dimensions, material gradations, exit velocities derived from the hydraulic model study, and flood routing results. The conclusions were that erosion would not be significant enough to reach the confining fluviolacustrine layer.

It was most desirable to locate the stilling basin foundation for the RCC on the John Day Tuff as much as possible. The stilling basin configuration had several limiting physical parameters that dictated some of the initial configuration. The left hillside was located such that the basin floor width to the left needed to be minimized to avoid a costly excavation. The foundation basin floor width to the right side also needed to be minimized since the John Day Tuff dropped off sharply and was overlain by the less desirable alluvial fan and slopewash materials. Depth of excavation also was a limiting factor, especially if the length of the basin extended too far downstream because of the underlying artesian aquifer as well as the alluvial fan materials which had relatively high groundwater levels. These conditions were identified utilizing geologic trench explorations and piezometric data from both existing and new piezometers.

## HYDRAULIC MODEL STUDY

A physical hydraulic model study was built at a 1:36 scale starting with trial layouts based on rough estimates utilizing plunge pool and hydraulic jump stilling basin calculations. Chute walls and stilling basin walls were initially oversized to ensure containment of all flows. Side slopes and overall configurations were chosen based on geology, topography, and constructability considerations using RCC. A large plunge pool with a smaller secondary pool was first tested with limited success. Various configurations of baffles, dentates and flow deflectors were also tested to identify feasible options. The existing chute required modifications which included removal of an existing flow deflector and the end vanes as well as raising of the existing chute walls at various locations. The pool configurations and baffle dimensions between the pools were varied until the most suitable configuration was developed. A vertical face near the base of the first baffle created the most significant impact on the effectiveness of the energy dissipation of the first pool. The downstream channel configuration and size was also studied and refined by trial and error.

## SELECTED STILLING BASIN DESIGN CONFIGURATION

The final design included a three stage plunge pool with pools separated by baffles as shown in Figure 1. Each successive pool is smaller due to energy dissipation of the previous pool. The floor and the left side is located on John Day Tuff whereas the right side is mainly on alluvial fan deposits as well as on compacted backfill. The left side also has a catch berm located within the excavation and above the RCC walls to minimize the possibility of raveling materials, rocks, or boulders from entering the stilling basin.

Configuration, slopes and dimensions of the RCC stilling basin were simplified as much as possible to facilitate construction. Minimal conventional concrete is incorporated into the basin design in order to simplify construction and minimize costs. The pools were designed to be free draining after the spillway flows subside for public safety reasons as well as to minimize freeze thaw damage to the RCC.

The spillway exit channel was designed to minimize erosion up to 2,000 ft$^3$/s, which is higher than the 1,000 ft$^3$/s flow of record. For flows above that level, some erosion to the channel and downstream area would be anticipated but not to the extent of infringing on the confining fluviolacustrine layer.

A warning system will be installed including upstream basin monitoring and reservoir inflow monitoring to alert the downstream population when high releases are anticipated as well as for extremely remote events that could lead to dam failure.

The spillway modifications were started in July of 1996 and completed in January of 1997. Placement of approximately 19,000 yds$^3$ of RCC in the stilling basin took three weeks.

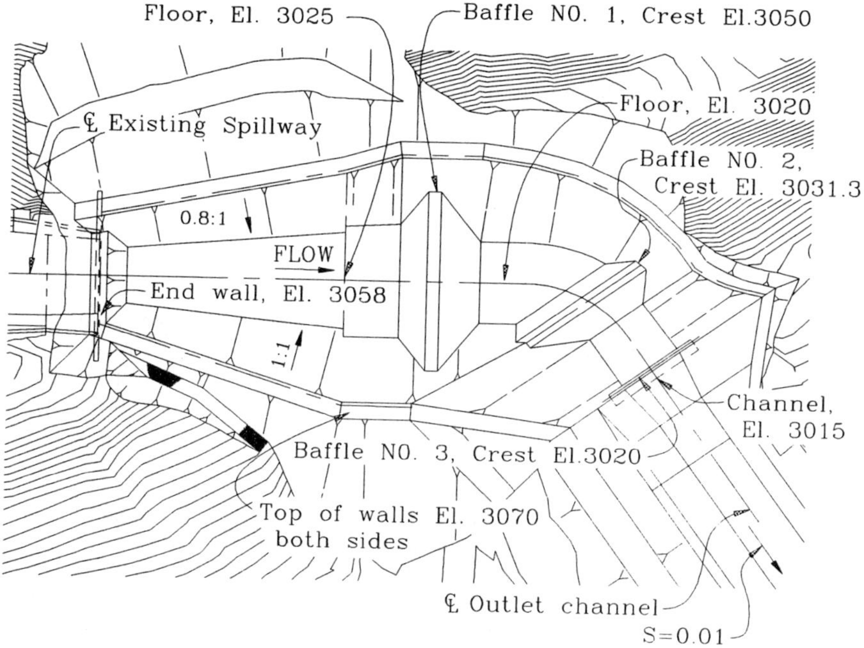

# OCHOCO DAM SPILLWAY MODIFICATION

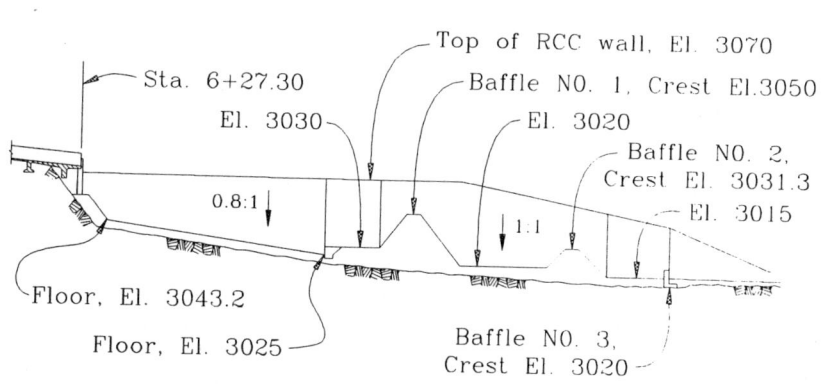

Plan view of stilling basin

Profile along centerline of stilling basin

FIGURE 1. CONFIGURATION OF THE STILLING BASIN DESIGN

# Overtopping Breaching of Noncohesive Embankment Dams

S. E. COLEMAN, R. C. JACK and B. W. MELVILLE
The University of Auckland, Auckland, New Zealand

ABSTRACT
Model studies of the overtopping failure of homogeneous embankment dams of cohesionless fill are presented. Observations are made in relation to breach channel profile, breach development, erosion processes and breach discharge. Equations are presented to describe discharge through the breach as a function of the breach crest length and the head on the breach crest as the breach develops.

BACKGROUND
The potential failure of a dam presents a significant hazard to downstream life and property. As a consequence, the development of civil protection plans in the event of a dam failure are increasingly being demanded for dams worldwide. Flood levels and discharges resulting from a dam failure are generally estimated for these plans either based on data from observed dam failures or using numerical models. The former data suffer from being rare and highly variable. Numerical models provide limited descriptions of dam-breach processes and development, calibration and verification of these models being based on observed dam failures or laboratory data on dam breaching which are virtually nonexistent (Singh, 1996).

As a consequence of dam-break flooding prediction being highly dependent on the adopted understanding of the physical processes involved in the breach development for the dam, the present investigations were undertaken to elucidate detailed descriptions of dam-breach processes and development for embankment dams. The dominant mode of failure for these structures, failure by overtopping (Schnitter, 1993), was investigated.

The results presented are relevant to earth dams, rockfill dams, fuse plugs, dykes and levees. These results can either be viewed as being indicative of breach processes and development in embankments of the present fill materials, or the results can be scaled based on physical modeling techniques to be appropriate to larger prototype dams of different fill materials. To this end, the proposed equations are in dimensionless form.

NONCOHESIVE EMBANKMENT EXPERIMENTS
Experiments were carried out for a series of four embankments of homogeneous fill. Each embankment had upstream and downstream slopes of 2.5:1 (H:V). Three of the embankments were 300 mm high, these embankments being constructed of a very fine

gravel of median sediment size $d_{50} = 2.4$ mm, a coarse sand of $d_{50} = 0.85$ mm, and a medium sand of $d_{50} = 0.29$ mm respectively. A 600 mm high embankment was constructed of the coarse sand. Each embankment was underlain by a toe drain to control the phreatic surface within the dam.

The experiments were conducted in a 1.52 m wide glass-sided flume (Figure 1). Breaching was initiated by raising the reservoir level upstream of the dam to a predetermined height above the level of a pilot channel formed in the dam crest. This channel was located at the flume side-wall to maximise possible lateral erosion for each embankment, the flume wall simulating the breach channel centreline.

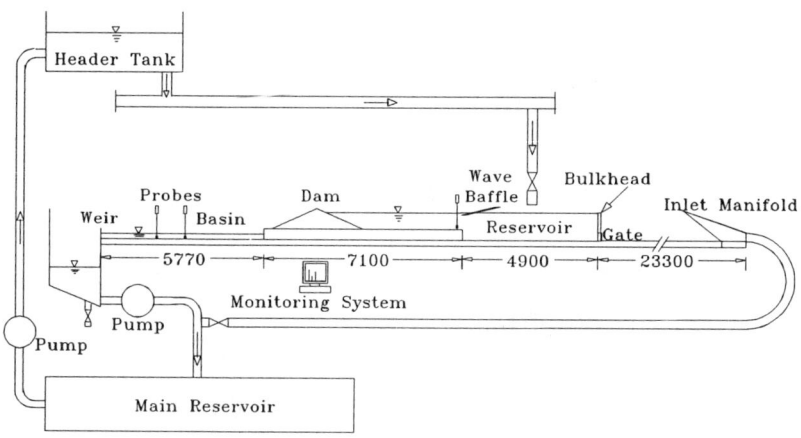

Figure 1. Experimental Setup.

With the reservoir maintained at a constant level by adjusting the supply flow, probes connected to a computer measured water levels in the reservoir upstream of the dam and in the weir basin downstream of the dam as the dam breach developed. A 120° V-notch weir enabled calculation of flow leaving the weir basin based on the measured water levels. Flow through the breach as this developed was then calculated by routing the flow leaving the weir basin back upstream to the downstream toe of the dam.

When a desired degree of embankment erosion had been achieved, flow into the reservoir was halted and the gate in the base of the bulkhead at the upstream end of the reservoir was opened. This resulted in a rapid cessation of flow through the dam breach with the lowering of the reservoir. Upon further lowering of the reservoir level, the surface profile of the eroded embankment was measured (in x, y and z coordinates) and automatically recorded by a surface profile measurement system connected to a computer.

The above experimental process for a given embankment was repeated a number of times, the degree of final erosion being different for each repetition. Each embankment was entirely rebuilt prior to retesting. Comparison of the measured water levels for the series of tests for each embankment confirmed the repeatability of the breaching process for the embankment and indicated the relative times at which the individual tests were halted. The picture of breach development with time was then built up from the embankment surface profiles measured over the series of tests for the embankment (Figure 2). Two series of tests were carried out for the 300 mm high embankment constructed of coarse sand.

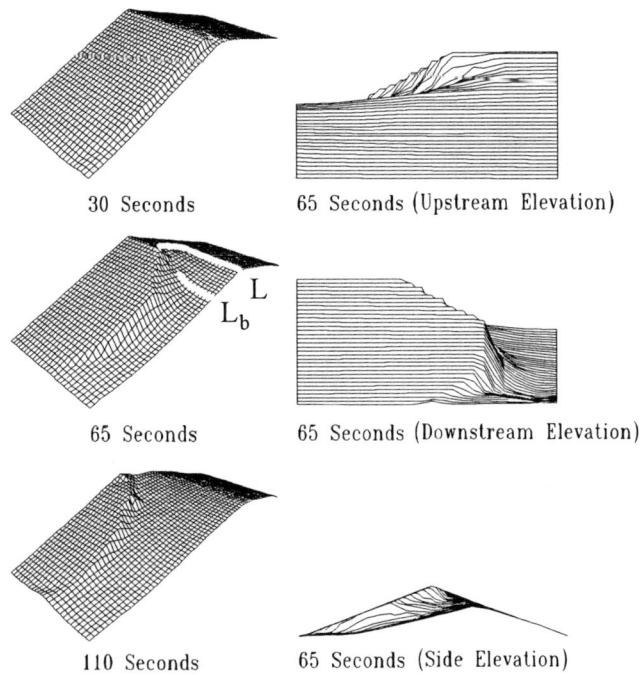

Figure 2. Breach Channel Profile and Breach Development for the 600 mm Coarse Sand Embankment.

## BREACH PROFILE AND BREACH DEVELOPMENT

Unless indicated, the comments relating to breach channel profile and breach development described here are common to all embankments tested and are illustrated in Figure 2 for the 600 mm high embankment.

## BREACH PROFILE

Previous researchers have primarily approximated breach cross-sectional shapes as simplified trapezoids or triangles. To a lesser extent, rectangular, parabolic, semicircular, cosine, and curved regime channel cross-sections have also been incorporated into breach development analyses.

The present results indicate that the breach channel is of an hourglass shape in plan. The channel is curved (with positive curvature) in cross-section below the waterline, and is convex (of negative curvature) in longitudinal profile with the channel apex located upstream of the embankment crest (Figure 2). The ratio of breach channel width to channel depth for a given breach discharge increases with increasing sediment size.

The apex of a given breach longitudinal profile defines the breach crest point for the profile. The breach crest, controlling breach discharge, is then curved in plan and elevation, the breach crest length, L, being greater than the width of the breach channel at the embankment crest, $L_b$ (Figure 2). This observation is important in regard to analysing both breach discharge and also previous attempts at quantifying this discharge.

Side-wall steepness above the waterline is a function of the sediment. Although the materials tested were cohesionless, capillary forces resulted in a pseudo-cohesive behaviour of the medium sand when wet. This resulted in overhangs forming above the waterline for this material. Near-vertical side-walls above the waterline were sustained by the coarse sand material in the same manner.

## BREACH DEVELOPMENT

Flow through the pilot channel initially erodes a small channel from the crest to the toe of the dam on the embankment downstream face. This channel widens and deepens with time, lateral erosion occurring at all points along the channel side-walls. Embankment material is eroded by the processes of tractive shear and turbulence. Breach channel side-wall erosion, due to side-wall undermining, exposes large amounts of unprotected material to these erosive processes.

Until vertical erosion of the breach is inhibited by the dam foundation, the hourglass shape of the channel in plan increases in curvature with time, this giving increasing rounding of the approach and exit channels. The breach channel slope decreases with time although there is also some upstream movement of the downstream toe of this slope to maintain slope values. Progressive erosion of the upstream face of the embankment by the accelerating approach flow results in upstream movement of the apex of a given breach longitudinal profile and lengthening of the breach crest with time.

Subsequent to the dam foundation inhibiting vertical erosion of the breach, the breach channel side-walls erode laterally accompanied by a lesser erosion of the exposed breach channel invert down to the dam foundation.

As the breach developed for the medium sand, the pseudo-cohesive property of this material resulted in erosion being more rapid at the downstream toe (a region of high turbulence) than on the downstream face of the embankment. A step then formed in the breach channel slope as the downstream toe of the breach channel migrated upstream. This step had an associated downstream roller which increased the local rate of vertical erosion and the size of the step. The step, however, quickly became unstable, the breach longitudinal profile consequently flattening out.

Scaling of capillary action forces would be necessary to scale the present results to larger embankments in regard to both steps in breach channel slope and also slopes of breach channel side-walls above the waterline.

## BREACH HYDRAULICS

Breach discharge rate, Q, is commonly estimated using the broad-crested weir equation, $Q = C_d L h^{3/2}$, where $C_d$ is the discharge coefficient, L is the crest length, and h is the hydraulic head on the crest. In utilising this equation with the breach width at the embankment crest, $L_b$, used to define L (ie $Q = C_d L_b h^{3/2}$), previous researchers have adopted values of $C_d$ ranging from 1.3 to 1.7 $m^{1/2}/s$. The present experiments indicate an equivalent value of $C_d$ of 1.3 $m^{1/2}/s$, where, owing to the curved nature of the breach crest, h is taken as the height of the reservoir surface above the apex point on the breach channel centreline for the present analyses.

With the breach crest (controlling discharge) located upstream of the embankment crest, and $L > L_b$, the present experiments indicate a more appropriate form of the broad-crested weir equation of

$$Q' = 0.24 L' h' \tag{1}$$

where $Q' = Q/[gH^5]^{1/2}$, $h' = h/H$, $L' = L/H$, and H is the height of the reservoir surface above the dam foundation. Equation (1) can be simplified to $Q = 0.75 L h^{3/2}$, indicating a broad-crested weir discharge coefficient of $C_d = 0.75$ $m^{1/2}/s$.

For the present experiments, Q, can be directly related to the head on the crest, h, by

$$Q' = 2.72 h' \tag{2}$$

where the discharge rate is that for the full breach cross-section, ie double that tested in the present experiments. This relation is dependent on breach dimensions remaining in the same proportions as the breach develops, ie $L \propto h$, and is valid only until vertical erosion of the breach becomes inhibited by the dam foundation. Eq. 2 provides a more general fit to the present experimental data (Figure 3) than Eq. 1. Adopting the relation $L = 5.68h$, which was found to adequately describe the present experimental results, Eq. 2 can be simplified to $Q = 8.52 h^{5/2}$ for the full breach cross-section.

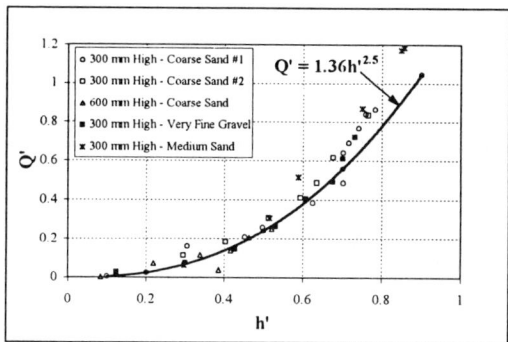

Figure 3. Breach Discharge Rate as a Function of Head on the Breach Crest for the Present Experiments (where Q is that for a half-breach cross-section).

## CONCLUSIONS

The present results indicate the curved three-dimensional nature of the breach channel, with cross-sectional shape varying with distance down the embankment face. In addition, the curved breach crest is shown to be located upstream of the embankment crest and is found to be longer than the breach width at the position of the embankment crest.

Embankment material is found to be eroded by the processes of tractive shear and turbulence. Tractive shear-dominated erosion (for coarser-grained materials) produces breach long sections with smooth downstream slopes: turbulence-dominated erosion (for finer-grained materials) produces vertical steps in these profiles. The process of breach development is greatly influenced by the effect of the dam foundation on the vertical erosion of the breach.

Breach discharge rate, Q, can be estimated as a function of the breach crest length, L, and the hydraulic head on the crest, h, using the dimensionless expression $Q'=0.24L'h'$. Alternatively, Q can be evaluated using $Q'=2.72h'$, where the discharge rate is that for the full breach cross-section, ie double that tested in the present experiments. This latter relation, which provides a more general fit to the present experimental data than the former equation, is valid only until vertical erosion of the breach becomes inhibited by the dam foundation.

## ACKNOWLEDGMENTS

The writers would like to thank G. Webby for his comments in relation to this study.

## REFERENCES

Schnitter, N. J. (1993). "Dam failures due to overtopping." Proc. Int. Workshop On Dam Safety Evaluation, Grindewald, Switzerland, Vol. 1, 13-19.

Singh, V. P. (1996). "Dam breach modeling technology." Kluwer Academic Publishers, Dordrecht, The Netherlands.

# Predicting Embankment Dam Breach Parameters - A Needs Assessment

TONY L. WAHL
U. S. Bureau of Reclamation, Denver, Colorado, USA

## ABSTRACT

Simulation of embankment dam breach events and the resulting floods are crucial to characterizing and reducing threats due to potential dam failures. Development of effective emergency action plans requires accurate prediction of inundation levels and the time of flood wave arrival at a given location. If population centers are located well downstream of a dam, details of the breaching process have little effect on the result; travel time, attenuation, and other routing effects predominate. However, in a growing number of cases, the location of population centers near a dam makes accurate prediction of breach parameters (e.g., breach width, depth, rate of development) crucial to the analysis. If breach parameters cannot be predicted with reasonable accuracy, increased conservatism with associated increased costs are required. This paper examines existing empirical procedures and numerical models used to predict breach parameters, reviews new technologies relevant to dam breaches, and outlines a program for development of an improved numerical model for the simulation of embankment dam breach events.

## BACKGROUND

Today there are numerous tools available for analyzing dam failures and the resulting floods. Wurbs (1987) compared state-of-the-art models, including the National Weather Service (NWS) Dam-Break Flood Forecasting Model (DAMBRK), the U. S. Army Corps of Engineers Hydrologic Engineering Center Flood Hydrograph Package (HEC-1), and the NWS Simplified Dam-Break Flood Forecasting Model (SMPDBK). DAMBRK, the most widely used dam failure analysis model, is being upgraded and will be replaced in the future by FLDWAV (Fread, 1993), which combines capabilities of DAMBRK and the NWS's DWOPER model. FLDWAV's improvements relative to DAMBRK include capability for modeling interconnected river systems, flows over or through levees, automatic Manning roughness calibration, improved numerical stability, and input and output improvements. FLDWAV's treatment of the actual dam-breach process is essentially identical to that of DAMBRK.

The two primary tasks in the hydraulic analysis of a dam breach are the prediction of the reservoir outflow hydrograph and the routing of that hydrograph through the downstream valley. Predicting the outflow hydrograph can be further subdivided into predicting the breach characteristics (e.g., shape, depth, width, rate of breach formation), and routing the reservoir storage and inflow through the breach. DAMBRK, FLDWAV, and other similar models treat the routing tasks—through the breach and through the downstream valley—in much greater detail than the breaching process. In fact, most models do not directly simulate the breach. Rather, the user of the model independently determines the ultimate breach parameters (i.e., dimensions of the fully developed breach), and the time required for breach formation. These

Table I. — Embankment breach models (V. Singh and Scarlatos, 1988; Wurbs, 1987).

| Model and Year | Sediment Transport | Breach Morphology | Parameters |
|---|---|---|---|
| Cristofano, 1965 | Empirical formula | Constant breach width | Angle of repose, others |
| Harris & Wagner, 1967; BRDAM, 1977 | Schoklitsch formula | Parabolic breach shape | Breach dimensions, sediment properties |
| Lou, 1981; Ponce & Tsivoglou, 1981 | Meyer-Peter Müller formula | Regime type relation | Critical shear stress, sediment, tailwater |
| BREACH, 1985 | Meyer-Peter Müller, modified by Smart | Rectangular, triangular, trapezoidal | Critical shear, sediment, tailwater, dry slope stability |
| BEED, 1985 | Einstein-Brown formula | Rectangular or trapezoidal | Sediment, tailwater, saturated slope stability |
| FLOW SIM 1 and FLOW SIM 2 | Linear erosion or Schoklitsch formula | Rectangular, triangular, trapezoidal | Breach dimensions, sediment properties |

parameters are provided as input to the routing model, and the model then simulates the development of the breach in a progressive fashion, usually a linear increase in breach dimensions over the span of the breach formation time. There is presently little research to support or refute the assumption of linear breach development.

Ultimate breach parameters are typically estimated using case study-based predictive equations that relate breach parameters to gross characteristics of the dam and reservoir, such as dam height and storage volume (MacDonald and Langridge-Monopolis, 1984; Froehlich, 1987, 1995; Reclamation, 1988; Von Thun and Gillette, 1990). Such relations have high uncertainty due to scatter in the available case study data, especially with respect to breach formation time and breach side slope angles. These relations also are based on a database of dam failure case studies that includes few examples of large dams or large reservoirs.

The National Weather Service BREACH model (Fread, 1985) and other similar numerical models do attempt to simulate the breach formation process in greater detail, but are not widely used at the present time. Table I shows physically-based dam breach models and their characteristics (V. Singh and Scarlatos, 1988; Wurbs, 1987). Several of these models are dependent on calibration coefficients that have not been effectively generalized. The greatest weakness of the existing models is their reliance on tractive-stress based erosion models that do not reflect the predominant mechanisms of headcut erosion, geotechnical slope failure, and lateral embankment erosion observed in case studies and physical model studies.

## BREACH PARAMETER DEFINITIONS

Embankment dam breaches are typically assumed to be approximately trapezoidal in shape; the breach geometry can be described in terms of a breach height, average breach width, and breach side slope angle. The slope of the breach invert in the flow direction is assumed to be horizontal. These parameters describe the breach geometry to the extent needed to compute flowrates through the breach, assuming discharge characteristics of a broad-crested weir.

Past research has been focused on the use of a single time parameter, termed the *breach formation time* or *time of failure*. Definitions of these terms have varied in the case study investigations, and the case study data exhibit large scatter, perhaps due to variable interpretations of breach formation time among dam failure eyewitnesses. DAMBRK's definition for the time parameter is:

*The time of failure as used in DAMBRK is the duration of time between the first breaching of the upstream face of the dam until the breach is fully formed. For overtopping failures the beginning of breach formation is after the downstream face of the dam has eroded away and the resulting crevasse has progressed back across the width of the dam crest to reach the upstream face.*

The second part of this definition describes a breach initiation phase that precedes the breach formation phase. For purposes of estimating available warning and evacuation time in the event of a dam failure, the *breach initiation time* should be defined as follows:

*The breach initiation time begins with the first flow over or through a dam that is of enough significance to warrant warning, evacuation, or heightened awareness of the potential for dam failure, and ends at the start of the breach formation phase.*

There is presently little guidance for the prediction of breach initiation times, and breach initiation time is not a factor in the DAMBRK analysis. However, it is likely that the breach initiation phase could be of significant duration in many cases (e.g., dams overtopped by only a small amount, flow over resistant abutment areas, etc.), and even if the breach initiation phase is short, it can still have a significant impact on loss-of-life from a dam failure. BREACH and other similar models do simulate the breach initiation phase, but their accuracy is questionable because they do not simulate headcutting and other observed embankment erosion mechanisms.

## IMPORTANCE OF BREACH PARAMETERS

Variation of breach parameters can affect peak discharge and inundation levels, as well as warning and evacuation time. The effect on peak discharges was examined by K. Singh and Snorrason (1984). For small reservoirs (those which experience significant drawdown before the breach is fully developed), changes in breach formation time can dramatically affect peak outflow. Variations of breach width can also produce large changes in peak outflow, especially for large reservoirs. Variations of breach height have a relatively small effect on peak outflow.

Petrascheck and Sydler (1984) demonstrated the sensitivity of peak flow, inundation levels, and flood arrival time to changes in breach width and breach formation time. For locations near a dam, both parameters can have a dramatic influence. For locations well downstream of a dam, timing of the flood wave peak can change significantly with changes in breach formation time, but peak discharge and inundation levels are insensitive to changes in breach parameters.

Warning and evacuation time can dramatically influence the loss-of-life from dam failure. When establishing hazard classifications, preparing emergency action plans, or designing early warning systems, good estimates of warning time are crucial. Warning time is the sum of the breach initiation time, breach formation time, and flood wave travel time from the dam to a population center. Case history-based procedures developed by the Bureau of Reclamation indicate that loss-of-life can vary from 0.02% of the population-at-risk with more than 90 minutes of warning time, to 50% of the population-at-risk when warning time is less than 15 minutes (Brown and Graham, 1988). More recent work by DeKay and McClelland (1991) shows similar extreme sensitivity to warning time.

## EMPIRICAL BREACH PARAMETER PREDICTION EQUATIONS

A review of literature performed by this author (Wahl, 1996), revealed numerous breach parameter prediction equations developed since about 1984. Individual equations have been based on analyses of case study compilations, generally comprising about 20 to 60 dams. The database compiled by this author contains 108 dam failures, and was used to analyze and

# PREDICTING EMBANKMENT DAM BREACH

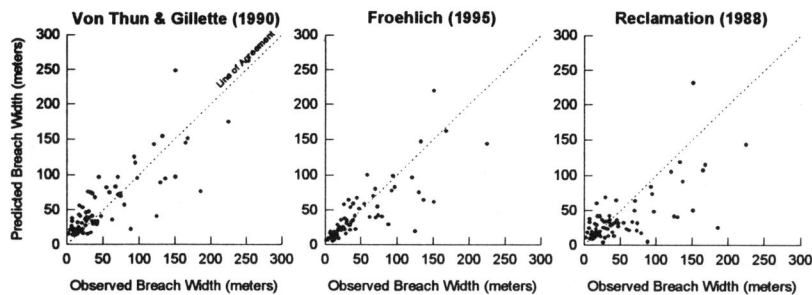

Figure 1. — Comparison of breach width prediction equations to case study data.

compare the various prediction equations. Figure 1 shows predicted and observed breach widths using three of the available breach-width prediction equations. There are some dramatic prediction errors evident with all three relations. Figure 2 compares predicted and observed breach formation times, using several available prediction equations. Again, large prediction errors are evident. It is also significant to note the large variation of reported breach formation times for some dams. This is evidence of potential problems in distinguishing between breach initiation and breach formation phases.

## MECHANICS OF EMBANKMENT BREACH

Laboratory testing and field observations of embankment dam failures have shown that headcutting is the predominant mode of failure for cohesive embankments or rockfill embankments with a cohesive core (Dodge, 1988; Powledge et al., 1989). In overtopping failures, headcutting typically initiates near the toe of the dam and advances upstream until the crest of the dam is breached; in some cases a headcut may initiate at the knickpoint present at the downstream edge of the crest. As the headcut advances upstream it will widen and assume a semi-circular shape which improves stability of the headcut through arching of the soil mass. In some cases, multiple stairstep headcuts form on the downstream face of the dam.

Headcut initiation takes place when the protective cover on an embankment fails, allowing localized erosion that creates an overfall. Factors affecting the initiation of headcutting include embankment slope, vegetation type and quality or riprap type and size, cover discontinuities, flow concentrations, flow velocities, and unit discharges. Headcut initiation can be modeled using tractive stress-based approaches (Temple and Hanson, 1994).

The key erosion zone once a headcut has formed is at the base of the headcut overfall. As material is eroded from this area, support for the above soil mass is removed, leading to sudden collapse of the soil block. Tailwater conditions at the base of the overfall and aeration of the nappe are key factors in headcut advance. The Agricultural Research Service has developed empirically-based procedures for estimating headcut advance rates in earth spillways, using a model that compares energy dissipation rate at an overfall to a headcut erodibility index for the material (Temple and Moore, 1994). These procedures are being incorporated into the SITES model used by the Natural Resources Conservation Service for design and analysis of earth spillways (Temple et al., 1994). These procedures have not yet been applied to steep slopes or breaching of earth embankments, but do hold promise for such applications.

## A NEW BREACH MODEL

Clearly, when population centers are close to a dam, accurate prediction of breach parameters is necessary to make reliable estimates of warning and evacuation time, peak outflow and inundation zones, and loss-of-life. Available models cannot fully address the needs for many of these cases. Great benefit could be obtained from development of an improved breach simulation model that is based on the observed erosion mechanics. The model should address the following issues:

- For a given set of conditions, will a dam breach?
- How much time is required to initiate a breach?
- How will the breach develop once it is initiated? Define the geometry of the breach during its development. Define total time for breach development and ultimate breach geometry.

Reclamation is now pursuing a cooperative effort with several agencies to develop such a model. The initial focus will be on the more tractable problem of breaches caused by overtopping, although the model should eventually be applicable to more complex piping and seepage-induced failures.

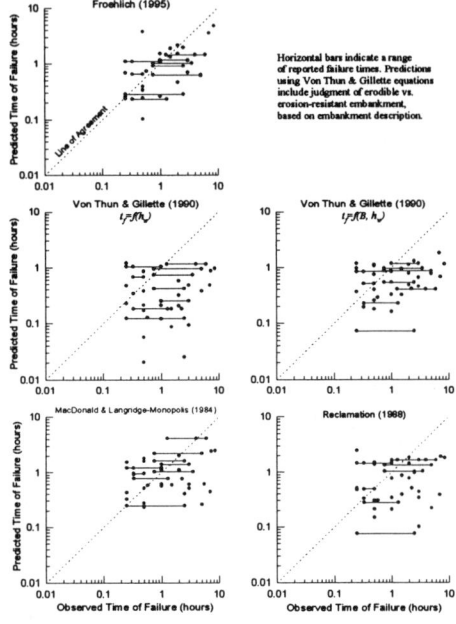

Figure 2. — Predicted vs. observed breach formation times using several prediction equations.

Initial efforts will be focused on the breach initiation phase of the problem. Recent research by Reclamation and others has improved the capability to assess riprap stability, erosion of vegetated surfaces, and headcut advance. Work on a generalized numerical model with unsteady flow and sediment transport capability will also be pursued. This will provide the basis for development of a more specialized dam breach model.

Recent advances in technology for analyzing headcut erosion, riprap stability, erosion of vegetated surfaces, and high energy erosion of resistant earth materials have all come about through extensive large-scale physical modeling efforts and collection of case study data from prototype structures. Similarly, large-scale physical modeling will be required to address complicating factors in embankment breaching processes, such as:

- Variable embankment and foundation configurations, materials, and densities of fill
- Effect of discontinuities, singularities, and flow concentrations
- Presence and depth of tailwater on the downstream slope
- Unique embankment features such as toe drains, blanket drains, erodible filter zones, etc.

## LITERATURE CITED

Brown, C. A., and W. J. Graham, 1988, "Assessing the Threat to Life from Dam Failure," *Water Resources Bulletin*, Vol. 24, No. 6, December, 1988.

Dekay, M. L., and G. H. McClelland, 1991, "Setting Decision Thresholds for Dam Failure Warnings: A Practical Theory-Based Approach," CRJP Technical Report No. 328, Center for Research on Judgment and Policy, University of Colorado, Boulder, CO.

Dodge, R. A., 1988, *Overtopping Flow on Low Embankment Dams — Summary Report of Model Tests*, REC-ERC-88-3, U. S. Bureau of Reclamation, Denver, Colorado, August 1988, 28 p.

Fread, D.L., 1984, "DAMBRK: The NWS Dam-Break Flood Forecasting Model," National Weather Service, Office of Hydrology, Silver Spring, Maryland.

Fread, D. L., 1985 (revised 1991), "BREACH: An Erosion Model for Earthen Dam Failures," *NWS Report*, National Oceanic and Atmospheric Administration, Silver Spring, Maryland.

Fread, D. L., 1993, "NWS FLDWAV Model: The Replacement of DAMBRK for Dam-Break Flood Prediction," ASDSO Conference, Kansas City, Missouri, Sept. 26-29, 1993, p. 177-184.

Froehlich, D. C., 1987, "Embankment-Dam Breach Parameters," in: *Hydraulic Engineering*, Proceedings of the 1987 ASCE National Conference on Hydraulic Engineering, Williamsburg, Virginia, August 3-7, 1987, p. 570-575.

Froehlich, D. C., 1995, "Embankment Dam Breach Parameters Revisited," in: *Water Resources Engineering*, 1995 ASCE Conference, San Antonio, Texas, August 14-18, 1995, p. 887-891.

MacDonald, T. C., and J. Langridge-Monopolis, 1984, "Breaching Characteristics of Dam Failures," *Journal of Hydraulic Engineering*, Vol. 110, No. 5, p. 567-586.

Petrascheck, A. W., and P. A. Sydler, 1984, "Routing of Dam Break Floods," *International Water Power and Dam Construction*, Vol. 36, p. 29-32.

Powledge, G. R., D. C. Ralston, P. Miller, Y. H. Chen, P. E. Clopper, and D. M. Temple, 1989b, "Mechanics of Overflow Erosion on Embankments. II: Hydraulic and Design Considerations," *Journal of Hydraulic Engineering*, Vol. 115, No. 8, August 1989, p. 1056-1075.

Singh, K. P., and A. Snorrason, 1984, "Sensitivity of Outflow Peaks and Flood Stages to the Selection of Dam Breach Parameters and Simulation Models," *Journal of Hydrology*, Vol. 68, p. 295-310.

Singh, V. P., and P. D. Scarlatos, 1988, "Analysis of Gradual Earth-Dam Failure," *Journal of Hydraulic Engineering*, Vol. 114, No. 1, p. 21-42.

Temple, D. M., and G. J. Hanson, 1994, "Headcut Development in Vegetated Earth Spillways," *Applied Engineering in Agriculture*, Vol. 10, No. 5, American Society of Agricultural Engineers.

Temple, D. M., and J. S. Moore, 1994, "Headcut Advance Prediction for Earth Spillways," presented at the 1994 ASAE International Winter Meeting, Paper No. 94-2540, Atlanta, Georgia, December 13-16, 1994, 19 p.

Temple, D. M., H. H. Richardson, J. A. Brevard, and G. J. Hanson, 1994, "The New DAMS2," presented at the 1994 ASAE International Winter Meeting, Paper No. 94-2544, Atlanta, Georgia, December 13-16, 1994, 7 p.

U. S. Bureau of Reclamation, 1988, "Downstream Hazard Classification Guidelines," ACER Technical Memorandum No. 11, Denver, Colorado, December 1988, 57 p.

Von Thun, J. L., and D. R. Gillette, 1990, "Guidance on Breach Parameters," unpublished internal document, U. S. Bureau of Reclamation, Denver, Colorado, March 13, 1990, 17 p.

Wahl, T. L., 1996, "Prediction of Embankment Dam Breach Parameters: Literature Review and Needs Assessment," USBR, Water Resources Research Laboratory, PAP-735, Denver, CO.

Wurbs, R. A., 1987, "Dam-Breach Flood Wave Models," *Journal of Hydraulic Engineering*, Vol. 113, No. 1, p. 29-46.

# Performance Parameter Evaluation of Potential Hydraulic Structure Failure Modes

JAY N. STATELER
U.S. Bureau of Reclamation, Denver, Colorado, USA

To promote efficient and effective monitoring for dam safety purposes, the Bureau of Reclamation has recently begun developing and documenting performance parameters for each of its dams. Fundamentally, the performance parameter evaluation addresses the question: "What should be done to properly look after the dam in the future, from a dam safety perspective, given what we know today?" The process used involves: (1) identifying the most likely failure modes for the dam, (2) identifying the key parameters to monitor that will provide the best indication of the possible development of each of the identified failure modes, (3) defining an instrumented and visual monitoring program to gather the necessary information and data, (4) defining the ranges of expected performance, and (5) defining the action to be taken in the event of unexpected performance. The goal of the work is to prevent uncontrolled reservoir releases that could cause loss of life or significant economic losses in areas downstream of a dam. This is the definition of "failure" used in the work and in this paper. The identification of potential failure modes for a dam is done in light of: (1) information and analyses that are currently available concerning the dam and damsite, (2) the current state-of-the-art in dam design and evaluation, and (3) the record and available knowledge regarding past dam failures. This paper will present a general discussion of failure modes of concern relative to the hydraulic structures associated with dams. The performance parameters methodology will be employed in a "global" sense. Static, earthquake, and flood loading conditions will be discussed, in that order.

## STATIC LOADING CONDITIONS

STATIC FAILURE MODE 1
A disruption/crack/break in a hydraulic structure could lead to the

introduction of water under high pressure into downstream areas of a dam, potentially causing failure of the dam. This failure mode is typically only of concern relative to the outlet works for an embankment dam (upstream of the gate chamber), where the potential exists that the high pressure water could result in piping of material and eventual breaching of the dam. The flow would most likely follow along the wall(s) of the hydraulic structure to a downstream exit point, as the hydraulic structure would serve to concentrate the erosive energy of the flow and would provide some measure of "roof" support (due to arching effects) for a developing "pipe."

The best defense against this potential failure mode is a design that: (1) effectively avoids potential problems associated with corrosion, deterioration, and unusual settlements (absolute and differential) that could give rise to a disruption/crack/break; (2) incorporates a downstream filter zone in the embankment (that would prevent piping of embankment materials); and (3) locates the gate chamber as far upstream as possible. For existing structures having some potential for development of this failure mode, monitoring should include: (1) monitoring at relatively frequent intervals for evidence of new seepage, or changed seepage conditions, in the vicinity of the downstream end of the hydraulic structure, with particular emphasis on evidence of material transport by any seepage noted, and (2) inspecting the interior of the hydraulic structure on an appropriate frequency to detect evidence of corrosion, deterioration, cracking, etc.

## STATIC FAILURE MODE 2

A crack, joint offset, or improperly filtered drain in the hydraulic structure could create an uncontrolled exit location for seepage flowing through erodible portions of the dam or foundation. Typically this failure mode is only a concern for embankment dams. If the seepage is transporting material, then failure of the dam is possible due to piping that leads to eventual breaching of the dam. Since the opening size of the exit may not be readily expandable as development of the failure mode progresses (since the hydraulic structure typically would be concrete, or possibly steel or iron), for catastrophic failure of the dam to occur, it would likely require development of another exit location shortly before dam failure (to allow a point of exit for higher seepage flow quantities).

The best defense against this potential failure mode is a design that effectively avoids the problem by preventing the development of cracks and joint offsets/displacements. This means effectively addressing the potential for corrosion, deterioration, large settlements, differential settlements, etc. during design of the hydraulic structure. Also designs should incorporate filters or filter zones around drain pipes and in other areas, where possible, that would prevent piping of embankment materials into the hydraulic structures. For existing structures having some potential for development of this failure mode, monitoring should include: (1) monitoring for evidence of seepage into hydraulic structures, with particular emphasis on evidence of material transport by the seepage, (2) inspecting the interior of the hydraulic structure on an appropriate frequency to detect cracks, displaced joints, etc. where seepage could enter, and (3) inspecting the embankment adjacent to and above the hydraulic structure for sinkholes, unusual settlements, etc.

## STATIC FAILURE MODE 3

Deterioration with time of a hydraulic conduit, perhaps in conjunction with inadequate design, could lead to structural collapse of the conduit and collapse of overlying material into the conduit. Though theoretically possible relative to a concrete dam, this failure mode basically is only relevant for embankment dams. Actual failure of the embankment dam could occur due to: (1) loss of freeboard at the dam crest, (2) transverse cracking of the disrupted embankment, leading to seepage erosion, or (3) injection of large amounts of high pressure water into the embankment, leading to seepage erosion and piping (same as Static Failure Mode 1).

The best defense against this potential failure mode is a design that avoids the problem. Where problems exist remediation is possible. For several Reclamation dams, a steel pipe has been inserted in an existing conduit, and the annulus grouted, to address deteriorating conditions. To detect circumstances requiring remediation, inspection of the interior of the hydraulic structure should be performed on an appropriate frequency.

## STATIC FAILURE MODE 4

Failure of a gate or valve can lead to uncontrolled release of the reservoir. To be a dam safety concern, the release would need to exceed the safe downstream channel capacity (i.e. there potentially could be an adverse impact on people located downstream of the dam). The reason for failure could be related to design, maintenance, and/or misoperation. The best defense against this potential failure mode incorporates the following:

> 1. An appropriate conceptual design such that redundant flow control devices are provided where possible, and that the hydraulic structure is divided into enough separate flow paths such that failure of one gate/valve would not be sufficient to exceed downstream channel capacity, where possible.

2. An appropriate structural and mechanical design of gates and valves, including appropriate checks/reviews of designs, both as they are completed and in the future (as new information/experience becomes available).

3. A program for routinely inspecting and exercising the gates/valves, and for providing up-to-date instructions and adequate training to personnel operating gates and valves.

## EARTHQUAKE LOADING CONDITIONS

The failure modes would be the same as those described above for static loading conditions; however, the earthquake loading could initiate development of the failure mode by causing damage/disruption. The best defense against earthquake-related potential failure modes is a design that effectively addresses the design seismic loadings. Since it is impossible to accurately forecast the exact seismic loading that may actually occur, and the structural response to the loading, an immediate post-earthquake inspection of the hydraulic structure (and entire dam) should be made when significant seismic shaking of the damsite occurs. At a minimum, this would serve to verify expected satisfactory performance. To be effective, a routine monitoring program would also need to be in place to adequately document pre-earthquake conditions for comparison with post-earthquake observations.

It is possible that the dam and/or hydraulic structures could suffer some ill-effects from an earthquake (associated with seepage performance) that will not show up until higher reservoir elevations are subsequently reached and/or until a substantial amount of time has passed. Therefore heightened awareness and monitoring may be appropriate for some time following an earthquake, with particular attention paid when the reservoir rises to elevations that have not been experienced since the earthquake.

## FLOOD LOADING CONDITIONS

### FLOODING FAILURE MODE 1
Gates could fail to operate or are not operated as expected. The problem could be due to: (1) stuck, damaged, or inoperable gates, (2) lack of power or backup power, (3) loss of access to the site (and no remote operation capability), and/or (4) operator error. Typically the problem relates to gates that are closed when they should not be, which leads to overtopping of the dam when it otherwise need not occur.

The best defense against this failure mode incorporates the following:

1. An appropriate structural and mechanical design of gates and valves, including appropriate checks/reviews of designs, both as they are completed and in the future (as new information/experience becomes available).

2. Redundant, independent power sources for equipment, and "guaranteed" access to the site and/or remote operational capability and backup capability.

3. A program for regular inspection of the gates/valves, power sources, and site access situation, regular exercising of the gates/valves, and providing up-to-date instructions and adequate training to personnel operating gates and valves.

## FLOODING FAILURE MODE 2
Flood discharges could lead to erosion at the toe of the dam, that eventually undermines the dam and leads to breaching failure. Because concrete dams and their foundations generally are quite erosion-resistant, this failure mode is most relevant to embankment dams. The best defense against this potential failure mode involves: (1) appropriately located spillway and outlet works stilling basins, and (2) visual monitoring during flood discharges and post-flood inspections to look for evidence of potential problems that could result in dam failure at higher flow rates and/or longer flow durations.

## FLOODING FAILURE MODE 3
Flood discharges can lead to erosion at the end of the hydraulic structure that, by headward erosion, eventually undermines and fails the structure and results in uncontrolled release of the reservoir. The best defense against this potential failure mode involves: (1) an adequate design for energy dissipation and/or erosion protection using scale model testing, as appropriate, and (2) visual monitoring during flood discharges and post-flood inspections to look for evidence of potential problems that could result in hydraulic structure failure at higher flow rates.

## FLOODING FAILURE MODE 4
Flow through the hydraulic structure during a flood event can cause erosion and damage to the structure (itself) due to cavitation, stagnation, or other irregular flow patterns. Dam failure could occur due to: (1) Scenario A: headward erosion of the hydraulic structure, which eventually leads to uncontrolled reservoir release, (2) Scenario B: undermining of the support

for the dam afforded by the foundation material around the hydraulic structure (which is being eroded away), which eventually leads to breaching of the dam, (3) Scenario C: erosion at the toe of the dam, due to flows no longer being confined within the hydraulic structure as they were intended to be, that eventually undermines the dam and leads to breaching failure, or (4) Scenario D: overtopping and breaching of the dam due to reduced flow capacity of the damaged hydraulic structure. Possible mechanisms associated with this failure mode include:

1. Cavitation occurs in a tunnel or conduit, leading to Scenarios B or D. Remediation work typically involves constructing air slots.

2. Stagnation occurring in an overflow spillway, leading to Scenarios A, C, or D. Stagnation occurs when the upstream end of a floor slab displaces upward relative to the downstream end of the floor slab immediately upstream. If water can penetrate the joint between the two slabs and get beneath the downstream slab, it would be highly pressurized and may be able to jack the downstream slab out.

3. Cavitation occurs in an overflow spillway, due to significant offsets at abutting floor slabs (the upstream end of the downstream slab being too low, relative to the downstream end of the upstream slab). Dam failure could occur by Scenarios A, C, or D.

4. Significant offsets develop in the spillway chute or stilling basin walls that lead to unusual flow patterns and failure of the chute walls and/or stilling basin. Alternatively flows could simply overtop the walls (with or without significant offsets along the walls). Failure of the dam could occur by Scenarios A or C.

5. Abraded or otherwise deteriorated concrete in the spillway floor slabs leads to erosion and/or slab removal, which causes toppling of the spillway walls. Dam failure could occur by Scenarios A or C.

The best defense against this potential failure mode is the same as for Flooding Failure Mode 3 above.

## SUMMARY

The performance parameter process provides a cost-effective means to identify failure modes at a dam that pose "non-negligible" risk, and to develop an appropriate monitoring program to address the concerns.

# ABRASION/EROSION IN STILLING BASINS

Leslie Hanna and Elisabeth Cohen
Bureau of Reclamation, Denver, Colorado, USA

## INTRODUCTION

Many stilling basins have experienced damage caused by rock, gravel, and sand brought into the basin by back flow over the stilling basin end sill. Normal operation of a hydraulic jump energy dissipation basin can cause a reverse flow eddy over the basin end sill and lower apron as shown in figure 1. This counter-rotating eddy is driven by a high velocity jet rising off the basin floor near the end of the basin. Riprap placed on the apron downstream of the basin end sill is typically designed to be stable under this condition. However, small material can be transported into the basin and trapped where turbulent flow continually moves the material about the surface, eroding the concrete. The cost for these repairs, in terms of time, effort and money, can be significant. If a means to reduce the reverse flow can be found, large savings can be obtained. One possible solution that is currently being studied at the Bureau of Reclamation's Water Resources Research Laboratory (WRRL) is to install flow deflectors in the basin to improve inter-basin flow conditions and minimize upstream velocities over the basin end sill (figure 2).

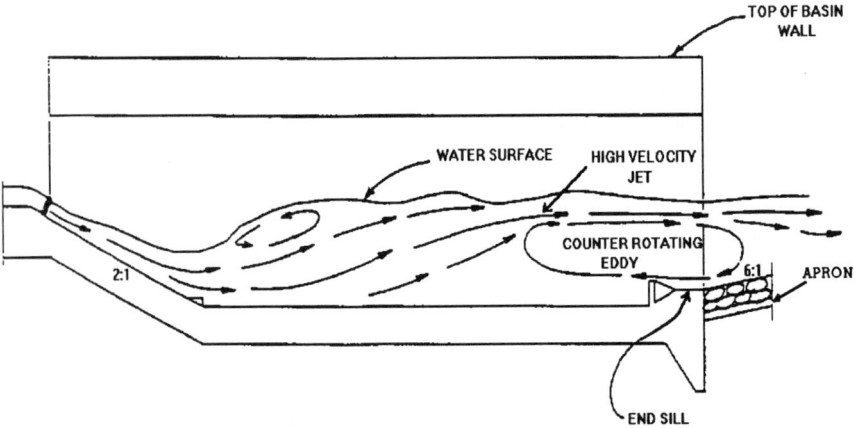

**Figure 1.** Counter-rotating flow eddy over basin end sill and lower apron.

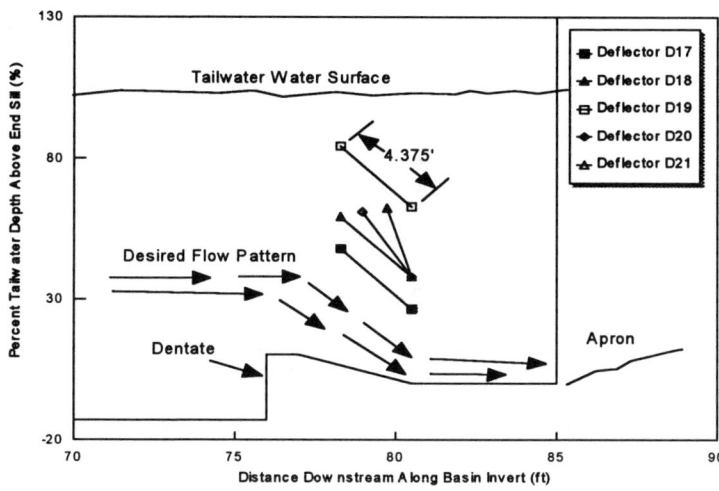

**Figure 2.** Deflector locations with respect to tailwater depth above the basin end sill.

## EXPERIENCES

Many stilling basins have experienced abrasion damage, as exemplified below at several Reclamation Dams. Often abrasion has progressed to depths exposing reinforcement and requiring repair of concrete by sawcutting, sandblasting, and concrete replacement with polymer concrete, or silica fume concrete.

*Vallecito Dam* - Vallecito Dam, completed in 1941, experienced abrasion/erosion damage in the outlet works stilling basin in the 1980's. The repairs involved a silica fume concrete with high slump and 9000 lbs/in$^2$ strengths. The work was completed in 1991. The spillway chute has since experienced more erosion indicating this is a continuing problem.

*Ridgway Dam* - An underwater inspection of the Ridgway Dam outlet works stilling basin revealed that the concrete floor was severely eroded, with the reinforcing bars exposed. The region will have this work repaired using a two phase process. The first phase is to construct bypass capacity to dewater the stilling basin, remove all materials, and determine the extent of repairs needed. The second phase the following year will be to make the repairs. The work scope is not determined but when all is complete the cost may be between $200,000 and $1,000,000.

*Taylor Draw Dam* - In 1991, about $200,000 was spent to repair abrasion damage to the Taylor Draw Dam outlet works stilling basin. After just one operating season, an inspection revealed that abrasion damage had again occurred. After repairs were complete the second time, a study conducted by WRRL demonstrated that the installation of flow deflectors improved the basin's flow distribution significantly, greatly reducing the potential for movement of material into the basin. The deflectors have been in place for 4 years with no further repairs to the basin concrete required.

## THE MODEL

A physical model is being used to investigate hydraulic conditions in Type II stilling basins and to study the affect of deflector positioning and inclination on flow patterns over the basin end sill. The study will be used to optimize and generalize flow deflector designs based on basin geometry and operating conditions. The Ridgway Dam outlet works and its Type II twin bay stilling basin are being used for the model investigations. The model includes the 42-in high pressure slide gates discharging into 2:1 sloping chutes and 12 ft wide bays. The basin is 85 ft long. Froude scaling was used to model the outlet works at a 1:10.5 scale. The downstream riprap apron topography was modeled on a 6:1 slope with moveable bed material to simulate the abrasion source. Unit discharges (q) (corresponding to 40-, 60-, 80-, and 100-percent gate openings for the Ridgway Dam outlet works), and percent of tailwater depth were used to describe flow conditions. Velocity measurements were determined using a sontek acoustic flow meter and were measured at the downstream end of the basin end sill in the center of the bay. Bottom velocities were measured 5.25-in above the basin end sill. All velocities are described in terms of average velocities. Tailwater was set according to the tailwater curve generated for the Ridgway Dam outlet works operations.

## INVESTIGATIONS

Flow conditions over the basin end sill were characterized with profiles representing average velocities (negative values represent velocities in the upstream direction) mapped along the vertical axis in the center of the bay for unit discharges of 29 ft$^3$/s/ft (40% gate), 41 ft$^3$/s/ft (60% gate), 52 ft$^3$/s/ft (80% gate), and 60 ft$^3$/s/ft (100% gate) as shown in figure 3. The vertical axis shows the relative depth in percent of total tailwater depth over the basin end sill. Initial investigations determined that the most effective position along the length of the basin was to locate the deflector directly above the downstream slope of the basin dentates. Once this was established, the most effective position along the vertical axis was investigated. Figure 3 shows that as values of unit q increase, the thickness of the high velocity (downstream) jet increases, thereby lowering the transition point between upstream and downstream velocities above the basin end sill. The effectiveness of the flow deflector is dependent on the vertical location of the deflector with respect to this transition point and its ability to trap and redirect a large enough portion of the high velocity jet (immediately above the transition point) to improve flow conditions. With this in mind, three vertical locations and several deflector angles were investigated.

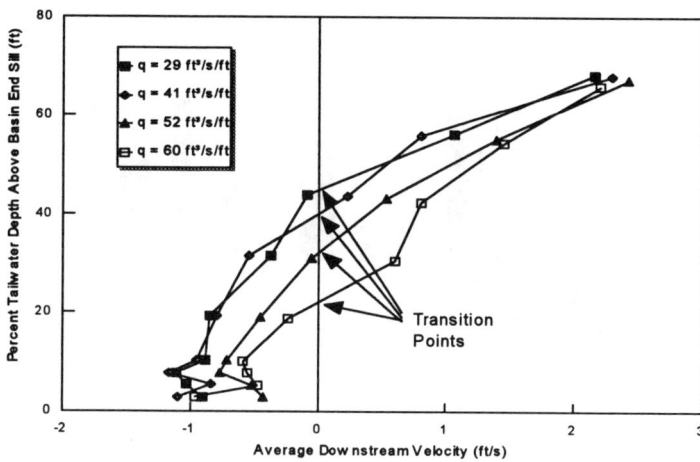

**Figure 3**. Velocity profiles measured along the vertical axis above the basin end sill.

All of the deflectors tested were 4.375 ft deep and were located as shown in figure 2. Deflectors D17 through D19 were positioned at an angle of 60 degrees, and deflectors D20 and D21 were positioned at 70 and 80 degrees, respectively.

Figure 4 shows bottom velocities measured near the basin end sill for deflector positions D17 through D21 for each flow tested. The results of these investigations show that the performance of each deflector varies over the range of flows. When the deflector was positioned low in the basin and just above the transition points of the higher flows (i.e. D17) the deflector performed well at the high flows. However, it became ineffective as the flow was decreased because the transition point moved above the location of the deflector. As a result, at the lower flows, the deflector missed a major portion of the high velocity jet because it was positioned below it. A similar problem occurred when the deflector was positioned too high (i.e. D19). Although the deflector was in good position (just above the transition point) to redirect the jet at the lower flows; as the flow was increased, the transition point moved too far below the deflector for it to remain effective The solution was to position the deflector (D18) between the locations of deflectors D17 and D19 where it would be less sensitive to the movement of the transition point. This produced positive downstream velocities (average) throughout the range of flows.

Next the angle of the deflector was varied. Deflectors D20 and D21 were installed at the same location as D18 except with the angle increased to 70 and 80 degrees respectively.

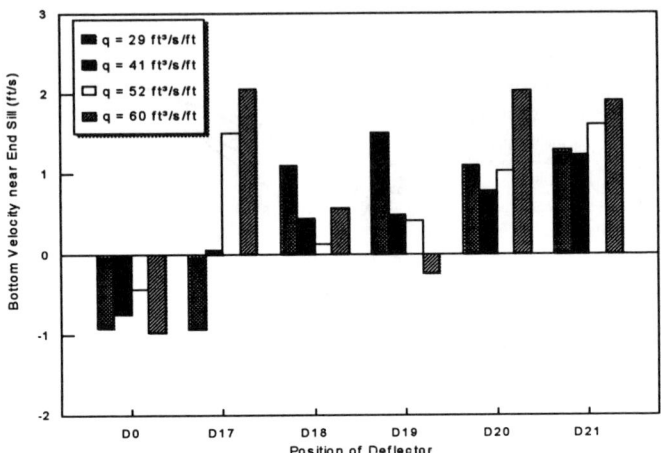

**Figure 4.** Bottom velocities (average) measured for each deflector position. (D0 indicates no deflector was installed.)

Figure 4 demonstrates that flow conditions were improved as the angle was increased and the best overall results, throughout the range of flows, occurred with deflector D21 installed.

Table 1 shows the velocity range within one standard deviation (67 percent confidence level) for the bottom velocities measured for deflector D21 and with no deflector (D0) installed. The table demonstrates that with deflector D21 installed, velocities over the basin end sill act predominately in the positive or downstream direction. Without a deflector, the velocities predominantly act in the upstream direction.

Table 1. Bottom velocities within one standard deviation.

| Deflector Position | Velocity Range Within One Standard Deviation (ft/s) | | | |
|---|---|---|---|---|
| | q = 29 ft³/s/ft | q = 41 ft³/s/ft | q = 52 ft³/s/ft | q = 60 ft³/s/ft |
| D21 | -.14 to 2.67 | -.08 to 2.56 | .05 to 3.13 | .14 to 3.62 |
| D0 | -2.1 to .239 | -2.22 to .02 | -1.62 to .43 | -1.7 to .51 |

Each of these investigations was conducted with the tailwater depth set at a specific level according to the tailwater curve for Ridgway Dam outlets works operations. Future investigations will determine the best deflector positioning relative to fluctuations in tailwater depth.

## CONCLUSIONS

Deflectors have been designed and installed at Taylor Draw Dam with marked improvements in stilling basin flow patterns; and based on the model study, performance of the deflectors show the potential for significant savings by reducing damage caused by abrasion.

The results of the Ridgway Dam hydraulic model study indicate that the effectiveness of the deflector depends on the basin discharge and on the deflector's relative position and sensitivity to the movement of the transition point throughout the range of operations. The study showed the deflector was most effective when it was located between 38 percent and 69 percent of the average tailwater depth over the full operating range, and positioned at an angle of 80 degrees.

Further investigations will determine if the deflector location can be generalized over large ranges of tailwater depth. If the variation of the tailwater (i.e. the operating range) is greater than 200 percent, a single deflector may not be effective. The structural design of the deflectors will depend on the material used, the overall width of the stilling basin, and the angle of the deflector. Future work may also involve determining the maximum basin width that the deflector design will be effective.

Further work at the Bureau of Reclamation Water Resources Research Laboratories will include generalizing flow deflector designs for Type III stilling basins.

## REFERENCES

[1] Dodge, Russ, "Hydraulic Study of Taylor Draw Dam Outlet Works," U.S. Department of the Interior, Bureau of Reclamation Report R-92-10, March 1992.

# Folsom Spillway Gate Failure and Lessons Learned

ROBERT V. TODD
United States Bureau of Reclamation
Denver, Colorado, USA

## INTRODUCTION

Spillway Tainter (radial), gate No. 3 failed at Folsom Dam, California, on July 17, 1995, in the early morning. The failure resulted in a discharge of 1,132 m$^3$/s into the lower American River. This paper will address how the forensic team formed to investigate this failure functioned, and arrived at a failure theory.

## BACKGROUND

The gate that failed was one of five service spillway gates 12.8 m wide by 15.2m high. There are also three other emergency spillway gates, with the same width, but 15.9 m high. The dam and gates were designed and contructed by the US Army Corps of Engineers (Corps) between 1948 and 1956. The Corps transferred the dam to the Bureau of Reclamation (Reclamation) for operation and maintenance in 1956.

An action team was first established comprising personnel from the Corps and Reclamation to design reinforcements for the remaining seven gates, and to design a new gate to replace the failed one. Shortly afterwards, the Corps and Reclamation agreed that, because of the importance of determining the actual failure mode a forensic team (TEAM) should be formed. This turned into a multi-disciplinary, multi-agency team. The primary agencies involved were Reclamation, Corps, California Department of Water Resources, California Department of Transportation (Caltran)), US Air Force McClellen Base, California (USAF), Osaka University, and Woodward-Clyde Corporation.

## INITIAL INVESTIGATIONS

After the No. 3 spillway bay was stoplogged, a Contractor secured the gate in its failed position and built scaffolding platforms so that the TEAM could nspect it, and allow photographs to be taken to record its post-failure conition. The gate was then cut into four main pieces and moved to a yard where the pieces were laid out. A thorough examination of the remaining gates by a climbing team of engineers drawn from the Reclamation and the Corps identified corrosion of bolts and nuts, and also of some of the structural memers, especially where ponding of water had occurred. Both Caltran and USAF performed numerous destructive and non-destructive tests on members of the failed gate, and nuts and bolts to determine their material characteristics.

The TEAM had a wide array of resources and expertise in the engineering field to draw upon, and these assets were utilized where applicable. At the end of each meeting, the TEAM leader would assign tasks which had been identified during the meeting to members who had the appropriate background and expertise. Each task would be assigned a completion date, which would be agreed upon by the TEAM. The USAF, with their extensive experience in forensic work of crashed aircraft, were assigned the task of determining a failure sequence. They examined all of the failed gate pieces, paying special attention to the various structural joints.

Because of the relatively light construction of the Folsom gates, compared to other Tainter gates of a similar size, it was thought that the gates would have a tendency to vibrate when being used. Dr. Ishii of Osaka University, who has had a lifetime of studying dynamics of gates, inspected the failed gate and wrote a report indicating a strong possibility of failure due to flow induced vibrations.

## INITIAL FINDINGS

Based on observing corrosion on the gates, led to the assumption that corrosion had played a major role in the failure. Dr. Ishii's report highlighted the possibility of a failure due to flow-induced vibrations. The tests performed by Caltran and USAF indicated that all materials of the failed gate met the required specifications as referenced on the original Contractor's construction drawings.

USAF failure sequence indicated that the failure was initiated through an overload of the strut bracing bolts at the joint of the diagonal closest to the trunnion at its attachment to the third strut, see Fig. 1. This occurred on the right side of the gate, looking downstream. After this joint failed, the other diagonal connections failed, then the two lower struts, with loss of bracing about their weak axii, buckled downwards. This resulted in the loss of structural integrity of the right arm system. The consequences were total failure of the gate, with only the right hoist chain preventing the gate from being swept down the spillway. The failure scenario was based on inspection of the failed gate. Theoretical analyses was performed to validate this scenario. A key area of the analyses was the diagonal brace joint, the strength of which was dependent upon four 19-mm-diameter bolts. In order to determine the structural capacity of this joint, Caltran designed a test fixture, using actual components from the gate. The joint failed at a maximum tensile load of 361 kN.

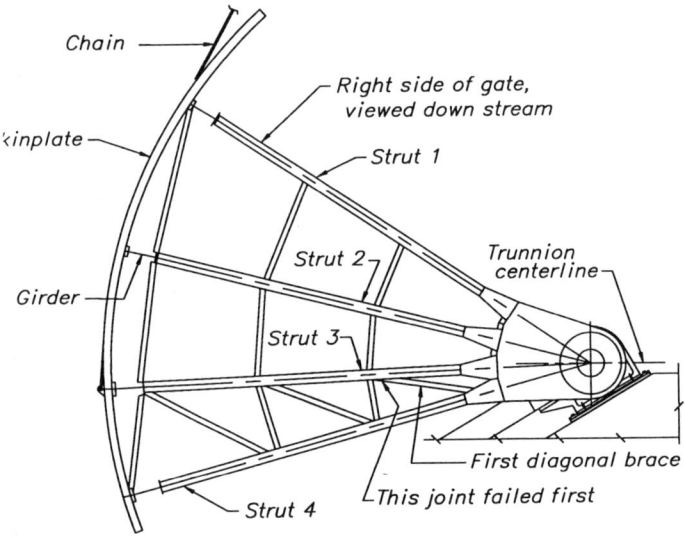

Fig. 1 - Initial failure at diagonal brace joint

## THEORETICAL ANALYSIS

Both Reclamation and USAF performed theoretical analysis using 2- and 3-dimensional (D) finite element models (FEM). The condition at failure was simulated in the models: gate being raised; opening 0.8 m; reservoir surface at elevation 141 m; head of water on gate was 13.3 m. The greatest loading applied to the gate, discounting vibration, occurs when the gate is just being raised from the sill. Under these conditions there are the following loads: static water load; dead weight of the gate; friction from the side seals; chain loads; and trunnion friction moments. The only unknowns are the trunnion friction moments which are dependent upon the coefficient of friction between the steel pin and the bronze bushing. By limiting the load carrying capability of the bracing joints, it was determined that the gate failed with a trunnion friction coefficient value of 0.22 to 0.28. The variation in the value was dependent upon the method of structural representation of the gate in the model and degree of element complexity.

It was not until December 1995, that the original calculations were found. The gate had been designed to the guidelines of the 1950's, but did not include the loading imposed by the trunnion friction moments. This omission, because of the large pin diameter of 0.81 m, resulted in an under-designed gate with minimal reserve structural capacity.

## DYNAMIC TESTING OF ORIGINAL GATE

To obtain additional parameters for the theoretical analysis of the gate, Dr. Ishii performed a modal analysis of gate No. 2. Data were obtained with the gate raised 15.2 cm, but no water loading on the gate. These data confirmed Dr. Ishii's preliminary analysis that the gate could be self-excited and fail under conditions of low-gate openings with high-reservoir heads. The TEAM at this juncture had conflicting information as to the failure loading, was it due to vibration, or to trunnion friction, or a combination of these loadings.

## DETERMINATION OF TRUNNION FRICTION

By strain-gauging two of the original gates, and using the data from the FEM, values of trunnion friction were determined. However, because the gates were being reinforced, they were all stoplogged off, and the resulting bearing pressure on the pins due to gate weight alone, was minimal. The analysis was found to be inconclusive as bearing pressures needed to be similar to those at the time of failure, that is at least 28 times higher. A friction testing fixture was made from one of gate No. 3's trunnions and the coefficient of friction was determined to be around 0.2 at bearing pressures equivalent to those at the time of gate failure.

## NEW EVIDENCE OF WALL PLATE MARKINGS

Early in 1996, a thorough visual inspection of the wall plates of all of the gates was made by a climbing team. This inspection determined that circular markings, which had been visible to Dr. Ishii through binoculars and long distance camera lenses, were due to grinding. Dr. Ishii had assumed the circular markings were due to gate vibration and scribed by bolt heads on the side seal plates. However, further information was still needed on the dynamics of the gates and whether there had been any vibrational loadings on the gate.

## TESTS ON REINFORCED GATES

After the gates had been reinforced in the Spring of 1996, additional testing was performed. Gate No. 2 was strain-gauged and static readings were taken of the stresses in two of the struts on either side of the gate. This was to obtain data to determine which of the FEM most closely predicted actual gate stresses. Accelerometers were also attached to the gate, and responses were obtained for the gate being raised to 1.2 m with a head of water of 11.5 m. It was found that the Reclamation's 3-D FEM using the ADINA program gave the closest correlation with actual gate stresses. The trunnion friction coefficients from this model, with a pin bearing stress of 11.9 MPa, were determined to be 0.24 and 0.14 for left and right trunnions, respectively. The accelerometer responses were minimal and no dynamic stresses were discernable. The dynamic characteristics of the original and reinforced gates were determined from the 3-D FEM to be very similar. This indicated that the original gate most likely did not vibrate in a self-excited mode. Based on these findings, the TEAM concluded that vibration did not play a key role in the gate failure.

## FINAL CONCLUSION OF FAILURE LOADING

The failure loading was determined to be trunnion friction, which increased over time due to corrosion on the pin. It had been overlooked in the initial design, and therefore the strut bracing and struts had not been sized adequately. With failure of the diagonal bracing, the two lower strut members failed around their weak axes. A more detailed explanation of the failure, and information on the various tests performed appear in the Forensic Report (1996).

## LESSONS LEARNED IN USING A FORENSIC TEAM

The lessons learned by forming a forensic team to investigate the failure of the Folsom Tainter gate are as follows:

1. Having a multi-disciplinary team was a major asset. The different capabilities and resources of the TEAM were used in performing a series of tasks. Initially, there was a need for a lot of tasks to be

addressed at the same time. With the resources available, these tasks were addressed in a timely manner.

2. When major issues were being discussed and findings of tasks presented, the TEAM was not narrowly focussed but brought a multiplicity of ideas and questions to the table.

3. Good leadership is required for a team to be successful. Chuck Howard, from the Reclamation Sacramento Regional Office, fulfilled this role. He oversaw the tasks definitions, scopes, and assignments, and kept the TEAM focussed.

4. Synergy, or group energy, was present at nearly all of the meetings. This is an environment which is dynamic and results in extra output and ideas.

5. Differences of opinion arise under team conditions; however, if these differences are addressed professionally without personal animosity, the outcome invariably strengthens the team.

There were a number of important lessons to be learned from the Folsom Tainter gate failure investigation, and they are identified as follows:

1. Because a gate operates satisfactorily over a number of years, it does not necessarily follow that it will continue to do so. There may be an inherent structural weakness, and the gate could fail without notice.

2. It is necessary, especially for larger spillway Tainter gates, that the gate design calculations are available for review. A review should be performed any time that there is a major inspection of a gate to ensure that the design takes into consideration current design standards.

3. In the case of large gates with trunnions requiring greasing, it is advisable to provide an automatic greasing system.

## REFERENCE

Spillway Tainter Gate No. 3 Failure, Folsom Dam (1996). Forensic Report, Bureau of Reclamation, Mid-Pacific Regional Office, Sacramento, California.

# RESERVOIR SEDIMENTATION IN CHINA
## --EXPERIENCES AND LESSONS
by

ZHOU ZHIDE and YANG XIAOQING
International Research and Training Center on Erosion and Sedimentation
P.O. Box 366 Beijing, China 100044

## ABSTRACT

Experiences and lessons on reservoir sedimentation based on the practice of more than 80,000 reservoirs in China are introduced. Selection of a rational operating rule for a reservoir is the first priority in dealing with reservoir sedimentation. The operating rule of impounding the clear and discharging the muddy is extensively discussed. The necessity of installation of outlets with sufficient discharge capacity to keep a long-term capacity of a reservoir is presented.

## INTRODUCTION

Since 1949 more than 80,000 dams have been built in China, among them 18,000 are large dams. Reservoirs behind these dams have experienced different situations in reservoir sedimentation. Reservoirs on sediment-laden rivers have experienced serious sedimentation and its induced problems. Many lessons and experiences dealing with sediment problems have been accumulated.

At present 150 or so dams higher than 30 m are being built in China, including the Three Gorges Project on the Yangtze River and the Xiaolangdi Project on the Yellow River. During the planning and design of these dams the existing experiences and lessons in reservoir sedimentation have been extensively utilized to make these projects successful.

## UNDERESTIMATION OF SERIOUSNESS OF SEDIMENT PROBLEMS

In the 1950s reservoir sedimentation was not always given its appropriate position in the planning and design of hydroprojects. Underestimation of the seriousness of sediment problems led to severe impacts on the management of the projects.

## SELECTION OF DAM SITE

Although reservoir sedimentation is not always an important factor in the selection of dam site, it may be of significant importance in some cases. Take the Liujiaxia Project for instance(Yu,1988).

Liujiaxia Project, commissioned in 1974, is a multi-purpose hydroproject built on the upper Yellow River. In order to fully utilize the water of Taohe River(annual runoff 5.17 billion m³ compared with 26.3 billion m³ of the Yellow River) the dam site was finally set 1.5 km downstream of the confluence of the Yellow and Taohe Rivers(Fig.1).

Compared with the reservoir capacity of 5.74 billion m³, the annual sediment load of the Yellow river, 89.4 million tons, is not large. Until now the pivot point of the delta in the main reservoir is still far away from the dam. Thus, sedimentation induced by the sediment influx of the Yellow River has not caused any trouble to the project so far.

The major problem is the formation of a river mouth bar near the confluence in the main river by the sediment carried by the Taohe River. As the storage of the Taohe arm is only 2% of the total storage capacity and the annual sediment load of the Taohe is quite large, 28.6 million tons, the inactive storage in the Taohe arm was filled up and the river mouth bar rose to the minimum pool level by 1980. Proximity of the river mouth bar resulted in a rapid increase in the amount of sediment passing through the turbines. The maximum annual amount of sediment passing through power unit 2 was as large as 11.6 million tons of sediment in 1978. Abrasion of turbine blades has been very serious. In June 1980, when more flow was required to meet an abrupt increase in power demand, the pool level in front of the dam suddenly dropped by a large amount because the river mouth bar impeded the flow of water to the dam from the upstream part of the main reservoir.

If the dam site of the Liujiaxia Project had been selected above the confluence of the Yellow and Taohe Rivers, sediment problems of the Liujiaxia Project would have not been so serious at the initial stage of the project. The decrease of the benefit of the project would have been compensated as remedial measures to use the water of the Taohe River could have been found.

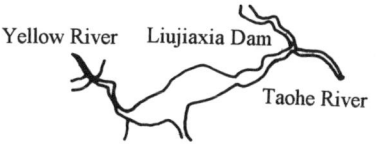

Fig. 1 Plan of Liujiaxia Project

## SELECTION OF OPERATING RULE

In the 1950s when a huge number of dams were built in China, all of the reservoirs were designed to adopt the operating rule of impoundment. However, such an operating rule is mainly appropriate for reservoirs built on clear rivers. For reservoirs built on sediment-laden rivers it induces serious reservoir sedimentation. Take the Sanmenxia Project for instance(Long and Li, 1995).

Sanmenxia Project, commissioned in 1960, is the first large multi-purpose hydroproject built on the Yellow River. It was originally designed under the operating rule of impoundment. The discharge capacity of outlets at pool level 315 m was 3,084 m³/s(the average annual discharge at Sanmenxia is 1,580 m$^3$/s). The original reservoir capacity was 35.4 billion m$^3$ at normal pool level 350 m.

From September 1960 to March 1962 1.7 billion m$^3$ of reservoir capacity below 335 m were occupied by sediment deposits. Meanwhile, upstream extension of backwater deposits was very serious and would cause severe impact on the City of Xi'an.

In order to relieve the serious situation of reservoir sedimentation the operating rule had been changed twice and reconstruction program had been carried out to enlarge the discharge capacity of outlets before 1973. At the third stage of operation when the operating rule of impounding the clear and discharging the muddy(I&D) is adopted, the equilibrium state of reservoir sedimentation has been reached, a long-term storage capacity of 5.9 billion m$^3$ is kept (Table 1).

Table 1 Three operational stages of Sanmenxia Project

| Stage | Time period | Operating rule | Discharge capacity (m$^3$/s) | Trap efficiency(%) |
|---|---|---|---|---|
| 1 | Sep.1960 to Mar.1962 | Impoundment | 3,084 | 92.9 |
| 2 | Mar.1962 to Oct.1973 | Flood detention & sediment flushing | 6,064 | 20.0 |
| 3 | Oct.1973 to present | I & D | 9,064 | 0 |

## MEASURES FOR MITIGATING RESERVOIR SEDIMENTATION

The first priority for mitigating reservoir sedimentation is to select a rational operating rule for a reservoir. Then an overall plan for sediment management should be developed. Commonly there are two basic approaches: 1) preventive measures to reduce sediment inflow to a reservoir, and 2) remedial measures to reduce sediment deposits in a reservoir. Combination of measures from both approaches may be selected to optimize a sediment management plan.

Potential preventive measures include watershed management, soil conservation, warping, sediment bypassing, debris or settling basins, and constructing new upstream dams. Potential remedial measures include density current venting, sediment flushing, dredging and excavating, siphonage, and dam raising. All these measures may be used for small reservoirs, but some of them may be not suitable for large reservoirs, such as siphonage and dredging.

## SEDIMENTATION AND OPERATING RULES OF RESERVOIRS

Although reservoir sedimentation is a universal problem, the extent of sedimentation of various reservoirs are quite different. Reservoirs may be classified into three groups based on the diagram shown in Fig.2, using ratio of reservoir capacity to annual water runoff, $\psi$, ratio of reservoir capacity to annual sediment load, $\phi$, and annual average sediment concentration of the river as parameters. The first group is reservoirs built on clear rivers with low annual average sediment concentration less than 1 kg/m$^3$. The second group includes reservoirs built on rivers with medium annual average sediment concentration in the range of 1 to 10 kg/m$^3$. The third group is reservoirs built on heavily sediment-laden rivers with annual average sediment concentration greater than 10 kg/m$^3$ (Fig.2).

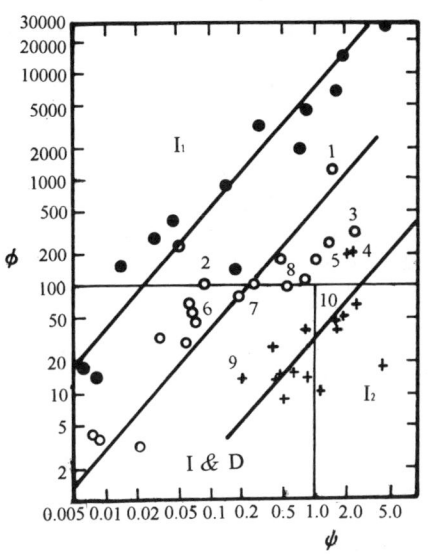

- Rivers of low concentration (<1 kg/m$^3$)
- Rivers of medium concentration (1-10 kg/m$^3$)
- Rivers of high concentration (>10 kg/m$^3$)

$\psi=V/W$  $\phi=V/W_s$

V-Reservoir storage capacity
W-Annual runoff
$W_s$-Annual sediment load (in volume)
1-High Aswan
2-Three Gorges Project
3-Lake Mead
4-Elephant Butte
5-Boulder
6-Low Aswan
7-Tarbela
8-Liujiaxia
9-Sanmenxia
10-Guanting

Fig. 2 Diagram showing various groups of reservoirs and various regions of operating rules

According to the situation of existing reservoirs, $\phi$ is an important index showing the extent of seriousness of reservoir sedimentation. Reservoir sedimentation is not a problem for reservoirs with $\phi$ larger than 100 (most reservoirs in group 1). Reservoir sedimentation is serious for reservoirs with $\phi$ less than 100 (most reservoirs in group 3).

The extent of sedimentation in a reservoir is closely related to the adopted operating rule of the reservoir. Three basic types of operating rules have been adopted in China: 1) impoundment, 2) I&D, and 3) detention. The first two types are the most often adopted. In Fig.2 three regions showing the adopted operating rules of reservoirs are depicted. $I_1$ region ($\phi$ >100) shows reservoirs without serious sedimentation problem, for those reservoirs adopting the operating rule of impoundment is natural. $I_2$ region ($\phi$ <100, $\psi$ 1) shows reservoirs with serious sedimentation problem. However, it is difficult to adopt I&D operating rule for those reservoirs as water is not enough to fully fill the reservoir after discharging muddy water during the flood season. Therefore, no other choice is possible other than impoundment operating rule. For reservoirs in the region of I&D($\phi$<100 and $\psi$<1.0) I&D operating rule is the first choice.

From the data of Sanmenxia Project (Table 1) it is clear that the influence of operating rule is decisive.

## LONG-TERM CAPACITY OF A RESERVOIR

When a new equilibrium alluvial channel in a reservoir is finally developed the remaining storage capacity of the reservoir is called the long-term capacity of the reservoir.

The strategy of reservoir sediment management is to maintain the long-term capacity of a reservoir as large as possible.

The storage of a reservoir consists of two parts, namely that over the flood plains and that of the main channel (Fig.3). The storage over the flood plains will be gradually lost to deposits carried by overflow of floods. However, a part of the storage of the main channel may be preserved through rational operation of the reservoir by lowering the pool level in the flood season. This part of the storage is the long-term capacity of a reservoir.

One of the prerequisite conditions to keep an adequate long-term capacity of a reservoir is to install sluicing outlets at a rational elevation with sufficient capacity.

## LOW-LEVEL OUTLETS

The design of low-level outlets installed in a dam should conform with the adopted operating rule of a project. For projects adopting the operating rule of impoundment the design of outlets should meet the demand for regulating runoff, unless density current venting should be considered. For projects adopting I&D operating rule regulating both runoff and sediment load is a must in design of outlets, while regulating sediment load plays a key role in the design of outlets. Practice in China shows sufficiently large discharge capacity of low-level outlets is the prerequisite of successful execution of I&D operating rule. Some empirical formula for determining the discharge capacity of outlets for projects adopting I&D operating rule have been developed.

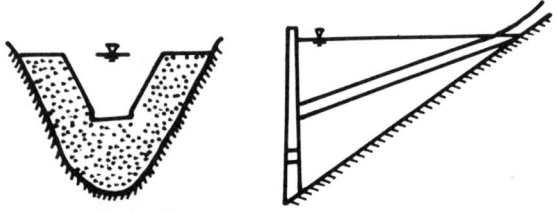

Fig.3 Sketch of long-term capacity of a reservoir

## CONCLUSIONS

1. Although reservoir sedimentation is a universal problem, its extent of seriousness is different for various reservoirs. For reservoirs built on clear rivers sedimentation is not a big problem, while for reservoirs built on sediment-laden rivers it is severe and should be dealt with seriously and cautiously.

2. The first priority in mitigating reservoir sedimentation is to select a rational operating rule for a reservoir. Fig.2 is recommended for selection of an optimum operating rule.

3. The operating rule of impounding the clear and discharging the muddy (I&D) is useful for reservoirs with the ratio of storage capacity of a reservoir to the annual sediment load being less than 100.

4. To guarantee the execution of the I&D operating rule and to keep a long-term capacity of a reservoir outlets with sufficient discharge capacity is of necessity.

## REFERENCES

Long Yuqian and Li Songheng, 1995, Management of Sediment in the Sanmenxia Reservoir, Advances in Hydro-Sciences and -Engineering, Vol.II, Beijing, China.

Yu Guanglin, 1988, Practice of Sediment Flushing at Low Water Level before Flood Season in Liujiaxia Reservoir, Yellow River, No.3 (in Chinese).

# Socioeconomic Analysis of Reservoir Sedimentation

ROLLIN H. HOTCHKISS AND FRANK H. BOLLMAN
University of Nebraska and Agricultural Industries, respectively, USA

## ABSTRACT

Current methods of economic analysis for reservoir sedimentation are short-sighted, narrowly focused, and do not allow for flexibility when operating a dam. The short-sighted nature of benefit-cost analyses is due to the use of a constant discount rate to determine the present values of future benefits and costs. Such a practice implies that any benefits more than about 100 years into the future are not worth any increase in initial project cost. Present analyses are narrow-minded because the physical and socio-economic implications of reservoir sedimentation are not considered upstream or downstream from a dam, nor are non-monetary items considered. Flexibility is denied dam operators because sediment management techniques, such as low-level outlet operation, cannot be justified using the constant and narrowly focused constant discount rate. It is recommended that discount rates be lowered or re-evaluated on a regular basis for all projects, that existing methods for socio-economic impacts be examined for reservoir sedimentation analysis, and that low-level outlet gates and similar appurtenances be seriously considered as a part of all dam designs and operations.

## INTRODUCTION

The discounted present value method is commonly used for the economic analysis of proposed dams and reservoirs. In this method, costs are compared to benefits throughout the anticipated design life and then discounted to a common point in time, such as the present. The discount, or interest rate, plays a pivotal role in assessing the cost and benefits over the design life. If benefits exceed costs, the project is deemed to be economically justified.

This method of accounting is flawed when considering sediment and sediment management because 1) a constant discount rate approach essentially dictates a short design life, 2) uncertainties in social and political futures require flexibility for dam operators to respond to changing conditions, and 3) current practice neglects most of the costs associated with sedimentation and does not account for benefits derived from sediment management.

The purpose of this paper is to examine these three shortcomings of present practices and propose improvement to current economic analysis procedures that will make it possible to justify sediment management techniques to ensure that dams are sustainable into the indefinite future.

## SHORTCOMINGS OF PRESENT VALUE DISCOUNTING

With limited global resources, dams should be constructed to operate indefinitely into the future. The design life of a dam is functionally defined as that time before major maintenance or expenditures make the project inoperational (Barkdoll and Odgaard, 1995). In practice, however, design life seems to be derived from a narrowly-defined economic analysis. For example, most dams worldwide are given a 50- to 100-year design life. Assigning relatively short design lives has to be questioned when, as a result of not incorporating adequate sediment management methods, additional costs are incurred to site replacement structures and the associated needless loss of benefits to future generations if alternative economically feasible sources of energy and water are not available. The too-short half-century to century-long design life is the direct result of basing economic analysis on the mechanics of discounting to present worth.

A discount rate seeks to express the future benefits of a dam project in terms of present values. The present value of gains minus the present value of costs equals the present net worth of the project. Projects are judged "economically feasible" when the benefit-cost ratio is 1:1 or greater. The interest (discount) rate used to discount future benefits of water projects significantly affects the resulting benefit-cost ratio (Hotchkiss and Bollman, 1996).

Most construction costs are incurred early in a project, whereas most benefits of sediment management occur in the future, when the dam is operating. As the discount rate is increased, the present value of any sediment management benefits mainly accruing in later years is greatly diminished. Table 1 illustrates the dominant nature of discount rate and why it essentially eliminates the consideration of sediment management at dams. For example, a one dollar benefit 50 years in the future when discounted to the present at a five percent rate is 8.7 cents. That same benefit if realized 75 years into the future is worth a present value of only 2.6 cents, or only one-third the benefit from the 50-year level.

Table 1. Effect of Different Discount Rates on the Present Worth (in cents) of a Future Benefit of One Dollar

|  | 3% | 5% | 8% | 10% | 12% |
|---|---|---|---|---|---|
|  | all values below are in cents | | | | |
| 50 years hence | 22.81 | 8.72 | 2.13 | 0.85 | 0.35 |
| 75 years hence | 10.89 | 2.58 | 0.31 | 0.08 | 0.02 |

The controlling aspect of discount rate is further illustrated using an example from Linsley and Franzini (1964). They demonstrate that only a moderate increase in discount rate affects annual cost more than a large difference in design life. For example, if a design life is increased from 45 to 100 years while the interest rate is increased from three to four percent, the increase in annual cost due to the higher discount rate is greater than the reduction in annual cost due to the longer design life.

Using a discount rate method to reduce costs and benefits to a present value favors short-term projects with substantial benefits in the near future at minimal investment. This approach cannot recognize the value of sediment management that would increase design life indefinitely. For example, assume that incoming sediments form a delta within the reservoir, far upstream from the dam. Once the delta face reaches the dam, sediment may be passed downstream through low-level sluice gates, effectively maintaining a storage volume indefinitely into the future (Fan and Morris, 1992). The construction of low-level outlets for this purpose, however, increases initial construction cost while benefits may not be realized for several years or even decades after the dam is constructed. While managing sediment for long-term sustainability is

## INCOMPLETE ANALYSIS

Besides discouraging sustainability because of short design life recommendations, current economic analysis procedures are also very narrow in scope. Most analyses performed in the past have focused on the narrow concept of "finances" instead of the larger socio-economic context in which dams operate (Hitzhusen, 1996). For example, the classic accounting method yields an estimate of how much sediment will be deposited in the planned reservoir during the economic life of the project. Given this estimate of sediment storage volume, the water storage requirements of the dam, and hence the dam height and cost, may be determined. This very narrow approach excludes many extended extenuating physical and socioeconomic consequences of reservoir sedimentation.

Reservoir sedimentation impacts not only the reservoir area of a project but far upstream and downstream as well. Table 2 summarizes some of the secondary and tertiary impacts of sedimentation, many of which are community-impacting and long lasting. For example, what are the costs associated with the bed and bank erosion downstream from a dam? Are those costs included in the benefit-cost analysis of the dam? If they were, then the benefits of passing sediment downstream to prevent erosion could also be included to offset the costs for low-level outlets. Initial capital cost would increase in order to lengthen the benefit stream and prevent downstream damages. The impacts are called economic locational externalities because they occur away from the primary site of disturbance.

Table 2. Sediment-related consequences of dam construction

| Primary impact | Secondary impact | Tertiary impact |
|---|---|---|
| Upstream deposition | tributary aggradation | see all others |
|  | increased groundwater levels | increased soil moisture in root zone |
|  |  | flooded homes |
|  | decreased navigational clearance |  |
|  | increased flood frequency |  |
|  | deposition at diversions |  |
|  | altered geomorphology |  |
|  | uncontrolled wetland creation |  |
| Downstream scour | armoring of bed | change in habitat |
|  | bank instability |  |
|  | tributary degradation |  |
|  | undercut diversions |  |
|  | increased bridge scour |  |
|  | lower groundwater levels | loss of riparian vegetation |
|  |  | agricultural impacts |
|  | decreased turbidity | aquatic habitat changes |
|  | geomorphic changes |  |
| Reservoir deposition | reduction in all benefits |  |
|  | reduced useful life |  |
|  | degraded water quality | decreased dissolved oxygen |
|  |  | interstitial deposition |
|  |  | contaminant concentration |

Even more overlooked than the physical impacts in Table 2 are the many and varied socioeconomic impacts of reservoir sedimentation. A dam and its reservoir are only a part of a larger watershed and the society which it benefits. Any watershed or reservoir measure should be evaluated as part of a sediment management system which includes not only the reservoir proper, but also environmental, political, social, economic, financial, technological, and other issues (Bruk, 1992). Several authors have proposed schemes for identifying and classifying "benefits" and "costs" of physical and social externalities (Bollman, 1992, and Hitzhusen et al. (1984). Non monetary issues may be quantified using economic principles such as hedonic pricing, option and bequest values, foregone benefits to future users, and existence values (Hitzhusen, 1996), or by simply defining meaningful quantities such as miles of gravel salmon spawning reaches that have been sedimented. Examples of how to account for some of the externalities and non monetary values are shown in Table 3 (from Hitzhusen, 1996).

Table 3. Benefits of Reservoir Sediment Management Described with Economic Measures (from Hitzhusen, 1996)

| Benefits of Reduced Sedimentation | Measures of Economic Benefits |
|---|---|
| Increased flood control | Reduced downstream economic loss |
| Increased water supply for agriculture, industry, business, and households | Marginal value product of water and willingness to pay of householders and recreators |
| Increased hydroelectric power | Foregone higher costs of alternative electric sources, e.g., oil |
| Increased recreation | Travel cost and contingent valuation measures |
| Increased adjacent reservoir property values | Hedonic pricing and market pricing of property |

## UNCERTAINTY

Two major sources of uncertainty exist when considering a long-term dam and reservoir: uncertainty in estimates of physical processes, and uncertainty in future land and water use and climate. Sediment transport and deposition are complex processes that at the present time cannot be estimated with a high degree of accuracy. Each estimate of soil erosion, sediment transport, deposition and resuspension, and sediment consolidation is subject to error. Estimates of reservoir life based on incomplete or inaccurate information can lead to underdesigned facilities without the ability to move sediment out of the reservoir. For example, of 23 reservoirs in India where sediment measurements were taken following construction, 21 experienced a higher-than-predicted deposition rate, and 8 reservoirs were filling in at a rate more than three times the predicted value (Central Water Commission, 1996). When estimates are greatly underpredicted, or when facilities for management are not included, expensive retrofitting becomes necessary, such as at the Sanmenxia project in China (Long, 1996), where final costs greatly exceeded the initial costs that would have included the retrofitted facilities.

Uncertainty in watershed land use and climate for projects designed for indefinite use must also be accounted for in design. Often the construction of a dam results in increased population and land use pressure from the planned uses of energy, water, and land that the reservoir makes possible. Subsequent tree clearing upstream and resulting soil erosion increases sediment delivery to the reservoir at a rate greatly exceeding that of the planning assumption. Such was the case at Anchicaya Reservoir in Colombia, where deforestation following construction of the dam reduced the project to a run-of-the-river benefit in only twelve years (Hitzhusen, 1996).

## RECOMMENDATIONS

Current economic analysis practices for reservoir sedimentation generally overlook economic life horizons beyond 100 years, are too narrowly focused, and are too inflexible. Changes are necessary in the traditional discounted present value method to allow dams to be economically justified for generations instead of decades. The recommendations are presented in three areas: discount rate procedures, socio-economic impact analysis, and uncertainty allowances.

### DISCOUNT RATE PROCEDURES

Alternatives to current schemes which penalize long-term projects must become available. Possible alternatives include incremental planning and variable discount rates. Incremental planning makes review and revision of the dam operating procedure an integral part of the economic operation of a dam. Over time, the initial project benefits and costs are monitored, not only to see if the projected benefits and costs are on track, but to explore if benefits can be increased or costs can be reduced, thus acting to ensure that the stream of projected net benefits is actually attained, or preferably, increased. This is equivalent to "resetting" the benefit/cost analysis to the present at set intervals in the future. Such a procedure could take into account changes in land use, land policy, pricing structures, and other socio-economic factors over time and overcome the disadvantages of a once-only analysis.

Variable discount rates might also be considered. For example, the discount rate may be significantly decreased or even set to zero for times past the traditional planning horizon, such as 50 or 100 years. Such a practice would give more "weight" to future benefits and avoidance costs and would result in a more evenly distributed stream of benefits over the life of the project, which could be extended practically indefinitely.

### SOCIO-ECONOMIC IMPACT ANALYSIS

Current methods of comparing benefits to costs using only on-site factors can be defined as a financial analysis. A socio-economic analysis is much broader and more complete because it includes externalities and factors that are usually not described in monetary terms. Hitzhusen et al. (1984) suggest a rational method for including benefits and costs associated with sedimentation for a wide variety of inputs. They propose accumulating costs and benefits across the income classes represented by the people served by the project. This method recognizes a wider variety of benefits and costs associated with sediment management than traditional methods and distributes them back to the people served, thus building a connection between a purely financial analysis and a broader socio-economic analysis.

## UNCERTAINTY ALLOWANCES

The uncertainty in estimates of sediment production, transport, and deposition coupled with the uncertainty of future land use argues that dams should be constructed with maximum flexibility for sediment management. For example, including outlets and maintaining them in a functional state with the idea of managing sediment can be an effective tool for managing sediment. In a case study describing the Sefid Rud Dam in Iran, Hassanzadeh (1995) relates how low-level outlets have extended the life of the dam. He emphasized that the installation of sediment flushing facilities should be considered at the planning stages of dams because retrofitting is difficult. It must be recognized that a more flexible operating plan may yield initially smaller net revenues than a flexible plan. Flexibility raises initial costs with the prospect of being able to manage sediment far into the future and sustain project benefits indefinitely. It will not only avoid high future costs associated with retrofitting, but will also ensure the flow of net social benefits at a level exceeding the planned net benefits.

## REFERENCES

Barkdoll, B.D., and Odgaard, A. J. 1995. Economic Aspects of Sedimentation Management. Proceedings, 6th International Symposium on River Sedimentation, p. 1155-1167, New Delhi, India, Nov. 7-11. New Delhi: Oxford and IBH Publishing Co.

Bollman, F. H. 1992. Institutional and Socio-Economic Analysis: Essential Aids to the Diagnosis of Sediment Management Problems. Discussion paper for organizational meeting of the International Coordinating Committee on Reservoir Sedimentation, ISMES in Seriate, Bergamo, Italy, December 21-22.

Bruk, S. 1992. The Management Challenge in Sediment Research. 5th International Symposium on River Sedimentation, V. 1, p. 39-46, Karlsruhe, Germany

Central Water Commission. 1996. Experience in Sedimentation of Indian Reservoirs . In Proceedings of International Conference on Reservoir Sedimentation, V. 1, p. 53-71, Ft. Collins, Colorado, September 9-13.

Fan, J., and Morris, G. 1992. Reservoir Sedimentation II: Reservoir Desiltation and Long-Term Storage Capacity. ASCE Journal of Hydraulic Engineering 118(3): 370-384.

Hassanzadeh, Y. 1995. The Removal of Reservoir Sediment. Water International 20: 151-154.

Hitzhusen, F. 1996. Economic Analysis of Reservoir Sedimentation. In Proceedings of International Conference on Reservoir Sedimentation, V. 1, p. 35-52, Ft. Collins, Colorado, September 9-13.

Hitzhusen, F., Macgregor, B., and Southgate, D. 1984. Private and Social Cost-Benefit perspectives and a Case Application on Reservoir Sedimentation Management. Water International 9(4): 181-184.

Hotchkiss, R. and Bollman, F. 1996. Reservoir Sedimentation: An Entreaty for Improved Analysis of Economic, Environmental and Social Effects and Consequences. In Proceedings of International Conference on Reservoir Sedimentation, V. 2, p. 885-890, Ft. Collins, Colorado, September 9-13.

Linsley R., and Franzini, J. 1964. Water Resources Engineering. New York, New York: McGraw-Hill Book Co.

Long, Yu-Qian. 1966. Sedimentation in the San-men-xia Reservoir, Yellow River. In Proceedings of International Conference on Reservoir Sedimentation, V. 3, p. 1293-1328, Ft. Collins, Colorado, September 9-13.

# Sedimentation in Sierra Nevada Reservoirs

RICHARD KATTELMANN
University of California
Sierra Nevada Aquatic Research Lab
Mammoth Lakes, California

ABSTRACT
Water resources in the Sierra Nevada are managed with a complex network of storage, diversion, and conveyance facilities. The hundreds of dams throughout the range have greatly altered sediment transport. Compilation of sedimentation records from reservoir surveys suggests that sediment yields from most catchments in the mountain range are generally less than 100 $m^3$ $km^{-2}$ $yr^{-1}$ (0.2 AF $mi^{-2}$ $yr^{-1}$). Although larger dams seem to have sufficient space to accumulate sediment for centuries, some smaller reservoirs have lost much of their capacity in just a few years or even in a single storm. Operation of these smaller structures has required removal of the accumulated sediment at major expense. Several options for coping with sediment problems in small reservoirs are being explored. Erosion control as part of a comprehensive watershed management strategy may limit the sediment supply in the long term. Redesign of structures to allow sediments to pass through the dams during high flows may offer the best hope for both operations and the stream system.

INTRODUCTION
Continuing public concern about environmental problems in the Sierra Nevada of California led to a Congressionally-mandated evaluation of the state of the environment of the entire mountain range. This Sierra Nevada Ecosystem Project attempted to identify the major ecological and resource problems in different parts of the Sierra Nevada and document the status of knowledge about the region. Water resources and aquatic habitat were central issues throughout the project, and more than two dozen of the chapters of the final report are concerned with aquatic resources. The extensive development of water resources in all river basins of the range has greatly altered flows of water and sediment out of the mountains. The interruption of sediment transport by dams affects both aquatic habitat and operation of the hydraulic facilities. This paper provides a brief overview of sediment deposition in reservoirs of the Sierra Nevada.

The Sierra Nevada is the principal mountain range of California and extends roughly northwest-southeast for more than 600 km (375 mi) along the eastern edge of the state and is about 100 km (60 mi) wide, on the average. The Mediterranean climate of the Sierra Nevada results in a strongly seasonal precipitation pattern with about half of the average annual precipitation occurring in winter and another third in late autumn. Streamflow generated below about 1500 m (4900 ft) is usually directly associated with storms, while streamflow above 2500 m (8200 ft) is almost entirely a product of spring snowmelt. Between these approximate bounds, streamflow is generated both by warmer storms in winter and by snowmelt during spring. Peak flows transport most of the sediment to reservoirs and are caused by different mechanisms in different parts of the Sierra Nevada. Most of the mountain range has a dense vegetation cover with little exposed soil.

## ESTIMATES OF SEDIMENT YIELDS

Natural surface erosion is generally regarded as small in the Sierra Nevada because of high infiltration capacity of the soils, predominance of snowmelt as a water input to soils, rarity of overland flow, predominance of subsurface flow, and relatively continuous vegetation cover. In the Sierra Nevada, the greatest potential for overland flow to occur appears to be below the snow zone in woodland-grassland communities between 300 and 900 m (1000 to 3000 ft) (Helley 1966).

There have been relatively few measurements of sediment transport and deposition in the Sierra Nevada. Suspended sediment is sampled at only a few stations throughout the mountain range and is a poor measure of total sediment load in mountain streams where most material is transported as bedload. The best means of determining sediment yields over long time periods is with repeated bathymetric surveys of reservoirs (e.g., Rausch and Heinemann 1984). Comparison of the bottom topography after a span of a few years allows calculation of the change in volume of sediment over the time interval (e.g., Rausch and Heinemann 1984). Most of the information for the Sierra Nevada came from a Soil Conservation Service study in the 1940s (Brown and Thorp 1947).

Estimates of average annual sediment yields in the Sierra Nevada were compiled from all available sources for part of the Sierra Nevada Ecosystem Project (Kattelmann 1996). These values provide order-of-magnitude approximations of sediment yield. The numbers should be considered uncertain and may contain some serious errors resulting from the original measurements, assumption of inappropriate densities if reported as mass rather than volume, and conversion from some unusual units. The period of measurement varies greatly between basins, resulting in different sediment delivery regimes depending on the inclusion of floods. Some of the compiled values were based on total basin area above the reservoir or measurement site, and others were based only on the sediment contributing area not regulated by upstream reservoirs and lakes. Most reported values are less than 100 $m^3$ $km^{-2}$ $yr^{-1}$ (0.2 AF $mi^{-2}$ $yr^{-1}$), which is the simple average of the compiled values. For comparison, the Colorado River basin produces about 300 $m^3$ $km^{-2}$ $yr^{-1}$ (0.6 AF $mi^{-2}$ $yr^{-1}$) and the Columbia River yields about 30 $m^3$ $km^{-2}$ $yr^{-1}$ (0.06 AF $mi^{-2}$ $yr^{-1}$) (Holeman 1968).

Very few measurements of reservoir sedimentation have been reported in the past two decades. The one-time measurements in isolation do not provide sufficient information or provide much confidence in using the values to infer differences between basins or over time. Comparison of modern sedimentation rates with those summarized by Brown and Thorp (1947) would be very useful in determining whether more intensive land management has altered sediment yields at the basin scale. Given the extensive water development in most river basins of the Sierra Nevada, changes in sediment delivery should be considered throughout each basin. Each dam in the network affects the channel below it. With the presence of many dams upstream, contributing areas for sediment are often much smaller than contributing areas for water (in the absence of exports out of the basin).

## SEDIMENT PROBLEMS IN SIERRA NEVADA RESERVOIRS

The generic sediment problem with impoundments is that they interrupt the natural transport of solids downstream. The change in sediment delivery can have geomorphic implications that, in turn, alter aquatic and riparian habitat. Reduction in gravel supply has greatly reduced spawning habitat in many streams and rivers of the Sierra Nevada. In addition, the accumulation of sediment behind dams reduces their useful storage volume and can interfere with operation of outlets and gates. All but the smallest dams in the Sierra Nevada had initial capacities of a few hundred thousand $m^3$ (more than a thousand acre-feet) and many started with tens or hundreds of millions of $m^3$ (tens or hundreds of thousands of acre-feet) of initial storage volume. Therefore, most reservoirs in the Sierra Nevada are unlikely to become even half filled with sediment for a few hundred years. Nevertheless, there are dozens of small diversion dams and forebays that can become filled with sediment in just a few years.

Unusually large floods can completely fill smaller diversion works, as occurred at Log Cabin dam on Oregon Creek and Hour House dam on the Middle Yuba in 1986 (Kondolf and Matthews 1993). Assuming that the dam is to remain in operation, the accumulated sediment must be removed. The methods of removal can have an assortment of impacts. Dredging, trucking, and disposal of the sediments in a stable location has been a costly approach to the problem. Ralston Afterbay on the Middle Fork American River has had sediment removed on six occasions between its completion in 1966 and 1986 (Georgetown Ranger District 1992). Location of suitable sites for long-term storage of removed sediments within a short distance from the reservoir has been difficult. The small forebays on Southern California Edison's Bishop Creek system have also required dredging of accumulated sediments. Estimates of the costs of dredging and transportation depend on access and distance to a disposal site, and have ranged from about \$26 $m^{-3}$ (\$20 $yd^{-3}$) to about \$3,500 $m^{-3}$ (\$2,700 $yd^{-3}$). A new dredging technology has recently been introduced and tested at Cresta Reservoir on the North Fork Feather River (Harrison and Weinrib 1996). This vortex slurry pump, called the "EDDY Pump", creates a powerful vortex that loosens and entrains solids in water being drawn into a suction tube. This system does not create turbidity at the dredging site and can redeposit the slurry underwater without excessive re-suspension.

Sluicing offers another option for removal of accumulated sediments from reservoirs (e.g., Scheuerlein 1990). Opening sluice gates or an outlet tunnel allows water levels to fall and sediment to be resuspended and flushed out with the water. Problems have arisen when sluicing has been conducted during summer months, at times when flows are inadequate to disperse the redeposited sediment. Accidental releases of sediment occurred on the Middle Yuba River from Hour House reservoir in 1986 and from Poe dam on the North Fork of the Feather River in 1988. More than $1 million was spent excavating sand out of the channel below Hour House dam, but a flood during the early stages of the North Fork Feather clean-up conveniently flushed all the excess sediment out of the channel (Ramey and Beck 1990; Kondolf and Matthews 1993).

When sediment is flushed out of reservoirs at low flows, it will be redeposited close to the dam. However, when it is introduced at higher flows, it will usually be carried downstream and dispersed. Engineering approaches to letting sediments pass through dams during high flows are being considered at several sites. The Pacific Gas and Electric Company (1994) is designing pass-through systems to retrofit two of its dams on the North Fork Feather River. Such pass-through systems could allow reservoir operations to interfere less with natural sediment transport and could have geomorphic benefits with regard to channel degradation below dams (Kondolf and Matthews 1993). Most small dams do not need to store water during peak flows.

## EROSION CONTROL AND WATERSHED MANAGEMENT

Some human activities disrupt natural geomorphic processes and accelerate erosion or destabilize hillslopes. When soil loss and sediment delivery occur at unusually high rates in response to some human disturbance, the utility of facilities for water storage and diversion and hydroelectric production can be reduced. For example, recent bathymetric surveys of Pardee Reservoir on the Mokelumne River suggest that the average annual rate of sediment deposition has more than doubled since the previous survey in 1943 when it was 150 $m^3$ $km^{-2}$ (0.3 AF $mi^{-2}$) (East Bay Municipal Utility District 1995). Parts of the Mokelumne River basin have been extensively roaded and logged in the past few decades, and there has been considerable concern about apparent increases in sediment yield.

Activities that purposefully move soil, such as construction of roads, and that reduce vegetative cover and root strength have the greatest potential for increasing erosion. When such actions occur in and near stream channels, acceleration of sediment delivery is likely. Destruction of riparian vegetation often leads to massive streambank erosion. Roads located near streams have been frequently identified as one of the worst contributors to reservoir sedimentation (e.g., Anderson, 1974). The North Fork Feather River is a commonly cited example of a basin with highly accelerated erosion. A comprehensive evaluation of sediment sources in the basin found that about 90 percent of the erosion and about 80 percent of the sediment yield is induced by human activities (Soil Conservation Service 1989). Most of the accelerated sediment delivery is caused by bank erosion where riparian vegetation has been eliminated by overgrazing and erosion from road cut and fill slopes (Soil Conservation Service 1989). A cooperative effort between local land owners, public agencies, Pacific Gas and Electric Company, and private individuals is attempting to reduce erosion throughout the basin. The Pacific Gas and Electric Company is involved because it operates two small reservoirs in the canyon of the North Fork Feather River. Sediment is rapidly filling the reservoirs, interfering with operation of the control gates on the dams, and accelerating turbine wear (Pacific Gas and Electric Company 1994). Watershed management has great potential to reduce sediment delivery within this basin because many of the sources are obvious and treatable.

## CONCLUSIONS

Although the large storage reservoirs of the Sierra Nevada appear to have capacities large enough to hold sediments delivered for several hundred years at current yields, operations of many smaller reservoirs are currently constrained by sediment accumulation. As dredging and transportation of stored sediment becomes expensive, other sediment management alternatives are being sought. Where erosion has been accelerated by human activities, control of near-channel sources of sediment can be effectively reduce long-term sediment supply. Changes in both structures and operations to allow sediments to pass through during high flow can allow the streams to function more naturally and improve the overall operational efficiency of hydropower and other diversion facilities.

## ACKNOWLEDGMENTS

Work contributing to this paper was supported by the Sierra Nevada Ecosystem Project as authorized by the U.S. Congress (HR5503) through a cost-reimbursable agreement (PSW-93-001-CRA) between the USDA-Forest Service, Pacific Southwest Research Station and the Regents of the University of California, Centers for Water and Wildland Resources. Much of the material presented here is abstracted from volume 2, chapter 30 of that report (Kattelmann 1996) and an article in the Proceedings of the International Conference on Reservoir Sedimentation, Fort Collins, Colorado. Continuing studies of the hydrology of the Sierra Nevada funded through NASA's Earth Observing System program through a grant to Jeff Dozier at the University of California at Santa Barbara also contributed to this paper.

## REFERENCES

Anderson, H. W. 1974. Sediment deposition in reservoirs associated with rural roads, forest fires, and catchment attributes. International Association of Hydrological Sciences Publication 113, 87-94.

Brown, C. B., and Thorp, E. M. 1947. Reservoir sedimentation in the Sacramento-San Joaquin drainage basins. Special Report 10, USDA-Soil Conservation Service, Sedimentation Section, Office of Research, Washington, D.C.

Dendy, F. E., and Champion, W. A. 1978. Sediment deposition in U.S. reservoirs, summary of data reported through 1975. Miscellaneous publication 1362, USDA-Agricultural Research Service, Oxford, MS.

East Bay Municipal Utility District. 1995. Letter from R. C. Nuzum to Mokelumne River Joint Ownership Protocol Group, October 10, 1995. Oakland, CA.

Georgetown Ranger District. 1992. Placer County Water Agency sediment storage site environmental assessment. USDA-Forest Service, Eldorado National Forest, Placerville, CA.

Harrison, L. L., and Weinrib, H. P. 1996. EDDY pump dredging demonstration at Cresta Reservoir. Proceedings of the North American Water and Environment Conference, ASCE, New York, NY.

Harrison, L. L., Lee, W. H., and Tu, S. 1995. Sediment pass-through, an alternative to reservoir dredging. Waterpower '95 Conference, ASCE, New York, NY, 2236-2245.

Helley, E. J. 1966. Sediment transport in the Chowchilla River basin: Mariposa, Madera, and Merced Counties, California. Ph.D. dissertation, University of California, Berkeley.

Holeman, J. E. 1968. The sediment yield of major rivers throughout the world. Water Resources Research 4:737-747.

Kattelmann, R. 1996. Hydrology and water resources. Sierra Nevada Ecosystem Project: Final Report to Congress, chapter 30, volume 2, Centers for Water and Wildland Resources, University of California, Davis.

Kondolf, G. M., and Matthews, W. V. G. 1993. Management of coarse sediment on regulated rivers. Report 80, Water Resources Center, University of California, Davis.

Pacific Gas and Electric Company. 1994. Project description: Rock Creek-Cresta sediment management project. San Francisco, CA.

Ramey, M. P., and Beck, S. M. 1990. Flushing flow evaluation: the North Fork of the Feather River below Poe Dam. Environment, Health, and Safety Report 009.4-89.9. Pacific Gas and Electric Company, San Ramon, CA.

Rausch, D. L., and Heinemann, H. G. 1984. Measurement of reservoir sedimentation. in Erosion and sediment yield: some methods of measurement and modeling, edited by R. F. Hadley and D. E. Walling, Geo Books, Norwich, England, 179-200.

Scheuerlein, H. 1990. Removal of sediment deposits in reservoirs by means of flushing. Hydrology in Mountainous Regions, II - Artificial Reservoirs; Water and Slopes, publication 194, International Association of Hydrological Sciences, Wallingford, England.

Soil Conservation Service. 1989. East Branch North Fork Feather River erosion inventory report, Plumas County, California. USDA-Soil Conservation Service, River Basin Planning Staff, Davis, CA.

# ABOUT DELIVERY RATIO: HOW DOES IT CHANGE IN TIME AND SPACE?

### GIAMPAOLO DI SILVIO and ANDREA MARION
University of Padua (Italy)

ABSTRACT
The mitigation of the sediment yield into reservoirs can be achieved by sediment source control. However, the variation of the sediment input in the upper part of the river network produces a significant variation of the sediment delivery to the reservoir after some time. An estimate of this response time as a function of some integral characteristic river parameters is obtained using a simplified river model with uniform sediments.

INTRODUCTION
Measures for mitigating reservoir sedimentation can be classified into the following four categories, according to their objective:
- removal of the sediments already deposited in the reservoir;
- reduction of the sedimentation rate while the sediments are flowing through the reservoir;
- interception of the sediments before they enter the reservoir;
- decrease of the sediment production from the watershed.

These approaches differ basically on when and where the control on the sediment transport is exerted. The last category is obviously the one which tackles the problem at its very roots and it is often considered the best choice when the sediment yield is very large.

Substantial decrease of the sediment production can be obtained for both surface erosion and mass movement. As far as surface erosion is concerned, mitigation measures are essentially based on the stabilizing effect of vegetation. In natural watersheds the mitigation can be achieved by upgrading grassland and forests, while in cultivated areas it can be achieved by improving land use and agricultural techniques. Erosion by mass movement usually occurs in the upper part of natural basins as landslides or debris-flow events. Such erosion can instead be reduced by appropriate structures (checkdams, retaining walls) on the slopes and on the extreme branches of the hydrographic network.

It should be stressed, however, that a decrease of sediment production does not correspond to an equal decrease of sediment yield, i.e. of the amount of sediments reaching the reservoir. The ratio between sediment production and sediment yield is generally called delivery ratio. The definition of delivery ratio usually includes the sediments that are permanently intercepted on the watershed slopes, as well as those temporarily stored along the fluvial system. The delivery ratio associated with the permanent interception should be estimated on the basis of the slope morphology and it is independent from the fluvial dynamics. By contrast, the delivery ratio in the river depends on the reduction of the net amount of sediments reaching the hydrographic network. Indeed, the river configuration is modified by any variation on the sediment input and the resulting sediment transport is continuously varying.

The morphological dynamics of natural rivers adapt to changes occurring at its boundaries on a time scale that is sometimes very long compared to the requirements of the designer. Moreover, the response time depends on the grain-size composition of the sediment input. This paper addresses the problem of evaluating the delivery ratio of river systems, i.e. the ratio between the net sediment input in the river and the sediment yield in the reservoir.

## DELIVERY RATIO IN A RIVER SYSTEM

A rough estimate of the delivery ratio in time and space can be obtained by the following schematic model. Let the river system under consideration be represented by a single prismatic channel, fed at its upper end and having the geometric, hydraulic and sedimentologic features of the downstream section of the real river uniformly distributed all along its length.

A characteristic length, $L_c$, and a characteristic time, $T_c$, are defined as follows:

$$L_c = \frac{H}{S_0} \quad (1a) \qquad T_c = \frac{H}{US_0} \quad (1b)$$

where H is a typical water depth for uniform flow, U is the relevant uniform velocity and $S_0$ is the bed elevation gradient. If the river length (L) is large enough compared to the characteristic length ($L_c$), and the flood period ($T_f$) is large enough compared to the characteristic time ($T_c$), the flow in the river can be considered as a sequence of steady, uniform conditions during the whole hydrological cycle (Di Silvio and Marin, 1996).

It also assumed that the river is in the long-term equilibrium conditions, i.e. the annual sediment input from the watershed is equal, for each grain-size fraction, to the annual sediment transport ($W_{0i}$). The annual transport can be expressed as the time integral of any fraction-by-fraction transport formula over the hydrological cycle. The formula by Engelund and Hansen for total transport is used here:

$$W_{0i} = \beta_i \frac{S_0^n}{b^p d_i^q} \left(\frac{d_i}{\sum \beta_i d_i}\right)^s f(Q) \qquad (2)$$

where $\beta_i$ is the bottom percentage of the i-th class, $d_i$ is the relevant grain size, b is the river width, Q is the water discharge and f is a function of the flow duration curve. Application of Eq. 2 to several rivers gave the following approximate values of the exponents: n=2, p=1, q=1.2, s=0.8.

If a small perturbation $w_0$ of the sediment supply is introduced, for example a reduction caused by mitigation measures in the upper part of the watershed, it propagates along the river. The annual sediment transport for each class, $W_i(x,t)$ can still be calculated with Eq. 2, provided the bed composition $\beta_i(x,t)$ and the local slope S (x,t) are known. Writing the balance equations for the stream and the mixing layer of the bottom leads to:

$$\frac{1}{b}\frac{\partial w_i}{\partial x} + \frac{\partial y_i}{\partial t} = 0 \qquad (3)$$

$$b\delta\frac{\partial \beta_i}{\partial t} = -\frac{\partial w_i}{\partial x} + \beta_i\frac{\partial w}{\partial x} \qquad (4)$$

$$w_i = W_i - W_{0i} \quad ; \quad \sum w_i = w = W - W_0 \quad ; \quad \sum y_i = y$$

where $y_i(x,t)$ represents the contribution of the i-th class to the bottom erosion y(x,t), $\delta$ is the thickness of the bottom mixing layer (assumed constant), W(x,t) is the total sediment transport and w(x,t) is the transport perturbation. The set of Eq. 2-4 can be solved for $y_i$, $\beta_i$ and $W_i$.

If the bottom composition $\beta_i$ is assumed constant in space and time(which is the case for a reasonably uniform material) and small perturbations ($w_0 << W_0$) are considered, Eq. 2 can be linearized to:

$$w_i(x,t) = W_{0i}\frac{n}{S_0}\frac{\partial y}{\partial x} \qquad (5)$$

By eliminating the variable $y_i$, the set of equations (2)-(5) provides the same diffusion equation for the fraction transport:

$$\frac{\partial w_i}{\partial t} = D\frac{\partial^2 w_i}{\partial x^2} \quad ; \quad D = \frac{nW_0}{S_0 b} = n\Psi\frac{L_c^2}{T_c} \qquad (6)$$

where $\Psi = W_0/Q_0$ is the sediment concentration in the flow associated to the steady equivalent discharge, $Q_0$=UHb, which delivers the same annual transport. It is worth noticing that the depth, H, and the velocity, U, associated to $Q_0$ are to be taken as the typical values used in (1a) and (1b). Eq. 7 is an extension to a river with a time-dependent water discharge of the equation found by de Vries (1975) for the case of uniform material.

The same equation, with the same diffusion coefficient, applies to the total transport fluctuation $w=\Sigma w_i$. The diffusion coefficient D is shown to depend on the overall sediment transport in the river at equilibrium. The solution of (6) provides the following expression of the delivery ratio for any cross-section of the river:

$$\frac{w_i}{w_{0i}}\left(=\frac{w}{w_0}\right) = \mathrm{erfc}\left(\frac{L}{\sqrt{4DT}}\right) = \mathrm{erfc}\left(\frac{1}{\sqrt{4n\Psi}}\cdot\frac{L}{L_c}\cdot\sqrt{\frac{Tc}{T}}\right) \quad (7)$$

where L is the distance of the cross section from the river "source", T is the time elapsed from the mitigation works, i.e. the reduction of the sediment input. The delivery ratio as a function of T and L is plotted in Fig. 1. It decreases with the river length and increases with the time: namely, in any section of the river the delivery ratio tends asymptotically to 1 as time increases. The response time of a river, however, tends to be very long, being proportional to the square of the river length, i.e. more or less proportional to the watershed area, and inversely proportional to the diffusion coefficient.

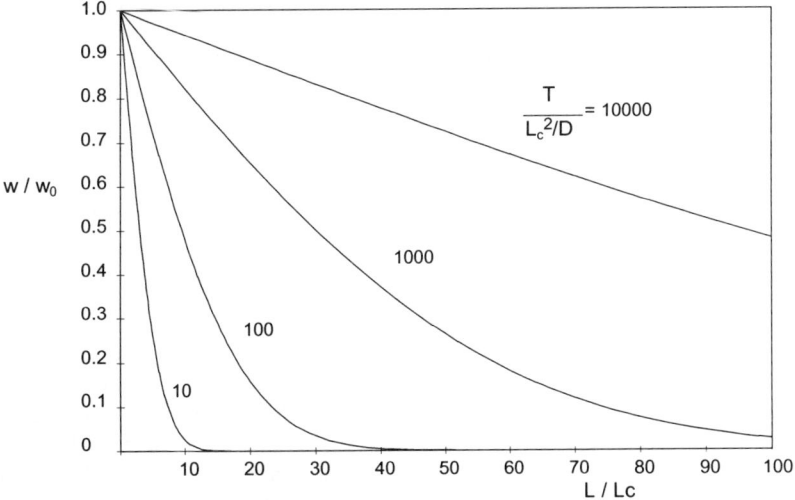

Figure 1.    Delivery ratio as a function of L and T

In order to give an example, in dimensional terms, of the response time, the number of years necessary for a 10% delivery ratio of the initial perturbation ($w/w_0=0.1$) is shown in Fig. 2 as a function of D and L. Some points are also included in the figure, corresponding to four different locations along the Po River (the longest river in Italy) and a tributary (the Mallero river at Sondrio). As it appears, the response time is of several decades, up to about 300 years for the entire river. It should be noted that moving along the river, the response time increases less than proportionally with the distance, because the diffusion coefficient increases as well. The time required for detecting significant benefits in terms of sediment yield from erosion-control works on the slopes is generally very long.

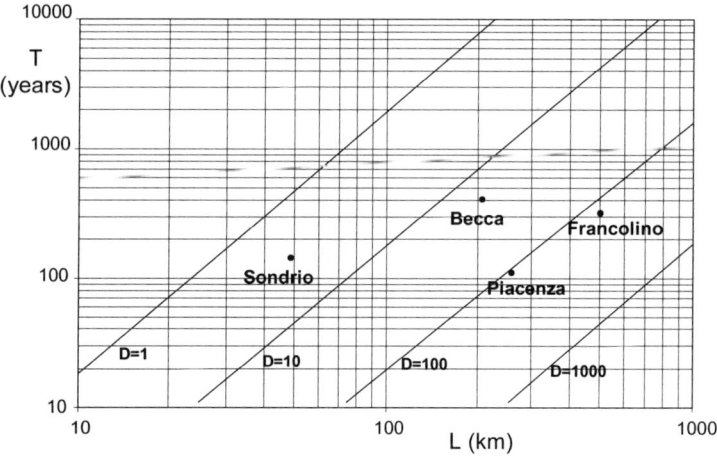

Figure 2. Response time of a river as a function of its length and diffusion coefficient for a delivery ratio of 10%.

## EFFECTS OF GRADATION

The solution found above (Eq. 7) shows that the delivery ratio is the same for the total transport and for the transport of any sediment grain size. This is true when the hypothesis that $\beta_i$ remains constant in time and space holds. Such assumption is valid only when sediments are almost uniform ($d_{16}/d_{84}>0.5$). If the hypothesis is relaxed, the response of the fines appears to be much faster than the response of the coarse fraction. Unfortunately, the solution of the set of equations for non-uniform material is not as direct as for uniform material. Numerical solutions, however, indicate that the propagation of the perturbations of each fraction is affected by the overall mixture composition, with a rather complex interaction (full coupling). As a general result, the fine fractions are subject to a smaller damping of the initial perturbation than the coarse ones. Therefore, the incidence of the transport of the fine fractions on the total transport increases with the distance from the source. By contrast, the coarse fractions are dominant in determining the bottom profile in the upper part of the river.

## CONCLUSIONS

The analytical evaluation of the delivery ratio in the simple case of a prismatic river with relatively uniform sediments shows that the response time of fluvial systems is very long (decades or centuries). The presence of non-uniform sediments increases the complexity of the problem, due to the fact that the dynamics of the different grain-size fractions are fully coupled. For a correct evaluation of the delivery ratio in real cases, a numerical solution of a morphological model capable of handling several sediment fractions (as Eqs. 1 to 4) is recommended. In any case, with the exception of the finest fractions, the benefits in terms of reservoir sedimentation to be expected from a decrease of sediment production tend to be relatively small and much delayed in time.

## ACKNOWLEDGMENTS

This work has been funded by the GNDCI (National Group for the Defense from Hydrogeologic Catastrophes) of the Italian National Research Council.

## REFERENCES

Di Silvio and Marin (1996). *Analytical approach to river morphodynamics: effects of space- and time-irregularities and grainsize non-uniformity.* FRIMAR EC-Program, Technical Report N.2, Padua (Italy), p.28.

Vries, M. de (1975). *A morphological time scale for rivers.* Proc. IAHR, Sao Paulo (Brazil), vol. 2, pp.17-23. (also Delft Hydraulics, Publ. N. 112)

# Reservoir Sediment Management Related To Sediment Quality and Contamination

### BERNHARD J. WESTRICH
Institute of Hydraulics, University of Stuttgart
70556 Stuttgart, Germany

**Abstract**
The layout and design of hydropower intake structures, spillways and bottom outlets is important to ensure the function of structures and installations endangered by sedimentation. Contamination of reservoir sediments is becoming an environmental issue requiring immission oriented sediment management strategies which will be addressed. Different structural and operational options for sedimentation mitigation are described such as flow guiding elements, levee raising, sediment flushing by water level lowering, flushing enhancement by pilot channel and sediment dredging. Practical problems will be illustrated by case studies and specific solutions found by physical and numerical modelling will be presented.

**Introduction**
Sedimentation is a big issue in planning and designing hydropower structures as well as maintaining and operating water reservoirs. Sedimentation has various implications for the safety and function of the hydraulic facilities and the reservoir life time as well as for the sediment budget, reservoir ecology and water quality. Contamination of reservoir sediments is an important aspect in reservoir maintenance because of contaminant remobilization risk and potential pollution of aquatic habitat by sediment removal. Mitigation of sedimentation is a challenging engineering task because of the complex interaction of reservoir hydrology, hydraulics, sediment quantity as well as hydraulic structures and their operation. Suitable site specific countermeasures are to be found which allow for sustainable reservoir water use under economic and ecological conditions.

**Sedimentation aspects**
The main goal in sediment managing is to operate the reservoir in such a way as to minimize the long term sediment trapping effect and to approach the natural sediment transport dynamics as well as possible. Different options are available such as: structural measures (design of hydropower intake, bottom outlet, movable gates; flow guiding elements, raising the levees), operational measures (sediment sluicing and flushing) and sediment dredging. The loss of coarse sediment fraction in the tailwater can be compensated to some extent by well performed flushing operation.

However, the removal of sand and gravel for feeding the tailwater river bed with coarser substrate is inevitably associated with the erosion of fine sediments causing undisired impact on fish and benthal microbiota due to increase of suspended load and clogging of interstitial habitat. Some hydraulic and sediment related aspects which have to be considered are reported by Wilcock et al. (1996)

## Chemical aspects

Physico-chemical sediment properties have to be investigated before designing a reservoir sediment management concept with correspponding actions. Therefore, sediment samples have to be taken for grain size analysis, fall velocity measurement and, in particular erodibility and soil mechanics tests (bulk density, shear resistance) are necessary to determine sediment parameters for numerical or physical modeling. Furthermore, information is required on sediment and porewater chemistry. Hence, in-situ measurements of dissolved components (peeper technique) and laboratory tests (leaching, speciation) must be conducted for assessment of adsorbed contaminant remobilization (heavy metal, organic pollutants etc.) and oxygen consumption rate due to bio-chemical reaction of released pore water.

Based on sediment contaminant inventory tests, the impact on the aquatic environment has to be quantified for technical specification of sediment slucing, flushing or dredging. Release of toxic, bioavailable matter and/or oxygen consuming components from resuspended sediments should be avoided. Clogging of interstitial habitat by resedimentation of fine sediments has to be minimized. If harmless dredged material is deposited on the reservoir banks, technical measures against mobilization by flood events have to be taken; e. g. impounding by small dams or covering by coarse material.

## Storage reservoirs

Water power storage reservoirs with a periodic discharge regime (days or weeks) show flow dynamics allowing enhancement of sediment removal. During the time period of electric power peak demand, the augmentation of discharge by reservoir water level lowering has a considerable effect on sediment removal. Hence, the sedimentation during low water demand period can be compensated during short period of peak power demand. The flow-through of suspended sediment can be enhanced by an appropriate design of flow cross sections and the use of guiding structures, e. g. submersible spur dikes. Examples are shown by two hydropower stations located in side arms of the upper river Danube. The upstream one has a design discharge of 74 $m^3/s$ and a head of 6,2 m. The original total volume of the reservoir in 1953 accounts to $1,5 \cdot 10^6$ $m^3$. The storage capacity is 0,8 Mio $m^3$ where half of it was lost by sedimentation. The diversion structure is a multiple gate weir preventing bed load from entering the power channel. The reservoir was originally excavated at a horizontal bed level of 484 m s l. Over more than 40 years of operation a huge sediment volume with local deposition layers up to 2 m has accumulated resulting in a change of the flow-through pattern and a deterioration of the flow approaching the turbines. (Fig. 1). The sediment management is aiming at an engineering solution with the following aspects:

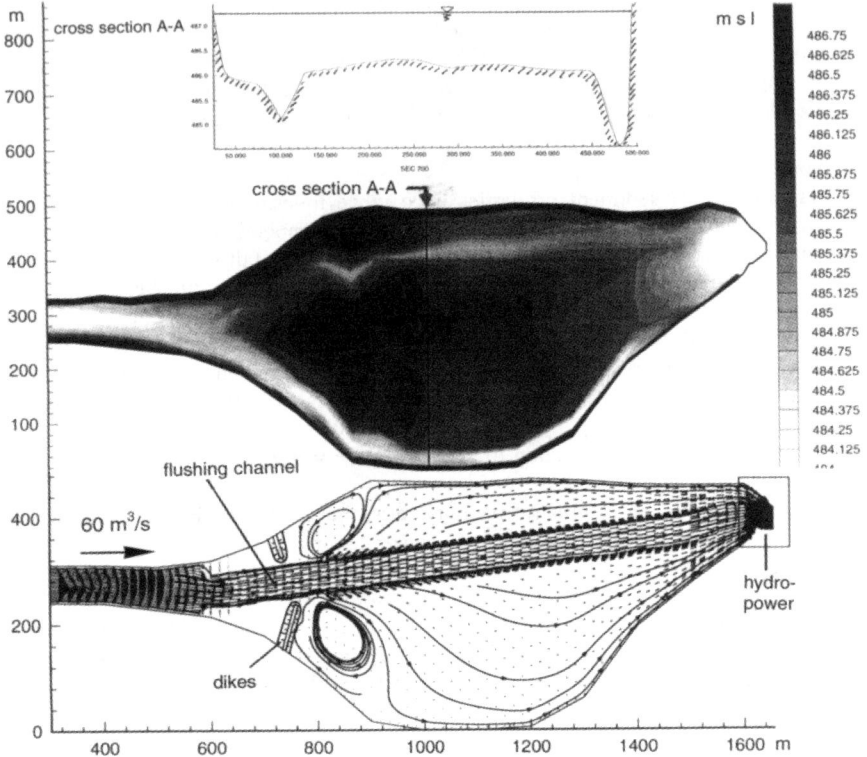

Fig. 1: (a) Reservoir bed profil (A-A) recorded in 1995
(b) Isolines of bed topography due to sedimentation from 1953 to 1995
(c) Streamlines at low water level, flushing channel, guiding dikes

- Providing a short circuit flow between inflow section and hydropower station
- Designing a flushing channel with maximum transport capacity and long term morphological stability.
- Sediment flushing enhanced by submersible dikes.
- Minimizing sedimentation in the shallow water area

The channel was optimized by extensive calculations using a 2-d suspended sediment transport model (Al-Zoubi, 1996) with the final design of 40 m width, 600 m length and submersible spur dikes on both sides near the entrance (Fig. 1) The channel provides a velocity enhancement during the water level drawdown period of about 2 hours a day and hence, results in an averaged overall sedimentation reduction of about 20%.

## River reservoirs

In most cases river run-off power station are being operated with constant head water level. Without navigation the reservoir water level can be drawn down during flood events for transport capacity enhancement. To perform a sediment flushing operation the discharge beyond which the water level is to be lowered, the duration and the additional released water volume, respectively has to be specified to achieve the required sediment removal. Field monitoring by continuous on-line water and sediment control measurements have to be performed to ensure that the site specific criteria in terms of suspended matter concentration and/or dissolved oxygen depletion and other quality related parameters are not exceeded.

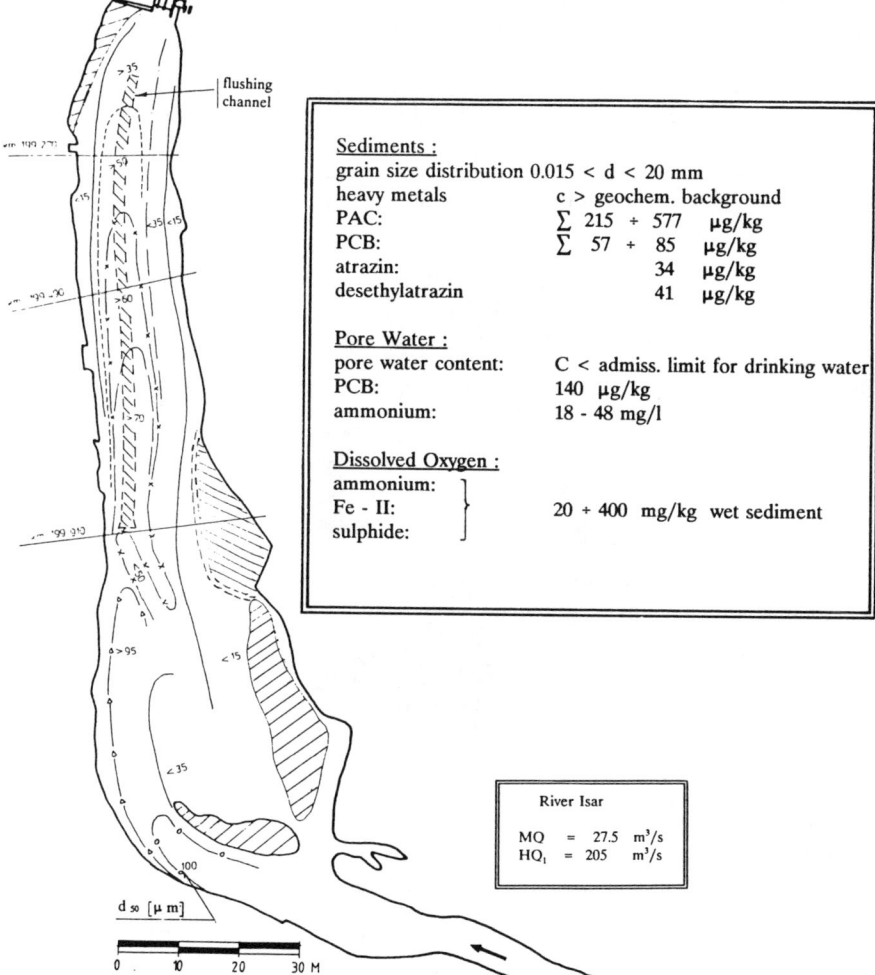

Fig.2: River Isar reservoir with sediment grain size distribution /2/, sediment and pore water analysis /3/ and pilot flushing channel.

A pilot project on sediment flushing by water level drawdown in a river reservoir is shown in Fig. 2. Prior to the first sediment flushing operation in 1991, a hydraulic (Westrich et al, 1992) and ecological feasibility study (Müller, 1989; Knell, 1988) was performed to ensure that a significant amount of sand and gravel can be removed and used for tailwater river bed feeding. Resuspension of fine sediments from shallow water zones should not occur because of the environmetal damage in downstream river sections and flood plains. An engineering solution was found as following: pilot channel dredging, water level lowering by 2 m as soon as the flood peak discharge exceeds about 300 m³/s. The following constraints were given by the local water authority: maximum augmentation of suspended sediment concentration by 5 g/l, dissolved oxygen content not less than 8 mg/l. Flushing operations have been repeated several times since, two times in 1993 and twice in 1995 (Scheuerlein et al, 1996). The pilot channel has proven to be efficient and useful for initiating sediment flushing. It has changed, of course its width and shape, but the sediment budget has substantially improved. Meanwhile bed load material (16 mm gravel) observed by tracer measurements has been moved to the tailwater. However, further structural measures (guiding dikes etc.) are envisaged to improve the removal effeciency and stabilize the flushing channel morphology.

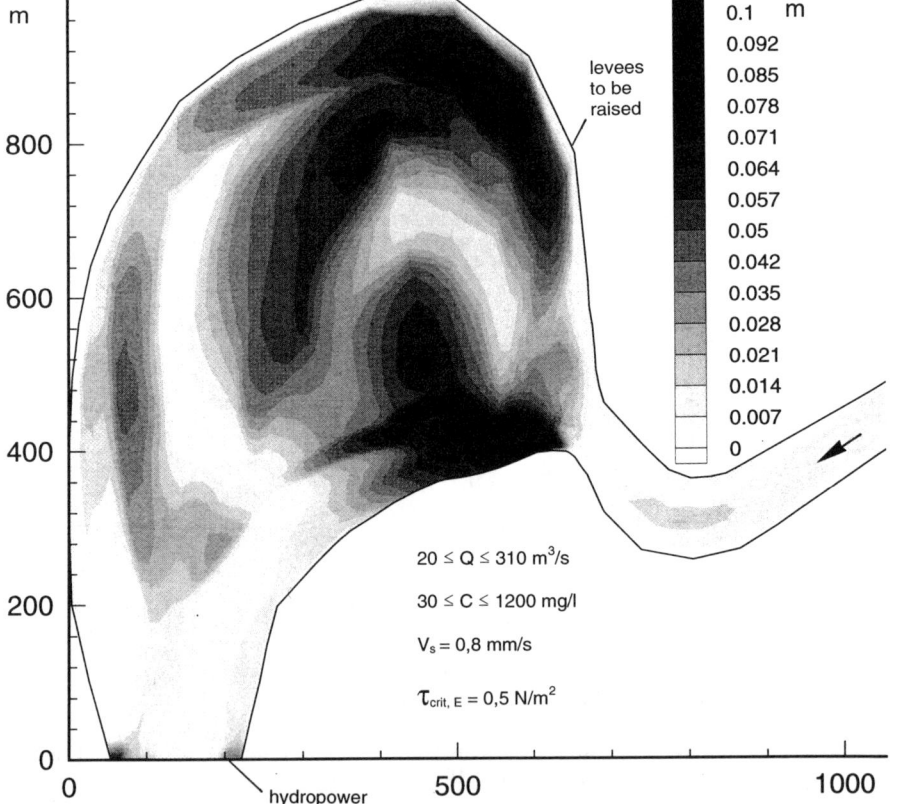

Fig. 3: Distribution of computed net sediment deposition from 1988 to 1992

Besides hydropower intake itself there are several other issues related to increasing risk of overtopping embankment. Even if there is no sedimentation problem for the power house the sediment volume can cause an elevation of the design water level. If flushing and dredging is not acceptable because of inviromental concern, raising the levees can be an ultimate measure.

When choosing suitable countermeasures the prediction of the quasi- equilibrium of reservoir bed level with respect to the water level elevation for the design flood is a key issue. For sedimentation prediction a numerical model can be applied to provide reliable information on the morphological bed evolution as shown for the river Iller reservoir Lautrach where the sedimentation problem will be solved by raising the levees. The numerical model (Fig.3) shows the spatial distribution of the sedimentation rate over the last four years indicating that the hydropower intake on the left is almost uneffected by the process. However, for safety reasons the levees will have to be raised in some sections. It shows the application of numerical models to design the reconstruction of levees to over come sedimentation problems.

## Conclusions

Sediment flushing is a very effective measure for sedimentation mitigation. The flushing efficiency however, is strongly dependent on the peak and duration of the discharge as well as the erodibility of the deposits. If natural flow cross sections do not allow for efficient flushing the use of structural elements for sediment transport capacity enhancement can be recommended and, the reservoir operation rules can be adapted. For managing contaminated sediments immission oriented investigations on environmental impact are required. Criteria for assessing sediment/water quality interaction are urgently needed for ecologically oriented sediment management strategies and for numerical morphology and quality modelling.

## References:

1) Westrich, B., Al-Zoubi, S., Müller, J. (1992): "Planning and designing a channel for reservoir sediment flushing", Inter. Symp.on Sedimentation, Karlsruhe Germany
2) Müller, J. (1989): "Investigation on sediment core samples from river Isar reservoir", Institute of General, Applied and Engin. Geol., Techn. Univ. München (internal report, in german)
3) Knell, B. (1988): " Organic pollutant analysis of river Isar sediments", Memmingen, Germany (internal report, in german)
4) Al-Zoubi, S., Westrich, B. (1996): "Sedimentation of lakes and river reservoir, Sediment problems in conduits and canals", (in german), ETH Zürich
5) Scheuerlein, H., Müller, J., Luff, H.: (1996):Experience with sediment flushing in the reservoir Bad Tölz", (in german) ETH Zürich Report Nr. 143. part 2, pp. 45-64.
6) Wilcock, P. R, Kondolf, M. G., Barta, Alan F.: "Specification of sediment maintenance flows for a large gravel-bed river", Water Res. Res., vol. 32, No. 9, 1996
7) Al-Zoubi, S.: "2-d finite element model for reservoir sedimentation prediction", PhD. thesis Universität Stuttgart (to be submitted in 1997).

# Double-mass curve used on sedimentology
## - A case study -

**Newton de Oliveira Carvalho**
*ELETROBRÁS, Centrais Elétricas Brasileiras S/A*
*Rua da Quitanda, 196 - 22° andar*
*Phone (021) 211-5353 - Fax (021) 516-4462*
*20091-000 - Rio de Janeiro, RJ - Brazil*

**ABSTRACT** The great spatial and temporal variability of sediment yield in the rivers, due to accelerated erosion, represents a major problem for the reservoirs. It is always necessary to make a prediction of the reservoir sedimentation in order to get its probable useful lifetime. This paper shows use of double-mass curves in a case study to get the correct increasing rate in sediment yield of a basin, or a location on a river for a good evaluation of the reservoir sedimentation.

## INTRODUCTION

The reservoir's sedimentation process affects slowly but effectively the reservoir capacity of discharge regularization used for hydroelectricity generation and other users. So, in designs of dams with reservoir regularization purpose, it is of primordial importance to carry out a prediction of its sedimentation considering the sediment distribution along the lake. Usually the studies are done by using an average annual inflow sediment load to the reservoir. Usually, the reservoir sedimentation evaluation is made considering a 100 year's horizon. It is necessary to take into account a coefficient for the increasing of the erosion in the basin through the increasing of the human activities. To take into account this, an usual procedure is to double the previous computed average annual sediment load. The usually adopted prediction time is 100 years, which is twice the economical useful lifetime of the development. So, the conclusions are made based on the new height-area-volume curves of the prediction for 100 years.

However, this procedure it is not very good, being more convenient to consider real rates of increase of the quantity of sediment throughout the considered time. This paper shows a case study taking care of this procedure, using annual average sediment discharge and the double-mass curve.

## REGION AND PROPOSED STUDY

A sedimentologic study of the Doradas river has been done by the author [Carvalho, 1994] in order to evaluate the sedimentation and the useful lifetime of the Doradas reservoir, to be formed by the Las Cuevas dam, located on a site with a watershed area of 160.3 km². Doradas river is formed by San Agaton and San Buenas creeks (Fig. 1) and is a tributary of Uribante river, flowing to Apure river, Orinoco basin, South-West region of Venezuela. The source areas of the creeks are the Andes mountains, region of sandstone formation and high precipitation (4 000 mm yr$^{-1}$). Located at the influence of San Cristobal city, the rural area has been populated slowly and is presenting an accelerated erosion.

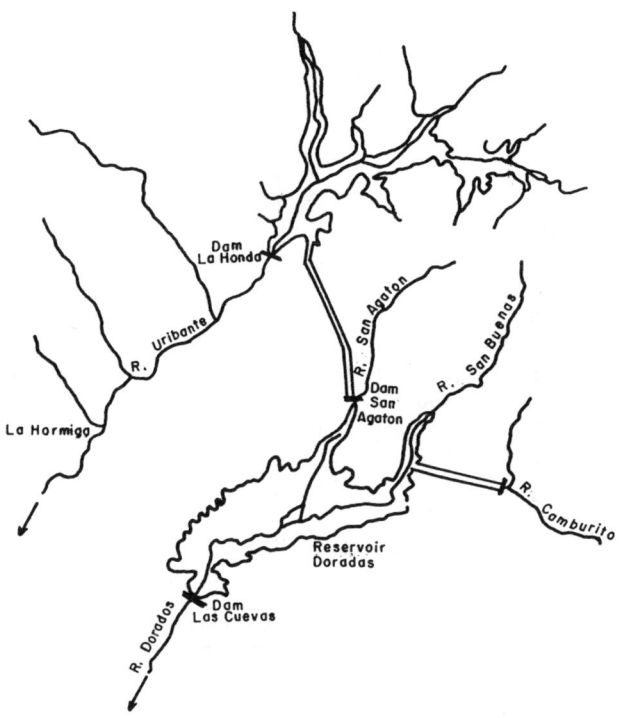

FIG. 1 Doradas reservoir position

The dam, to be constructed for a hydroelectric power plant of La Colorada (920 MW), it is part of the complex of Uribante-Caparo of CADAFE. In spite of the few available data of Doradas basin, the reservoir sedimentation has been evaluated as of 350 years, considering a sediment load equal to twice the present amount, taking into account the gradual increasing of sediment yield rate due to accelerated erosion. For next phase of design it is better to use the sediment rate as showed ahead in this article.

## SEDIMENT YIELD AND REGIONALIZATION

Table 1 presents, in the first five columns, the data gathered at the Environment Ministry of Venezuela. The total sediment yield values were got by adding 30% on suspended sediment values to take into account the bed load.

Table 1 - Sediment yield and degradation height in the Alpes basin of Venezuela (data gathered at the Environment Ministery)

| N. | River | Station | Drain. Area ($km^2$) | Susp.Sed. Yield ($t\ km^{-2}\ yr^{-1}$) | Total Sed. Yield ($t\ km^{-2}\ yr^{-1}$) | Degrad. height ($mm\ yr^{-1}$) |
|---|---|---|---|---|---|---|
| 1 | Azuero | Pte. Junin | 371 | 811 | 1 054 | 0.66 |
| 2 | Quiimimari | Buenos Aires | 1 113 | 1 213 | 1 577 | 0.98 |
| 3 | Frio | Pte. Frio | 1 450 | 2 609 | 2 690 | 1.68 |
| 4 | Uribante | Pte. Uribante | 2 606 | 1 919 | 2 495 | 1.56 |
| 5 | Chururu | Pte. Chururu | 140 | 300 | 390 | 0.24 |
| 6 | Doradas | Pte. Doradas | 582 | 2 474 | 3 216 | 2.01 |
| 7 | Paguey | El Paso | 810 | 525 | 682 | 0.42 |
| 8 | La Yuca | Pte. La Yuca | 265 | 989 | 1 286 | 0.80 |
| 9 | Masparro | Pte. Masparro | 495 | 616 | 800 | 0.50 |
| 10 | Bocono | Bocono | 450 | 944 | 1 227 | 0.77 |
| 11 | Bono | Peña Larga | 1 580 | 1 873 | 2 435 | 1.52 |
| 12 | Tucupido | Pte. Tucupido | 440 | 841 | 1 093 | 0.68 |
| 13 | Portuguesa | Pte. Portuguesa | 810 | 1 852 | 2 408 | 1.50 |
| 14 | Las Marias | Pte. Las Marias | 325 | 738 | 959 | 0.60 |
| 15 | Morador | Pte. Morador | 610 | 631 | 820 | 0.51 |
| 16 | Guache | Puente Viejo | 300 | 1 300 | 1 690 | 1.05 |
| 17 | Acarigua | Pte. Acarigua | 970 | 2 021 | 2 627 | 1.64 |

Fig. 2 presents the values of total sediment yield against the drainage area of the stations using the data of Table 1. The fitted curves represents sediment yield regionalizations and were got by visual process. By the analysis of the lines it can be seen that the Doradas basin is located in the superior line which corresponds to a higher sediment yield in the region.

The total sediment for Las Cuevas dam, which forms the Doradas reservoir, is of about $4\ 000\ t\ km^{-2}\ yr^{-1}$.

## INCREASING EROSION ON THE BASIN

Table 2 shows data of suspended sediment on the station of Doradas river at Puente Doradas (drainage area = $582\ km^2$), downstream Las Cuevas dam, of the period 1968 to 1974.

Using its data it is proposed to demonstrate that there is having a gradual increase in sediment yield from the sediment measurements in the Doradas river, as an evidence of an increasing erosion in the basin. Certainly the human activity in the region is the main cause. This can be demonstrated by a study of the increased anthropic effects.

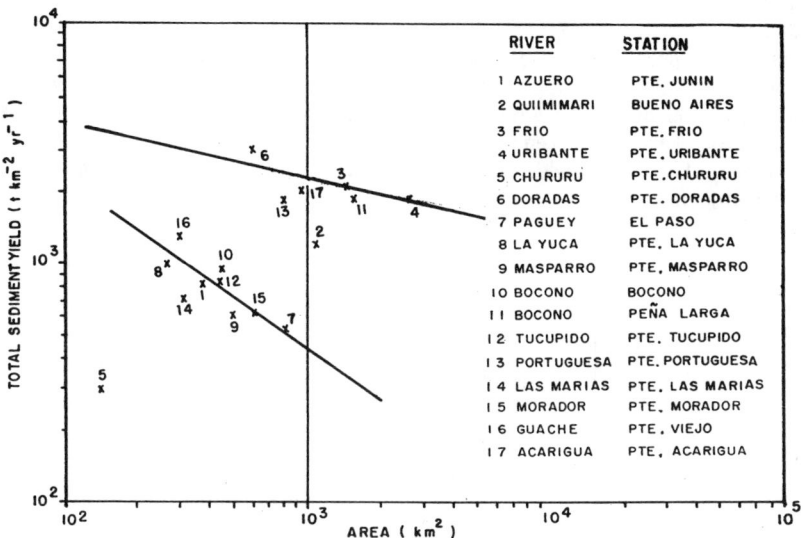

FIG. 2 Regionalization of total sediment yield x drainage area in the Andes (Venezuela)

Table 2 - Suspended sediment on the Doradas river at Puente Doradas, drainage area of 582 km² (data gathered at the Environment Ministery of Venezuela)

| ANO | Average Water Disc. ($m^3 s^{-1}$) | Σ Average Water Disc. ($m^3 s^{-1}$) | Annual Susp Sed. Yield ($t\ km^{-2}\ yr^{-1}$) | Σ Annual Susp Sed. Yield ($t\ km^{-2}\ yr^{-1}$) | Concent. Média ($mg\ l^{-1}$) |
|---|---|---|---|---|---|
| 1968 | 36.17 | 36.17 | 1 795.9 | 1 795.9 | 916.3 |
| 1969 | 29.63 | 65.80 | 905.3 | 2 701.2 | 563.9 |
| 1970 | 36.72 | 102.52 | 2 329.0 | 5 030.2 | 1 170.6 |
| 1971 | 53.51 | 156.03 | 4 284.6 | 9 314.8 | 1 477.7 |
| 1972 | 72.56 | 228.59 | 4 680.2 | 13 995.0 | 1 190.4 |
| 1973 | 39.92 | 268.51 | 1 672,3 | 15 667.3 | 773.1 |
| 1974 | 37.29 | 305.08 | 1 750.6 | 17 417.9 | 866.3 |
| Aver. | 43.58 | | 2 488.3 | | |

The double-mass curve of the case study is obtained by putting columns of sums of average water discharge against sums of suspended sediment yield (Fig. 3).

The double-mass curves is used for studying trends in sediment yield and for detecting the effect of watershed practices on sediment yield [Searcy and Hardison, 1960]. Analysing the breaks in the line there can be seen three years of higher sediment discharge (1970/71/72) due to floods in the same period. From 1972/73 to 1973/74 it is possible to note a normalization of the sediment discharge. A second analysis shows that the last period of 1973/74 presents a higher tendency of sediment discharge in relation to 1968/69, in spite of the smaller values of 1974 compared to that of 1968. This can be seen by the inclination of the straight lines forming the angle indicated on Fig. 3. It is evident that the Doradas is having a gradual degradation due the soil erosion.

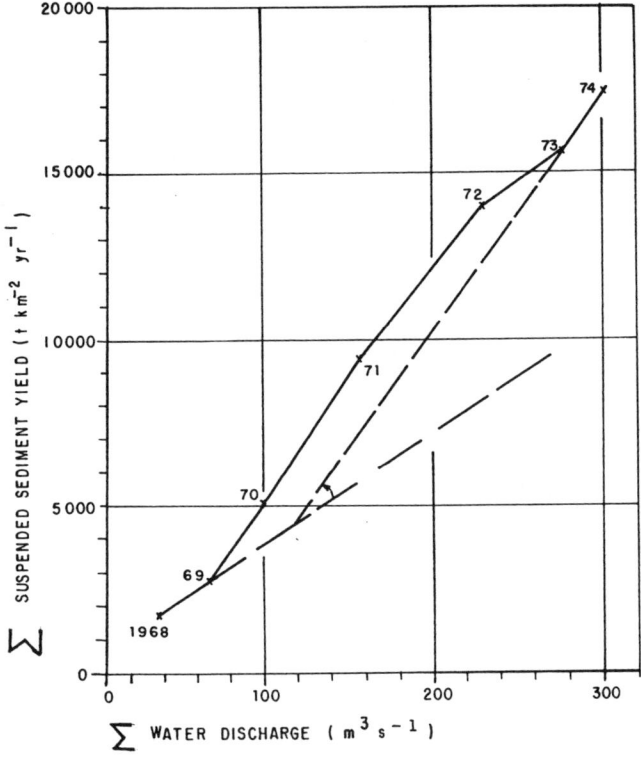

Fig. 3 - Sediment double-mass curve for Doradas river

The rates between the values are computed as:

from 1973/74:
$$r_1 = \frac{1,672.3 + 1,750.6}{39.92 + 37,29} = 44.34 \tag{1}$$

from 1968/69:
$$r_2 = \frac{1,795.9 + 905.3}{36.17 + 29.63} = 41.05 \tag{2}$$

The above computations show that the sediment yield on the biennium 1973/74 is higher than on the biennium 1968/69. The rate $E_c$ (crescent erosion) is computed as:

$$E_c = (r_1 - r_2)/r_1 = (44.34 - 41.05) / 44.34 = 0.074 \tag{3}$$

This means that there was an increase of 7.4% in the sediment yield during seven years, from 1968 to 1974. Let's use $R_i$ to denote annual rate of increase. So, $(1+R_i)^7=1.074\%$. Then, $R_i=1.025\%$. In ten years: $(1+R_i)^{10}-1=10.74\%$. In 100 years: $(1+R_i)^{100}-1=177.3\%$, and in 1996 it will be $(1+R_i)^{29}-1=34.45$, higher than that of 1968.

## CONCLUSIONS

As shown, several stations are located in Andes region, having high sediment yield and also undesirable values; among them are the stations in the basins of Doradas, Uribante, Quiimimari, Portuguesa, Acarigua and Bocono, as shown on Fig. 2.

It is obvious that these basins deserve protection cares, beginning with a better study of the sedimentologic aspects, geology, precipitation, forest cover and others.

The Doradas basin presents a high degradation height of 2 mm yr$^{-1}$, the highest value in Table 1. Certainly it is necessary to implement protection work on the basin in order that the erosion does not reach higher values. It is convenient to remember that the actual value (1996) of the degradation height is 2.6 mm yr$^{-1}$, considering the increase of the sediment yield, as demonstrated by the double-mass curve analysis.

The high values of sediment yield are greatly influenced by the sandstone rocks in the region, presenting easy disintegration, by the high precipitation (4 000 mm), and the steep slopes of the mountains. The sediment control in the basin is a primary factor, beginning by the remained forest preservation and also by the reforestation of some areas. The protection of the roads and also of the slopes, which present erosion sources, is another important aspect.

In the point of view of the prediction of the reservoir sedimentation, the present study shows that the adopted value of twice the quantity of sediment may be accepted a hundred years. The double-mass curve for Doradas station was figured for a short period of time of seven years. For better prediction it is necessary to gather recent values which could present a great modification in the basin if the human activity has increased.

## REFERENCES

CARVALHO, Newton de Oliveira (Abril.1994). Estudos sedimentológicos para o reservatório Doradas. Convênio CADAFE/Eletrobrás/Furnas, do Brasil. Sibéria, Venezuela.

SEARCY, James K. and HARDISON, Clayton H. (1960). Double-Mass Curves. US Geological Survey, WSP 1541-B. Washington, DC.

# THE DEVELOPMENT OF EQUILIBRIUM PROFILES FOR FLUSHING CHANNELS

ROBERT H.A. JANSSEN and HSIEH WEN SHEN
Graduate Student and Professor
University of California at Berkeley, Berkeley, USA

## ABSTRACT

Hydraulic flushing and sediment-pass-through have been proposed as ways of preserving reservoir storage capacity. Both require full reservoir drawdown using low-level outlets, leading to incision of a flushing channel in the sediment bed. In wide reservoirs, the flushing channel slope and width vary along its length. Three stages of channel widening and incision during reservoir drawdown were identified from a physical model. Initially, when the water surface slope is small, little erosion and channel widening occurs. Then, when the flow through the reservoir achieves riverine conditions, the flushing channel widens and incises rapidly. Finally, when the flow through the reservoir is confined to the flushing channel, the channel incises more slowly into the sediment bed, and widens intermittently by bank failure processes.

## INTRODUCTION

Mahmood [1987] has reported that although the world's total reservoir storage capacity increased 25 times between 1950 and 1970, that this storage capacity is being lost due to sedimentation at a rate of 1% per year, or about 50 km$^3$ per year, and is costing an estimated $6 billion per year to replace. Mikhalev [1971], Mahmood [1987], and Fan and Morris [1992], among others, have proposed *hydraulic flushing* or *sediment-pass-through* (SPT) operations as a means of maintaining reservoir storage capacity. These operations involve drawing down the water level in a reservoir using a low level outlet in order to promote erosion of sediment in the reservoir (flushing), and/or to allow any incoming sediment to be transported through the reservoir (SPT).

Based on field experiences, Gvelsiani and Shamal'tzel [1971], Mahmood [1987], and Wu [1989], concluded that without full reservoir drawdown, sediment erosion and transport is very limited. Hotchkiss and Parker [1988] modeled reservoir drawdown in a narrow flume, measuring equilibrium bed slopes under various drawdown scenarios. Chang et al. [1996] used a one-dimensional flow and sediment transport model to develop operating rules for SPT to ensure a long-term sediment mass balance in a series of reservoirs.

In narrow, gorge-type reservoirs, sediment transport and erosion can occur over the full width of the reservoir bed. However, in wider reservoirs a *flushing channel* is incised into the sediment bed as the reservoir level is drawn down, and most of the sediment transport and erosion takes place within this channel. The sediment transport capacity of the flushing channel depends on its width and slope. Lai [1994] presented a comprehensive set of results for reservoir drawdown experiments conducted in a wide flume, with steady, clear-water inflow. He found that over time, the flushing channel reached an equilibrium width, and that the flushing channel characteristics were very similar to those observed in the field. The study presented here looks at the development of the flushing channel slope and width during the drawdown procedure.

## DESCRIPTION OF PROCESSES

When an outlet that is located lower than the sediment level at a dam is first opened, the sediment in its immediate vicinity is eroded and discharged, forming a flushing cone. If the reservoir is not drawn down, the extent of this erosion will be limited. However, if the reservoir is drawn down, both progressive and retrogressive erosion may occur along the length of the reservoir. *Progressive erosion* occurs as the backwater effects in the upstream reaches of the reservoir are reduced, resulting in an increased energy slope and increased sediment transport capacity. As the reservoir is drawn down further, the extent of progressive erosion moves downstream.

*Retrogressive erosion* starts at the downstream end of the reservoir when the flow is no longer controlled by the outlet, but rather by the local open channel, or riverine, conditions. This results in an increase in energy slope, increased transport capacity, and incision of a flushing channel into the sediment bed. The extent of the retrogressive erosion will move upstream as the flushing channel widens and incises further into sediment bed. As the reservoir is drawn down, the rate of widening and incision decreases, until an equilibrium state is reached. The incision of a flushing channel into the sediment bed by retrogressive erosion is the subject of this paper.

## PHYSICAL MODEL

A physical model was set up to investigate the flushing channel formation due to retrogressive erosion in a reservoir wide enough to allow the flushing channel to widen without constraint. To start understanding the basic principles involved in flushing channel formation, the experiments presented here were performed using uniformly sized, non-cohesive sediment, being transported as bed load. Future work will have to take into account sediment gradation, suspended load, and cohesion effects. In order to demonstrate the effect of reservoir drawdown on the development of the flushing channel, the results of a simple case are presented here: Steady inflow to the reservoir and no incoming sediment load.

## EXPERIMENTAL SET-UP

The physical model is located in a 2.4 m wide, 1.5 m deep, and 50 m long concrete, non-recirculating flume at the Richmond Field Station of the University of California at Berkeley. A sketch of the longitudinal section of the experimental set-up is shown in Figure 1. A 15 cm wide sluice gate, located in the center of the dam, was opened to draw down the reservoir. Uniformly ground walnut shells were used as sediment, with a mean size of 1.25 mm, and a specific gravity of 1.39. The sediment bed was prepared using a template attached to a carriage running along rails on the top of the flume walls. The sediment surface was initially flat, at an elevation of 10 cm above the sluice gate sill. The flume was then slowly filled to about 16.5 cm above the sluice gate sill (depth of water about 6.5 cm).

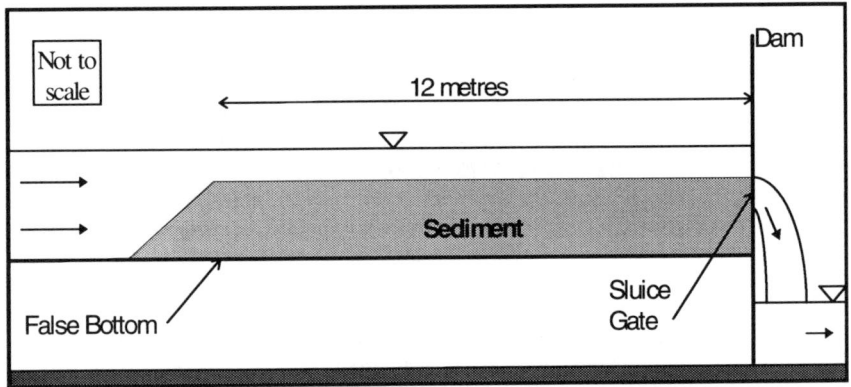

**FIGURE 1: Sketch of Longitudinal Section of Model**

Before the start of the drawdown the inflow and outflow were balanced, and a flushing cone was formed in the sediment immediately upstream of the sluice gate, reaching a stable size within a few minutes. The drawdown operation was started by opening the sluice gate from its initial position of about 1.6 cm open, to 10 cm open. The outflow from the sluice gate was sampled at discrete intervals. Water surface elevations were measured using both ultrasonic sensors and point gauges, while bed surface elevations during the run were measured using a disc gauge. The width of the flushing channel was measured visually using a ruler suspended just above the water surface. At the end of each run, when the flume had been drained, the bed surface was surveyed using a point gauge attached to the carriage.

## RESULTS

The discharge through the sluice gate over time is plotted in Figure 2. The flow reaches a maximum immediately after the sluice gate is fully opened, and then drops as the reservoir is drawn down. Also shown in Figure 2 are plots of the flushing channel width over time at two locations. The flushing channel starts to widen at 3.5 m before it does at 5.5 m (which is further upstream), and reaches a larger width at 3.5 m. This is consistent with retrogressive erosion, which starts at the dam and proceeds upstream. From the plot of the final channel widths in Figure 3 it can be seen that the channel width decreases away from the dam. A section further upstream from the dam will start to widen later (as seen on Figure 2), when the flow through the reservoir is lower, and therefore achieves a smaller width.

From Figure 2, three different stages in the flushing channel width development can be identified. Initially, the channel does not widen very much. This is followed by a short period of very rapid widening of the flushing channel. Then, after a sudden change in the rate of widening, the flushing channel continues to widen at a more gradual rate.

The water surface profiles plotted in Figure 4 help to explain the development of the flushing channel. Initially, the water surface is still above the sediment surface, and the water surface slope is relatively flat. The resulting low bed shear stresses are not sufficient to cause erosion. At around 10 minutes, however, the water surface has dropped to just above the initial sediment surface, and the water surface slopes are much steeper. The resulting increased shear stress leads to high erosion rates, and rapid widening of the flushing channel.

As the reservoir is drawn down, the discharge through the dam decreases, while the flushing channel widens and deepens. Eventually, the flushing channel will have sufficient capacity to convey all the flow, and the water surface will drop below the top of the flushing channel banks. It was observed that the rate of channel widening suddenly decreases at this time. The flow through the reservoir is then confined to the flushing channel, and the channel widens through intermittent bank failures as it incises into the sediment bed.

## CONCLUSIONS

The width of a flushing channel is determined by both the rapid widening while the reservoir is drawn down, and by the intermittent bank failure process when the flow through the reservoir is confined to the flushing channel. For analysis of hydraulic flushing or SPT, both these processes have to be taken into account when evaluating the flushing channel width and slope, and hence the sediment transport capacity. Further work is required to investigate how the width and slope of the flushing channel are affected by incoming flow and sediment load.

## FUTURE WORK

The experimental work presented here will be extended to cover a wider range of steady and unsteady inflows, as well as incoming sediment loads. Further down the line, other factors affecting the equilibrium flushing channel will be taken into account, such as repeated drawdown operations, different sluice gate settings, and non-uniform sediment sizes.

## REFERENCES

**Chang, H.H., Harrison, L.L., Lee, W., and Tu, S.** [1996], "Numerical modeling for sediment-pass-through reservoirs", *Journal of Hydraulic Engineering*, ASCE, Vol. 122, No. 7, July.

**Gvelesiani, L.G. and Shamal'tzel, N.P.** [1971], "Studies of storage work silting of hydroelectric power plants in mountain rivers and silt deposition fighting", *Proceedings of the 14th Congress of the International Association for Hydraulic Research*, Paris.

**Hotchkiss, R. and Parker, G.** [1988], "Reservoir Sediment Sluicing - Laboratory Study", *Proceedings of the ASCE National Conference on Hydraulic Engineering*, Colorado Springs, Colorado. S.R. Abt and J. Gessler, Editors.

**Lai, J.S.** [1994], "Hydraulic flushing for reservoir desiltation", Ph.D. dissertation submitted to the University of California at Berkeley, Berkeley, CA.

**Mikhalev, M.A.** [1971], "Control of silting in reservoirs on mountain rivers", *Proceedings of the 14th Congress of the International Association for Hydraulic Research*, Paris.

**Wu, C.M.** [1989], "Hydraulic properties of reservoir desilting", *Proceedings of the 23rd Congress of the International Association for Hydraulic Research*, Ottawa.

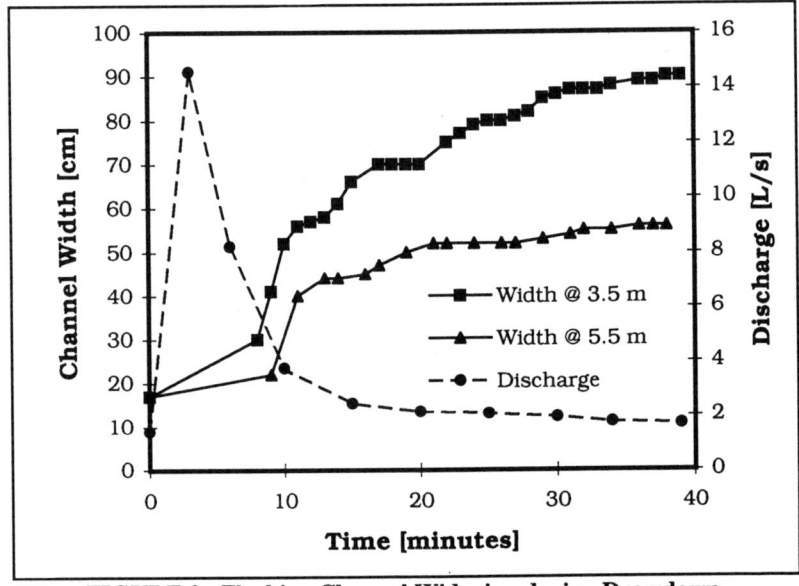

**FIGURE 2: Flushing Channel Widening during Drawdown**

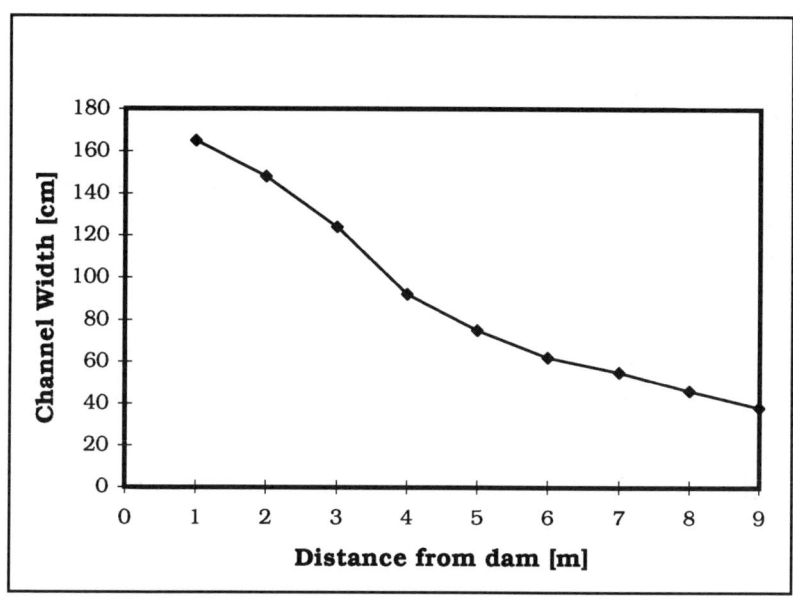

**FIGURE 3: Final Flushing Channel Widths**

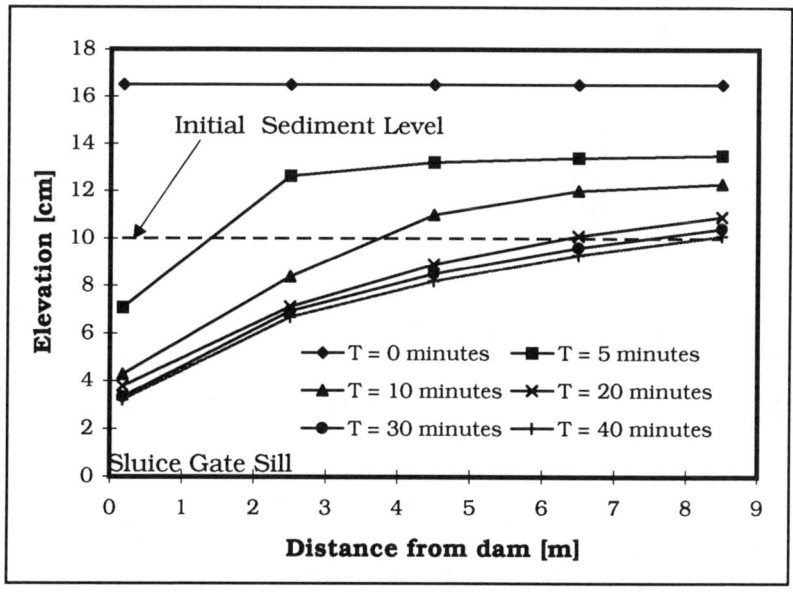

**FIGURE 4: Water Surface Profiles during Drawdown**

# MODELING THE IMPACTS OF RESERVOIR EMPTYING

A. PETITJEAN, F. MAUREL, J.P. BOUCHARD, J.C. GALLAND
Electricité de France, Laboratoire National d'Hydraulique,
6 Quai Watier
78400 Chatou, France

## ABSTRACT

Reservoir emptying is compulsory in France, for safety reasons. It requires an impact study to foresee the consequences of the emptying for all the potential users of water and to mitigate these consequences. A study methodology has been developed at EDF to answer these requirements. It includes estimate of the amount of deposited sediments, measures of physical and mechanical properties of the sediment, prediction of erosion during emptying and design of remedial devices. After the description of the physical phenomena involved in such operations, an example on the Durance river highlights the methodology.

## 1. INTRODUCTION

### 1.1. RESERVOIR DAM MANAGEMENT AND REGULATION

Electricité de France (EDF) manages 75 % of the surface water in France. So reservoir sedimentation is a major problem for EDF. Moreover for large dams (height superior to 20 m and capacity superior to $15.\ 10^6\ m^3$), emptying is compulsory every 10 years to enable inspection in order to prevent potential dam-break.

Due to the quantities of sediment trapped in the reservoirs, flushing or emptying may have negative consequences. In particular, sediment of high concentration may affect the water quality and even the ecological environment of downstream reaches. For this reason, an impact study is assessed to foresee the consequences of the emptying for all the potential users of water and to mitigate these consequences, if required. Preparation of such an operation thus requires a large amount of communication and scientific works and has to be started upto 28 months before the targeted date.

In many cases, the answer to these problems is searched in limiting as much as possible the sediment release downstream the dam. The methodology and tools for that purpose are being set up, corresponding to 4 different stages : evaluating the amount of sediment settled down within the reservoir, measuring sediment properties, predicting the amount of sediment that could leave when emptying and, finally, designing structures to limit sediment release. This last point was treated by Galland et al. [1]

### 1.2. PHYSICAL PHENOMENA

The hydrodynamic action on sediments during a lowering of reservoir level has been described in particular by J. S. Lai and H. W. Shen [2] on the basis of physical modeling and field observations. Our monitoring of a number of drainings of French dams enables us to confirm their findings. As the level begins to drop, the high velocities near the bottom gate lead to a limited erosion of a cone in the sediments.

As the level continues to lower, velocities increase in the upstream part of the reservoir, resulting in erosion of bottom sediments when the shear stress exceeds the critical shear stress. If the level drops rapidly, erosion occurs throughout the reservoir, and retrogressive erosion may develop beginning in the downstream part.

The second cause of resuspension of sediments has been described less frequently: it is the mud banks which rapidly emerge above the water, becoming unstable and tending to slide. The proportion of this mud from the banks in overall resuspended sediment increases when discharge in the river is low, and is therefore higher during emptying than during flushing.

The process of mud slides is very complex, as it involves phenomena of soil mechanics: loading of the deposits due to emersion, dissipation of pore pressure leading to consolidation of the deposits, under-pressure at the foot of the bank as sediments are drained away and a carving effect as the river bottom is hollowed out. Moreover it may be strongly coupled to hydrodynamic effects.

If the water level drops slowly enough, erosion begins first in the upstream part of the reservoir. The suspended solids progress downstream and may be redeposited in the part where velocities are low. A layer of fresh mud is then formed and is rapidly resuspended at the end of emptying. This process is highly detrimental to the environment, as concentrations of suspended solids may be very high, and for a short time, exceed 100 g/l.

Downstream of the dam, the suspended matter is often responsible for a deterioration in water quality. Cohesive sediments have the capacity to fix pollutants transported by the river. When the reservoir is stratified, reducing conditions prevail at the bottom of the reservoir; resuspension of the deposits will then lead to a release of toxic elements, consumption of dissolved oxygen and a freeing of $NH_3$.

## 2. MEASUREMENT OF THE AMOUNT OF SEDIMENT IN A RESERVOIR

The estimate of the amount of sediment deposited within a reservoir requires the deployment of specific techniques when no regular monitoring has been performed. Four different methods were tested on the Rophemel reservoir : bi-frequencies (210 kHz and 33 kHz) bathymetry, high pressure jet, seismic reflection and core sampling. On this example, the best choice was to perform a 210 kHz bathymetry survey for the definition of the roof of the sediment layer and then a core sampling campaign to obtain the thickness of the sediment layer. The seismic reflection should nevertheless be tested further because it still has an interesting potential.

Core samples are also of great help for the prediction of the amount of matter that will be eroded, as they allow for the measurements of the mud properties. This prediction can be done with very simple tools as the MOBILI model presented hereafter.

## 3. CALCULATION OF THE AMOUNT OF SEDIMENT THAT MAY BE ERODED

To date, few attempts have been made to simulate all the phenomena described before. The coupling of soil mechanics, solid transport and hydrodynamics is indeed difficult to represent, in particular because of the absence of a soil mechanics model adapted to failure of soft soil like mud. This lead us to treat the problem using a global approach which consists in considering that the channel created in the reservoir obeys the same laws as do natural rivers in equilibrium with respect to their alluvial deposits [2].

## 3.1. A MORPHOLOGICAL APPROACH : MOBILI

MOBILI is based on the morphological laws of regime theory, and gives the general characteristics of the final state reached by the reservoir at the end of the operation, assuming that hydrodynamic and soil mechanics equilibriums are both reached. By comparing this final state with the initial state, one can determine the total volume of sediments mobilized during the operation. No time information is available, however, so that forecasting the evolution of suspended solids concentrations over time is very difficult.

Geometry is discretized by cross sections and the reservoir is divided into reaches in which the sediment properties are supposed to be constant. In each reach, the general characteristics of the river bed must be evaluated. This means the bed slope, the width of the channel, the water depth and the bank slope in a state of equilibrium. This last parameter is totally independent of the other three and can be evaluated with soil mechanics measurements of cohesion, for instance. Three unknowns remain, and three equations are therefore needed. The first hypothesis is uniform flow conditions in each reach; the second stipulates that no solid transport occurs at the entrance to the reservoir at the end of the operation; and the third assumes morphological equilibrium.

In the case of a reservoir with cohesive sediment deposits, these three hypotheses give us three equations:

1 - Manning Strickler law

2 - Sediment transport equilibrium is reached when the shear stress applied by water on the river bed equals the critical shear stress for erosion ($\tau_c$).

3 - The last relation is a morphological one. Yalin has given a complete review of the numerous forms of regime theory [4]. The governing principle and the resulting relation differ depending on the author. We have chosen the principle proposed by Ramette [5]: the bed is dug with a minimum expenditure of energy. This leads us to maximize the Froude number. Assuming that the river profile is rectangular, and using the Manning Strickler law, one finds a specific ratio of width to depth $\frac{l}{h} = 18$ that Ramette has verified as being consistent with natural river morphology data.

These three equations have an analytical solution:

$$j = \left\{ \frac{18.\left(\frac{10}{9}\right)^2 \cdot \left(\frac{\tau_c}{\rho g}\right)^{8/3}}{n.Q} \right\}^{6/13} \qquad h = \left\{ \frac{\left(\frac{10}{9}\right)^{1/3} \cdot n^2 \cdot Q^2 \cdot \left(\frac{\rho g}{\tau_c}\right)}{18^2} \right\}^{3/13}, \qquad l = 18h$$

where j is the slope of the energy grade line.

These relationships are used by MOBILI to define the final state reached by a reservoir after emptying. The reservoir is described by a 1D procedure. The free surface is calculated from the dam, where the level is imposed, to the upstream part of the reservoir, using the value of j. This makes it possible to evaluate the bed level, since we know h. The cross section of the eroded channel is defined as a rectangular channel of width l, centered on the bottom of the initial sediment. Finally, bank failure is taken into account by means of stability slopes on both sides of the river. This operation is summarized on Figure 1.

The parameters of the model are: the critical shear stress for erosion $\tau_c$, the Manning coefficient n, and the discharge Q. The remaining data needed to apply the model are the geometry of the deposit and the stability slopes of the banks.

Let us now examine the assumptions underlying the MOBILI code. The first is that a state of equilibrium is reached at the end of the operation, and that morphological laws can therefore be applied. This theory was developed for natural rivers whose evolution is slow. Its extension to draining of reservoirs has not yet been proved feasible. This is particularly true for flushing during a flood, which is a very rapid operation (2 or 3 days): the result given by the model should be considered to be the equilibrium state of the reservoir after several flood episodes at discharge Q. It gives an idea of the state of sedimentation that can be maintained in the reservoir when managed by flushing.

The second hypothesis relates to modeling of mud, whose properties are supposed to be uniform in each reach. This is possible only if a core sampling survey has been carried out before. Some measurements are then needed, such as concentrations or mechanical properties like cohesion or yield value. Certain empirical laws given by Migniot [6] can be used to determine a representative critical shear stress and a stability bed slope. Where possible, flume studies can give a more reliable value of $\tau_c$.

### 3.2. APPLICATION TO THE ESCALE RESERVOIR

A validation of MOBILI was performed on the Escale reservoir on the Durance river in southeast France. The reservoir was impounded for the first time in 1963, with an initial capacity of 15 million m$^3$. Since, it has filled with cohesive sediments which reached a volume of 10.6 million m$^3$ in 1981 and 11.6 million in 1993. In October 1993 and January 1994, two severe floods occurred (a 10-year flood followed by a 100-year flood), flushing the reservoir to a certain extent. Following this, the total volume of remaining sediment was found to be 10 million m$^3$.

The aim of our study was to determine whether MOBILI could forecast the volume that might be flushed during these floods. This was the first time the software was used for this purpose and that its validation was possible thanks to the large amount of data available on the reservoir:

- a topographic survey of the reservoir before impounding in 1963 gave the bed rock position.
- core sampling had been performed for the purposes of an earlier study on the sedimentation of the reservoir. They gave the distribution of sand and mud layers. Other measurements consisted in grain size, mass concentration and some mechanical properties like yield value. In addition, the critical shear stress for erosion had been estimated by means of flume tests.
- two hydrographic surveys, before and after the floods, allowed for good assessment of the flushing effect of the floods.

The reservoir was divided into 4 reaches in which critical shear stress, bank slope and Manning coefficient had to be found.

The best way to evaluate the stability bank slope was to measure it on profiles of the reservoir before the flood.

The Manning roughness coefficient was 0.012, typical for this kind of mud; the critical shear stress determined with the flume tests ranged between 7,7 and 22 Pa.

The total amount of resuspended matter estimated by MOBILI was 1.5 million m$^3$. This is very close to the measured value of 1.57 million m$^3$ obtained by comparing the two hydrographic surveys. The result is shown on figure 2.

This test demonstrates that the code is indeed able to forecast the sedimentary impact of a flood in a reservoir. Globally, the error with respect to eroded volumes is 5%, but this low value is due to the compensation of errors in opposite directions in the different reaches. If this had not been the case and all errors had been compounded, the total relative error would have been 30%.

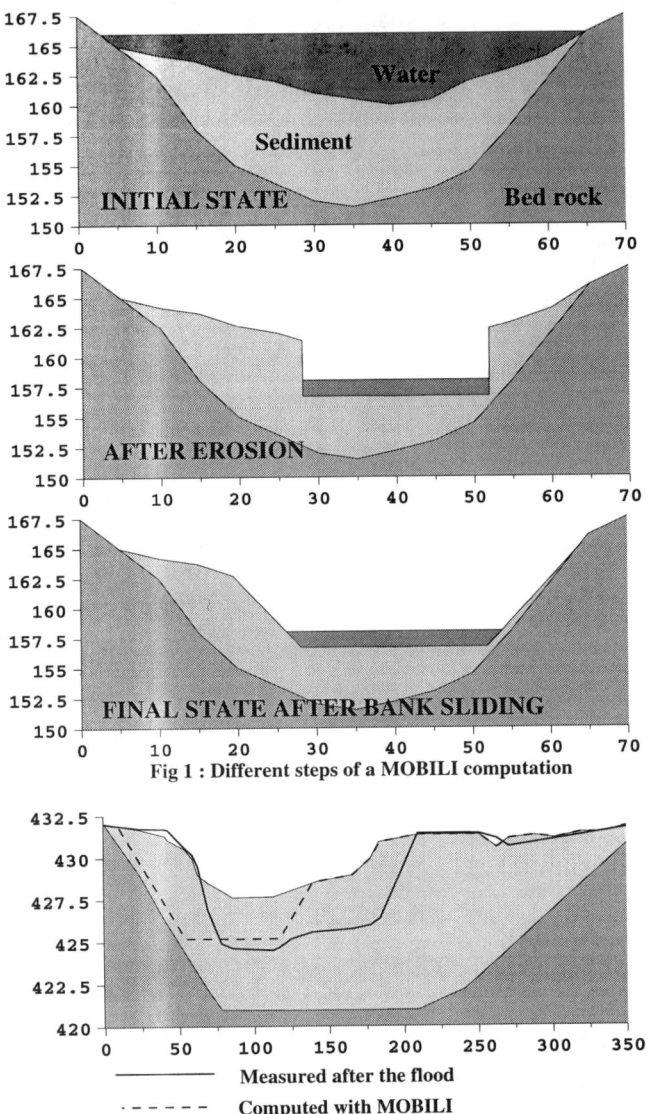

Fig 1 : Different steps of a MOBILI computation

Fig. 2 : Comparison between the computed geometry and the measured cross section after the flood in the Durance river

## 4. CONCLUSIONS AND PERSPECTIVES

In this paper we presented MOBILI which is based on the morphological laws of regime theory, and gives the general characteristics of the final state reached by the reservoir at the end of an operation such as draining or flushing.

We have shown the importance of mud slippages from reservoir banks in such operations. Taking these phenomena into account is indispensable for good management of sediments in reservoirs.

The second lesson is the importance of field measurements prior to such studies. This presentation, focusing more on modeling of the phenomena, inadequately highlights the difficulty of defining model parameters and the amount of measurements needed to define them correctly. Sediments are often extremely heterogeneous, and meas ents taken at only a few points are of little significance. Validation of the models described here cannot be dissociated from validation of the entire set of data which served to describe the reservoir and the sediment deposits.

It is therefore the entire study methodology which is important, including the choice of variables to be measured, the importance of taking heterogeneity into account, and the modeling itself which must continue to be validated if such studies are to increase in reliability.

## REFERENCES

[1] GALLAND J.C., BOUCHARD J.-P., MAUREL F.
Limiting Sedimentological Impact of Reservoir Emptying. 2nd ICCORES Conference, Fort Collins,USA, 1996

[2] LAI J.S., SHEN H.W.
Flushing sediments through reservoirs.
Journal of Hydraulic Research, p. 237, N. 2, Vol. 34, 1996.

[3] BOUCHARD J.-P., MAUREL F., PETITJEAN A.
Numerical modeling of resuspension in reservoirs. 2nd ICCORES Conference, Fort Collins,USA, 1996

[4] YALIN M.
River Mechanics, ed. Pergamon Press, 1992.

[5] RAMETTE M.
Cours d'eau sauvages, cours d'eau aménagés. EDF report HE-40/79/10b, (in french), 1979.

[6] MIGNIOT C.
Tassement et rhéologie des vases (1) and (2) (in french).
La Houille Blanche n° 1 et n°2, 1989.

# Desilting Reservoir Sediment Deposits by Drawdown Flushing

### JIHN-SUNG LAI
Assistant Research Fellow, Hydraulic Research Laboratory,
National Taiwan University, Taipei 106 - 17, Taiwan, R. O. C.

### HSIEH WEN SHEN
Professor, Department of Civil Engineering,
University of California at Berkeley, Berkeley CA 94720, USA.

### ABSTRACT
Reservoir desiltation techniques through hydraulic flushing have been employed effectively in many field cases. It usually results in strong degradation in the reservoir if the reservoir water level can be drawn down significantly. A diffusion model and an unsteady mobile-bed model were established to simulate one-dimensional degradation flushing processes during drawdown operations. Three sets of data collected in a 0.5m-wide flume were used to calibrate and verify the models. The simulated results of the bed elevation changes were in good agreement with the measured data.

### INTRODUCTION
Generally speaking, there are three control measures of reducing reservoir sedimentation, including (1)reducing sediment yield from watershed, (2)dredging, (3)hydraulic flushing or density current venting. Among them, hydraulic flushing techniques have been employed successfully in many field cases around the world to preserve the useful reservoir storage [1].

Incoming sediments can be deposited first in the main channel and extend laterally across the width of the impounded reservoir. Hydraulic flushing is to remove sediments by scouring sediment deposits and/or passing incoming sediment through a reservoir. It may occur when the desilting outlet is opened to release sediment and water from the reservoir. In practice, hydraulic flushing is proved to be effective if the reservoir water level can be drawn down significantly. If sufficient water level drawdown near the dam can be generated, flow in the impounded reaches will imitate original riverine patterns to create a flushing channel by cutting in the deposits. Thereby, strong erosion on reservoir deposits may take place, which is called retrogressive erosion.

There were several models developed to simulate sediment transport during flushing. White (1990) developed numerical models by using the backwater profile calculation to study the feasibility and effectiveness of flushing operations. Coping with different sediment transport equations, one-dimensional diffusion models were developed to simulate bed profile variations. They could be solved analytically for constant diffusivity and simple boundary conditions, or numerically for changeable diffusivity and complex boundary conditions [2]. Lai and Shen (1995) developed a one-dimensional unsteady mobile-bed model by adopting the approximate Riemann solver to simulate degradation flushing processes in a reservoir. In this study, experimental data used were collected by Peng and Niu (1987) in a 0.5m-wide flume. The diffusion model and the unsteady mobile-bed model were employed to investigate degradation processes during flushing in a reservoir. The simultaed results of bed evolution from the models were then compared with measured data to show good agreements.

## GOVERNING EQUATIONS

In a rectangular channel without lateral inflow, one-dimensional unsteady flow equations can be written as:

$$\frac{\partial h}{\partial t} + \frac{\partial (hu)}{\partial x} = 0 \tag{1}$$

$$\frac{\partial (hu)}{\partial t} + \frac{\partial}{\partial x}(hu^2 + gh^2/2) = gh(S_o - S_f) \tag{2}$$

and the sediment continuity equation is:

$$\frac{\partial z_b}{\partial t} + \frac{1}{(1-p)} \frac{\partial q_s}{\partial x} = 0 \tag{3}$$

Equation (1) and (2) present mass and momentum conservation for water flow. $h$ is the water depth; $u$ is the mean velocity; $g$ is the gravitational acceleration; $S_o$ is the bottom slope; $S_f$ is the friction slope; $z_b$ is the bed elevation; $p$ is the porosity of bed material; $q_s$ is the volumetric sediment transport rate per unit width. The unknowns are $u$, $h$, $z_b$, $S_f$ and $q_s$, and two additional relationships of flow resistance and sediment transport rate are required. The friction slope can be expressed by Manning formula or associated with the hydraulic radius, representative roughness height (i.e. median diameter), and velocity [3].

Based on the concept of energy balance in sediment transport, the total streamwise energy generated in a stream reach per unit time (i.e. available stream power(Bagnold, 1966)) may be divided into two components: the energy moving sediment and the energy overcoming flow resistance. The volumetric sediment transport rate per unit width can be written:

$$q_s = \frac{q}{\theta(s-1)}(J - J_c) \tag{4}$$

where $q$ is discharge per unit width; $\theta$ is a coefficient; $s$ is the specific gravity of the sediment; $J$ is the energy gradient; $J_c$ is the critical energy gradient.

## EXPERIMENTAL DATA

To examine the validity of the models for the simulation of the flushing processes during drawdown operations, three sets of experimental data collected by Peng and Niu [4] were used. The experiments were conducted in a 0.5m-wide flume covered by sand ($d_{50}$ =0.45 mm). A 0.5m-wide gate was installed in the flume and could be opened suddenly to draw down water level rapidly near the gate. Table 1 lists the experimental conditions. The cumulative volume of flushed sediment deposits varying with time and the bed profile along the flume at specific time were recorded.

Table 1  Experimental data collected in a 0.5m-wide flume (Peng and Niu [4])

| Original Run No. | Discharge ($m^3$/s) | Initial Deposit Depth at Dam (m) | Sand Feeding Rate ( ton/(m·s) ) | Length of Sand Bed (m) | Initial Bed Slope (%) | Running Time (min) |
|---|---|---|---|---|---|---|
| a | 0.0114 | 0.071 | 0. | 8.73 | 0. | 46 |
| b | 0.0200 | 0.071 | 0. | 8.73 | 0. | 35 |
| c | 0.0200 | 0.064 | $2.16 \times 10^{-5}$ | 8.60 | 0.18 | 40 |

## NUMERICAL SOLUTIONS

In this section, the numerical treatments for the diffusion model and unsteady mobile-bed model were discribed in the following.

### DIFFUSION MODEL

Based on the assumption of steady uniform flow condition with no lateral sediment inflow, a diffusion equation can be derived from the sediment continuity equation combined with Eq.(4), and it is written as:

$$\frac{\partial z_b}{\partial t} - K \frac{\partial^2 z_b}{\partial x^2} = 0 \ ; \quad \text{where} \quad K = \frac{q}{\theta(1-p)(s-1)} \quad (5)$$

where $K$ is the coefficient of diffusivity. Essentially, Eq.(5) is a second order parabolic type of partial differential equation which can be solved by an analytical procedure with a constant diffusion coefficient $K$ and simple boundary/initial conditions. For practical purposes, the numerical method is more powerful to deal with the complex boundary/initial conditions and a changeable diffusion coefficient $K$, especially for flushing operation [2,4].

Regarding numerical methods for solving Eq.(5), the Crank-Nicolson implicit finite difference scheme was employed to calculate the numerical solutions of the bed profile $z_b(x,t)$. The Crank-Nicolson implicit scheme is unconditionally stable and can achieve the second-order accuracy.

Two boundary conditions are needed to close the system. The downstream boundary condition is the bed elevation at dam, which is the Dirichlet condition, $z_b = z_b(0,t)$. At the upstream end of reservoir, the Neumann boundary (derivative type) condition, $\partial z_b(L,t) / \partial x = f(t)$, is applied; $L$ is the length of the reservoir. Derived from Eq.(5), the upstream boundary condition can be expressed as:

$$\left.\frac{\partial z_b}{\partial x}\right|_{x=L} = \frac{\theta(s-1)G_L}{\gamma_s q} + J_c \qquad (6)$$

where $G_L$ is the upstream sediment supply in ton/s per unit width, and $\gamma_s$ is the specific weight of sediment. The initial condition is given by the bed profile of the previous sediment deposits.

## UNSTEADY MOBILE-BED MODEL
Eq.(1) and Eq.(2) can be rewritten in a conservative vector form:

$$q_t + [f(q)]_x = b(q) \qquad (7)$$

where $q = [h, hu]^T$, $f(q) = [hu, hu^2 + gh^2/2]^T$ and $b(q) = [0, gh(S_o - S_f)]^T$. Equation (7) is a hyperbolic system. With the simulation domain being discretized by finite difference method, the solution at the interface between elements (or grids) is a Riemann problem. The Riemann problem is an initial value problem of Eq.(7) without source term $b(q)$. $f(q)$ is a normal outward flux with respect to one side of the element and $q$ at time $t=0$ are assumed to be known. The quantities $q_L$ and $q_R$ are the piecewise constant properties on the left and right, respectively, of the element interface. The outward normal flux $f(q)$ can be obtained by solving the initial Riemann problem. For the approximate solution of the Riemann problem, the approach used is to generalize the Riemann solver of the linear case to the nonlinear case. This approximate solution can be obtained by using flux-difference splitting method to solve h and u. Then, the bed elevations for advanced time step can be calculated in Eq.(3).

In the flow part, two boundary conditions applied in the model are the given hydrograph at upstream and the water stage at sluice gate. The boundary conditions for sediment continuity equation are the sediment supply rate at the upstream end and the sediment discharge derivative equal to zero at the downstream end. The initial condition is given by the initial values of $u$, $h$ and $z_b$ at each element.

## RESULTS OF SIMULATION
### Diffusion Model
The Crank-Nicolson implicit finite difference scheme is unconditionally stable; therefore, there is no limitation of the time interval. For the better resolution of bed elevations, equal space and time intervals are selected to be 0.2 m and 10 seconds, respectively. Given the upstream and downstream boundary conditions, bed elevations at each node can be obtained by solving a tridiagonal matrix system. To determine the value of the coefficient $\theta$ for diffusivity in Eq.(5), the criteria for calibration is to evaluate whether the calculated results such as bed profile were able to simulate the experimental results with reasonable accuracy. Run $a$ is first selected to calibrate the coefficient $\theta$. The calibrated $\theta$ will then be applied to other runs.

To obtain a better fit with measured data, it is found the value of coefficient $\theta$ is 1.73 (i.e. $K = 0.028$ by Eq.(5) with $p = 0.5$ and $s = 2.65$) and the critical energy slope $J_c$ is 0.0023 for Run $a$. As shown in Figure 1(a), with initially a zero bed slope the simulated bed profile of Run $a$ is plotted to compare the measured data at 90 second. Additional simulated bed profiles are also plotted at the 30 second and 2,000 second to illustrate the progress of bed erosion. Using the calibrated resistance coefficient and critical energy slope, the diffusion model was applied to predict the bed profile for other runs. The calculated results are shown in Figures 1(b) and 1(c) for Run $b$ and $c$, respectively.

## Unsteady Mobile-Bed Model

The unsteady mobile-bed model was developed to solve the bed elevation and water surface profiles at each computation time step. The time step is restricted to the Courant condition, and therefore the time interval is selected to be 0.2 second with the nonuniform grid size. The comparison between diffusion model and unsteady mobile-bed model will be discussed for the simulation in drawdown flushing.

Applying the sediment transport equation in Eq. (4) to obtain the sediment discharge flux at the interface of elements, the unsteady mobile-bed model was used to simulate the bed profiles. As shown in Figures 2(a) through 2(c), the bed profiles simulated by these two models are plotted together, and the simulated bed profiles for each run are close to each other. Thus, both models can simulate hydraulic flushing performed in a flume by drawdown operations.

## CONCLUSION

A one-dimensional diffusion model and an unsteady mobile-bed model were established to simulate degradation flushing processes during drawdown flushing. For both models, the simulated results of the bed elevation changes were in good agreement with the measured data from the experiments conducted in a 0.5m-wide flume. However, more studies are needed when the models are applied to the cases with lateral erosion, complicated reservoir geometry, or nonuniform sediments.

## REFERENCE

1. Bruk, S.(1985). *Methods of computing sedimentation in lakes and reservoirs*, UNESCO, Paris, 224p.
2. Lai, J.S. and Shen, H.W. (1996). "Flushing sedment through reservoirs," *Jour. of Hydraulic Research*, IAHR, Vol. 34, no.2, pp.237-255.
3. Lai, J.S. and Shen, H.W. (1995). "Degradation flushing processes in reservoir," *XXVIth IAHR Congress*, London, UK, Sept. 11-15, 1995.
4. Peng, R., and Niu J. (1987). "Numerical model for headward erosion on bed load." *J. Sediment Research*, Vol 3 (In Chinese).
5. White, W.R.(1990). "Reservoir sedimentation and flushing." In: *Hydrology in Mountainous Regions. II-Artificial Reservoirs; Water and Slopes, Proc. of two Lausanne Symposia*, IAHS, Publ., no.194, pp.129-139.
6. Zhao, D.H., Shen, H.W., Tabios III, G.Q., Lai, J.S., and Tan, W.Y. (1994). "A finite volume two-dimensional unsteady flow model for river basins." *J. Hydr. Div.*, ASCE, Vol.120(7).

# DESILTING RESERVOIR SEDIMENT DEPOSITS

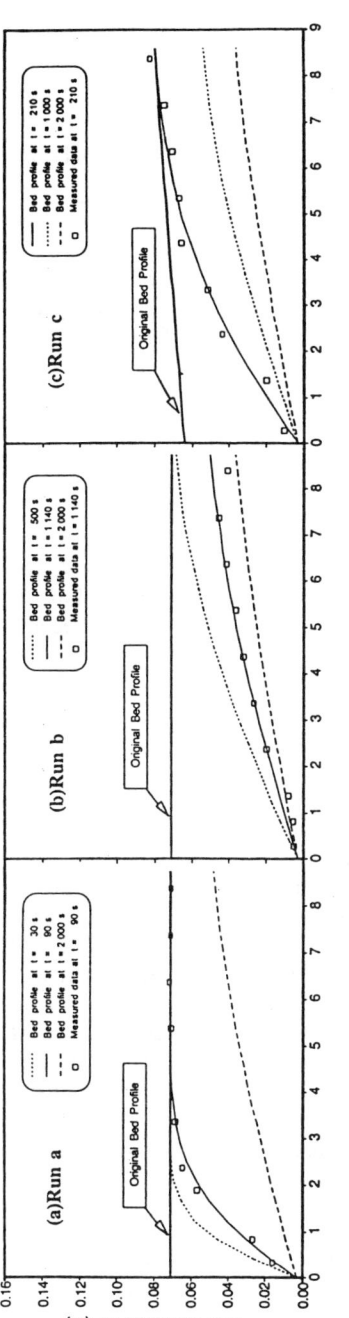

Figure 1 Bed profiles simulated by diffusion model for (a)Run a, (b)Run b and (c)Run c

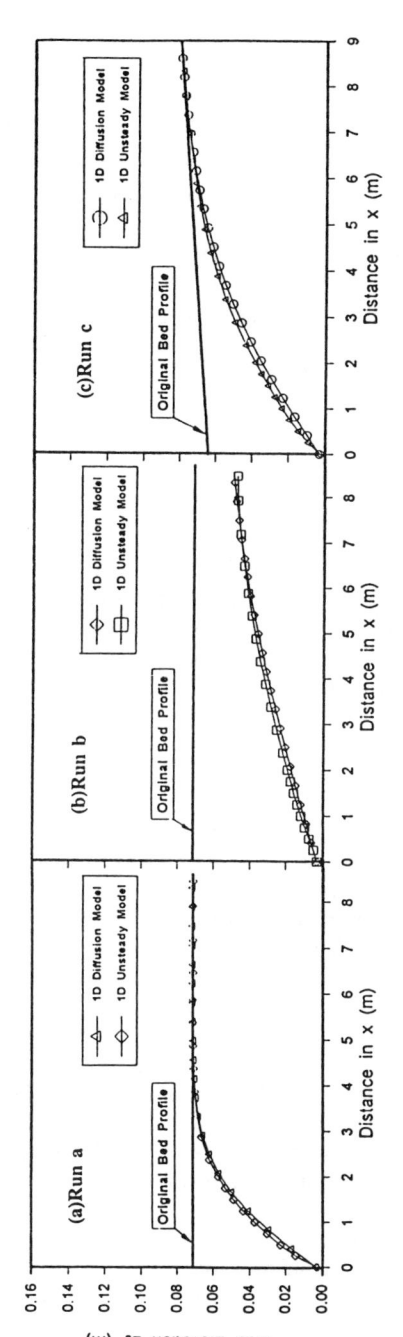

Figure 2 Comparison of simulated results from diffusion model and unsteady mobile-bed model for (a)Run a, (b)Run b and (c)Run c

# NON-EQUILIBRIUM SEDIMENT TRANSPORT MODELING OF SANMENXIA RESERVOIR

ALBERT MOLINAS
Assistant Professor
Department of Civil Engineering
Colorado State University
Fort Collins, CO 80523

BAOSHENG WU
Ph.D. Student
Department of Civil Engineering
Colorado State University
Fort Collins, CO 80523

## ABSTRACT

The Generalized Stream Tube model for Alluvial River Simulation (GSTARS) is modified to accommodate the non-equilibrium sediment transport and the computation of sediment transport capacity by size fractions. An application of the modified model in the Sanmenxia reservoir is presented. The comparison shows that the computed and measured channel bottom and water surface profiles after three flood events are in good agreement.

## INTRODUCTION

Sanmenxia Reservoir (Fig. 1) is located in the lower part of the middle Yellow River in China. It is a multipurpose hydro-project mainly used for flood control. The drainage area above the dam amounts to 688,000 km$^2$. The Yellow River cuts through an extensive loess plateau with an area 580,00 km$^2$, of which 439,000 km$^2$ suffer from severe soil erosion. The sediments yield of the gullied loess area amounts to more than 10,000 tons/km$^2$ annually. The yearly average sediment load at Sanmenxia station amounts to 1.6 billion tons, 86% of which comes during the flood season from July to October. A suspended sediment concentration of 911 kg/m$^3$ was measured at Sanmenxia station during a flood in 1977. Since 1974, the reservoir has been operated by storing only relative clear water in non-flood season and disposing the muddy water during flood season from July to October. As a result, the deposition in the reservoir has been kept in a state of equilibrium. A usable capacity in the reservoir has been preserved for regulation of water and sediment.

Since the sediment concentration along the Sanmenxia reservoir can be very high, and the sediment inflow varies to a great extent from flood to flood, the sediment transport in the reservoir is usually in a highly non-equilibrium state (Molinas, Yang, and Wu

1994). Moreover, the sediment size gradation also varies dramatically from flood to flood depending on the locality covered by the rainfall. Therefore, the sediment transport of different size groups is highly variable. In contrast to the sediment transport in fairly clear water without too much wash load, the transport capacity in the Yellow River is evidently affected by the changes of sediment fall velocity and the specific weight due to high concentration (Molinas, Yang, and Wu, 1996).

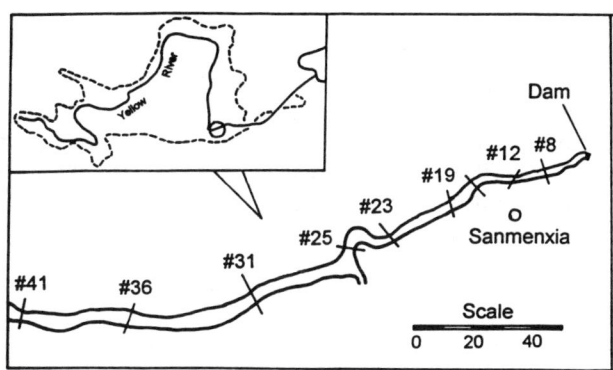

Fig. 1  Sketch of Sanmenxia Reservoir

The Generalized Stream Tube model for Alluvial River Simulation (GSTARS) was developed by Molinas and Yang (1986) for solving river engineering problems. This model is a semi-two dimensional model using stream tubes. The basic equations used in the model include energy equation, momentum equation, sediment continuity equation, and four optional sediment transport equations. In this paper, the GSTARS model is modified to accommodate the characteristics of the Yellow River including non-equilibrium sediment transport, and the computation of sediment transport capacity by size fractions. An application of the modified model in the Sanmenxia reservoir is presented.

## NON-EQUILIBRIUM SEDIMENT TRANSPORT

For suspended sediment transport, the actual sediment discharge is affected by diffusion. The diffusion process affects essentially the suspended load during deposition and entrainment because certain sediments, especially the fines, require considerable time or distance in settling or in attaining their transport capacity. Therefore, suspended load may not adjust immediately to flow conditions. It requires a certain development distance to reach the equilibrium concentration. In mathematical modeling of sediment transport, sediment load is usually computed at each discrete cross section. This computed sediment load needs to be corrected if the development length for sediment transport is longer than the grid size $\Delta x$.

From diffusion theory of sediment transport, the non-equilibrium suspended sediment transport equation can be expressed as

$$\frac{dC_t}{dx} = -\frac{\alpha\omega}{q}(C_t - C_*) \tag{1}$$

where $C_t$, $C_*$ are the suspended sediment concentration and transport capacity, respectively, $q$ is the unit discharge, $\omega$ is the sediment fall velocity, and $\alpha$ is the non-equilibrium suspended sediment transport coefficient. The relationship given by (1) is first order linear differential equation. Assuming that the sediment transport capacity varies linearly with respect to distance along the channel, the solution of (1) can be obtained (Han 1980, Han and He 1990)

$$C_{ti} = C_{*i} + (C_{ti-1} - C_{*i-1})\exp\left(-\frac{\alpha\omega\Delta x}{q}\right) \\ + (C_{*i-1} - C_{*i})\frac{q}{\alpha\omega\Delta x}\left[1 - \exp\left(-\frac{\alpha\omega\Delta x}{q}\right)\right] \tag{2}$$

where $i$ is the cross section index counted from upstream to downstream. For non-uniform sediment mixtures, this equation is assumed to be applicable to each size fraction.

## SEDIMENT TRANSPORT CAPACITY BY SIZE FRACTIONS

The computation of sediment transport capacity by size fractions is one of the key elements in numerical simulation for non-uniform sediment mixtures. In this paper, a method identified as the transport capacity fraction method is used for routing sediment mixtures. The total sediment concentration is first computed by using a total sediment transport equation. The computed total concentration is then divided into size fraction concentrations by a transport capacity-size distribution function. This method is expressed mathematically as

$$C_{*k} = p_{ck}C_*, \quad \sum_k p_{ck} = 1 \tag{3}$$

where $C_{*k}$ is the sediment transport capacity of size fraction $k$, and $p_{ck}$ is the sediment transport capacity fraction by weight. The total sediment concentration $C_*$ is computed from the universal stream power transport equation for natural rivers developed by

Molinas and Wu (1996)

$$C_*(PPM) = \frac{1430(0.86 + \sqrt{\psi})\psi^{1.5}}{0.016 + \psi} \quad (4)$$

where $\psi$ is the universal stream power, which is expressed as

$$\psi = \frac{V^3}{\frac{\gamma_s - \gamma_m}{\gamma_m} g D \omega_m \left[\log\left(\frac{D}{d_{50}}\right)\right]^2} \quad (5)$$

where $D$ is the average flow depth, $d_{50}$ is the median diameter of bed material, $g$ is the gravitational acceleration, $V$ is the average flow velocity, $\gamma_s$ and $\gamma_m$ are specific weight of sediment and water-sediment mixtures, and $\omega_m$ is the fall velocity of $d_{50}$ in water-sediment mixtures.

The transport capacity size fraction is computed by Molinas and Wu (1995) equation

$$p_{ck} = \frac{p_{bk}\left[\left(\frac{d_k}{d_{50}}\right)^\alpha + \zeta\left(\frac{d_k}{d_{50}}\right)^\beta\right]}{\sum_k \left\{p_{bk}\left[\left(\frac{d_k}{d_{50}}\right)^\alpha + \zeta\left(\frac{d_k}{d_{50}}\right)^\beta\right]\right\}} \quad (6)$$

in which

$$\left.\begin{aligned}\alpha &= -2.9\exp\left[-1000.0\left(\frac{V}{V_*}\right)^{2.0}\left(\frac{D}{d_{50}}\right)^{-2.0}\right] \\ \beta &= 0.2\sigma_g \\ \zeta &= 2.8 F_r^{-1.2}\sigma_g^{-3.0}\end{aligned}\right\} \quad (7)$$

where $d_k$ is the particle size of size fraction $k$, $F_r$ is the Froude number ($F_r = V/\sqrt{gD}$), $p_{bk}$ is the $k$th size fraction of bed material, $V_*$ is the shear velocity, and $\sigma_g$ is the standard deviation of bed material size ($\sigma_g = \sqrt{d_{84}/d_{16}}$). This equation considers the sheltering and exposure effects on a given size fraction due to the presence of other sizes in the bed material.

## MODELING OF SANMENXIA RESERVOIR

The modified GSTARS model is used to simulate the non-equilibrium and non-uniform sediment transport in Sanmenxia reservoir for a time period of 50 days from July 1, 1974 to August 18, 1974. The water and sediment inflows are shown in Fig. 2. During flood events the reservoir is operated at low water level to facilitate the passage and sluicing of sediment. The sediment deposited in the backwater zone during the low flow season (when the reservoir level was kept high) is sluiced by high sediment transport capacity of flows combined with high drawdowns during the flooding season. Fig. 3 is a comparison of the computed channel bottom and water surface profiles with the measured profiles. The initial channel bottom profile is also shown in Fig. 3. It can be seen that the computed channel bottom and water surface profiles after three flood events are in good agreement with the measured profiles. Comparing channel bottom profiles, it can be seen that the sluicing of deposited sediment in the backwater zone 60 km to 120 km upstream from the dam is successfully simulated.

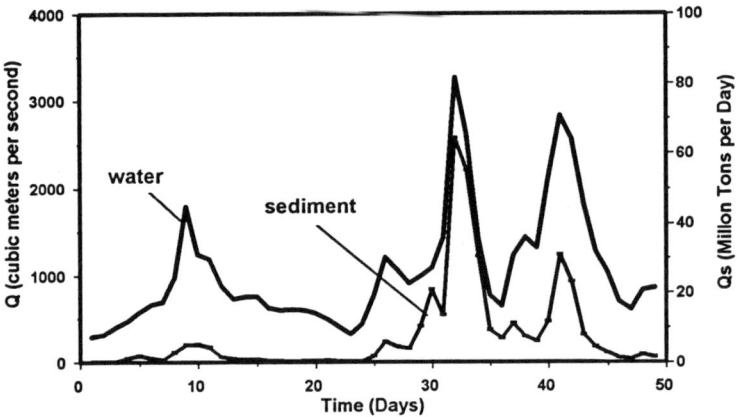

Fig. 2   Water and Sediment Hydrographs

## CONCLUSIONS

The Generalized Stream Tube model for Alluvial River Simulation (GSTARS) is modified to accommodate non-equilibrium sediment transport and the computation of sediment transport capacity by size fractions. The modified model also includes newly developed sediment transport equation for natural rivers. An application of the model to the Sanmenxia reservoir is presented. The comparison shows that the computed and measured channel bottom and water surface profiles after three flood events are in good agreement.

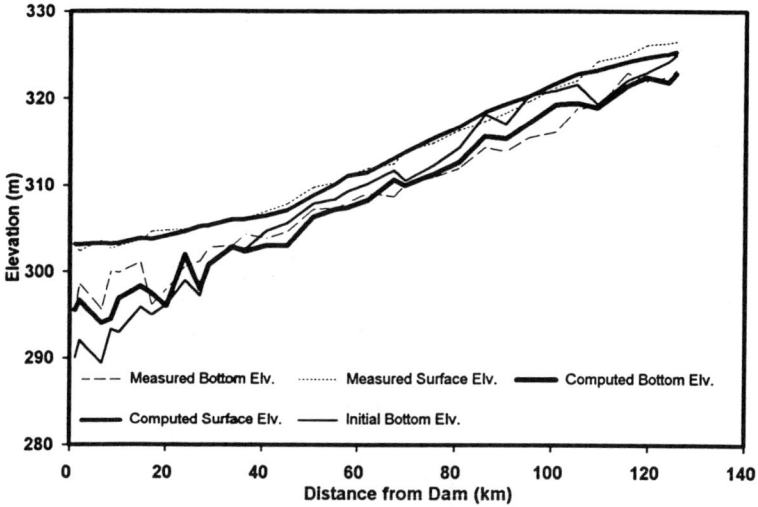

**Fig. 3** Comparison of Computed and Measured Channel bottom and Water Surface Profiles

## REFERENCES

Han, Q. (1980). "A Study on the Non-equilibrium Transportation of Suspended." Proc. 1st Int. Symp. on River Sedimentation, Beijing, China, pp.793-802.

Han, Q, and He, M. (1990). "A Mathematical Model for Reservoir Sedimentation and Fluvial Processes." Int. J. of Sediment Res., IRTCES, Vol. 5, No. 2, pp.43-84.

Molinas, A., and Yang, C. T. (1986). "Computer Program User's Manual for GSTARS (Generalized Stream Tube model for Alluvial River Simulation)," U.S. Department of the Interior, Bureau of Reclamation, Engineering and Research Center, Denver, Colorado, 142pp.

Molinas, A., Yang, C. T., and Wu, B. (1994). "Application of GSTARS to the Yellow River Morphological Studies," Colorado State University, Fort Collins, Colorado, 23pp.

Molinas, A., and Wu, B. (1995). "Transport of Nonuniform Sediment Mixtures," J. Hydr. Engrg., ASCE, (Submitted, August 1995).

Molinas, A., and Wu, B. (1996). "Sediment Transport in Natural Rivers," J. Hydr. Research, IAHR, (Submitted, October 1996).

Yang, C. T., Molinas, A., and Wu, B. (1996). "Sediment Transport in the Yellow River," J. Hydr. Engrg., ASCE, Vol.122, No.5, pp.237-244.

# DESILTING OF WATER RESERVOIRS IN ALGERIA BY DREDGING
## STUDY CASE : THE FERGUG RESERVOIR

**MOHAMED ERRIH** and **HABIB BENDAHOU**
Asst. Professors, Institute of Hydraulics, University of Sciences & Technology of Oran
P.O. Box 1505, El-Mnaouer Oran, Algeria

### ABSTRACT
Close to 37 reservoirs (middle and great size) are now in operation in Algeria. Some of them are in need of silt clearance. Different techniques of desiltation have been used in Algeria, however, the dredging method remains the most employed. The objective of this paper is to examine the dredging operation undertaken in the Fergug reservoir. This operation is called *Prototype Project*. The informations obtained will serve as a database for the desilting of other reservoirs in operation. The dredge that was used is a De Groot Nijkerk, Dutch brand. The desilting project was at the start assigned to the Dutch Company H.A.M. The dredging have begun in September, 1989 and stopped in March, 1990 (6 months). After one year stop, the Algerian Company of Maritime Works (SO.NA.TRA.M) has resumed the dredging works from March, 1991 until March, 1992 (12 months). The useful capacity has passed from 2 (siltation of 90%) to 9 millions of cubic meters (siltation of 50%). However, the desilting procedures by dredging of reservoirs are often very costly, specially for developing countries : the desilted cubic meter cost 14.53 Algerian Dinars ($ ½), including equipement (materials) and about 52% of water is utilized for 48% of extracted silt. So, the problem must be attacked at the origin to preserve the lands against erosion and to reduce the sediment transport, making then the reservoir last longer.

### INTRODUCTION
The complex process of erosion-solids transport-sedimentation is characterized by a large space-time irregularity suited to the hydroclimatical, physical, geomorphological and socieconomical conditions of Algeria. The flooded wadi carries a very important mass of suspended solid particles. Water show a muddy and colloid aspect. The wadi carries on its bed voluminous bodies making the hydrometrical measurements of bed loads difficult and indeed impossible (Errih *et al.*, 1992). The recorded water velocity often reaches 5 m/s (Medinger, 1960). Even the measurement of water velocity by current meters is a difficult task in the flooding period, particularly during the passage of flood waves : breakage of suspension cables and propellers of current meters. The sediment concentration is relatively high and very variable. The concentrations are generally maximal during the first of September-October with wadi beds initially dry during the summer season.

The specific erosion rate in Algeria is over 2000 tons/km$^2$/year on average with a maximum of 4000 to 5000 tons/km$^2$/year respectively in the region of Dahra and on the catchment of Agrioun wadi which supplies the Ighil-Emda reservoir (Demmak, 1982). The wadies of the Mediterranean Basin pour out 200 millions tons of sediments annually on average into the sea. We can a priori say that the values of sediments transport of wadi are smaller than the real values, because the bed load is not measured. The small wadi carry a relatively important sediment volume despite a weak flood flow. Table 1 shows the annual values of liquid and solid volumes of some experimental micro-catchments for the year 1984/85.

**Table 1. Values of liquid and solid volumes of small wadi. Legende: S=catchment area (ha); Vw= volume of flow (thousand cubic meters); Vs=sediment transport (tons); C=sediment concentration (g/l); T=specific erosion rate (tons/km²/year).**

| Experimental cachment | S (ha) | Vw(10³ m³) | Vs(tons) | C (g/l) | T(tons/ km²/year) | Remarks |
|---|---|---|---|---|---|---|
| BS21 - 1 | 169 | 43.3 | 1659 | 38.4 | 981.7 | Hill |
| BS21 - 2 | 84 | 10.1 | 274.3 | 27.3 | 326.5 | Hill |
| Souaghi - 1 | 27 | 0.9 | 116 | 124 | 429.7 | Glacis |
| Souaghi - 2 | 27 | 1.8 | 323.3 | 180 | 1197 | Glacis |
| Bad Land 1 | 8 | 8.2 | 327.4 | 45.3 | 4655 | Bad Land |
| Bad Land 2 | 18 | 12.1 | 861.4 | 71.4 | 4786 | Bad Land |

The reduction of useful capacity by siltation are estimated to be nearly 20 millions m³/year, for the whole Algerian hydraulic structures. The lifetime average of reservoirs does not exceed 30 years. Table 2 gives the evolution of siltation of some small and middle sized reservoirs. Figure 1 shows the storage potentialities evolution and the siltation state of the reservoirs in operation, for a period going until the year 2010. The total yearly water capacity is clearly below the maximum storage capacity. This is due to the drought which lasts since 1980 on the one hand, and the bad water resources management on the other hand (Errih, 1993). The importance of the siltation of reservoirs in Algeria has been, since the fifties, the subject of several studies (Duquennois, 1955; Thévenin, 1960; Belbachir, 1980; Bellouni, 1980; Demmak, 1980; Errih *et al.*, 1992).

**Table 2. Evolution of reservoir siltation of small and middle size. Legende : d=before dredging (evacuated volume 8 millions of cubic meters from 1968 to 1972);ad=after dam surelevation (1977).**

| Reservoir | year of priming | capacity (10⁶ m³) | | | siltation / year (10⁶ m³) | erosion rate(%) |
|---|---|---|---|---|---|---|
| | | initial | 1967 | 1977 | | |
| Fodda | 1932 | 228 | 139 | 109 | 2.6 | 52 |
| Hamiz | 1935 | 21.5 | 12 | 17.3d | 0.33 | 51 |
| Ghrib | 1939 | 280 | 185 | 150 | 3.4 | 46 |
| Ksob | 1940 | 11.6 | 3 | 29 | 0.27 | 84ad |
| FoumEl Guerza | 1950 | 47 | 35 | 27 | 0.74 | 43 |

Varied techniques of desilting have been applied. Method such as extraction by siphon device, flushing by current density have been developped to recuperate capacity. However, the dredging is the most utilized. The flushing by current density technique was used for the first time in the 1953 in the Iril-Emda reservoir, built upon the Agrioun wadi, for electricity production. The extraction by siphon device was thought for the first time by Jandin who has developped and utilized this method between 1892 and 1894 to evacuate the sediments from the Djidiouia reservoir (Hannoyer,1974). About 1.4 million of silt has been extracted in period of three years.

**Table 3. Volume of extracted silt from some reservoirs in Algeria from the year 1958 to 1968**

| Period | 1958-1961 | 1962-1964 | 1965-1966 | 1967-1968 |
|---|---|---|---|---|
| Reservoir | Cheurfas | Sig | Fergug | Hamiz |
| Capacity of extracted silt (10⁶ m³) | 10 | 1 | 3 | 1.2 |

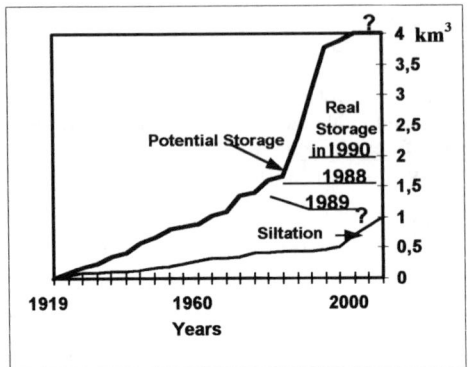

**Figure 1 Evolution of potential storage and siltation**

## THE ALGERIAN EXPERIENCE OF DESILTING BY DREDGING

It was only in the year 1950 that we started seriously taking care of the dredging of reservoirs in Algeria (Thévenin, 1960). Generally, the dredging necessitated the excavation of a channel in the sediments to facilitated the flow of density currents towards the bottom outlets. The work performed by the Lucien Demay dredge (suction-cutter dredge) is shown in Table 3 in chronological order from the year 1958 to 1968. The Lucien Demay dredge is in theory able to dredge 1 cubic meter of silt (density of about 1.6), in situ, using about 5 cubic meters of clear water. When the suction and discharge pipes get plugged, the dredge consumes more water, up to nine times of extracted silt (Bellouni, 1980; Belbachir, 1980). During the dredging period in the Hamiz reservoir (5½ months), 8300 hours of pumping have been really performed and 730 hours of interruption have been recorded. According to Bellouni (1980), the dredging has been interrupted because of mechanical and piping system failures, plugs in the pipes (25% of the stops), the changing of places, extension or shortening of the floating pipeline.

## THE DREDGING OF FERGUG RESERVOIR

### SITUATION AND CHARACTERISTICS OF THE RESERVOIR

This second desilting of the Fergug reservoir, after the one performed in 1965, has been called *Prototype Project* . The informations collected will serve as a database for desilting other reservoirs in operation. The Fergug reservoir is located upon the Habra wadi, in the North West of Algeria, in the foot of the Atlas chain, at about 10 km upstream of the city of Muhammadia. The reservoir is supplied by El-Hamman, Fergug and Thaghzut wadies. It was built in the year 1871, and has a height of 50m and a lenght at the top of 500m. It has initially a useful capacity of 18 millions of cubic meters. At a bathymetrical survey made in 1973, the useful capacity was only 12.71 millions of cubic meters. The cachment area of the Habra wadi, upstream of the reservoir, is 420 km². The mean annual precipitation is 430mm. The stream flow to the reservoir is 56 millions of cubic meters per year on average. The reservoir performs the following functions : regulation of the Fergug and El-Hammam wadies; drinking water supplying to the cities of Oran, Arzew and Muhammadia; industrial water supplying to the industrial area of Arzew and irrigation of the Habra land.

### LEVEL OF SILTATION AND SEDIMENTS' NATURE

In order to get an idea of the siltation level, it was proceeded in May, 1983 to a bathymetrical survey of the reservoir. At 700m upstream of the reservoir the level was 100.5m and we could still measure depths of 5 to 7m. These deepened at the intake structure. Upstream of this limit, the mean depth of water is 4m. The silt ceiling reaches in the affluents, at 2.5km upstream of reservoir, the free surface. The useful capacity was reduced to 2 millions of cubic meters, giving a siltation of about 90%. Twelve borings, spread uniformly over the reservoir area, have been executed in order to probe the depth of the sediments and investigate their geotechnical properties. The borings which have depths from 6 to 28m were bounded by the alluvial layer of the old wadi bed. The sediments of thickness varying from 3 to 25m, generally, appear in the form of grey and brown coherent soils : 85% are sandy-loamy clays; 5% of the soils are sandy-gravelly clays and 10% are loamy-gravelly sands. The natural water content of the sediments decreases with the depth : 1) at depths from 0 to 10m, the water content is high and the consistency index has values between 0.4 to 0.6 (liquid to very soft consistency); 2) at depths from 10 to 20m, the consistency index takes values between 0.6 to 0.75 (very soft to soft consistency); 3) at depths under 20m, values from 0.75 to 1 (soft soils to consistent) are recorded for the consistency index.

**EQUIPEMENT UTILIZED**
Two different techniques of desilting the Fergug reservoir have been proposed at the start : 1) mechanical process using bucket-line dredge after a complete reservoir drawdown and using trucks for transporting the silt; 2) hydraulic process using a suction-cutter dredge working underwater, and the silt transport by a pipe. The hydraulic desilting have been chosen because of economical considerations. It cost 22.28 Algerian Dinars ( ~$1) to desilt 1 cubic meter by the machanical method while it cost only 11.19 Algerian Dinars (AD) using the hydraulic method.
The dredging project realization was assigned to H.A.M. a Dutch Company. The works have started in September, 1989 and ended in March, 1990. After one year stop, the Algerian Company of Maritime Works (SO.NA.TRA.M) has resumed the works which lasted until March, 1992.
The *Rezug Yusef* dredge, a De Groot Nijkerk Dutch brand, is a stationary suction-cutter dredge with a floating crane for moving the anchors, assembling and disassembling of the floating pipeline and the fuel suppling of the dredge. The dredge hull is made of six metallic caissons, and is 52.5m long and 8.9m wide. The pump's motor power is 1250 Horsepowers (Hp). The jet pump's and the cutter head's powers are respectively 195 and 150 Hp. The mud pump utilized for the suction and evacuation of the silt is a centrifugal pump (three blades) with a maximum flow rate of mixture of 1600 l/s and an suction head of 28m using a 700mm diameter pipe. The suction and discharge pipes have respectively a diameter of 650mm and 700mm. The floating pipeline of 700mm diameter constitutes the link between the suction dredge and the exposed pipeline. The steel wall thickness is 16mm. Each pipe's part (two different lengths of 6 and 12m each) is carried by floats made of two coupled bodies, having 8.3m length and 1.05m diameter. The floating pipeline's total length is 530m. The steel made submersible pipeline is put counterpart to the exposed pipeline on the left shore of the lake. It is made of steel duct having 12m length and 700mm diameter, joined by rubber elbows. The wall thickness is 12mm. The total length of the submersible pipeline is 945.6m. The explosed pipeline is made of 700mm diameter and 12m long steel ducts. The wall thickness is 12mm. The elbows are obtained by using plastic joints and steel junctions. The exposed pipeline's length is 731.2m (Benchehaima and Bounouara, 1992).

**DREDGING PROCESS**
The maximum and mean dredging depths are 16m and from 8 to 10m respectively. The sediments are dredged in parallel trenchs, following a central axis of cross dredging. The average cross dredging speed is on the order of 0.5m/s. The dredged layer thickness is about 3m. The cutter head work is made easy by using three water jets : one applied on the left, another on the right and the third in the cutter head itself. When the soils are loose, they are sucked by the suction duct towards the pump. Cavitation phenomena is avoided by using a security valve. The suction duct and the pump are equiped with a device for removing gas contained in the silt. The mud

pump furnishes the suction pressure as well as the discharge pressure to eject the silt towards zone passing through the floating and the exposed pipelines. Ventilation ducts are set at the end of the floating pipeline, providing air intake in case of pressure decrease. But, since the dumping zone is located lower than the suction dredge, we have to throttle the discharge pipe, so it can give enough pressure. The dredged materials are, generally, soft and deep which makes the use of piles as pivots of cross dredging impossible, because these piles will have very large sizes (Stigler and Burges, 1980; Guetarni, 1989). For this reason the solution called *X tree* or *Christmas tree* was used, in which the back of the suction dredge is fixed with three axis. The strain of the anchorage cables is leapt constant with hoists. The two lateral anchors are hung up to a float by a rope in order to not sink in the silt.

**Table 4. Dredging results obtained by H.A.M and SO.NA.TRA.M**

|  | H.A.M | SO.NA.TRA.M |
|---|---|---|
| Data of strated works | September 9, 1989 | March 16, 1991 |
| Data of ended works | March 4, 1990 | March 31, 1992 |
| Number of working days | 135 (2700 hours) | 150 (1650 hours) |
| Number of days of interruption | 41 | 195 |
| Total of silt production ($m^3$) | 4,818,032 | 2,817,670 |
| Daily mean production ($m^3$ per day) | 35,689 | 18,784 |
| Hourly mean production ($m^3$ per hour) | 1,784 | 1,707 |
| Volume of utilized water ($m^3$) | 5,115,551 | 3,061,600 |
| Mean rate /silt (%) | 48.5 | 48 |

## DREDGING EVOLUTION STATE

A daily report is established at the end of each working day, giving the effective number of hours of work and the silt volume extracted. From this data, the daily dredge efficiency and the quantity of water used are determined. Table 4 shown comparatively the differents results obtained by H.A.M and SO.NA.TRA.M. 44% of the work-stops for H.A.M were due to mechanical problems (hoist motors and tugboat self-starter failures), 14% were due to disassembling and assembling the submersible pipeline, because the dredging of Fergug wadi was done following that of El-Hammam wadi and finally 42% because of holidays. 58% of the stops for SO.NA.TRA.M were due to mechanical problems (such as tugboat self-starter and hoist motors failures, shearing of pump's shaft, oil leakage), 27% were caused by water shortage for the dredging, 10% because of a plug in the discharge pipe and finally 5% because of holidays.

From Table 4 we can say that :

- H.A.M has produced in 6 months twice of that produces by SO.NA.TRA.M. The reasons are as follow : 1) H.A.M had 2 teams worked in total 20 hours a day, while SO.NA.TRA.M had only one team that worked 11 hours a day; 2) SO.NA.TRA.M used the equipement that was already been used by H.A.M, which was more sensitive to mechanical failures risks.

- The two companies hourly output is almost the same and largely exceeds the hourly output mentioned in the dredge performance (minimum of 800 cubic meters per hour).

The estimated project period was 36 months. It was reduced to 18 months. The dredging was interrupted for a year because of water shortage due to successive years of drought. The water storage was to be utilized in priority for water supply. The dredging has given the following results :

- Extracted silt volume      7,635,703 cubic meters
- Water quantity used (that is $1.07 m^3$ of water for $1 m^3$ of silt)      8,177,151 cubic meters
- Average monthly fuel quantity used      120,000 litres

- Cost of the desilted cubic meter (equipement not included)      1.55 AD
- Cost of the desilted cubic meter
 (equipement and installation included without amortization)      14.53 AD
- Amelioration of the water quality : reduction of chemicals quantity used in the water treatment.

## CONCLUSION

From the results of the dredging works performed in the Fergug reservoir, we may conclude that, technically, the dredging is an efficient mean to recuperate capacity . But, in spite of this advantage and of the acquired experience and the innovations realized in this area in Algeria, the dredging is often very costly, specially for developing countries. This is more noticeable in drought period. Thus, we must perform the dredging preferably during the wet phase of the hydrological cycle. We estimate that the initial storage capacity of Algerian reservoirs will be reduced by 2/3 at the end of century. It is imperative to undertake appropriate and urgent measures to preserve our hydraulic potential. So, the problem of siltation of reservoirs must be attacked upstream to preserve land against erosion, on one hand, and to reduce the sediments transport, one the other hand, extending then, certainly, the lifetime of the reservoirs.

## REFERENCES

**Belbachir, K.** 1980, 'Desilting of Hamiz dam', *International Seminar of Experts on Reservoir Desiltation*, Tunis, Communication 8, p.10

**Bellouni, M,** 1980, 'I. Main Courses of action undertaken by Algeria for the desilting of dams in operation, II. Study of the Hamiz dam dredging, III.The Lucien Demay dredge and the desilting of the Hamiz dam',*International Seminar of Experts on Reservoir Desiltation*,Tunis,Com.13,p.25

**Benchehaima, A. and Bounouara, A.** 1992, *'Expériences algériennes de dévasement des barrages par dragage (cas du Fergug)'*, Projet de fin d'études ingéniorat, Institut d'Hydraulique de l'USTO, Oran, p.78

**Demmak, A.** 1980, 'The algerian experiences as regards the control of dam siltation', *Internatioanl Seminar of Experts on Reservoirs Desiltation*, Tunis, Com. 10, p.8

**Demmak, A.** 1982, *'Contribution à l'étude de l 'érosion et des transports solides en Algérie septentrionale'*, Doctor Thesis, Université Pierre Marie-Curie, Paris, 5-20

**Duquennois, H.** 1955, *'Lutte contre la sédimentation des barrages réservoirs'*, Electricité et Gaz d'Algérie, Compte rendu N°2, publication interne, Alger, p.28

**Errih, M., Bekhti, B., Benamar, B., Benshila, B., Boudjenane, N., Ladouani, A. and Sad-Chemloul, N.** 1992, 'Problem of siltation of small reservoirs in Algeria', *Proceedings of the 7th international conference in transport and sedimentation of solid particles,*. **224**, 9-11 June 1992,Wroclaw, Poland, 174-180

**Errih, M.** 1993, 'MIYAH : programme de calcul de régularisation de l'écoulement superficiel au moyen de barrages-réservoirs', *Actes des Deuxièmes Journées Tunisiennes de Géologie Appliquée*, 17-19 mai 1993, Sfax, 590-600

**Guetarni, A.** 1989, *'Brochure technique du barrage Fergug'*, Agence nationale des barrages, publication interne, Alger,p.5

**Hannoyer, J.** 1974, 'Nouvelle méthode de dévasement des barrages-réservoirs', *Annales de l'Institut Technique du Batiment et des Travaux Publics*, **314**, 146-153

**Medinger, G.** 1960, 'Transport solide des oueds algériens', *Annuaire Hydrologique de l'Algérie*, année 1958/59, Alger,5-31

**Stigler, C. and Burges, A.** 1980, 'Le dévasement des réservoirs par dragage', Discours au séminaire SO.NA.TRA.M/DRAGAGE, Alger, 3 avril 1980, p.5

**Thévenin, J.** 1960, 'La sédimentation des barrages-réservoirs en Algérie et les moyens mis en oeuvre pour préserver les capacités', *Annales de l'Institut Techniques du Batiment et des Travaux Publics*, **156**, 25-35

# SEDIMENT MANAGEMENT IN BUFFER BASINS AT HYDROPOWER PLANTS

## H. SCHEUERLEIN AND F. EIBL
Univ.-Prof. Dr.Ing., Head
Dipl.-Ing., Research Associate
Water Resources Institute
University of Innsbruck
Austria

**ABSTRACT**

The paper deals with sediment deposition in buffer basins at peak energy producing hydropower plants. A principle superimposing a spiral-shaped forced vortex with the spiral flow in a bend is introduced. The principle is used to stimulate sediment deposition at defined locations in buffer basins. The basic hydraulic laws are analysed and the governing parameters are identified. Experience with model studies for two basins where the principle has been applied succesfully is used as basis for further fundamental investigation.

## INTRODUCTION

Hydropower plants used for peak energy production are operated only several hours a day according to the actual energy demand. In the Austrian Alps numerous relatively small hydropower plants generating peak energy are equipped with daily storage basins to compensate the discrepancy between water supply and energy demand. Usually the water is withdrawn at constant rates from rivers by means of diversion dams. The buffer basins have limited capacity (0,1-0,2 Mio $m^3$, depth 3-6 m ) which makes them rather sensitive to capacity losses due to sediment depositions. The basin inflow diverted from natural rivers carries sediment as bedload as well as in suspension. The hedload sediment is deposited totally in the basin, the suspended material only partly. As the operation of the basin is characterized by continuously changing water depth ($0 < y \leq y_{max}$) the deposits of both bedload and suspended material spread all over the basin. The removal of the rather uniformly distributed sediment causes technical problems and creates undesired costs. On the other hand, sediment control is indispensable as otherwise the sediment might enter the power intake which is particularly dangerous as far as coarse material consisting of quartz particles is concerned as it would ruin the turbines within short time. The finer particles (entering the basin in suspension) might rather be tolerated as they are mainly composed of clay. Favourable sediment management has to consider the following objectives:

- Separation of bedload and suspended load should take place close to the diversion outfall into the basin.
- Provision should be made for the bed material to settle and accumulate at a place which has sufficient distance from the power intake and easy access for removal.
- Suspended material should be kept from settling down before it reaches the power intake where it may leave the basin together with the turbine water.

The ambitious demands expressed above cannot be met easily. By means of an extensive investigation involving also physical modelling, however, a solution equally simple and effective could be found. Although the whole research might deserve more attention, this paper will concentrate only on one detail of the total concept namely the method of separating and accumulating the coarse material by using a principle based on forced vortex and secondary current.

## FUNDAMENTAL HYDRAULIC PRINCIPLES

The hydraulic principle described in this paragraph must be understood as a composition of two well-known features: the forced vortex and the spiral flow induced by secondary current in channel bends.

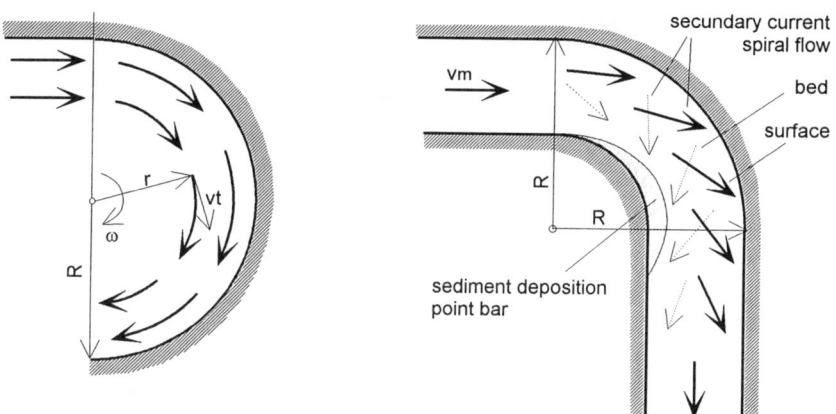

FIG. 1  Forced vortex       FIG. 2  Flow pattern in a bend

The forced vortex (Fig. 1) is characterized by constant angular velocity $\omega$ and the tangential velocity $v_t$ decreasing linearly with decreasing distance r from the center of the vortex.

$$v_t = \omega\, r \tag{1}$$

The typical feature of flow in open channel bends is characterized by the well-known spiral flow with the velocity vector at the water surface directed towards the concave bank and the velocity vector close to the bed directed towards the convex bank (Fig. 2). Spiral flow originates from superposition of the main flow following the longitudinal slope of the river due to gravity, and the circular secondary current perpendicular to the main flow direction caused by centrifugal and inertia forces due to the bend. The main flow component can be described as

$$v_m = \sqrt{\frac{8g}{f}} \sqrt{R_h S_t} \quad (2)$$

with $v_m$ standing for the mean flow velocity in longitudinal direction. $R_h$ is the hydraulic radius, $S_t$ is the energy gradient in longitudinal direction, and f is the Darcy-Weisbach resistance coefficient. The circular secondary current can be described by the balance

$$y\,g\,S_c = \int_{z=0}^{z=y} \frac{v_z^2}{r} \cdot dz \quad (3)$$

where y is the flow depth, g the acceleration due to gravity, $S_c$ is the energy gradient perpendicular to the main flow direction, z is a vertical coordinate starting from the bed, $v_z$ is the flow velocity in longitudinal direction (main flow) at z, and r is the radius of the bend.

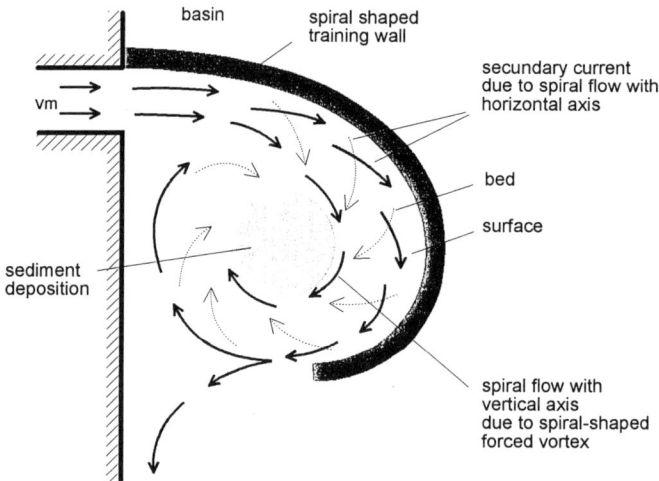

FIG. 3  Double spiral sand trap

Fig. 3 shows the combined application of the two previously described principles (forced vortex and spiral flow) at the entrance of a conduit into a basin with a training wall of steadily increasing curvature as auxilliary device. The intention of the measure is to provide favourable conditions for the sediment entering the basin as bedload to settle in the center region formed by the training wall whereas the sediment in suspension should be carried on. The efficiency of the measure depends significantly on the strength of the secondary current in particular of the flow close to the bed. From equation 3 it can be seen that the secondary current is governed by the energy gradient responsible for the cross flow ($S_c$) which is proportional to the square of the initial flow velocity and indirect proportional to the radius of the curved alignment. Since the initial flow velocity necessarily decreases as soon as it enters the basin due to friction and flow resistance, the only way to avoid fading of the secondary current is to gradually increase the curvature. This can be achieved by shaping the alignment of the training wall as a spiral curve of the type

$$\frac{1}{r} = \frac{s}{a^2} \qquad (4)$$

where s is the length of the arc under consideration, and a is a proportionality factor. The spiral curve described with equation 4 has an asymptotic center point. The flow determined by the shape of the training wall has the characteristics of a forced vortex with the tangential flow velocity being proportional to the radius, i.e. decreasing steadily towards the center of the spiral.

## DOUBLE SPIRAL SAND TRAP

The principle presented above can be used to concentrate sediment deposits at a distinct location. Sediment entering the basin as bedload is caught by the spiral flow with horizontal axis which develops along the curved training wall due to secondary current. As the sediment follows the flow close to the bed it gradually moves towards the center of another spiral flow with vertical axis due to forced vortex where it finally settles and accumulates.

The principle has been applied successfully in two buffer basins which are part of a complex hydropower system in western Austria (Fig. 4 to 7). Shape and dimensions of the training walls and certain auxilliary structures have been designed and optimized by means of physical models. One of the measures has already been realized with rather low budget. The performance of the measure is reported to be excellent and fully to the expectations of the client. The second application will be completed within a couple of months.

## CONCLUSIONS

The technique to combine spiral flow due to secondary current with a spiral-shaped forced vortex in order to stimulate sediment deposition at a clearly defined location has proved to be both feasible and effective. The fundamental hydraulic principles have been analyzed and the governing parameters have already been identified. The principle has been applied successfully at two buffer basins for hydropower plants. The necessary measures and auxilliary devices have been optimized by means of hydraulic model tests. In order to develop generally applicable design criteria and to identify the limits of the principle in practice a fundamental research project is in progress.

## ACKNOWLEDGEMENT

The authors are grateful to the Vorarlberger Kraftwerke AG, Bregenz for providing the possibility to carry out the model studies mentioned in this paper and for the readiness to adopt and to realize a rather unconventional solution.

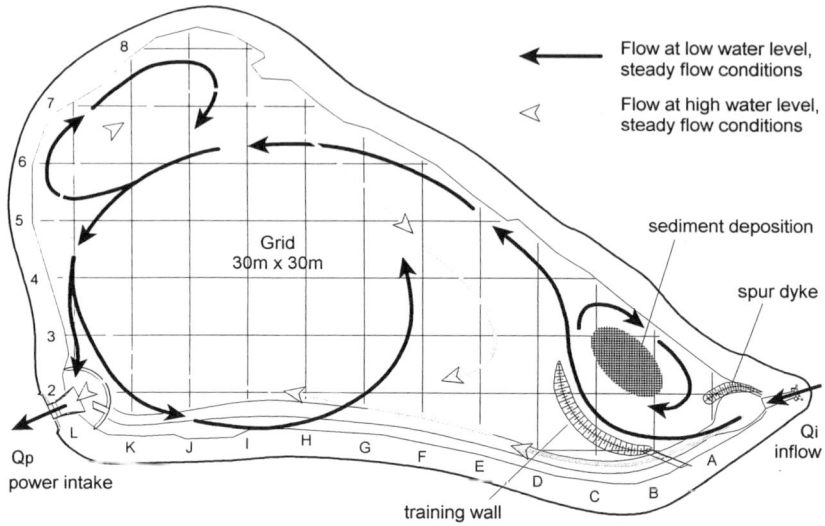

FIG. 4  Buffer basin Ach in West Austria with double spiral sand trap

FIG. 5  Double spiral sand trap, Ach basin model

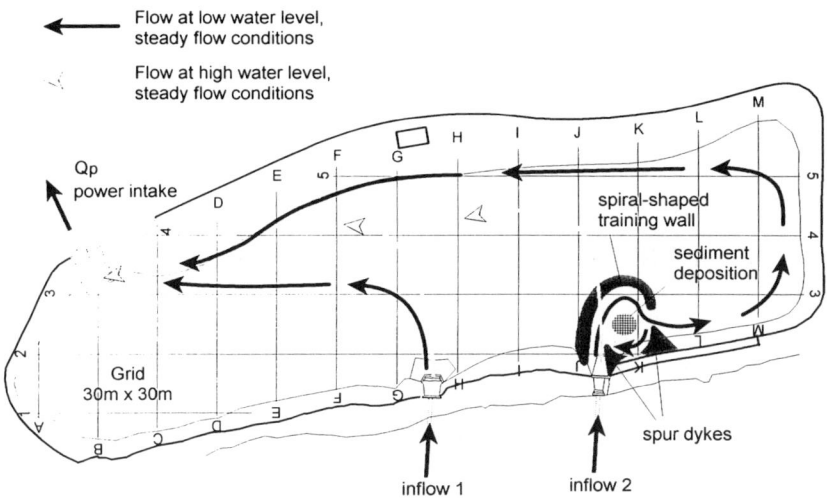

FIG. 6  Buffer basin Andelsbuch in West Austria with double spiral sand trap

FIG. 7  Double spiral sand trap, Andelsbuch basin model

# SEDIMENT MANAGEMENT AT AN ALPINE RESERVOIR

H. KNOBLAUCH, G. HEIGERTH, T. DUM
Institut for Hydraulic Structures and Water Resources Management
Technical University of Graz
Austria

## ABSTRACT

The discharge of the biggest Austrian glacier, called Pasterze, is stored in the Margaritze reservoir. This reservoir and the two reservoirs of the powerstation Glockner-Kaprun are connected with a pressure tunnel. Because of the abrasion of the glacier the sediment rate of the Margaritze-inflow is very high. The reservoir has little sediment storage capacity because of the small gross storage. Flushing by opening the bottom outlet is very problematic, therefore we were looking for the possibilities to transfer sediment to the much bigger reservoirs at Glockner-Kaprun using the pressure tunnel.
The solution presented in this paper is the sediment management of suspended sediment load including the following steps:
A sand trap, that is designed to create the flow processes required to eliminate from the delivery flow all the grain sizes above a certain diameter, and only these.
To transfer the suspended sediment through the 11.6 km long pressure tunnel we had to consider different studies for practicable service instructions. Basic facts of suspended load transport and the investigation of various operating conditions combined with sediment ratio-measurement on both ends of the pressure tunnel made sure, that sediments of different fractions can pass the tunnel without any risk of deposition.

## INTRODUCTION

The discharge of the biggest Austrian glacier, the Pasterze, is stored in the Margaritze and from there transported into the alpine reservoirs Wasserfallboden and Mooserboden of the Glockner-Kaprun power station through the 11.6 km long Möll tunnel. The mean diameter is D=3.05 m with a flow rate between 10 and 20 $m^3/s$. (Fig. 1). Because of the glacial abrasion this water has a very high solid matter content.
Since the content of 3.2 million $m^3$ of the Margaritze reservoir is rather small (comparing to both reservoirs in the Kaprun valley with a content of each 85 million $m^3$) the space for sediments is small, too. The siltation has already led to major problems at the bottom-outlet and has also necessitated the rebuilding. The flushing required for keeping the

bottom-outlet free are complicated for several reasons, so that the management (Tauernkraftwerke AG) has looked for other possibilities to avoid the silting.

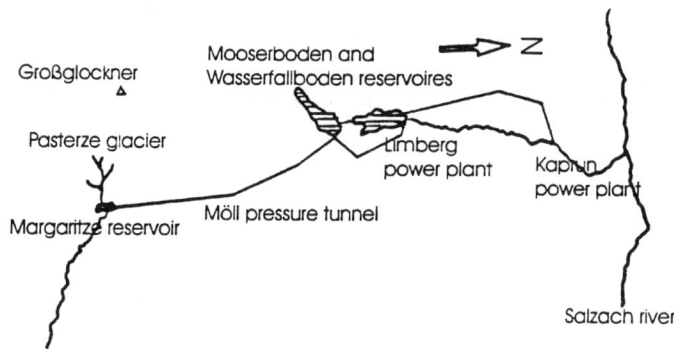

Fig. 1: System of the power plant Glockner-Kaprun

First the mud conveyed by special mud pumps was transported to a special sand-trap to eliminate from the delivery flow all the grain sizes above a certain diameter, and only these. Various considerations led to the study of a longitudinal-type sand trap. Using the results of a laboratory test on a hydraulic model (section model), the fundamental dimensions for the prototype test were determined. The results obtained led to structural measures that ensure a high sand-removal efficiency.

After passing the sand-trap the conveyed mud was pumped to the intake structure of the pressure tunnel and removed into the annual reservoir of the Kaprun valley. The present study aimed at the basic possibility to transport a part of the sediments silted in the reservoir without major sedimentations at the tunnel itself or in the power conduit of the Limberg power plant. In this study no attention was paid to the sediment's effects on the mechanical parts of plants such as turbines, valves or draining pipes.

## DEVELOPMENT OF THE SAND TRAP

Our studies were intended to find a method for making only the particle sizes with a diameter d greater than 1.0 mm settle (this diameter having been found by theoretical studies). Sizes smaller than that had to remain in the delivery flow. A conventional sand trap design was provided with a ground step, where well-controlled sand removal was to start. In order to estimate a first sand-catching region, maximum possible sand-catching lengths were determined for the 1.0-mm size. Using the assumption for the sand-catching length [HUBER/SCHMIDT, 1973], as well as the settling velocity for the individual particle, the following fundamental idea was adopted for the hydraulic design, whereas the following areas can be distinguished over the length development of the sand trap (Fig. 2).

**(a) Inlet area**

**(b) Stilling section:** This about 3-m long section is intended to bring about a certain preliminary sizing in the particle distribution over the cross-sectional area. This section terminates in a bottom step 10 or 20 cm high.

**(c) Sand removal area:** This space is separated from the so-called sand-catching chamber underneath by a grating. Normal bar spacing is 20 mm. The gratings were designed to allow the bars to be arranged either lengthwise or across the flow direction.

**(d) Outlet:** A hand-operated gate was provided for flow control.

## MODEL TESTS IN THE LABORATORY

The sand trap designed on the basis of theoretical studies was simulated as a section model to scale 1:1 and installed in the Department's glass channel. Channel width, b, was 25 cm. The purpose of the model test was the functional check and adjustment of the areas of the facility as described above under a), b) and c). The measuring scheme provided for the variation of the following parameters: Height of ground step (elevation of pull-off edge), shape of step, flow velocity and length of grating. Variation and combination of these parameters led to the development of the optimum sand-trap design for the prototype.

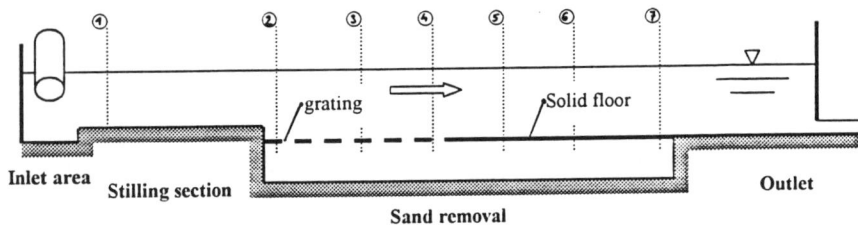

Fig. 2: Section model showing measuring planes and the length development

A measuring scheme was realised for a range of flow rates (20 l/s < Q < 47 l/s) within certain limits as defined by the pumping facility, for both types of sand trap, in order to obtain optimum results by statistical studies.

For the feasible variations, the following measurements were taken: measurement of velocity fields, visual check (eye, photography, video camera) of behaviour for a certain rate of sand transport.

For testing sand trap behaviour when supplied with sand, three different size ranges of model particles were used:
- (1) Quartz sand      $2.0 < d < 4$ mm
- (2) Ceramic sand     $0.3 < d < 0.7$ mm
- (3) Fine silt        $d < 0.1$ mm

## IN SITU PROTOTYPE TESTS

The type of sand trap that resulted from our model studies was thoroughly tested for effectiveness during a six-day test scheme.

Two sand trap widths were also adopted for the prototype (Type 1: b=40cm; Type 2: b=70 cm). For a mean flow of $Q = 83$ l/s, this gave a water depth within the sand-catching area of approx. 50 cm (Type 1) and 30 cm (Type 2), respectively.

Concentrations and grain-size distributions were measured at the following locations: downstream of sand intake in the reservoir, in the sand-catching chamber (upstream end, middle, downstream end), at the outlet of sand trap.

The aim of the prototype tests was to prove the satisfactory performance of the sand trap on the basis of statistical evaluation of measuring records.

Measured data was evaluated mainly for suspended-solids concentration, particle-size distribution, and quantitative balance.

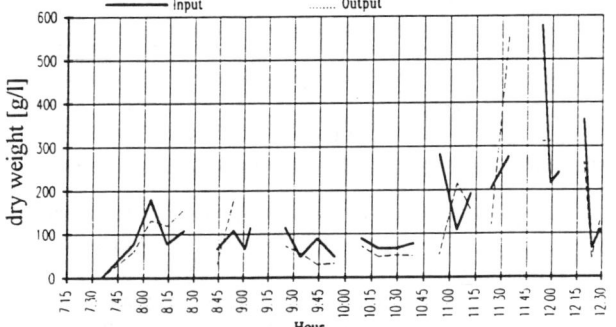

| Input [t] | Output [t] |
|---|---|
| 17.7 | 18.2 |
| 8.4 | 10.4 |
| 10.6 | 7.0 |
| 12.6 | 9.3 |
| 15.1 | 13.6 |
| 11.5 | 16.3 |
| 15.7 | 15.7 |
| 7.6 | 6.1 |
| 99.2 | 96.6 |

Fig. 3: Characteristic of sand trap performance

Evaluation of suspended-sediment concentration gave values of between $c_T = 6$ % and $c_T = 20$ % of solids. We noted that these values showed a typical pattern dependent on pump operation (The suspended-sediment concentration depends on the depth of the pump in the deposited sediment). The main criterion for demonstrating the good performance of the sand trap was the particle-size distribution curve. This showed that the adopted design was excellently suited to accomplish the intended purpose of the sand trap (removal from the delivery flow of all particles with a diameter greater than 1.0 mm). The third fundamental quantity was obtained by evaluation of the quantitative balances: Out of the total suspended load, a percentage of 1.3% (by volume, wet) was retained in the sand chamber. The sand trap was designed for a "safe" removal of particles with a diameter equal to and larger than 1.0 mm. This requirement implied that some particles smaller than 1.0 mm were also allowed to settle. The amount of silt unintentionally removed is of a magnitude of about 1% of the sediment load (wet) carried along.

Evaluation of measured data led to the following design elements to be included in the final proposal (for a flow, Q, of approximately 83 l/s):

- stone trap at the entrance to intercept major particles,
- stilling area with a length, L, of approx. 2.5 m,
- step with pull-off edge with a height, h, of 10 cm,
- sand-catching chamber with a channel width, b, of 40 cm,
- length of bottom grating for the sand chamber should be approx. 4.0 m,

## SEDIMENT TRANSPORT THROUGH THE MÖLL TUNNEL

A special sludge pump stationed on a raft delivered the mud-water-mixture (55 to 75 l/sec) through a approximately 150 m long pipe into the coarse sand trap which was stationed above the inflow of the transfer tunnel. The desilted mud-water-mixture was delivered through another pipe to the inflow area of the tunnel. There it is mixed with water of the Margaritze reservoir and is transported through the Möll tunnel to the Möll pumping station, from where it is pumped to the Mooserboden reservoir or is flowing to the Wasserfallboden reservoir in case that the Limberg power station is working (Fig. 1)

The sampling points were at the raft, before, in and after the desilting, in the Möll pumping station and in the Limberg power station. At all points both concentration and grain-size distribution were punctually determined, a recorder had been installed at one measuring point only in the Möll pumping station, which continually recorded the concentration of the solids.

The sediment samples of the Margaritze reservoir proved that the major part is in the silt grain range. But the tests showed that also coarser grains are sometimes transported, so that e.g. the samples of the sand trap contained maximum grains of up to 50 mm.

A quantity balance was prepared by comparing the concentration measurements at the desilting outlet to the Möll pumping station and to the particular flowing through.

On comparing the grain-size distribution between the desilting outflow and Möll pumping gear it was proved that this showed only smaller deviations. In the case of higher concentrated sediment-water-mixtures of the desilting outflow samples of 1 l were taken, which were filtered and analyzed. In the case of the highly diluted mixtures in the Möll pumping station samples of 100 l were taken.

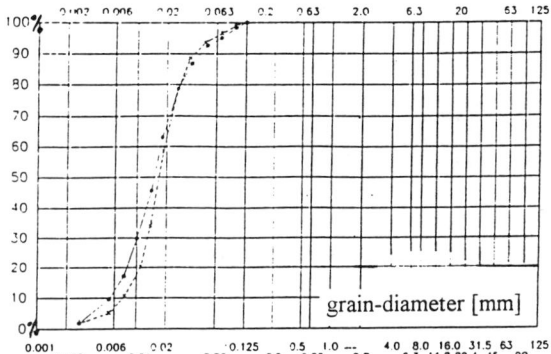

Fig. 4: Grain-size distribution between the desilting outlet and the Möll pumping station

It was proved that all sediment load of the Möll tunnel (inlet) have almost the same form and that they arrive with an approximately same grading curve at the sampling points at the Möll pumping station (outlet). Problems can be caused if a sediment cloud arrived in the Mooserboden reservoir is again retracted by starting the turbines of the Limburg power plant and if it is additionally loaded by the Möll tunnel.

## SUMMARY

The above article deals with the management of sediment at an alpine reservoir. With the use of a sand trap all particles exceeding a certain diameter, and only these were removed from the delivery flow. For this purpose measurements were taken both on a section model in the laboratory and on a prototype.

The desilted mud-water-mixture was delivered through another pipe to the inflow area of the tunnel and was transported through the Möll tunnel to the Möll pumping station, from where it was pumped to two bigger reservoires.

The results of the tests have demonstrated the effectiveness of the algorithm adopted and the excellent performance of the sand trap and the sediment transport through the Möll pressure tunnel without sedimentation.

## REFERENCES

Bollrich, et al (1989). "Technische Hydromechanik" Bd.2, VEB Verlag f. Bauwesen Berlin

Buhrke, Kecke, Richter (1988). "Strömungsförderer", VEB Verlag Technik Berlin

Graf, W.H. (1971). "Hydraulics of Sediment Transport", McGraw-Hill

Huber, Schmidt: „Bemessung von Entsanderanlagen", Schweizer Bauzeitung 40/1973

Kasanski, M. (1981). "Feststofftransport in Rohrleitungen - Gegenüberstellungen von Untersuchungen im Modell und in der Natur", Franziusinstitut TU-Hannover, Heft 52

Raudkivi, A.J. (1982). "Grundlagen des Sedimenttransportes", Springer Verlag

Yalin, M.S.: „Mechanics of Sediment Transport", Pergamon, 1977

# SOLUTION OF SEDIMENT PROBLEMS FOR HYDRO CASCADE ON MOUNTAIN RIVER

## ALEXANDER M. PROUDOVSKY, VICTOR B. RODIONOV, VLADIMIR O. SARANCHEV, MIKHAIL JA. GILDENBLAT,
Center for Hydraulic Reserches, NIIES, Moscow, Russia

### ABSTRACT

The solutions of problems related to sedimentation in the Nizhne-Cherek hydro cascade reservoirs and to protection of the cascade structures and equipment against sediment are presented. The prognosis of changes in river processes connected to cascade construction were made using numerical modelling. Physical modelling was used to investigate anti-sediment structures. As a result of the study some new solutions are offered.

The paper presented for this Congress (topic Da) "Hydraulic Aspects of Designing Nizhne-Cherek Hydro Cascade" gives characteristics of the Cherek River and describes the principal traits the hydro cascade on this river. The problems connected with changes in the sediment transport caused by construction of the cascade and turbine equipment protection are of importance in designing the cascade. The Cherek River is a typical mountain river with high gradient. The river bed ranges from pebbles to boulders. The sediment flow is distributed very non-uniformly within the year with 94% to 97% of annual sediment transport occurring during the two summer flood months. The mean annual suspended sediment load is about one million tons, the maximum suspended sediments concentration recorded in field investigations is about 20 kg/m$^3$. The mean monthly sediment concentration as a function of mean monthly water discharge for various sediment particle sizes are presented in Fig 1. It is clear from this figure, that in low water periods (when water discharge does not exceed 30-40 m$^3$/s) the river carries relatively fine sediments.

For several reasons, the downstream Aushiger hydro is the first-priority project. The reservoir of $1 \cdot 10^6$ m$^3$ total capacity is proposed 6 km downstream of the head cascade reservoir with total storage of $4.5 \cdot 10^6$ m$^3$ available to feed this hydro. The Aushiger hydro will be supplied with water from the downstream reservoir till the Soviet hydro is commissioned. After that the downstream reservoir will function as a spare one in case of stoppage of the Soviet hydro. The above data are indicative of the fact that the annual sediment inflow is comparable with the storage capacity of both resevoirs and the siltation time of reservoirs is expected to be relatively short.

Because the downstream reservoir will be used as the main source of the Aushiger hydro water supply only for a short time period, problems of taking necessary

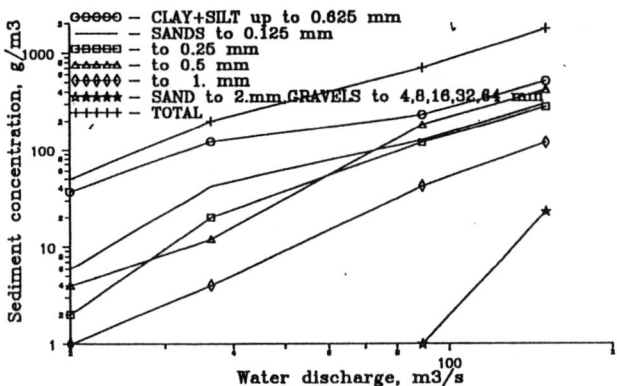

Fig.1 Water–Sediment inflow relationship.

measures to prevent entrainment into the headrace of coarse sediments, which can adversely effect the turbines, and of securing this reservoir storage capacity for daily flow control, are not so keen. The chronology of downstream reservoir sedimentation and the arrival of different size sediment to the intake of the Aushiger hydro at the first stage of cascade construction was studied from this point of view. The results of calculations using program HEC-6 [1] are presented in Figs 2 and 3.

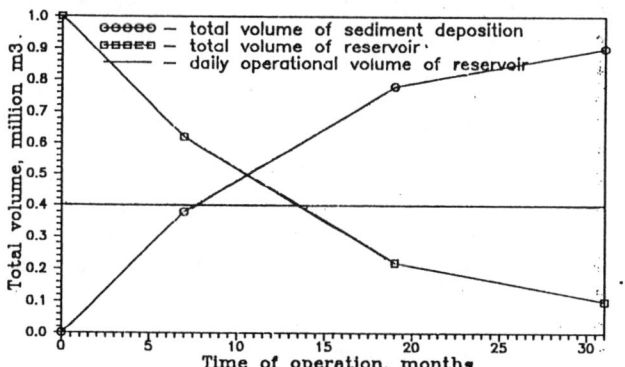

Fig.2 Chronology of spare reservoir total volume lowering as a result of sediment deposition before Soviet hydrostation operation beginning.

As seen from Fig 2, the downstream reservoir's ability to provide capacity sufficient for daily control of the Aushiger hydro discharges in low water seasons (required capacity is about $0.4 \cdot 10^6$ m$^3$) may be maintained only for 1...2 years. of natural water and sediment inflow.

Assuming that only sediments of larger that 0.25 mm size incoming the Aushiger hydro headrace system present abrasion-wear problem for the turbines, it may be expected that this problem will not arise if the time interval between start-up of the Aushiger and Soviet hydros does not exceed 2-3 years. It is expected that the Soviet

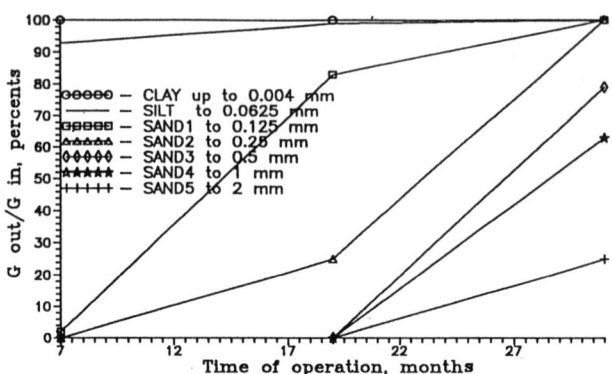

Fig.3 Chronology of relative sediment discharge $G_{out}/G_{in}$ for spare reservoir with water discharge 100–150 m$^3$/s.

hydro will be commissioned after 2 years of Aushiger operation. If this period is attained, the sediments will cause no difficulties for the Aushiger hydro operation. But in any case, to avoid this problem it is necessary to create the head reservoir as soon as possible after the Aushiger hydro commissioning. The head dam is the first-priority object.

After creation of the upstream (head) reservoir the water supply of the cascade will be from this reservoir. Until the head reservoir is fully silted up, only fine sediments will pass downstream. In low water periods the total river flow net obligatory releases (about 2 m$^3$/s) will be passed through the head dam. Only in flood period with flows exceeding 75 m$^3$/s considerable amounts of water will be spilled through the dam of the head reservoir. At the same time to eliminate entrainment of coarse sediments to turbines and silting-up of the headrace, the major part of relatively coarse sediments flowing into the reservoir have to be discharged downstream of the head dam. The main part of the storage necessary for daily flow control is placed in the Soviet hydro headrace. For that reason the loss of the head reservoir working storage does not present any risk and there is no need to pass flood discharges with low water levels in the reservoir. The calculation of head reservoir sedimentation was fulfilled when designing the Nizhne-Cherek hydro cascade. The results of this calculation presented in Figs 4 and 5 show the following: when passing the flood discharges through the reservoir with normal water level, no settling of clay particles within the reservoir takes place. Reservoir is sedimented-up with silts and sands. More coarse particles are moving above the delta.

Life-span of the reservoir is about 13 years. After that sands problematic for turbines begin to approach the headrace intake. Thus, there is no doubt about the necessity to undertake certain measures preventing the entrainment of relatively coarse sediments into the headrace about 10 years after the head reservoir operation has begun. However these measures may be taken after several years of reservoir operation.

In the couse of designing of the hydro cascade, settling basins of different types at the entry to the headrace have been considered. Among conventional settling basins preference was given to a three-chamber basin with regular flushing ensuring settling

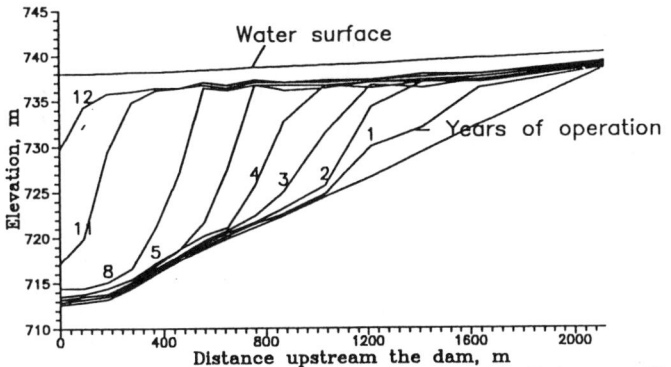

Fig.4 Change by time of upstream reservoir thalweg profile.

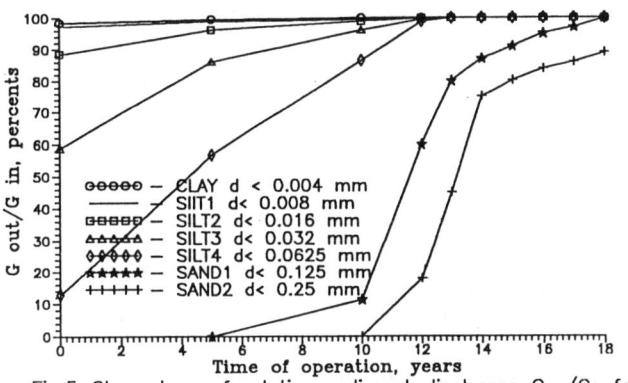

Fig.5 Chronology of relative sediment discharge $G_{out}/G_{in}$ for head reservoir with water discharges 100–150 m$^3$/s.

of particles coarser than 0.25 mm. But the cost of such a settling basin appeared to be very high. Moreover there are certain difficulties caused mainly by hydroabrasion of flushing galleries and equipment. For this reason it was considered advisable to apply another solution based on a suction system [2,3]. According to this solution the sediments are intercepted and diverted by steel tubes due to the head difference created by the dam. In the case under consideration a trench will be provided on the bank separated from the reservoir by a protection embankment (Fig 6).

At the entry to the trench the first row of tubular suction devices is located. To intercept bed sediments across whole trench the tubes must be installed to ensure the merging of cones, forming in deposits. It is necessary in this case to install two tubes in each row. The openings of suction devices are placed much below the level of sediment deposition. The horizontal tube sections are located on the berm along the trench. This berm is above the dead storage level, but below the normal water level. On the upstream face of main dam the suction tubes pass trough diaphragms descending to the tailrace channel of the main spillway. Suction devices are fixed on

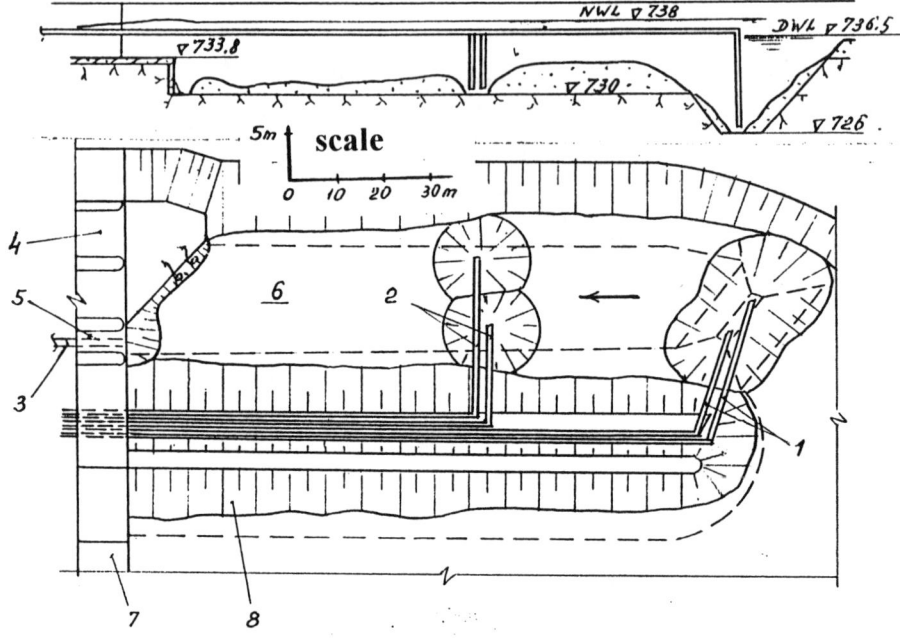

Fig 6. Anti-sediment structure based on suction system 1 - 1st row of suction devices, 2 - 2nd row of suction devices, 3 - low-level sluice, 4 - headrace intake, 5 - icepass, 6 - settling canal, 7 - main spillway, 8 - protection embankment.

pontoons with lifting gears facilitating maintenance and replacement of the tubes. The flow through the tube begins when the suction device is submerged. The discharge in the
tube may be regulated by lifting the downstream end of the tube or by air admission into the vacuum zone. When tubes start to function, the cones with nearly natural slopes are formed in the bed deposits in front of each suction device. Along the upstream slopes of the cones the bulk of incoming bed load approaches the suction devices and is transported by water flow through the tubes to the downstream pool. Suspended sediments may miss the first row of suction devices. But due to the fact that the discharge and particle sizes of sediments downstream of the first suction devices row are much lower than at the entrance to the trench, the water depths here are larger and flow velocities are respectively lower. Relatively large sediment particles, which have passed of first row of suction devices, settle down to the bed. To divert these sediments from the headrace intake, the second row of suction devices is installed. The functioning of these devices is the same as those in the first row.
The headrace intake is located at the end of the trench. Upstream of this intake there is a slanting threshold diverting the rest of abrasion-prone sediments to low-level sluices. Above the sluice an icepass is provided. The proposed structure was studied on physical model. Geometrical scale of the model was 1:35.

As mentioned above, after commissioning the Soviet hydro water to the Nizhne-Cherek cascade will be supplied from the head reservoir. Thus, the water flow at river stretch between the upstream dam and Aushiger hydro tailrace exit to the river will be much less that in natural conditions even during floods. At the same time practically all natural inflow of coarse sediments will be discharged through the head dam. It will cause settling of sediments in the river channel downstream of the head dam and within the spare reservoir. For this reason the regulating storage of the spare reservoir may be lost. To secure this storage it is proposed to pass flood waters through the downstream dam spillway with the gates fully open. When passing the flood the water levels upstream of the dam of the spare reservoir will change according to water discharges. Filling of the spare reservoir will take place after the flood peak. To estimate the minimum capacity of the regulating storage of the spare reservoir, assessment of the deposition process between two dams was made using the recorded hydrograph for several years of high water period.

According to results of this assessment, inspite of a very high sediment concentration, live storage of the spare reservoir is not less that $0.44 \cdot 10^6$ m$^3$ during the whole period under consideration (Fig. 7). It is enough for daily storage at the Aushiger hydro when the Soviet hydro is stopped.

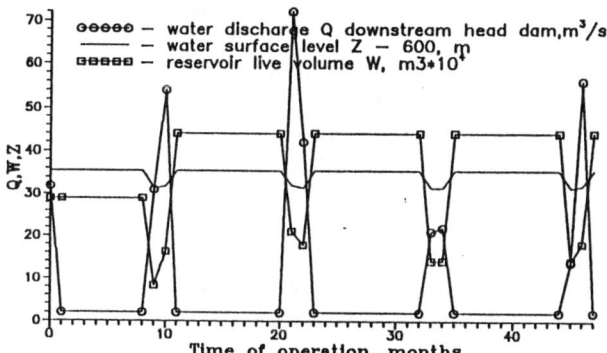

Fig.7 Change by time of spare reservoir live volume and water surface level after Soviet hydro operation beginning.

REFERENCES

1. U.S.Army Corps of Engineers, Hydrologic Engineering Center, "HEC-6 Scour and Deposition in Rivers Reservoirs", User's Manual, 1993, Davis, USA.
2. Hotchkiss R.H. and Xi Huang, "Hydrosuction Sediment Removal from Reservoirs", International Coordinating Committee on Reservoirs Sedimentation, St-Petersburg, 1994.
3. Proudovsky A.M. and Rodionov V.B. "The Problems Concerning Reservoir Sedimentation and Sediment Passage through New Hydroprojects on Nile River", Proceedings of the International Conference on Reservoir Sedimentation, Fort Collins, USA, 1996, V.3, pp. 1777-1796.

# Effects of Reservoir Regulation on Ice Jam Thickness

## JON E. ZUFELT
Research Hydraulic Engineer, USACRREL, Hanover, NH   USA

### ABSTRACT
Hydropower operations alter the natural levels of discharge in a river. In a seasonal sense, the effect of hydropower regulation is to average the flow, cutting off the very high and very low periods of discharge that may result in flooding or drought conditions. Peaking operations, however, may reverse this trend, resulting in flows that are much higher or lower than the natural flow levels for that time of the year. During winter, natural discharge levels are typically low and regulation for hydroelectric generation may result in brief periods of abnormally high and possibly low discharge under ice-covered conditions. Large variations in discharge over the hydropower cycling period may result in ice movement or grounding. Therefore, the range of discharge fluctuation is often limited during ice formation and breakup periods when the ice cover is most likely to move. This paper looks at the effects of these unsteady discharge fluctuations on the resulting ice cover thickness through the use of a numerical model. Two reservoir configurations are presented, which help examine the effects of hydropower regulation on the ice cover thickness in the reaches upstream and downstream from a hydropower facility.

### BACKGROUND
Hydropower peaking refers to the scheduling of discharge so as to provide energy when needed most, typically on a diurnal cycle of morning and evening when electrical demands are highest. At facilities with small storage reservoirs, peaking allows the most efficient use of the often limited inflow volumes available during the winter months. Peaking is also used at facilities with very large reservoirs where inflow volume is less of a constraint. The peak discharge is sometimes given as a percent or multiplier of the base or average daily inflow, the flow that would exist under run-of-river discharge conditions. The off-peak discharge must compensate for the above-average flows during peaking and is typically set at some minimum, environmentally acceptable level. The overall discharge range, therefore, can be rather large, with the ratio of peak to off-peak discharge as high as 6 to 1.

Ice covers may present problems for peaking operations. Stable ice cover formation is best accomplished at low-velocity, steady flows. Once stabilized, i.e., frozen in place, the cover can withstand higher flows without movement or failure. The extreme ranges of discharge resulting from peaking operations can either inhibit the formation of an ice cover or cause an existing cover to fail and move downstream. Continued generation and failure of the ice cover downstream of a hydropower facility results in greater overall ice volume contributing to the accumulation. Larger ice volumes generally translate into higher water levels and increased risk of flooding during the spring breakup period. Higher water levels may reach far enough upstream to increase the tailwater level of the hydropower facility and thus reduce generating capacity and efficiency. Smaller volume reservoirs often experience a stage variation due to the draining and filling of the reservoir over the peaking cycle. In extreme cases, continued downstream movement of the ice accumulation into the reservoir results in a loss of storage volume and reduced peaking capacity.

As a result, limits on the peak discharge, or range of discharge fluctuation, are often instituted when an ice cover is present or at least during the likely formation and breakup periods. Limits are developed through experience or by the use of a numerical model to estimate the ice thickness and the resulting water levels expected (or allowable) for different levels of peak discharge. The equilibrium ice thickness represents the minimum thickness required to achieve a stress balance within a static ice accumulation under steady, uniform flow conditions. The downstream-acting forces of accumulation weight and water shear stress are balanced by the resisting shear stress at the banks and the internal accumulation strength. Equilibrium thickness can be directly obtained by:

$$\eta_{eq} = \frac{BS}{2\mu(1-s_i)}\left(1+\left[1+\frac{f_i u^2 \mu(1-s_i)}{2BS^2 s_i g}\right]^{1/2}\right) \qquad (1)$$

where B is the river width, S is the slope of the water surface (which is equal to the bed slope for a uniform flow condition), $f_i$ is the Darcy-Weisbach friction factor for flow along the underside of an ice accumulation, u is the average water velocity, $s_i$ is the specific gravity of ice, g is the acceleration due to gravity, and $\mu$ is a coefficient describing the strength properties of the ice. Equation 1 is typically utilized with a steady discharge equal to the peak value of the cycling period, which would represent the worst case water levels. The peaking discharge hydrograph, however, results in unsteady, non-uniform levels of water velocity, depth, ice thickness, and ice velocity. When an ice accumulation fails, it moves downstream and may reform at a thickness quite different from that calculated by equation 1 with the resulting water levels being either over- or under-estimated. In many cases, a more detailed numerical model is necessary to provide accurate water level determinations.

## NUMERICAL MODEL DESCRIPTION

Zufelt and Ettema (1996) describe a fully coupled, one-dimensional finite difference model for simulating water and ice motion. The model solves for the water depth beneath the ice accumulation, d; water velocity, u; ice thickness, η; and ice velocity, υ. The model is based on the conservation of mass and momentum equations for water flow:

$$\frac{\partial d}{\partial t} + u\frac{\partial d}{\partial x} + d\frac{\partial u}{\partial x} = 0 \tag{2}$$

$$\frac{\partial u}{\partial t} + u\frac{\partial u}{\partial x} + g\frac{\partial d}{\partial x} + gs_i\frac{\partial \eta}{\partial x} - gS_o + \frac{f_b u^2 (B+2d)}{8Bd}\left[1 + \frac{f_i}{f_b}\frac{B}{(B+2d)}\left(\frac{u-\upsilon}{u}\right)^2\right] = 0 \tag{3}$$

and ice movement:

$$\frac{\partial \eta}{\partial t} + \upsilon\frac{\partial \eta}{\partial x} + \eta\frac{\partial \upsilon}{\partial x} = 0 \tag{4}$$

$$\frac{\partial \upsilon}{\partial t} + \upsilon\frac{\partial \upsilon}{\partial x} + gs_i\frac{\partial \eta}{\partial x} + g\frac{\partial d}{\partial x} + g(1-s_i)K_p(1-p)\frac{\partial \eta}{\partial x} + \frac{g(1-s_i)}{B}k_0\lambda K_p(1-p)\eta - gS_o - \frac{f_i}{8s_i\eta}(u-\upsilon)^2 = 0 \tag{5}$$

where x and t are the space and time coordinates, $S_o$ is the bed slope, $f_b$ is the Darcy-Weisbach resistance coefficient for flow along the bed and banks, $K_p$ is the passive pressure coefficient, p is the accumulation porosity, $k_0$ is the coefficient of lateral pressure, and λ is the friction coefficient of ice on ice. Equations 2–5 as presented above assume a uniform rectangular channel with a constant bed slope. Upstream and downstream boundary conditions are required for both the ice and water variables. The system of equations is solved using a Preissmann four-point implicit finite difference scheme.

## EXAMPLE APPLICATIONS

Two examples are presented below to demonstrate the effects of unsteady peaking discharge rates on the thickness of ice accumulations and resulting water levels. The first case examines the thickness downstream of a reservoir, while the second looks at the thickness upstream of a small reservoir where stage fluctuates over the cycling period.

## CASE 1—DOWNSTREAM EFFECTS

For this first example, a uniform rectangular channel with a width of 150 m and bed slope of 0.0005 extends for a distance of 30 km downstream from a peaking facility into the backwater of another dam. The initial conditions of depth, water velocity, and ice thickness were determined by running the steady off-peak discharge through the model until conditions stabilized. The model was then run using upstream boundary conditions of the peaking discharge hydrograph and an equilibrium thickness ice cover. The downstream boundary conditions were specified as constant water depth and zero ice velocity. The equivalent ice strength coefficient, $\mu$, is 1.27, the Darcy-Weisbach friction coefficients for the bed and ice are 0.10 and 0.14, respectively, and the accumulation porosity is 0.4. The computational length and time steps are 250 m and 1 minute, respectively.

The peaking discharge hydrograph rises from an off-peak level of 150 $m^3/s$ to 450 $m^3/s$ over 30 minutes, holds at 450 $m^3/s$ for 3 hours, and then returns to the off-peak level. Two evenly spaced peaks occur in each 24-hour period. The equivalent average discharge (based on volume) for this peaking hydrograph is 237.5 $m^3/s$. Maximum ice thickness and water elevations are attained following two 24-hour cycling periods. Figure 1 presents the water discharge with time at the upstream end, mid-reach, and downstream end. Also plotted is the equivalent average discharge. The high-frequency fluctuations on the rising limbs of the mid-reach and downstream hydrographs are attributable to the destabilization and movement of the ice cover. As the ice cover comes into motion, the resistance to flow decreases and the local water velocity, and, hence, discharge, increases. The attenuation of the discharge peak with distance downstream is clearly evident from the figure. The time of peak discharge and time between peaking cycles are brief enough so that the discharge at downstream locations does not reach the peak value or fully recede to the off-peak value.

The effects of the variable discharge and the attenuation of the discharge peak on ice thicknesses downstream from a dam are demonstrated in Figure 2. The thickness profiles attained with steady discharges equal to the peak value of 450 $m^3/s$ and the mean value of 237.5 $m^3/s$ show a constant thickness until reaching the reservoir backwater, where the thickness then decreases with increasing water depth. The fully coupled model utilizes the unsteady peaking flow hydrograph and shows the effects of the partial terms in equations 2–5. The unsteady discharge and ice motion in the upstream reaches result in thicknesses greater than those expected with a steady discharge of 450 $m^3/s$. Conversely, the attenuation of the discharge wave as it enters the reservoir backwater results in thicknesses that are less than those expected with a steady discharge of 450 $m^3/s$. The resulting water levels calculated by using a steady peak discharge of 450 $m^3/s$ are lower in the upstream and higher in the downstream reaches than those calculated by the fully coupled model.

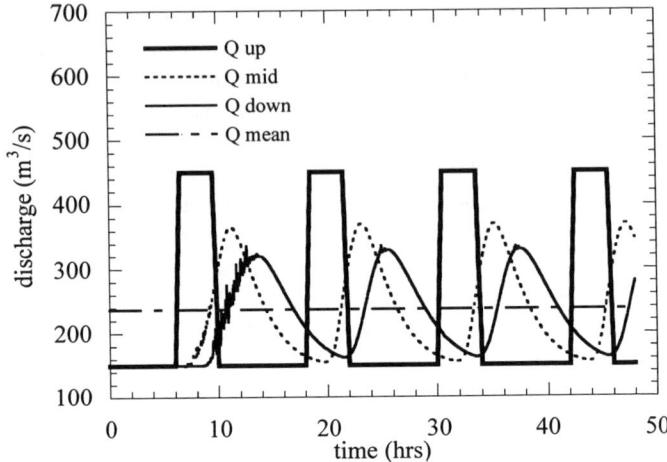

Figure 1. Discharge hydrographs for case 1.

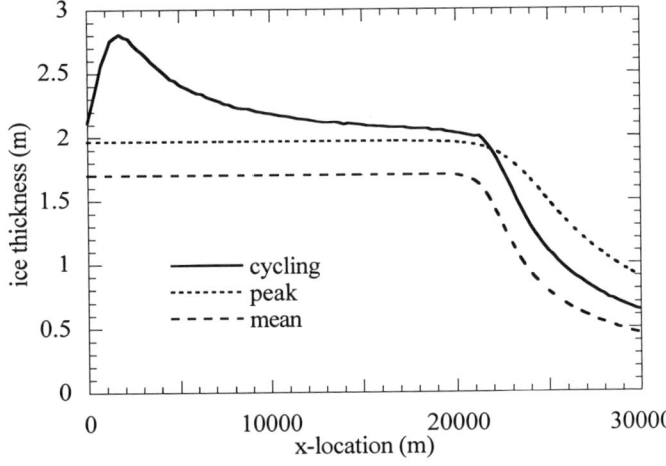

Figure 2. Ice thickness profiles for case 1.

## CASE 2—UPSTREAM EFFECTS

The second example examines a smaller reservoir where a constant inflow is just sufficient to sustain the peaking discharge at a hydropower facility. As a result, the stage within the reservoir fluctuates over the cycling period. The modeled reach extends 10 km upstream from a peaking hydropower dam, with all other parameters equivalent to those specified in case 1. In this case, however, the upstream discharge is held constant at 237.5 m$^3$/s and the peaking discharge hydrograph of case 1 is used as the downstream discharge. The initial conditions are for reservoir stages at their minimum levels (at the start of the filling portion of the cycle), with equilibrium ice thicknesses corresponding to a flow of 237.5 m$^3$/s. Figure 3 presents discharge and stage hydrographs for upstream, mid-reach, and downstream (dam) locations. The effects of peaking on reservoir stages is clearly evident from the figure. The reduced depths at the dam during the peak flow portion of the discharge hydrograph result in significant increases in water velocity and ice thickness. The thickness at the dam increases from an initial value of 0.8 m to 2.15 m after two cycling periods.

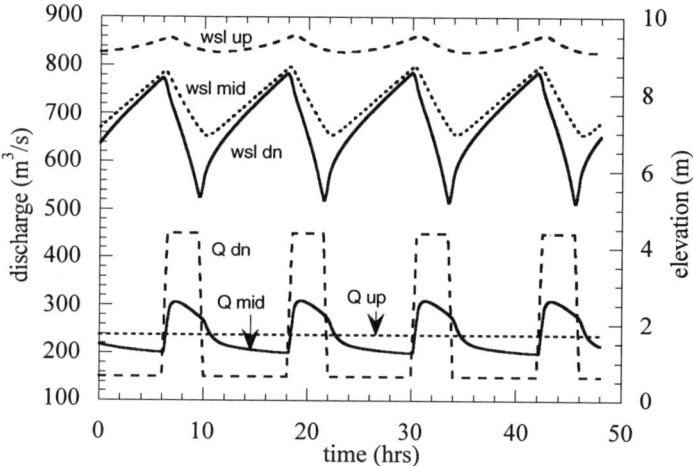

Figure 3. Water surface level and discharge hydrographs for case 2.

## CONCLUSIONS

The discharge immediately upstream or downstream of a peaking hydropower facility is highly unsteady. As the change in discharge (from off-peak to peak flow) increases or the time to achieve this change decreases, the partial terms with respect to time and space in equations 2–5 increase, thereby increasing ice thickness. Equilibrium formulations, such as equation 1, do not include these partial terms, and will thus underpredict the ice thickness. The peaking flow hydrograph, however, will be attenuated as it travels downstream, potentially reducing the peak and the steepness of the discharge wave at locations a considerable distance from the dam. In many cases, a fully coupled model that includes ice motion, such as the one presented herein, is necessary to accurately determine ice thickness and water level expected upstream or downstream of a peaking hydropower facility.

## REFERENCES

Zufelt, J.E. and Ettema, R., 1996. "Ice Jam Dynamics." IIHR Technical Report No. 380, Iowa Institute of Hydraulic Research, The University of Iowa, Iowa City, IA.

# ICE EFFECTS ON RIPRAP: SMALL-SCALE TESTS

D.S. SODHI, S. BORLAND, J.M. STANLEY, and C.J. DONNELLY
U.S. Army Cold Regions Research and Engineering Laboratory
72 Lyme Road, Hanover, NH, 03755-1290 USA

ABSTRACT
We conducted model tests to simulate interaction between floating ice sheets and sloping banks protected with riprap stones. Two series of tests were conducted, representing ice action against model riprap bank protection when the ice sheet moves perpendicular and at an angle of 45° to the shoreline. The first series of tests simulates ice shoving action, while the second series of tests incorporated both shoving and shearing actions of ice in equal proportion. We conducted 35 tests during the first series and 53 tests during the second series. The results indicate that the size of maximum stone ($D_{100}$) should be about 2.5 times the ice thickness to avoid damage from ice action either perpendicular or at an angle of 45° to the shoreline. The data on the probability of riprap failure indicate that the likelihood of riprap damage increases with the slope of a riprap protected bank.

INTRODUCTION
The ice-related mechanisms that can damage riprap shore protection can be categorized as: (a) the "plucking" action by rising and falling water levels, (b) the shoving and (c) shearing action of moving ice sheets, and (d) the ice-affected hydraulics of a river. This study deals with the effects of typical direct shoving and shearing action of moving ice sheets on the riprap protection of a bank. Sodhi et al. (1996) review the literature on this subject.

Because the dominant forces involved during ice action on riprap are from gravity, friction, and ice flexural failure, we followed Froude similarity principles (Ashton 1986) to scale various parameters and variables. Froude scaling calls for reducing all the lengths by a geometric scaling factor. It also calls for reduction of flexural strength by the same factor. Taking various constraints of the testing facility into account, we established the geometric scaling factor to be 10. Assuming the flexural strength of full-scale ice to be 700 kPa (100 psi), we needed the strength of the model ice to be 70 ± 20 kPa (10 ± 3 psi).

EXPERIMENTAL SETUP AND PROCEDURE
Figure 1 shows a schematic drawing of a two-tiered frame, which supported the model riprap bank for tests with direct ice shoving action. We could set the slope of the model riprap bank to 2V:3H, 1V:2H, or 1V:3H. Figure 2 is a sketch of the model

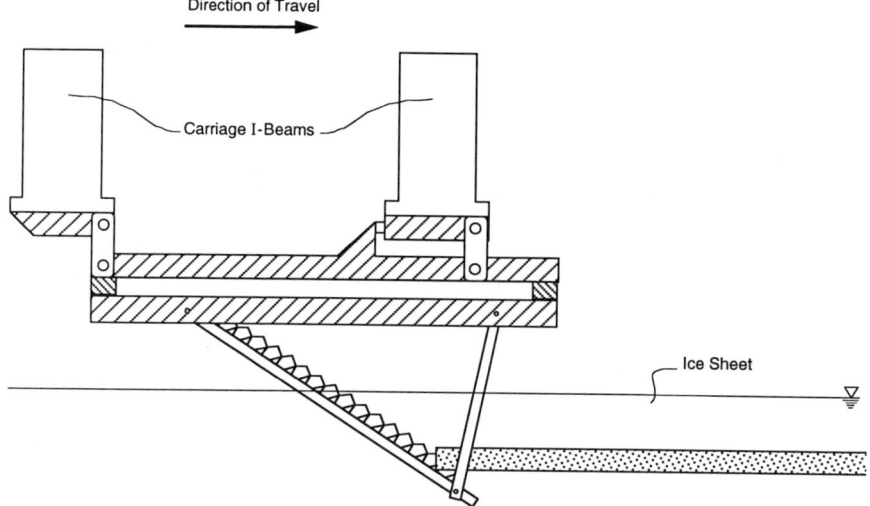

Figure 1. Setup for tests for direct ice action on the model riprap bank.

riprap bank used for the 45° ice action, which consisted of two sloping wooden surfaces. Each surface was arranged symmetrically on both sides of the center line at an angle of 45° in the plan view, and had a slope of 2V:3H, 1V:2H or 1V:3H in the vertical planes perpendicular to the 45° lines. During both series of tests, we placed wooden boards around the sloping surfaces to contain the sand, filter fabric, and riprap stones placed on them. We first placed sand to a depth of 38 mm (1.5 in.) on the sloping wooden surfaces, and the aluminum angles attached to the surfaces provided stability to the sand. To simulate current practice for shore protection, we covered the sand with a filter fabric, and then placed a mix of rocks on top of the filter fabric. In the second series of tests, we performed tests with two different rock mixes on the two sloping surfaces with a given ice sheet.

Sodhi et al. (1996) describe the gradation and sizes of stones used in these series of tests. We used a metamorphic rock, which has a density of 2600 kg/m³ (160 lb/ft³), as model riprap. The shape of an individual rock was typically blocky to slightly elongated with angular edges. We manually separated the stones into bins of different sizes because sieves and handling equipment for stones this large were unavailable. We first placed the large size rock by hand, distributing them evenly on the model bank surface, and then successively placed smaller size rocks to make the thickness of the riprap blanket approximately 1.5 to 2 times the $D_{100}$ stone, as is done in current practice.

We attached the entire assembly of the model riprap bank to a steel frame under the carriage that spans the test basin in the Ice Engineering Facility of the Laboratory. During a test, we moved the carriage at a speed of 40 cm s$^{-1}$ along the length of the basin. The relative movement between a stationary model ice sheet and the moving riprap bank produced an interaction in which the model ice sheet deflected up and broke into small blocks after failing in bending. After some amount of ride up, the broken ice pieces piled up on the model riprap.

The steel frame, under which the riprap model was attached, had five load cells installed at appropriate places to measure horizontal and vertical forces during the ice–riprap interaction. Because of the excessive weight of the rocks and the sand during the second series of tests, we had to install extra braces and clamps to support the model, and these extra supports prevented us from measuring the vertical force. However, we could still measure the horizontal force during the tests.

We grew model ice sheets in the test basin from a solution of urea (1% by weight) in water. We started the growth of a model ice sheet by a wet seeding process, which enabled us to obtain ice sheets with small size grains. We allowed the ice sheet to grow at a temperature of –20°C (–4°F) in the basin for a certain length of time so it

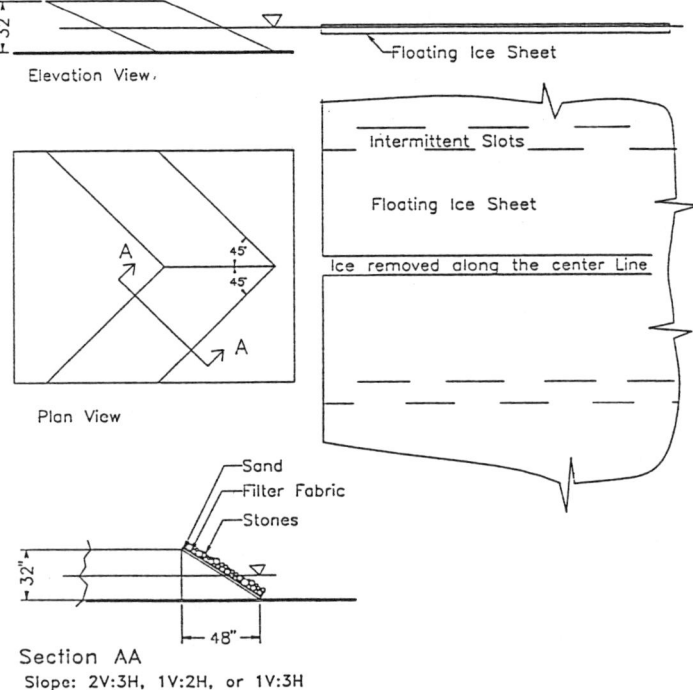

Figure 2. Model riprap bank and ice sheet for modeling oblique ice action. Section AA shows the placement of sand, filter fabric and riprap stones.

could attain the desired thickness. After the ice growth process, we increased the ambient temperature in the basin to 0°C (32°F) for a certain duration of time to allow the model ice to decrease its flexural strength through a tempering process. To characterize a model ice sheet, we measured the characteristic length and the flexural strength. Sodhi et al. (1996) have described in detail the procedures to measure the properties of a model ice sheet. When the flexural strength was within the range of desired strength, we conducted a riprap test.

During the tests of direct ice action, we cut slots in the ice sheet that were equal to the width of the model riprap bank along the length of the model basin. This stopped interference between the model side walls and the neighboring ice sheet. For the second series of tests, in which ice action was a combination of shoving and shearing, we cut two slots in the center, separated by about 30 cm (12 in.), and removed the ice between them (Fig. 2). This prevented the ice from interacting with the stones placed near the ridge line of the model. After some experimentation, we cut two more rows of staggered, intermittent slots at the edge of the model riprap bank, as shown in Figure 2. This confined and stabilized a certain width of the model ice sheet so that it alone interacted with the model riprap. Once we established this procedure, we continued the same pattern of cutting slots in the model ice sheets. Thus, we had a reproducible means of comparing the results of tests with various stone sizes and ice thicknesses.

RESULTS
During the experimental program, we changed the slope of the model riprap bank, the size and the mix of the rocks, and the thickness of the model ice sheets. Sodhi et al. (1996) present the results of the first series of tests in terms of measured horizontal and vertical forces, outcome of interaction as pile-up or ride-up events, and damage to the model riprap bank.

Observations made during both series of tests showed that a model ice sheet would initially ride up on the model riprap bank, and then pile up there. As the interaction proceeded, the ice sheet was forced to go between the piled-up ice and the model riprap. It was during these events that the model riprap was damaged. At times, the ice sheet pushed the stones out of their location and carried them with it to the surface of the piled up ice. After a test, we took the whole assembly of the model riprap bank and the piled-up ice outside the basin to melt the ice. After the ice melted and exposed the bank, we observed the damage to the riprap protection.

We considered the model riprap protection to have failed if the ice took out the stones and exposed a certain area of filter fabric. We noted the severity of damage by designating numbers ranging from 0 to 3 according to the area of filter fabric exposed by the ice action. We designated the severity of damage as follows: (a) 0 for no damage to the riprap, i.e., insignificant movement of rocks, (b) 1 for the exposed area of filter fabric being less than 20 cm$^2$ (3 in.$^2$), (c) 2 for the exposed area of filter

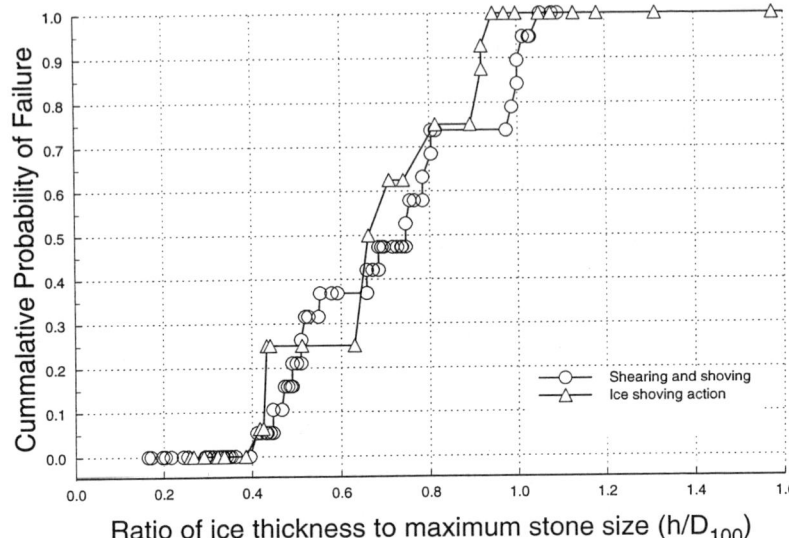

Figure 3. Cumulative probability of riprap failure versus $h/D_{100}$.

fabric being in the range of 20 cm² (3 in.²) and 65 cm² (10 in.²), and (d) 3 for the exposed area of filter fabric being greater than 65 cm² (10 in.²).

From the data obtained from tests, we tabulated the ratio of ice thickness ($h$) to the maximum rock size ($D_{100}$) and the corresponding riprap failure, as indicated by a damage severity level equal to or greater than 2. After sorting the values of $h/D_{100}$ in ascending order, we noticed that the model riprap bank did not fail for values of $h/D_{100}$ less than 0.4 and failed all the time for the values of $h/D_{100}$ greater than 1. In between these two values of $h/D_{100}$, the occurrence of riprap failure was somewhat random. If there were $m$ failures and $n$ non-failures in the set of data for random failures, we generated a cumulative probability of failure by adding $1/m$ for each failure and 0 for each non-failure to the cumulative probability value at each ascending value of $h/D_{100}$ ratio.

Figure 3 shows a plot of the resulting cumulative probability of riprap failure versus $h/D_{100}$ ratio for both series of tests. The plot of cumulative probability of riprap failure does not indicate the effect of the slope of the riprap bank. When we compared the probability of failure obtained from the set of data on random failures $m/(m+n)$ for each slope, we found that it was lower for the slope of 1V:3H than that for steeper slopes of 1V:2H and 2V:3H, as shown in Table 1 for shearing and shoving action. This indicates that the likelihood of riprap damage increases with the slope of a riprap bank.

Table 1. Data on failure and non-failure events during shearing and shoving tests with three slopes.

| Slope | Number of Failure events (m) | Number of non-failure events (n) | Probability of failure m/(m+n) |
|---|---|---|---|
| 1V:3H | 4 | 50 | .077 |
| 1V:2H | 11 | 15 | 0.44 |
| 1V:1.5H | 8 | 17 | 0.32 |

## CONCLUSION

We conducted two series of model tests to simulate the interaction between ice and riprap for an ice sheet moving perpendicular to and at an angle of 45° to the shoreline. We changed the stone size and mix, the slope angle of the model riprap, and the ice thickness during the test program. The damage incurred during these two series of tests told us that the maximum size of riprap stones to sustain no damage should be about 2.5 times the ice thickness. The data on the probability of riprap failure indicate that a riprap bank with lower slope has less likelihood to be damaged during an ice action.

## ACKNOWLEDGMENTS

The authors are grateful to C. Schelewa for his help in conducting the tests, and J.L. Wuebben and K. Carey for suggestions. This work was completed with funding from the Civil Works Program, U.S. Army Corps of Engineers.

## REFERENCES

Ashton, G. D. (ed.) (1986) *River and Lake Ice Engineering*, Water Resources Publications, Littleton, Colorado.

Sodhi, D. S., Borland, S. L. and Stanley, J. M. (1996) Ice action on riprap: Small-scale tests. CRREL Report 96-12, U. S. Army Cold Regions Research and Engineering Laboratory, Hanover, N.H. 03755-1290.

# Ice Jam Mitigation for Small Streams

## JAMES H. LEVER
U.S. Army Cold Regions Research and Engineering Laboratory, Hanover, NH

## ABSTRACT

Small streams can cause severe ice-jam flooding. Ice booms can mitigate freezeup ice jams for low cost and environmental impact provided suitable low-velocity pools are available. Low-cost breakup ice-control structures also exist, although work remains to quantify their effectiveness. Ice weakening could provide effective breakup ice-jam mitigation at very low cost and environmental impact. However, large natural variability in ice hydraulic conditions, lack of suitable theory and incomplete field data make it difficult to quantify their effectiveness.

## 1. INTRODUCTION

Ice jams are stationary accumulations of ice that restrict flow. Both freezeup and breakup ice jams occur, classified according to the principal processes that form them (USACE 1994, Beltaos 1995). Freezeup ice jams occur during prolonged cold spells, as frazil ice produced in highly turbulent river reaches collects along slower reaches. Although the process may take several weeks, freezeup jams can progressively block a significant portion of the channel and cause local flooding despite low winter flows. Snowmelt or rainfall can trigger breakup of an ice-covered river and the formation of breakup ice jams. The process is more dynamic than freezeup. Sudden fracture and movement of an ice cover can yield powerful surges of water and ice. These ice runs may stop abruptly at river constrictions or intact ice covers on slower reaches to form thick ice accumulations; flooding may result within a matter of minutes.

Ice-jam mitigation for small streams (say, 20- to 60-m-wide unregulated rivers) must deal with typically high average slopes ($> 0.001$) and shallow channels (1–3 m). Rapids may remain open through prolonged cold spells to generate prodigious volumes of frazil ice and consequently thick ice accumulations. Breakup surge velocities can exceed 5 m/s, and small channel depths at the jam sites can yield ice jams that are firmly grounded to the riverbed.

Small streams account for the majority of ice-jam sites brought to the attention of the Corps of Engineers. While flood damages can be significant locally, they are usually insufficient to justify the cost of conventional flood-control structures such as dams and levees. Environmental concerns (including ecological, recreational and aesthetic concerns) may also limit the appeal of conventional structures.

Here, we discuss two categories of ice-jam mitigation techniques for small streams: ice-control structures and ice weakening. These techniques attempt to utilize knowledge of ice-jam processes to mitigate ice-jam flooding for low cost and environmental impact. Field examples are referenced that provide data on cost (1994 dollars) and performance. We also discuss the difficulty in assessing the effectiveness of these techniques quantitatively. For more thorough descriptions of ice-jam mitigation, see Belore et al. (1990), USACE (1994) and Beltaos (1995).

## 2. ICE CONTROL STRUCTURES

Specialized ice-control structures (ICSs) have been developed to mitigate ice jam flooding (Tuthill 1995). An ice boom, consisting of floating elements tied to bed or bank anchors with wire rope, is often used to mitigate freezeup ice jams. The boom collects ice that would otherwise accumulate downstream, and the resulting ice cover insulates the water to reduce local ice production. Surface velocities must be below about 0.8 m/s for a boom to collect ice (Foltyn and Tuthill, 1996).

In 1982 the Corps installed an ice boom on the 160-m-wide Allegheny River to mitigate ice-jam flooding at Oil City, PA. It cost $7,900/m of river width. Flow is reduced in the Allegheny during freezeup to initiate ice-cover formation at the boom. In 1989 the Corps built a 1.5-m-high weir on the 110-m-wide Oil Creek to complete its ice-jam mitigation project for Oil City. Although intended to create a low-velocity pool for an ice boom, the weir alone initiates early ice-cover formation. The Oil Creek structure cost $21,000/m of river width. Oil City has not experienced serious ice-jam flooding since these structures were built. CRREL designed an ice boom for the 80-m-wide Salmon River in Idaho, and analyses show that it prevented ice-jam flooding in Salmon, ID in 1990-91 (White 1992, Zufelt and Bilello 1992).

Ice booms are seasonal structures, have low environmental impact, and have low to moderate cost. These advantages disappear if a conventional weir is needed to reduce local flow velocities. At CRREL, we have developed a seasonally installed, 0.9-m high "tension weir." It consists of rubberized fabric supported by wire mesh and anchored into the banks with wire rope. It cost about $3,600/m of river width and performed well during three seasons of field trials, forming ice covers without use of an ice boom (Figure 1). Although it requires seasonal installation and removal, the tension weir appears to be an attractive alternative to conventional weirs in small streams where low cost and environmental impact are important.

Figure 1. Ice cover formation on CRREL tension weir in Ompompanoosuc River, Vermont.

Ice booms and weirs can help mitigate breakup ice jams by reducing downstream ice accumulations and delaying ice runs until downstream ice has moved out. However, structures with greater ice-retention capacity have been built to mitigate breakup ice jams. An ICS on the Ste. Anne River in St. Raymond, Quebec, consists of a low-head weir with ice-retaining piers. The piers retain the ice cover formed behind the weir, and the pool stores ice arriving from upstream. The Corps of Engineers designed a similar structure for the 46-m-wide Cazenovia Creek in West Seneca, NY (Gooch and Deck 1990). It wasn't built because the community could not provide its portion of the project cost, about $56,000/m-width of river.

Figure 2. Credit River ICS arrests a breakup ice run. (Photo courtesy C. Worte.)

An ICS on the Credit River in Mississauga, Ontario, controls ice breakup without use of a weir (Belore et al. 1990). The structure consists of concrete piers spaced at 2-m intervals across a 27-m-wide channel adjacent to a natural floodplain (Figure 2). The piers arrest a breakup ice run to form an ice jam. Large boulders extend across the floodplain to restrain ice pieces as flow bypasses the structure. The structure cost $23,000/m of river width, and no flood damages have occurred since its construction in 1988.

CRREL has developed a low-cost ICS to control breakup ice jams on small streams (Lever et al. 1996). It consists of massive sloped blocks, partially buried in riprap, placed across the river adjacent to a natural floodplain. The blocks will arrest a breakup ice run and form a stable, partially grounded ice jam. Trees or boulders on the floodplain retain ice pieces in the river channel while allowing flow to bypass the structure. Large gaps between blocks allow easy fish and canoe passage. Built in 1994 in Hardwick, Vermont, the ICS cost $3,600/m-width and has performed well during the four breakup events experienced to date (Figure 3).

Figure 3. Breakup ice run pushes ice above sloped-block ICS.

## 3. ICE-WEAKENING TECHNIQUES

Ice weakening involves reducing the strength of the ice cover at an ice-jam site to allow a breakup ice run to move through without jamming. Operations are normally conducted a few weeks before the threat of breakup.

Mechanical weakening achieves an immediate reduction in ice-cover strength. One technique is to cut a pattern of slots through the ice along the thalweg using a commercially available trencher. The slots allow easy fracture of the ice cover as an ice run moves through. Results by Jolicoeur et al. (1984) for the 37-m-wide Beaurivage River near Quebec City indicate that diagonal patterns achieve good results for minimum cutting time. A 2.5-tonne, four-wheel-drive trencher should cut a 30-m-wide pattern of diagonal slots in 0.5-m-thick ice at a rate of 600 pattern-m/day and a cost of about $1/pattern-m (plus mobilization costs).

Ice breaking is another mechanical ice-weakening technique. By breaking a broad channel through the ice along the thalweg, the reach should offer very low resistance to a moving ice run. A newly developed, 18-tonne amphibious excavator (Figure 4) appears to be suitable for use in small streams (Haehnel 1995). It is capable of breaking a 30-m-wide channel through 0.5-m-thick ice at a rate of 300 m/day and a cost of about $6/pattern-m (plus mobilization costs).

Figure 4. Amphibious excavator used to break a channel in an ice cover.

Thermal ice weakening accelerates the deterioration of ice-cover strength that occurs naturally during the breakup season. One method involves drilling an extensive pattern of holes through the ice along the thalweg. If average temperatures remain above freezing, the holes widen significantly within a week. The City of Oconto, WI, has successfully employed this technique for several years, using a small tractor with an attached post-hole digger (Haehnel 1996). Equivalent rented equipment should drill a 30-m-wide grid of 20-cm-dia. holes on 3-m spacings through 0.5-m-thick ice at a rate of 300 pattern-m/day and a cost of $0.6/pattern-m.

Another thermal ice-weakening technique involves applying an environmentally acceptable dark material on the ice ("dusting") to enhance solar absorption. The dark material initiates melting of the ice surface to create a water layer that continues the absorption process. The State of Alaska has used aircraft to apply sand on the Yukon River for many years to mitigate breakup ice jams. Haehnel et al. (1996) showed that a hydroseeder can efficiently apply mulched leaves to a small stream provided there is reasonable road access. The hydroseeder can treat a 30-m-wide swath at a rate of 2,000 pattern-m/day and a cost of $2/pattern-m.

## 4. ASSESSING EFFECTIVENESS

To assess the effectiveness of ice-jam mitigation, we must determine the change in ice-jam severity caused by a particular mitigation technique. Because ice-jam severity depends on spatial and temporal variations in ice volume, ice strength, runoff and channel morphology, this is a difficult task.

Until recently, all numerical models calculated ice-jam thickness based on a static force equilibrium (Zufelt and Ettema 1996). These models could not predict the onset of breakup, the sequence of ice runs, the likelihood or location of ice jams, or ice-jam thickness when momentum effects are important. Furthermore, key model parameters (e.g., ice friction and cohesion, ice-jam roughness) have large uncertainties due to lack of good field data. Recent efforts to extend ice-jam theory to include the movement and arrest of ice runs (e.g., Shen et al. 1990, Zufelt and Ettema 1996, Hopkins et al. 1996) are very encouraging. However, these models are still under development, and field data suitable for their validation are scarce.

Physical models have been used extensively to develop ice-control structures (Beltaos 1995, Lever et al. 1996) but not, as yet, to develop or assess ice-weakening techniques. Physical models can be expensive relative to numerical models and more restricted in spatial extent and parametric scope; they also suffer from a lack of field data to validate results. However, they incorporate the physics of the ice-jam process provided model distortions are kept within reasonable limits.

The complexity and variability of ice-jam processes also make field assessment of mitigation techniques difficult. Except for specific freezeup jams (Zufelt and Bilello 1992), we cannot yet predict a river's ice-jam behavior in the absence of the mitigation technique. Comparison of pre- and post-technique stage-frequency distributions could determine effectiveness. However, complete long-term data records are very scarce, and data uncertainties in short-term records can obscure the results unless the technique yields a strong influence on ice-jam behavior.

What are the practical alternatives? One approach is to be conservative: build a robust ICS that handles (as confirmed by model tests) severe ice-hydraulic conditions. But conservatism increases costs and can thereby limit applicability in small streams. Fortunately, breakup ICSs have shown a trend towards lower cost as knowledge of the breakup process has improved. The Credit River ICS uses in-channel ice storage and flow bypass on a natural floodplain to reduce costs. The sloped-block ICS goes further by using a few massive elements to trigger the natural ice-jamming process. The effectiveness of these structures can be assessed over a reasonable period because they produce a dramatic effect: arrest and retention of breakup ice runs.

Assessment is more difficult for ice-weakening techniques because their effect is more subtle. Numerous questions arise: How weak does an ice cover need to be to pass an ice run? What spatial extent of ice cover should be weakened? What channel morphologies will arrest ice runs even in the absence of an ice cover? If thermal weakening is used, how is enhanced solar absorption linked to reduced strength? Despite these questions, agencies have used ice weakening because in principle it could work, and it is inexpensive and expedient. These advantages justify research into the effectiveness of ice weakening, and we are attempting to document the performance of ice-weakening efforts at several sites.

## CONCLUSIONS

Ice-jam mitigation for small streams must deal with difficult ice-hydraulic conditions and requirements for low cost and environmental impact. Ice booms are well suited to mitigate freezeup ice jams, provided suitable low-velocity pools are available. Breakup ICSs have shown progress towards lower cost and environmental impact, and their field performance should yield acceptable proof of effectiveness within a few years. Ice-weakening techniques could, in principle, provide breakup ice-jam mitigation in small streams for very low cost and environmental impact. However, the complex ice-jam process, lack of applicable theory and incomplete field data makes assessment of their effectiveness a long-term undertaking.

## REFERENCES

Belore, H.S., B.C. Burrell and S. Beltaos (1990). Ice jam mitigation. Can. J. Civil Engineering, Vol. 17, pp. 675-685.

Beltaos, S., ed. (1995). River ice jams. Water Resources Publications, Highlands Ranch, CO, 372 p.

Foltyn, E.P. and A.M. Tuthill (1996). Design of ice booms. Cold Regions Technical Digest 96-1, CRREL, Hanover, NH.

Gooch, G.E. and D.S. Deck (1990). Model study of the Cazenovia Creek ice control structure. Special Report 90-29, CRREL, Hanover, NH, 31 p.

Haehnel, R.B. (1995). Breaking river ice to prevent ice jams. Ice Engineering Information Exchange Bulletin No. 11, CRREL, Hanover, NH.

Haehnel, R.B. (1996). Drilling holes in ice to reduce ice jam potential. Ice Engineering Information Exchange Bulletin No. 14, CRREL, Hanover, NH.

Haehnel, R.B., C.H. Clark and S. Taylor (1996). Dusting river ice with leaf mulch to aid in ice deterioration. Special Report 96-7, CRREL, Hanover, NH.

Hopkins, M.A., S.F. Daly and J.H. Lever (1996). Three-dimensional simulation of river ice jams. 8th Int. Conf. on Cold Regions Engineering (ASCE), Fairbanks, AK, p. 582–593.

Jolicoeur, L., B. Michel and J. Labbe (1984). Cutting trenches in an ice cover to prevent ice jams. Workshop on the Hydraulics of River Ice, Fredericton, New Brunswick, p. 127–136.

Lever, J.H., G. Gooch, A. Tuthill and C. Clark (1996). A new, low-cost ice control structure. Part 1: Concept development. Part 2: Construction and performance. 8th Int. Conf. on Cold Regions Engineering (ASCE), Fairbanks, AK, p. 617–639.

Shen, H.T., H. Shen and S.M. Tsai (1990). Dynamic transport of river ice. Journal of Hydraulic Research (IAHR), Vol. 28 (6), p. 659–671.

Tuthill, A.M. (1995). Structural ice control: Review of existing methods. Special Report 779, CRREL, Hanover, NH.

U.S. Army Corps of Engineers (1994). Ice jam flooding: Causes and possible solutions. Engineering and Design Pamphlet EP1110-2-11, Washington, D.C.

White, K.D. (1992). Salmon River experimental ice boom: 1989-90 and 1990-91 winter seasons. Special Report 92-20, CRREL, Hanover, NH.

Zufelt, J.E. and M.A. Bilello (1992). Effects of severe freezing periods and discharge on the formation of ice jams at Salmon, Idaho. CRREL Report 92-14, CRREL, Hanover, NH.

Zufelt, J.E. and R. Ettema (1996). Ice jam dynamics. Iowa Institute of Hydraulic Research, Technical Report No. 380, University of Iowa, Iowa City, IA.

# SIMULATION OF DYNAMIC RIVER ICE TRANSPORT AND JAMMING

SHUNAN LU and HUNG TAO SHEN
Clarkson University
Potsdam, NY 13699-5710
USA

RANDY D. CRISSMAN
New York Power Authority
Niagara Falls, NY 14302
USA

## ABSTRACT

Winter operations of hydropower generating stations can be significantly affected by the possibility of ice jam formations in rivers. A numerical model for dynamic river ice transport is used to study the ice jamming processes in the upper Niagara River. This study is intended as an example to demonstrate the use of the model to analyze ice jamming processes for hydropower project.

## INTRODUCTION

The formation of river ice jams can have significant negative impacts on hydropower productions. River ice jam formation is a very dynamic and highly two-dimensional phenomenon. The classical static ice jam theory (Pariset and Hausser 1961, Uzuner and Kennedy 1976, Beltaos 1983), which can determine the one-dimensional ice jam thickness profile, cannot describe the ice jamming process. Shen et al. (1993) developed an analytical formulation and a two-dimensional numerical model for dynamic river ice transport. The model has been used to study ice transport and ice jam dynamics in rivers (Shen and Lu 1996, Shen et al. 1997). This model is refined and applied to the upper Niagara River (Fig. 1). The model provides a useful tool for investigating possible methods to alleviat the impact of ice runs on hydropower productions.

The Niagara Power Project of the New York Power Authority (NYPA) and the Sir Adam Beck (SAB) generating stations of Ontario Hydro (OH) divert water from the Grass Island Pool (GIP) of the upper Niagara River just upstream of Niagara Falls (Crissman 1990). The combined generating capacity of the US and Canadian stations is about 4,500 megawatts. The winter operation of these generating stations is at times hampered by ice transported into the river from Lake Erie. Ice jams and subsequent flooding of the low lying areas along the upper Niagara River sometimes occur. Ice also causes power generation losses, because water diversions are reduced to provide additional flow to transport ice over Niagara Falls and reduce the potential for ice jamming.

## THE NUMERICAL MODEL

Detailed descriptions of the model formulation and implementation were given by Shen et al. (1993, 1997a). The governing equations for the model of the river hydraulics are the shallow water equations of motion including the effects of surface ice. The ice transport model includes the mass and momentum balances of the river ice under the influences of wind and water drag, gravity, internal ice resistance, and bank and bed friction. The internal ice resistance is calculated from a viscous-plastic constitutive law (Hibler 1986). The ice pressure is determined from the granular formulation used in classical ice jam theory (Shen et al. 1993). The hydrodynamic equations are solved by a finite-element method using the lumping technique and leap-frog time integration. The ice dynamics are simulated with a Lagrangian discrete-parcel method with smoothed-particle hydrodynamics. Ice parcels are allowed to move tangentially along a solid boundary. The boundary friction, which opposes the ice movement, is calculated using a dynamic Coulomb yield criterion (Shen et al. 1997a).

The water level at Fort Erie, diversion flows at the hydropower intakes, and water discharge past the International Control Structure at the downstream boundary are the prescribed hydrodynamic boundary conditions. The ice discharge entering the river from Lake Erie is the boundary condition for the ice dynamic simulation.

## MODEL CALIBRATION

The Manning's bed resistance coefficients along the river were determined by calibrating the model with the observed discharge and water levels during the ice-free period prior to two major lake ice runs, that occurred on February 23-27, 1975 and January 23-26, 1992. The calibrated Manning's coefficients varied between 0.015 and 0.028 along the river. The ice transport model was calibrated for the February 24-27, 1975 lake ice run. This ice run event was selected for calibrating the model because of the availability of detailed field data. In addition to discharge and water levels, a series of charts showing the observed ice conditions in the GIP were available and were compared to the simulated ice transport pattern. The calibrated parameters for the ice dynamic simulation model are summarized in Table 1. These parameters are the same as those for a large scale model (Su et al. 1996) of ice transport in the GIP. The value of the internal friction angle is consistent with the value commonly used in river ice jam studies. The wind drag coefficients on water and ice surfaces are consistent with values previously used in a model for ice transport in Lake Erie (Chieh et al. 1993).

The viscous-plastic constitutive law relates the internal ice stress to the strain rate of the ice flow. During the ice stoppage and jamming process the ice mass behaves more like a solid and its stress state becomes independent of the strain rate. The viscous-plastic law approximates the rigid behavior of a stationary ice mass by a state of very slow flow (Hibler 1986). To overcome this inherent deficiency in the viscous-plastic law, empirical parcel stoppage criteria are used in the numerical model to simulate ice stoppage and jamming. These empirical criteria for the upper Niagara River model are:

- Stage I: When $|\vec{V_i}| < 0.15 m/s$ and $t_i > 3.5m$, where $\vec{V_i}$ = ice velocity and $t_i$ = ice thickness, the ice parcel is forced to stop temporarily for one time step. The parcel is allowed to move at the next time step under the influence of the dynamic forces.
- Stage II: When $|\vec{V_i}| < 0.10 m/s$ and $t_i > 3.5m$, the ice parcel is forced to stop permanently.

Table 1: Parameters Used in the Ice Dynamic Simulation

| Parameter | Description | Value | |
|---|---|---|---|
| $N_{max}$ | Maximum ice concentration | 60% | |
| $\phi$ | Internal friction angle of ice | 46° | |
| $\tan\phi_B$ | Bank-to-ice friction coefficient | 1.04 | |
| $\tan\phi_b$ | Bed-to-ice friction coefficient | 1.04 | |
| $j$ | Empirical constant | 15 | |
| $e$ | Principal axes ratio of yield curve | 2 | |
| $n_i$ | Single layer Manning's coefficient of surface ice | 0.025 | |
| $\frac{\rho_a \gamma^2}{\rho}$ | Wind-on-water stress coefficient | $1.5 \times 10^{-6} W^{-0.2}$ | $W \leq 1.0 m/sec$ |
| | | $0.6 \times 10^{-6} W^{0.5}$ | $1.0 m/sec < W < 15.0 m/sec$ |
| | | $3.12 \times 10^{-6}$ | $W \geq 15.0 m/sec$ |
| $C_a$ | Wind-on-ice drag coefficient | 0.0015 | |

## SIMULATION OF NIAGARA RIVER ICE DYNAMICS

The calibrated model was used to simulate the January 1992 lake ice run. This ice event was selected to illustrate the ice jam dynamics in the upper Niagara River since it was representative of severe ice jamming processes related to power generation losses and the potential for flooding. Moreover, large scale physical and numerical model simulations of this event were conducted for the GIP (Alden Research Lab. 1996, Su et al. 1996). These simulations provide additional data for further validation of the model.

Fig. 1: The Upper Niagara River

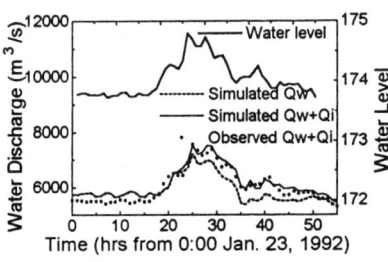

Fig. 2: Observed Water Level and Comparison of Discharges at Fort Erie

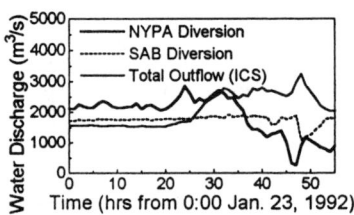

Fig. 3: Diversions and Discharge Past the International Control Structure

# DYNAMIC RIVER ICE TRANSPORT AND JAMMING

The simulation of the January 1992 ice event spanned the 55-hour period from 0000 hours on January 23 to 0700 hours on January 25. The results of the simulation are presented in charts referenced to the begining of the simulation. The ice-affected period for this ice run was between 2200 hours on January 23 (Hour 22) to 0700 hours on January 25 (Hour 55). Table 2 summarizes the observed and simulated occurrences of the major ice events during this ice run. An ice stoppage occurred near the NYPA intakes shortly after the arrival of the lake ice due to the congestion of the ice in the narrow passage between the intake wall and the ice island formed in the shallow area of the Grass Island Pool. Fig. 2 shows the observed water level and the comparison of simulated and estimated prototype discharges at Fort Erie. The estimated prototype total discharge was determined by a one-dimensional unsteady flow simulation, which approximates the ice-water flow by a water flow of the same mass flow rate (Crissman et al. 1993). Fig. 3 shows the observed diversion at NYPA and OH intakes and the discharge past the International Constrol Structure. Fig. 4 shows the recorded wind velocity during the ice run. The wind direction angle is measured counterclockwise from true North. Fig. 5 shows the boundary ice discharge at Fort Erie and the ice discharge in the Tonawanda and Chippawa Channels at the upstream end of Grand Island. A comparison of figures 4 and 5 shows the effect of wind on the lake ice discharge as well as its distribution into the two channels. The increasing ice accumulation and the accompanying reduction in water discharge in the Tonawanda Channel contributed to the reduction of the ice discharge in that channel. Fig. 6 compares observed and simulated water levels along the river. Fig. 7 shows a set of ice velocity distribution plots corresponding to the occurrences of the major ice events listed in Table 2. The simulation results presented indicate that the calibrated model can correctly simulate the ice run dynamics in the upper Niagara River. The model is being used to study the effectiveness of operational and structural mitigating measures for alleviating ice impacts on hydropower operations on the Niagara River (Shen et al. 1997a).

Table 2: Comparison of Chronology of Simulated and Observed Ice Events

| Event | Observed | Simulated |
|---|---|---|
| Ice arriving at NYPA Intakes from the East and Little West Channels | ~ 1000 Jan. 24 (Hour 34) | 0945 Jan. 24 (Hour 33.75) |
| Initiation of the ice island in GIP | After 1100 Jan. 24 (After Hour 35) | 1200 Jan. 24 (Hour 36) |
| Ice arriving at GIP from the West Channel | 1500 Jan. 24 (Hour 39) | 1500 Jan. 24 (Hour 39) |
| Ice Jam occurring in the vicinity of NYPA intakes | 2025 ~ 2041 Jan. 24 (~ Hour 44.5) | ~ 2030 Jan. 24 (Hour 44.5) |
| Surface ice coverage in the West Channel (observation time) | 60~80% (2030~2109 Jan.24) (Hour 44.5~45.15) | ~80% (2000~2100 Jan. 24) (Hour 44~45) |
| Ice Stoppage beneath the North Grand Island Bridge | After 0015 Jan. 25 (After Hour 48.25) | 0030 Jan. 25 (Hour 48.5) |

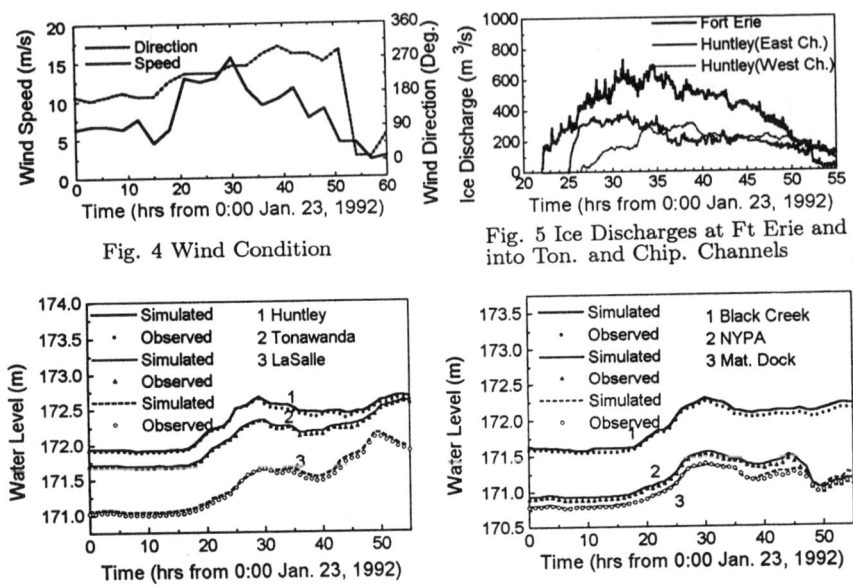

Fig. 4 Wind Condition

Fig. 5 Ice Discharges at Ft Erie and into Ton. and Chip. Channels

Fig. 6 Comparison of Observed and Simulated Water Levels along the River

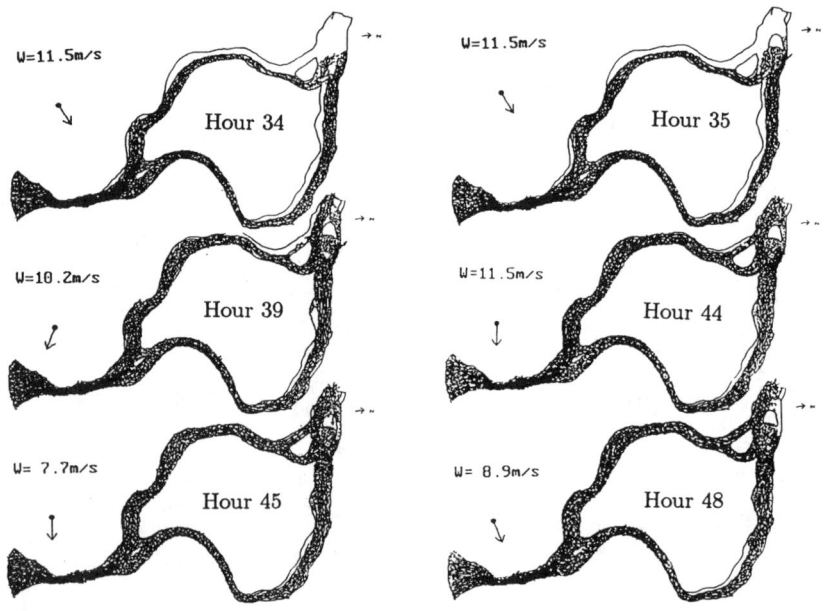

Fig. 7 Simulated Ice Distributions in the Upper Niagara River, Jan. 1992 Lake Ice Run

## SUMMARY AND CONCLUSIONS

A two-dimensional numerical model for dynamic transport of river ice was developed and applied to the upper Niagara River to study the ice jamming process in relation to hydropower operations. Comparisons of the simulation results with field data show that the model is capable of describing the ice transport and jamming processes accurately. The model can be used to assist in the design of effective mitigation measures to improve the hydropower prodution on rivers which are affected by ice in the winter.

## ACKNOWLEGEMENTS

This study was supported by the New York Power Authority through Contract No. 029494-91.

# References

[1] Alden Research Laboratory (1996) "Physical model studies of ice concerns at the Niagara Power Project Intake," Report submitted to the New York Power Authority.

[2] Chieh, S.H., Wake, A., and Rumer, R.R. (1983) " Ice forecasting model for Lake Erie," *J. of Waterway, Port Coastal and Ocean Engineering*, ASCE, 109(4), 392-415.

[3] Crissman, R.D. (1990) "An on-line early warning system for ice jams and stoppage on the upper Niagara River," *J. of Hyd. Res.*, Vol. 28, No. 6, 719-728.

[4] Crissman, R.D., Chin, C. L., Yu, W., Mizamura, K., and Corbu, I (1993) "Uncertainties in flow modeling and forecasting for Niagara River," *J. Hyd. Engrg.*, ASCE, 119(11) 1231-1250.

[5] Hibler, W.B. III (1986) "Ice dynamics," *The Geophysics of Sea Ice*, N. Untersteiner, ED., Plenum Press, New York, 577-640.

[6] Shen, H.T., and Lu, S. (1996) "Dynamics of river ice jam release," *Proc.,8th Int'l Conf. Cold Regions Eng.*, ASCE, Fairbanks, 594-605.

[7] Shen, H.T., Lu, S. and Crissman, R.D. (1997) "Numerical simulation of ice transport over the Lake Erie-Niagara River ice boom," *Cold Regions Sc. and Tech.*, In Press.

[8] Shen, H.T., Lu, S. and Liu, L. (1997a) "Numerical model studies of ice transport and jamming in the upper Niagara River," *Report No.97-1*, CEE Dept., Clarkson University, Potsdam, NY.

[9] Shen, H.T. and Chen, Y.C., Wake, A. and Crissman, R.D.(1993). "Lagrangian discrete parcel simulation of river ice dynamics," *Int'l J. of Offshore and Polar Eng.*, 3(4), 328-332.

[10] Su, J., Shen, H.T., and Crissman, R.D. (1996) "A numerical model study on ice transport in the vicinity of Niagara River hydropower intakes," *Proc., Int'l Symp. on Ice*, Vol. 3, IAHR, Beijing.

# Ice Retention with Artificial Islands on the St. Marys River

### ANDREW M. TUTHILL and KEVIN L. CAREY
Cold Regions Research and Engineering Laboratory, Hanover, NH, USA

## ABSTRACT
For the past two decades, a navigation ice boom has alleviated ice problems at the head of the Little Rapids Cut, a channel constriction on the St. Marys River near Sault Ste. Marie, Michigan. This study assesses the feasibility of replacing portions of the ice boom with artificial islands constructed of quarried stone.

## INTRODUCTION
The St. Marys River ice boom (SMRIB) is located at the head of the Little Rapids Cut, 2 miles south of the Soo Locks near Sault Ste. Marie, Michigan (Fig. 1). The boom is made up of 1-ft x 2-ft x20-ft timbers and has two arms with a central navigation opening (Fig. 2). Since its first installation 20 years ago, the boom has reduced the rate of ice cover progression in the Little Rapids Cut and minimized ice interference with the Sugar Island Ferry, which provides the 250 residents of Sugar Island with their only transportation link with the mainland.

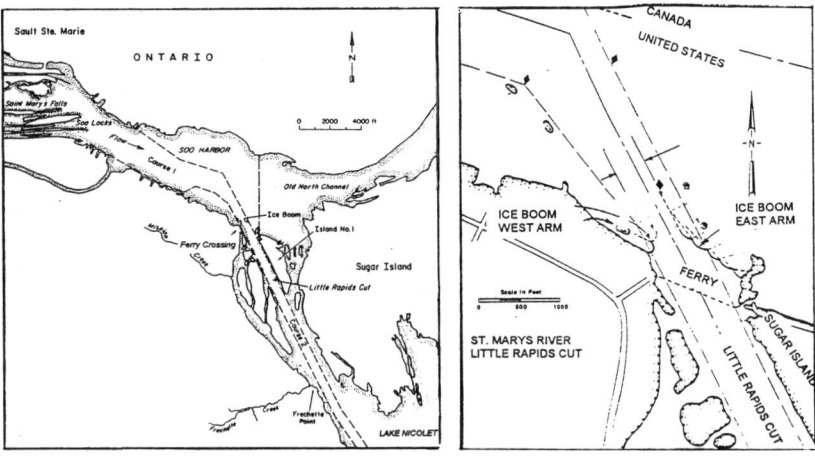

Fig. 1  Area map.          Fig. 2.  Local site map.

The Soo Area Office of the Detroit District of the US Army Corps of Engineers is responsible for the boom's annual installation, removal, and maintenance. Although proven effective, the ongoing costs and risks associated with boom operation and maintenance led the Detroit District to contract with the Cold Regions Research and Engineering Laboratory (CRREL) to examine options for replacing the boom, or portions of it, with permanent structures, such as rock islands. Tuthill and Carey (1996) presents the results of this study.

This paper examines the relationship between the ice processes on the St. Marys River and ice boom performance, addressing the issues of ice cover formation in Soo Harbor, ice bridge formation at the entrance to the Little Rapids Cut, and ice-edge progression within the cut. The study estimates the effect on ice cover progression that would result from replacing the existing east arm of the boom with a line of artificial islands. A preferred alternative that would incorporate the islands and the anchoring system of the boom spans is presented.

## BACKGROUND

The St. Marys River is 45 miles long, connecting Lake Superior and Lake Huron. Lake Superior's average winter outflow is 73,000 cfs, 70% of which passes through the Little Rapids Cut. Under normal winter conditions, sufficient ice cover develops in Soo Harbor between late December and mid-January to form an ice bridge at the head of the cut. Undisturbed, this ice bridge stabilizes the ice cover in the harbor and prevents ice from moving into the cut. The ice bridge is occasionally disturbed by warming and wind or by vessel traffic. This can destabilize the ice field in Soo Harbor and bring large amounts of broken ice into the cut. As a result of the Great Lakes-St. Lawrence Seaway Navigation Season Extension Program which began in 1971, the ice bridge at the head of Little Rapids Cut was continuously breached by ship traffic. The increased ice movement into the cut frequently halted or interfered with ongoing ferry operations between Sugar Island and the mainland, blocked the navigation channel, and increased the frequency of ice jams in the lower Little Rapids Cut.

As a result, in 1974, the Detroit District of the Corps of Engineers contracted Acres American Incorporated to conduct a physical hydraulic model study to develop concepts for some form of physical ice restraint (Acres American Inc., 1975; Cowley et al., 1977). Goals were to help stabilize the ice cover in Soo Harbor, moderate ice problems in the cut, and generally aid winter navigation. The study recommended an ice boom with an opening for navigation, which was designed by the Detroit District with assistance from CRREL. The boom was first installed in December of 1975 and, because of its effectiveness, has been installed annually up to the present time. Perham (1978) documents the boom's performance during its first two seasons.

## ICE REGIME ON THE ST. MARYS RIVER AND ICE BOOM PERFORMANCE

A main objective of the current study was to estimate the change in ice discharge into the Little Rapids Cut that would result from replacing the east arm of the boom with a line of artificial islands. The study investigated the performance history of the boom, focusing on important influences on ice discharge into the cut, such as air temperature, wind, and vessel activity.

The following description of the ice formation process is based on review of the SMRIB annual report series maintained by the Detroit District. The most important period in terms of ice discharge into the Little Rapids Cut occurs between mid-December and mid-to-late January, when the ice cover is forming in Soo Harbor. Much of the Soo Harbor area has surface water velocities at or below 1 ft/s, allowing the formation of thermally grown sheet ice, especially along the northern shore (Fig. 1). Frazil and thin plate ice form in the remaining open water area between St. Marys Falls and the entrance to the Little Rapids Cut. When this drifting ice reaches the cut entrance, it is either retained behind the two arms of the boom, or passes through the 375-ft-wide navigation opening. Initially, this ice collects against the sheet ice cover on Lake Nicolet, at the downstream end of the cut. Additional ice adds to a 1.5-to 3-ft-thick accumulation that progresses upstream within the cut.

If the air temperature is consistently low, the loose accumulations of frazil and small floes quickly consolidate upstream of the boom arms into stable ice fields. In an average winter, these ice fields have formed by early January. The formation of a stable ice bridge across the entire opening occurs, on average, three weeks later, and usually marks the start of a stable midwinter period. Lacking a major thaw, or frequent vessel activity, ice discharge into the cut and ice cover progression are minimal after this time. Thaw periods can delay the formation of a stable ice cover at the entrance of the Little Rapids Cut and weaken the ice cover in Soo Harbor. The thaws are typically followed by strong northwesterly winds that can result in heavy ice discharges into the cut.

The overall coldness of a winter, measured in maximum accumulated freezing degree days (AFDD), appears to be the single most important factor influencing the maximum upstream position of the ice cover in the Little Rapids Cut. Fig. 3 shows a clear relationship between maximum AFDDs and the maximum upstream ice edge position for the winters following the Winter Navigation Demonstration Program. The winter of 1995 is an exception, which may be explained by the timing of warm and cold spells. Note that during the Winter Navigation Program (1971-79), the ice edge frequently reached the ferry crossing over a wide range of maximum AFDDs.

Two approaches were used to estimate the position of the upstream ice edge with the east boom arm replaced by islands. The first method was based on the assumption that, with the islands alone (and no east boom), ice discharge into the cut would increase by 50 % during the early part of the ice formation period. The second method was based on test results from the 1975 Acres American physical model study. Both methods relied on a derived relationship between cumulative ice volume in the accumulation, and maximum ice edge position, based on a calculated equilibrium ice accumulation thickness multiplied by channel width and reach length.

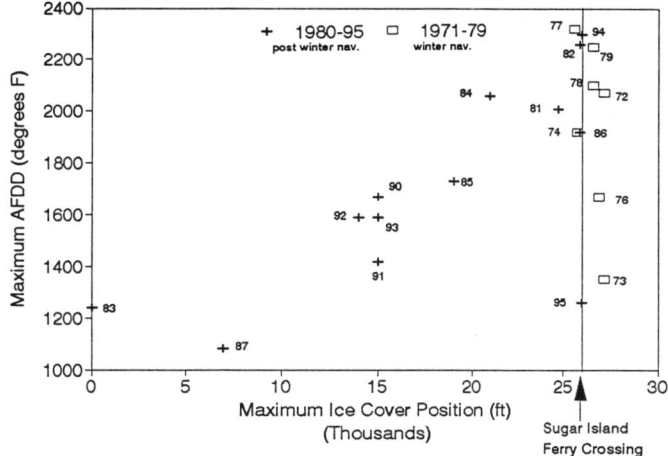

Fig. 3. Maximum upstream ice edge position vs. accumulated freezing degree days.

The first method assumes that the greatest increase in ice discharge into the cut (as a result of the elimination of the east boom) will occur in the early ice formation period, when the floes are too small to arch between the islands. The early ice formation period is defined as the time interval that begins with the first appearance of ice in Soo Harbor, and ends with the formation of a stable ice cover on the east side of the cut entrance. The method assumes that ice discharge into the cut during this period will increase above existing (with-boom) levels by roughly 50%. The approximate dates that a stable ice field existed upstream of the east arm of the boom are known from the SMRIB annual reports. The resulting calculated maximum ice edge positions (Prediction Method 1) are plotted for each year of record, along with the observed maximum ice edge positions in Fig. 4. By this estimating method, the ice edge would have progressed past the ferry crossing 11 times during the 19 years of record, compared to the 6 times under existing conditions.

The 1975 Acres American physical model study tested an alternative quite similar to the original boom, with a navigation opening of 250 ft. The study also tested an alternative consisting of a 2000-ft-long line of 10-ft-diameter cells along the east side of the navigation channel, in the approximate location of the proposed line of islands, with a 550-ft-wide navigation opening. The tests found that the total ice discharge into the cut was 30% greater under the line-of-cells alternative. If each winter's total ice accumulation volume is augmented by 30% (Prediction Method 2), the ice edge would have progressed past the ferry crossing 15 times in the 19 years of record instead of the observed 6 times (Fig. 4).

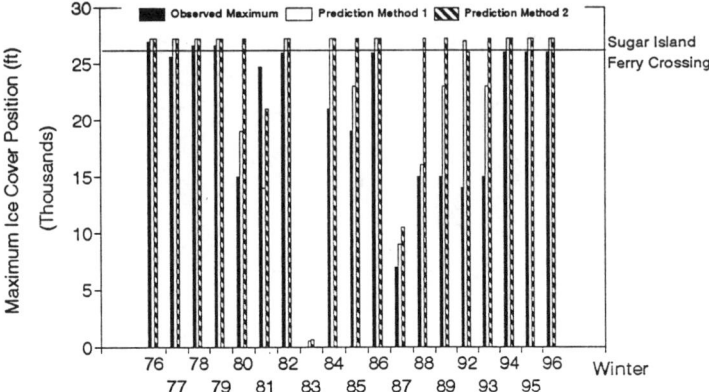

Fig. 4. Observed vs. predicted maximum ice edge positions. Method 1 assumes a 50% increase in ice discharge during the early part of the ice formation season. Method 2 is based on test results from the Acres American (1975) Study.

## PREFERRED ALTERNATIVE

This study recommends the design alternative portrayed in Fig 5. Six new ice islands would be constructed along the east side of the channel. Islands 1, 2, 3, and 4 could serve as above-water boom attachment points for the first three boom spans. The west end of span 5 would connect to existing underwater anchors. The junction of spans 5 and 4 would attach to existing underwater anchor A2E and to existing Island 2E. The construction of new islands 6 and 7, between the two existing islands, would further stabilize the ice cover on the east side of the channel, and help protect boom spans 4 and 5 from impacts by large floes.

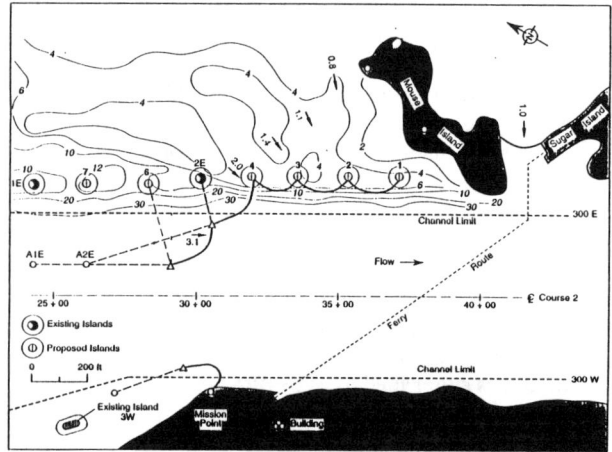

Fig. 5. Map showing the preferred alternative.

Although this alternative calls for boom spans, their installation and removal would be simplified by using the islands as above-water anchoring points.

CONCLUSIONS

This study identified the ice formation period as the time when ice first appears on the Soo Harbor until a stable arch forms across the head of the Little Rapids Cut. During this period, ice edge progression in the Little Rapids Cut is the most rapid and ice retention at the head of the cut is therefore the most critical. Serious ice problems result when the ice edge reaches the Sugar Island Ferry crossing. This has happened during six winters since the boom's first installation in 1975.

This study recommends taking a conservative and incremental approach to the replacement of the east boom with islands. In spite of the costs and risks involved with the current boom, its performance record is good. The current design is the result of 20 years of observations of performance and analysis of failure events. With the existing structure in place, the ice edge in the cut reaches the ferry on a fairly frequent basis, so any design alternative that increases ice discharge into the cut beyond current levels may be unacceptable. The authors believe that a solution that widens the navigation opening from 375 to 650 ft would result in a significant increase in the ice discharge into the cut, and an increased frequency of the ice edge reaching the Sugar Island Ferry crossing.

The preferred design alternative is conservative. It preserves the present navigation opening width and the eastern boom spans. As experience is gained under this new configuration, it may be possible to delete certain boom spans if the islands appear to be retaining the ice sufficiently.

REFERENCES

Acres American Inc. (1975) "Model Study of the Little Rapids Cut Area of the St. Marys River, Michigan," Final Report. Acres American Inc., Buffalo, NY.
Cowley, J.E., J.W. Hayden, and W.W. Willis (1977) "A Model Study of the St. Marys River Ice Navigation," Canadian Journal of Civil Engineering, Vol. 4, 1977, pp. 380-391.
Perham, R.E. (1978) "Ice and Ship Effects on the St. Marys River Ice Booms," Canadian Journal of Civil Engineering, Vol. 5, 1978, pp. 222-230.
Tuthill, A. M., and K. L. Carey (1996) "St. Marys River Ice Boom Replacement Project: Ice Island Design Options," Contract report to the U. S. Army Engineer District, Detroit. U S Army Cold Regions Research and Engineering Laboratory, Hanover, NH. July 1996.

# ICE CONTROL AT LOCKS AND DAMS

F. DONALD HAYNES
USA Cold Regions research and Engineering Laboratory, Hanover, NH, USA

## ABSTRACT

Locks and Dams have problems with ice every winter, especially those in the north. The most severe problem is ice accumulation in the miter gate recess. The second most severe problem around locks is ice in the upper approach. Another severe icing problem is water leaking past J-seals and subsequently freezing on cold surfaces, such as trunnion arms and adjacent concrete walls. In this paper, solutions to some of the most severe problems are presented., such as bubblers or some type of heater.

Key Words: Lock and dam icing, Ice control, Air bubblers

## INTRODUCTION

Locks and Dams across the Nation have problems with ice every winter, especially those in the northern states. Haynes et al. (1993) identified 12 serious ice problems around locks, the most severe of which is ice accumulation in the miter gate recess. Ice can grow out from the cold concrete wall a foot or two. This prevents complete opening of the miter gate and is a hazard to tows. The second most severe problem around locks is ice in the upper approach. This ice can prevent miter gates from being opened, it can make exiting the lock difficult for upbound tows, and it can be pushed into the lock chamber by downbound tows. Another severe icing problem is water leaking past J-seals and subsequently freezing on cold surfaces, such as trunnion arms and adjacent concrete walls. This can render tainter gates inoperable. In addition icing can affect personnel safety. In this paper, solutions to some of the most severe problems are presented.

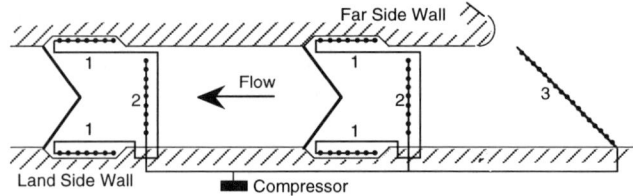

Figure 1. Schematic of a lock bubbler system. The various diffuser lines are (1) recess flusher, (2) air screen, and (3) deflector air screen.

# ICE CONTROL AT LOCKS AND DAMS

## AIR BUBBLER SYSTEMS

The first complete air bubbler system to keep ice out of the lock chamber was designed at CRREL and installed at Starved Rock Lock and Dam in 1985. Because of the success of this bubbler system (Fig. 1), many such systems have been installed at locks and dams in northern states. Details of a bubbler system are given in Engineer Manual 1110-8-1 (FR) (US Army Corps of Engineers 1990). The air supply and diffuser pipes are anchored to the walls and sills of a lock chamber by straps and bolts. A diffuser line is placed in each miter gate recess (1) to flush ice out of this area so that miter gates can be fully opened. Another diffuser line, called an air screen (2), is one that spans the lock chamber. One air screen is located just upstream of the upper gates to help prevent ice from being pushed into the lock chamber by a downbound tow. Another air screen is located just upstream of the lower miter gates to create an area of open water so that the recess flushers can move ice into this area freely; then the gates can be fully opened. The third type of air diffuser line is a deflector screen (3) placed at an angle across the upper approach and used to direct moving ice toward the dam. This line can be used in conjunction with a submersible tainter gate, located next to the lock, that passes ice downstream.

Typically, a 700-kPa (100-psi), 21-m$^3$/min (750 cfm) compressor is used to supply air to submerged diffuser pipes, which have 10-mm ($^3/_8$-in.) diameter orifices spaced about 2.5 m (8 ft) apart. Current designs use one 125-hp electric-motor-driven compressor located about half-way between the ends of the lock chamber. This is practical because only the diffuser lines at one end of the lock chamber are used at a time.

## ELECTRIC HEATERS

During cold spells, ice can readily grow on the upstream faces of tainter gates. At many projects, tainter gates need to be moved periodically to control the upper pool level. To free the gate from the upper pool ice sheet, heaters have been installed at Lock and Dam 2 on the Upper Mississippi River. These heaters consist of U-type tubular heaters placed inside of watertight cavities. Each heater has a 492 W/m (150 W/ft) capacity. Other heaters have been placed on the trunnion arms and on the support channels.

Laboratory tests on the ice buildup process attributable to water level changes on cold surfaces (Haynes et al. 1994) led to the design of a heater panel for preventing ice growth or shedding ice if it had formed. The design consists of two thin aluminum plates separated by aluminum spacers, between which a self-regulating heat cable is placed. The overall dimensions of the panel are 1.22×2.44 m×31.75 mm (4 ft×8 ft ×1$^1/_4$ in.) thick. Aluminum was chosen because it is lightweight, it does not rust, and it is a good conductor of heat. Self-regulating heat cable was chosen because it is waterproof and because it draws less current as it heats up. Because of this, the cable does not require a thermostatic control. With 14.6 m (48 ft) of heat cable in one panel at a spacing of 1 ft, the total power is 1920 W. In using the self-regulating heat cable, an important fabrication requirement is to waterproof all junctions and the end of the cable. One advantage of the heater panel is that it can be made in any size, any shape, and any power density.

Figure 2. Heater panel in a miter gate recess at Starved Rock Lock and Dam.

The panels were specifically designed to be mounted by Tapcon screws on the concrete walls of a miter gate recess. One of these panels was installed and field tested at Starved Rock Lock and Dam (Fig. 2). It is very effective in keeping ice off the concrete wall. If seven of these panels were used to cover the entire miter gate recess, power to them could be cycled from one panel to another by a programmable controller. Cycling the power means that only one panel would be on at a time and the total power would remain at 1920 W. The reason the power can be cycled from one panel to another about every 40 minutes is that only a thin layer of ice, about 1 mm ($^1/_{32}$ in.), has to be melted at the interface to keep the panel free of ice. Some ice growth is permissible on the panels in a lock chamber. Massive ice growth on the panels will prevent opening of the miter gates. Once this thin layer is melted, the remaining mass of ice will fall off and can be moved away by the recess flushers.

## RADIANT HEATERS

Ice can form on cold metal surfaces because of water spray, mist, water level changes, freezing rain, or refrozen melt water. Sometimes this ice accumulates in very inaccessible, but critical, areas, such as the gear rack on the end of a roller gate. Figure 3 shows two 3000-W radiant heaters just above the gears at Lock and Dam 16 on the Upper Mississippi River. These heaters are suspended by a wire rope cable, and help remove ice from the gears. They are located about 5 m (15 ft) above the water level. However, at times personnel still have to be lowered down by a sling to chisel ice off of the gears so the gate can be moved. This task is one of the most dangerous jobs around locks and dams.

Laboratory tests were conducted at CRREL with infrared heaters to determine the effects of three variables on heater performance: wind speed, distance between heaters and ice, and ambient air temperature. The results are given by Haynes et al. (1995).

Figure 3. Infrared heaters in the form of a T above gears on a roller gate.

## SIDE SEAL HEATERS

A side seal rub plate is a steel plate, cast in the concrete pier wall, that mirrors the edge of the curved tainter gate skin plate (Fig. 4). A rubber J-seal, bolted to the gate, is in contact with this rub plate over the entire length of the skin plate. In this design, a heater cavity is formed by a channel welded to the back of a rub plate. A larger channel is welded to the rub plate around the heater cavity, and insulation is placed between the two channels to focus the heat toward the rub plate. One pair of U-type heating elements with a capacity of about 656 W/m (200 W/ft) is placed in the heater cavity. Other designs have used pipes to form heater cavities, around which insulation is placed. Side seal heaters prevent ice formation on the rub plate and also help to keep the J-seals flexible in cold weather to maintain a watertight seal with the rub plate.

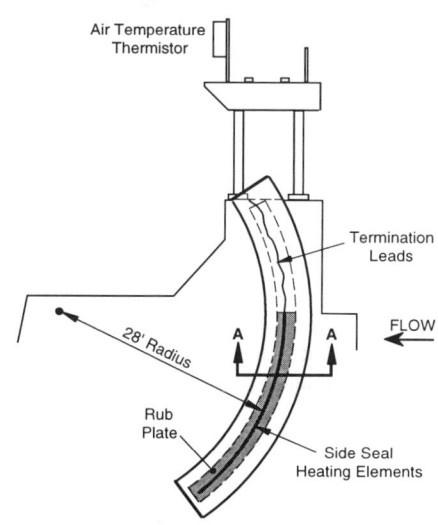

Figure 4. Side seal heater location for a tainter gate.

A primary requirement in the use of electrical heating elements is replaceability. All side seal heater designs should provide for easy replacement. In most cases the heating elements are placed in cavities filled with air, and these should be moisture-proof.

Figure 5. Heaters placed in a walkway ramp prevents ice buildup.

## OTHER ICE CONTROL TECHNIQUES

There are many opportunities for using heat to prevent icing around locks and dams. While this can be done as a retrofit in a rehabilitation, it is usually easier to incorporate heaters in new construction. A good example of using heat is shown in Figure 5. One-half of this sloped walkway ramp at Starved Rock Lock and Dam has electrical heaters buried in it, and one-half does not. Keeping half of this ramp ice-free in the winter provides considerable safety for personnel.

Hot air heaters, such as salamanders, are used at many projects because they provide a high quantity of concentrated heat, and they are portable. Electrical heaters have been considered for use in keeping floating guidewalls ice-free. Heaters have been placed in concrete mats around roller rail supports for miter gate strut arms at Marseilles Lock and Dam on the Illinois River.

## CONCLUSIONS

Several ice control methods have been presented in this report. Implementation of these methods and their economic feasibility is a function of the severity and frequency of the ice problem they address. Personnel at each lock and dam can determine if bubblers or some type of heater can be used to solve ice problems in critical areas. An advantage of using electrical heaters is that their operation can be automated. The prioritization of ice problems to be solved, consistent with funding constraints, has to be made for each lock and dam individually.

## REFERENCES

**US Army Corps of Engineers** (1990) Winter navigation on inland waterways. EM 1110-8-1 (FR), Office of the Chief of Engineers, Washington, D.C.

**Haynes, F.D., R. Haehnel, and L. Zabilansky** (1993) Icing problems at Corps Projects. Technical Report REMR-HY-10, US Army Engineer Waterways Experiment Station, Vicksburg, MS.

**Haynes, F.D., R. Haehnel, and L. Zabilansky** (1994) Panel heaters used to control ice growth caused by fluctuating water levels. The REMR Bulletin, vol. 11, no. 1, US Army Engineer Waterways Experiment Station, Vicksburg, MS.

**Haynes, F.D., R. Haehnel, and C.H. Clark** (1995) Use of radiant heaters to prevent icing. The REMR Bulletin, vol. 12, no. 2, US Army Engineer Waterways Experiment Station, Vicksburg, MS.

# ICE PROBLEMS ON HYDROPOWER PROJECT WŁOCŁAWEK (LOWER VISTULA)

WOJCIECH MAJEWSKI
Technical University of Gdańsk, Gdańsk, Poland

## ABSTRACT

Hydropower project Włocławek was designed as one of 8 run-off-river dams on the lower Vistula. It was commissioned in 1970. Unfortunately till today it remains as a single project, thus causing many problems during ice formation and transport. The paper includes the description of the project and results of some studies which were carried out during 25 years of its operation. In 1982 due to very unfavorable hydrological and meteorological conditions large ice jam formed on the reservoir resulting in severe flood along the upper part of the reservoir. Conclusions drawn from the experience of design, construction and exploitation of the project concerning ice management are presented.

## INTRODUCTION

Vistula is the largest Polish river of the length 1041 km. It flows from south to north across the whole country and discharges to the Baltic Sea. Due to the differences in time of river freeze-up and ice-run between southern and northern section of Vistula severe ice problems appeared in the past, especially along the lower Vistula. In the beginning of winter in December large amounts of frazil ice formed, thus leading to frazil-ice jams. Similar situation existed in spring during ice-run. Use of icebreakers was usually necessary to form channel in the northern section of the river to allow ice-run from the south, which starts usually 10 to 14 days earlier then in the north. Lower Vistula is 391 km long river section which was prone to the development of ice jams and winter floods.

In 1970 the hydraulic project Włocławek on the lower Vistula was completed and put into operation. It has been designed as the first hydraulic project of the Lower Vistula Cascade (LVC) consisting of 8 similar run-off-river reservoirs. The main aim of LVC was the production of electric energy improvement of navigation, and flood protection. Unfortunately the economic situation of the country did not allow to continue this program and Włocławek project remained till today as a single dam and reservoir on the lower Vistula with all consequences of this situation.

## WŁOCŁAWEK PROJECT

Włocławek project consists of earth dam (20 m high, 650 m long), 10 bay spillway section (20 m each) equipped with steel vertical gates, hydraulic powerplant of the capacity 162 MW (6 Kaplan units) operating mainly as peak power, and navigation lock (115 m long, 12 m wide). The project forms the run-off-river reservoir of the total volume 400 mln m$^3$ and the length about 50 km which extends between two cities Włocławek and Płock (Fig.1). The useful volume of the reservoir is 55 mln m$^3$ and the head at the dam is 11.3 m. Maximum depths are near the dam reach 15 m. The average discharge in dam cross-section is 930 m$^3$/s, and the maximum discharge observed during the operation of the project was 6900 m$^3$/s. Discharge through the powerplant at normal water elevation in the reservoir (57.30 m) is 2190 m$^3$/s (Majewski, 1996)

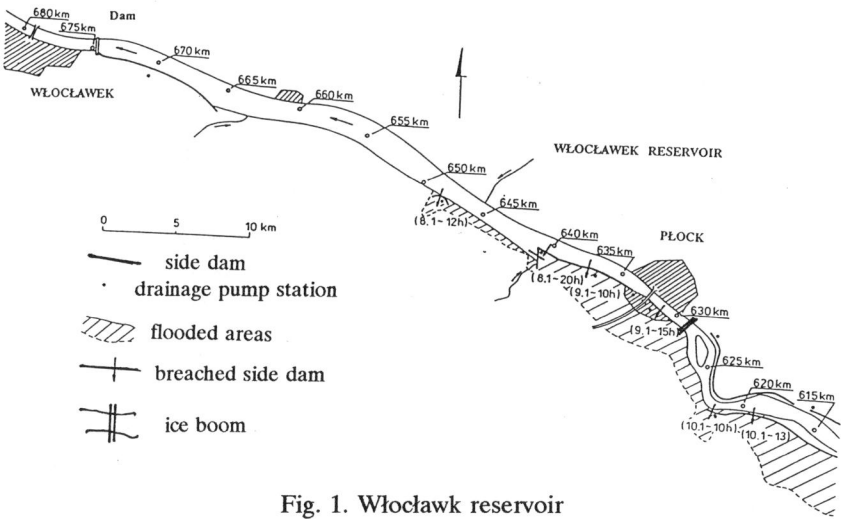

Fig. 1. Włocławk reservoir

The spillway section is equipped with steel vertical gates which can be lowered by 2.20 m in order to discharge water and ice floes. The total discharge with fully lowered gates is 1300 m$^3$/s. For higher discharge it is necessary to raise the gates to obtain total discharge of 7500 m$^3$/s (Fig.2). The formation of the reservoir resulted in increased sedimentation at the rate 1.5 to 1.8 mln m$^3$ per year. This includes bed load (30%) and suspended load (70%). Deposition of bed load in the upstream part of the reservoir formed shoals and islands which facilitated the formation of bottom ice, deposition of frazil ice and initiation of ice jams. The main problems which appeared on the reservoir and during dam operation were as follows (Majewski, 1984):
- formation of hanging dams due to frazil ice inflow to the reservoir already covered with solid ice causing increased water elevations in the upstream part and thus creating the danger of breaching of flood dikes,

- discharge of ice floes through the spillway,
- operation of spillway gates and navigation lock gates during cold weather,
- ice breaking on the reservoir and on the river downstream from the dam.

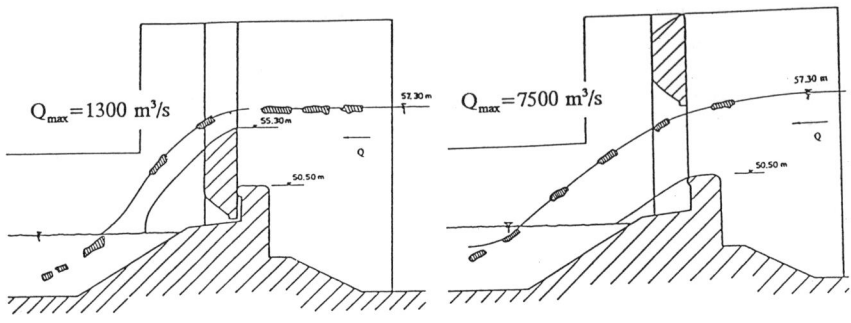

Fig. 2. Operation of the gates

## ICE CONDITIONS ON THE LOWER VISTULA

Lower Vistula has a very nonuniform character along its course. In some sections it has natural character, meandering and braided. In some parts the river was trained for navigation. In many areas the river is shallow, with shoals and sandy islands. This situation was favorable for the formation of hanging dams. Similar situation was during spring ice-run. In the past numerous ice jams formed, which were accompanied by severe floods inundating agricultural land and affecting the cities located along the river.

Construction of Włocławek dam resulted in important changes in hydraulic and ice regimes. The duration of various ice forms in this part of the river was on the average about 80 days per year. The total amount of days with ice phenomena did not change, however, the duration of particular ice forms changed considerably after dam construction. Time of various ice forms before and after dam construction is as follows:
- frazil ice movement 48/8 days,
- floating ice 9/6 days,
- solid ice cover 25/64 days.

Solid ice cover on the reservoir forms now much faster than on the natural river, and lasts much longer. Solid ice cover on the reservoir forms due to the extension of border ice, freezing of frazil ice, or juxtaposition of ice floes. During one winter one, two or even three ice periods may happen due to consecutive periods of cold and warm weather. Solid ice cover on the reservoir formed usually earlier than on the river upstream from the reservoir where large amounts of frazil ice were produced. Frazil ice discharged into the reservoir, moved under solid ice cover and formed hanging dams, which diminished flow cross-sections, increased flow resistance and raised water elevations, especially in the upstream part of reservoir in the region of Płock.

In general Włocławek project was not designed properly to withstand severe winter conditions and operation during low air temperatures. After severe experience, changes in the design and exploitation were introduced. Several extreme situations indicated that water elevations in Płock were up to 2 m higher with ice cover on the reservoir, than for free surface flow. It appeared also that gates on the spillway have no heating or deicing system, which presented severe problems during gates operation in winter conditions.

## OPERATION OF THE PROJECT IN WINTER CONDITIONS
The main problem of the operation of the reservoir in severe winter conditions was the possibility of gates maneuvering to control ice and water discharge. During severe winter 1982 when air temperatures dropped down at night time even to -20 °C the gates which had no heating were frozen to the pears and the sill of the weir. Moreover, the leaking water formed large hanging masses of ice. In such situation opening or closing of the gates was very difficult, time consuming, and sometimes impossible. Similar situation existed on the navigation lock.

The next problem was the passage of ice through the weir at low water discharge which resulted in low downstream water elevation. Large floes of ice dived from the top of spillway into stilling basin and damaged concrete bottom and floor blocks. Additional disadvantage was caused by the concentration of ice and water discharge downstream of the dam along comparatively short spillway section (250 m) in relation to the whole width of the reservoir (1500 m) downstream from the dam. The water discharge over lowered gates was limited to 1300 m$^3$/s. When the inflow to the reservoir exceeded this value it was necessary to fully open the gates. This procedure required first closing of discharge when gates were lifted and then water was discharged through the opening between spillway sill and the bottom of the gate. This operation lasted few hours. A very dangerous moment was when ice floes began to dive below the gate which caused the possibility to damage the sealing of the gate.

## WINTER FLOOD ON THE RESERVOIR IN 1982
This flood was the result of the coincidence of extreme hydrological and meteorological conditions which occurred in Poland. In December 1981 solid ice cover formed over the whole reservoir with discharge below 1000 m$^3$/s. At the end of December high air temperatures and rains caused rapid increase of discharge and ice break-up along the whole length of Vistula River. Ice floes were flowing into reservoir and were transferred in the downstream direction through the spillway. On January 6 there was a sudden drop of air temperature from +8 °C during daytime to -20 °C during night. This situation was accompanied by strong winds (up to 20 m/s) blowing in the upstream

direction. This way ice-run was stopped and very rough solid ice cover consisting of refrozen ice floes formed immediately over the whole reservoir. Vistula upstream from the reservoir remained ice-free and produced enormous amounts of frazil ice which was flowing into reservoir and deposited underneath already existing solid ice cover. Large hanging dams formed and blocked flow cross-sections. Maximum discharge which lasted only one day reached 3900 m$^3$/s. Two days later maximum water elevation in Płock was reached resulting in the overtopping of flood dikes which were breached in 5 places causing inundation of 10 000 ha of land and 2230 farms (Fig.1.). The changes of discharge and water elevations are shown in Fig.3. Water elevation with ice cover exceeded free surface elevation, for the same discharge, by nearly 2.5 m.

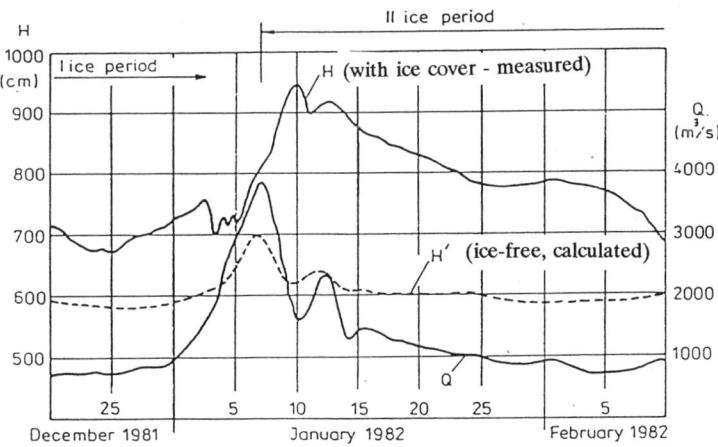

Fig. 3. Discharge and water elevations in Płock (winter 1982)

During this flood 100 mln m$^3$ of ice were deposited in the reservoir. One of the measures which was usually used on the reservoir, was the formation of ice free channel by icebreakers and discharge of floating ice downstream. Two difficulties were observed. First there were problems in crushing ice cover by icebreakers when large amounts of frazil ice were deposited under ice cover. Some help in the operation of icebreakers was the use of explosives which weakened the ice cover. The second difficulty was connected with discharging of crushed ice downstream. Close to the dam, because of large cross-sectional area, flow velocities are small, not exceeding 0.1 to 0.2 m/s. In case of the wind blowing in the upstream direction crushed ice was stopped and refroze thus forming again solid rough ice cover and the work of icebreakers was simply wasted.

## ENGINEERING MEASURES AFTER 1982 WINTER FLOOD

It has been assumed that the crest of the dike should be 1.0 m higher than water surface elevation during last flood. Dredging in the upstream part of the reservoir is indispensable to remove all obstacles which could stop floating ice. In order to improve the operation of weir gates during low air temperatures special aeration system in front of each gate was installed. Moving air along the upstream face of the gate prevent the formation of ice which adhered to the gate and the piers. The main problem of the 1982 winter flood was the inflow of large amounts of frazil ice into reservoir where solid ice cover already existed. Frazil ice was produced along the river section upstream from the reservoir. The best measure to decrease the amount of frazil ice was the formation of solid ice cover along river reach upstream from the reservoir. This was achieved by means of floating ice booms which stopped ice floes, ice pans and frazil slush moving downstream (Fig.1). This way solid ice cover formed very fast upstream from the reservoir, insulated the water from heat loss and thus prevented the production of frazil ice.

## CONCLUSIONS

Winter operation of reservoirs and hydraulic structures is much more difficult than in summer during free surface flow (Majewski, 1986). There is danger of the formation of ice jams of various kinds which may considerably increase water elevations in comparison with free surface flow. This situation may result in overtopping of flood dikes, breaching them and inundation of land and farms or cities. In order to avoid dangerous situations the following conditions must be fulfilled:
- operation of gates and hydraulic installations must be possible even during low air temperatures,
- upstream part of the reservoir should be clear of any obstacles which might stop floating ice and initiate ice jam,
- spillway should allow to discharge ice downstream without damaging hydraulic structure
- formation of ice-free channel by icebreakers along the reservoir is very often a good solution to lower water elevations in the upstream part of the reservoir,
- restriction of the production of frazil ice along the river upstream from the reservoir may be achieved by the installation of ice booms which speed up the formation of ice cover on the river.

## REFERENCES

Majewski, W.,(1984) Backwater Profiles on Hydroelectric Reservoir with Ice Cover, IAHR Ice Symposium 1984, Hamburg, Vol.1.

Majewski, W.,(1986) Ice Problems at Hydropower Installations and Hydraulic Structures, IAHR Symposium on Ice, Iowa City, Vol.3

Majewski, W., (1996) Ice Problems during 25 Years of Operation of Hydraulic Project Włocławek on Vistula River, 13th Ice Symposium, Beijing, Vol.1.

# Model Scale Measurements of Ice Loads on a Submerged Turret Loading (STL) Tanker

S. Løset and Ø. Kanestrøm

*Norwegian University of Science and Technology, NTNU, Norway*

**Abstract**

A recently developed Submerged Turret Loading concept (STL) for loading oil offshore is in use on several oil fields in the North Sea. In Arctic waters, the presence of floating ice provides several additional challenges mainly connected to loads from level ice, broken ice and pressure ridges. Model scale tests with the STL have been performed in the Hamburgische Schiffbau-Versuchsanstalt GmbH (HSVA) ice tank in Hamburg. The model ship had a conventional icebreaking bow and a length between perpendiculars of 6.535 m. The testing was performed at a scale of 1:36. A wedge in front of the chain table cleared ice from the STL-buoy area. The tidal current tests indicated that the STL-system has good weather vaning qualities in ice.

## 1. INTRODUCTION

Model tests and experiences from full scale operation of the STL/STP (Submerged Turret Loading/Submerged Turret Production, hereinafter abbreviated STL) concept in open water have proven the performance of this concept with regard to dynamic behaviour and weather vaning qualities (Fig. 1). In the Arctic, floating ice provides several additional challenges. A major concern is the weather vaning capability of the ship when there is a substantial shift in the ice drift direction. In principle, the STL can minimise the applied ice forces by yawing to a heading coincident with the advancing ice while breaking the ice in bending with a conventional icebreaker bow. The tidal current may cause such a shift in the ice drift.

A more extensive attempt to assess these issues is the present experiments in model scale. The operational performance and forces exerted on the ship and mooring system were explored when the system was towed through the anchoring system against level ice, broken ice and pressure ridges.

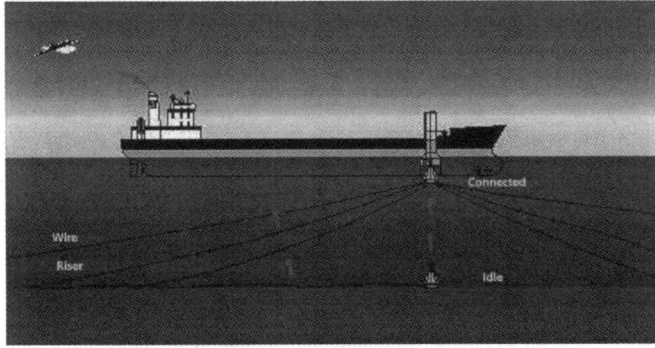

Fig. 1.  Sketch of the STL in use, open water.

## 2. EXPERIMENTAL PROCEDURES

The experiments were conducted in the large ice tank at HSVA. The tank is 78 m long, 10 m wide and 2.5 m deep. The basin is equipped with a motor driven towing carriage. A beam was mounted on the main carriage and extended about 3.5 m beyond the aft end of the main carriage. At the far end, the beam made a 90° bend downwards through the wake behind the model ship. The part of the beam that extended through the water was fixed to an underwater platform 0.83 m below the water surface, thus modelling a water depth of 30 m. Further, the underwater platform ('sea bed') was hooked on rails mounted on the tank wall about 0.5 m below the water surface (see Fig. 2a). In this way the underwater platform was pushed by the main carriage.

The model ship had a conventional icebreaking bow with a length between perpendiculars of 6.535 m ($Lpp$ = 6.535 m). This corresponds to 234 m in full scale (scale factor $\lambda = 36$) and approximately 120 kDWT. The breadth and draught of the model were $B = 1.079$ m and $T = 0.418$ m, respectively, corresponding to $B = 38.8$ m and $T = 15.0$ m in full scale. The model was placed towards the left tank wall side (Fig. 3). This was to ensure that the model could turn 180° to starboard without hitting the tank wall.

The model of the STL ship was connected to the underwater platform through 8 mooring lines that were fixed to the turret (see Fig. 2). The characteristics of the mooring system were symmetrically modelled (Løset et al., 1997a). The tension in each line was measured through a load cell that provided the load-time traces. A triaxial cell measured the forces in $x$, $y$ and $z$-directions at the turret. The data logging provided records from the whole run of each test as well as $x$ and $y$-positions of the ship (measured by potentiometers).

(a)    (b)

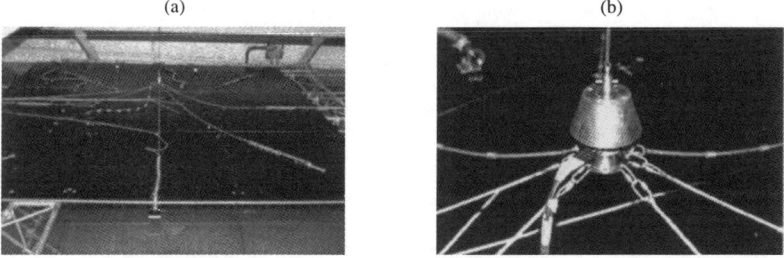

Fig. 2. Photos of (a) the mooring lines fixed to the underwater platform, and (b) buoy with the triaxial load cell.

(a)    (b)

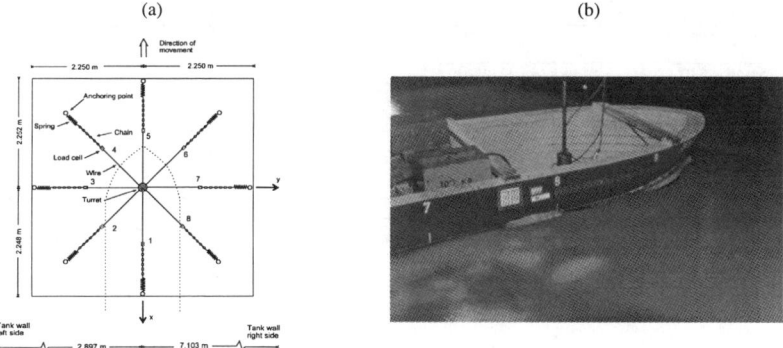

Fig. 3. (a) Plan view (seen from above) of the test set-up with major dimensions and co-ordinates indicated, and (b) photo of the ship during testing in level ice.

The physical model scale (*ms*) tests in ice are assumed to be scaled to full scale (*fs*) by the theory of geometric similarity (see e.g. Timco, 1984; Evers and Jochmann, 1993). This similarity is satisfied when all linear dimensions of the model is equal to a constant ($1/\lambda$) times the corresponding linear dimensions of the full scale:

$$L_{fs} = \lambda L_{ms} \qquad (1)$$

Froude scaling will be used for scaling the model results. The application of Froude scaling assumes that the level ice properties scale properly.

## 3. RESULTS

The model tests were carried out in two phases where the actual test runs of Phase I were conducted in July 1996 and the tests of Phase II were continued and completed in November 1996. The tests included level ice, broken ice and pressure ridges. A full scale ice thickness (*h*) of 1.2 m corresponds to a model ice thickness of 33 mm. The model ice was of fine-grained columnar type and grown from a sodium chloride solution (about 0.65 % concentration). The procedures and preparation of the HSVA model ice are thoroughly described by Evers and Jochmann (1993).

The forces on the STL tanker in three directions (*x*, *y* and *z*) were measured by the triaxial load cell located on the buoy. From the triaxial cell, the resultant force ($F_{tot}$) was calculated according to

$$F_{tot} = \sqrt{F_x^2 + F_y^2 + F_z^2} \qquad (2)$$

where $F_x$, $F_y$ and $F_z$ are the forces in the *x*, *y* and *z* -direction, respectively. Further, the run provided force-time traces in all mooring lines.

For each ice sheet, the effective run length was about 50 m. This test span was typically divided into three equal test sections, each of length 200 seconds. In general, less than 5 % of the run was used to accelerate the ship and advance the model into stable ice conditions. Further, some oscillations occurred when the run was stopped or the speed was shifted to another value. Hence the central 90 % of the time-series records were used to compute average forces and angles. The observations showed substantial variations of load with time, but a rather stable average value.

The ice generally broke about 6 times the ice thickness (0.24 m-0.26 m) out from the bow. The fluctuation of the total force is probably due to this sequential breaking and advancing of the model. Table 1 shows parts of the test matrix.

Table 1. Parts of the test matrix.

| Sheet | Test | Description | Model scale | | Full scale | |
|---|---|---|---|---|---|---|
| | | | v (m/s) | h (mm) | v (m/s) | h (m) |
| Ice Sheet 1, 16.07.96, $\sigma_{f,ms}$ = 35 kPa ($\sigma_{f,fs}$ = 1260 kPa) | | | | | | |
| 1 | 1010 | Breaking new level ice, forward[1] | 0.05 | 30 | 0.30 | 1.1 |
| 1 | 1020 | Breaking new level ice, forward | 0.075 | 30 | 0.45 | 1.1 |
| 1 | 1030 | Breaking new level ice, forward | 0.125 | 30 | 0.75 | 1.1 |
| Ice Sheet 2, 19.07.96, $\sigma_{f,ms}$ = 20 kPa ($\sigma_{f,fs}$ = 720 kPa) | | | | | | |
| 2 | 2010 | Breaking new level ice, forward | 0.05 | 42 | 0.30 | 1.5 |
| 2 | 2020 | Breaking new level ice, forward | 0.075 | 42 | 0.45 | 1.5 |
| 2 | 2030 | Breaking new level ice, forward | 0.125 | 42 | 0.75 | 1.5 |
| Ice Sheet 4, 12.11.96, $\sigma_{f,ms}$ = 35 kPa ($\sigma_{f,fs}$ = 1260 kPa) | | | | | | |
| 4 | 4010 | The tanker starts at an angle of 30° in the ice[2] | 0.05 | 45 | 0.30 | 1.6 |
| 4 | 4020 | Breaking new level ice, forward | 0.05 | 45 | 0.30 | 1.6 |
| 4 | 4030 | Breaking new level ice, forward | 0.075 | 44 | 0.45 | 1.6 |
| 4 | 4040 | Breaking new level ice, forward | 0.125 | 43 | 0.75 | 1.5 |

[1] Each run lasted about 200 sec. This means that test series 1010-1030, 2010-2030 and 4010-4040 used one towing tank length each.
[2] Constant speed kept until the tanker was 'in line'. Three equal time steps for Tests 4020-4040.

Table 2 shows the model scale results from Test series 1000, 2000 and 4000. In broken ice, the average total forces in full scale did not exceed 2500 kN and the single mooring line loads were not above 1200 kN. The forces measured in the broken ice and tidal current tests were less than 50 % of the level ice forces. The tidal current tests indicated that the STL-system has good weather vaning qualities in ice. In most cases, the wedge in front of the chain table effectively cleared ice from the STL-buoy area.

A frequency analysis was performed to find the characteristic load frequencies when breaking level ice. A fast Fourier transform (FFT) was employed to compute the spectra (Newland, 1984). The last column in Table 2 shows the results from these computations. The frequency marked (*f*, Hz) will typically be the ice breaking frequency because this will give the largest variation of the loads and be cyclic.

Table 2. Level ice forces (N), model scale (mean values with ± indicating the empirical standard deviation), and marked frequency peaks $f$ from the FFT analysis.

| Test | Line 1 | Line 2 | Line 3 | Line 4 | Line 5 | Line 6 | Line 7 | Line 8 | $F_{tot}$ | $f$ (Hz) |
|---|---|---|---|---|---|---|---|---|---|---|
| 1010 | 0.1 ± 0.1 | 1.0 ± 1.7 | 4.2 ± 4.7 | 1.8 ± 2.1 | 11.6 ± 5.6 | 13.5 ± 7.6 | 0.4 ± 0.3 | 2.7 ± 3.4 | 24.8 ± 12.1 | 0.36 |
| 1020 | 0.0 ± 0.0 | 3.5 ± 4.9 | 11.4 ± 9.3 | 6.7 ± 7.3 | 15.6 ± 5.1 | 9.8 ± 8.6 | 0.2 ± 0.2 | 1.5 ± 3.2 | 36.2 ± 14.3 | 0.26 |
| 1030 | 0.1 ± 0.1 | 2.5 ± 3.2 | 13.6 ± 7.6 | 7.7 ± 6.3 | 19.4 ± 7.0 | 12.9 ± 10.0 | 0.2 ± 0.0 | 0.5 ± 1.7 | 41.1 ± 13.9 | 0.22, 0.27 |
| 2010 | 0.0 ± 0.1 | 1.7 ± 3.9 | 23.4 ± 19.7 | 20.8 ± 19.3 | 46.0 ± 14.5 | 30.4 ± 19.0 | 2.7 ± 4.6 | 2.5 ± 6.3 | 105.8 ±27.4 | 0.19, 0.22 |
| 2020 | 0.0 ± 0.0 | 1.2 ± 2.9 | 32.9 ± 10.1 | 29.1 ± 16.8 | 49.7 ± 11.5 | 21.5 ± 14.0 | 0.0 ± 0.0 | 0.0 ± 0.0 | 104.7 ±30.9 | 0.26 |
| 2030 | 0.0 ± 0.0 | 1.4 ± 3.6 | 35.2 ± 12.9 | 38.2 ± 20.1 | 54.6 ± 12.4 | 27.6 ± 15.0 | 0.0 ± 0.0 | 0.0 ± 0.0 | 123.3 ±34.4 | 0.16-0.27 |
| 4010 | 0.0 ± 0.0 | 0.0 ± 0.0 | 0.0 ± 0.0 | 1.0 ± 3.7 | 82.2 ± 48.3 | 122.0±61.6 | 112.0±51.8 | 32.9 ± 18.9 | 240.0±152.1 | - |
| 4020 | 0.0 ± 0.0 | 0.3 ± 1.7 | 8.9 ± 16.0 | 2.2 ± 5.1 | 39.5 ± 22.5 | 44.8 ± 31.6 | 32.3 ± 37.1 | 6.8 ± 13.3 | 200.0 ±50.2 | 0.11, 0.14 |
| 4030 | 0.0 ± 0.0 | 1.5 ± 4.3 | 22.6 ± 25.2 | 8.1 ± 11.5 | 54.1 ± 17.8 | 44.5 ± 37.0 | 33.9 ± 37.0 | 5.5 ± 9.5 | 215.0 ±46.4 | 0.20, 0.24 |
| 4040 | 0.7 ± 5.2 | 0.1 ± 1.1 | 4.3 ± 9.2 | 0.9 ± 3.1 | 28.6 ± 20.1 | 32.9 ± 12.9 | 7.7 ± 14.6 | 0.7 ± 3.0 | 143.0 ±50.0 | 0.24-0.26 |

## 4. APPLICATION

The forces are scaled by $\lambda^3$ where $\lambda = 36$. Further, the difference in density between sea water ($\rho_{w,fs} = 1025$ kg/m$^3$) and the model tank water ($\rho_{w,ms} = 1005$ kg/m$^3$) affects the scaling. Hence the forces are scaled according to the following equation:

$$F_{fs} = \frac{\rho_{w,fs}}{\rho_{w,ms}} \lambda^3 \cdot F_{ms} \qquad (3)$$

The scaled results are used when evaluating the feasibility of the STL system. The results are shown in Figs. 4 and 5. Tests 1010-1030 are referring to an ice sheet thickness $h = 1.08$ m, a flexural strength $\sigma_f = 1260$ kPa and speeds of 0.3, 0.45 and 0.75 m/s, respectively.

For Tests 2010-2030, the mean ice sheet thickness in full scale was 1.51 m with a flexural strength $\sigma_f = 720$ kPa and speeds of 0.3, 0.45 and 0.75 m/s, respectively. The maximum average total force was 5868 kN (Test 2030) with an average line tension in the front line of 2600 kN. The corresponding maximum loads were 8929 kN and 3507 kN.

Fig. 4 shows almost a linear relationship between the total force and the ship speed. This corresponds well with common theory on ice resistance in level ice at low speeds (Løset, 1995).

Test 4010 is a 30° test and 4020-4040 are straight forward tests with respectively 0.3, 0.45 and 0.75 m/s speeds in full scale. The mean ice thickness was 1.59 m, and the flexural strength $\sigma_f = 1260$ kPa. Fig. 5a seems to show a reduction of the total force with increasing speed. However, for this particular ice sheet, the reduction of $F_{tot}$ is rather due to a slight spatial decrease of the ice thickness where the two higher speeds were run as shown in Table 1.

Tests 4020-4040 gave mean total forces from 7709 kN to 11421 kN and for the 30° test the extreme total force was 34125 kN. For Tests 4020-4040, the largest extreme force was 14729 kN. However, we should bear in mind that such severe ice conditions as in the 4000 test series are unlikely in the eastern Barents Sea (Løset et al., 1997b). For instance, in the spring when the thickest ice is encountered, the flexural strength will not be at its maximum (Shkhinek et al., 1997).

Fig. 5b shows a photo taken by the underwater video camera of the wedge clearing ice floes in front of the chain table and mooring lines. The photo does also show submerged ice floes along the hull.

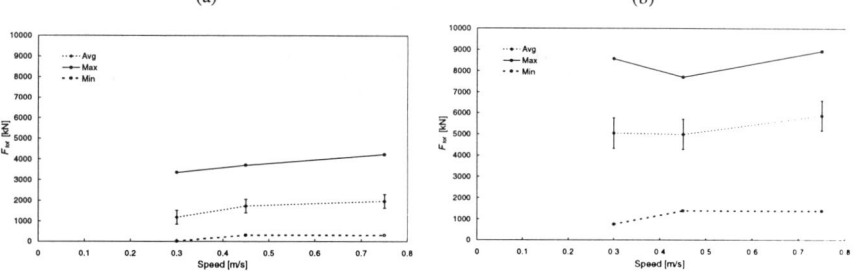

Fig. 4. $F_{tot}$ vs. speed in full scale for (a) Tests 1010-1030, and (b) Tests 2010-2030.

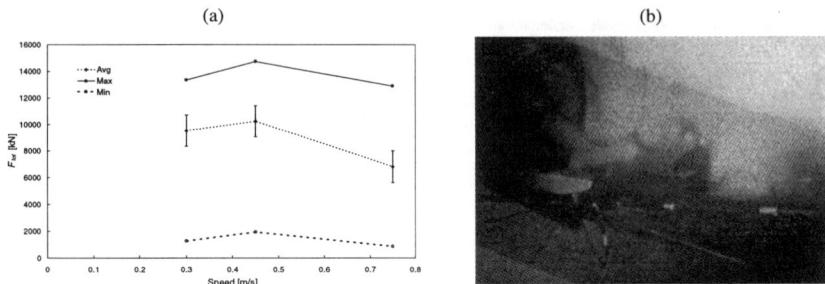

Fig. 5. $F_{tot}$ vs. speed at full scale for (a) Tests 4020-4040. (b) Photo of the wedge clearing ice floes in front of the chain table and mooring lines.

The dynamic response of the moored STL system is examined in the following. For shallow waters, the damping from the mooring lines is low (Faltinsen, 1990). Therefore, special attention should be made for the ice breaking frequency $(f_b)$:

$$f_b = \frac{V}{l} \qquad (4)$$

The damped natural period $T_d$ of a mechanical system can be expressed as follows

$$T_d = \frac{2\pi}{\sqrt{1-\xi^2}} \sqrt{\frac{m}{k}} \qquad (5)$$

where $m$ is mass, $k$ stiffness, and $\xi$ is given by

$$\xi = \frac{c}{c_{cr}} \qquad (6)$$

Here, $c$ is the damping and $c_{cr}$ the critical damping expressed as (Bergan et al., 1986):

$$c_{cr} = 2m\frac{2\pi}{T} \qquad (7)$$

where $T$ is the natural period. For a water depth of 40 m, $\xi$ will be in the range of 0.02-0.03. Hence the damped natural period reads

$$T_d = 0.155\sqrt{m} \qquad (8)$$

where $m$ is in tons, $T_d$ in seconds. Here we have used a total stiffness of the mooring system $k = 1686$ kN/m (Løset et al., 1997a). From the model scale frequencies reported in Table 2, the model scale periods scaled by $\lambda^{1/2}$, we find that the ice breaking periods range from 16.9 to 54.5 seconds, full scale. Thus, the ice breaking period may coincide with the natural period in surge for ships with masses (including added mass) in the range of

$$11\,888 \text{ tons} < m < 123\,632 \text{ tons}$$

An ordinary tanker falls into this range. The size of the ice breaking force component should also be considered. Heavier ice forces will obviously create larger amplitudes and loads on the system than small ice forces. In the cases with predominant (characteristic) periods less than 20-25 seconds, the dynamic ice breaking force will be taken by the inertia of the ship. Only the mean force variation caused by yaw motion or spatial changes in the ice parameters will be transferred directly to the mooring system (Sandvik, 1990).

## 5. CONCLUSIONS

A number of small scale tests with the STL have been performed in the HSVA ice tank in Hamburg. The model ship had a conventional icebreaking bow and a length between perpendiculars of 6.535 m. The testing was performed at a scale of 1:36. The purpose of the tests was to study the feasibility of the STL concept in level ice, broken ice, and pressure ridges. The most important results are as follows (all forces refers to full scale):

- In level ice, 1.5 m full scale ice thickness and 720 kPa flexural ice strength, the maximum average total force was 5868 kN with an average line tension in the front line of 2600 kN. The corresponding maximum loads were 8929 kN and 3507 kN. The measured forces (scaled) are therefore within the capacity of the present STL mooring system.
- Tests 4020-4040 gave mean total forces from 7709 kN to 11421 kN and the largest extreme force was 14729 kN. For the eastern Barents Sea, we should bear in mind that such severe ice conditions as in the 4000 test series are unlikely.
- In broken ice, the average total forces did not exceed 2500 kN and the single mooring line loads were not above 1200 kN.
- The wedge in front of the chain table cleared ice from the STL-buoy area.
- The tidal current tests indicated that the STL-system has good weather vaning qualities in ice.

*Acknowledgement.* The authors would like to thank APL for their support to the project. We also wish to thank staff at HSVA for their professional execution of the test programme. Finally, we would like to thank the Commission of the European Communities (DG XII, Science, Research and Development) for giving us the opportunity to study the STL concept in ice.

## REFERENCES

Bergan, P.G., P.K. Larsen and E. Mollestad (1986): Svingning av konstruksjoner. The Norwegian Institute of Technology, Department of Structural Engineering, Tapir, 278 p. (in Norwegian).

Evers, K.U. and P. Jochmann (1993): An Advanced Technique to Improve the Mechanical Properties of Model Ice Developed at the HSVA Ice Tank. The 12th International Conference on Port and Ocean Engineering under Arctic Conditions (POAC), Hamburg, 17.- 20. August 1993, Volume II, pp. 877-888.

Faltinsen, O.M. (1990): Sea Loads on Ships and Offshore Structures. Cambridge University Press, 328 p.

Løset, S. (1995): Ice Loads on the STL Tanker. The Norwegian Institute of Technology, Department of Structural Engineering, Report No. R-9-95, Trondheim, 26 p.

Løset, S., Ø. Kanestrøm, T. Pytte, K.U. Evers, P. Jochmann and P.C. Sandvik (1997a): Model Tests in Ice of a Submerged Turret Loading (STL) Concept. Proceedings of the 16th International Conference on Offshore Mechanics and Arctic Engineering, Yokohama, 13-18 April 1997 (in press).

Løset, S., K. Shkhinek, P. Strass, O.T. Gudmestad, E.B. Michalenko and T. Kärnä (1997b): Ice Conditions in the Barents and Kara Seas. Proceedings of the 16th International Conference on Offshore Mechanics and Arctic Engineering, Yokohama, 13-18 April 1997, (in press).

Newland, D.E. (1984): An Introduction to Random Vibrations and Spectral Analysis. Sec. Edition, Longman Scientific & Technical, England, 377 p.

Sandvik, P.C. (1990) Feasibility of Floating Production Concept for Arctic Conditions. ESARC Report No. 19. Task 89/02, SINTEF Report No. SF 604040.06, 116 p.

Shkhinek, K., T. Kärnä, O.T. Gudmestad, S. Løset, A. Bolshev, S. Mischenko, E. Chasovskih, E. Lehmus and P. Strass (1997): Potential Structures for the Russian Arctic Offshore. Proceedings of the 16th International Conference on Offshore Mechanics and Arctic Engineering, Yokohama, 13-18 April 1997 (in press).

Timco, G. W. (1984): EG/ADS: A New Type of Model Ice for Refrigerated Towing Tanks. Cold Regions Science and Technology, Vol. 12, pp. 175-195.

# Solution of the Saint Venant Equations through the use of Riemann based methods

A J CROSSLEY and N G WRIGHT
Dept. of Civil Engineering, University of Nottingham,
Nottingham, UK

## ABSTRACT

A study has been made on the use of Riemann based methods to solve the Saint Venant equations of open channel flow. Different Riemann solvers have been applied to the dam-break problem including explicit and implicit formulations. It was observed that for this particular test case, the method that best dealt with increasing reservoir/tailwater depth ratio's (leading to higher Froude numbers) was that due to Roe. For smaller depth ratio's the methods tested performed equally well. Comparisons between explicit and implicit versions of Roe's scheme showed that whilst higher CFL numbers were obtainable using the implicit method, there was a cost of having to use a more complex numerical algorithm and smearing of the solution occurred at steep gradients.

## INTRODUCTION

Computational Fluid Dynamics or CFD is an invaluable tool for hydraulic engineers. Its use enables detailed predictions to be made about the flow for a particular channel under certain conditions, without the need to build a test rig or take expensive field measurements. However a number of problems are encountered when applying numerical techniques to this area. Firstly natural channels or rivers tend to have highly irregular geometries, leading to a complex mathematical model and difficulties in producing a suitable computational grid. Mixed regions of flow can also cause difficulties for some numerical methods e.g. at hydraulic jumps. Many numerical methods are restricted by a stability condition for the maximum time step. This often results in the need to use small time steps which can be very computationally expensive.

The commercial software currently available for performing 1-d simulations is quite advanced and satisfactory in many cases where there is little cross sectional or vertical variation within the flow. However if this is not the case then the software can become unstable around critical flow areas. This is usually dealt with by introducing false 'numerical' diffusion that smear out transitions.

The purpose of the study is to investigate efficient computational methods for solving the Saint Venant equations, with a view to developing a model that will be able to predict the flow conditions for any channel, without the need for extensive computing capabilities. To this end a class of methods known as Riemann solvers are being considered for open channel flow calculations. Riemann solvers were originally developed for solving compressible gas dynamics problems, and have a number of desirable properties which make them suitable for the application under consideration. In particular the techniques being studied fall in to the category of Total Variational Diminishing (TVD) schemes meaning that non-physical quantities such as negative depths will not be produced by the calculations. The construction of Riemann based methods is via a Finite Volume discretisation. The Finite Volume method is proving to be increasingly popular within the CFD community, and has several advantages over the more traditional Finite Difference methods such as the Lax-Wendroff, MacCormack or Preissmann schemes typically used in hydraulic simulations. Whereas Finite Difference methods are based upon approximating the differential forms of equations, Finite Volume techniques discretise the integral form. This ensures that the numerical approximation is conservative and that the method is valid across any discontinuities in the flow variables. In addition methods based on Finite Volumes are more applicable to channels with irregular geometries.

## MATHEMATICAL MODEL

The full Saint Venant equations form a two-dimensional system of partial differential equations with source terms for the bed friction and bed slope. The equations represent mass and momentum conservation along the channel and may be derived from the Navier-Stokes equations with the principle assumptions being that the fluid is inviscid, the pressure variation with depth is hydrostatic, and that the vertical component of the acceleration is negligible. For flow in a prismatic rectangular channel the 1-d form of the equations (see [1], [2] or [3] for details) may be written as

$$\frac{\partial A}{\partial t} + \frac{\partial Q}{\partial x} = 0$$

$$\frac{\partial Q}{\partial t} + \frac{\partial}{\partial x}\left(\frac{Q^2}{A} + \frac{gA^2}{2b}\right) = gA(S_o - S_f)$$

(1)

where $A$ is the cross sectional area, $Q$ is the discharge, $b$ is the width of the channel, $S_o$ is the bed slope and $S_f$ represents the friction slope which can be defined by either the Manning or Chezy formulae. There are other representations of the Saint Venant or Shallow Water equations. This particular formulation was chosen as the equations are written in conservative form and can be classified as conservation laws for which Riemann based methods were designed.

## NUMERICAL METHOD

In a mathematical context the equations can be thought of as a system of non-linear first order hyperbolic conservation laws with source terms. Considerable effort has been put into solving such equations numerically, especially within the field of aerodynamics, and it is from this field that most of the methods used in open channel flow originate. In vector form (1) can be expressed as

$$\mathbf{U}_t + \mathbf{F}_x = \mathbf{R} \qquad (2)$$

where in this instance

$$\mathbf{U} = \begin{pmatrix} A \\ Q \end{pmatrix}, \quad \mathbf{F} = \begin{pmatrix} Q \\ \dfrac{Q^2}{A} + \dfrac{gA^2}{2b} \end{pmatrix}, \quad \mathbf{R} = \begin{pmatrix} 0 \\ gA(S_o - S_f) \end{pmatrix}.$$

Generally a system such as (2) is solved numerically by discretising it into the form

$$\frac{\mathbf{U}_i^{n+1} - \mathbf{U}_i^n}{\Delta t} + \frac{1}{\Delta x}\left[\theta\delta^-(\mathbf{F}_{i+1/2}^{*n+1}) + (1-\theta)\delta^-(\mathbf{F}_{i-1/2}^{*n})\right] = \theta\mathbf{R}_i^{n+1} + (1-\theta)\mathbf{R}_i^n$$

where $\delta^-(\mathbf{F}_{i+1/2}^*) = \mathbf{F}_{i+1/2}^* - \mathbf{F}_{i-1/2}^*$ and $\mathbf{F}_{i+1/2}^*$ is an approximation to the flux function $\mathbf{F}$. The superscript and subscript denote the time-level and the position of the approximation respectively. The implicitness parameter, $\theta$, is chosen within the range $0 \leq \theta \leq 1$, and is 0 for an explicit scheme and 1 for a fully implicit method. Different numerical methods are represented by different choices for the approximate flux function. It is assumed that within each computational cell the values of $\mathbf{U}$ represent a cell average. The problem is to determine an expression for the flux values to enable the update to be performed. This can be done by solving a series of Riemann problems.

### The Riemann problem

A Riemann problem is defined as an initial value problem (IVP) of the form

$$u_t + F_x = 0$$

with the initial conditions

$$u(x,0) = \begin{cases} u_l & x < x' \\ u_r & x > x' \end{cases}$$

where the initial values maybe discontinuous across the point $x'$, which is normally taken to be the origin, and $u$ and $F$ may either be scalars or vectors. In terms of the numerical method a Riemann problem is defined at every interface between adjoining cells, using the cell values $\mathbf{U}$ to define the initial conditions. If $\mathbf{U}_l$ and $\mathbf{U}_r$ are set as the cell values then a first order method is produced. If a more complex formula is used, for example by including flux limiters [8] then a second order discretisation can be obtained. The solution to the Riemann problem is then used to calculate the flux values across the interface.

## RESULTS

A popular test case for checking open channel flow models is the dam-break problem, which is simple to code and tests the models ability to cope with flows where discontinuities are present. The problem entails solving the homogeneous Saint Venant equations with the initial conditions

$$h(x,0) = \begin{cases} h_l & \text{if } x < 0 \\ h_r & \text{if } x > 0 \end{cases}$$

$$Q(x,0) = 0$$

where $h$ is the depth of the flow given by $h = A/b$ for a rectangular channel. An analytical solution can be found for this problem enabling evaluations and comparisons to be made of different numerical methods. This test case has been used to evaluate a number of methods, as detailed below.

An implicit scheme based on the Roe Riemann solver [4] was tested on the dam-break problem. By setting the parameter $\theta$ to be either 0 or 1, the method could be made to be either explicit or implicit. The advantage of using an implicit method over an explicit scheme was demonstrated through the methods ability to generate stable solutions with CFL numbers greater than one. The disadvantages of implicit methods over explicit methods is that they require more computational effort and tend to give more diffuse (smeared) solutions near discontinuities. This was seen to be the case. Some sample results are shown in figures 1 and 2 where the Superbee [8] flux limiter was used to obtain second order spatial accuracy.

Another Riemann solver known as the Toro TR solver [9] was applied to the same problem, this time using an explicit formulation. In cases where the ratio $h_l : h_r$ was less that 5 : 1 this method gave very similar results to the explicit version of Roe's method. However for higher ratios of $h_l : h_r$ the method failed and unstable solutions were obtained. This corresponded to supercritical regions developing within the flow.

In order to verify the observations a comparison was made with software developed by Dr P A Sleigh at the University of Leeds. The package included explicit formulations of both Toro's TR and Roe's Riemann solver in addition to Toro's exact method [9]. The results were comparable to those seen previously, and there was little difference between the solutions produced using the exact and approximate Riemann solvers. Both of Toro's methods appeared to break down when supercritical regions were present.

## CONCLUSIONS

In the studies performed so far three different Riemann solvers have been compared for the dam-break problem. Whilst comparable results were obtained for sub critical conditions, only the Roe Riemann solver was seen to produce satisfactory solutions when super critical regions were present in the flow. Other Riemann solvers have been developed (for example those Osher and Solomon [6] and Harten, Lax and van Leer [5]) and their relative merits have yet to be established here. A method based on an implicit version of Roe's scheme was seen to produce stable solutions at CFL numbers greater than the stability limit of the corresponding explicit scheme. However this was at the cost of an increase in complexity of the algorithm and the smearing of any sharp gradients present.

These studies indicate that use of a Riemann based approach is feasible. However, careful consideration must be given to the implementation of the solvers and boundary conditions. Work is currently underway to adapt the work to arbitrary geometries, which is essential if the techniques are to be used in practical situations. Further the model is being modified to include adaptivity. Eventually the aim of the project is to combine the 1-d and 2-d models in some fashion so that an effectively 2-d model can be developed. The intention is that the algorithm will switch between the two models depending upon the flow conditions.

## REFERENCES

[1] Abbott M : Computational Hydraulics, Pitman Publishing Ltd, 1979.

[2] Chow V : Open Channel Hydraulics, McGraw Hill, 1959.

[3] Cunge J, Holly F and Verwey A : Practical aspects of computational river hydraulics, Pitman Publishing Ltd, 1980.

[4] Garcia-Navarro P, Priestley A and Alcrudo F : An implicit method for water flow modeling and channels and pipes, J. Hydraul. Res., **32**, 721–742, 1994.

[5] Harten A, Lax P and van Leer B : On upstream differencing and Godunov type schemes for hyperbolic conservation laws, SIAM Review, **25**, 35–61, 1983.

[6] Osher S and Solomon F : Upwind difference schemes for hyperbolic systems of conservation laws, Mathematics of Computation, **38**, 339–374, 1982.

[7] Roe P : Approximate Riemann solvers, parameter vectors and difference schemes, J. Comput. Phys., **43**, 357–372, 1981.

[8] Sweby P, High resolution schemes using flux limiters for hyperbolic conservation laws. SIAM J. numer. Analysis, **21**, 9995–1011, 1984.

[9] Toro E : Riemann problems and the WAF method for solving the two-dimensional shallow water equations, Phil. Trans. R. Soc. Lond. A, **338**, 43–68, 1992.

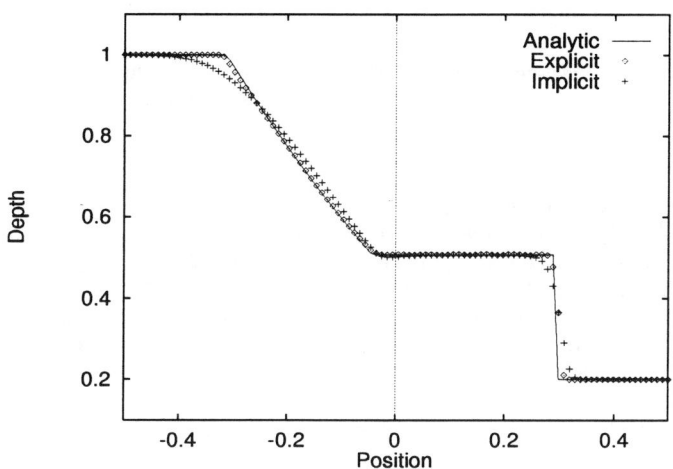

Figure 1: Depth profile $h_l : h_r = 5 : 1$

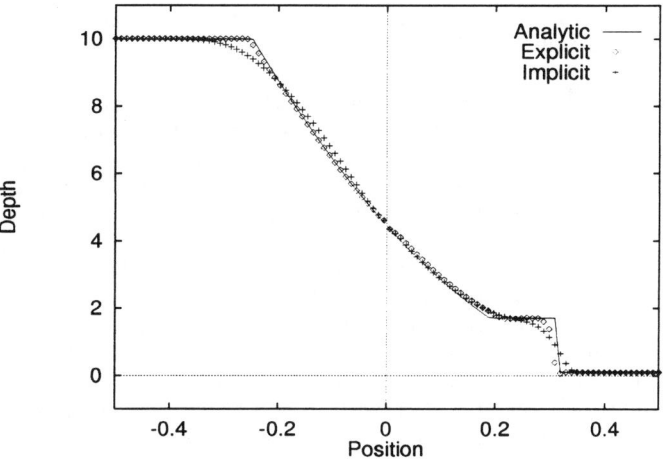

Figure 2: Depth profile $h_l : h_r = 100 : 1$

# Time Accurate Simulation of Free Surface Flow Problems

Weixing Yuan, Norbert Riedel, Rudolf Schilling
University of Technology, Munich, Germany

## ABSTRACT

This paper deals with the treatment of time accurate calculations of inviscid incompressible flows with free surface by using dual time stepping based on an artificial compressibility formulation. The flow field and the free surface location are calculated by coupling the free surface kinematic and dynamic equations with the equations of motion for the flow under free surface. Second-order time-differencing scheme and the cell-centered finite volume scheme based on flux difference splitting are used for these calculations.

KEY WORDS : free surface, unsteady flow, water turbine

## INTRODUCTION

Currently the water level just behind a draft tube exit of water turbine is used for water turbine efficiency calculation if no static pressure is measured near the draft tube exit. Measurements of T. Toyokura [1] confirmed that the water level changes in open channel near draft tube outlet and it is not equivalent to the static pressure at draft tube outlet. The investigation of Ch. Schneider [2] shows that there is an interaction between draft tube and tailwater flow. It becomes particularly a more important problem considering low head water turbines. This problem is being solved by a numerical simulation technique. This paper describes a method to compute the inviscid incompressible flows with free surface in a time accuracy manner.

A significant difficulty for incompressible flow calculation occurs since the continuity equation is given not in a time evolution form, but in the form of a divergence-free constraint. One possibility to couple changes of the velocity field with the pressure field while satisfying the continuity equation is the artificial compressibility method introduced by Chorin [3]. The basic idea is to add a pseudotime derivative of pressure to the continuity equation, which directly couples the pressure and velocity field.

Another difficulty is to represent and follow the free surface. The VOF method [4] has been used whenever the grid is stationary. However both applying the free surface boundary conditions and tracking the free surface in time are difficult, because the grid lines do not represent the free surface and they will be present between many grid lines [5]. Another efficient alternative, which uses a water level height function, has been adopted in this paper because of its accuracy in representing the free surface. The free surface elevation is determined by the free surface velocity.

To solve free surface problem time-accurately a dual time stepping based on the artificial compressibility method is used in this paper. To simplify the description, the 2-D Euler equations are discussed here to obtain a method, which allows for straight forward extension to solve a 3-D problem.

## GOVERNING EQUATIONS

For a inviscid incompressible fluid flowing under the influence of gravity, the continuity and Euler equations can be expressed as

$$\nabla \cdot \mathbf{v} = 0,  \tag{1}$$

$$\frac{D V}{Dt} + \nabla p = 0, \tag{2}$$

where v is the velocity component. In this expression, the density and the hydrostatic component have been absorbed in the pressure term p. All components are nondimensionalized by reference length $L_{ref}$ and reference speed $U_{ref}$.
According to Chorin [3], after the transformation from the physical coordinate system to the computational coordinate system defined by $\xi = \xi(x,z;t)$, $\eta = \eta(x,z;t)$, $\tau = t$, the equations of motion for the two-dimensional incompressible flow can be written in a unified artificial compressibility formulation

$$\mathbf{I}_{m1} \frac{\partial Q}{\partial t^*} + \mathbf{I}_{m2} \frac{\partial Q}{\partial \tau} + \frac{\partial F}{\partial \xi} + \frac{\partial G}{\partial \eta} = 0, \tag{3}$$

where $\mathbf{I}_{m1}$ andd $\mathbf{I}_{m2}$ are modified identity matrices and t* the pseudotime. The vectors are

$Q = J^{-1} \{p, u, w\}^T$,
$F = J^{-1} \{\beta^2(U-\xi_t), uU + \xi_x p, wU + \xi_z p\}^T$,
$G = J^{-1} \{\beta^2(V-\eta_t), uW + \eta_x p, wW + \eta_z p\}^T$,

where $U = \xi_t + \xi_x u + \xi_z w$ and $W = \eta_t + \eta_x u + \eta_z w$. J is the Jacobian determinant of the transformation. Note that the terms $\xi_t$ and $\eta_t$ which come out from the velocity of grid movement are not equal to zero for a dynamic grid system. In the formulation ß is referred to as an artificial speed of sound. It may also be interpreted as a relaxation parameter for the pressure calculation. By setting different modified identity matrices the three typical artificial compressibility methods can be recovered :

Case 1: $\mathbf{I}_{m1}$ = diag.[0,0,0], $\mathbf{I}_{m2}$ = diag.[1,1,1];
Case 2: $\mathbf{I}_{m1}$ = diag.[1,0,0], $\mathbf{I}_{m2}$ = diag.[0,1,1];
Case 3: $\mathbf{I}_{m1}$ = diag.[1,1,1], $\mathbf{I}_{m2}$ = diag.[0,1,1].

Case 1 is the original Chorin form in generalized coordinates. It may be used for unsteady flows if the parameter ß is set to be very large so that the equations become "almost" accurate. However, the value for ß must be chosen with care when using the upwind-differencing. Because of this constraint, we consider alternative iteration procedures. By introducing upwind-differencing the pseudotime step $\Delta t^*$ can be chosen very large for case 2 introduced by S. E. Rogers [6] and for case 3 by C. L. Merkle [7]. The two approaches ensure that the physical time step will be limited by the physical constraints of the problem and not by the numerical stability. In case 3 the physical time term can be understood as the source term in the modified residual. So the original unsteady problem is converted into a number of steady state calculations in pseudotime t* and it allows to apply some well-developed hyperbolic algorithms.

The set of equations are solved taking into account the dynamic and kinematic conditions of the free surface. The first states that the static pressure on a free surface is constant; i.e., $p = p_b + z/Fr$. The Froude number is defined as $Fr=U^2_{ref}/gL_{ref}$. The second states that the free surface is a material surface; i.e., once a fluid particle is on the free surface, it remains on the surface forever: $w = dh/dt = h_t + uh_x$. Where h(x,t) is the free surface location and h is single valued. It does not allow the breaking waves.

## NUMERICAL SOLUTION

The free surface and the bulk flow solutions are coupled by first computing the bulk flow at each pseudotime step, and then using the bulk flow free surface velocities to calculate the movement of the free surface. After the free surface elevation is updated, its new values are used as a boundary condition for the pressure on the bulk flow for the next subiteration. Several subiterations (pseudotime stepping) are required at each time step. For rapid convergence, both the global physical time stepping and the local pseudotime stepping are used.

## BULK FLOW SOLUTION

Equation (3) can be in a full implicit form with second-order in time

$$\frac{1}{J}|^{n+1}[\frac{I_{m1}}{\Delta t^*}+1.5\frac{I_{m2}}{\Delta \tau}]\Delta q^{n+1,m+1}+\delta_\xi(A\,\Delta q^{n+1,m+1})+\delta_\eta(B\,\Delta q^{n+1,m+1})=$$

$$=-\frac{I_{m2}}{\Delta \tau}(1.5\,Q^{n+1,m}-2\,Q^n+0.5\,Q^{n-1})-[\delta_\xi F^{n+1,m}+\delta_\xi G^{n+1,m}], \qquad (4)$$

where n means the index of time step and m the index of subiteration. A and B are the jacobian matrices of the flux F and G, and $q=\{p,u,w\}^T$, $q^{n+1,m+1}=q^{n+1,m}+\Delta q^{n+1,m+1}$. When $\Delta q^{n+1,m+1}$ goes to zero, the right-hand-side of the equation (4) goes to zero, and $q^{n+1,m+1}$ to $q^{n+1}$, resulting in the solutions of the (n+1)-th time step. A cell-centered finite volume scheme [8] based on the flux-difference split form of upwind-differencing is applied in this paper. However, the similarity transformation of the Jacobian matrix of [9] is used for matrix A with the average of $q_R$ and $q_L$ at cell face, which are calculated using variable extrapolation (MUSCL) based on a second-order upwind scheme with a Minmod limiter. Note that the jacobian matrix A includes the artificial compressibility parameter ß, which affects not only the continuity equation, but also the momentum equations. This indicates that the extremely large values of ß may cause large errors in the differencing of the momentum equations.

## FREE SURFACE SOLUTION

The kinematic condition on the free surface is used to calculate the movement of grids on free surfaces. Due to $h_\tau = h_t + x_\tau h_x$, one obtains $h_\tau = w - u h_x + x_\tau h_x$ in the computational domain. The movement of the free surface is calculated implicitly. The time term $h_\tau$ can be approximated using a second-order backward formulation. So the height of the free surface can be calculated as $h^{n+1,m+1}=(h_\tau^{n+1,m+1}\Delta\tau+2h^n-0.5h^{n-1})/1.5$. Throughout the interior of the free surface the derivative $h_x$ is computed using a second-order centered difference stencil. A dissipation term $0.5|u-x_\tau|\Delta x \delta_x^2 h$ is added to prevent odd-even decoupling at adjacent nodes which may lead to unphysical oscillatory solutions, where $\delta_x^2$ represents the central second difference operator.

## APPLICATION OF BOUNDARY CONDITIONS

The boundary conditions used in this paper is based on characteristics method. The number of positive or negative eigenvalues of the matrix A determines the number of waves propagating information from the interior of the computational domain to the boundary. In this study an inflow total pressure or a uniform velocity profile is assumed at the inflow and the static pressure is specified at the outflow, while other components are extrapolated from the computational domain to the boundary.

The wall boundary $\zeta=\zeta_{wb}$ and the free surface $\zeta=\eta_{fs}$ are material surfaces, so we obtain $D\zeta/Dt=0$; i.e. $\zeta_t + u\zeta_x + w\zeta_y = 0$. This formulation can be used for calculating the flux terms. Because of $\zeta_t = -\zeta_x x_\tau - \zeta_y y_\tau$, one can use $(u-x_\tau)\zeta_x + (w-y_\tau)\zeta_y = 0$ to correct the velocity components for calculating the matrices A and B.

## RESULTS

The calculations are performed in a flexible coordinate system and the grids are algebraically generated. The grids near the free surface and/or near the moving plate move in accordance with the movement of free surface and/or moving plate, while others remain stationary. The computation was initiated with a horizontal water surface. In the following tests the ß is set to be 1 or 2 and the CFL number for local pseudotime step is set to be $10^3$. 9 subiterations are applied for the last three tests.

## ONE-DIMENSIONAL VALIDATION

To verify the time-accuracy of the above mentioned three approaches of artificial compressibility formulation, the flow through an inviscid 1-D channel with an oscillating back pressure, described in [6,7], has been calculated. The numerical treatment is similar to that of [6].

Fig.1 shows the velocities of analytical solution and the computed results for case 1 and case 3 considering a frequency $\omega = 10$. Because of the constraint of the value for ß, the case 1 does not provide accurate results, even if 300 subiterations were applied. About 10 to 30 subiterations were required for case 3 to reach convergence; i.e., RHS of equation 4 being less than $10^{-5}$. Further tests have shown that the case 2 and case 3 of the formulation have the same time-accuracy property. However, the convergence of case 3 is slightly faster than that of case 2. Therfore, case 3 of the formulation will be selected to compute the following free surface problems.

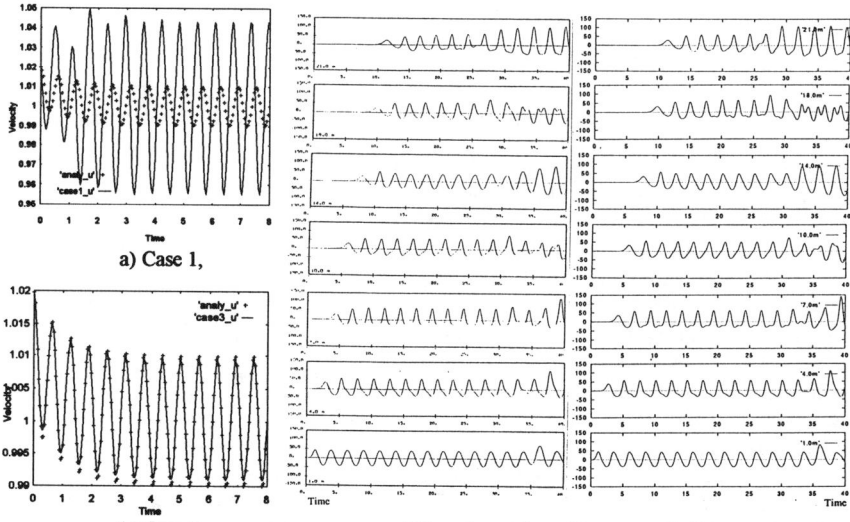

a) Case 1,

b) Case 3.

Fig.1 Solutions for 1-D channel.

a) Experimental profiles,   b) Computed profiles.

Fig.2 Wave profiles of the 2-D wave channel.

## TWO-DIMENSIONAL WAVE CHANNEL

As an initial test case for the unsteady flow with free surface a two-dimensional wave channel problem is calculated, which has been studied experimentally by Chapalain [10] in a 35.54 m long and 40 cm deep tank with a piston-type wavemaker. The amplitude of the motion of the wavemaker is 7.8 cm and the period is 2.5 s.

This problem has been successfully solved by using dual time stepping. Fig. 2 shows the comparison of computed and measured wave profiles at seven different x locations as a function of time. Note that, no damping term has been used in the calculation, which means that the wave tank ends at a perfectly reflecting wall. 355× 10 cells are generated in the whole region. The calculation has been performed for 16 periods with 72 physical time steps per period and took about 90 minutes CPU time on a hp700 workstation, while 16000 time steps were required in work [5] and each time step took 33 seconds on a Cray X-MP by using case 1 for the formulation.

## BODY/FLUID INTERACTION

When a vertical plate is uniformly accelerated from rest into initially stationary fluid of finite depth with a free surface, a thin jet occurs and rises rapidly up the plate. As in [11] the acceleration of moving plate was set to 2g in this paper and the time step $\Delta\tau$ is set to 0.01. The simulation was run for 30 time steps. Fig.3 shows the free surface elevation at three different times. Fig.4 shows the free surface elevation at the plate versus time. The numerical solution coincides very well with the analytical solution.

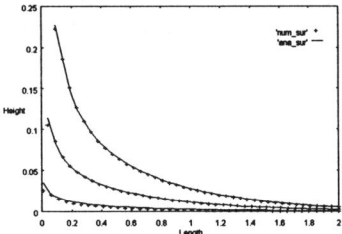
Fig.3 Free surface elevation.

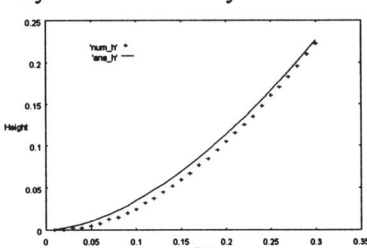
Fig.4 Free surface height at moving plate.

## OPEN CHANNEL NEAR A TURBINE DRAFT TUBE

According to the experimental investigations of [2] a significant interaction of draft tube and tailwater flow is existing. The work [12] has also confirmed that the combination of draft tube and open channel affects the efficiency and output of water power plants. These imply that the tailwater geometry can been optimized. Obviously, the numerical simulation of the upstream and downstream of the draft tube outlet is needed for the optimum adaption of a turbine of the water power plant. To validate the developed code, a test case of T. Toyokura [1] has been recalculated.

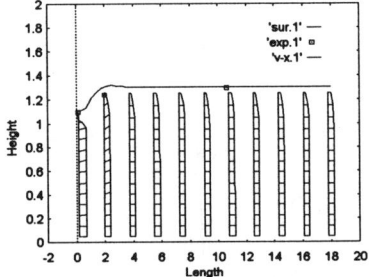
Fig.5 Free surface and velocity profile behind a draft tube.

The test case has been studied experimentally by T. Toyokura in a 1.89 m long and 105 mm wide open channel. The inflow velocity is 1m/s. The cumputational domain comprises a mesh of 50x14 cells. The results have shown that there is a discrepancy between the pressure at the draft tube exit and the pressure given by the water level just behind the draft tube. This coincides with the experimental results of [1]. Fig.5 shows the measured and computed free surface location with velocity profiles. The water level at the open channel entrance is depressed against higher water level downstream the open channel due to the velocity head at the draft tube exit. This is consistent with the measurements of [1,12].

## CONCLUSION

An implicit dual time stepping method based on the artificial compressibility formulation has been used for solving the unsteady incompressible and free surface flow problem with primitive variable form. Due to the upwind-differencing scheme based on flux-difference splitting, very large values of CFL in pseudotime can be used to obtain rapid convergence. The free surface is represented accurately in terms of a flexible general curvilinear coordinate system. The computed results match the analytical values and experiment data very well.

The value of artificial speed of sound ß affects the stability and the pseudotime step $\Delta t^*$ influences the convergence of the calculation. It implies that an optimum conbination between ß and $\Delta t^*$ can improve the subiteration convergence. These numerical simulations are being extended to 3-D inviscid and viscid flow problem.

## REFERENCE

1. T. Toyokura, et al, Flow in Open Channel Near Draft Tube Outlet of Low Head Turbine. IAHR and AIRH Symposium 1990, Belgrade Yugoslavia.
2. Ch. Schneider, Y. Mochkaai, W. Knapp, R. Schilling, Head Recovery in the Tailwater of Low Head Hydro Power Stations, 27th Congress of the International Association for Hydraulic Research, San Francisco, 10-15 August, 1997, to appear.
3. A J.Chorin, A Numerical Method for Solving Incompressible Viscous Problems. J. Compute. Phys. 2, pp. 12-26, 1967.
4. C. W. Hirt, et al, Volume of Fluid (VOF) Method for the Dynamics of Free Boundary. J. Comput. Phys. 39, pp. 201-225, 1981.
5. M. Beddhu, et al, A Time Accurate Calculation Procedure for Flows with Free Surface Using a Modified Artificial Compressibility Formulation. Applied Mathematics and Computation, Vol. 65, pp. 33-48, 1994.
6. S. E. Rogers, et al, Upwind Differencing Scheme for the Time-Accurate Incompressible Navier-Stokes Equations. AIAA Journal, Vol. 28, No.2, February 1990.
7. C. L. Merkle, et al, Time-Accurate Unsteady Incompressible Flow Algorithms Based on Artificial Compressibility. AIAA Paper 87-1137, 1987.
8. N. Riedel, R. Schilling, Water Turbine Runner Design Including Rotor- Stator Interaction. Sixth International Symposium on Computational Fluid Dynamics, Lake Tahoe, Nevada, Sept. 4-8, 1995.
9. D. L. Whitfield, et al, Numerical Solution of the Two-Dimensional Time-Dependent Incompressible Euler Equations. Engineering and Industrial Research Station Report MSSU-EIRS-ERC-93-14, Mississippi State University, Mississippi State, MS, 1994.
10. G. Chapalain, R. Cointe, A. Temperville, Observed and Modeled Resonantly Interacting Progressive Water-Waves, Coastal Engineering, 16, pp 267-300, 1992.
11. D. I. M. Forehand et al, Numerical Prediction of Extreme Free-Surface Flows Caused by Body/Fluid Interaction. Sixth International Symposium on Computational Fluid Dynamics, vol. 1, Lake Tahoe, Nevada, September 4-8, 1995.
12. E. Kita, et al, Tailwater Level and Net Head in Low Head Power Plant. The 3rd Japan-China Joint Conference on Fluid Machinery, Osaka, April 23-25, 1990.

# STUDY ON HIGHLY-ACCURATE NUMERICAL METHOD FOR ADVECTION TERM

KOJI ASAI[1], TOSHIMITSU KOMATSU[2], KOICHIRO OHGUSHI[3] and KESAYOSHI HADANO[4]

1) Environmental Hydraulics Section, Chugoku National Industrial Research Institute, Kure 737-01, Japan
2) Dept.of Civil Eng., Kyushu University, Fukuoka 812-81, Japan
3) Dept.of Civil Eng., Saga University, Saga 840, Japan
4) Dept.of Civil Eng., Yamaguchi University, Ube 755, Japan

## ABSTRACT

A new computational scheme is proposed in order to compute an advection term accurately. This scheme is the implicit version of the HORNET and named the Implicit HORNET scheme. The Implicit HORNET scheme is based on the split operator approach and can be easily applied to a multi dimensional problem. The several model computations are carried out to verify the high performance of this scheme.

## INTRODUCTION

Water pollution in an enclosed sea, river, lake, underground water, etc. is a very severe problem for the human living environment. It is necessary to examine prevention of the water pollution in order to make the comfortable water environment, and counterplans for improving the water quality are required if the water pollution occurs.

A precise prediction of transport of contaminants is required to examine the prevention of the water pollution or the improvement of the water quality. The governing equation describing the phenomena of transport is the partial differential equation called the advection-diffusion equation. It is very difficult to solve this equation analytically unless boundary and initial conditions are quite simple. Therefore, a numerical approach using a computer is a popular and useful method.

Recently, Asai et al. presented the HORNET scheme [1]. This scheme can be easily treated since it is an explicit scheme and it has almost the same accuracy as the 6-point scheme [2] or the SOWMAC scheme [3] . However, it is not easy to apply the HORNET scheme to multidimensional problems because this scheme has been developed for one dimensional problem only. In this study, we have attempted to improve the HORNET scheme and change it into an implicit scheme to make it possible to apply to multidimensional problems.

## THE IMPLICIT HORNET SCHEME

### THE HORNET SCHEME
The one dimensional advection-diffusion equation is written as,

$$\frac{\partial \Phi}{\partial t} + u \frac{\partial \Phi}{\partial x} = D \frac{\partial^2 \Phi}{\partial x^2} \tag{1}$$

where $\Phi$ is concentration of contaminant, $u$ is a velocity in $x$-coordinate direction, $D$ is a physical diffusion coefficient. The HORNET scheme is the numerical method for solving Eq. (1) . The final expression of the HORNET scheme is given as follows :

$$\Phi_i^{n+1} = (\beta+K)\Phi_{i+1}^n + (1-\alpha\theta-2\beta-2K)\Phi_i^n + (\alpha\theta+\beta+K)\Phi_{i-1}^n - \alpha(1-\theta)\Phi_i^{n-1} + \alpha(1-\theta)\Phi_{i-1}^{n-1} \tag{2}$$

where $K = -\frac{1}{2}\left[\alpha-\alpha^2(3-2\theta)\right]$, $\theta = \frac{2\alpha^2-3\alpha+1-12\beta}{3(\alpha^2-\alpha-2\beta)}$, $\alpha = \frac{u\Delta t}{\Delta x}$, $\beta = \frac{D\Delta t}{\Delta x^2}$

$\alpha$ is the Courant number, $\beta$ is the diffusion number, $\Delta x$ and $\Delta t$ are the computational grid intervals for space and time respectively.

Highly-accurate results can be obtained with the HORNET scheme. However, the direct application of the HORNET scheme to multi dimensional problem could not give fine results. Extension of the original concept of the HORNET scheme to a multi dimensional problem was not straightforward.

### DERIVATION OF NEW SCHEME
The new scheme is based on the split operator approach. The split operator approach is a useful numerical procedure to solve the advection-diffusion equation. The advection term and the diffusion term are computed separately in the same time step in the split operator approach. The basic equation to be solved is the pure advection equation and given as follows :

$$\frac{\partial \Phi}{\partial t} + u \frac{\partial \Phi}{\partial x} = 0 \tag{3}$$

For simplicity, the velocity $u$ is assumed to be constant. Eq. (3) is discretized in the implicit form as follows :

$$\frac{\Phi_i^{n+1} - \Phi_i^n}{\Delta t} + u\left[\theta \frac{\Phi_i^{n+1} - \Phi_{i-1}^{n+1}}{\Delta x} + (1-\theta)\frac{\Phi_i^n - \Phi_{i-1}^n}{\Delta x}\right] = 0 \tag{4}$$

Eq. (4) is the same form as the generalization of the Crank-Nicolson method. It is important to estimate the truncation error terms of Eq. (4) in order to compose a high order scheme. By using the Taylor seise analysis, we can obtain the truncation error terms as follows :

$$\frac{\partial \Phi}{\partial t} + u\frac{\partial \Phi}{\partial x} = \frac{1}{2!\Delta t}\left[-\frac{\partial^2 \Phi}{\partial t^2}\Delta t^2 - 2\alpha\theta\frac{\partial^2 \Phi}{\partial t \partial x}\Delta t \Delta x + \alpha\frac{\partial^2 \Phi}{\partial x^2}\Delta x^2\right]$$

$$+ \frac{1}{3!\Delta t}\left[-\frac{\partial^3 \Phi}{\partial t^3}\Delta t^3 - 3\alpha\theta\frac{\partial^3 \Phi}{\partial t^2 \partial x}\Delta t^3 \Delta x + 3\alpha\theta\frac{\partial^3 \Phi}{\partial t \partial x^2}\Delta t \Delta x^2 - \alpha\frac{\partial^3 \Phi}{\partial x^3}\Delta x^3\right]$$

$$+ \frac{1}{4!\Delta t}\left[-\frac{\partial^4 \Phi}{\partial t^4}\Delta t^4 - 4\alpha\theta\frac{\partial^4 \Phi}{\partial t^3 \partial x}\Delta t^3 \Delta x + 6\alpha\theta\frac{\partial^4 \Phi}{\partial t^2 \partial x^2}\Delta t^2 \Delta x^2\right.$$

$$\left. - 4\alpha\theta\frac{\partial^4 \Phi}{\partial t \partial x^3}\Delta t \Delta x^3 + \alpha\frac{\partial^4 \Phi}{\partial x^4}\Delta x^4\right] + H.O.T. \tag{5}$$

The terms of the right hand side of Eq. (5) represent the truncation error terms and induce the numerical error. In this paper, we call the truncation error term the numerical diffusion term. The numerical diffusion terms include the time derivative and the cross derivative with respect to $t$ and $x$. It is convenient to transform them into the spatial derivative only. By differentiating Eq. (3) with respect to $t$ and $x$, the following relationships are obtained.

$$\frac{\partial^2 \Phi}{\partial t^2} = u^2\frac{\partial^2 \Phi}{\partial x^2}, \quad \frac{\partial^2 \Phi}{\partial t \partial x} = -u\frac{\partial^2 \Phi}{\partial x^2}, \quad \frac{\partial^3 \Phi}{\partial t^3} = -u^3\frac{\partial^3 \Phi}{\partial x^3}, \quad \frac{\partial^3 \Phi}{\partial t^2 \partial x} = u^2\frac{\partial^3 \Phi}{\partial x^3}, \quad \frac{\partial^3 \Phi}{\partial t \partial x^2} = -u\frac{\partial^3 \Phi}{\partial x^3}$$

$$\frac{\partial^4 \Phi}{\partial t^4} = u^4\frac{\partial^4 \Phi}{\partial x^4}, \quad \frac{\partial^4 \Phi}{\partial t^3 \partial x} = -u^3\frac{\partial^4 \Phi}{\partial x^4}, \quad \frac{\partial^4 \Phi}{\partial t^2 \partial x^2} = u^2\frac{\partial^4 \Phi}{\partial x^4}, \quad \frac{\partial^4 \Phi}{\partial t \partial x^3} = -u\frac{\partial^4 \Phi}{\partial x^4}$$

Substitution of these relationships into Eq. (5) results in the following :

$$\frac{\partial \Phi}{\partial t} + u\frac{\partial \Phi}{\partial x} = \left[-\alpha^2 + 2\alpha^2\theta + \alpha\right]\frac{\Delta x^2}{2!\Delta t}\frac{\partial^2 \Phi}{\partial x^2} + \left[\alpha^3 - 3\alpha^3\theta - 3\alpha^2\theta - \alpha\right]\frac{\Delta x^3}{3!\Delta t}\frac{\partial^3 \Phi}{\partial x^3}$$

$$+ \left[-\alpha^4 + 4\alpha^4\theta + 6\alpha^3\theta + 4\alpha^2\theta + \alpha\right]\frac{\Delta x^4}{4!\Delta t}\frac{\partial^4 \Phi}{\partial x^4} + H.O.T. \tag{6}$$

The numerical diffusion terms should be vanished to get a high-accuracy calculated result. The first term and the second them of the right hand side can be eliminated by introducing the new artificial terms. The third term of the right hand side can be canceled by adopting the value of $\theta$ that reduces the coefficient of this term to zero.

In order to discretize the third derivative, four grid points for space are required at least. However, the number of used grid points for space should be under three to develop a compact scheme. Therefore, the third derivative is represented by the cross derivative with respect to $t$ and $x$. By adding the second and third artificial numerical diffusion terms to Eq. (3), the following equation is obtained.

$$\frac{\partial \Phi}{\partial t} + u\frac{\partial \Phi}{\partial x} = K_2 \frac{\partial^2 \Phi}{\partial x^2} + K_3 \frac{\partial^3 \Phi}{\partial t \partial x^2} \tag{7}$$

Eq. (7) is discretized as follows :

$$\frac{\Phi_i^{n+1} - \Phi_i^n}{\Delta t} + u\left[\theta \frac{\Phi_i^{n+1} - \Phi_{i-1}^{n+1}}{\Delta x} + (1-\theta)\frac{\Phi_i^n - \Phi_{i-1}^n}{\Delta x}\right] = K_2 \frac{\Phi_{i+1}^n - 2\Phi_i^n + \Phi_{i-1}^n}{\Delta x^2}$$

$$+ K_3 \frac{\Phi_{i+1}^{n+1} - 2\Phi_i^{n+1} + \Phi_{i-1}^{n+1} - \Phi_{i+1}^n + 2\Phi_i^n - \Phi_{i-1}^n}{\Delta t \, \Delta x^2} \tag{8}$$

The numerical diffusion terms of Eq. (7) are as follows :

$$\frac{\partial \Phi}{\partial t} + u\frac{\partial \Phi}{\partial x} = \left[-\alpha^2 + 2\alpha^2\theta + \alpha + 2\beta\right]\frac{\Delta x^2}{2!\,\Delta t}\frac{\partial^2 \Phi}{\partial x^2} + \left[-\alpha^2 + 3\alpha^2\theta + 3\alpha\theta + 1 + 6\gamma\right]\frac{\Delta x^2}{3!}\frac{\partial^3 \Phi}{\partial t \partial x^2}$$

$$+ \left[-\alpha^4 + 4\alpha^4\theta + 6\alpha^3\theta + 4\alpha^2\theta + 12\alpha^2\gamma + \alpha + 2\beta\right]\frac{\Delta x^4}{4!\,\Delta t}\frac{\partial^4 \Phi}{\partial x^4} + \text{H.O.T.} \tag{9}$$

where $\quad \beta = \frac{K_2 \Delta t}{\Delta x^2}, \quad \gamma = \frac{K_3 \Delta t}{\Delta t \, \Delta x^2}$ .

To obtain Eq. (9), the following relationships are used.

$$\frac{\partial^3 \Phi}{\partial t^3} = u^2 \frac{\partial^3 \Phi}{\partial t \partial x^2}, \quad \frac{\partial^3 \Phi}{\partial t^2 \partial x} = -u\frac{\partial^3 \Phi}{\partial t \partial x^2}, \quad \frac{\partial^3 \Phi}{\partial x^3} = -\frac{1}{u}\frac{\partial^3 \Phi}{\partial t \partial x^2}$$

$\beta$ and $\gamma$ are decided so as to reduce the values of the coefficients of the first and second terms of the right hand side of Eq. (9) to zero and expressed as follows :

$$\beta = \frac{-\alpha + \alpha^2 - 2\alpha^2\theta}{2} \tag{10} \qquad \gamma = \frac{(1+\alpha)(-1 + \alpha - 3\alpha\theta)}{6} \tag{11}$$

$\theta$ can be decided from the coefficient of the third term of the right hand side of Eq. (9). Substituting Eq. (10) and (11) into the coefficient and putting the value of the coefficient to zero, we can obtain the following relation.

$$\alpha^2(\alpha^2 - 1)(2\theta - 1) = 0 \tag{12}$$

From Eq. (12), $\theta$ is decided to be 1/2. The final expression of the scheme is given as follows :

$$[-\alpha\theta-\gamma]\Phi_{i-1}^{n+1}+[1+\alpha\theta+2\gamma]\Phi_i^{n+1}+[-\gamma]\Phi_{i+1}^{n+1}$$
$$=[\alpha(1-\theta)+\beta-\gamma]\Phi_{i-1}^n+[1-\alpha(1-\theta)-2\beta+2\gamma]\Phi_i^n+[\beta-\gamma]\Phi_{i+1}^n \tag{13}$$

if the flow velocity $u$ is negative the scheme is expressed as following:

$$[-\gamma]\Phi_{i-1}^{n+1}+[1+|\alpha|\theta+2\gamma]\Phi_i^{n+1}+[-|\alpha|\theta-\gamma]\Phi_{i+1}^{n+1}$$
$$=[\beta-\gamma]\Phi_{i-1}^n+[1-|\alpha|(1-\theta)-2\beta+2\gamma]\Phi_i^n+[|\alpha|(1-\theta)+\beta-\gamma]\Phi_{i+1}^n \tag{14}$$

$$\beta=\frac{-|\alpha|+\alpha^2-2\alpha^2\theta}{2}, \qquad \gamma=\frac{(1+|\alpha|)(-1+|\alpha|-3|\alpha|\theta)}{6}$$

Eq. (13) or (14) is the implicit version of the HORNET scheme and called the Implicit HORNET scheme.

## MODEL COMPUTATIONS AND DISCUSSION

### ONE DIMENSIONAL MODEL COMPUTATION

Model computations of the one-dimensional pure advection were made to test the Implicit HORNET scheme. The initial distribution is the superposition of the Gaussian distribution (the standard deviation is 1.5m and the peak value is 1 at x=150m), the semi-ellipse distribution (the peak value is 1 at x=125m and the half width of x-axis is 10m) and the rectangle distribution (the height is 1 and the width is 10m). This distribution is transported downstream for 100sec under the uniform flow velocity of 0.5m/sec. The spatial grid size $\Delta x$ is 1.0m and the time increment $\Delta t$ is 0.2sec.

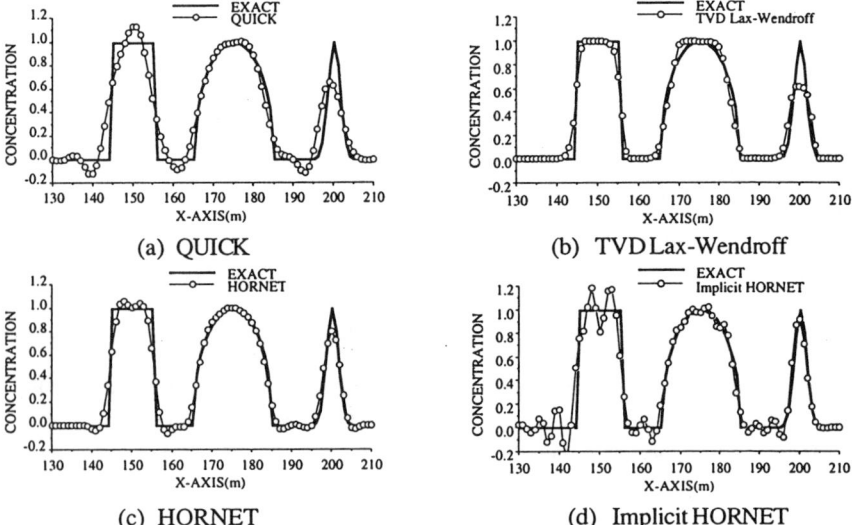

Fig.1 Computational results

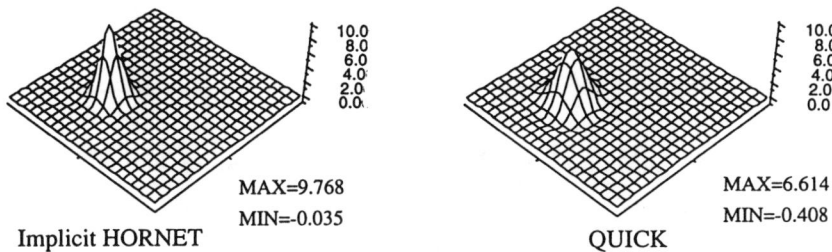

Fig.2 Computational results for 2-d advection

Fig.1 shows the computational results. The results by the TVD Lax-Wendroff scheme, the QUICK scheme and the HORNET scheme are also shown in Fig.1 for comparison. The QUICK scheme yields the over shoots and the under shoots. The TVD Lax-Wendroff scheme gives no spurious oscillation. However, it gives the flattened local extreme. This feature may make the results of the numerical diffusion simulation unreliable. The HORNET scheme provides the best solution from among them. The spurious oscillation is shown in the result by the Implicit HORNET. It should be noted that this spurious oscillation can be suppressed by regulating the value of $\theta$ slightly.

## TWO DIMENSIONAL MODEL COMPUTATION

The application of the Implicit HORNET scheme to multidimensional advection problems can be easily made by using the time splitting method. The rotation of a cone-shaped scalar field is used to realize the two dimensional advection. The initial condition is the Gaussian distribution of the standard deviation 200m. The initial distribution is centered at $x$=1400m, $y$=1400m and its peak value is 10. The spatial grid sizes $\Delta x$, $\Delta y$ are 100m and the time increment $\Delta t$ is 50sec. The Gaussian distribution is advected around under the angular velocity of $2\pi/12000$ sec$^{-1}$ for 12000sec (one revolution).

Fig.2 shows the computational solutions by the Implicit HORNET scheme and the QUICK scheme. The implicit HORNET scheme produces the better result. On the other hand, the result by the QUICK scheme shows the peak clipping error and the relatively large under shoots. It is found that the Implicit HORNET scheme can be applied to multidimensional problems easily and accurately.

## REFERENCES

(1) K.ASAI et. al. : Development of a simple and high-accurate scheme for 1-d diffusion simulation, Proc. XXVIth Cong.IAHR, Vol.2, pp.34-39 (1995)

(2) T.KOMATSU et. al. : Numerical calculation of pollutant transport in one and two dimensions, J.H.H.E, Vol.3, No.2, pp.15-30 (1985)

(3) T.KOMATSU et. al. : Refined Numerical Scheme for Advective Transport in Diffusion Simulation, J. Hydr. Engrg., ASCE, Vol.123, No.1, pp.41-50 (1997)

# Implicit TVD Methods For Modelling Discontinuous Channel Flows

A.I.DELIS AND C.P.SKEELS
Department of Mathematics, University of the West of England,
Bristol, United Kingdom

## ABSTRACT
The Saint Venant equations for modelling flow in open channels are solved in this paper using two implicit total variation diminishing (TVD) schemes. The applicability and performance of second and third order accurate implicit TVD schemes are investigated for the computation of free–surface flows. Both of the unsteady computational models are compared in their prediction of extreme steady state solutions, such as hydraulic jumps.

## 1. INTRODUCTION
In this paper numerical schemes are presented for solving the Saint Venant equations, which model the one–dimensional open channel flow. Derivation of these equations can be found in [1]. Several classical finite difference schemes are commonly used in industry such as Preissmann scheme [1], but they are highly inaccurate in modelling discontinuous flow. The Saint Venant hyperbolic equations yield discontinuous solutions, which can be difficult to represent accurately without the use of a shock–capturing method.

Total Variation Diminishing (TVD) schemes where introduced by Harten [4]. Their main property is that they are second–order accurate (except at extrema) and oscillation free across discontinuities. Explicit TVD methods have been widely reported [2] due to their simplicity. Their main disadvantage lies in the CFL number restriction on the time step. Implicit schemes have been proved to be unconditionally stable, even when a linearization technique is applied to solve a non–linear hyperbolic equation [7]. This paper investigates the more difficult case of implicit schemes modelling channel flow that is represented by a non–linear system of equations.

## 2. GOVERNING EQUATIONS
The one–dimensional approach to the unsteady flow of water is governed by the St.Venant equations:

$$\frac{\partial \mathbf{U}}{\partial t} + \frac{\partial \mathbf{F(U)}}{\partial x} = \mathbf{G}, \tag{1}$$

where

$$\mathbf{U} = (A, Q)^\tau, \qquad \mathbf{F(U)} = \left(Q, \frac{Q^2}{A} + gI\right)^\tau, \qquad \mathbf{G} = (0, gA(S_0 - S_f))^\tau, \qquad (2)$$

with $A$ = wetted cross-sectional area; $Q$ = flow rate; $I$ = hydrostatic pressure force; $g$ = acceleration due to gravity; $S_0$ = bed slope; $S_f$ = friction slope. The friction slope $S_f$ is defined in terms of Manning's roughness coefficient. For rectangular channels $A = Bh$, where $B$ = constant channel width and $h$ = water depth. In this case, the Jacobian matrix of the flux function $\mathbf{F}$ is given by

$$J = \frac{\partial \mathbf{F(U)}}{\partial \mathbf{U}} = \begin{bmatrix} 0 & 1 \\ c^2 - u^2 & 2u \end{bmatrix}, \qquad (3)$$

where velocity $u = Q/A$, and celerity $c = \sqrt{gh}$. The Jacobian matrix has eigenvalues $a^{1,2} = u \pm c$ and corresponding eigenvectors $\mathbf{e}^{1,2} = (1, u \pm c)^\tau$. Let $R$ denote the matrix whose columns are the eigenvectors of $J$.

Let $\Delta x$ be the grid size in space such that $x = i\Delta x$ and let the time increment be denoted by $\Delta t$ such that $t = n\Delta t$. Let $a_{i+1/2}^{1,2}$, $R_{i+1/2}$, $R_{i+1/2}^{-1}$, denote the quantities $a$, $R$, $R^{-1}$ evaluated at a symmetric average of $\mathbf{U}_i$, $\mathbf{U}_{i+1}$ [2] (where $\mathbf{U}_i$, $\mathbf{U}_{i+1}$ denote approximations of $\mathbf{U}$ at the left and right of the computational cell). These averages are chosen to guarantee the physically valid discontinuities in the solution. The difference between the characteristic variables (wave-strengths) are

$$\alpha_{i+1/2}^{1,2} = R_{i+1/2}^{-1}(\mathbf{U}_{i+1} - \mathbf{U}_i) = R_{i+1/2}\Delta_{i+1/2}\mathbf{U}.$$

## 3. IMPLICIT TVD SCHEMES

A general form of an implicit numerical scheme can be written as

$$\mathbf{U}_i^{n+1} + \lambda\theta(\tilde{\mathbf{F}}_{i+1/2}^{n+1} - \tilde{\mathbf{F}}_{i-1/2}^{n+1}) - \theta\Delta t \mathbf{G}_i^{n+1} = \mathbf{U}_i^n - \lambda(1-\theta)(\tilde{\mathbf{F}}_{i+1/2}^n - \tilde{\mathbf{F}}_{i-1/2}^n) + (1-\theta)\Delta t \mathbf{G}_i^n \qquad (4)$$

where $\lambda = \Delta t / \Delta x$ and $0 \leq \theta \leq 1$. For $\theta \neq 0$, the scheme is implicit. $\tilde{\mathbf{F}}_{i\pm 1/2}^n$ are the numerical flux vectors; the spatial accuracy of the scheme depends on their form. The numerical flux for the general scheme (4), can be written in the form

$$\tilde{\mathbf{F}}_{i\pm 1/2} = \frac{1}{2}\left[\mathbf{F}_i + \mathbf{F}_{i\pm 1} + R_{i\pm 1/2}\mathbf{D}_{i\pm 1/2}\right], \qquad (5)$$

where $\mathbf{F}_i = \mathbf{F}(\mathbf{U}_i)$ and $\mathbf{D}_{i\pm 1/2}$ is the scheme dependent vector function. In this paper, two schemes are applied to open channel flow. Both these schemes previously have been used in modelling gas dynamics [7] and achieve high order accuracy while retaining TVD properties.

### NON-MUSCL APPROACH

The elements of $\mathbf{D}_{i+1/2}$, denoted by $d_{i+1/2}^k$, have the following form for the second-order upwind scheme, as presented in [7]:

$$(d^k_{i+1/2})^u = \frac{1}{2}\psi(a_{i+1/2})(L^k_{i+1} + L^k_i) - \psi(a^k_{i+1/2} + \gamma^k_{i+1/2})\alpha^k_{i+1/2}, \qquad k = 1, 2.$$

where

$$\gamma^k_{i+1/2} = \begin{cases} \frac{1}{2}\psi(a_{i+1/2})(L^k_{i+1} - L^k_i)/\alpha^k_{i+1/2} & \alpha^k_{i+1/2} \neq 0 \\ 0 & \alpha^k_{i+1/2} = 0 \end{cases}.$$

The flux limiter functions $L^k_i$ control the second–order terms, so that a smooth non–oscillatory result, even in the presence of discontinuities, is guaranteed. The limiter function $L^k_i$ is expressed in terms of wave–strengths as

$$L^k_i = \mathrm{minmod}(\alpha^k_{i+1/2}, \alpha^k_{i-1/2}),$$

where

$$\mathrm{minmod}(x, y) = \mathrm{sgn}(x)\max\{0, \min[|x|, y\,\mathrm{sgn}(x)]\}.$$

Other limiters can be found in the literature [7].

The function $\psi$ is the entropy correction to the eigenvalues $|a^k_{i+1/2}|$ and it is employed in each scheme to ensure compatibility while guaranteeing satisfaction of the energy inequality.

## MUSCL APPROACH

The MUSCL–TVD interpolation formula provides a one parameter family of second–order schemes and one third–order scheme. TVD properties are guaranteed for the geometric approach using a slope rather than flux limiter. This limits the gradients of the dependent variables. The MUSCL scheme replaces the arguments, $\mathbf{U}_i$ and $\mathbf{U}_{i+1}$, of the numerical fluxes by $\mathbf{U^L}_{i+1/2}$ and $\mathbf{U^R}_{i+1/2}$, where $\mathbf{U^L}_{i+1/2}$ and $\mathbf{U^R}_{i+1/2}$ are calculated as follows

$$\begin{aligned}
\mathbf{U^L}_{i+1/2} &= \mathbf{U}_i + \frac{1}{4}\left[(1-m)\Delta^+_{i-1/2} + (1+m)\Delta^-_{i+1/2}\right], \\
\mathbf{U^R}_{i+1/2} &= \mathbf{U}_{i+1} + \frac{1}{4}\left[(1-m)\Delta^-_{i+3/2} + (1+m)\Delta^+_{i+1/2}\right], \\
\Delta^-_{i+1/2} &= \mathrm{minmod}(\Delta_{i+1/2}\mathbf{U}, \beta\Delta_{i-1/2}\mathbf{U}), \\
\Delta^+_{i+1/2} &= \mathrm{minmod}(\Delta_{i+1/2}\mathbf{U}, \beta\Delta_{i+3/2}\mathbf{U}),
\end{aligned}$$

where $\beta$ is a compression parameter whose value is in the range

$$1 \leq \beta \leq \frac{3-m}{1-m}, \quad m \neq 1.$$

The spatial order of accuracy is determined by the chosen value of $m$. In this study $m = 1/3$ is used, giving third–order accuracy [7]. The second order numerical flux is analogous to (5), but evaluated at $\mathbf{U^L}_{i+1/2}$ and $\mathbf{U^R}_{i+1/2}$ and the elements of $\mathbf{D}_{i+1/2}$ are

$$d^k_{i+1/2} = -\psi(a^k_{i+1/2})\alpha^k_{i+1/2}, \qquad k = 1, 2.$$

Here $a_{i+1/2}^k$, $R_{i+1/2}$ and $\alpha_{i+1/2}^k$ are evaluated as before, but the arguments $\mathbf{U}_i$ and $\mathbf{U}_{i+1}$ have been replaced by $\mathbf{U}^L{}_{i+1/2}$ and $\mathbf{U}^R{}_{i+1/2}$ respectively.

## CONSERVATIVE LINEARIZED FORM

Solving for $\mathbf{U}^{n+1}$ in (4) we normally need to solve a set of nonlinear algebraic equations iteratively. To avoid this costly procedure, the implicit operator is linearized and solved. Following the same procedure as in [3] and [7], the conservative linearised form can be written in the form of a block tridiagonal system

$$A_1 \delta \mathbf{U}_{i-1} + A_2 \delta \mathbf{U}_i + A_3 \delta \mathbf{U}_{i+1} = -\lambda(\tilde{\mathbf{F}}_{i+1/2}^n - \tilde{\mathbf{F}}_{i-1/2}^n) + \Delta t \mathbf{G}_i^n, \tag{6}$$

where

$$A_1 = -\frac{\lambda\theta}{2}(J_{i-1} + B_{i-1/2})^n$$

$$A_2 = I + \frac{\lambda\theta}{2}(B_{i-1/2} + B_{i+1/2})^n + \Delta t J_G,$$

$$A_3 = \frac{\lambda\theta}{2}(J_{i+1} - B_{i+1/2})^n,$$

with $B_{i+1/2} = \left[R \cdot \text{diag}\psi(a_{i+1/2}^k) \cdot R^{-1}\right]_{i+1/2}^n$, $\delta \mathbf{U}_i = \mathbf{U}_i^{n+1} - \mathbf{U}_i^n$ and $J_G$ is the Jacobian matrix of the source term $\mathbf{G}_i^n$. The above scheme is only first order accurate in time, but second order in space for the upwind scheme and third order in space for MUSCL. Setting $\theta = 1$ the method is unconditionally stable and well suited for the calculation of steady flows by solving the unsteady equations.

## 4. NUMERICAL RESULTS

In order to make realistic comparisons between the schemes the full system of equations (1) is solved. In [5] and [6], test problems and their analytical solutions are presented for discontinuous flows in rectangular channels with variable bottom slopes. The first problem, which we shall call Problem 1, has a subcritical inflow and outflow with a supercritical central section. The second, which we shall call Problem 2, has a subcritical inflow and supercritical outflow. Both examples have rectangular channels 10m wide, but it is 100m long for Problem 1 and 1km long for Problem 2. Both use an inflow discharge of 20m$^3$s$^{-1}$ and Manning's roughness coefficient is constant at 0.03 for Problem 1 and 0.02 for Problem 2. A grid of 51 points in space has been used for the numerical simulations for each problem. The time increment is calculated to maintain a predetermined CFL number at the end of each time step and the flow conditions are computed until convergence is achieved. The type of flow in the two examples is determined by their variable bed slopes, $S_0$, which are given as functions of the exact water depths and their derivatives [5, 6]. The analytical solutions are plotted with a solid line in all figures.

Problem 1 was run at a CFL number of 30 for both schemes. The value of $\beta = 1$ produced the best results for the MUSCL scheme. The solutions produce a similar

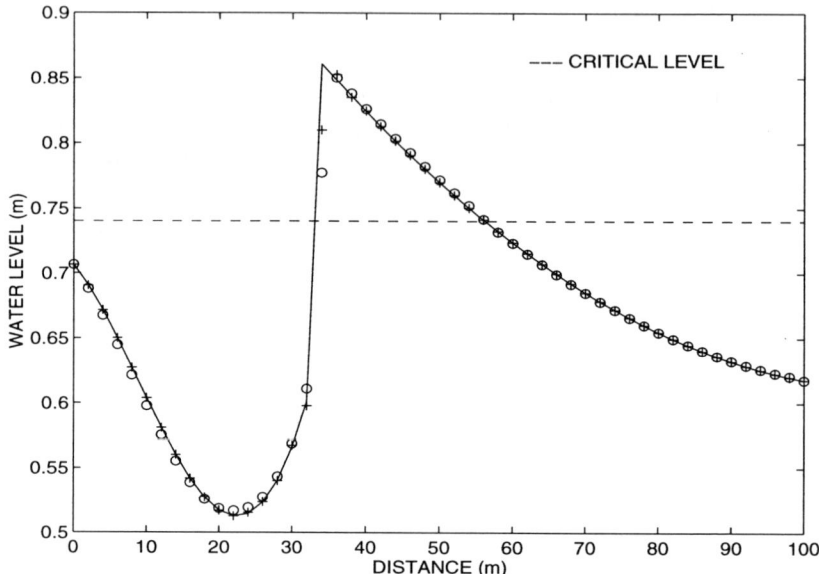

Figure 1: Water surface profile for Problem1, Upwind(o), MUSCL(+).

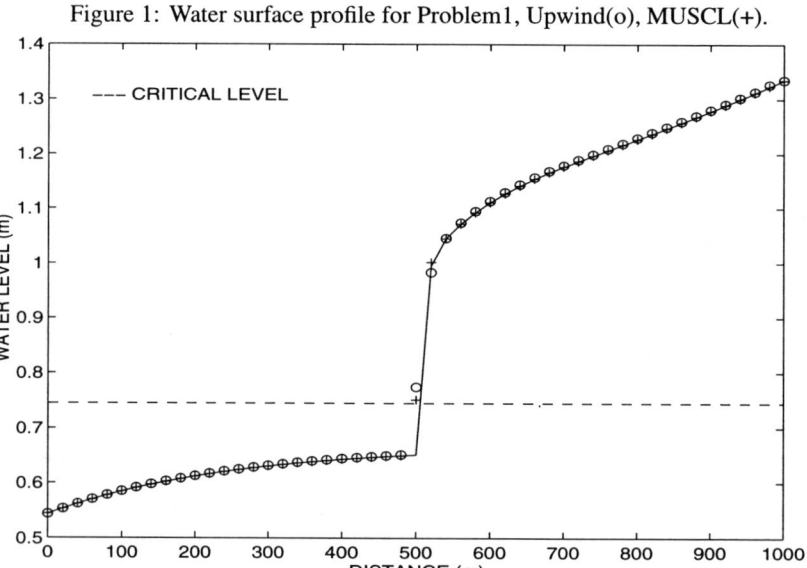

Figure 2: Water surface profile for Problem 2, Upwind(o), MUSCL(+).

water surface profile, though the upwind one is less accurate in the smooth region than the third order MUSCL. Nevertheless both methods accurately predict the hydraulic jump, as shown in Figure 1. For the 51 grid points, the implicit schemes save almost 60% of the CPU time required by the explicit versions of the schemes.

Problem 2 was run at a CFL number of 10 for both schemes. For this example the value of $\beta = 4$ produced the best results for the MUSCL scheme. Again, the solutions produce a similar water surface profile and the third order MUSCL scheme was predictably more accurate than the second order non–MUSCL scheme in the smooth region. Both methods accurately predict the hydraulic jump, as shown in Figure 2. For this problem the implicit schemes save almost 30% of the CPU time necessary by the explicit versions.

## 5. CONCLUSIONS

This paper presents high–resolution implicit TVD methods for modelling open channel flow calculations with hydraulic jumps. Using the same conservative linearized formulation, two implicit schemes were compared for their accuracy in modelling discontinuous steady state flows against analytical solutions. The numerical solutions accurately predict the water profile, including a smooth transition from subcritical to supercritical flow. The hydraulic jumps are sharply captured with no over or undershoots. The most accurate scheme is the third order MUSCL. The CFL stability restriction that applies to explicit schemes is greatly relaxed; higher CFL values, and therefore much larger time steps, can still produce very accurate results in substantially less computation time.

## REFERENCES

[1] J.A.Cunge, F.M.Holly and A.Verwey (1980), *Practical Aspects of Computational River Hydraulics*, Pitman, London, U.K.

[2] A.I.Delis and C.P.Skeels (1996), *TVD Schemes for open channel flow*, (submitted for publication).

[3] P.Garcia–Navarro, F.Alcrudo and A.Priestley (1995), *An implicit method for water flow modelling in channel and pipes*, J. Hydraulics Research, 32, 721–741.

[4] A.Harten (1983), *High–resolution schemes for hyperbolic conservation laws*, J. Comput. Phys., 49, 357–393.

[5] I.MacDonald, M.J.Baines and N.K.Nichols (1994), *Analysis and computation of steady open channel flow using a singular perturbation problem*, Numerical Analysis Report 7/94, Dept. of Mathematics, University of Reading, U.K.

[6] I.MacDonald, M.J.Baines and P.G.Samuels (1995) *Steady open channel test problems with analytical solutions*, Numerical Analysis Report 3/95, Dept. of Mathematics, University of Reading, U.K.

[7] H.C.Yee (1989), *A class of high–resolution explicit and implicit shock capturing methods*, NASA–TM101088.

# Numerical Simulation of Large Eddy Structure in Hydraulic Jump

YOSUKE YAMASHIKI[1], TAMOTSU TAKAHASHI[2],
PODALYRO A. DE SOUZA[3], and KIKUO TAMADA[4]
1)Researcher, International Lake Environment Committee, Kusatsu, Japan
2) Professor, Disaster Prevention Research Institute, Kyoto University, Uji, Japan
3)General Director, CTH - Centro Tecnologico de Hidraulica e Recursos Hidricos - DAEE/EPUSP
3),4)Professor, Polytechnic School, University of Sao Paulo, Sao Paulo, Brazil

## ABSTRACT
The principal goal of this study is the development of a vertical two dimensional mathematical model for estimating large-eddy structure and bed erosion downstream of a spillway. The formation of large eddies in the hydraulic jump was visually analyzed using a high velocity video camera. A mathematical model was developed using Marker And Cell Method, which involves SGS (Sub Grid Scale) model and PSI-Cell (Particle Source In Cell) method. The transportation process of sediment and air bubbles were simulated in Lagrangean form, by introducing air bubbles and sediment markers. Schematic movement characteristics of air bubbles were simulated by this numerical model.

## INTRODUCTION
The determination of a mathematical model to analyze large eddy formation and air bubble entrainment, as well as related erosion downstream of a spillway, if it is adequately established, will offer great advance both for technical and environmental study. In Brazil more than 93 % of total electricity is generated by hydraulic power plants and downstream erosion and cavitation of spillway are one of the most severe problem faced in the maintenance of dams. Many physical models have been developed for experimental study for erosion estimation and prediction. Recently, super-saturation of both oxygen and nitrogen downstream of spillway also became a most severe environmental problem. Adequate numerical models are needed for these environmental phenomenon, however no solver efficiently simulates this air-water mixed flow with free surface.

The spillways in Brazil, especially in Sao Paulo (SP) states, are characterized by huge discharge and relative low head,which is mainly applied into upstream of Parana River Basin. Application of hydraulic jump as a main source of energy dissipation are common by introducing USBR I to III types dissipator in SP.

Recently, large eddy formation in the hydraulic jump has been studied by several researchers. The growth process of vortices in the hydraulic jump was visually observed and analyzed using the high velocity video camera by Long et al.(1991)[1]. They observed it two-dimensionally for each Froude number and concluded that the large eddy is composed of several principal vortices, which are generated in the initial part of the jump and which travel downstream, and of small dependent vortices. The rotational direction of all the main vortices are coincidences. Their study suggested the necessity of mathematical modeling of a hydraulic jump with a time-dependent dynamic solution.

For the numerical solution of a hydraulic jump, the consideration of air entrainment effects is essential. The aeration characteristics of a free hydraulic jump have been studied experimentally (Rajaratnan 1962, Schroder 1963, Rusch and Leutheusser 1972). The average air bubble concentration in vertical section of hydraulic jump has been experimentally studied by Khalifa (1979).

Numerical solution of radial free hydraulic jump with steady condition using the Strip Integral Method, considering the air entrainment effect, was studied by Kharifa et al (1992). This study indicate that, with the steady condition, it is already possible to establish a mathematical model of hydraulic jump with high precision. Submerged hydraulic jump has been simulated numerically using the standard k-ε model (Long et al 1990, Fabio et al 1993).

The use of k-ε turbulent model for numerical simulation of free hydraulic jump may be useful for the steady solution. However, application for free hydraulic jump may have difficulties in determination of bundary condition and adequate constant variables.

In this study, the dynamic solution of the hydraulic jump, as well as jet entrainment was studied to obtain better information about the growing process of large eddies. The Marker And Cell method was modified, combining SGS model and PSI-CELL method[6),7)]. A lagrangean solution for the air bubble transport, assuming that diameters of air bubbles are initially uniform and determined by surrounding water pressure, was applied. The mechanism of picking-up and transport process of bed material in the macro turbulence zone was simulated using the lagrangean model based on Tchen's equation[2),6)], assuming that diameter of bed materials is uniform.

The computer code was written in C language, using reference "103 Y0/VIFMAC" of the computer library of Kyoto University, created by Takemoto[3)] and developed by Tetsuro Sakai et al[4)].

## GOVERNING EQUATIONS

The system of governing equations are the grid-filtered time-dependent two-dimensional compressible (with low Mach number) mixed flow Navier-Stokes, liquid phase continuity equations. SGS model for only liquid phase is introduced[5)]. The effect of Lagrangean sediment marker is considered using PSI-CELL method.,[6),7)]:

Continuity Equation for Water $\frac{\partial}{\partial t}f_l\rho + \frac{\partial}{\partial x_j}f_l\rho \overline{u_j} = 0$ (1)

$$\frac{D(f_l\rho \overline{u_i})}{Dt} = \frac{\partial}{\partial x_j}(1+f_a+2.5f_s)\mu\left(\frac{\partial \overline{u_i}}{\partial x_j}+\frac{\partial \overline{u_j}}{\partial x_i}-\frac{2}{3}\frac{\partial \overline{u_k}}{\partial x_k}\delta_{ij}\right) - \frac{\partial \overline{p}}{\partial x_i} + f_l\rho g_i - \nabla\{(f_l\rho R_{ij})+(f_l\rho L_{ij})+(f_l\rho C_{ij})\}+S_{pi} \quad (2)$$

Reynolds Term $R_{ij} = -2KS_{ij}$   $S_{ij} = \frac{1}{2}\left(\frac{\partial \overline{u_j}}{\partial x_i}+\frac{\partial \overline{u_i}}{\partial x_j}\right)$   $K = (C_s\Delta)^2[2S_{ij}S_{ij}]^{1/2}$ (3)

$$\Delta^2 = (\Delta x^2 + \Delta y^2)/2 \quad (4)$$

Leonards Term $L_{ij} = \frac{\Delta_k^2 \partial^2(\overline{u_i}\,\overline{u_j})}{24\partial x_k \partial x_k}$, Cross Term $C_{ij} = -\frac{\Delta_k^2 \overline{u_i}}{24}\frac{\partial^2 \overline{u_j}}{\partial x_k \partial x_k} - \frac{\Delta_k^2 \overline{u_j}}{24}\frac{\partial^2 \overline{u_i}}{\partial x_k \partial x_k}$ (5)

$S_{pi} = -\frac{\partial M_{pi}}{\partial t}$   $\overline{M_{pi}} = V^{-1}\left(\sum_{k=0}^{k=N_s} m_{pk}u_{pik}\right)$ (6)

where $f_L, f_g, f_s$ are volumetric ratio of Liquid(water), *Gas(air)*, and *Solid(sediment)* phases, respectively, i, j, k =1,2, $\rho$ is water density, $u_i$ is velocity component of water in i direction, p is water pressure ; g is gravity force (i=2), *Rij, Cij, Lij* are Reynolds, Cross, Leonards term, respectively, determined following Kano et al[8)], Cs is Smagolinsky[9)] coefficient, $\mu$ is water viscosity, $\Delta x$, $\Delta y$ are horizontal and vertical grid size, V : Cell volume (grid area for 2-D calculation), $S_{pi}$ is negative production term for flow field by particle movement, $m_{pk}$ specific mass of particle k, $u_{pik}$ is i direction velocity component of particle k.

In the present stage, Smagolinsky type SGS model is used (Cs = 0.20) However considering the difficulty in determining coefficient Cs, dynamical modeling of SGS coefficient for multi phase flow may be beneficial. Considering the effect of Lagrangean air-bubble and sediment, GAL-LES may be one of the adequate solution[10]. However, these models were not applied in the present study.

The "QUICKEST" third-order finite different scheme[11], is used in this study. For the calculation of near-surface cell velocity, the first order central differential scheme was used. Another surface cell velocity, which can not be assumed by differential equation, was estimated by copying the near-cell velocity, and subtracting the gravity effect.

The Lagrangean sediment transport equation, which treats both bed load and suspended load is as follows[6]:

$$\rho(\frac{\sigma}{\rho}+C_M)A_3 d^3 \frac{d\bar{u}_p}{dt} = \varepsilon \left(-\frac{1}{2}C_D \rho |\bar{u}_r| \bar{u}_r A_2 d^2 + \rho(1+C_M)A_3 d^3 \frac{d\bar{u}_f}{dt}\right) + \rho(\frac{\sigma}{\rho}-1)A_3 d^3 (g_i - \mu_f g_j) \quad (7)$$

where $C_D$ is friction coefficient, $\bar{u}_p$ is sediment velocity vector, $\bar{u}_f$ is water velocity vector, $\bar{u}_r = \bar{u}_p - \bar{u}_f$, $A_2 = \pi/4$, $A_3 = \pi/6$, d is particle diameter, $C_M$ is virtual mass coefficient (=0.5), $\sigma$ is density of sediment, $\rho$ is water density. And $\varepsilon$ is covering coefficient, determined only when sediment is at bottom and another sediment is shading upstream. $\mu_f$ is static friction coefficient only works only when sediment is at bottom. This term was determined according to Nakagawa et al.[12] For the simulation of successive saltation movement, we followed Gotoh's method[6].

The Lagrangean air bubble transport equation is introduced similarly as:

$$\rho \beta A_3 a^3 \frac{d\bar{u}_a}{dt} = -\frac{1}{2}C_{Da}\rho|\bar{u}_{ra}|\bar{u}_{ra} A_2 a^2 + \rho \beta A_3 a^3 \frac{D\bar{u}_f}{Dt} - \rho g A_3 a^3 \quad (8)$$

where $\beta$ is virtual mass, given as 0.5, $\rho$ is water density, $a$ is air bubble diameter, $C_{Da}$ : Friction Coefficient of Air Bubble, given as 2.6. The Adams-Bashforth method was used for time integration.

The air entrainment process was treated as follows: The entrainment from the water surface was treated by distributing air bubble marker particles in the sub-cells of the surface cells with no water marker cell. These distributed air bubble marker particles are transported according to the surface cell velocity using the same type of transport equation as the sediment. In this case, the surface air drag coefficient is assumed to 1.0. The air entrainment by the air large bubble captation is simulated by distributing the air bubble marker particles in the inner air cells. The air bubble pairing and dividing can not be simulated in this study.

## EXPERIMENTAL STUDY

An experimental study was realized in Ujigawa Hydraulic Laboratory in Kyoto University using a rectanglar channel 8m long and 0.5m wide. The maximum discharge was 20 l/s and the hydraulic jumps of three different Froude number were studied in terms of the velocity distribution and of the downstream erosion with 2 mm diameter of uniform bed material, sheeted from 50 cm downstream of the beginning of hydraulic jump. Figure 1 shows successive images of the frontal part of hydraulic jump with Fr.=5.7, captured by high velocity video camera. Positions of neutral particles A, B and C in each frame are shown using circlar, square, and trianglar white markers, respectively. Orbits of these neutral particles are traced and presented in Figure 2, where characteristics of the large eddy- the rotation directions were the same even for small dependent eddy - are presented. The velocity distribution of hydraulic jump, obtained by analyzing frames of high velocity video camera, are presented in Figure 3. Rapid change of the velocity field is observed for each 0.1 second. Air bubble entrainment was observed from the frontal part of hydraulic jump, as described by Long et al[1]. The Buoyancy effect of air bubbles seems to be very small in the frontal part of hydraulic jump, however in the rear part, this effect seems to become larger in proportion to large eddy decay and velocity decrease. Characteristics of sediment lifted-up process by large eddies was observed downstream. Bottom sediment transport occurred only when large eddy passed near the movable bed.

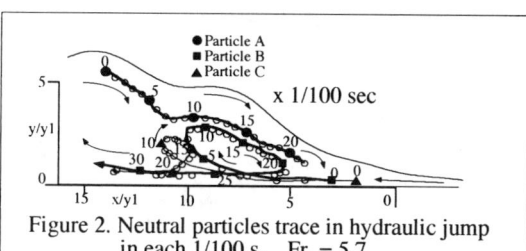

Figure 2. Neutral particles trace in hydraulic jump in each 1/100 s.   Fr. = 5.7

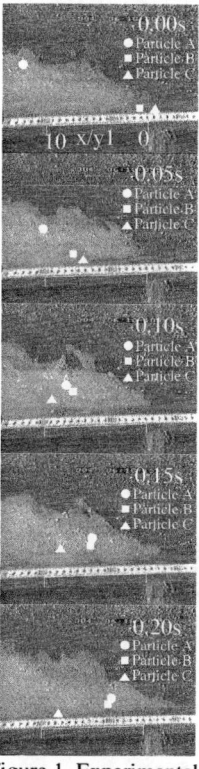

Figure 1. Experimental study using high velocity video camera (Fr.=5.7)

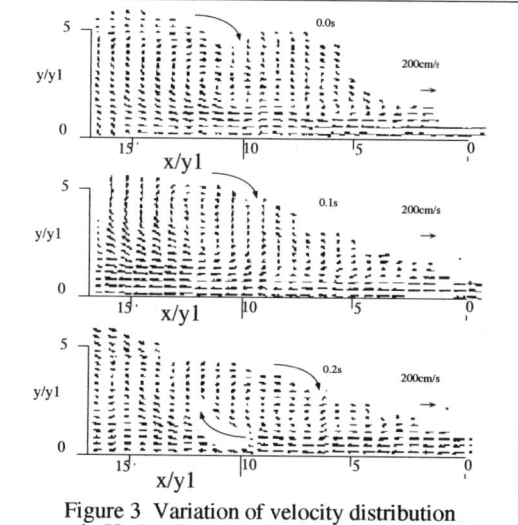

Figure 3 Variation of velocity distribution in Hydraulic Jump. (each 0.1 s) Fr = 5.7

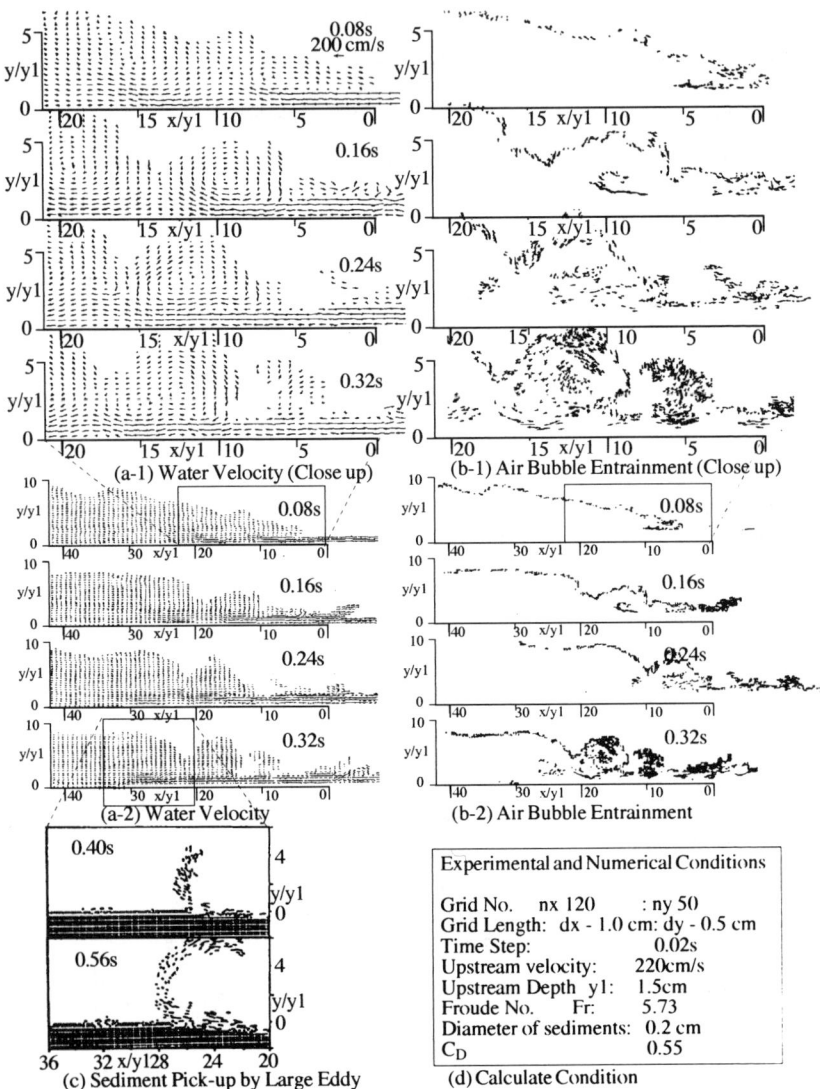

Figure 4.(a)(b)(c)(d) Numerical Simulation of Hydraulic Jump with Fr. = 5.7

## NUMERICAL RESULTS

Numerical simulation was realized only for two dimensional condition. Large eddy growth and air bubble entraimnent, as well as sediment lifted-up process in hydraulic jump is illustrated in Figure 4. In 4(a), the dynamic eddy calculation focussed on upper regime is presented. Each vortices is more separated in the simulated in comparison with experimental study. It may because of the difference between the two dimensional eddy structure in simulation and three dimensional structure in the experimental study. Comparing with the experimental data and video analysis (Figure 1 to 3), dynamic growth of large eddy was described by the numerical simulation. Due to the high Reynolds number of upstream region near the outlet of the jet, the velocity value become unstable near this region. Application of QUICKEST scheme and SGS model offered more stable solution in this case for most part of the hydraulic jump, except surface cell where the application of QUICKEST is not available. Future modification of this model should be focussed on the modification of differencial scheme near the surface cells.

## CONCLUSION

Two dimensional large eddy structure of hydraulic jump are studied both numerically and experimentally. The air-bubble entrainment process and transport process, as well as the vortices generation and transportation were visually analyzed by high-velocity video camera. Schematic movement characteristics were simulated by the two-dimensional MAC method introducing lagrangean air bubbles. The interaction between air bubbles and eddy structures was not clarified in this study, hence, not included in the model. Sediment picking-up process by large eddy was modeled introducing lagrangean sediment marker. This sediment picking-up process was observed in the video analysis, however, was not calibrated in detail in this study. Future study on the modeling of hydraulic jump should be focussed on the effect of air bubbles for liquid phases. Three dimensional model should be used for future study, however, it should be accompanied with adequate treatment for free surface and with better experimental data.

ACKNOWLEDGEMENT

The authors wish to thank Professor T. Sakai and H. Gotoh in Kyoto Univ. for many helpful suggestions including computational strategies, Dr. Jayme. Pinto, Eng. Rene Zaidan, Celso Aoki, Winston Kanashiro of University of Sao Paulo for many helpful suggestions, Prof. Y. Satofuka and other master students in DPRI for helping experimental study, Dr. K.Shimizu and Mr. S.Pareek in RCEQC - Kyoto Univ. for grammatical correction.

REFERENCES

1) DEJIANG Long et al. ; Structure of Flow in Hydraulic Jumps, J.of Hyd. Reser. IAHR, Vol.29, 1991, No.2.
2) Tchen, C., M. Equation of the motion for a particle suspended in an homogeneous field. Ph.D. Dissertation, Delft University, 1947. 3) Yukimasa TAKEMOTO et al. Numerical Solution program for viscous incompressible fluid flow with the Marker and Cell Method, 103 Y0/VIFMAC, Bulletin of Large Computer Center - Kyoto Univ. 1983. 4) T. Sakai, T. Mizutani, H. Tanaka and Y.Tada, Numerical Simulation of breaking wave on slope, Bulletin of 34th Coastal Engineering Symposium, JSCE, 1987, pp71 - 75. 5) Y. Murai and Y. Matsumoto, Flow Structure of Bubble Plume in Rectangular Tank, Proc. Of the 5th int. Symp on C.F.D., Vol.2, (1993), pp.285 - 290. 6) Hitoshi GOTOH, Study of sediment particle dynamics and its application for movable bed, Doctor Theses submitted to Kyoto University, 1992. 7)Crowe C.T., Sharma M.P. And Stock D.E.: The Particle-Source-In Cell(PSI-CELL) Model for Gas-Droplet Flows, Jour. Fluids Eng., pp.325-332, 1977. 8)M.KANO, T.KOBAYASHI, and T. ISHIHARA, Prediction of Turbulent Flow in Two-Dimensional Channel with Turbulence Promoters, Mechanical Engineering Research, No.50.449. 9) Smagolinsky, J., Monthly Weather Rev., 93(1963),pp.99. 10) Nadaoka, Yagi, and Nihei, Development of LES turbulent model for high concentrated solid-liquid mixture flow, Bulletin of Hydraulic Engineering, JSCE, No.38, 1994. 11) Y. Takemoto et al. A Computer Code for Time-dependent, Viscous, Incompressible Fluid Flows Using the Third-order Upwind Finite-difference Scheme called "QUICK. 132 y0/QMAC2D, J. of SuperComputer in Kyoto Univ. Vol.17, No,6, 1984. 12) Hiroji NAKAGAWA, T. TSUJIMOTO, Y.HOSOKAWA, Stochastic sediment movement in movable bed, DPRI Annual Bulletin, Kyoto Univ. No.22, B-2, pp.553-573, 1978

# Allowance for Secondary Flows in the 3 D Numerical Flow Models of Piping Systems

R. KLASINC, T. DUM, H. KNOBLAUCH, R. REITBAUER
Institut for Hydraulic Structures and Water Resources Management
Technical University of Graz
Austria

ABSTRACT

The subject of this paper is turbulent pipe flow, a phenomenon of great importance in the hydraulic structures of hydro power schemes. The aim of our studies was to test and adapt a three-dimensional numerical program for general use in hydraulics. A program has been developed that is based on the differential equations by Navier-Stokes, using the finite volume method for the discretisation of the flow field.
The numerical results were compared with the hydraulic parameters measured on the physical model. Two geometries were investigated: a sudden expansion of a pipe and a T-junction. A great number of pressure and velocity values were measured at different points, using the LDA system. For the sudden expansion, good agreement with the calculation using the k - $\varepsilon$ model was obtained. As to the T-junction, a strong secondary flow was seen to be present. In order to improve the results of the numerical calculation, the k - $\varepsilon$ model was substituted by a Reynolds stress model (RSM), which brought results that were considerably closer to the physical conditions.
This study has shown that even for complex flows, appropriate turbulent models are of practical use.

## INTRODUCTION

For optimising the hydraulic design of closed pipe systems for hydro power schemes as well as cooling and irrigation systems, it is necessary to conduct a great number of comparative investigations into the hydraulic parameters of the individual piping components. According to the current practice of pipe hydraulics studies, flows are considered as turbulent one-dimensional processes with a square law relationship between hydraulic loss and mean fluid velocity. The total head loss in a hydraulic system is equal to the sum of friction loss through the whole system and the shape-dependent losses caused by the individual conduit components. Naturally, this one-dimensional view is acceptable only for simple geometries. As soon as more complex cases are concerned, three-dimensional treatment becomes imperative. The aim of the studies dealt with in this article was to determine the practical use of numerical models in the calculations for hydraulic conduit systems. Studies should focus on conduit components such as profile extensions, bends, and junctions. We ought to find adequate numerical models permitting low-cost variant studies for a great number of different geometries. In addition, such a numerical model would help to determine local dissipation and forces acting on the walls. Attention should furthermore be given to the areas of subatmospheric pressure to avoid cavitation risks.

Another method of studying turbulent flow is by the use of physical flow models. These are generally reduced-scale reproductions of prototypes, built to allow laboratory study. Sometimes only part of the object is built in the laboratory, where flow mechanisms and their effects can be studied under controlled boundary conditions.

A model is considered similar to a real object where geometric, kinematic, and dynamic similarity is present. A basic requirement for obtaining similarity is the identification of the forces acting on the mechanisms of flow. Different model laws depending on the type of forces involved are used. However, there are certain limits to the use of the model laws for hydraulic tests. This is due to practical reasons. Flows that are much determined by viscosity forces will better be dealt with by use of the Reynolds model. The requirement that the velocity scale be inversely proportional to the longitudinal scale leads to high flow velocities in a reduced-scale model.

There is advantage in using numerical models in such cases. In view of the rapid development of computer technology, physical tests can increasingly be substituted by numerical solutions. An essential criterion for the use of numerical models is their behaviour in the case of secondary flows, and flow through a pipe bend involves substantial secondary flows (Fig. 1).

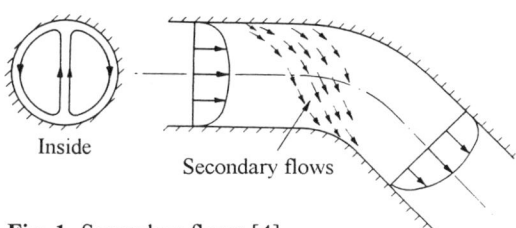

**Fig. 1** Secondary flows [4]

We conducted comparative studies between different numerical turbulence models, between the conventional k - ε model on the one hand and the newly introduced Reynolds Stress Model (RSM) on the other hand. Practical testing of the two models showed that in the case of secondary flow, calculations using the RSM gave better agreement with the physical conditions.

## PHYSICAL MODEL
### MODEL CONSTRUCTION

Both models were constructed in plexiglass piping with an inside diameter (D1, D) of 140 mm. The sudden expansion in Model 1 was reproduced by a plexiglass pipe with an inside diameter (D2) of 172 mm. Wall thickness was 5 mm. (Fig. 2 and 3)

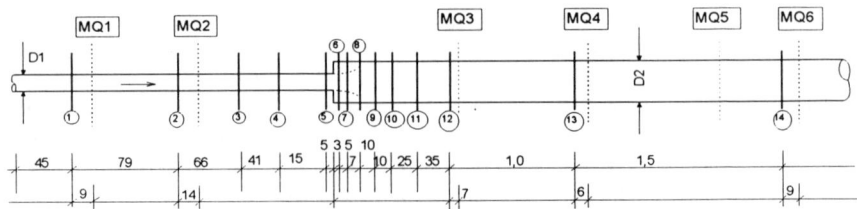

**Fig. 2** Model 1; cross sections for measuring differential pressure (MQ1, ...), cross sections for measuring by LDA (1, ..., 14).

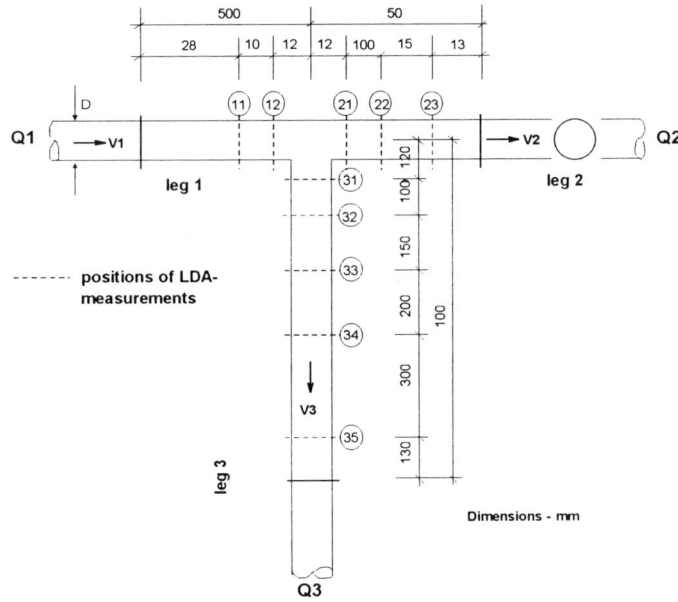

**Fig. 3** Model 2; cross sections for measuring by LDA.

## APPLIED MEASURING TECHNIQUES

Magnetic-inductive transducers were used for measuring discharge. Pressure data was registered by differential pressure transducers. A computer-controlled data acquisition device averaged the fluctuating data (3000 values in 20 seconds).

Velocity was measured using an LDA device with an argon-ion laser, which allowed measurement of two-dimensional and even three-dimensional flows. The test section was enclosed in a water-filled rectangular plexiglass box with access to the laser beam being provided through its bottom. This arrangement is shown in Figure 4 and was designed so as to allow two-dimensional coincident velocity measurement of axial and radial flows and to minimise refraction effects on the input laser beams at the cylindrical pipe surface.

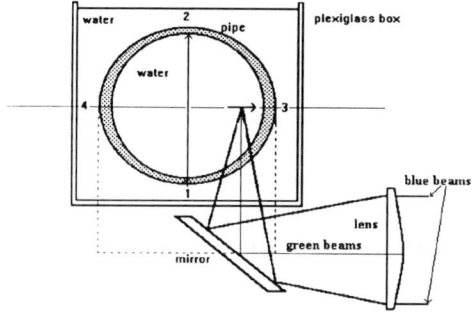

**Fig. 4** Experimental arrangement for the measurement with a two-dimensional LDA system

## NUMERICAL MODEL

The processes of flow are determined by the variables of velocity, pressure, density, and temperature. These are described using five conservation equations. The turbulence movement is described using methods of statistics. By putting the hydraulic parameters, averaged over time, into the conservation equations, the number of unknown quantities is increased. By introducing the so-called k-ε model, the number of differential equations is increased to 7. The turbulent kinetic energy, k, stands for the intensity of turbulence. The rate of dissipation, ε, is a measure of the kinetic energy converted into heat and sound. The finite volume method is used as the discretisation method. Apart from the k-ε model, another turbulence model is implemented in program FIRE [1], i.e. the Reynolds Stress Model (RSM) [2]. The program provides for the option to use one of the two turbulence models. Unlike the k-ε model, the RSM dissolves the fluctuating quantities of velocity and pressure into their spatial components. This fluctuating movement leads to additional stresses (first derived by Reynolds) at the volume element. The Reynolds Stress Model differs from the k-ε model in calculating the components of these stresses for all the three directions:

$$\sigma_x = -\rho \cdot \overline{u^2}; \quad \tau_{xy} = -\rho \cdot \overline{u \cdot v}; \quad \tau_{xz} = -\rho \cdot \overline{u \cdot w}; \quad \sigma_y = -\rho \cdot \overline{v^2}; \ldots$$

In this context, we refer to the tensor (i.e. the matrix) of the Reynolds stresses. With account being taken of its symmetry, this possesses six components, which have also to be modelled for the Reynolds Stress Model, that is, six additional differential equations have to be solved. This naturally implies substantial additional computing work. But such a model is particularly suited to allow for the anisotropy of turbulence, which is important especially near the walls and in zones where the flow tends to separate from the walls. [5]

Flow processes in which secondary flows are important can in this way be calculated more accurately than by use of the k-ε model. In addition, the Reynolds Stress Model yields fluctuation velocities $\overline{u^2}$, $\overline{u \cdot v}$, $\overline{u \cdot w}$, $\overline{v^2}$ ....
Values obtained from LDA measurement can also be used to determine the fluctuation quantities of the velocities for the physical model. These rms values (rms - root mean square; $\sqrt{\overline{u^2}}$ corresponds to rms in the u direction) can directly be compared with the results from the Reynolds Stress Model. A serious disadvantage of the Reynolds Stress Model is the substantial extra computing work involved and a more problematic computing process with respect to reaching convergence of the solutions.
The equations for treating flow behaviour near the walls correspond to the formulas used for the k-ε model. [3]

## RESULTS

For sudden pipe expansions (Fig. 2), the results obtained from the numerical k-ε model were very close to the results measured on the physical model, so it was not necessary to carry out further analyses using RSM. In order to compare the results from the two numerical models, we tested flow through a T-junction (Fig. 5 and 6). This paper only presents the case of flow in one of the legs of the junction (90°).

**Fig. 5** Comparison of the kinetic energy of the two numerical models,
a) RSM model and b) k-ε model

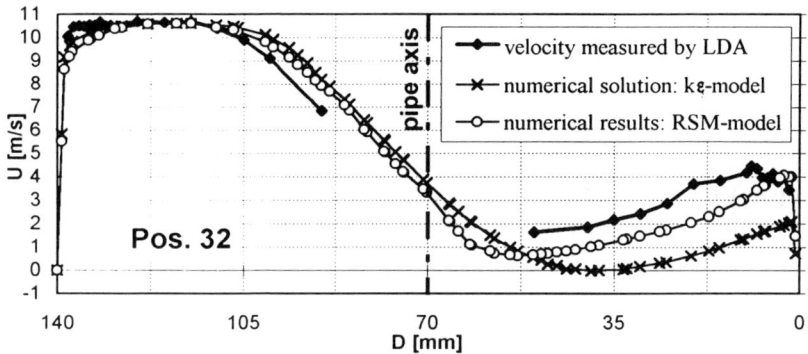

Fig. 6 Comparison of numerical (RSM and k-ε model) and physical model (LDA measurement), leg 3, position 32 (Fig. 3)

## CONCLUSION

In order to test the accuracy of numerical flow models allowing for secondary flows, the well-tried k-ε model for turbulent flows was compared with the newly introduced Reynolds Stress Model (RSM). Where the proportion of secondary flows was low, agreement between the results of the k-ε model and the measured results was satisfactory. A test using a T-junction showed, however, that the RSM was better suited to yield reliable numerical results. We plan to apply the RSM to study complex geometries as are found in the hydraulic structures of hydro power schemes.

## ACKNOWLEDGEMENTS

The authors wish to express their thanks to the Verbundgesellschaft electricity board Austria for the ready support they gave to our studies.

## REFERENCES

[1] AVL (1995). "FIRE - Instruction Manual," Version 6.2, Graz
[2] Basara, B., Plimon, A., Bachler, G. (1996). *On the Simulation of Turbulent Flow and Heat Transfer in an Intake Manifold.* 6. Int. Conf. Flow Modelling, Florida
[3] Dum, T. (1996). *Verifikation eines numerischen Strömungsmodells anhand physikalischer Modelle*, Dissertation, TU-Graz
[4] Miller, D. S. (1990). *Internal Flow Systems.* BHRA (Information Services)
[5] Oertel, H., LAURIEN, E. (1965). *Numerische Strömungsmechanik.* Springer-Verlag, Berlin Heidelberg New York

# NUMERICAL SIMULATION OF FLOW IN A CHANNEL WITH A WAVY SIDEWALL

**Thomas Molls and Gang Zhao**
Southern Illinois University, Carbondale, IL 62901-6603

## Abstract

Supercritical flow in a channel with a wavy sidewall is numerically simulated by solving the 2D depth-averaged equations using a finite-difference model DASH (Molls and Chaudhry 1995, Molls et. al. 1995). Time differencing is accomplished using a second-order accurate Beam and Warming approximation; while, the spatial derivatives are approximated by second-order accurate central differencing. The equations are solved in transformed computational coordinates using an alternating-direction-implicit (ADI) scheme. The turbulent and depth-averaging stresses are neglected. Vector and contour plots are used to qualitatively depict the flow. Near the straight sidewall, the numerical results are in close accordance with experimental data obtained by Mizumura (1995).

## Introduction

Most channels (natural and man-made) have irregular boundaries. Obviously, boundary shape can greatly affect the velocity and water surface profile in an open-channel. Particularly challenging is the simulation of supercritical flow in a channel with a wavy sidewall. In this case, the flow is complicated due to the formation of cross waves and the behavior of a numerical model under these severe conditions is especially revealing.

## 2D Depth-Averaged Equations

Assuming a hydrostatic pressure distribution and neglecting the effective and depth-averaging stresses, the governing equations in conservative vector form are as follows (Chaudhry 1993, Molls and Chaudhry 1995):

$$\frac{\partial \mathbf{Q}}{\partial t} + \frac{\partial \mathbf{E}}{\partial x} + \frac{\partial \mathbf{F}}{\partial y} = \mathbf{G}$$

where

$$\mathbf{Q} = \begin{pmatrix} h \\ hU \\ hV \end{pmatrix} \quad \mathbf{E} = \begin{pmatrix} hU \\ hU^2 + gh^2/2 \\ hUV \end{pmatrix} \quad \mathbf{F} = \begin{pmatrix} hV \\ hUV \\ hV^2 + gh^2/2 \end{pmatrix} \quad \mathbf{G} = gh \begin{pmatrix} 0 \\ S_{ox} - S_{fx} \\ S_{oy} - S_{fy} \end{pmatrix}$$

where $S_o$ = channel bottom slope and $S_f$ = friction slope. Usually, $S_f$ is quantified using the Manning or Chezy formula.

## Experimental Verification Data

The numerical model was applied to a channel with one straight sidewall and one wavy sidewall made of smooth steel (sine wave with amplitude of 0.025 m and period of 1 m). The channel was 10 m long and 0.3 m wide with a bottom slope of 0.0129. The flow through the channel was 0.0176 m³/s corresponding to an average Froude number of 1.57. The water depth in the channel, over a 3 m segment, was experimentally obtained by Mizumura (1995).

## Model Parameters

The computations were performed on a 91x21 grid with $\Delta x$=0.067 m and $\Delta y$=variable (see Fig. 1). At the inflow boundary, the water depth and velocity were assumed constant. An average inlet water depth was calculated from the experimental data. Since Mizumura (1995) did not obtain velocity data, an average inlet velocity was determined ($U_{up}$=Q/A) and the inlet V velocity was set to zero. These boundary conditions assured proper inflow into the channel. At the outlet, all variables were extrapolated from interior points. Free-slip conditions were applied at the sidewall boundaries. The model was run with a Chezy C of 100.

## Results

The computed solution is qualitatively depicted in Figs. 2 and 3. Fig. 2 shows velocity vectors and streamlines in the channel. The vector plot indicates a surprisingly low velocity near the straight sidewall. Away from the straight sidewall, the velocities are larger and highly skewed. The streamlines highlight the two-dimensional nature of the flow. Fig. 3 and Fig. 4, a contour plot and a carpet plot of the water depth, clearly depict the cross waves in the channel.

Figures 5 and 6 compare the computed water depth with experimental data (Mizumura 1995) near the straight sidewall and near the channel centerline, respectively. At the inlet, to insure proper inflow, the velocity and water depth were assumed constant across the channel; therefore, the solution was allowed to develop for two meters before comparison with the experimental data. Also, since the outlet boundary was extrapolated, no comparisons were made near this boundary. In general, the agreement between the computed and experimental results is favorable. Near the straight sidewall (Fig. 5), the computed and experimental water depths are nearly in phase; however, the computed solution slightly underpredicts the experimental wave peaks and overpredicts the troughs. Near the channel centerline (Fig. 6), agreement between the numerical results and experimental data is somewhat compromised - here, the phase lag is more pronounced.

As expected, including channel resistance (i.e. Chezy C) improves the agreement between the numerical and experimental data. Including resistance tends to increase the predicted water depth and shorten the wave period. Also, near the channel centerline (Fig. 6), the model is more sensitive to the inclusion of resistance.

## Summary and Conclusions

In this paper, the 2D depth-averaged equations were applied to supercritical flow in a channel with a wavy sidewall. The model results were depicted using vector, contour, and carpet plots to clearly indicating the nature of the flow. Furthermore, the computed solution was compared with experimental data obtained by Mizumura (1995). Along the straight sidewall, the computed and experimental water depths were in phase; however, the model did not precisely predict the wave peaks and troughs. Near the channel centerline, the agreement between numerical and experimental data is somewhat compromised. As expected, including channel resistance (Chezy C) improves the agreement between the numerical model and experimental data.

## References

Chaudhry, M.H. (1993). *Open-channel flow*. Prentice Hall, Englewood Cliffs, N.J.

Mizumura, K. (1995). "Free-surface profile of open-channel flow with wavy boundary.", *J. Hydr. Engrg.*, ASCE, 121(7), 533-539.

Molls, T.R., and Chaudhry, M.H. (1995). "Depth-averaged open-channel flow model.", *J. Hydr. Engrg.*, ASCE, 121(6), 453-465.

Molls, T.R., Chaudhry, M.H., and Khan, K.W. (1995). "Numerical simulation of two-dimensional flow near a spur-dike.", *Advances in Water Resources*, 18(4), 227-236.

FLOW IN CHANNEL WITH WAVY SIDEWALL 243

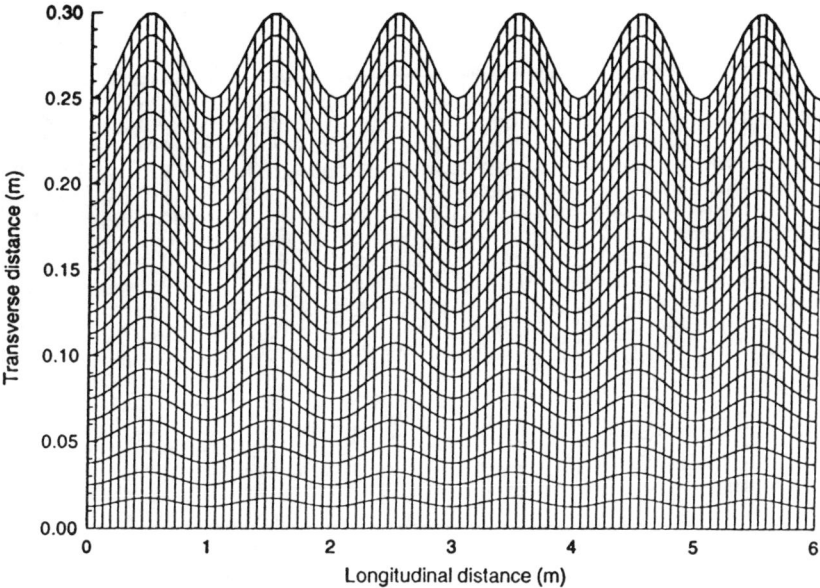

Figure 1 - Computational grid (note: grid not to scale)

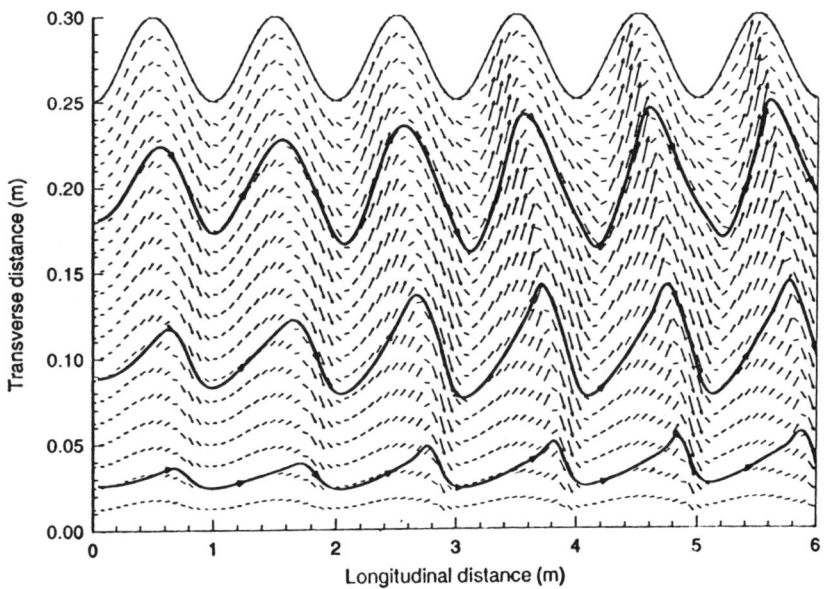

Figure 2 - Velocity vector plot

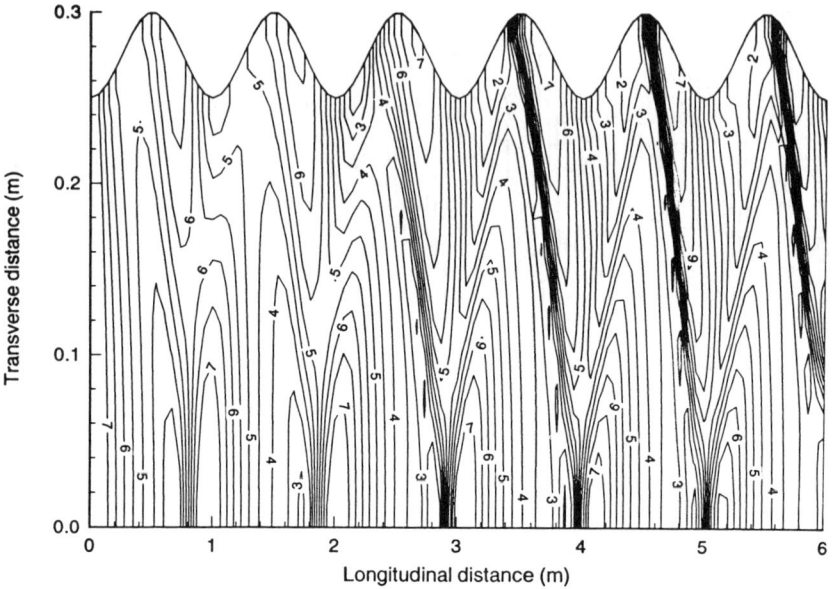

**Figure 3** - Water depth contour plot (contours in cm)

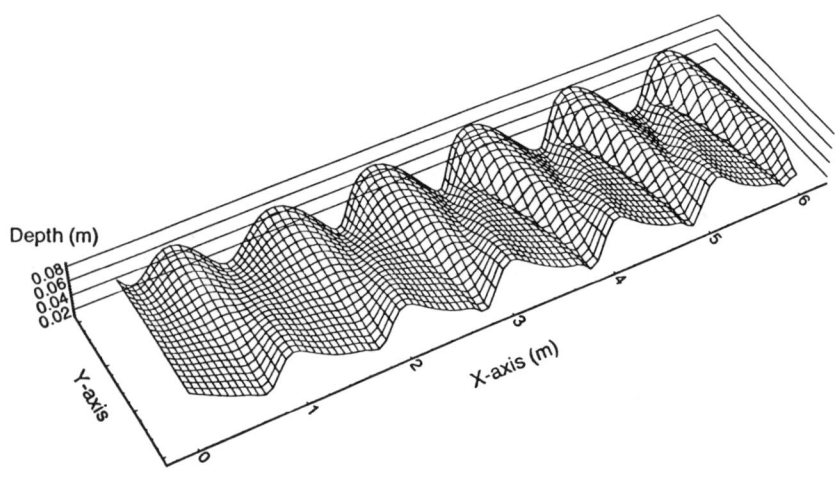

**Figure 4** - Water depth carpet plot

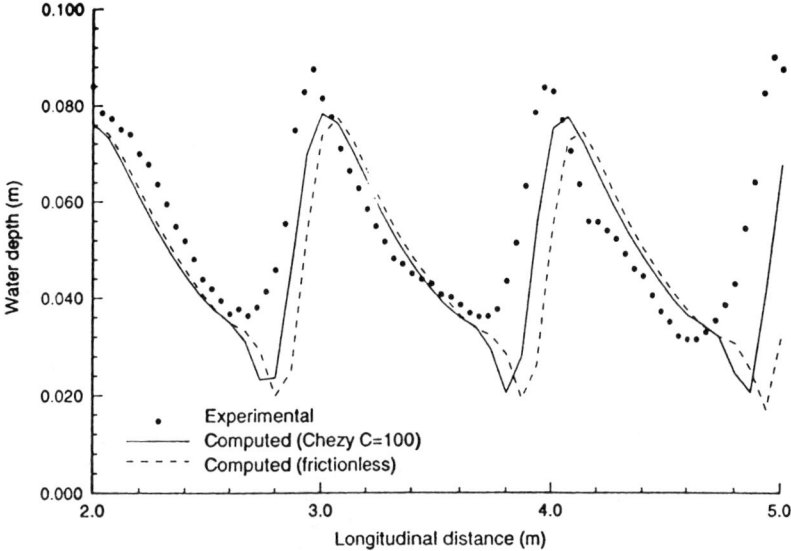

**Figure 5** - Water depth 15 cm from straight sidewall

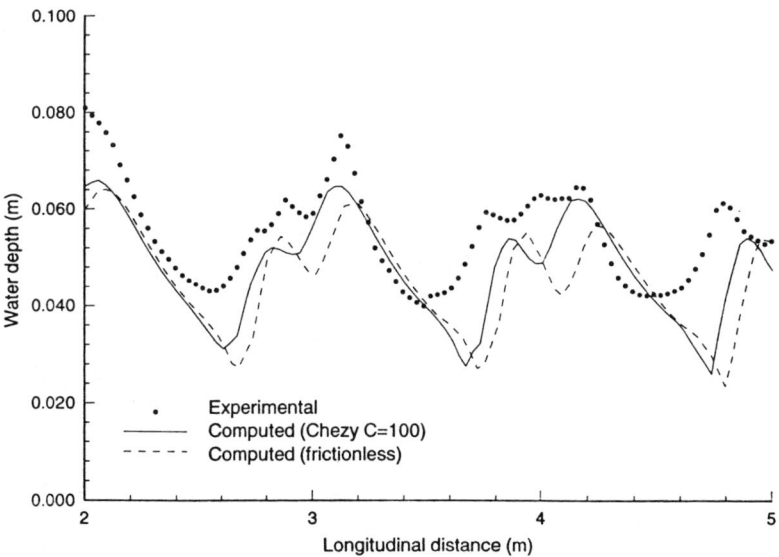

**Figure 6** - Water depth 105 cm from straight sidewall

# 3-D NUMERICAL SOLUTION OF FLOW IN SINE-GENERATED MEANDERING COMPOUND CHANNEL

## H.S.JIN[*], S.EGASHIRA[*] and B.Y.LIU[+]

[*] Faculty of Science & Engineering, Ritsumeikan Univ., Kusatsu, Shiga 525, Japan
[+] The NEWJEC, Inc., 20-19 Shimanouchi 1-chome, Chuo-ku, Osaka 542, Japan

### ABSTRACT

Three-dimensional solutions for flows in a single as well as a compound sine-generated meandering channel are obtained numerically. A sigma transformation in vertical co-ordinate is employed for following the dynamic free water surface. The results on main flow shifting, secondary current pattern, water surface super-elevation and so forth agree closely with the measured data of flume tests. The secondary current direction in most locations of compound channel is opposite to that in single channel.

### 1. INTRODUCTION

Fluid flows should be described in three-dimension, especially local phenomena concerned, such as curved channel flow and bed deformation, flow around hydraulic structure, compound channel flow, etc. During the last decade, computational fluid dynamics has brought about extraordinary progress because a number of researchers have devoted a good deal of effort to the subject and computer technology has made rapid advances, thus making it realistic to simulate more complex flow phenomena.

A full three-dimensional description requires the determination of free water surface from kinematic and dynamic boundary conditions and more sophisticated computation. To reduce this computational difficulty, some investigators, such as Lardner and Cekirage (1988), have proposed a mode splitting approach, i.e., coupled the shallow-water and full 3D models to obtain the free-surface elevation and the 3D flow field. Several other 3D modelings have been reported in the literatures, e.g., Casulli and Cheng (1992), among others. Herein a three-dimensional numerical model, based on boundary-fitted orthogonal curvilinear co-ordinates (Chau and Jin, 1995) in horizontal plane and vertical sigma-transformation, is developed. The method does not incorporate mode splitting, but instead solves directly the primitive three-dimensional governing equations. These equations are discretized and solved by an implicit finite difference scheme. The continuity equation, incorporated with the relationship between velocity components and water surface level based on the momentum equations, is solved firstly to obtain the advanced water surface level. Then the momentum equations are solved to get the advanced velocity with the new water surface. The method has been applied to simulate single and compound sine-generated meandering channel flows. The results are compared with the measurements.

## 2. FORMULATION OF THREE-DIMENSIONAL MODELING

For modeling water flow with irregular physical boundaries, a boundary-fitted orthogonal curvilinear co-ordinate system is much preferable. However, it is not so easy to generate numerically the body-fitted co-ordinates in three-dimension, especially for those with unsteady boundaries, including bank boundary and free water surface, e.g., tidal flow, flow in alluvial river, etc. In this study, shallow water problem is concerned, numerically-generated boundary-fitted orthogonal curvilinear co-ordinates ($\xi$ and $\eta$) (Chau and Jin, 1995) and a sigma co-ordinate ($\theta$) are employed for horizontal plane as shown in fig.1(c) and vertical direction, respectively. The sigma co-ordinate is defined as $\theta = (z - Z_b)/h$, in which z is the vertical Cartesian co-ordinate ranging from $z=Z_b$ at floor ($\theta=0$) and $z=\zeta$ at water surface ($\theta=1$), and $h = \zeta - Z_b$ is the water depth. By using the sigma approach, water column is divided into the same number of layers which are independent of the water depth. The main advantage of such system is the fact that it fits to both the moving free surface and the bottom topography.

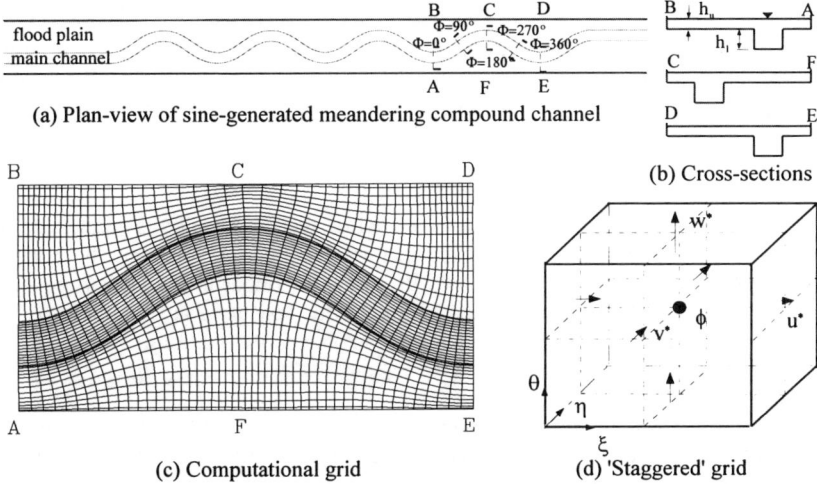

Fig.1 Sine-generated meandering compound channel and computational grid

### 2.1 GOVERNING EQUATIONS

For a shallow water problem, the vertical accelerations are usually neglectable in the vertical momentum equation, which leads to the hydrostatic pressure equation. Thus the governing equations for three-dimensional flow in ($\xi$, $\eta$, $\theta$, t) — system with the hydrostatic pressure and Boussinesq's approximation are expressed as follows:

$$\frac{\partial h}{\partial t} + \frac{1}{g_{11}g_{22}}\left[\frac{\partial(u^* h g_{22})}{\partial \xi} + \frac{\partial(v^* h g_{11})}{\partial \eta}\right] + \frac{\partial w^*}{\partial \theta} = 0 \qquad (1)$$

$$\frac{1}{h}\frac{\partial hu^*}{\partial t} + \frac{1}{hg_{11}g_{22}}[\frac{\partial(g_{22}hu^{*2})}{\partial \xi} + \frac{\partial(g_{11}hu^*v^*)}{\partial \eta}] + \frac{\partial u^*w^*}{h\partial \theta} + \frac{u^*v^*}{g_{11}g_{22}}\frac{\partial g_{11}}{\partial \eta} - \frac{v^{*2}}{g_{11}g_{22}}\frac{\partial g_{22}}{\partial \xi}$$
$$= -\frac{g}{g_{11}}\frac{\partial \zeta}{\partial \xi} + \frac{1}{g_{11}g_{22}}[\frac{\partial(g_{22}\sigma_1)}{\partial \xi} + \frac{\partial(g_{11}\tau_{21})}{\partial \eta}] + \frac{\partial \tau_{31}}{h\partial \theta} + \frac{\tau_{12}}{g_{11}g_{22}}\frac{\partial g_{11}}{\partial \eta} - \frac{\sigma_2}{g_{11}g_{22}}\frac{\partial g_{22}}{\partial \xi} + fv^* \quad (2)$$

$$\frac{1}{h}\frac{\partial hv^*}{\partial t} + \frac{1}{hg_{11}g_{22}}[\frac{\partial(g_{22}hu^*v^*)}{\partial \xi} + \frac{\partial(g_{11}hv^{*2})}{\partial \eta}] + \frac{\partial v^*w^*}{h\partial \theta} + \frac{u^*v^*}{g_{11}g_{22}}\frac{\partial g_{22}}{\partial \xi} - \frac{u^{*2}}{g_{11}g_{22}}\frac{\partial g_{11}}{\partial \eta}$$
$$= -\frac{1}{g_{22}}\frac{\partial \zeta}{\partial \eta} + \frac{1}{g_{11}g_{22}}[\frac{\partial(g_{22}\tau_{12})}{\partial \xi} + \frac{\partial(g_{11}\sigma_2)}{\partial \eta}] + \frac{\partial \tau_{32}}{h\partial \theta} + \frac{\tau_{21}}{g_{11}g_{22}}\frac{\partial g_{22}}{\partial \xi} - \frac{\sigma_1}{g_{11}g_{22}}\frac{\partial g_{11}}{\partial \eta} - fu^* \quad (3)$$

in which, $u^*$, $v^*$, $w^*$ are the $\xi$, $\eta$, $\theta$ components of flow velocity, respectively, and these are defined as: $u^* = (ux_\xi + vy_\xi)/g_{11}$, $v^* = (ux_\eta + vy_\eta)/g_{22}$ and $w^* = hd\theta/dt$. Herein u and v are the velocity components in Cartesian co-ordinates. g and f are the gravitational acceleration and Coriolis coefficient, respectively. $\sigma_i$ and $\tau_{ij}$ are the effective normal and shear stress which are evaluated with the Boussinesq's approximation (Rodi, 1980), and simply the effective viscosity ($v_t$) is calculated by empirical formula, i.e., $v_t \sim hu_*$ (Fischer et al., 1979). $u_* = [C_f(u^{*2} + v^{*2})]^{1/2}$ is the friction velocity. $C_f$ is the resistance coefficient at the bed surface which is evaluated in terms of the velocity near the bottom: $C_f = 1/[8.5 + 2.5\ln(z'/k_s)]^2$ in which $k_s$ is the roughness height of the bed surface and $z'$ is the distance away from the bed with an order of $k_s$. $g_{11}$ and $g_{22}$ are the orthogonal curvilinear co-ordinate transformation relationship (Chau and Jin, 1995).

## 2.2 INITIAL AND BOUNDARY CONDITIONS

At the open boundaries, the water surface level or flow rate which is converted into velocity with a common empirical formula is specified. At the wall boundaries, impenetrable condition is employed, i.e., the velocity normal to the bank is equal to zero and the tangential component, conforming to the resistance law, is not zero. The initial conditions correspond to the initial value in boundary conditions.

## 2.3 NUMERICAL METHOD

A "staggered" grid arrangement is employed as shown in fig.1(d), i.e., velocity component is stored at corresponding control volume face. The above governing equations are discretized and solved by a specially developed finite difference scheme. At first, the continuity equation is solved, in which the advanced velocity components are implicitly expressed in terms of the water surface level based on the momentum equations as: $\tilde{u}^*_{i+\frac{1}{2},j,k} = A^u_{i,j,k}(\zeta^{n+1}_{i+1,j} - \zeta^{n+1}_{i,j}) + B^u_{i,j,k}$, $\tilde{v}^*_{i,j+\frac{1}{2},k} = A^v_{i,j,k}(\zeta^{n+1}_{i,j+1} - \zeta^{n+1}_{i,j}) + B^v_{i,j,k}$. Nonlinear terms in eq.(1) are linearized as: $hu^* = (\zeta - Z_b)u^* = \zeta^{n+1}u^{*n} + (\zeta^n - Z_b)\tilde{u}^* - \zeta^n u^{*n}$. $A^u_{i,j,k}, \ldots, B^v_{i,j,k}$ are the coefficients. The index 'n' denotes the time level of discretized hydrodynamic variables. i, j and k are the node numbers in $\xi$, $\eta$ and $\theta$ direction, respectively. Integrating vertically the discretized continuity equation, an algebraic equation set for water surface level is formulated as:

## MEANDERING COMPOUND CHANNEL

$$a_{P_{i,j}}\zeta_{i,j}^{n+1} = a_{E_{i,j}}\zeta_{i+1,j}^{n+1} + a_{W_{i,j}}\zeta_{i-1,j}^{n+1} + a_{N_{i,j}}\zeta_{i,j+1}^{n+1} + a_{S_{i,j}}\zeta_{i,j-1}^{n+1} + b_{i,j} \quad (4)$$

where $a_{P_{i,j}}$, $a_{E_{i,j}}$, $a_{W_{i,j}}$, $a_{N_{i,j}}$, $a_{S_{i,j}}$ and $b_{i,j}$ are the coefficients. Solving the above equation set by Tri-Diagonal Matrix Algorithm (TDMA) with Alternating Direction Iteration, the water surface elevation $\zeta^{n+1}$ at advanced time is determined. Then the momentum equations are discretized with the advanced $\zeta^{n+1}$ using backward, upwind and central finite differences for the time derivation, advection and diffusion terms, respectively. These result in an implicit numerical algorithm. The algebraic equation sets are solved iteratively by TDMA. We thus obtain the solution of horizontal velocities at advanced time. Finally the vertical velocity component can be found by integrating eq.(1) in a control volume. Thus, one time step calculation is completed. Executing the procedure step by step, the water surface level and velocity are solved.

### 3. FLOW IN SINE-GENERATED MEANDERING CHANNEL

The present method is employed to simulate the flows in compound channel as well as in single one, respectively. Computed results are compared with flume data reported by Liu (1991). The flume is 12m long and 1m wide with a central located main channel. The main channel is 20cm wide and 3cm deep, and its centreline is taken as sine-generated-curve, $\theta = \theta_{max}\sin(2\pi s/L)$, in which the maximum angular deviation ($\theta_{max}$) of the centreline is $35°$ and the wave length L along the centreline is 2.2m. Four and a half waves are formed, as shown in Fig.1(a) and (b). The flume slope is 0.009.

### 3.1 FLOW IN A SINGLE CROSS-SECTION CHANNEL

A single sine-generated meandering channel flow in which the flow discharge is 0.87 l/s and average water depth about 1.4 cm is simulated by the present method. The computed and measured contours of longitudinal velocity at $\Phi=0°$, $90°$ and $180°$ sections and water surface level are shown in figs.2 and 3, respectively. Fig.4 shows the pattern of secondary currents at $\Phi=90°$ and $180°$ sections.

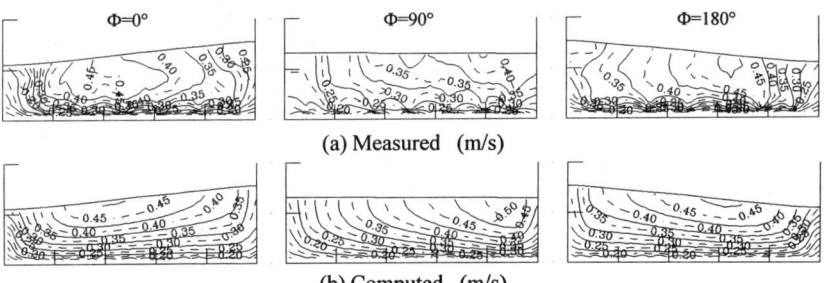

Fig.2 Main flow contour in sine-generated meandering channel

The present method reproduces correctly water surface super-elevation as well as the location of maximum main flow velocity region in such a sine-generated meandering channel. The maximum main flow region is located near the inner bank at bend flex (at $\Phi=0°$, $180°$ or $360°$) in such a sine-generated meandering channel with flat bed. With attention focused on the secondary currents around the bend flex, it is the apparent nature that the flow points to the outer bank near surface and to the inner bank near bottom, respectively. A marked relief of water surface forms near the outer bank. Around Φ=90° and Φ=270°, water surface is calm a little. Secondary currents, therefore, are weak and tend to the same direction. These show that the simulated results mimic well the measurements in both magnitude and shape of water surface, velocities of main flow and secondary currents.

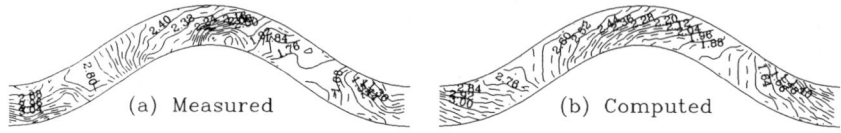

Fig.3 Water surface level contour in sine-generated curved channel (cm)

## 3.2 COMPOUND CHANNEL FLOW

Data for a compound channel flow with a central located sine-generated main channel were obtained from the flume test (Liu, 1991), in which the flow discharge was 7.15 l/s and mean water depth in main channel about 4.4 cm. The computed water surface level and

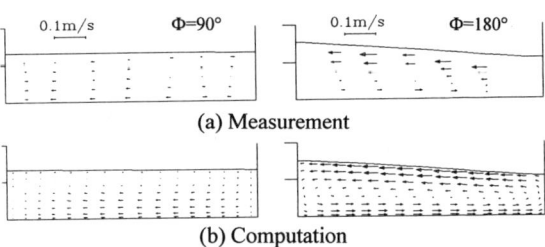

Fig.4 Secondary current in sine-generated bend channel

secondary currents in main channel region are compared with the experimental measurements in figs.5 and 6, respectively. The simulated results agree closely with the measured data. The corner vortex in the main channel is reproduced fairly well by the present method. Both of the experimental and computed results show that compound channel flow is characterized by such natures, uneven water surface over the main channel, corner-induced secondary currents, accelerated or decelerated secondary currents in inbank region, cross currents over the main channel and so forth. In addition, it is noticeable that the vortex vector of secondary currents tends to downstream at Φ=90° and to upstream at Φ=180°, respectively, and is inverse to that of the single meandering channel (see figs.4 and 6). It does not conform to the result predicted by means of the depth-integrated model with a streamline method (Jin et al., 1995). Around the concave region of bend flex (at Φ=0°, 180° or 360°), an independent secondary current forms near the water surface.

## 4. CONCLUSIONS

Based on the primitive 3-D governing equations, a three-dimensional numerical model for free water surface flows is developed with a rigorous implicit finite difference scheme. Numerically generated boundary-fitted orthogonal curvilinear co-ordinates in horizontal plane and vertical sigma-transformation are useful for treating the irregular and dynamic boundaries. The numerical 3D solutions for flows in a single sine-generated meandering channel as well as in a compound one mimic the measured data in flume tests quite well. The directions of secondary current vortex can be predicted well for compound channel flow by present method. They are not reproduced correctly by means of 2-D depth-integrated streamline method.

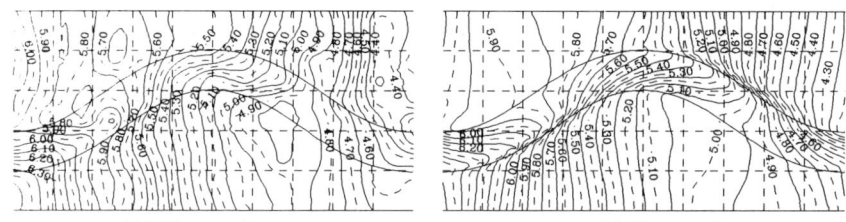

(a) Measured    (b) Computed
Fig.5 Water surface level in sine-generated compound channel (cm)

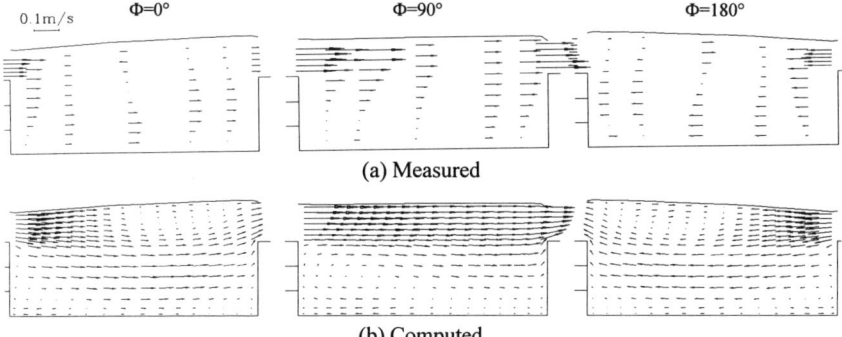

Fig.6 Secondary currents in sine-generated compound channel flow

## REFERENCES

Casulli, V. and Cheng, R.T. (1992). Semi-implicit finite difference methods for three-dimensional shallow water flow. I.J. for Numerical Method in Fluids, 15, 629-648.

Chau, K.W. and Jin, H.S. (1995). Numerical solution of two-layered, two-dimensional tidal flow in boundary-fitted orthogonal curvilinear co-ordinate system. International Journal for Numerical Method in Fluids, 21(11), 1087-1107.

Fischer, H.B., List, E.J., Koh, R.C.Y., Imberger, J. and Brooks, N.K. (1979). *Mixing in inland and coastal waters.* Academic Press, Inc.

Jin, H.S., Egashira, S. and Liu, B.Y. (1995). Characteristics of meandering compound channel flow evaluated with two-layered, 2-d method. Annual Journal of Hydraulic Engineering, JSCE, 40, 717-724.

Lardner, R.W. and Cekirge, H.M. (1988). A new algorithm for three-dimensional tidal and storm surge computations. Applied Mathematical Modelling, 12, 471-481.

Liu, B.Y. (1991). Study on sediment transport and bed evolution in compound channels. Doctoral Dissertation, Kyoto University, Japan.

Rodi, W. (1980). *Turbulence models and their application in hydraulics, state-of-the-art.* IAHR Publication, DELFT, The Netherlands.

# 3-D TURBULENT FLOW SIMULATION FOR THE TAIL WATER CHANNEL

## GUAN JIANYONG    CHEN BIHONG
(Dalian University of Technology, Dalian, China, 116023)

**ABSTRACT**
In this paper, the simulation of 3-D turbulent flow with free surface in the tail water channel is presented. The outflow from the turbine is discharged through the draft tube into the inlet of the tail water channel. The governing equations are the three dimensional incompressible N-S equations and K-$\varepsilon$ turbulent model equations which are discretized by control volume method to predict the mean flow and turbulence. The method uses to predict the free water surface profile is based on the concept of a fractional volume of fluid[4] (VOF). This model is used to simulate the turbulent flow field in the tail water channel of Fengman Hydropower Plant in China.

## INTRODUCTION

In hydraulic engineering, turbulent flow is one of the complicated problems. Several studies have been made to describe 2-D turbulent flow[1][2][3]. However, most practical problems are 3-D flow field. Then a three dimensional K-$\varepsilon$ turbulent model is needed to be developed to improve the simulation of engineering practice. As a part of the tail race river of hydropower plant, the tailwater channel flow field plays an important role in gaining energy. In this paper, the inlet flow discharged from the draft tube of a hydropower plant is subcritical, since its Froude number is smaller than 1.

The 3-D N-S equations and their performance in the tail water channel flow are described in this paper. VOF method is used here to treat the free surface boundary configurations. The finite volume method is used to discretize the governing equations.

## GOVERNING EQUATIONS

The equation of continuity, momentum equations and the K-$\varepsilon$ turbulent modeling equations for three dimensional, incompressible turbulent flow are as follows:

$$\frac{\partial U}{\partial X} + \frac{\partial V}{\partial Y} + \frac{\partial W}{\partial Z} = 0 \tag{1}$$

$$\frac{\partial U}{\partial t} + \frac{\partial UU}{\partial X} + \frac{\partial UV}{\partial Y} + \frac{\partial UW}{\partial Z} = -\frac{1}{\rho}\frac{\partial P}{\partial X} + \frac{\partial}{\partial X}(\tau_{XX}) + \frac{\partial}{\partial Y}(\tau_{XY}) + \frac{\partial}{\partial Z}(\tau_{XZ}) \tag{2}$$

$$\frac{\partial V}{\partial t} + \frac{\partial UV}{\partial X} + \frac{\partial VV}{\partial Y} + \frac{\partial VW}{\partial Z} = -\frac{1}{\rho}\frac{\partial P}{\partial Y} + \frac{\partial}{\partial X}(\tau_{YX}) + \frac{\partial}{\partial Y}(\tau_{YY}) + \frac{\partial}{\partial Z}(\tau_{YZ}) \tag{3}$$

$$\frac{\partial W}{\partial t} + \frac{\partial UW}{\partial X} + \frac{\partial VW}{\partial Y} + \frac{\partial WW}{\partial Z} = g_Z - \frac{1}{\rho}\frac{\partial P}{\partial Z} + \frac{\partial}{\partial X}(\tau_{ZX}) + \frac{\partial}{\partial Y}(\tau_{ZY}) + \frac{\partial}{\partial Z}(\tau_{ZZ}) \tag{4}$$

$$\tau_{XX} = 2v_t \frac{\partial U}{\partial X} - \frac{2}{3}K \quad (5)$$

$$\tau_{YY} = 2v_t \frac{\partial V}{\partial Y} - \frac{2}{3}K \quad (6)$$

$$\tau_{ZZ} = 2v_t \frac{\partial W}{\partial Z} - \frac{2}{3}K \quad (7)$$

$$\tau_{XY} = \tau_{YX} = v_t(\frac{\partial U}{\partial Y} + \frac{\partial V}{\partial X}) \quad (8)$$

$$\tau_{XZ} = \tau_{ZX} = v_t(\frac{\partial U}{\partial Z} + \frac{\partial W}{\partial X}) \quad (9)$$

$$\tau_{YZ} = \tau_{ZY} = v_t(\frac{\partial V}{\partial Z} + \frac{\partial W}{\partial Y}) \quad (10)$$

$$v_t = C_\mu \frac{K^2}{\varepsilon} \quad (11)$$

$$\frac{\partial K}{\partial t} + \frac{\partial UK}{\partial X} + \frac{\partial VK}{\partial Y} + \frac{\partial WK}{\partial Z} = \frac{\partial}{\partial X}(\frac{v_t}{\sigma_K}\frac{\partial K}{\partial X}) + \frac{\partial}{\partial Y}(\frac{v_t}{\sigma_K}\frac{\partial K}{\partial Y}) + \frac{\partial}{\partial Z}(\frac{v_t}{\sigma_K}\frac{\partial K}{\partial Z}) + G - \varepsilon \quad (12)$$

$$\frac{\partial \varepsilon}{\partial t} + \frac{\partial U\varepsilon}{\partial X} + \frac{\partial V\varepsilon}{\partial Y} + \frac{\partial W\varepsilon}{\partial Z} = \frac{\partial}{\partial X}(\frac{v_t}{\sigma_\varepsilon}\frac{\partial \varepsilon}{\partial X}) + \frac{\partial}{\partial Y}(\frac{v_t}{\sigma_\varepsilon}\frac{\partial \varepsilon}{\partial Y}) + \frac{\partial}{\partial Z}(\frac{v_t}{\sigma_\varepsilon}\frac{\partial \varepsilon}{\partial Z}) + C_1\frac{\varepsilon}{K}G - C_2\frac{\varepsilon^2}{K} \quad (13)$$

$$G = v_t\{(\frac{\partial U}{\partial Y} + \frac{\partial V}{\partial X})^2 + (\frac{\partial V}{\partial Z} + \frac{\partial W}{\partial Y})^2 + (\frac{\partial W}{\partial X} + \frac{\partial U}{\partial Z})^2 + 2[(\frac{\partial U}{\partial X})^2 + (\frac{\partial V}{\partial Y})^2 + (\frac{\partial W}{\partial Z})^2]\} \quad (14)$$

$$\frac{\partial F}{\partial t} + \frac{\partial FU}{\partial X} + \frac{\partial FV}{\partial Y} + \frac{\partial FW}{\partial Z} = 0 \quad (15)$$

Where U, V, W are the components of velocity in the X, Y, Z directions respectively; P is the pressure; $K, \varepsilon$ are the turbulent kinetic energy and its dissipation rate respectively; $v_t$ is the turbulent eddy viscosity; G is the turbulent production term; t is time; F is the volume of fluid function. The values of constants $C_1$, $C_2$, $C_\mu$, $\sigma_k$ and $\sigma_\varepsilon$ used in the model are listed in Table 1.

Table 1. Values of the constants in the K-$\varepsilon$ model

| $C_\mu$ | $C_1$ | $C_2$ | $\sigma_k$ | $\sigma_\varepsilon$ |
|---|---|---|---|---|
| 0.09 | 1.43 | 1.92 | 1.00 | 1.30 |

## DISCRETIZATION OF THE MATHEMATICAL EQUATIONS

The partial differential equations are discretized by using finite volume method. A staggered grid system is used as shown in Fig 1. Scalar quantities such as P, $K, \varepsilon$, $v_t$ are stored at the intersections of grid lines, velocities are stored at the control volume surfaces as shown in Fig 1, Fig 2 and Fig 3.

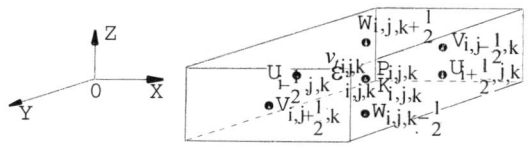

Fig 1. 3-D staggered grid and variables distribution

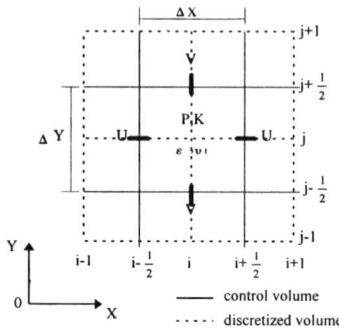

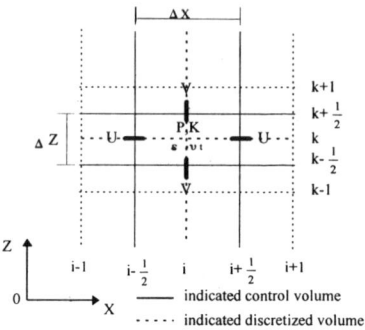

Staggered grid and variables distribution in X-Y plane
Fig 2.

Staggered grid and variables distribution in X-Z plane
Fig 3

## BOUNDARY CONDITION

### (1) Upstream boundary conditions

The velocity and turbulent quantities are specified at the inlet. The distribution of velocity U at the inlet

$$U_{in} = \frac{Q}{LH} \quad (16)$$

Where Q is the inlet discharge; L,H are the inlet width and height respectively; the velocities V,W are assumed to be zero, as in :

$$W_{i,j,k} = V_{i,j,k} = 0 \quad (17)$$

The turbulent kinetic energy K, dissipation rate $\varepsilon$ and the eddy viscosity $\upsilon_t$ can be obtained from Eqs. (18)(19)(20)

$$K_{i,j,k} = 0.004 U_{in}^2 \quad (18) \qquad \varepsilon_{i,j,k} = \frac{U_{in}^3}{0.4H} \quad (19) \qquad \upsilon_{t i,j,k} = \frac{C_\mu K_{i,j,k}^2}{\varepsilon_{i,j,k}} \quad (20)$$

### (2) Downstream boundary conditions

At the downstream end the water head is specified. The longitudinal gradients of the other variables are assumed to be zero as in:

$$\frac{\partial U}{\partial X} = \frac{\partial V}{\partial X} = \frac{\partial W}{\partial X} = \frac{\partial K}{\partial X} = \frac{\partial \varepsilon}{\partial X} = 0 \quad (21)$$

### (3) Solid boundary

On the wall boundary, the wall function method proposed by Launder and Spalding is used and may be expressed as :

$$\frac{U_p}{U^*} = \frac{1}{\kappa} \ln(E \frac{U^* Y_p}{\nu}) \quad (22)$$

Where $U_p$ is the resultant velocity paralleled to the wall; $U^*$ is the resultant friction velocity; k is the VonKarman constant; E is a roughness parameter.
Imposing local equilibrium conditions near the wall yields:

$$K = \frac{U^{*2}}{\sqrt{C_\mu}} \quad (23) \qquad\qquad \varepsilon = \frac{U^{*3}}{\kappa Y_p} \quad (24)$$

**(4) The free water surface boundary**

On the water surface, it is assumed that the first derivatives of $K, \varepsilon$ with respect to Z are approximately zero, and the pressure P is zero.

$$\frac{\partial K}{\partial Z} = \frac{\partial \varepsilon}{\partial Z} = 0 \quad (25)$$

The stability conditions:

$$\Delta t \leq \min_{i,j,k}(\frac{\Delta X}{|U_{i,j,k}|}, \frac{\Delta Y}{|V_{i,j,k}|}, \frac{\Delta Z}{|W_{i,j,k}|}) \quad (26)$$

A linear stability analysis shows that:

$$\Delta t \leq \min_{i,j,k}(\frac{1}{2v_t} \frac{\Delta X^2 \Delta Y^2 \Delta Z^2}{\Delta X^2 \Delta Y^2 + \Delta Y^2 \Delta Z^2 + \Delta Z^2 \Delta X^2}) \quad (27)$$

## APPLICATION OF THE MODEL

This model and computer program are applied to calculate 3-D flow field in the tail water channel of Fengman Hydroelectric Power Plant. The general layout of tail water channel is shown in Fig. 4 and Fig. 5. The inlet discharge is 500m³/s. The length of the channel in X direction is 72m and the height H and width L of the inlet are 5.5m and 64m respectively. The inlet herein is the exit of draft tube of turbogenerator set. Froude number at the inlet is 0.156, so it is a subcritical flow in the tail water channel. Space steps for calculation $\Delta X \times \Delta Y \times \Delta Z = 2m \times 4m \times 0.5m$ and time step = 0.6 sec. The computational results are shown in Fig. 6, 7, 8, 9, 10, 11 respectively.

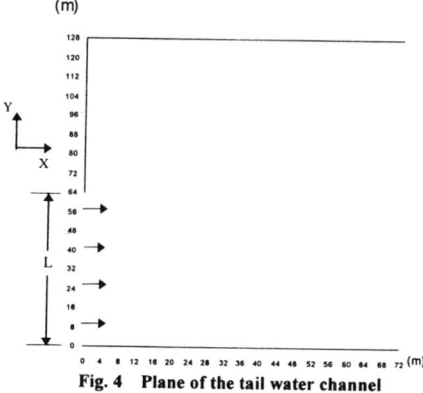

**Fig. 4** Plane of the tail water channel

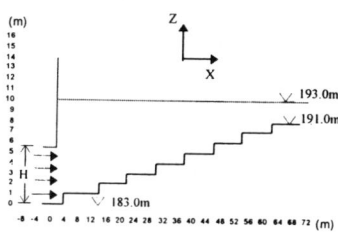

**Fig. 5** Longitudinal profile

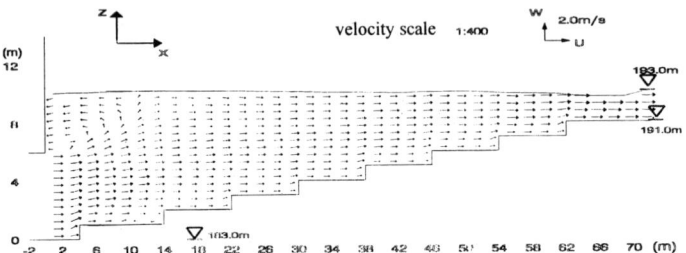

Fig.8 Velocity vector distribution at Y=24m

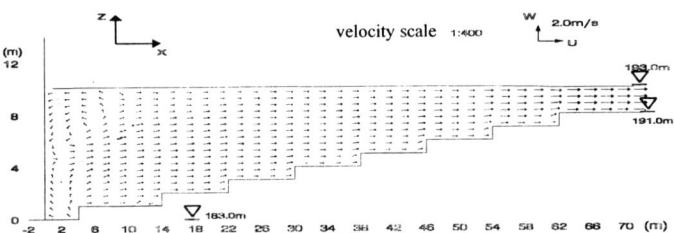

Fig.9 Velocity vector distribution at Y=72m

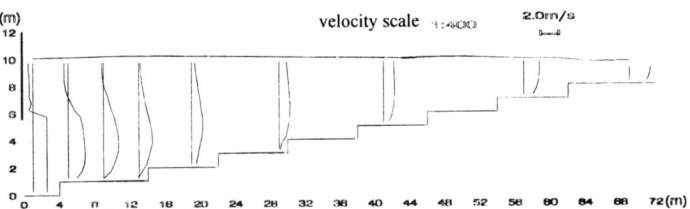

Fig.10 U Velocity profiles at Y=24m

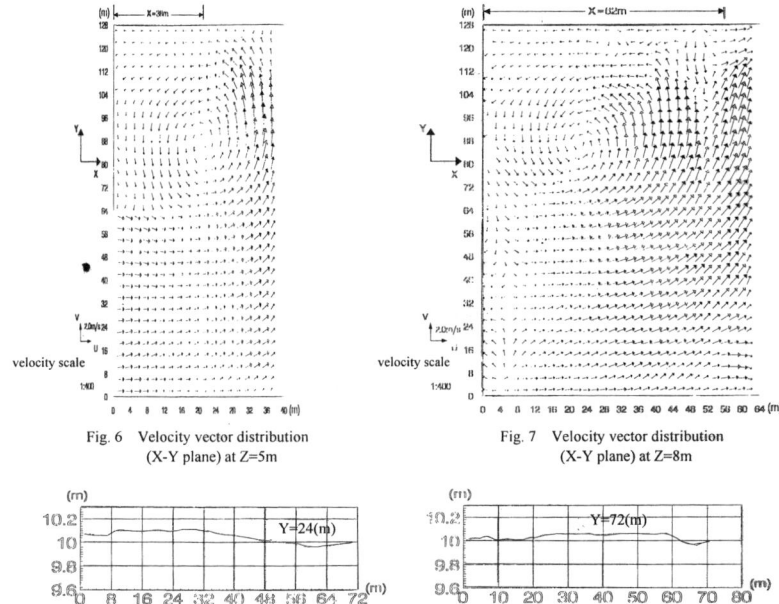

Fig. 6 Velocity vector distribution (X-Y plane) at Z=5m

Fig. 7 Velocity vector distribution (X-Y plane) at Z=8m

Fig. 11 Water surface profiles (X-Z plane) at Y=24m, 72m respectively

## CONCLUSION

The three dimensional K-$\varepsilon$ turbulent model has been used to predict the characteristics of flow in tailwater channel. This method might be a useful alternative in the solution of other similar free surface turbulent flow. The model can predict the water surface profile and mean velocity field as well as the pressure field in the open channel.

## REFERENCES

[1] Nallasamy, M., Turbulence Models and their Applications to the Prediction of Internal Flows, A Review, Com. & fluid, Vol. 15, No. 2, 1987.

[2] Peter M. Steffler and Yee-chung Jin, Depth Averaged and Moment Equations for Moderatedly Shallow Free Surface Flow, J. Hydraulic Research, Vol. 31, No. 1, 1993.

[3] Long. D, et al, Structure of Flow in Hydraulic Jumps, J. Hydraulic Research, Vol.29, No. 2, 1991.

[4] Hirt, C. W., Nichols, B. D., Volume of Fluid Method for the Dynamics of Free Boundaries, J. of Computational Physics 39, 1981.

# RITTER'S DAMBREAK WAVE REVISITED

## GUIDO LAUBER and WILLI H. HAGER
VAW, ETH-Zentrum
CH-8092 Zurich, Switzerland

The classic Ritter solution for the dambreak wave is compared with new observations collected in a smooth rectangular and horizontal channel. Whereas deviations are large at the wave fronts general agreement was found in the wave center.

## INTRODUCTION

Ritter has presented the first solution to the dambreak wave. He was not aware of the shallow water equations of the De Saint-Venant and the subsequent wave theory provided by Boussinesq. A historical account on these early contributions to the wave formation in open channels is given by Hager and Chervet (1996). The characteristic equations in a rectangular prismatic channel may be written as (Liggett, 1994)

$$\frac{d(v \pm 2c)}{dt} = g(S_o - S_f) \qquad (1)$$

$$\frac{dx}{dt} = v \pm c \qquad (2)$$

where v=cross-sectional velocity, $c=(gh)^{1/2}$=wave celerity in shallow water with g=gravitational acceleration and h=flow depth, t=time, $S_o$=bottom slope, $S_f$=friction slope, and x=streamwise coordinate, with origin at the dam section. Ritter considered a frictionless fluid in the horizontal channel, for which $S_o - S_f = 0$. With $h(v=0) = h_o$ as the initial water depth, the solution of (1) is

$$V = 2(1 - Y^{1/2}) \qquad (3)$$

where $Y = h/h_o$, and $V = v/(gh_o)^{1/2}$. Further, with $X = x/h_o$ and $T = (g/h_o)^{1/2}t$ as dimensionless location and time, measured from the instant of rupture, (2) yields for

$$\frac{X}{T} = 2 - 3Y^{1/2}. \qquad (4)$$

Solving for Y gives with m=X/T and Q=VY=relative discharge

$$Y = \left[\frac{1}{3}(2-m)\right]^2, \qquad (5)$$

$$V = \frac{2}{3}(1+m),\tag{6}$$

$$Q = \frac{2}{27}(1+m)(2-m)^2.\tag{7}$$

At the dam section (m=0) the flow depth is thus $Y_0=4/9$, the velocity $V_0=2/3$ and the discharge $Q_0=8/27$. Further, the free surface is a parabola defined by the *negative front* $Y(m=-1)=Y_-=1$, where $V_-=0$ and $Q_-=0$, and the *positive front* $Y(m=2)=Y_+=0$ where $V_+=2$ and $Q_+=0$. The corresponding front velocities are, from (4), $(X/T)_-=-1$, and $(X/T)_+=+2$.

Ritter's dambreak wave is still a rough basis of the dambreak phenomenon, and is normally adopted as the boundary condition of numerical approaches. The present project was initiated to provided new data on dambreak waves in sloping channels. The horizontal channel was also considered and these results are compared subsequently with the Ritter solution.

## EXPERIMENTS

The experiments were conducted in a rectangular channel 500mm wide and 14m long. The right channel wall was of glass, the bottom and the left side of black PVC. The boundary roughness height was practically zero, and all flows were in the turbulent smooth regime. Preliminary experiments indicated negligible scale effect provided the initial flow depth $h_0$ is larger than 300mm.

The axial flow depth and the average axial velocity were measured with a Video system connected to a suitable computer. The removal of the dam was simulated with a plane gate connected to a pressurized air system. The acceleration of the gate was 2g and it was also verified in preliminary experiments that faster removal did not change the wave formation. Fig.1 shows the experimental set up and the gate (Reinauer and Lauber 1996).

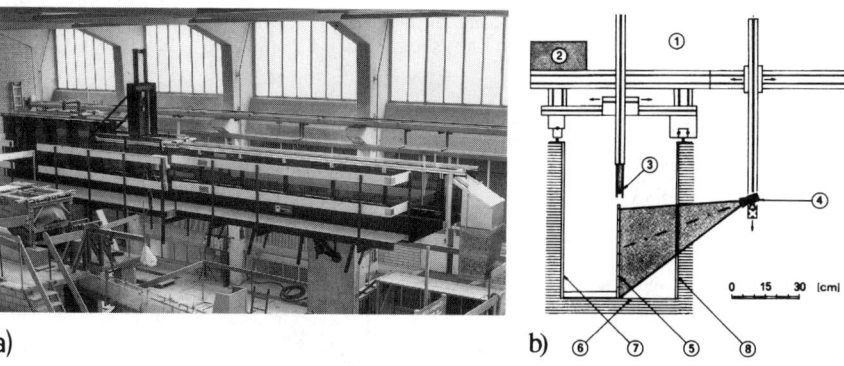

**Fig.1** Experimental stand (a) Side view, (b) section with ① trolley, ② light source, ③ line light, ④ video camera, ⑤ observation plane, ⑥ channel bottom, ⑦ channel wall, ⑧ glass front.

The Video camera is a SONY interline-transfer CCD XC-77RR. The exposure time can be varied between 1/60 and 1/10000 s, and 50 frames per s can be made. The data are interpreted with a KS 400 program provided by KONTRON Electronik GmbH. A detailed description of the system is given by Lauber and Hager (1995).

The accuracy of our observations is as follows: for flow depths 1mm, i.e. ±1% if 100mm is considered a typical flow depth. Perturbations of flow were larger in the front than in the rear wave portions. Velocities could be determined to ±0.1m/s, i.e. ±5% if 2ms$^{-1}$ is considered typical for the horizontal channel. The velocity was difficult to determine in the first instant of dambreak, due to highly accelerating flow and strongly inclined streamlines. Also, difficulties arose both at the wave front and at the wave end. It should be noted that the channel was completely removed from water splashes and drops prior to each experiment, such that the dambreak wave propagated in an absolutely dry smooth channel.

## EXPERIMENTAL RESULTS

In all plots, the quantities were non-dimensionalized with length $h_0$, time $(h_0/g)^{1/2}$, velocity $(gh_0)^{1/2}$ and unit discharge $(gh_0^3)^{1/2}$ such that $X=x/h_0$, $T=(g/h_0)^{1/2}t$, $Y=h/h_0$, $V=V/(gh_0)^{1/2}$ and $Q=q/(gh_0^3)^{1/2}$. The basin length was $\lambda_s=L_B/h_0=11.05$ in all experiments discussed subsequently. Fig.2a) shows non-dimensional flow depth Y as a function of $m=X/T$, for various $T-\Delta T \times n$, with one time increment $\Delta T=(g/h_0)^{1/2}0.5s$. The full curve is Ritter's solution (5). Deviations between observations and (5) are large in the first time increments, for both X/T<0 and >0. The free surface is curved differently in the reservoir zone with the negative front propagating faster than X/T=−1. This is due to streamline curvature effects. The positive wave is much slower, however, because of the so-called initial wave phenomenon described by Lauber and Hager (1997). The dynamic wave with a propagation speed $(X/T)_F=+2$ starts only at time $T=2^{1/2}$ and overtakes the initial wave at time T=3. Except for the wave front regions, the Ritter solution adequately describes the free surface profile until the negative wave has reached the upstream end (T≅8). Note that all curves pass through $Y_0=4/9$ at the dam section, in agreement with the Ritter solution. The front location is about $(X/T)_F=1.30$. During the emptying process, the free surface is nearly horizontal in the reservoir and tends towards the Ritter solution downstream from the dam section.

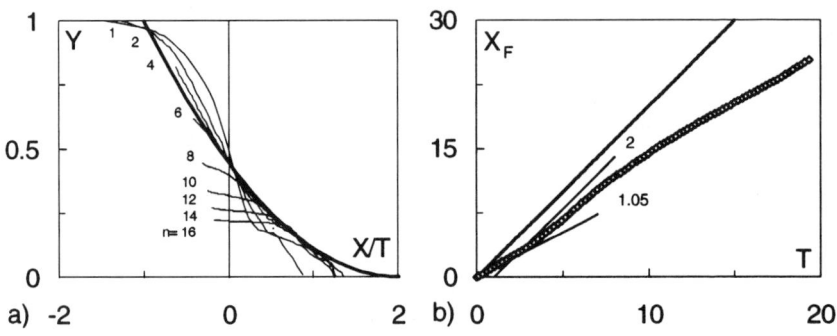

**Fig.2**  a) Free surface profile Y(X,T) and b) Positive front propagation $X_F(T)$. (—) Ritter solution, (—) experiments.

Fig.2b) refers to the positive front propagation $X_F(T)$ with $X_F=1.05T$ for the *initial wave*, and $X_F=2(T-2^{1/2})$ for the *dynamic wave*. The front velocity decelerates for large T due to viscosity (Lauber and Hager 1997).

The velocity distribution $V(X,T)$ is shown in Fig.3a). Deviations from the Ritter solution are seen to be large again for small time both at the positive and negative wave fronts. In the central wave portion, the agreement is satisfactory, however, and Ritter's solution describes adequately the flow phenomena. At the positive front, deviations are significant, due to (1) the initial wave for n=1 and 2, where $X_F$ is much smaller than predicted, (2) the reduction of velocity for larger n and $(X/T)_F=1.30$ for the wave location. During the emptying process, the negative front shift to the right side, because X remains constant for increasing time T. Close to the dam section, Ritter's solution is again satisfactory.

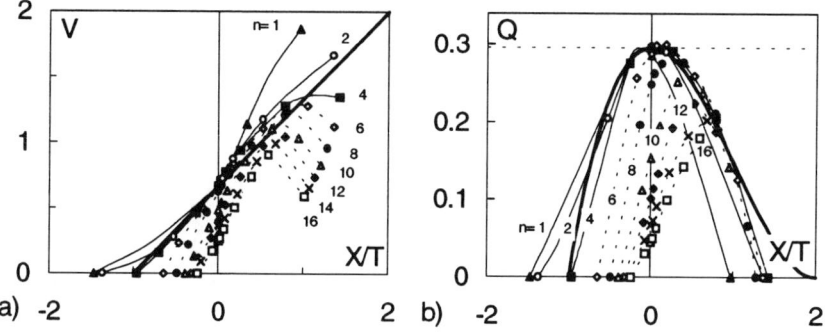

Fig.3  a) Velocity $V(X,T)$ and b) discharge $Q(X,T)$. Notation Fig.2.

The discharge distribution $Q(X,T)$ as shown in Fig.3b) follows Ritter's solution in the central wave portion. Deviations are again notable for small time, and at the two propagation zones. For 2<n<6 the maximum discharge $Q=8/27$ at the dam section is accurately predicted. As a general conclusion, it may thus be stated that the Ritter solution is unable to describe a dambreak wave in the initial phase and at both propagation ends.

## CONCLUSIONS

The first solution to the dambreak wave in a horizontal frictionless and prismatic rectangular channel was provided by Ritter. A comparison with experiments in a smooth channel indicates significant deviations both at the negative and positive wave fronts. Whereas the negative wave propagates with a dimensionless velocity of $-1.4$ compared to $-1.0$ according to Ritter, the positive front has a relative propagation velocity of only $+1.30$, compared to $+2.0$ according to Ritter. Significant differences are also found in the initial wave formation, mainly due to streamline curvature effects. In the central wave portion, general agreement can be noted between our observations and Ritter's solution.

The effects of bottom slope and length of reservoir are reported in a subsequent paper. The authors are currently investigating the effect of bottom roughness on the dambreak wave.

## ACKNOWLEDGMENTS

We would like to acknowledge the generous assistance and support of A. Chervet, VAW.

## REFERENCES

- Hager, W.H., Chervet, A. (1996). History of dambreak wave. *Wasser Energie Luft* **88**(3/4): 49-54 (in German).
- Lauber, G., Hager, W.H. (1995). Optical sensing of extremely unsteady open channel flows. *Wasser Energie Luft* **87**(11/12): 275-278 (in German).
- Lauber, G., Hager, W.H. (1997). Positive front of dambreak wave. *27 IAHR Congress* San Francisco, submitted.
- Liggett, J.A. (1994). *Fluid mechanics*. McGraw-Hill: New York.
- Reinauer, R., Lauber, G. (1996). Steep channels for hydraulic modelling. *Schweizer Ingenieur und Architekt* **114**(8): 121-124 (in German).

# SIMULATION OF WAVES GENERATED BY LANDSLIDES IN VAIONT DAM

## MASANORI MICHIUE AND OSAMU HINOKIDANI
Department of Civil Engineering
Tottori University, Tottori, Japan

### 1. INTRODUCTION

Almost 30 yeas ago, in 1963, a large scale landslide into the reservoir occurred at the Vaiont Dam in Italy. This landslide produced a giant wave in the dam reservoir. The wave dashed against the opposite bank and reached a height of 270m above the water surface of reservoir. This wave washed over the dam causing great damage to downstream area, and 2,125 people were killed in this disaster[1].

Dams are among the largest man−made structures in existence, bestowing multiple benefits to modern societies. In this rare case when a dam broke, the destruction caused both upstream and downstream was terrible. In order to prevent this kind of destruction, strict safety inspections of a dam itself are required. In addition to this, it is also important to carefully consider safety factors for this kind of waves such as Vaiont dam. In this paper, a two−dimensional numerical simulation model based on shallow water equations integrated with the water depth have developed and is applied to Vaiont landslide.

### 2. GOVERNING EQUATIONS AND NUMERICAL PROCEDURE

#### 2.1 GOVERNING EQUATIONS

In this study, we can use the following two−dimensional depth−integrated shallow water equations written in a vector form to simulate waves in a reservoir.

$$\frac{\partial U}{\partial t} + \frac{\partial E}{\partial x} + \frac{\partial F}{\partial y} = C \tag{1}$$

$$U_{ij}^{n+1} = \frac{1}{2}(U_{ij}^n + \overline{U}_{ij})$$
$$- \frac{\Delta t}{2\Delta x}\{(\overline{E}_{i+1j} - \overline{E}_{ij}) - (\overline{Q}_{xi+1j} - \overline{Q}_{xij})\}$$
$$- \frac{\Delta t}{2\Delta y}\{(\overline{E}_{ij+1} - \overline{E}_{ij}) - (\overline{Q}_{xij+1} - \overline{Q}_{xij})\} + \frac{1}{2}\overline{C}_{ij}\Delta t \qquad (8)$$

$$Q_{xij} = \frac{k}{8}(U_{i+1j}^n - 2U_{ij}^n + U_{i-1j}^n) \qquad (9)$$

$$Q_{yij} = \frac{k}{8}(U_{ij+1}^n - 2U_{ij}^n + U_{ij-1}^n) \qquad (10)$$

in which Q is an artificial viscosity term, k is a constant, and n indicates a time step. In this paper, in the prediction step, the backward difference scheme is used and the forward difference scheme is used in the correction step. The computation scheme in the simulation is shown in Fig.1.

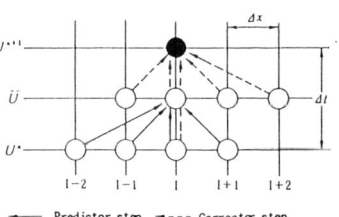

Fig.1 MacCormack scheme.

## 2.2 INITIAL AND BOUNDARY CONDITIONS

Flow velocities of u and v at initial conditions are zero and water level are constants follows :
$u_{i,j} = v_{i,j} = 0$,
$z_{i,j} + h_{i,j} =$ const. (11)
Fig.2 shows the boundary conditions. At the wall boundary, the artificial viscosity term is zero, and the velocities and water depth are expressed by following equations.
boundary condition for x direction ;
$u_{i-1} = v_{i-1} = 0$,
$h_{i-2} + z_{i-2} = h_{i-1} + z_{i-1}$ (12)
boundary condition for y direction ;
$u_{j-1} = v_{j-1} = 0$, $h_{j-2} + z_{j-2} = h_{j-1} + z_{j-1}$ (13)

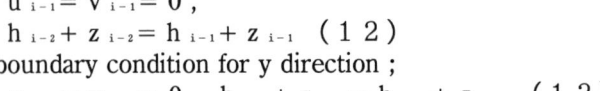

Fig.2 Boundary conditions.

## 2.3 MODELING OF INFLOW SEDIMENTS

$$U = \begin{pmatrix} h \\ uh \\ vh \end{pmatrix}, \quad E = \begin{pmatrix} uh \\ u^2h + \frac{1}{2}gh^2 \\ uvh \end{pmatrix}, \quad F = \begin{pmatrix} vh \\ uvh \\ v^2h + \frac{1}{2}gh^2 \end{pmatrix} \quad (2)$$

$$C = \begin{pmatrix} 0 \\ gh(S_{ox} - S_{fx}) + \frac{\partial}{\partial x}\left\{\varepsilon \frac{\partial(uh)}{\partial x}\right\} + \frac{\partial}{\partial y}\left\{\varepsilon \frac{\partial(uh)}{\partial y}\right\} \\ gh(S_{oy} - S_{fy}) + \frac{\partial}{\partial x}\left\{\varepsilon \frac{\partial(vh)}{\partial x}\right\} + \frac{\partial}{\partial y}\left\{\varepsilon \frac{\partial(vh)}{\partial y}\right\} \end{pmatrix} \quad (3)$$

$$S_{fx} = \frac{n^2 u \sqrt{(u^2 + v^2)}}{h^{4/3}}, \quad S_{fy} = \frac{n^2 v \sqrt{(u^2 + v^2)}}{h^{4/3}} \quad (4)$$

$$S_{ox} = -\frac{\partial z}{\partial x}, \quad S_{oy} = -\frac{\partial z}{\partial y} \quad (5)$$

in which h is water depth, z is bed level, u and v are depth-averaged velocities in x and y direction respectively, g is the acceleration due to gravity, $S_{ox}$ and $S_{oy}$ are bed slopes in x and y direction respectively, $S_{fx}$ and $S_{fy}$ are energy gradients in x and y direction respectively, $\varepsilon$ is a eddy viscosity coefficient in the horizontal direction, n is Manning coefficient, and $\varepsilon$ is estimated by the following equation :

$$\varepsilon = 0.01 \Delta^{4/3} \quad \text{(unit:cm)} \quad (6)$$

in which $\Delta$ is a mesh size.

In this study, MacCormack's scheme is used for horizontal and time differences. In the MacCormack scheme, governing equations are divided into two steps (prediction step and correction step), and in each step, for the advection term and the pressure term in the horizontal direction, the forward difference scheme and the backward difference scheme are used separately. For other terms such as the diffusion term, the central difference scheme is used in the horizontal direction. This MacCormack scheme has a second order of accuracy and in general, artificial viscosity terms are added for the stabilization. Eqs.(1)–(5) can be written by MacCormack scheme as follows :

prediction step :

$$\overline{U}_{ij} = U_{ij}^n - \frac{\Delta t}{\Delta x}\{(E_{ij}^n - E_{i-1j}^n) - (Q_{xij}^n - Q_{xi-1j}^n)\} \\ - \frac{\Delta t}{\Delta y}\{(F_{ij}^n - F_{ij-1}^n) - (Q_{yij}^n - Q_{yij-1}^n)\} + C_{ij}\Delta t \quad (7)$$

correction step :

The phenomenon we are aiming to reproduce is the change in water level when sediment flows into the reservoir area. The accumulation effect due to this inflow of sediment must be reflected on river bed. In order to model this, the phenomenon of inflowing sediment is expressed by changing over time the calculated river bed height in accordance with the amount of inflow sediment as shown in Fig.3.

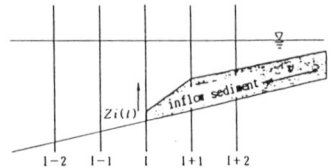

Fig.3 Modeling of inflow sediment.

## 3. SIMULATION OF VAIONT LANDSLIDE

The Vaiont dam is 265m high and was built in 1960. On oct. 9, 1963, a landslide occurred on the left bank of the reservoir which measured about 1.8km in length.

The landslide moved a quantity of sediment 250m deep, a distance of 400m towards the right bank at a speed of 20m/s − 30m/s[1]. The amount of sediment was about 270 million cubic meters.

The landslide pushed the reservoir water towards the right bank and created a wave which reached the bank at a height of 970m above the sea level, up to 270 meters above the normal water level of EL.700m. The massive wave overflowed the dam at a height of about 150m above the top of the dam. Fig.4 shows the topography before and after the disaster. The complete change in the reservoir shows just how large this disaster was.

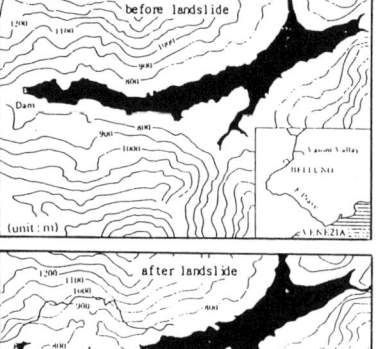
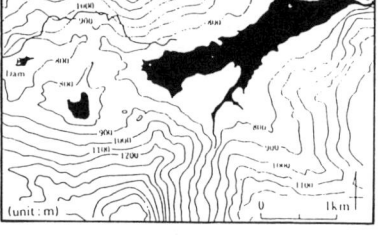

Fig.4 Topography of before and after disaster.

### 3.1 SIMULATION CONDITIONS

To recreate this kind of phenomenon, the calculated area was set for an area of 1800m wide across and 3750m long up− and down−stream, including the dam reservoir, as shown in Fig.7(a). The region inside was divided into a 25m × 25m grid.

Next, a model of sediment inflow is created using the amount of terrain movement before and after the landslide. It is estimated that the sediment proceeds at a speed at which the rise and fall in the river bed radiates out from the center of the slipped landslide area as shown in Fig.5. In this case, an estimated speed is 25m/s according to current research[1]. The calculation interval is 0.01 seconds and the calculated time is assumed to be 100 seconds.

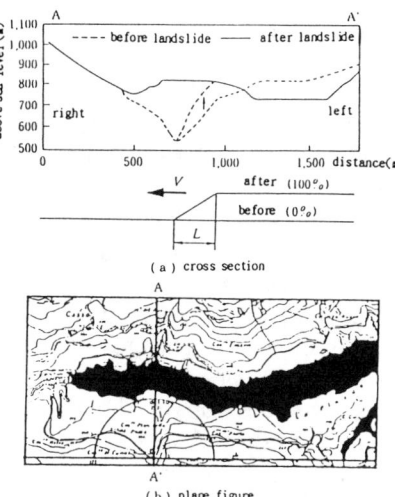

Fig.5 Modeling of landslide.

## 3.2 SIMULATION RESULTS

Fig.6 shows the calculated results at 10 seconds intervals for the cross-section of the landslide's center axis. It can be seen here how the wave moving upstream to the right bank returns to the left bank. The

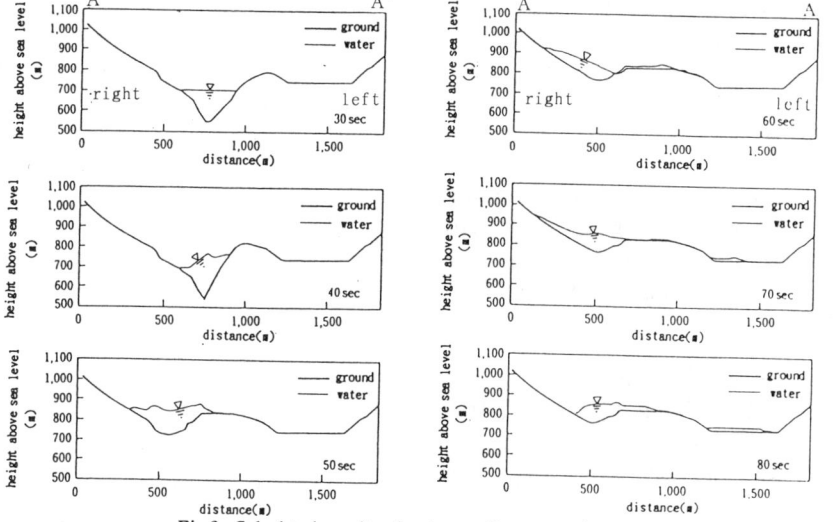

Fig.6 Calculated results of water profiles on section A-A'.

wave reaches a maximum height of 950m. This value accords fairly well with the actually measured one.

Fig.7 compares the maximum wave height above the sea level and positions of calculated results at the right bank with marks actually observed at the Vaiont dam site. From these figures, it can be seen the calculated results are in a good agreement with the observed one.

Fig.8 shows the changes over time in the depth of water which overflows the top of the dam. It thus shows how a wave of over 100m in height overflowed the dam in slightly more than 10 seconds.

## 4. CONCLUSIONS

In this paper, a two-dimensional numerical simulation model based on shallow water equations integrated with the water depth has developed and is applied to Vaiont landslide. According to comparison with simulation results and the observed one, it is found how well the calculation method introduced here recreates the change in reservoir water level which accompanied the inflow of sediment.

## REFERENCES

[1] G. Melidoro : Introductory Report, Proc. of the Meeting of the 1963 Vaiont Landslide, pp.31–42, 1986.

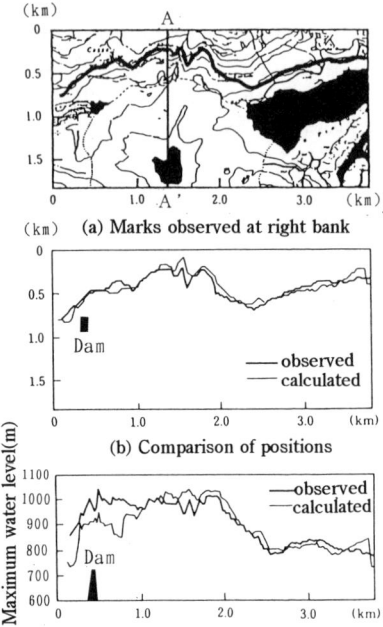

(a) Marks observed at right bank

(b) Comparison of positions

(c) Comparison of maximum wave height

Fig.7 Comparison between calsulated results and the observed one.

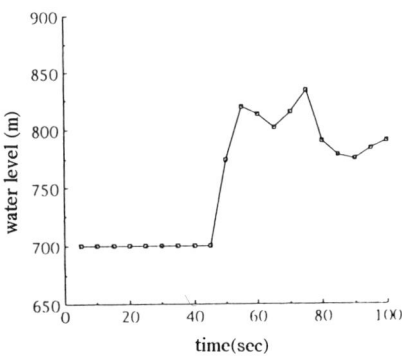

Fig.8 Time variation of wave height at the dam.

# AN EXTENDED DEPTH-AVERAGED TURBULENCE MODEL FOR FLOW CONSTRICTED BY COFFERDAMS

JIAN LIU* and AKIHIRO TOMINAGA**
*Overseas Civil Engineering Department, NEWJEC Inc., Chuo-ku, Osaka 542, Japan
**Department of Civil Engineering, Nagoya Institute of Technology, Nagoya 466, Japan

## ABSTRACT

An extended depth-averaged k-ε turbulence model is proposed for simulating the flow in a channel constricted by cofferdams. In the present paper, the coefficient $C_\mu$ is expressed as a function of shear strain rate scale S and vorticity scale Ω to eliminate the excessive production of turbulent kinetic energy around a stagnation point. The model constants $C_1$, $C_2$ in the ε-equation are modified as functional forms of time scales G/ε and Sk/ε to account for non-equilibrium effects. Three-dimensional effects are considered by modifying the friction coefficient in bottom shear stresses. The computed solutions are compared with experimental data. The agreement between the calculated and experimental results is quite satisfactory.

## INTRODUCTION

In the construction of hydraulic engineering works on large rivers, cofferdams are constructed for diversion works. For example, a staged cofferdam construction approach is being used in the Chinese Three Gorges Project, which is the biggest hydroelectric project in the world. The layout of cofferdams has significant influence on both channel navigation and channel closure. The change of the river stream due to a one-sided constriction of the river by cofferdams is characterized by constriction and expansion of the flow (Figure 1). The flow in the constricted section is characterized by separation and aeration, and full simulation of the flow is very difficult. The expansion of flow is accompanied by interaction of the transition zone with an eddy one and with the river bed. In a result of this interaction, bank and eddy boundary layers as well as the central body of the flow are formed. Generally, constriction and expansion of the flow are determined by the length of the longitudinal cofferdam, the layout of upstream and downstream transversal cofferdams (configuration of foundation pit) and the constriction degree (defined as the ratio of the constricted river width (B-b) to river width B). Liu (1991, 1995) has discussed the effects of the constriction degree, length of the longitudinal cofferdam and configuration of foundation pit on the flow by using the depth-averaged k-ε turbulence model of Rodi (1979) (hereafter referred to as original model). The computed results are generally in agreement with experimental data, but the predicted recirculating zone is smaller than the measured data. The original model originates in the standard k-ε turbulence model developed by Launder and Spalding (1974) (hereafter denoted as standard model). Thus, the two models have the same advantages and disadvantages. There are many difficulties in predicting the complex flows by using the standard k-ε model. For instance, the standard model is not suitable for predicting the flows with impinge, separation and streamline curvature, because the turbulent viscosity is over-predicted by the standard turbulent viscosity formulation in these flows. However, the standard model has the advantages of low computation cost and computational stability, so it is widely used for practical flow computations. In this paper, we will expand the performance of the original depth-averaged model by modifying model constants in the standard model and friction coefficient in bottom shear stresses, making it more compatible with practical flow conditions. Since the original model does perform well in certain flows, we will make the modified model reduce back to the original model, when simple flows are met. The extended model is examined against the flow constricted by cofferdams. In the process of computation, non-rectangular meshes proposed by Liu (1995) are used to treat the irregular geometry due to a trapezoidal foundation pit.

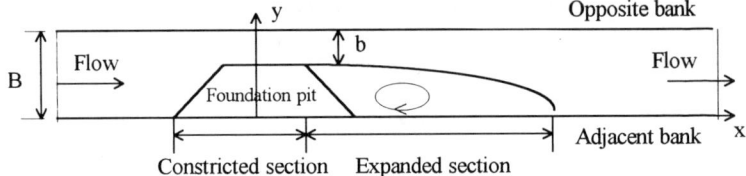

Figure 1    Schematic layout of cofferdams (plan view)

## MATHEMATICAL MODEL

### GOVERNING DIFFERENTIAL EQUATIONS

The steady, two-dimensional, depth-averaged equations are obtained by integrating the Navier-Stokes equations over the local water depth. The following assumptions are used in this process: hydrostatic pressure distribution, incompressible fluid, uniform velocity distribution in the vertical direction, small channel bottom slope, negligible wind shear at the water surface, and negligible Coriolis acceleration. The two-dimensional depth-averaged equations, along with the continuity, can be written as:

$$\frac{\partial(hu)}{\partial x}+\frac{\partial(hv)}{\partial y}=0 \quad (1)$$

$$\frac{\partial(uu)}{\partial x}+\frac{\partial(uv)}{\partial y}=-g\frac{\partial(z_b+h)}{\partial x}+\frac{1}{\rho h}\frac{\partial(h\tau_{xx})}{\partial x}+\frac{1}{\rho h}\frac{\partial(h\tau_{xy})}{\partial y}-\frac{\tau_{bx}}{\rho h} \quad (2)$$

$$\frac{\partial(uv)}{\partial x}+\frac{\partial(vv)}{\partial y}=-g\frac{\partial(z_b+h)}{\partial y}+\frac{1}{\rho h}\frac{\partial(h\tau_{xy})}{\partial x}+\frac{1}{\rho h}\frac{\partial(h\tau_{yy})}{\partial y}-\frac{\tau_{by}}{\rho h} \quad (3)$$

where u and v are the depth-averaged velocities in the streamwise direction(x) and the cross-stream direction(y), respectively; $\tau_{xx}$, $\tau_{xy}$, $\tau_{yy}$ the effective stresses; $\tau_{bx}$, $\tau_{by}$ the bottom shear stress components in the x- and y- directions, respectively; h the local water depth, $z_b$ the bottom elevation; the symbol $\rho$ denotes fluid density. Finally, g is the gravitational acceleration.

### DEPTH-AVERAGED TURBULENCE MODEL

To solve equations (1)-(3), the stress terms on the right-hand side of (2) and (3) must be expressed as explicit functions of the depth-averaged velocities and water level. The effective stresses ($\tau_{xx}$, $\tau_{xy}$, $\tau_{yy}$) consist of laminar viscous stresses, turbulent stresses and stresses due to depth-averaging, so it is difficult to quantify them. In practical river flows, laminar viscous stresses are very small and are negligible. The turbulent stresses are quantified based on the Boussinesq eddy-viscosity concept, assuming proportionality to the unit discharge gradients. The stresses due to depth-averaging are difficult to quantify and are neglected. Applying the

Boussinesq eddy-viscosity concept and substituting the depth-averaged velocities, the effective stresses become:

$$\tau_{xx} = \frac{2\rho v_t}{h}\left(\frac{\partial(hu)}{\partial x}\right), \quad \tau_{xy} = \frac{\rho v_t}{h}\left(\frac{\partial(hu)}{\partial y}+\frac{\partial(hv)}{\partial x}\right), \quad \tau_{yy} = \frac{2\rho v_t}{h}\left(\frac{\partial(hv)}{\partial y}\right) \quad (4)$$

where $v_t$ is turbulent kinematic viscosity and is determined from the following relation:

$$v_t = C_\mu k^2/\varepsilon \quad (5)$$

in which $C_\mu$ is an empirical constant; k and ε are the turbulent kinetic energy and its dissipation rate. The variation of k and ε is determined from the following semi-empirical transport equations (Rodi 1979):

$$\frac{\partial(uk)}{\partial x}+\frac{\partial(vk)}{\partial y}=\frac{\partial}{\partial x}\left(\frac{v_t}{\sigma_k}\frac{\partial k}{\partial x}\right)+\frac{\partial}{\partial y}\left(\frac{v_t}{\sigma_k}\frac{\partial k}{\partial y}\right)+G+P_{kv}-\varepsilon \quad (6)$$

$$\frac{\partial(u\varepsilon)}{\partial x}+\frac{\partial(v\varepsilon)}{\partial y}=\frac{\partial}{\partial x}\left(\frac{v_t}{\sigma_\varepsilon}\frac{\partial \varepsilon}{\partial x}\right)+\frac{\partial}{\partial y}\left(\frac{v_t}{\sigma_\varepsilon}\frac{\partial \varepsilon}{\partial y}\right)+C_1 G\frac{\varepsilon}{k}+P_{\varepsilon v}-C_2\frac{\varepsilon^2}{k} \quad (7)$$

where the coefficients $C_1$, $C_2$, $\sigma_k$, $\sigma_\varepsilon$ are the model coefficients; G is the production of turbulent kinetic energy due to the interaction of turbulent stresses with horizontal mean velocity gradients; and $P_{kv}$, $P_{\varepsilon v}$ are the source terms originating from non-uniformity of vertical profiles. These terms are calculated from the following formulae:

$$G = v_t\left[2\left(\frac{\partial u}{\partial x}\right)^2+2\left(\frac{\partial v}{\partial y}\right)^2+\left(\frac{\partial u}{\partial y}+\frac{\partial v}{\partial x}\right)^2\right] \quad (8)$$

$$P_{kv} = C_k[C_f(u^2+v^2)]^{3/2}/h, \quad P_{\varepsilon v} = C_\varepsilon [C_f(u^2+v^2)]^2/h^2 \quad (9)$$

in which $C_f$ is a friction coefficient; the coefficients $C_k$, $C_\varepsilon$ are constants, and are given by

$$C_k = 1/C_f^{1/2}, \quad C_\varepsilon = 3.6\, C_2\, C_\mu^{1/2}/\, C_f^{3/4}, \quad (10)$$

In the original model, the standard values of $C_\mu$, $C_1$, $C_2$, $\sigma_k$, $\sigma_\varepsilon$ come from the standard model of Launder and Spalding (1974). In standard k-ε turbulence model, the model coefficients are assumed as constant according to a set of experiments for simple turbulent flows, but these are not suitable for the flows such as separated flows. The coefficients $C_\mu$, $C_1$, $C_2$ were determined under the equilibrium flow condition, which means that production and dissipation of k balance each other and convective/diffusive effects are negligible. However, for complex flows, such as the flows with separation, recirculation and large streamline curvature, flow equilibrium does not exist. When a flow is not under equilibrium, besides the length and time scales dictated by the flow configuration, dynamic characteristics create extra length and time scales according to the local flow structure. In the standard model, the model coefficients do not respond to such structural changes because their values are fixed; variation of these coefficients must be considered when the global and local length and time scales are not comparable (Liu et al. 1996). In addition, the standard model overestimates production G of k around the frontal corner of square rib, as Murakami et al. (1996) pointed out. The sensitivity analysis indicates that for the coefficients of the k-ε model, only $C_\mu$, $C_1$, $C_2$ have significant effects on the flow field, and the coefficients $\sigma_k$, $\sigma_\varepsilon$ have insignificant effects on the flow field. Thus, $C_\mu$, $C_1$, $C_2$ are the only parameters modified in this paper. The modification of the coefficient $C_\mu$ by Murakami et al. (1996) is adopted in the present work, that is, $C_\mu$ is expressed as a function of shear strain rate scale S and vorticity scale Ω to avoid the excessive production of k around a stagnation point. According to the ideas of Shih (1995) and Liu et al.(1996), we have considered time scales G/ε and Sk/ε in the ε-equation by introducing functional forms for the production and dissipation terms, based on a realizability analysis. The mathematical expressions of the original and modified model parameters discussed above are summarized in Table 1. In Table 1, the function $f_\mu$, shear strain rate scale S and vorticity scale $\Omega$ for two dimensional incompressible flow are defined as

Table 1  Comparison of coefficients adopted by different models

| MODEL | $C_\mu$ | $C_1$ | $C_2$ | $\sigma_k$ | $\sigma_\varepsilon$ |
|---|---|---|---|---|---|
| original | 0.09 | 1.44 | 1.92 | 1.0 | 1.3 |
| modified | $0.09 f_\mu$ | 1.19+G/$\varepsilon$ | 17/(6+Sk/$\varepsilon$) | 1.0 | 1.3 |

$$f_\mu = \Omega/S, \quad \Omega/S \leq 1; \quad f_\mu = 1, \quad \Omega/S > 1 \tag{11}$$

$$S = \sqrt{2\left(\frac{\partial u}{\partial x}\right)^2 + 2\left(\frac{\partial v}{\partial y}\right)^2 + \left(\frac{\partial u}{\partial y} + \frac{\partial v}{\partial x}\right)^2}, \quad \Omega = \left|\frac{\partial u}{\partial y} - \frac{\partial v}{\partial x}\right| \tag{12}$$

## BOTTOM SHEAR STRESSES

The bottom shear stresses in the x- and y- directions are modeled according to the following formulae:

$$\tau_{bx} = \rho C_f u (u^2 + v^2)^{1/2}, \quad \tau_{by} = \rho C_f v (u^2 + v^2)^{1/2} \tag{13}$$

where the friction coefficient $C_f$ depends on the roughness of the river bottom. For hydraulically smooth river beds, $C_f$ is determined from the channel formula of Schlichting (1960):

$$C_f = 0.027 \left(\frac{v}{R\sqrt{u^2 + v^2}}\right)^{1/4} \tag{14}$$

and for rough beds from Manning's law:

$$C_f = n^2 g / h^{1/3} \tag{15}$$

in which R is the hydraulic radius of flow cross section, $v$ is the molecular kinematic viscosity, and $n$ stands for the Manning roughness factor. The friction coefficient $C_f$ is corrected for three-dimensional effects according to Tingsanchali and Maheswaran (1990):

$$C_{fc} = C_f \left(1 + \tan^2(2\alpha)\right)^{0.5} \tag{16}$$

where $\alpha$ is the turning angle between the main flow or surface streamline direction and a reference direction which is the upstream approaching direction in this study.

## NUMERICAL COMPUTATION

The governing equations are discretized by the finite control volume method in a staggered grid system in which grids near cofferdams are clustered. A Cartesian staggered grid system is employed so that the grid lines do not coincide with the surfaces of the skew cofferdams. Non-rectangular cells proposed by Liu (1995) are used for treatment of the skewed boundaries. The convective/diffusion terms in the governing equations are discretized by using a hybrid finite difference scheme (Patankar 1980). The SIMPLEC method proposed by Van Doormaal and Raithby (1984) is used for the coupling between water depth and velocities. The boundary conditions for every governing equation must be specified at all boundaries. Details about the treatment of the boundary conditions may be found in the paper of Liu (1995). Convergence criterion is set as the maximum normalized residual of all dependent variables less than $10^{-4}$.

## RESULTS AND DISCUSSION

The computation is conducted corresponding to the experimental case as shown in Table 2. The calculated values are compared with the experimental values and computed results by the original model (Liu 1995). Figure 2 shows comparisons of the calculated depth-averaged longitudinal velocities with the experimental data. The numerical prediction provides a very good estimation except for under-prediction near the upstream corner of the foundation pit (x/B= -0.4). The calculated and measured water depths are depicted in Figure 3. The calculated values are in good agreement with the experimental data, but the depression in the constricted section is not reproduced. The discrepancy on the profiles can also be the omission of physical mechanisms such as the role of the vertical component of the secondary flow and the instability near the longitudinal cofferdam. Figure 4 shows calculated and measured velocity vector distributions and streamlines. It should be mentioned that the scale of longitudinal and transversal directions is 1:2.5. The significant change of the velocity profiles occurs at the upstream corner of the foundation pit and the downstream side of the downstream transversal cofferdam. It is found that the original and extended models can reproduce flow separation at the upstream corner of the foundation pit. The calculated results by the two models have no big difference at the corner; this is probably due to the skewness of the upstream transversal cofferdam. The eddy zone downstream side of the downstream transversal cofferdam is underestimated in both models. However, the length and width of the recirculating zone by extended model is closer to measurements compared with the original model. It is also found that three dimensional correction of $C_{fc}$ has an insignificant effect on the velocity vector field near the longitudinal cofferdam.

Table 2   Experimental conditions

| Width of flume B | 30cm |
|---|---|
| Length of longitudinal cofferdam | 24cm |
| Length of foundation pit near to bank | 54cm |
| Width of foundation pit | 15cm |
| Angles between upstream and downstream transversal coffer-dams and adjacent bank | 45°  135° |
| Discharge Q | 3500cm³/s |
| Water depth of inlet | 8.7cm |

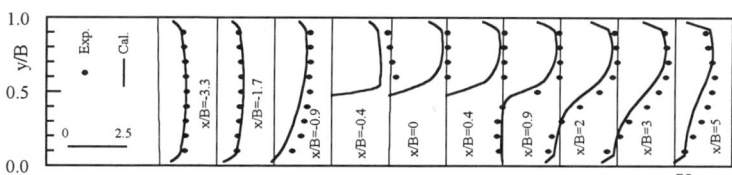

Figure 2   Comparison of computed longitudinal velocities with measurements

## CONCLUSIONS

The flow constricted by cofferdams is predicted by using an extended depth-averaged k-ε turbulence model. The extension of model coefficients and friction coefficient is successful. The extended model provides better results than the original model. The applicability of the extended model to the complex turbulent flowfields around the structures such as cofferdams seems to be quite good from the viewpoint of engineering applications.

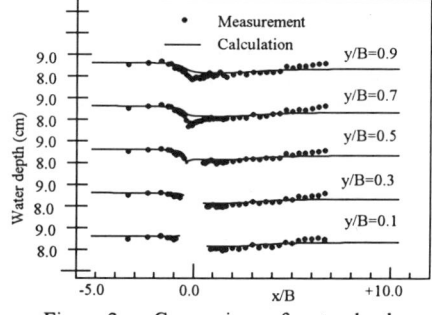

Figure 3   Comparison of water depths

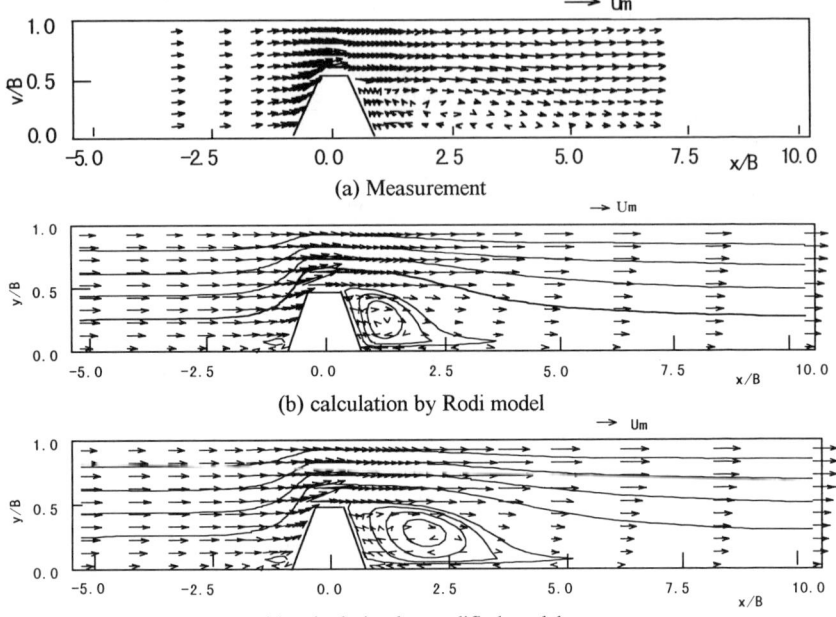

Figure 4   Velocity vector distributions

## REFERENCES

Liu, J.(1991), *Numerical Simulation of River Constriction by Cofferdams*, Master thesis, Wuhan University of Hydraulic and Electric Engineering, Dec. (in Chinese).

Liu, J.(1995), *Numerical Study on Flow in Open Channels with Hydraulic Structure*, Ph.D. thesis, Nagoya Institute of Technology, Dec..

Liu, J., Shyy, W., Vu, T. and Perng, C.-Y.(1996), Two-equation turbulence modeling for internal flows with non-equilibrium and rotating effects, *Flow Modeling and Turbulence Measurements VI*, Chen, Shih, Lienau & Kung (eds.), Balkema, Rotterdam, pp. 347-354.

Launder, B.E. and Spalding, D.B.(1974), The numerical calculation of turbulent flows, *Computer Methods in AppliedMechanics and Engineering*, Vol. 3, pp. 269-289.

Murakami, S., Mochida, A. and Iizuka, S.(1996), New trends in turbulence models for prediction of wind effects on structures, *5th Turbulent Flow Forum*, Oct.3,Tokyo, pp.1-52.

Patankar, S.V.(1980), *Numerical Heat Transfer and Fluid Flow*, Hemisphere, New York.

Rodi, W.(1979), *Turbulence Methods for Environmental Problems*, Hemisphere, New York.

Schlichting, H.(1960), *Boundary Layer Theory*, McGraw Hill Book Co., Inc., New York.

Shih, T.-H.(1995), Developments in computational modeling of turbulent flows, *Proceedings of the International Symposium on mathematical Modeling of Turbulent Flows*, Tokyo, Japan, Dec. 18-20, pp. 21-44.

Tingsanchali, T. and Maheswaran, S.(1990), 2-D Depth-averaged flow computation near groyne, *Journal of Hydraulic Engineering, ASCE*, Vol. 116, No.1, pp. 71-86.

Van Doormaal, J.P. and Raithby, G.D.(1984), Enhancement of the SIMPLE method for predicting incompressible fluid flows, *Numerical Heat Transfer*, Vol.7, pp. 147-163.

# 2-D FLOW AND SEDIMENT SIMULATION FOR THE FLOOD REGULATION OF A RESERVOIR WITH WATER INTAKE

| | |
|---|---|
| **Fayi Zhou** | University of Alberta, Canada |
| **Bihong Chen** | Dalian University of Technology, China |
| **Charles C.S.Song** | University of Minnesota, U.S.A. |

## ABSTRACT

In this paper, the flood regulation by operating the downstream sluice gates for a reservoir with a water intake is studied. The two-dimensional depth-averaged flow equations are solved by the boundary fitted finite volume method(FVM) based on MacCormack prediction-correction scheme. The bed deformation caused by both of the bed load and incoming suspended sediment is solved in a coupled way. A practical reservoir flood regulation operation is simulated.

## INTRODUCTION

The sediment transport and bed deformation near the water intake during the flood regulation operation of the reservoir must be studied in the design of sediment control of a water intake. In this paper, numerical approach is used to simulate a flood regulation operation of a reservoir in Taiwan. Body fitted mesh system is used to adapt the irregular boundary and the downstream gate operation is simulated. Both suspended load and bed load are considered. The model can handle the gradually and suddenly opening or closing of the sluice gates.

## GOVERNING EQUATIONS

The 2-D shallow water equations are written as follows:

$$\frac{\partial U}{\partial t} + \frac{\partial E}{\partial x} + \frac{\partial F}{\partial y} = S \tag{1}$$

where

$$U = [h, hU, hV, h\overline{C}]^T$$
$$E = [hU, hU^2 + gh^2/2, hUV, hU\overline{C}]^T$$
$$F = [hV, hUV, hV^2 + gh^2/2, hV\overline{C}]^T$$
$$S = [0, S_{fx}, S_{fy}, S_c]^T$$
$$S_{fx} = -gh\frac{\partial Z_b}{\partial x} - \frac{\tau_{bx}}{\rho}$$
$$S_{fy} = -gh\frac{\partial Z_b}{\partial y} - \frac{\tau_{by}}{\rho}$$
$$S_c = -\alpha\omega_s(\overline{C} - C_*) \tag{2}$$

where h is depth of water; $Z_b$ is the bed elevation; U, V are depth-averaged velocity components along x and y direction, respectively; $\tau_{bx}$ and $\tau_{by}$ are bottom shear stresses; $\overline{C}$ is the depth-averaged sediment concentration; $C_*$ is the critical sediment transport capacity of the flow which will be given later; $\alpha$ is the settling probability of the suspended grains at time t; $\omega_s$ is the fall velocity of the grain.

The bed load transport rate $q_b$ (ks/(s.m)) is based on Samov's equation (CSHE, 1992):

$$q_b = K_b \sqrt{D} (\frac{\overline{U}}{U_c/1.2})^3 (\overline{U} - \frac{U_c}{1.2})(\frac{D}{h})^{1/4} \tag{3}$$

where, $K_b$ is a coefficient(0.95); $D$ is the mean diameter of the grain(m); $h$ is the depth of the water(m); $\overline{U}$ is the mean velocity(m/s); and $U_c$ is the incipient velocity of the grain which can be expressed as

$$U_c = 1.34 \sqrt{\frac{\gamma_s - \gamma}{\gamma} gD} (\frac{h}{D})^{0.143} \tag{4}$$

Following formula (Q.Han, 1990) is used to calculate sediment transport capacity of the flow:

$$C_* = 0.000148 \gamma_s (\frac{U^3}{\frac{\gamma_s - \gamma}{\gamma} gh\omega_s})^{0.92} \tag{5}$$

where $\gamma_s$ and $\gamma$ are the specific weight of the sediment and water respectively. Here $C_*$ has the same unit with $\gamma_s$.

The equation for bed deformation is based on the Exner Equation as

$$\frac{\partial z_b}{\partial t} + \frac{1}{\gamma_s'} (\frac{\partial (q_{sx} + q_{bx})}{\partial x} + \frac{\partial (q_{sy} + q_{by})}{\partial y}) = 0 \tag{6}$$

where $\gamma_s'$ is the dry specific weight of the bed load material; $q_{sx}$ and $q_{sy}$ are the x-, y-components of the suspended sediment transport rate, respectively.

## NUMERICAL SCHEME

Integrating Eq.(1) over a control volume, we have

$$\frac{\partial \overline{U}}{\partial t} + \frac{1}{A} \iint_\Omega (\frac{\partial E}{\partial x} + \frac{\partial F}{\partial y}) ds = \frac{1}{A} \iint_\Omega S ds \tag{7}$$

where, $\overline{U}$ is the area-averaged value of $U$; $A$ is the area of control volume $\Omega$. Then we have

$$\frac{\partial \overline{U}}{\partial t} + \frac{1}{A} \oint_L \vec{H} \cdot \vec{n} \cdot dL = \overline{S} \tag{8}$$

Where, $\vec{H} = i\vec{E} + j\vec{F}$. The over bars denote the volume-averaged variables which are usually assumed to be the values on the center of a cell. $\vec{n}$ is the normal direction of the cell surface $dL$. Eq.(1) can be solved by MacCormack prediction-correction scheme as follows:

Prediction step:

$$\overline{U^{n+1}} = U^n - (\frac{\Delta t}{A}) \sum_{i=1}^{2} (H \cdot L_i^+ + H^- \cdot L_i^-) + \Delta t \overline{S^n} \tag{9}$$

Correction step:
$$U^{n+1} = \frac{1}{2}(\overline{U^{n+1}} + U^n) - (\frac{\Delta t}{A})\left[\sum_{i=1}^{2}(\overline{H^+ \cdot L_i^+} + \overline{H^+ \cdot L_i^-}) - \sum_{i=1}^{2}(\overline{H \cdot L_i^+} + \overline{H^- \cdot L_i^-})\right] + \frac{\Delta t}{A} \cdot \overline{S^{n+1}} \quad (10)$$

where, $L_i^+$ is the downstream and $L_i^-$ is the upstream length vector of the finite volume $\Omega$. $H^+$, $H$ and $H^-$ are the flux vectors at downstream, current and upstream finite volume respectively. In Eq.(9) and Eq.(10) the over bar means the predicted value.

## BOUNDARY CONDITIONS
### Upstream Boundary
The velocity condition at the inlet can be specified as:
$$\bar{u}(x_0, y, t) = Q(t) \cdot \frac{h(x_0, y)^{2/3}}{\sum B_i h(x_0, y)^{5/3}} \quad (11)$$

where $B_i$ is the width of $i$-th element of the inlet boundary. Such a velocity specification has taken into account the effect of depth variation on the velocity distribution along the transverse. For suspended sediment, the inlet depth-averaged concentration is given. For bed load transport, zero-gradient condition of bed load rate is assumed.

### Downstream boundary
Both sluice gates and overflow weir exist in this study and the correspondent formula are used. The treatment of sluice gate operation will not be described for the limit of space.

### Solid boundary
No flux can pass through the solid boundary. Along the tangential direction, the full slip boundary condition is used.

## MODEL APPLICATION
Liyutan reservoir located in the central area of Taiwan is a multi-purpose reservoir. The water intake of the power plant is located in the right bank, just upstream of dam, see Fig.1. Several operation cases are simulated and for the limit of space, only 2.33 year flood regulation is presented.

### The flow field under peak flow of 2.33 year flood
The upstream of the computation domain is 1.2km upstream of the dam section. Mesh distribution of the computation configuration is shown in Fig.1 (total grid number=NNX*NNY=75*27). Manning's coefficient is chosen between 0.025 and 0.050 depending on the bed topography and is varied at different locations. The result is shown in Fig.6 and Fig.7. Good agreement for the water surface elevation between computation and experiment at the backwater area of the reservoir is obtained. The velocity comparison is also satisfactory.

### Flood regulation - 2.33 year flood

During the flood regulation operation, the sluice gates are operated in the following ways: when the discharge is less than 800 m$^3$/s, the water intake works and the opening of downstream gate is adjusted to ensure the water head in front of the intake is kept at 604.5m (NWL). When the discharge is increased further, the intake gate is closed and the downstream gates are fully opened. In computation, the inlet suspended sediment concentration is 0.25%(in weight ratio). The mean diameters of the suspended sediment is 1mm and the correspondent fall velocity is 12cm/s. The mean diameter of the bed material is set to $D_{50}$=60 mm. Bed roughness n=0.035. Fig.2 shows the bed deformation at the end of flood. Fig.4a is the calculated hydrograph of upstream and downstream sections and Fig.4b is the free surface elevation variation with time under the 2.33 year flood regulation. The gates are assumed to be suddenly opened when this discharge exceeds 800 m$^3$/s. The sharp increase of the discharge and the sudden drop of water level can be seen from this figure. Fig.3 is the variation of suspended sediment concentration with time at several locations. It is found that there is hardly any coarse grain (D=1mm) in front of intake (the maximum C is only 0.010%) Therefore, basically there will be no coarser grain ($\geq$ 1mm) coming into the water intake under 2.33 year flood provided there is no fine sand deposit within the flush channel. Fig.5 shows the profiles of the cross sectional bed surface. From Fig.2 and Fig.5, it can be seen that under 2.33 year flood, the left part of the reservoir is in deposition while at the right part near the water intake, there is scour occurs. And just in front of the spillway part, there is slight deposition. This indicates that during the flood regulation operation, there is no much sediment deposition in front of the water intake. All of the phenomena mentioned above are qualitatively consistent with those observed form the physical experiment.

## CONCLUSIONS

The proposed 2-d model can be used to simulate the steady and unsteady open channel flows including both sub-critical and supercritical flow. One important feature of the model is its capacity of handling the gate operation which is often encountered during the flood regulation operation of the reservoir. In current model, only the single size of grain is considered. The error exist without considering the grading of the grain, the armoring and the sorting.

## REFERENCES

1. CSHE, *The hand book of Sedimentation*, Task Committee of Sedimentation, Chinese Society of Hydraulic Engineering, The Press of Chinese Environment Science, 1992.
2. Fayi Zhou, Bihong Chen(1994) " The bed deformation downstream of the hydropower plant", *the Proceeding of II IAHR symposium, Asia and pacific region*, August, 1994, Singapore.
3. Han Qiwei (1990), "The study of non-equilibrium sediment transport", *Proceeding of*

*International Symposium of Fluvial Sedimentation*, Beijing.
4. Zhao, D.H., Shen, H.W., "Finite-volume two-dimensional unsteady-flow model for river basins", *Journal of Hydraulic Engineering*, Vol.120, No.7, July,1994.

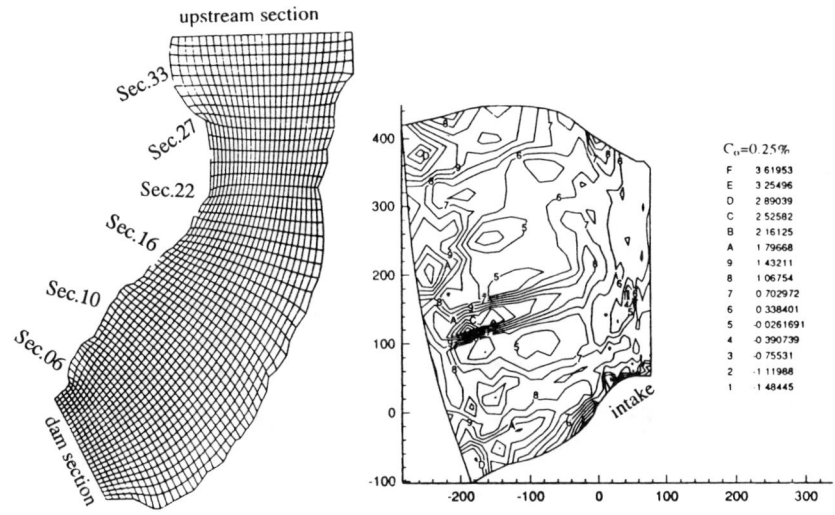

Fig.1 Part of 2-D mesh distribution  Fig.2 Bed deformation at the end of 2.33 year flood
(positive values represent deposition,
negative values represent scour)

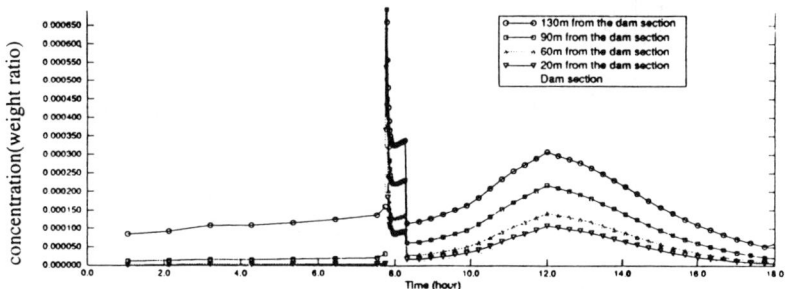

Fig.3 The sediment concentration variation with time (inlet $C_0=0.25\%$)

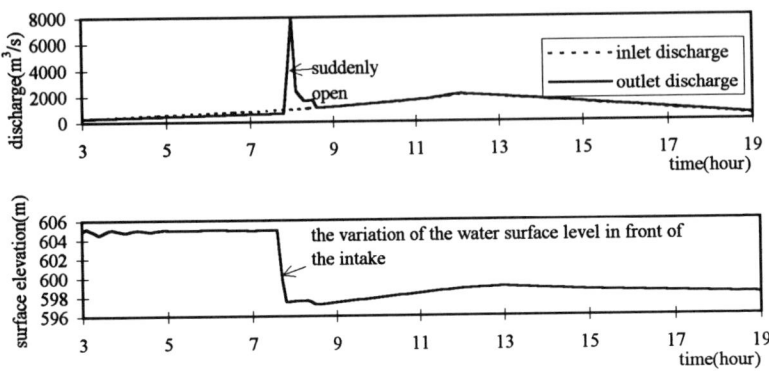

Fig.4 The regulation of 2.33 year flood

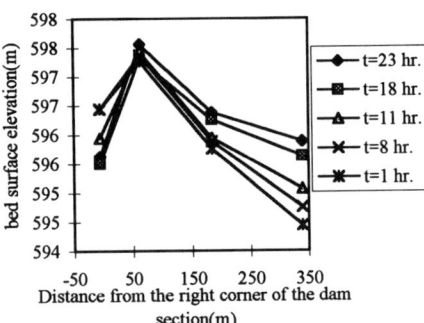

Fig.5a Bed deformation at the section 60m upstream of the dam section

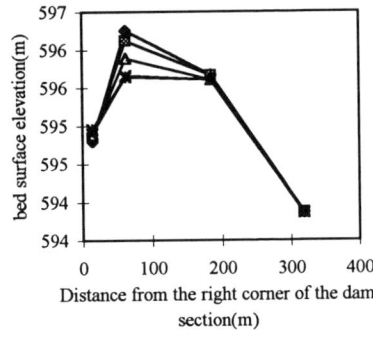

Fig.5b. Bed deformation at the section 20m upstream of the dam section

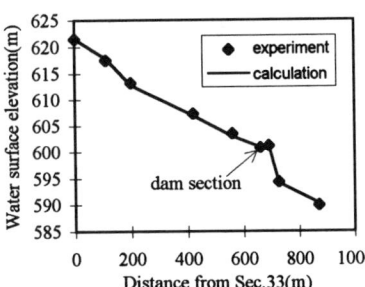

Fig.6 Free surface elevation along the central axis(steady flow for comparison)

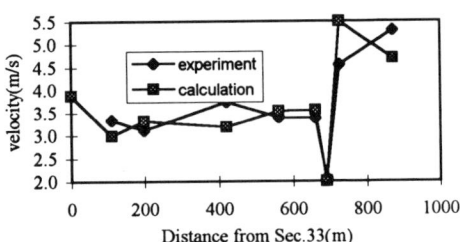

Fig.7 Velocity values along the central axis(steady flow for comparison)

# 3D numerical modelling of sediment deposition and bed changes in a tunnel-type sand trap

## NILS REIDAR B. OLSEN and HILDE MARIE KJELLESVIG
SINTEF Civil and Environmental Engineering, 7034 Trondheim, Norway

### INTRODUCTION
The hydraulic design of the sand trap is important in planning of hydropower plants using water from sediment-carrying rivers. It is necessary to ensure that the sand trap will function with sufficient efficiency. Physical modelling of sediment deposition in a sand trap is difficult because of practical problems when scaling small sediment particles. The alternative, 3D numerical modeling, has previously been used for a steady situation for both free surface flow (Olsen and Skoglund, 1994) and a tunnel-type sand trap (Olsen and Chandrashekhar, 1995). The calculated concentration profiles from these studies compared well with laboratory measurements.

The sand trap will fill up over time, and the geometry will change between each flushing event. It is therefore useful to calculate the trap efficiency over time using changing bed geometry. The present study focus on the numerical modeling of changes in the bed geometry of the sand trap as deposition occurs. The results have been assessed by a laboratory experiment.

### THE NUMERICAL MODEL
The computer program SSIIM was used in this study. The calculation of the water flow is based on the transient Reynolds averaged Navier-Stokes Equations. The k-$\epsilon$ model is used to resolve the Reynolds stress term. A control volume method is used for discretization, using an upstream implicit method. The SIMPLE method is used for the pressure coupling. The sediment concentration is calculated with the transient convection-diffusion equation, using an implicit Power-Law Scheme. The numerical methods are described in more detail by Patankar (1980), Rodi (1980), Melaaen (1992) and Olsen (1996). The sediment concentration at the bed is calculated by van Rijn's formula (1987):

$$c_{bed} = 0.015 \frac{d^{0.7}}{a} \frac{\left[\dfrac{\tau_0 - \tau_{critical}}{\tau_{critical}}\right]^{1.5}}{\left[\dfrac{(\rho_s - \rho_w)g}{\rho_w v^2}\right]^{0.1}} \tag{1}$$

The sediment particle diameter is denoted d, $\tau_0$ is the bed shear stress, $\tau_{critical}$ is the critical bed shear stress for movement of sediment particles, $\rho_w$ and $\rho_s$ are the density of water and sediment, v is the viscosity of the water, g is the acceleration of gravity and a is a reference level, set equal to the roughness height. Since the concentration is given in the center of each bed cell, the reference level did not always correspond to half the size of the bed cell. The concentration was then extrapolated using the Rouse equation (1937)

$$\frac{c_{cell}}{c_{van\ Rijn}} = \left[\frac{h-b}{h}\frac{a}{h-a}\right]^{\frac{w}{\kappa u_*}} \tag{2}$$

Here, h is the depth, b is half the height of the bed cell, a is the roughness height, w is the fall velocity of the particles, $u_*$ is the shear velocity and $\kappa$ is a constant equal to 0.4.

Two algorithms were necessary to take into account the effect of the sloping bed. The formula given by Brooks (1963) were used to decrease the shear stress at the bed as a function of the bed slope. Also, an avalanche routine was used to change the bed when the bed slope became larger then the angle of repose for the material (Olsen, 1996).

The procedure to find the sediment concentration at the bed was used in combination with sediment continuity to determine the bed changes.

## THE LABORATORY EXPERIMENT

The laboratory study was conducted in a plexiglass tunnel. Figure 1 shows the geometry of the model. A water discharge of 15 liters/second was used. Sediments were added at the upstream boundary of the model in a quantity of 80 kg during 88 minutes. A longitudinal bed level profile at the centerline was measured afterward. The sediments had a maximum size of 0.9 mm, a minimum size of 0.03 mm and a $d_{50}$ of 0.5 mm.

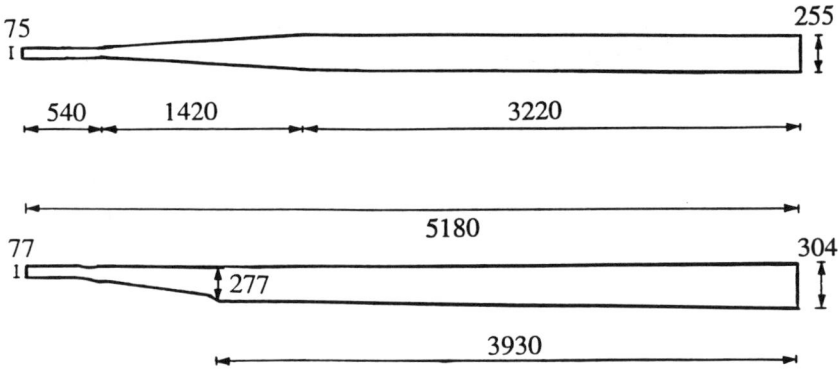

Figure 1. Sketch of the laboratory model geometry seen in plan (above) and from the side (below). Dimensions are given in mm.

## NUMERICAL MODEL RESULTS

A structured grid with 96x13x15 cells in the streamwise, cross-streamwise and vertical direction was used. The grid is shown in Fig. 2. The time step was 10 seconds, with 5 inner iterations pr. time step for the water flow calculation and 50 inner time steps for the sediment calculation.

Figure 2. Grid seen from above.

Fig. 3 shows a longitudinal profile with velocity vectors along the centerline of the tunnel at the entrance. There is a contraction in the entrance of the flume, giving a jet pointing in the downwards direction. This gives a recirculation zone at the tunnel roof in the entrance. The same recirculation zone was observed in the laboratory experiment. The corresponding jet at the entrance of the bed introduced high shear stress at the bed. This caused very little deposition at the entrance of the tunnel. The deposition pattern is shown from above in Fig. 4. At the upstream part of the deposit, the magnitude is higher closer to the side wall than at the center of the sand trap. This was also observed in the laboratory experiment.

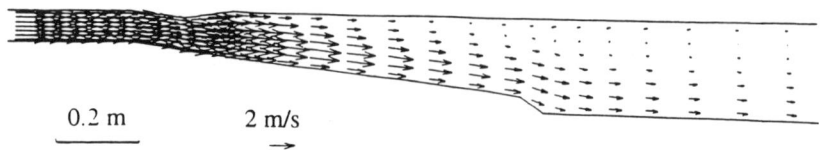

Figure 3. Longitudinal profile with velocity vectors along the centerline of the tunnel at the entrance.

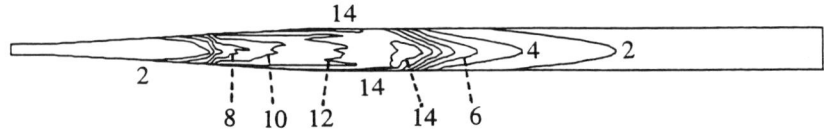

Figure 4. Map of deposited material shown with contours. The levels are given in cm.

Vertical concentration profiles along the centerline of the tunnel are shown in Fig. 5. The concentration is fairly uniformly distributed vertically at the entrance. Over the deposits, there is a higher concentration near the bed. The concentrations decrease sharply downstream the deposits. This is also in correspondence with observations from the laboratory experiment, where the deposit grew to a certain height and then moved forwards as a delta front.

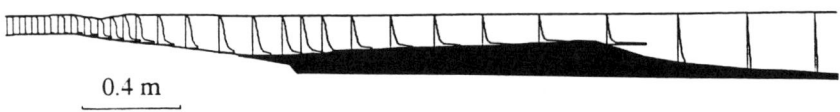

Figure 5. Concentrations in a longitudinal profile. The dark area is the deposited material.

The measured and calculated bed profile is shown in Figure 6. The numerical model is able to reproduce the main deposition pattern of the profile.

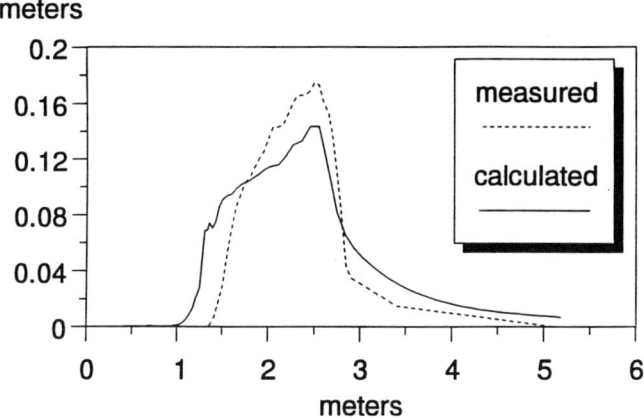

Figure 6. Longitudinal profile of calculated and measured deposition levels.

## CONCLUSIONS

The numerical model is able to predict the bed movements in the sand trap. The corresponding laboratory experiment shows good agreement with the results from the calculation.

## REFERENCES

Brooks, H. N. (1963), discussion of "Boundary Shear Stresses in Curved Trapezoidal Channels", by A. T. Ippen and P. A. Drinker, ASCE Journal of Hydraulic Engineering, Vol. 89, No. HY3.

Melaaen, M. C. (1992) "Calculation of fluid flows with staggered and nonstaggered curvilinear nonorthogonal grids - the theory", Numerical Heat Transfer, Part B, vol. 21, pp 1-19.

Olsen, N. R. B. and Skoglund, M. (1994) "Three-dimensional numerical modeling of water and sediment flow in a sand trap", IAHR Journal of Hydraulic Research, No. 6.

Olsen, N. R. B. and Chandrashekhar, J. (1995) "Calculation of water and sediment flow in desilting basins", 6th. International Symposium on River Sedimentation, New Delhi, India.

Olsen, N. R. B. (1996) "Three-dimensional numerical modeling of local scour", HYDROINFORMATICS-96, Zurich, 1996.

Patankar, S. V. (1980) "Numerical Heat Transfer and Fluid Flow", McGraw-Hill Book Company, New York.

van Rijn, L. C. (1987) "Mathematical modeling of morphological processes in the case of suspended sediment transport", Ph.D Thesis, Delft University of Technology.

Rodi, W. (1980) "Turbulence models and their application in hydraulics", IAHR State-of-the-art paper.

Rouse, H. (1937) "Modern Conceptions of the Mechanics of Fluid Turbulence", Transactions, ASCE, Vol. 102, Paper No. 1965.

# Increased output of old power plants
# - limitation and problems to be solved

### HERMOD BREKKE
Professor, The Norwegian University of Science and Technology, Trondheim, Norway

ABSTRACT

The paper includes a study of increasing the output of existing power plants.

By exchanging the existing runner in the turbine with a new runner with increased flow and output, cavitation may occur. The paper gives an analysis of the design parameters and the limitation of the increase in output of Francis turbines, which is the most commonly used turbine type. When the flow in the powerplant is increased the increased velocity in the conduit system often leads to problems with the governing stability of the hydro-power plant.

One solution is increasing the cross sections of the conduit system (tunnels and penstocks). However, by means of a pressure feedback system of the governor, stability can be obtained and thus save the cost of increasing the cross sections of the conduits.

The paper gives a brief description of the pressure feedback system with illustration by means of a block diagram.

## INCREASING OUTPUT OF A FRANCIS TURBINE
The initial step in the design of a Francis runner will always be the calculation of the main dimensions which is based upon design head $H_n$ and required design flow $Q_n$ and an available Net Positive Suction Head ($NPSH_A$).

The procedure for the calculation of main dimensions and shape of a runner with increased output is based on some very important parameters. The parameters in question are discussed in the following text.

## THE SPEED NUMBER
The speed number is the most important parameter when sudying increased power of a turbine. The connection between the specific speed $n_s$ and the speed number $^*\Omega$ which is referring to the flow at best efficiency point (BEP) will be:

$$^*\Omega = \omega \sqrt{^*Q} / (2gH_n)^{0.75} \qquad \text{where } \omega = n \cdot \pi/30$$

and $\quad n_s = K^{0.5}(2g)^{0.75} 30 / \pi \,^*\Omega$

Here $K = Q_n/{}^*Q$ where $Q_n$ = nominal flow and ${}^*Q$ = BEP flow.

The dimensionless speed number may also be expressed by the reduced dimensionless outlet blade velocity at the shroud $\underline{u}_2 = u_2/(2gH)^{0.5}$ and the outlet blade angle $\beta_2$.

$${}^*\Omega = \sqrt{\pi}\,\underline{u}_2^{3/2}\,\sqrt{\tan(\pi - \beta_2)} \tag{1}$$

or

$$\underline{u}_2 = u_2 / (2gH)^{0.5} = {}^*\Omega^{2/3} / (\pi \tan(\pi - \beta_2))^{1/3}$$

## THE NET POSITIVE SUCTION HEAD (NPSH)

The NPSH-value of a runner is depending on the outlet velocity at the shroud and the outlet angle of the blades ($\pi-\beta_2$). However, the blade curvature, the curvature of shroud and crown as well as the blade lean angle $\theta$ (described later) has also an influence on the cavitation performance.

For a given geometry of a runner, the following empirical formula for the *required* $NPSH_R$ may be established (note: $NPSH_R < NPSH_A$ must be fulfilled):

$$NPSH_R = u_2^2 / (2g)\left[aK^2 \tan^2(\pi - \beta_2) + b\right] \tag{2}$$

where a and b is depending of the geometry, and

$1 < a < 1.2$ and $0.05 < b < 0.15$ where $b = f({}^*\Omega)$

## THE REACTION RATIO

The reaction ratio describes the relative pressure drop from inlet to outlet of the runner, i.e. $(h_1-h_2)/H_n$. An increase of the reaction ratio is one possibility to avoid inlet cavitation.

By studying the equation for the absolute specific energy $E=gH$ and the relative specific energy (ROTHALPY), $I = gh + w^2/2 - u^2/2$, the following equations may be established:
*   Assuming friction free flow i.e. $I = I_1 = I_2$ = constant
    Then: $gH_n \cdot \eta_n = u_1c_{u1} - u_2c_{u2} = (E_1-I_1)-(E_2-I_2) = (gh_1+c_1/2)^2 - (gh_2+c_2/2)^2$
*   By assuming $c_{u2} = 0$ at best efficiency and further assuming the cross section of the runner to be equal at inlet and outlet i.e. $c_{m1} = c_{m2}$
*   Then the following equation for the reaction ratio yields:

$$\frac{h_1 - h_2}{H_n} = \frac{u_1 c_{u1}}{gH_n} - \frac{c_{u1}^2}{2gH_n} \tag{3}$$

The reaction ratio then depends on the inlet velocity vector diagram and the inlet blade angle as illustrated in fig. 1.

# INCREASED OUTPUT OF OLD POWERPLANTS

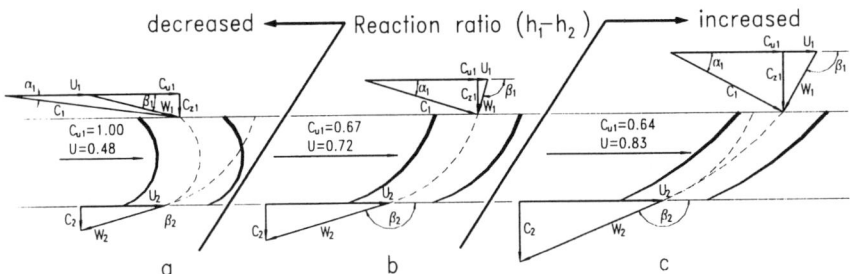

*Figure 1.* Blade cascades for different reaction ratios for $c_{u2}=0$ and $\underline{u}_1\underline{c}_{u1}$ = const.

It should also be emphasized that an increased inlet blade velocity $u_1$ obtained by increasing the inlet diameter automatically reduces the absolute tangential velocity component $c_{u1}$ if the efficiency is unchanged, and thus the reaction ratio increases.

For a given specific speed according to eq (1), an increased reaction ratio and increased inlet pressure may be obtained by changing the blade angles as shown in fig. 1.

## ADJUSTMENT OF INLET ANGLES

In order to maintain the hydraulic efficiency, the inlet angles must be adjusted if the outlet angle is changed (see fig. 1). By a study of the velocity vector diagram for the best efficiency point on an existing runner, denoted by asterix, and for increased outlet angles $(\pi - \beta_2)$, the following equation for the adjusted inlet angles can be derived:

$$\tan(\pi - \beta_1) = \tan(\pi - {}^*\beta_1) \frac{\tan(\pi - {}^*\beta_1)({}^*u_1 - {}^*c_{u1})}{\tan(\pi - {}^*\beta_1)(u_1 - c_{u1})} \tag{4}$$

## THE BLADE LEAN ANGLE

The blade lean is defined by the angle $\theta$ and geometry parameters as illustrated in fig. 2. The reduced pressure gradient $d\underline{h}/dy = d(h/H_n)/dy$ in a stream surface may be established by eq. (5) expressed by the reduced dimensionless meridonal velocity $\underline{c}_m = c_m/(2gH_n)^{0.5}$ and angular speed $\underline{\omega} = \omega/(2gH_n)^{0.5}$.

$$\frac{d\underline{h}}{dy} = 2\left\{ \begin{array}{l} \left[\left(\frac{1}{R} - \frac{\cos\delta\cos^3\beta}{r}\right)\frac{\underline{c}_m^2}{\sin^3\beta} - \frac{\underline{c}_m(\partial \underline{c}_m/\partial z)}{\tan\beta} + 2\underline{\omega}\underline{c}_m\cos\delta\right]\tan\theta \\ +\left(\frac{\sin\delta}{r\tan^2\beta} + \frac{1}{\rho}\right)\underline{c}_m^2 + \left(2\underline{\omega}\frac{\sin\delta}{\tan\beta}\right)\underline{c}_m + \underline{\omega}^2 r\sin\delta \end{array} \right\} \tag{5}$$

Eq. (5) is based upon the equilibrium of forces and is valid for a runner with infinite number of blades i.e. potential flow. In addition to eq. (5) the rothalpy equation (6) based on the inlet energy and the equation of continuity must be used.

$$\underline{h} = \underline{\omega}^2 r^2 - \underline{c}_m^2/\sin^2\beta + (1 - 2\underline{u}_1\underline{c}_{u1}) - J \tag{6}$$

The theory is 2-dimensional only, but is still useful in order to obtain a physical understanding and to improve the blade shape in the basic design followed by a full 3D Euler analysis by CFD codes, and finally followed by a viscous code.

In the rothalpy equation (6), J is the estimated losses along a streamline from inlet to the regarded point and $\underline{u}_1 = u_1/(2gH_n)^{0.5}$ and $\underline{c}_{u1} = c_{u1}/(2gH_n)^{0.5}$, which are the reduced circumferential blade velocity and the absolute flow velocity component in tangential (circumferential) direction, respectively.

In addition to eq. (5) and (6), the equation of continuity is used based on the distance between the streamlines, the blade angles and the thickness and the number of the blades in order to find the meridional velocity $\underline{c}_m$. (The real blades are used for the calculation of $\underline{c}_m$ only because the theory is based on an infinite number of blades.)

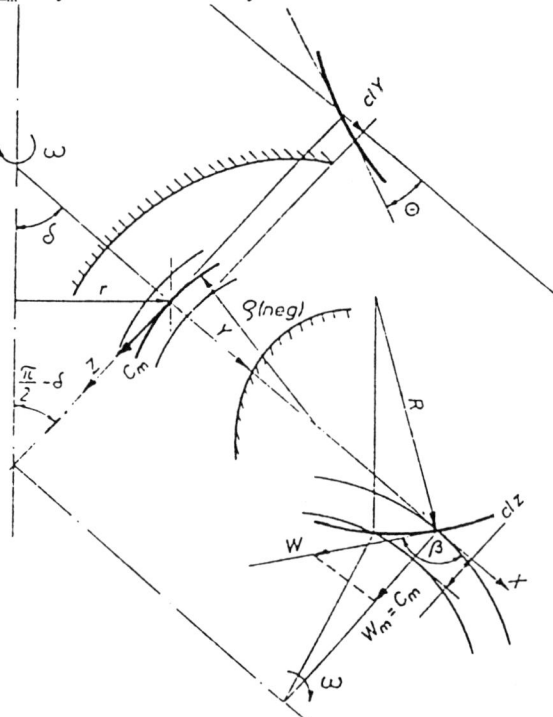

***Figure 2.*** *Definition of blade lean and the blade geometry.*

After the basic blade shape is adjusted by means of eq. (5) in order to obtain a suitable pressure gradient (d$\underline{h}$/dy), a 3D-CFD analysis is made for a final adjustment of the blades.

In fig. 3, the balanced pressure distribution on a geometrically balanced turbine blade (right) is compared with a blade with traditional design (left). Note that the low pressure zone at the inlet of the shroud has disappeared in the blade with balanced blade lean angles θ to the right. By using balanced blades of the new design increased range of flow and head can be allowed for the turbine without cavitation problems.

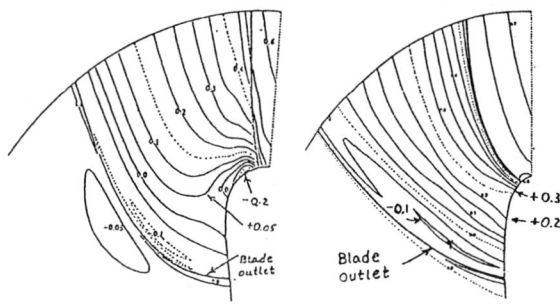

*Figure 3. Pressure distribution on suction side of a traditional runner (left) and on a runner of new design with balanced blade lean angle θ (right) (courtesy Kvaerner).*

## STABILITY PROBLEMS CAUSED BY INCREASED FLOW

Because of the increased velocity in the conduit system in a hydropower plant the governing stability of the turbines may be poor.

As a rule of the thumb the ratio between the inertia time constant of the generator and the inertia time constant of the penstock ($T_a/T_w$) must fulfill following require-ments

For high head plants: $T_a/T_w > 4$ for $0.4 < h_w < 0.7$ $\left( h_w = \dfrac{a^0 c}{2gH_n} \right)$

For low head plants: $T_a/T_w > 3$ for $h_w \geq 1.0$

For increased flow velocity, $T_w$ will increase and sufficient governing stability may not be obtained. One solution to solve this problem could be to increase the cross section of the conduits. This solution is very expensive, and fortunately a pressure feedback system has been developed during the recent 5 years in order to solve the stability problems in cases where the inertia the ratio $T_a / T_w$ is below the normal acceptable limit. In the last chapter of this paper a brief description of the principle of a pressure feedback system will be given:

## THE PRESSURE FEEDBACK SYSTEM

The arrangement is based upon a computer based governor system. In addition to the feedback signal from speed and load also a feedback signal from the pressure in front of the turbine is included in the new system.

The pressure feedback signal is fed via a special algorithm to the governor.

The system is based upon the principle that the effect from pressure rise in front of the turbine must be damped down by the governor in order to avoid amplification of pressure oscillations which in turn leads to power and speed oscillations.

The principle of a pressure feedback system is illustrated in the simplified block diagram in fig. 4. The diagram is valid for a Francis turbine where there is an influence from the speed on the flow caused by the turbine characteristic illustrated by $Q_n$.

A pressure feedback system as described was installed on a 350 MW high head Francis turbine governor for Svartisen powerplant in Norway. The stability was then obtained without any surge chamber, saving some US $ 5 Mill in excavating cost.

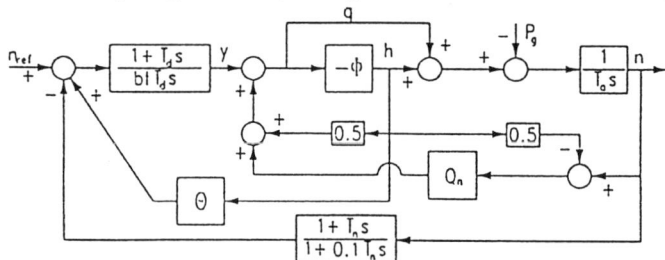

*Figure 4. Block diagram for a turbine governing system with a pressure feedback system.*

In the block diagram $\Theta$ represents the pressure feedback element for which the following equation yields:

$$\Theta(s) = \frac{KT_1 s}{(T_1 s + 1)(T_d s + 1)(0.5s + 1)} \qquad (7)$$

It has been proven theoretically that by tuning such a pressure feedback element, a stable speed governing can be obtained for an originally unstable power plant.

## CONCLUSION

By controlled parameters an improved design and increased output may be obtained for a traditionally designed Francis turbine runner. The governing stability may in many cases be solved by introducing a pressure feedback signal to a computer-based governor. By means of these two factors an increase of 10-20% may in general be obtained for power plants built 15-20 years ago.

## SYMBOLS

| | | | | |
|---|---|---|---|---|
| H | = | head energy m W.C.) | $T_a$ = | generator time constant (sec) |
| h | = | hydraulic pressure (m W.C.) | $T_w$ = | water conduit time constant (sec) |
| Q | = | turbine discharge (m³/s) | $h_w$ = | Allievi pipe characteristic |
| n | = | turbine speed (rpm) | $T_d$ = | integral time constant (sec) |
| ω | = | turbine angular vel. (rad/sec) | $T_n$ = | derivative time constant (sec) |
| s | = | Laplace operator (s = jω) | $T_1$ = | press. feedback time constant (sec) |

# Flow-Induced Multiple-Mode Vibrations of Lightly Damped Rectangular Trashrack Prisms

### K. KERENYI, H. DROBIR, T. STAUBLI, G. DORRER, N. HEJL
Institut für Konstruktiven Wasserbau, Technische Universität Wien, A-1040 Wien, Austria

### ABSTRACT
The experimental set-up and recent results of an investigation on flow-induced vibrations for a multiple-mode lightly damped rectangular prism are described. The identification of the fluid dynamic force coefficients are determined by means of a regression analysis accompanied by online numerical simulation, which allows minimization of errors. An application example of these procedures is given.

## 1. INTRODUCTION
Trashracks are important parts of hydro-power plants and outlet works of dams. If the function of the dam is flood control only, then the outlet works are large and there is little chance for objects smaller than the least dimension of the conduit (or control gate) to become lodged in the conduit. For such cases there is little need for narrow trashracks; large beams or columns are then installed at the entrance to the conduit to prevent large objects, such as trees from entering. However, for power intakes, large pieces of material entering the outlet works can become lodged in the shut-off value, the wicket gates, or even in the turbine runner. For power intakes it is common to space trashrack members to provide clear openings of not more than 5 to 10 centimeters.
In the design of any structure immersed in the flow, the possibility of vibration or response to dynamic loading should always be considered. In flowing, water, structures are inherently exposed to dynamic loading. It is necessary for the designer to insure that dynamic loads which do occur are within acceptable limits. Both ultimate strength of the structure and the possibility of failure through fatigue must be considered.

## 2. REVIEW OF TRASHRACK STUDIES
Individual trashrack bars can generally vibrate in directions in the plane (transverse) and normal to (streamwise) the plane of the rack. Vibrations in the transverse direction take place with the frequency of the shed vortices. However, in the streamwise direction, the bars will vibrate with twice the frequency of the shed vortices because the bar will pass through two complete cycles of upstream-to-

downstream movement as it passes through one complete cycle of motion from one side to the other. This, streamwise motion of the rack can be triggered by reduced velocities corresponding to the half of that triggering frequency of the transverse vibrations.

Syamalarao (1987) investigated trashracks with rectangular bar sections for different spacings and angles of incidence. For single bars, the data stem from the earlier work of Callander (1987), Saraiva and Ramos (1983), and others. Strouhal numbers varied from 0.06 for aspect ratios smaller than 2.8, to 0.16 for the larger aspect ratios. He found increasing Strouhal number with decreasing wake width behind the bar is smaller, which occurs when separation from the leading edge reattaches downstream to the bar. Callander observed a range of critical Strouhal numbers of 0.35 to 0.15 for bars with elongation ratio e/d between 0.5 and 1.5. At angles of incidence greater than zero, the flow separating at the leading edge reattaches only to one side of the bar, resulting in smaller value of Sh. The streamwise length e is more relevant than width d for controlling the critical vibration frequency. Syamalarao investigated the behavior of an elastically suspended rigid rack which could move in the direction normal to the rack (streamwise vibrations). Naudascher (1994) has formulated the range of streamwise vibrations for members having e/d ratios of 10. The streamwise mode is induced by vortices separating from alternating edges of the member. Vigander (1979) found that streamwise modes of vibration were present in the design of the Raccoon Mountain trashracks for which the bars were strongly stiffened by welded lateral bars. During the study of a ½-scale model composed of a series of 25cm diameter bars and suspended for testing in a lock culvert, several of the members failed from fatigue after only a few hours of testing. Subsequent shaking tests in air and response tests in water showed that the members vibrated in streamwise modes.

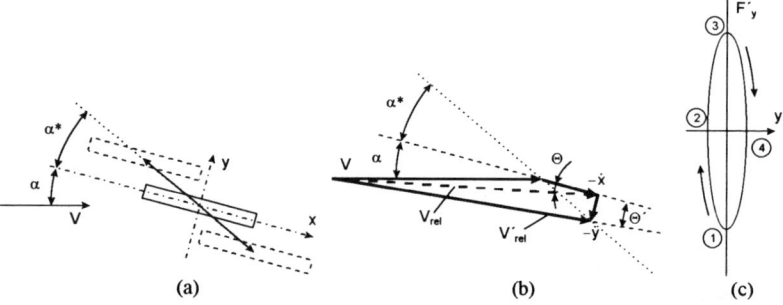

**Fig.1:** (a) Oblique bar movement; (b) corresponding relative velocity of approach $V_{rel}$; (c) corresponding force-displacement diagram

In a trashrack that vibrates in the streamwise mode, the trashrack bars are subjected to simultaneous transverse and streamwise forces. The coupling of streamwise and transverse bar movements leads to an increased streamwise excitation.

An example of how an additional degree of freedom may affect the flow-induced vibration of a structure is given in Fig.1. It illustrates the consequences of oblique bar vibrations obtained due to an error α* caused by coupling. During such vibrations, the velocity of the bar will have components $\dot{x}$ and $\dot{y}$ (Fig 1a,b) so that the instantaneous approach velocity and incident angle change from $\vec{V}_{rel}$ to $\vec{V}'_{rel} = \vec{V} - \dot{\vec{x}} - \dot{\vec{y}}$ and from Θ to Θ', respectively (Fig 1b). Assuming that the instantaneous hydrodynamic force component $F_y$ acting on the bar can be estimated with the aid of quasi-steady analysis, one obtains the force displacement diagram for the fluctating component $F'_y = F_y - \overline{F}_y$ as depicted in Fig. 1c ($\overline{F}_y$ = mean value of $F_y$). The area inside the loop of this diagram corresponds to the work effected on the bar per vibration cycle. For the conditions treated, this work is positive, thus contributing to the excitation of the bar. The argument shows that coupled movements in the y direction due to a misalignment α* do affect the flow-induced vibration in the x-direction.

## 3. EXPERIMENTAL SET-UP

As described above, the main significance of the choice of the experimental set-up lay in the coupling between two degrees of freedom. For this reason a free oscillator model was selected for this investigation. Emphasis was put on the identification of fluid dynamic forces. An innovative method of online measurement and numerical simulation was developed (Fig. 2). In the process of simulation it was necessary to identify a multi-degree equation of motion for various parameters (1).

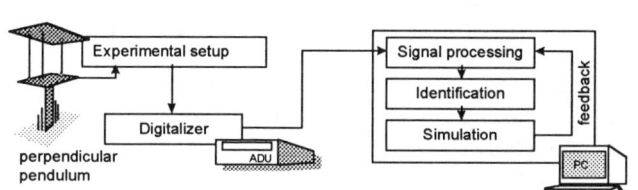

Fig. 2: Block diagram illustrating the dataflow

$$M \cdot \ddot{\vec{q}} + B \cdot \dot{\vec{q}} + C \cdot \vec{q} = \vec{F}(\ddot{\vec{q}}, \dot{\vec{q}}, \vec{q}, \alpha, V, \delta_x, \delta_y, f_{nx}, f_{ny}, Tu) \qquad (1)$$

A rectangular prism with an elongation ratio e/d = 10 (Fig. 3) was positioned in a laboratory free surface channel. The test section was 0.94 m wide by 16 m long, with a water depth of 0.38 m. The aluminum test prism had sharp edged corners and the surface was highly smoothened.

Of great importance in such studies is the approach-flow turbulence. For this reason, the tests were undertaken in a laboratory flume especially designed for possible control of turbulence level Tu. The flow was conditioned with honeycombs and filter mats in the inlet (Fig. 4). The intensity of free-stream turbulence was less than 1%. To guarantee constant velocity distribution, a special trumpet-shaped inlet was constructed.

The prism was free to oscillate both in the cross-flow and in the streamwise direction. The restoring forces were provided by linear coil springs. Thus, the prism acted as a linear mass oscillator with two perpendicular degrees of freedom. The free oscillator was constructed as a perpendicular pendulum (Fig. 5). This pendulum was suspended with three swing arms. In order to enable the perpendicular vibration, cardan joints were used. The stiffness and related natural frequencies could be varied by means of coil springs. The pendulum was mounted on a turntable to vary the flow incidence.

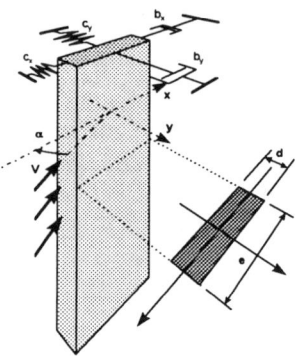

Fig. 3: Cross-secctional shape of the test bar (e/d=10).

One of the purposes of this investigation was to achieve the insight into the hydrodynamic coupling mechanisms between the two vibration modes and to determine the interaction inbetween the unsteady lift and drag forces. Vibration tests were run with reduced velocities $V_{rx}$ and $V_{ry}$ from 35 to 51 ($V_{ri}$ = $v/(d\ f_{ni})$ with v = streamwise velocity, $f_{ni}$ = natural frequency in air in either x- or y- direction). The flow incidence angle was varied between 30° and 45°. Furthermore, the Scruton number (Sc = $2m_r\delta$) was also changed in both directions between 8 and 180.

The response of the body-oscillator was measured by deflection cells as well as by accelerometers. The first derivative was used to determine the velocity. Fluid dynamic forces were measured with strain gauges. The strain gauges had be calibrated dynamically and statically. To vary the damping, which is a nonlinear function of amplitude, electromagnetic dampers were used in streamwise direction and hydraulic dampers in transverse direction. While beyond the scope of this paper, it is important to note that variation of damping can be employed to analyze stability of the hard oscillator.

A hot film probe was employed to identify the turbulence characteristics at the test section, namely the scale and the intensity. The longitudinal integral length scale of turbulence was computed from the

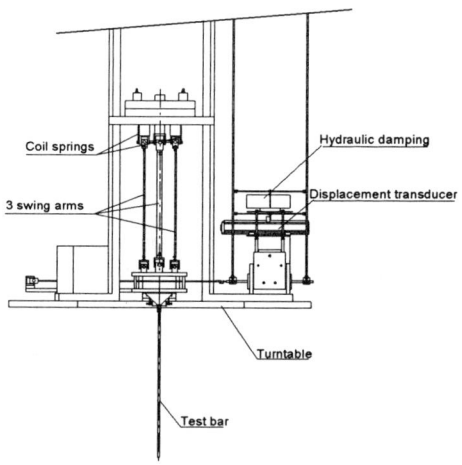

Fig. 5: Mounting device for test bar.

autocorrelation function of the streamwise fluctating velocity and the time mean free stream velocity. The integral length scale was used to determine the measurement time. Positioning the hot film probe into the wake permitted determination of the vortex shedding frequency.

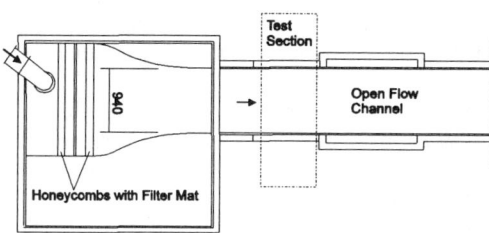

Fig. 4: Plan view of inlet and test flume.

As to the compilation of measurement data, a signal processing computer was linked to a second computer responsible for the identification of the fluid dynamic forces and the integration of the equation of motions. The sampling rate of the digitalizied measurement data was 75 Hz. Due to low frequency response of our measurement system, no aliasing occured (the maximum frequency encountered in the signals was smaller than 10 Hz). Digital filtering was necessery to eliminate signal noise. The cut of frequency of the Butterworth filter $5^{th}$ order was employed between 15 and 30 Hz.

## 4. MODELIZATION OF FLUID FORCES

The aim of this analysis is to identify the fluid force in the form of a nonlinear polynom being a function of displacement velocity and acceleration. Digitally acquired data are required for this analysis. The regression analysis used here, which is based on least square procedure (LSP), has the interest to select a small number of important independent variables when the number of variables is very large (Akaike 1970).

The difference between the mathematical model and the measured forces is called error (**s**). This is displayed in matrix form:

$$\mathbf{s} = \mathbf{Z} \cdot \mathbf{a} - \overline{\mathbf{F}} \qquad (2)$$

The residual sum of squares (**RSS**) must be minimized to calculate the coefficients **a**.

$$|\mathbf{RSS}| = |\mathbf{s}^T \cdot \mathbf{s}| \Rightarrow \min \qquad (3)$$

If one partially derives the terms of equation (3) to the vectorial coefficients, then the following application results in:

$$\mathbf{Z}^T \cdot \mathbf{s} = \mathbf{0} \qquad (4)$$

Inserting equation (2) in equation (4) results in the condition for **â**:

$$\hat{\mathbf{a}} = \mathbf{Z}^T \cdot \mathbf{Z}^{-1} \cdot \mathbf{Z}^T \cdot \overline{\mathbf{F}} \qquad (5)$$

One disadvantage of the LSP is that as the results increase in quality, they also increase in quantity such that the number of terms is too large. To counter this effect it is necessary to determine which terms are best. The selection of the best variables is accomplished as follows:
- Determine number (P) and type of variables
- Create estimate models with a *single* variable; calculate coefficients **a** and **RSS** (calculate P number of estimate models). Determine best model as one with the lowest **RSS** value.
- Create new estimate models with *two* variables on the condition that the best model from the previous step will be extended by a second term. Calculate **a** and **RSS** values again multiplied by P-1.
- Usually the model is extended one variable at a time, but calculation is terminated when the information criterion is greater than the information criterion in the previous step.

The criteria used here are known as the Akaike Information Criterion:

$$\text{AIC}^*(I_p) = \log\left|\hat{\text{RSS}}(I_p)\right| + n \cdot \frac{2 \cdot i_p}{T} \quad (6)$$

...and the Basic Information Criterion:

$$\text{BIC}^*(I_p) = \log\left|\hat{\text{RSS}}(I_p)\right| + n \cdot \frac{i_p \cdot \log(T)}{T} \quad (7)$$

## 5. COMPARISON OF MEASUREMENT AND MODELIZATION

The regression analysis explained in the previous section was used to identify the fluid dynamic coefficient functions determining the right side of the equation of motions (eqs.(*1*)) The following dimensionless equations of motion (*8*), (*9*) show the interaction between the fluid dynamical system and the structural system. This differential equation system is nonlinear due to fluid dynamic loading.

$$\ddot{\xi} + \frac{\delta_x}{\pi}\dot{\xi} + \xi = \frac{V_{rx}^2}{8\pi^2 m_r} C_x(\ddot{\xi}, \dot{\xi}, \xi, \ddot{\eta}, \dot{\eta}, \eta, A_\xi) \quad (8)$$

$$\ddot{\eta} + \frac{\delta_y}{\pi}\dot{\eta} + \eta = \frac{V_{ry}^2}{8\pi^2 m_r} C_y(\ddot{\xi}, \dot{\xi}, \xi, \ddot{\eta}, \dot{\eta}, \eta, A_\eta) \quad (9)$$

The following example demonstrates the identification and simulation procedure. The test was run with a reduced velocity $V_{rx} = 38.9$ and $V_{ry} = 37.5$ and a reduced mass $m_r = 200$ with an angle of flow incidence of 40°. The logarithmic decrement of the mechnical damping was $\delta_x = 0.164$ and $\delta_y = 0.037$. The natural frequencies of the mechanical system were $f_{nx} = 0.789$ and $f_{ny} = 0.801$. The cross-section of the test bar had a width $d = 0.005$ m and height $e = 0.05$ m, with a water depth $h = 0.38$ m.

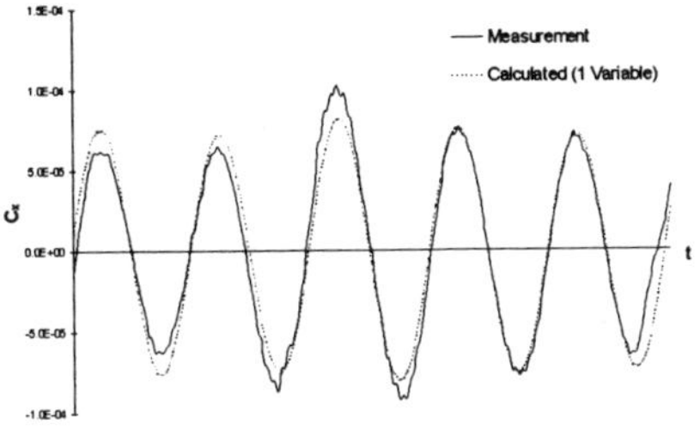

# LIGHTLY DAMPED RECTANGULAR PRISMS

$C_x = 1.317 \cdot \xi$

(a)

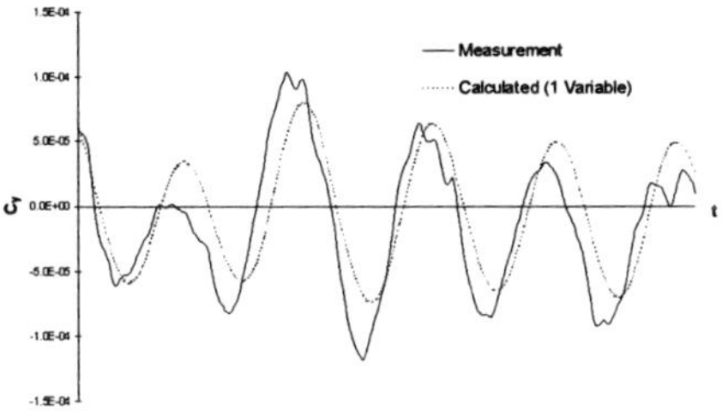

$C_y = 2.208 \cdot \eta$

(b)

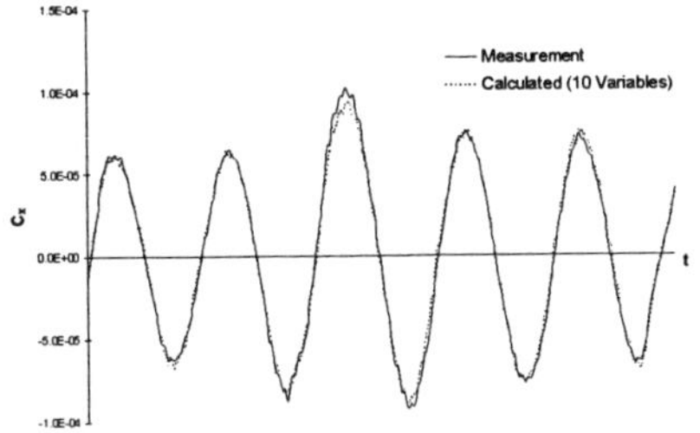

$C_x = 5.124 \cdot \xi + 0.119 \cdot \dot{\xi} + 5.005 \cdot \ddot{\xi} - 0.883 \cdot \eta - 0.535 \cdot \dot{\eta} - 0.621 \cdot \ddot{\eta} + - 0.533 \cdot \xi^2 \ddot{\xi} + 0.108 \cdot \xi \dot{\xi}^2 + 0.014 \cdot \xi^2 \dot{\xi} - 0.337 \cdot \xi^3$

(c)

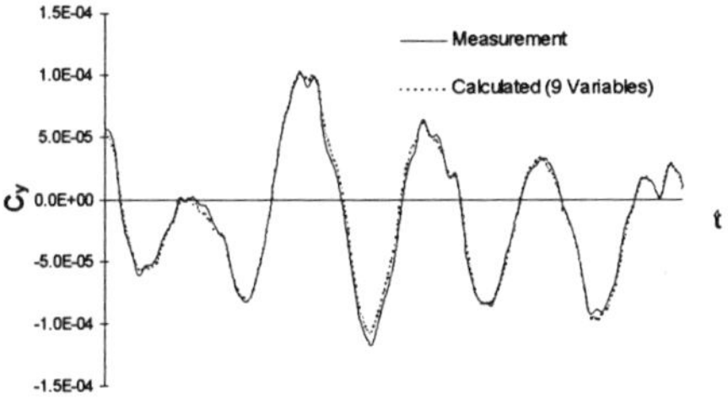

$C_y = -1.083 \cdot \xi + 0.080 \cdot \dot{\xi} - 0.721 \cdot \ddot{\xi} + 6.185 \cdot \eta + 0.585 \cdot \dot{\eta} + 5.419 \cdot \ddot{\eta} + + 0.571 \cdot \eta^2 \dot{\eta} + 0.057 \cdot \eta^3 + 0.584 \cdot \dot{\eta}^3$

(d)

**Fig. 6:** Stepwise procedures of selection of variables using AIC and BIC criteria. (a,c) Selecting one variable in streamwise and transverse mode. (b,d) AIC and BIC criterion taking a minimum value for 10 variables in streamwise and 9 variables in transverse direction.

The number P of variables was 16. Number T of samples was 1000. Only linear and cubic variables were allowed. Coupling variables were taken to be linear. Fig. 6 illustrates the selection of the best variables. Fig. 7 shows that the trend AIC function reaches a minimum of 10 variables in streamwise modes and uses 9 variables for a minimum in transverse mode. This figure also

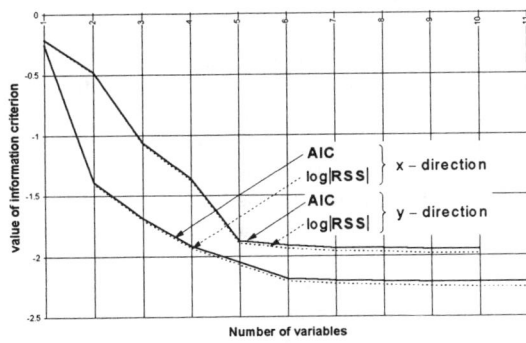

**Fig. 7:** Effect of AIC and BIC criterion

shows that **RSS** always decreases. After 5 selections the nonlinear polynomial fit hardly gets better.
For the simulation the nonlinear differential equation system had to be transformed in first-order form.

$$\begin{pmatrix} \dot{\xi}_1 \\ \dot{\xi}_2 \\ \dot{\eta}_1 \\ \dot{\eta}_2 \end{pmatrix} = \begin{pmatrix} \xi_2 \\ f(\xi_1,\xi_2,\eta_1,\eta_2,A_\xi,A_\eta) \\ \eta_2 \\ f(\xi_1,\xi_2,\eta_1,\eta_2,A_\xi,A_\eta) \end{pmatrix} \quad (10)$$

The integrator used for simulation were the classical seven-order Runge-Kutta-Fehlberg (RKF7) methods. It includes a stepwise control procedure based on a complete coverage of the leading term of the local truncation error. These methods require fewer evaluations per step than other Runge-Kutta methods of corresponding order, if the latter ones are also used with stepwise control. The numerical solution for our example is presented in Fig. 8. For a proper numerical simulation it is very important to identify the correct terms of the fluid dynamic force coefficient function.

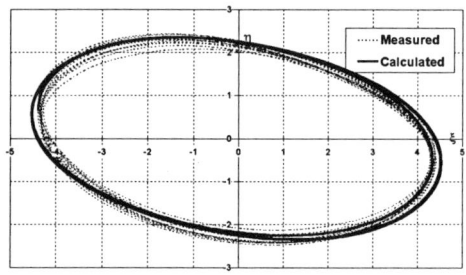

**Fig. 8:** Phase plot of nondimensional displacements; RKF7 integrator was used for simulation

## 6. CONCLUSION

This paper describes a new and innovative numerical procedure to efficiently determine fluid dynamic coefficients, which allow the design engineer to determine forces acting on vibrating bars in general flow conditions. However, it is important to point out that the investigations described here are ongoing and by no means complete. We found the following points to be of central importance: advantage of online simulation, the determination of fluid dynamic coefficient function, the nonlinear modelization and possible future use of more than one test bar.

Online simulation has the great advantage of allowing immediate check measurement data for relevance. Particulary during the development phase of the experiment, the online simulation proved to be of great importance. The engineer no longer has to work with fluid dynamic constants, rather with a fluid dynamic coefficient function. In other words, he or she gets a dynamic loading to calculate the response of the structure. The nonlinear modelization of the fluid dyanmic coefficient function is an attempt to obtain a better description of coupling mechanisms between the two vibration modes to determine the interaction between the unsteady lift and drag forces. This same modelization process is also necessary to calculate unstable limit cycles and to follow the unstable branch of the hard oscillator. Finally, as our investigations made use of only one test bar for trash racks, we can see a future research possibility in employing series of test bars.

## ACKNOWLEDGEMENT

The authors wish to extend their cordial thanks to D.I. H. Prechtl and Ing. F. Apeltauer for their invaluable help with the test facilities. They are also indebted to Mr. N. Massinger for his assistance with the measurements.

## APPENDIX I - REFERENCES

AKAIKE, H. (1970), Statistical Predictor Identification, Ann. Inst. Statist., 22 (1970), 203-217
CALLANDER, S. J. (1987), Flow-induced vibrations of cylinders with rectangular section,
　　Institute of Hydromechanics, University of Karlsruhe, Karlsruhe, Germany
NAUDASCHER, E. and Y. WANG (1993), Flow-induced vibrations of prismatic bodies and
　　grids of prisms, Journal of Fluids and Structures, Vol. 7
SRAIVA, J. G. and C. M. RAMOS (1983), Hydrodynamic loads and vibrations of trashrack
　　elements, 20th Congress IAHR, Moscow, Russia
SYAMALARAO, B. C., N. D. THANG, and E. NAUDASCHER (1987), Vibration of trashracks
　　in flow with different incidence angles, BHRA International Conference on
　　Flow-Induced Vibrations, Bowness-on-Windermere, England
VIGANDER, S. (1979), Trashrack vibration studies: Raccoon Mountain pumped-storage
　　project, Report No. 43-47, Tennessee Valley Authority, Knoxville, Tennessee

## APPENDIX II - NOTATION

| | | | |
|---|---|---|---|
| A | dimensionless coefficient | $Sh=f_n \cdot d/V$ | Strouhal number |
| a | coefficient | s | error |
| B | mechanical damping matrix | T | samples |
| C | mechanical stiffness matrix | Tu | turbulence intensity |
| d | width of the body | V | mean velocity of flow |
| e | streamwise length dimension of the bar | $V_r = V/f_n \cdot d^{-1}$ | reduced velocity |
| F | force | x | streamwise displacement |
| $f_n$ | natural frequency | y | transverse displacement |
| h | depth of flow | Z | matrix of variables |
| $I_P$ | Index set corresponding to variables | $\alpha$ | incidence angle of flow |
| M | mass matrix | $\alpha^*$ | error incidence angle due to coupling |
| $m_r = m/\rho \cdot d^{-2}$ | reduced mass | $\delta$ | logarithmic decrement of mechanical damping |
| n | degree of freedom | $\eta = y/d$ | nondimensional deflection in y direction |
| P | possible independent variables | $\Theta$ | instantaneous approach velocity incidence angle |
| q | displacement vector | $\Theta'$ | instantaneous approach velocity incidence angle due to coupling |
| | | $\xi = x/d$ | nondimensional deflection in x direction |

# THE ANALYSIS OF ROTOR-STATOR INTERACTION PHENOMENA IN A PUMP-TURBINE

Y. QIAN
Fuji Electric Co., Ltd.   Japan
C. ARAKAWA
The University of Tokyo   Japan

ABSTRACT
In this paper, a high-accuracy 3-D viscous numerical code, which has been developed previously by authors, is developed to analyze the flows through a pump turbine, which are included the rotating blades (runner) and stationary blades (guide vanes and stay vanes). The numerical analysis are performed with the multiple blades interaction between rotor-stator by a 3-D Navier-Stokes code. The motions of the separation flows are predicted by reducing the discharge rate from design to off-design operation conditions. In order to accurately capture the strong vortex, a new boundary condition of turbulent parameters on the wall is introduced in this simulation. The computational results are found to be in reasonable agreement with the experimental results. The special attentions of this work are focused on how can introduce the multiple blades techniques to analyze the performance of pump turbine, how can reduce the computational memory and CPU time, and how can apply the numerical techniques to hydraulic turbine and pump design.

1. INTRODUCTION
Several researchers have published numerous papers on the reverse flow phenomena using measurement techniques. In a series of combined analytical and experimental researches, Kurokawa(1994) provided experimental results of reverse flow patterns over a wide discharge range. These experimental results provided a detailed description and became well established as research and design tools in industry. On the other hand, following the development of computational hardware and Computational Fluid Dynamics(CFD), CFD techniques have been widely used to simulate the flow patterns and predict the performance of hydraulic turbines and are known as "numerical experiment". Qian and Arakawa(1991) successfully developed a 3-D numerical code to simulate the flow through a hydraulic turbine. The code considers the influence of viscosity with some turbulent models and can be successfully applied to simulate the phenomenon of the channel vortex in Francis turbines and steady block-block interaction(1994). Aschenbrenner(1996) have developed the Euler and Navier-Stokes codes to simulate the interaction phenomena in Francis type turbine. The fluid flow interactions in hydraulic turbine have significant influence on performance and efficiency. This work is a attempt to develop the multiple blades techniques to analyze the performance of pump turbine and predict the inlet reverse

flows of pump turbines under partial flow conditions in pump operation. The pump-turbine model analyzed is one of the test models being researched and developed.

It is well known that the k-ε turbulent model is not suitable for predicting strong vortex flows, for example, discuss for basic confined swirling flows by Hogg and Leschziner(1989). Qian(1996) checked the difference vortex flows through the draft tube against the k-ε turbulent model simulation. The shortcomings of the k-ε turbulent model are found in strong swirling flow. This phenomena confirms the fact that the k-ε turbulent models have a limitation in the case of strong vortex flows. On the other hand, if higher level turbulent models are selected, such as Reynolds stress model and LES, longer computational time is needed. At present, it is very difficult for industrial design to simulate the flow patterns, due to the huge CPU time required. Therefore, in this paper, the k-ε model is still selected to model the eddy viscosity in the Reynolds average Navier-Stokes equations, even though its shortcomings have been known. According to the above reason, Qian(1995) proposed a new ideal model to modify the turbulent parameters near the wall in the k-ε model. This new method is also introduced in this paper. The simulation results can be used with confidence as a design tool, and the process of the iterations converge smoothly despite the instability of the vortex progress during the iteration.

## 2. NUMERICAL METHOD AND BOUNDARY CONDITIONS

Pseudo-compressibility was adopted as in previous papers (Arakawa and Qian, 1991,1996). The Navier-Stokes equation is transformed into the linear Delta form using the Beam-Warming method(1976), and moreover the TVD scheme based on the upwind and approximately factored method was used. The turbulent model that the authors use for the simulation is the standard k-ε 2 equations turbulent model with wall function condition and a modified new wall boundary condition proposed by Qian(1995)

The computational domain includes the runner, guide vanes and stay vanes. The inlet boundary conditions are given by the discharge rates. For the boundary at the exit of guide vanes, the differentials of velocity components in the streamwise direction are set at 0 for the initial iteration, and then a non-reflective boundary condition proposed by Dulikravich(1993) is introduced and the velocities are corrected in order to conserve the mass flow rate to be the same as the inlet discharge. Although the flow inside the interface of rotor-stator is inherently unsteady due to blade row interaction, this work is based on the time-averaged flow field. Several previous researches, such as Saxer(1994),; Gallus(1995), have indicated that the steady multiple blades calculation approach predicts the time-averaged flow field very well. Therefore, the steady multiple blades calculation are performed when runner rotates through a constant angle value in the tangential direction. The mesh coincident lines are shown in Fig. 1. To accurately simulate the influence from the runner, interpolation is introduced to transfer information across the 3-D sliding boundary. In the patch boundary between the stationary blades and the rotating blades, a grid-point overlap exists at each other. The computational domains are separated to every channel of blade-to-blade. The boundary conditions on the periodical surfaces are exchanged in the iteration. All computations are carried out on two EWS computers. On the blade surfaces, impermeable conditions are specifying. The tangency condition is enforced by specified the contravariant velocity W=0 on the crown and band surfaces, and

V=0 on the blade surfaces. The contravariant velocities and their pressures are obtained by linear extrapolation from the interior points when required.

## 3. ANALYSIS RESULTS OF THE PUMP TURBINE

Fig. 2 shows the outline of the pump turbine model and velocity vectors through the runner, guide vanes and stay vanes. Fig. 3 shows the results of pressure contours at different relative locations against the position of the runner under design discharge rate (Q=Qd). The pressure fluctuation can be captured by analysis all computational results. The results of the time-averaged absolute velocity and pressure distributions at the inlet of the pump under design operation conditions are shown in Fig.4. $C_\theta$, $C_r$, $C_z$ are respectively the tangential, centrifugal, and axial velocities averaged between the rotating blades in the circumferential direction, and $C_p$ is the pressure coefficient, all of which are made non-dimensional using the spouting velocity corresponding to the net head $\sqrt{2gH}$. The symbols in the Fig.4 are experimental results. Fig. 5 shows the results of the velocity and pressure coefficients, which are defined in figure 4 under the partial discharge Q=0.7Qd. The axial velocity created a negative volume near the band side and the tangential velocity rise at the same region. The results show that the reverse flows are created near the band side of the pump inlet when the discharge is reduced to less than 70% of the design discharge. The reverse flow separation point can be accurately obtained by data analysis. The flow patterns are observed using computational graphic visualization on the computational display. Fig. 6, and 7 show the velocity vector at the Qd and 0.7Qd operation conditions. This result clearly shows that a vortex exists at the 0.7Qd near the band and extends a long way in the downstream direction along the band surface.

## 4. CONCLUSION

The numerical technologies of rotor-stator interaction have been introduced to analyze the performance of pump turbine. Because the steady multiple blades techniques are especially developed for this work, the computations can be performed on small size computers, such as desk-top EWS. These numerical techniques will became very important tool in the pump turbine design.

The pump reverse flow patterns are also simulated using numerical techniques. The recirculation flow phenomena, reverse flow starting conditions, and the characteristics of the pump turbine have been predicted in the paper. These results have already been adopted as a part of the design routine for daily use after introduction some modification scales.

The authors believe that this paper is only a first step for the analysis of rotor-stator interaction in pump turbines. In the near future, a further series of analyses will be reported. Several points will be studied, for example, measurement the rotor-stator interaction, analysis the pressure fluctuation in comparison with experiment, introducing a conservative algorithm to handle the unstable flows, and developing highly-accurate turbulent models which can be computed on small scale computers.

## REFERENCE

Arakawa, C., Qian, Y., and, Samejima, M.(1991) Turbulent flow Simulation of Francis Water Runner With Pseudo-Compressibility, GAMM-Conference on Numerical Methods in Fluid Mechanics, 259-268

Arakawa, C., Qian, Y. and Kubota, T.( (1996) Turbulent flow Simulation of Runner for Francis Hydraulic Turbines Using Pseudo-Compressibility, ASME Journal of Fluid Engineering, Vol. 118, pp.285-291

Aschenbrenner, T., Riedel, N. and Schilling, R.(1996) Fluid Flow Interactions in Hydraulic Machinery, XXVIII IAHR Symposium, Valencia, Spain.

Gallus, H, E,, Groillus, H.,and Lambertz, J., (1982) The Influence of Blade Number Ratio and Blade Row Spacing on Axial-Flow Compressor Stator Blade Dynamic Load and Stage Sound Pressure Level, ASME Journal of Engineering for Power ,Vol.104,pp.633-644

Gallus, H. E., Zeschky, J.. and Hah, C., (1995) Endwall and Unsteady Flow Phenomena in an Axial Turbine Stage, ASME Journal of Turbomachinery, Vol.117, pp562-570.

Kurokawa, J., Kitahora,T. and Jiang, J.(1994), Pumps Using Inlet Reverse Flow Model, XXVII IAHR Symposium, Beijing, China.

Qian,Y.,Arakawa,C.,and Kubota,T.,(1994), Numerical Flow Simulation on Channel Vortex in Francis Runner, XXVII, IAHR Symposium Beijing, China.

Qian, Y.,Arakawa,C.(1995), 3-D Numerical Analysis of Bulb Turbine Runner Performance with and Without Tip Clearance, FED-Vol.227, Numerical Simulation in Turbomachinery ASME.

Qian, Y., Suzuki, R., Arakawa, C., (1996) Analysis of the Performance of Bulb Turbine Using 3-D Viscous Numerical Techniques, XXVIII, IAHR, Symposium, Valencia, Spain.

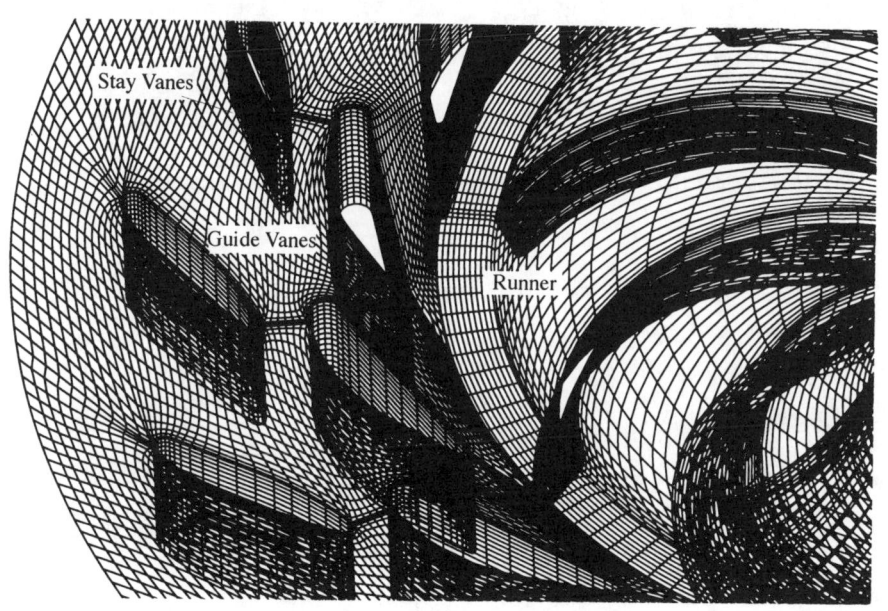

Fig.1 Computational grid for the pump turbine

Fig. 2 The outline of pump turbine, and calculated velocity through the runner, guide vanes and guide vanes

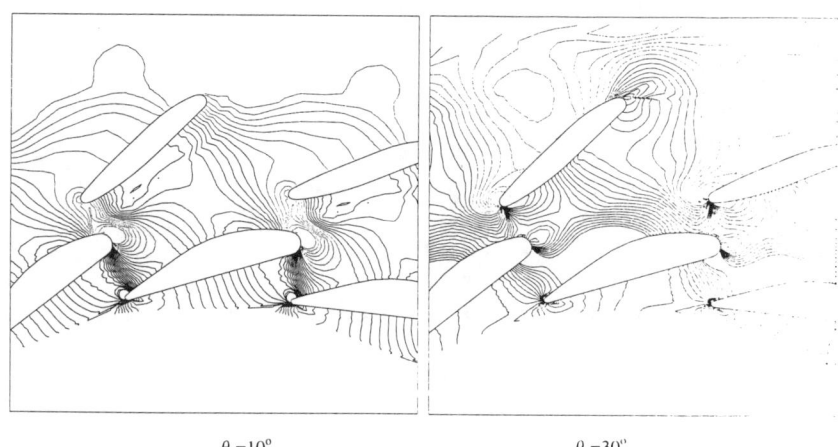

$\theta = 10°$          $\theta = 30°$

Fig.3 The pressure contours on the section of stay vanes and guide vanes

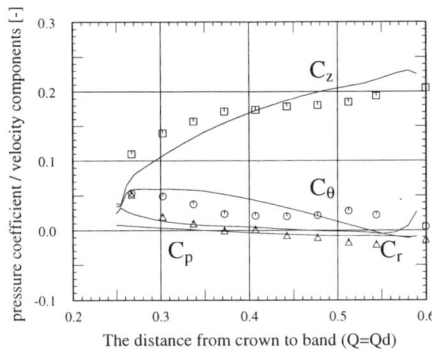 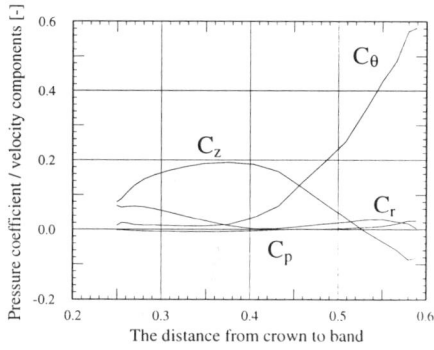

Fig.4 Velocity and pressure coefficient (Q=Qd)   Fig.5 Velocity and pressure coefficient (Q=0.7Qd)

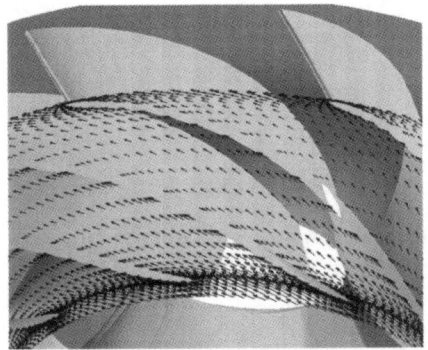 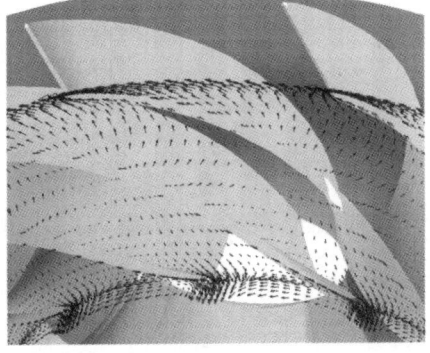

Fig.6 Velocity vectors at runner inlet (Q=Qd)   Fig.7 Velocity vectors at runner inlet (Q=0.7Qd)

# Head Recovery in the Tailwater of Low Head Hydro Power Stations

Ch. Schneider, Y. Mochkaai, W. Knapp, R. Schilling
University of Technology, Munich, Germany

## ABSTRACT

The utilization of low head hydro power today is based on tubular turbines with increasing specific speed. The high amount of kinetic fluid energy in this kind of turbines emphasizes the importance of flow deceleration downstream the runner to recover the corresponding velocity head. Theoretical and experimental investigations have shown that there is a usable flow deceleration even downstream the draft tube exit. The purpose of this paper is to provide information on how to optimize this process of flow deceleration in the tailwater.

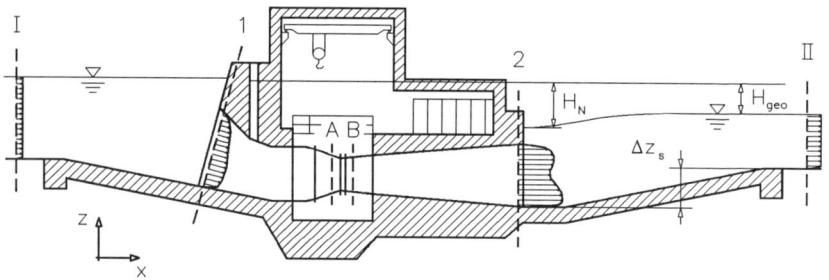

Fig.1: Net head $H_N$ and geodetic head $H_{geo}$ in a low head power station

## INTRODUCTION

Today's utilization of low head hydro power is restricted nearly exclusively to the application of tubular turbines, i.e. bulb-, pit- and S-turbines. These turbines are characterized by high values of the specific speed $n_q = nQ^{0.5}H^{-0.75}$. Optimum results are obtained with values of the unit speed $Q'_1$ that rise monotonously with the specific speed. The resulting economic benefit of smaller diameters at higher values of $n_q$ has established a continuing trend to increase the specific speed of turbines for low head applications. The values of the specific velocity head $h_c/H$, however, grow proportionally with the square of the unit flow $Q'_1$. Due to this fact the velocity head $h_{c,2}$ at the draft tube exit may reach values of 5% or more of the geodetic head, s. [6]. A portion of this exit velocity head is recovered in the tailwater, though. The continued deceleration of the flow results in a depression of the water level at the draft tube exit, thus increasing the net head.

## STATE OF THE ART

In a first approximation the tailwater can be compared with an abrupt area enlargement, also known as Carnot diffuser. With some simplifying assumptions the pressure recovery $\Delta p$ of such a diffuser can be estimated by means of the conservation of momentum, considering the mean flow velocities $\bar{c}_2$ and $\bar{c}_{II}$ at the respective cross sections, s. [2]:

$$\Delta p = \rho \bar{c}_{II}(\bar{c}_2 - \bar{c}_{II}) \tag{1}$$

The introduction of the draft tube area ratio $AR_{dt} = A_2/A_B$ and the tailwater area ratio $AR_{tw} = A_{II}/A_2$ allows to transform this equation to suit the terminology of hydraulic machinery:

$$\frac{\Delta h_{tw}}{H_{geo}} = \frac{16}{g\pi^2} \frac{1}{AR_{dt}^2} \frac{AR_{tw} - 1}{AR_{tw}^2} Q_1'^2 \tag{2}$$

This equation shows that at high values of $Q_1'$ the head recovery may reach a considerable level of even a few percent of $H_{geo}$. The experimental testing of ducts with abrupt area enlargement, however, usually results in higher values of $\Delta p$ than those obtained by equation 1. Archer, for instance, has explained this fact with the diffusing effect of the dead water besides the emerging jet, s. [1]. In fact, equation 1 yields only the smallest possible value of $\Delta p$ which still satisfies the conservation of momentum. In this case the difference of velocity head $\Delta(h_c)_{2-II}$ is completely dissipated. In an abrupt area enlargement the expansion of the jet is caused by the dissipative process of turbulent disintegration of the bounding surface which is very close to this worst case.

Therefore any item which supports the jet expansion in another way will increase the achieved values of $\Delta p$. In fact there are examples how to improve the pressure recovery in an abrupt area enlargement. Idelchik has described experiments with baffles at the entrance of a Carnot diffusers which resulted in an essential reduction of the head loss of about 40%, s. [5]. Other experiments with a combination of a conical diffuser and an abrupt area enlargement have shown a dependence of the optimum diffuser angle on the area ratio of the abrupt enlargement. Above this, theoretical considerations of Euteneuer have shown that also a swirling inlet flow may increase the head recovery in a Carnot diffuser, s. [4]. Thus, radial and circumferential velocity components obviously help to expand an emerging jet. As a water turbine draft tube provides both such velocity components to the tailwater, the expanding jet downstream the draft tube exit can be considered as an extended diffuser which increases the head recovery above the minimum value estimated with equation 1. Due to these facts the tailwater head recovery could be optimized by variation of the appropriate parameters.

## TEST FACILITIES

To get an idea of the influence of different tailwater parameters laboratory tests have been carried out with a suitable low head test rig. This test rig is equipped with a model bulb turbine of 320mm runner diameter and a free surface tailwater channel. The turbine was operated at about 2.8m head and a discharge of about 0.5 $m^3/s$. Two different draft tube designs were used, both with nearly the same area ratio $AR_{dt} \approx 4$. Variant 0 is a simple conical diffuser whereas variant 1 has a transition from circular to rectangular cross section and thus a rectangular exit cross section. The tailwater geometry was adapted to a very common design, s. fig. 5. The head recovery $\Delta h_{tw}$ was determined by reading a level gauge connected to pressure taps at the draft tube exit and a level gauge in the tailwater some distance downstream.

## INFLUENCE OF THE UNIT DISCHARGE

Using draft tube variant 1 the values of $\Delta h_{tw}$ were measured in a number of operating points $Q'_1 > Q'_{1,opt}$ at two different water levels in the tailwater. These different levels correspond to area ratios $AR_{tw}$ of about 3.6 and 9.0. Fig. 2 shows the results together with their fitting curves (solid lines) and the curves determined by applying equation 2 (dashed lines). It is obvious that there is a distinct advantage of the low area ratio $AR_{tw}$. The investigation confirms the quadratic dependence $\Delta h_{tw} = k\,Q'^2_1$. The measured factor $k_m$, however, is considerably higher than the value $k_t$ obtained by equation 2. The ratio $k_m/k_t$ is about 1.7, thus showing that the real head recovery exceeds the minimum value by about 70%.

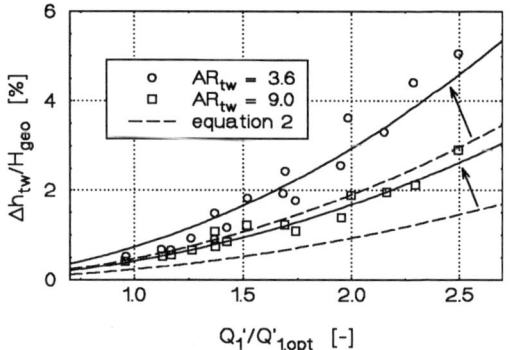

Fig.2: Influence of the unit discharge

## INFLUENCE OF THE DRAFT TUBE GEOMETRY

Similar measurements were carried out with draft tube variant 0 to enable a comparison of different draft tube geometries. Fig. 3 shows the values of $\Delta h_{tw}$ measured at high unit speed $n'_1 = 1.5\,n'_{1,opt}$ and high unit discharge $Q'_1 = 2.2\,Q'_{1,opt}$. The results obtained with draft tube 0 are distinctly lower than the values measured with variant 1. Additional measurements were carried out without modelling the tailwater flume bottom to reach the higher values of $AR_{tw}$ shown in fig. 3. The results of draft tube 0 coincide in substance with calculated values (solid lines) derived from equation 3, an extension of equation 2 to free surface flow and non-uniform inlet flow:

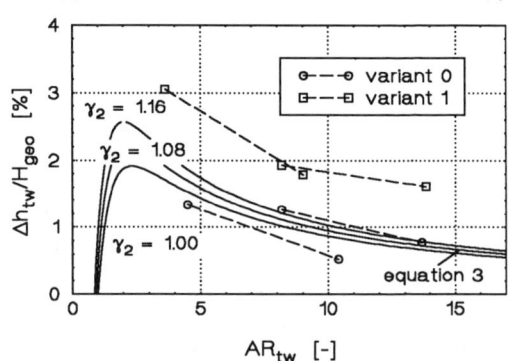

Fig.3: Influence of the draft tube geometry

$$\rho Q(\bar{c}_{II} - \gamma_2 \bar{c}_2) = \frac{1}{2}\rho g b \left(h_2^2 - h_{II}^2 - \Delta z_s(h_2 + h_{II})\right) \qquad \gamma = \frac{\int_A c^2 \, dA}{\bar{c}^2 A} \quad (3)$$

The pressure distribution on the slope of the flume bottom was assumed to be hydrostatic, based on the mean water level $(h_2 + h_{II})/2$. The consideration of a non-uniform inlet flow with momentum non-uniformity factors $\gamma_2 > 1$ results only in little differences of the calculated values.

## INTERACTION OF DRAFT TUBE AND TAILWATER FLOW

According to the results presented above obviously there is an interaction of draft tube and tailwater flow. Swirl and radial velocity components at the draft tube exit tend to expand the emerging jet with cross sections that correspond to the cross section of the draft tube exit. Thus the jet emerging from variant 0 expands with circular cross sections, s. fig. 4. As the tailwater cross section is rectangular with a high aspect ratio there is not much advantage of this jet expansion. The flow has to be adapted to this cross section in a turbulent and dissipative process. The jet leaving variant 1, however, expands more or less with rectangular cross sections which fit well into the rectangular tailwater provided by the test rig. Thus the bounding surfaces reattach very fast to the flume walls reducing turbulence and losses.

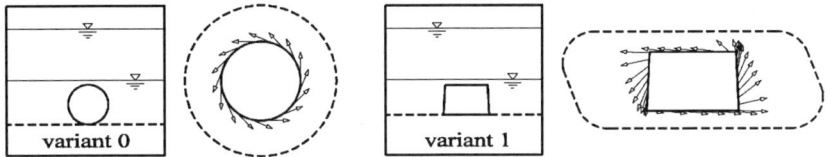

Fig.4: Influence of draft tube geometry on the jet expansion in the tailwater

## OPTIMIZATION OF THE TAILWATER GEOMETRY

Due to the advantage of low area ratios $AR_{tw}$ it appears reasonable to reduce the area ratio near to the draft tube by inserting guiding walls into the tailwater. According to equation 2 the maximum values of $\Delta h_{tw}$ are expected at $AR_{tw} \approx 2$ which corresponds nearly to a flush position of the guiding walls at the draft tube exit. With variant 0 in fact the values of $\Delta h_{tw}$ rise significantly at reduced values of the clear span $b/B$ whereas with variant 1 there is a distinct decrease. This finding again supports the idea that the tailwater geometry should be adapted to the draft tube geometry. Guiding walls with low clearance obstruct the expansion of the jet induced by variant 1 whereas they fit rather well to the circular jet cross section of variant 0.

Besides such inserted elements there are also some parameters of a common tailwater geometry which can be changed to optimize the head recovery. Fig. 6 shows the measured head recovery with varying the length of the horizontal bottom section, the slope of the inclined bottom portion and the draft tube submersion. Optimum results were achieved with low submersion, a horizontal portion of 1D length and a slope of 8°. The tests were carried out with draft tube variant 1.

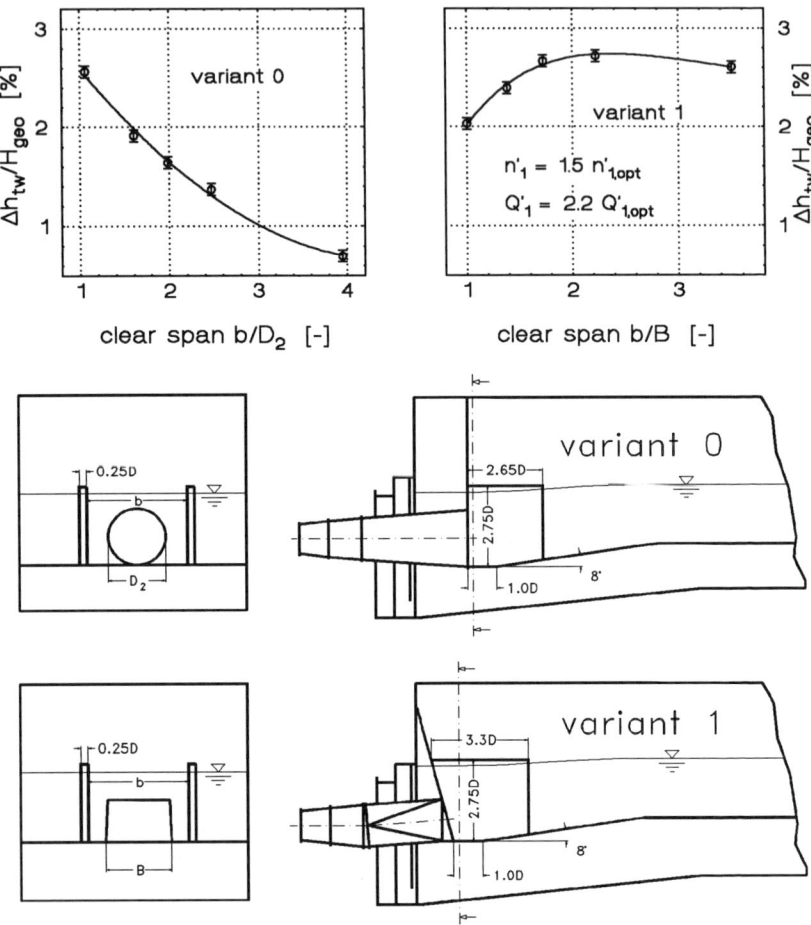

Fig.5: Influence of guiding walls on the head recovery

## CONCLUSION

The economy of low head hydro power stations is severely affected by high rates of exit loss which may reach values of 5% or more of the geodetic head. Theoretical and experimental investigations, however, have shown that a part of the exit velocity head is recovered in the tailwater, thus decreasing the water level at the turbine exit. Swirl and radial velocity components cause an expansion of the jet which emerges from the draft tube. Due to this fact the free jet downstream the diffuser exit acts like an extended diffuser with a free bounding surface. It has been shown that this process can be subject to optimization by an adaption of the tailwater and the draft tube design. According to the results presented there is still a considerable potential to improve the performance of low head power stations.

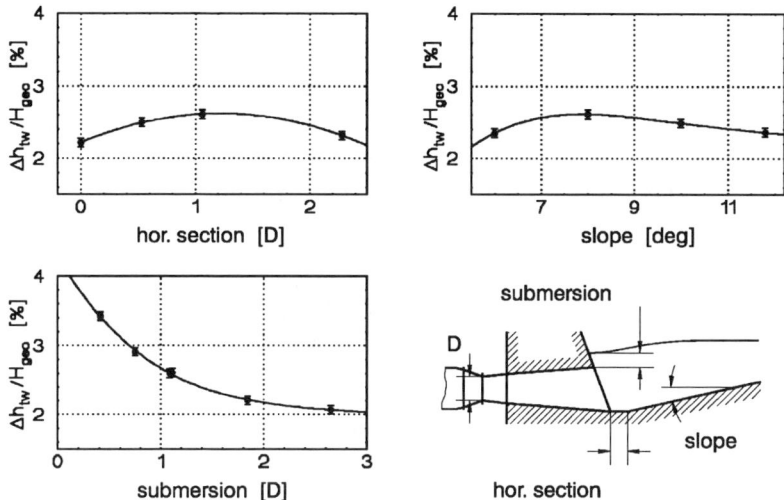

Fig.6: Influence of common tailwater parameters on the head recovery

## REFERENCES

[1] W.H. Archer: *Experimental Determination of Loss of Head Due to Sudden Enlargement in Circular Pipes*, Trans. ASCE, Vol. 76, 1913, S. 999

[2] R.P. Benedict, N.A. Carlucci, S.D. Swetz: *Flow losses in abrupt enlargements and contractions*, Transactions ASME, series A: Journal of engineering for power 88(1966), S. 73-80

[3] J. Chevalier, H. Giraud: *Études théoriques et expérimentales de l'évolution de l'énergie dans les ouvrages aval des installations hydroélectriques de basse chute*, La Houille Blanche, Nr. 2-3, p.155-166, 1968

[4] G.-A. Euteneuer: *Ein- und mehrstufige Carnot-Stossdiffusoren bei drallbehafteter Strömung*, Forsch. Ing.-Wes., Bd. 34 (1968), Nr.5, S. 154-158

[5] I.E. Idelchik: *Handbook of hydraulic resistance*, Springer-Verlag, 1986

[6] E. Kita, K. Kubota, Y. Kimoto: *Tailwater Level and Net Head in Low Head Power Plant*, Proceedings of the 3rd Japan-China Joint Conference on Fluid Machinery, Vol. 1, Osaka, 1990

[7] Ch. Schneider, W. Knapp, R. Schilling: *Optimised Tailwater and Draft Tube Design of Low Head Small Hydro*, Conf. Proc. Small Hydro, Water Power & Dam Construction Conf., Kuala Lumpur, October 1996

[8] Ch. Schneider: *Untersuchung der Wechselwirkung schnelläufiger Wasserturbinen mit dem Unterwasser*, PhD Thesis, Technical University of Munich, 1997

# EXPERIMENTAL EVALUATION OF THE COMPRESSIBLE PORTION OF A LIQUID FLOW CROSSING A CAVITATING CONTROL VALVE

FRATINO U. and PICCINNI A. F.
*Dipartimento di Ingegneria delle Acque - Politecnico di Bari*
*via E. Orabona, 4 - 70125 BARI (Italy)*

## ABSTRACT

This paper describes a method for evaluating, through experimental results, the compressible volumetric fraction released by a liquid mass crossing a cavitating control valve, by means of some remarks on the values of the pressure drops. It is known that the relationship, valid in fully developed flow with no cavitation occurrence, used for estimating the loss coefficient K falls when the cavitation phenomenon appears and the reason of this behaviour is to be find in the different capacity to dissipate of the arisen compressible portion of flow.

## INTRODUCTION

The determination of the performances of a control valve must necessarily consider a careful analysis of the hydraulic characteristics to assure its correct behaviour in every working condition. Under this point of view, the determination of the loss coefficient curve and of the cavitation characteristics are of great importance for a correct design and management.

However, the appearance of the cavitation phenomena invalidates the usual equations used to define the loss coefficient values. Just to evaluate, in an analytical way, the inconsistencies in the hydraulic behaviour of the control valves in developed cavitation conditions, a new approach to the problem has been defined through some considerations about the volumetric composition of the conveyed fluid.

This approach puts in evidence the discrepancies that arise when the influence of the compressible fraction of the fluid, originated by the local pressure drop, is not carefully evaluated.

For this purpose, by means of an energetic approach, the influence of the presence of an meaningful amount of the compressible fluid fraction on the expression of the loss coefficient has been evaluated and, in a following time, this compressible volumetric fraction has been derived.

## MATHEMATICAL APPROACH

In steady condition, the energy of a fluid crossing an upstream section, named *0-0*, in the temporal unit is equal to the sum of the energy of the fluid crossing a downstream

section *1-1* and of a contribution due to the internal energy U and to the mechanical one dissipated between the two observed sections.
In a fluid with elastic characteristics, this energy balance can be written as:

$$\int_{A_0}\left(p+\frac{\rho V^2}{2}+g\rho z+\rho U\right)VdA = \int_{A_1}\left(p+\frac{\rho V^2}{2}+g\rho z+\rho U\right)VdA + \Delta N \quad (1)$$

where $z$ is the geometrical level of the centre of sectional area $A$, $p$ the absolute pressure in the same point, $g$ the acceleration due to the gravity, $V$ the mean velocity, $U$ the specific internal energy and $\Delta N$ the lost mechanical power between the two section upstream and downstream the device location.

Taking in account of the continuity equation of the mass flow rate, it is possible to integrate the previous equation in the hypothesis that the change of the pressure $p$ in the sectional area and the change of density of the compressible fluid with the mean velocity $V$ can be neglected.

With these assumptions, the internal energy variation depends by the thermodynamic process to which the fluid crossing the two sections is subjected and, for a polytropic process with no addition of heat from the outside, this change is equal to:

$$U_1 - U_2 = \int_{p_1}^{p_0} pdv = \frac{p_0}{\rho_0} - \frac{p_1}{\rho_1} - \int_{p_1}^{p_0}\frac{dp}{\rho} = \frac{p_0}{\rho_0} - \frac{p_1}{\rho_1} - \frac{n}{n-1}\left(\frac{p_0}{\rho_0} - \frac{p_1}{\rho_1}\right) \quad (2)$$

where $n$ is the polytropic process exponent and $v$ is the specific volume of the fluid.
Defined as $\Delta H_i$ the head loss referring to an absolutely incompressible fluid, the integration of the equation (1), taking in account the equation (2), brings to:

$$\Delta H - \Delta H_i = \frac{n}{n-1}\frac{p_0}{g\rho_0}\left[1-\left(\frac{\rho_1}{\rho_0}\right)^{n-1}\right] \quad (3)$$

that defines an additional contribute to the dissipation rate, since, for the characteristics of the involved variables, this term is always a positive quantity to correlate only to the fraction of fluid with compressible behaviour.

Defined as *1-1* the section just downstream a cavitating hydraulic device location, the water-vapor mixture density $\rho_1$ in this section can be expressed by the relationship:

$$\rho_1 = X\rho_w + Y\rho_v \quad (4)$$

where $\rho_w$ and $\rho_v$ are the density of water and of vapor respectively, whereas $X$ and $Y$ are their volumetric ratios.

Assuming that the fluid density in the section *0-0* is equal to the water density, the equation (3) can be made dimensionless by means of the velocity head, calculated in the section *0-0*, in the following form:

$$\frac{\Delta H - \Delta H_i}{\frac{V_0^2}{2g}} = K - K_i = \frac{2np_0}{V_0^2 \rho_0 (n-1)}\left[1-\left(1-Y\left(1-\frac{\rho_v}{\rho_w}\right)\right)^{n-1}\right] \quad (5)$$

where $K$ and $K_i$ are the loss coefficients, calculated in the usual way, referred respectively to a water-vapor mixture and to an incompressible fluid.

So, through the equation (5), it is possible to justify the increase of the loss coefficient $K$, that can be verified when the appearance of a cavitation phenomenon causes a partial vaporisation of the fluid.

Referring to a generic polytropic process and known the experimental values of the hydraulic variables, the equation (5) permits to derive the volumetric fraction $Y$ of the fluid mixture:

$$Y = \frac{1 - \left[1 - \frac{(K - K_i)\rho_w V_0^2 (n-1)}{2np_0}\right]^{\frac{1}{n-1}}}{1 - \frac{\rho_v}{\rho_w}} \qquad (6)$$

The equation (6) has a general validity and, for an isothermal process, it can been reduced, solving a Taylor series, in the following way:

$$Y = \frac{1 - e^{-\frac{(K - K_i)\rho_w V_0^2}{2p_0}}}{1 - \frac{\rho_v}{\rho_w}} \qquad (7)$$

The definition of the thermodynamic process to which the compressible fluid portion is subjected is very hard to identify and the exponent of the polytropic transformation can vary between a minimum and a maximum value, respectively equal to one for an isothermal process and equal to 1,41 for an adiabatic one.

However, some experimental trials by Gubarev [4] with air have shown as the expected value for this variable is equal to 1,15, very close to the one typical of the isothermal process.

## EXPERIMENTAL SET-UP

The experimental tests were carried out in the laboratory of the Department of Water Engineering of Bari Polytechnic on different hydraulic devices, both in steady and in cavitation conditions.

The experimental set-up is composed by two parallel 100 mm and 200 mm diameter pipelines on which the hydraulic devices have been installed, supplied by an electropump assuring a 0,9 MPa pressure value and discharges over 100 l/s.

The hydraulic devices, used for the trials, were two control Larner Johnson ported valves of 100 mm and 200 mm diameters which present at their end a moving steel cylinder with sixteen parabolic shape openings [1, 2, 3].

The pressure taps used for evaluating the pressure values were located one diameter upstream and six diameters downstream the device location and the measures were performed by means of precision Bourdon tube pressure gauges. The discharge value has been evaluated by a measuring orifice plate located on the delivery pipe and by three calibrated volumetric tanks.

The experimental tests, performed to analyse the hydraulic behaviour of the devices, have been carried out at different upstream pressures and at different opening degrees, allowing to define completely their hydraulic characteristics both referring to the steady and the cavitation parameters [1, 2, 3].

## ANALYSIS OF THE RESULTS

The experimental determination of the loss coefficient K, calculated with the usual formulas referred to a fully developed flow and in lack of cavitation, has allowed to define several graphs as that in *fig. 1*. It represents the trend of the loss coefficient $K$ versus the Reynolds number $Re$ for the 200 mm Larner Johnson ported valve at an opening degree equal to 0,40. This graph presents three different areas: in the first one, the loss coefficient $K$ is in reverse proportion to $Re$, in the second one, it seems quite constant whereas, in the third one, it appears to increase in a strong way. The trend of $K$, in this last zone, is in agreement with the previous remarks for which the increase of the observed variable can be attributed to the rising of the cavitation with the consequent developing of cavitating bubbles that define compressible zones in the fluid mass.

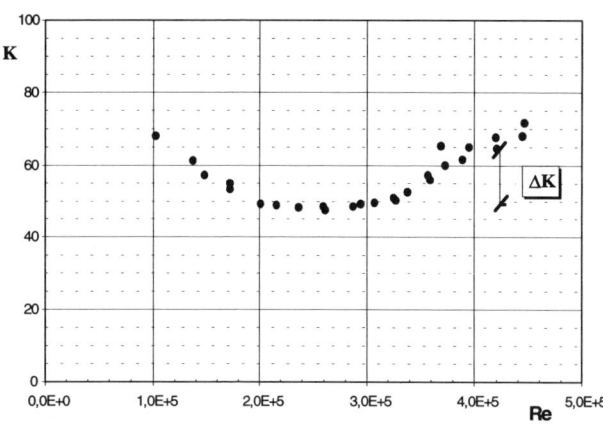

**Fig. 1** - *Loss coefficient vs. Re in 200 mm LJ valve ($\psi=0,40$)*

In the light of this remark the experimental results have been used to estimate, applying the equation *(6)*, the volumetric fraction of compressible fluid able to justify the growth of the pressure losses. In this way, several graphs that represent the compressible fraction $Y$ versus the ratio between the pressure drop $\Delta p$ and the upstream constant pressure $p_0$, have been obtained for different values of $n$ and, in *fig. 2*, one of these graph has been reported as example. The figure shows that the curve

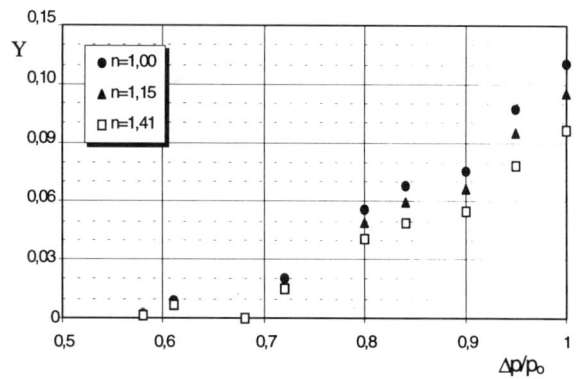

**Fig. 2** - *Volumetric ratio Y vs. $\Delta p/p_0$ in 100 mm LJ valve ($\psi=0,30$)*

obtained from the isothermal hypothesis presents the maximum values for $Y$, whereas the curve from the adiabatic one has the minimum values; however, in all cases, the fraction $Y$ aims to increase when the ratio $\Delta p/p_o$ increases.

The influence of the upstream pressure on the definition of the volumetric fraction $Y$ is suitable for a further comment. In fact, at the same opening degree $\psi$, the compressible volumetric fractions of the fluid mass versus the pressure ratio $\Delta p/p_o$ have been compared (fig. 3). In the figure, for the 100 mm Larner Johnson ported valve, at an opening degree of 0,10, the curves, calculated for a $n$ value equal to 1,15, are related to three different upstream pressure values and the trend of the curves, appears to be in agreement with some previous experimental remarks [1,2]. In fact, the dimensionless volume $Y$, at the same pressure ratio, is smaller connected with the greater upstream pressure values and this occurrence can be explained as the higher tendency proper of an hydraulic system to cavitate for lower upstream constant pressure. So, the pressure ratio $\Delta p/p_o$ can be interpreted as a cavitation parameter $\sigma$, whereas the volumetric ratio $Y$ defines, by means of a direct proportion, the potential number of vapour bubble implosions. Besides the three curves, when the pressure ratio increase, aim to growth in an asymptotic way.

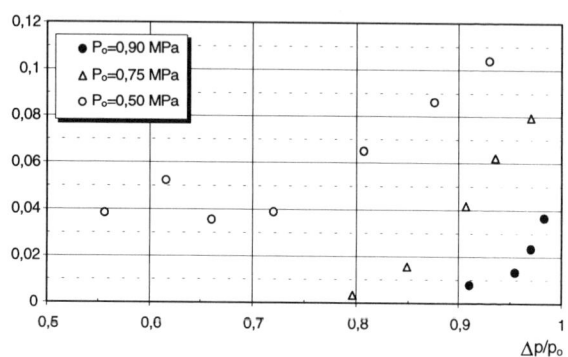

**Fig. 3** - *Volumetric ratio Y vs. $\Delta p/p_o$ for different upstream pressure values in 100 mm LJ valve ($\psi=0,10$)*

At least, in *fig. 4*, it is reported, for the 200 mm Larner Johnson ported valve, a graph that summarises the trends of the volumetric fraction $Y$ versus the

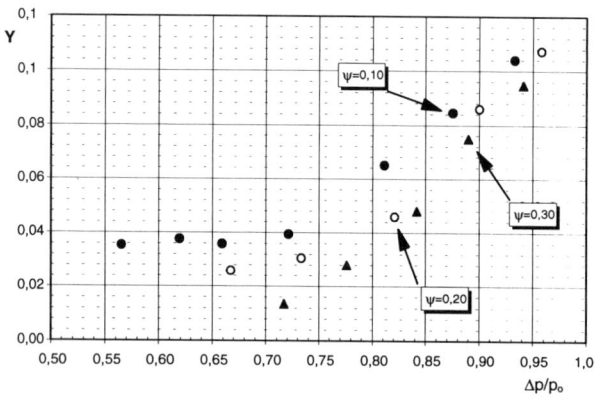

**Fig. 4** - *Volumetric ratio Y vs. $\Delta p/p_o$ at different opening degrees in 200 mm LJ valve*

pressure ratio $\Delta p/p_0$, varying the opening degree $\psi$ and at a 0,50 MPa constant upstream pressure. The graph confirms, for the same value of the ratio $\Delta p/p_0$, as the compressible volumetric fraction $Y$ increases when the opening degree decreases.
Besides the experimental points seem to be fitted by means of a single curve that can be considered as representative of the whole behaviour of the valve typology. The trend of this global curve depends, in a strong way, on the geometrical characteristics of the shear layer zone of the flow downstream the valve and, hence, on the path of the streamlines inside of the body of the device, in addition to the pressure history and to the physical characteristics of the flow.

## CONCLUSIONS

The research has allowed to quantify, through an equations of energy balance, the additional contribution to the pressure drop originated by an hydraulic device, that can be attributed to the appearance of a compressible portion in the fluid mass when the hydraulic device works under developed cavitation conditions.
This approach, by means of the results derived from a large laboratory investigation on two control valves in cavitating conditions, has made possible the numerical evaluation of the volume of the arisen compressible portion of the fluid mass crossing these valves.
The calculated values of the volumetric fraction of compressible portion present in the cavitating fluid mass confirm some experimental remarks, peculiar of the cavitation phenomena, and the compressible volumetric fraction, when the dissipation rate increases, fits a curve that is defined mainly by the geometrical characteristics of the separation zone of the flow downstream the valve in which the high velocity gradients create a large number of eddies.

## REFERENCES

[1] **Castorani A., De Martino G., Fratino U.**: "*Metodologie di valutazione delle caratteristiche cavitanti delle valvole di regolazione*" XXV Convegno di Idraulica e Costruzioni Idrauliche, Torino (Italy), September 1996.
[2] **Castorani A., De Martino G., Fratino U.**: "*Determination of critical cavitation limit in the pressure control devices*" Proceedings of the XVIII IAHR Symposium on Hydraulic Machinery and Cavitation, Valencia (Spain), September 1996.
[3] **Fratino U.**: "*Effetti cavitanti nelle valvole di regolazione*" PhD Thesis, Università di Napoli (Italy), February 1996.
[4] **Idelchik I. E.**: "*Handbook of Hydraulic Resistance*", 3rd Edition, CRC Press, Inc. 1994.

# Design of throttled surge tanks for high-head plants. Pressure wave transmission and reflection at a T-junction with an orifice in the lateral pipe.

### R. PRENNER and H. DROBIR
Institut für Konstruktiven Wasserbau, Technische Universität Wien, A-1040 Wien, Austria

### ABSTRACT

In pressure duct systems with surge tanks, the headrace tunnel cannot always be protected from pressure surges induced by turbine, pump and valve regulation, which in the case of superpositioned pressure waves can lead to massive damage to the tunnel lining. In order to ascertain the influence of the surge orifice on the transmission of pressure in the tunnel, a simple experiment with T-junction models was conducted in which various symmetrical and assymmetrical circular-orifice plates were used under various steady basic flows with single pressure waves. The suitability and limitation of application of a steady throttle headloss assumption in the common unsteady calculation was tested by a standard characteristics method using control calculations of the experiments.

## 1. INTRODUCTION

*General:* In complex water supply systems of high-head plants, control and regulation procedures produce time-dependent changes of flow and pressure in the conduits. Understanding of the unsteady hydraulic resistance behaviour of the restriction orifice in the surge tank is of great importance for the dynamic calculation according to elastic transient fluid theory, for it influences the pressure oscillation in the pressure duct system (Fig.1).

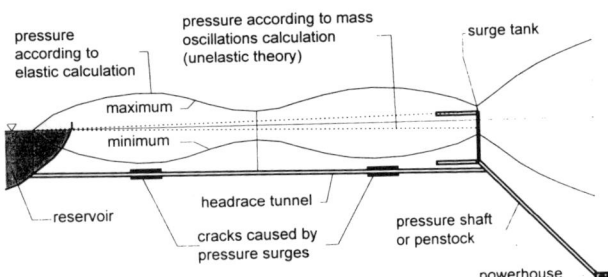

Fig 1: Pressure duct system of a high-head powerplant, maximum pressure surges stimulated the $2^{nd}$ harmonic oscillation of the headrace tunnel.

Because of the bifurcation and the non-linear throttle headloss, dangerous oscillation of surge pressure can arise and cause damage to the pressure tunnel system.

*Problem:* The present investigation is concerned with the behaviour of pressure waves during transmission through throttles in surge tanks. As there is still uncertainty regarding the extent of transmission and reflection of pressure waves in this construction, basic experiments were carried out with single pressure waves (to simplify the problem for analytical purposes) in a branch system with various circular orifice plates in the lateral pipe. Results were interpreted on the one hand by depicting, for practical purposes, the transmission coefficient in diagrams and, on the other, by applying a characteristics method [1, WYLIE and STREETER, 1993]. Here, a parabolic throttle resistance law was used to calculate numerically the validity for unsteady scenarios. The question was whether, or under what constraints, the quadratic law of resistance which applies to steady cases was also valid for unsteady cases.

## 2. PREVIOUS STUDIES - THEORETICAL APPROACH

Pressure wave transmission and reflection at throttle plates in duct systems are determined by the unsteady hydraulic resistance behaviour of the orifice plates. Many studies (e.g. JAEGER [2,1933], ZIENKIEWICZ and HAWKINS [3, 1954], SETH [4, 1973], BERNHART [5, 1977]) have developed methods of calculating water hammer behaviour at orifice plates of throttled surge tanks based on ALLIEVI's [6, 1903] equations. In all studies known to the authors, the calculations with steady headloss coefficients were developed only for the transmitted pressure wave in the pressure tunnel and included additional influences such as closure time and the wave reflection from the free surge tank water level. None of the studies isolated the pressure wave transmission and distribution in the junction with a throttle in the lateral pipe. Investigations of the transmission and reflection behaviour of presure waves on the throttle element itself have only been published for straight pipelines and shock tubes. As it is difficult to obtain the measurements, only few such studies have been published, i.e., by CONTRACTOR [7, 1965], TRENGROUSE, IMRIE and MALE [8,1966], FUNK, WOOD and CHAO [9,1972], and BERGER [10, 1978]. The results of the these studies can only be related to bifurcated systems with orifice plates in a qualitative sense, for interaction arises when pressure waves are distributed among the three pipes.

**Notations**

| | |
|---|---|
| $a_i$ | wave propagation velocities |
| $A_i$ | cross section areas of pipes |
| g | gravitational acceleration |
| h | piezometric head |
| $h_{f,orifice}$ | orifice head loss |
| $h_{jou}$ | Joukowsky head variation $a \cdot \Delta v/g$ |
| i | as subscript denotes number of pipe branches |
| $I_f$ | pipe friction gradient |

| | |
|---|---|
| r | reflection factor in the approach pipe |
| $s_2$ | transmission factors into the main branch |
| $s_3$ | transmission factors into the side branch |
| t | time, as a subscript denotes partial differentiation |
| v | velocity |
| $\Delta v$ | change in velocity |
| x | distance along pipe axis, as a subscript denotes partial differentiation |
| $\alpha$ | slope of pipe axis |
| $\xi_{orifice}$ | orifice head loss coefficient |
| $\mu$ | coefficient of contraction ($A_{contraction} / A_{orifice}$) |

For an unthrottled pipe branch, the transmission approach according to JAEGER [2, 1933] was applied to constant pipe cross-sections of $A_1=A_2=A_3$ and constant pressure wave speeds of $a_1=a_2=a_3$. The following simplified equation for the transmission coefficients for the branches yields (1).

$$s_2 = \frac{2 \cdot A_2}{A_1 + A_2 + A_3} = s_3 = \frac{2}{3} = 0.66 \qquad \ldots (1)$$

Headlosses in the pipe junction were generally taken into consideration by a headloss coefficient for steady flows of $\xi_{orifice}$, which is related to the kinetic energy (2).

$$h_{f,orifice} = \zeta_{orifice} \cdot \frac{v_{pipe}^2}{2 \cdot g} = \left(\frac{A_{pipe}}{\mu \cdot A_{orifice}} - 1\right)^2 \cdot \frac{v_{pipe}^2}{2 \cdot g} \qquad \ldots (2)$$

The reflection factor r (3) and the transmission factors $s_2$ (4) and $s_3$ (5) in a throttled pipe branch with the above assumptions, are given by

$$r = \frac{-3 \cdot a^2 + a \cdot \sqrt{(3 \cdot a)^2 + 8 \cdot \xi_{orifice} \cdot h_{jou} \cdot g}}{4 \cdot \xi_{orifice} \cdot h_{jou} \cdot g_3} \qquad \ldots (3)$$

$$s_2 = 1 - r \qquad \ldots (4)$$

$$s_3 = 2 \cdot r \qquad \ldots (5)$$

For the calculation of the unsteady pipe flow of the experiment, the momentum (6) and the continuity (7) equations were used, based on the assumptions of the one-dimensional fluid flow theory. These equations were solved numerically, under inclusion of boundary conditions, with an explicit characteristics method.

$$\frac{\partial h}{\partial x} + \frac{1}{g} \cdot \left(\frac{\partial v}{\partial t} + v \cdot \frac{\partial v}{\partial x}\right) + I_f = 0 \qquad \ldots (6)$$

$$A \cdot \frac{\partial v}{\partial x} + \frac{g \cdot A}{a^2} \cdot \left(\frac{\partial h}{\partial t} + v \cdot \frac{\partial h}{\partial x}\right) + \frac{\partial A}{\partial x} \cdot v + \frac{g \cdot A}{a^2} \cdot v \cdot \sin\alpha = 0 \qquad \ldots (7)$$

The influence of additional inertia forces of the orifice flow, the delayed scaling-off of the boundary layer, cavitation and unsteady friction effects in the orifice flow were not taken into consideration in this approach since experimental calculations indicate that they are of significance only for relatively small orifice area/pipe area cross-section ratios of 1:32 ($\zeta_{orifice} \geq 2500$) or less.

## 3. EXPERIMENTAL INVESTIGATIONS

The size of the experiment was essentially determined by the amount of room available in the laboratory. A 100-mm/diameter steel pipe system was rigidly fixed to a foundation plate with a special steel construction to prevent shifting of the T-section during pressure wave transmission. The single pressure waves were produced by the impact of a pendulum mallet on a rubber damped piston and introduced into the lower half of the penstock pipe.

The pipe lengths after the measuring points were chosen so that no reflection waves from the ends of the pipe branches would influence the behaviour of the transmitted single pressure waves. The pressure waves were recorded by high-resolution inductive pressure gauges (0-100 meter) and transferred by triggering via an A/D converter to a PC - measuring program. Measurement of the partial flows was carried out via two magnetic inductive flow meters (Fig.2).

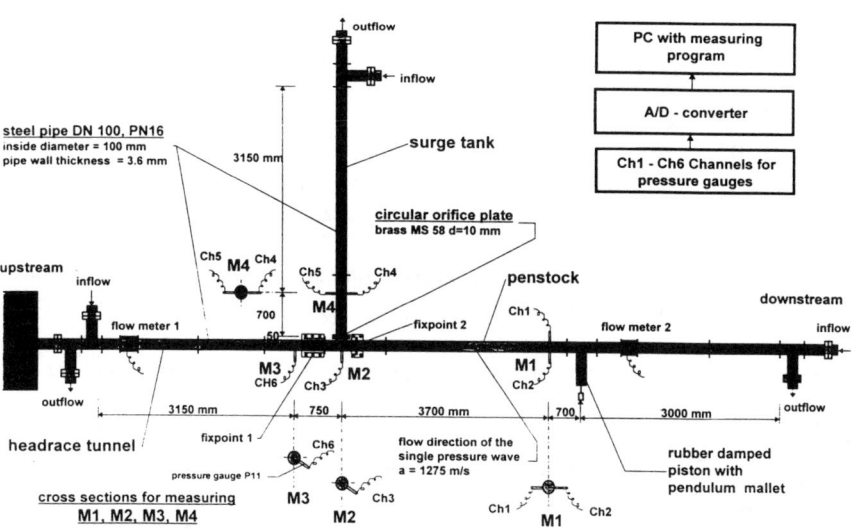

Fig. 2: General layout of experimental arrangement

In the experiment, 6 orifice plates with square edges and 2 orifice plates used once with beveled and once with sharp edges were investigated under various steady basic flow conditions (Fig. 3).

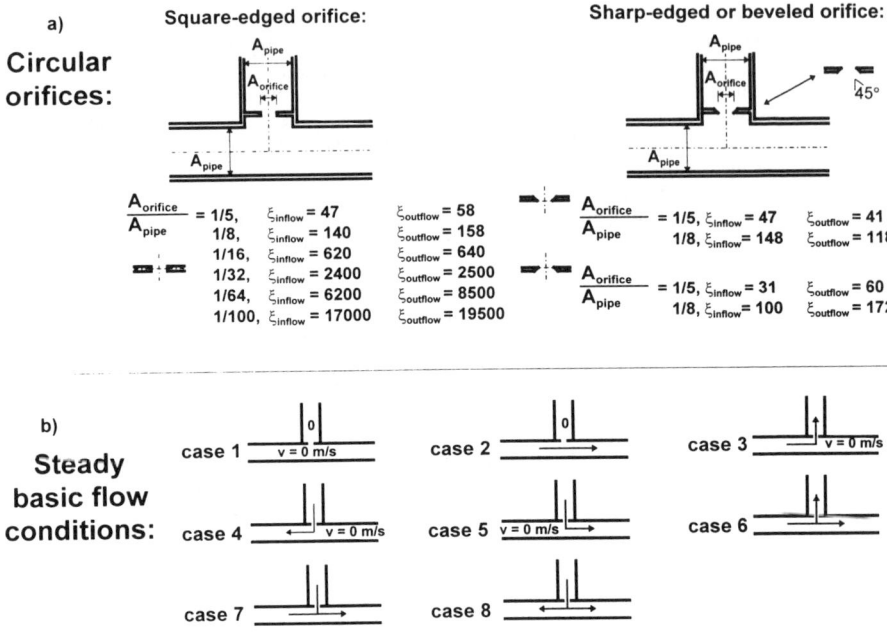

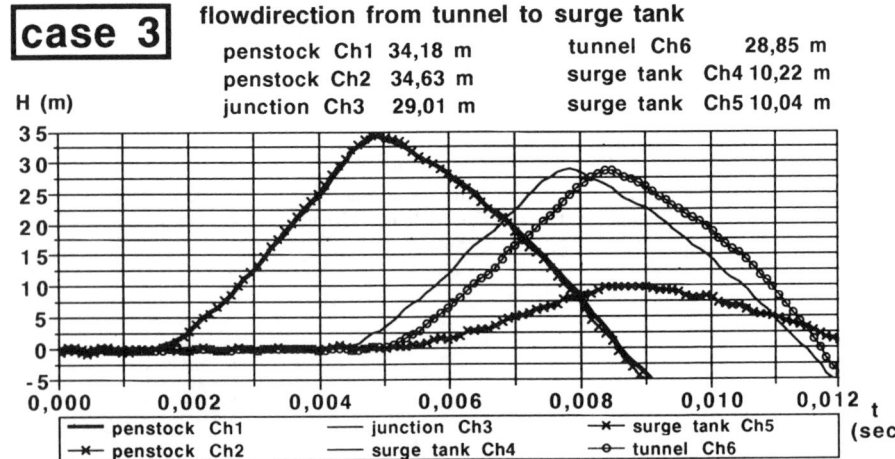

Fig. 3: Program of experimental investigations, a) types of orifices and throttle ratios
  b) basic flow combinations

The experiments covered a magnitude of maximum pressure waves from 5 m to 70 m. Because of the constant geometry, different gradients of the incident pressure wave resulted (Fig. 4).

Fig. 4: Test results of a single pressure wave experiments with 1/100 square orifice

For the interpretation of results, it was necessary to eliminate additional physical influences such as unsteady fluid friction and pseudo pressure transmission (induced by deformation of orifice plates and unavoidable elastic fixing in the foundation plate) in order to arrive at the pressure wave transmission factor itself. This was corrected by control calculations based on measurements at the unthrottled and the fully throttled branches, which deviated from the transmitted values by a maximum of 4%. A comparison of the transmission factors of the individual orifices with butt edges and without basic flow conditions can be seen in Fig. 5.

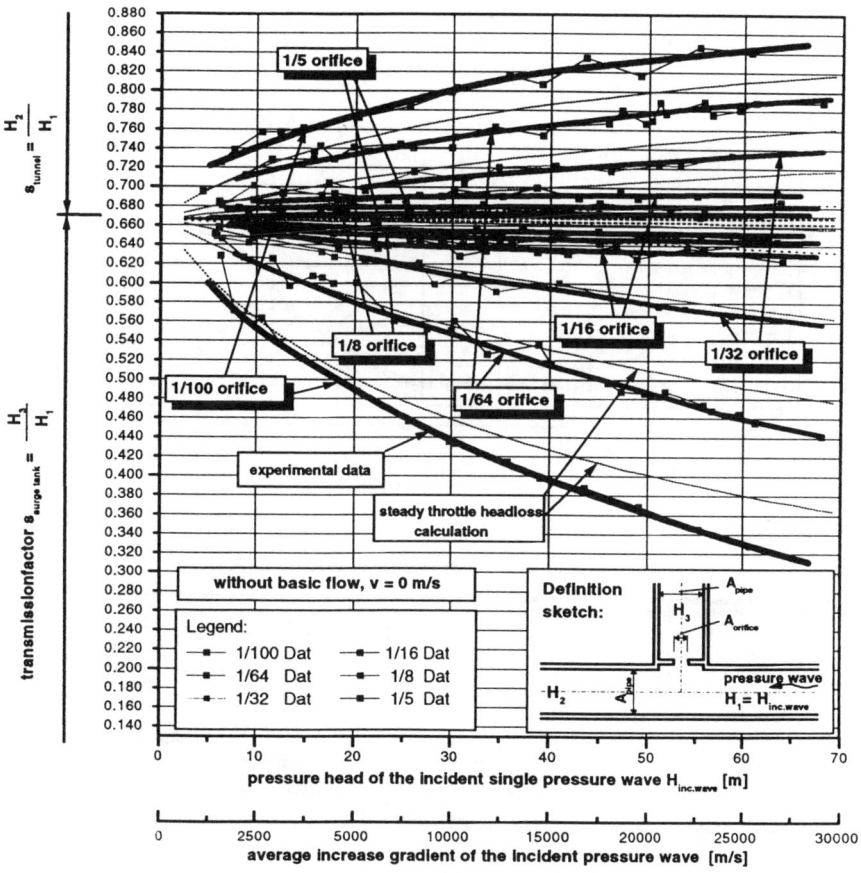

Fig. 5: Transmissionfactors in the surge tank and in the pressure tunnel, without basic flow ($v = 0$), all orifice plates with square edges

## 4. COMPARISON OF MEASURED TESTRESULTS AND RESULTS OF CALCULATIONS

The calculations of the experiments show that, for the orifices with area ratios of 1:5 to 1:16 commonly used in surge tanks, the quasi-steady headloss approach corresponds very satisfactorily with the results of the measurements (Fig. 6a). Results are not entirely satisfactory for area ratios of 1:32 to 1:64 (Fig. 6b). For ratios smaller than 1:50, the authors found it expedient to take into consideration the influence of the inertia of the high local acceleration in the vicinity of the orifice as well as the delayed scaling off of the boundary layer, although such small orifices are not relevant in the practice of surge tank design. In the experiments with superpositioned basic flows, these phenomena increased or decreased according to flow direction.

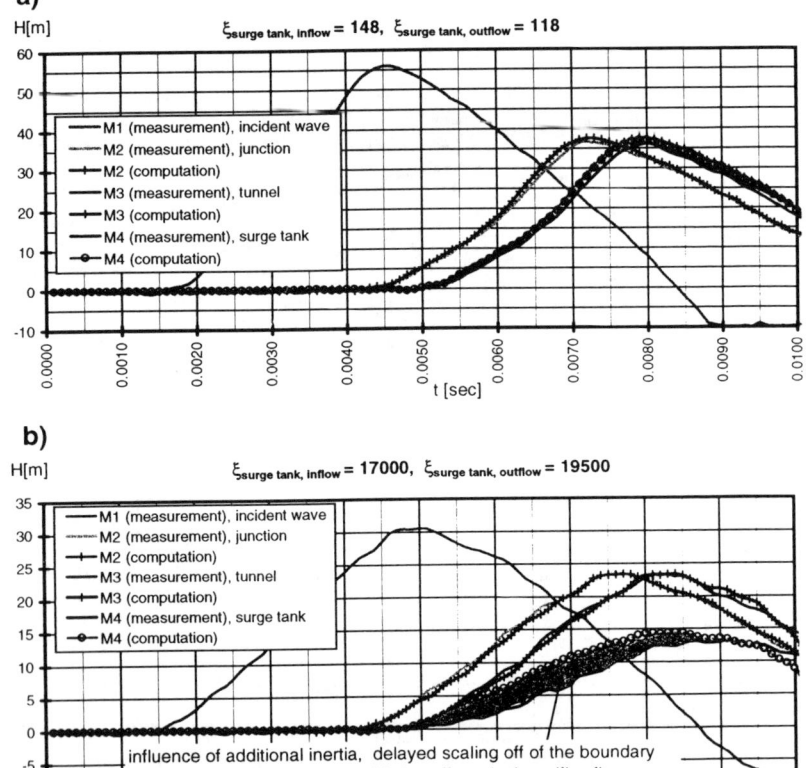

Fig. 6: Computed measurement for single pressure wave without basic flow (v=0)
a) 1/8 orifice with sharp edges in the surge tank, b) 1/100 orifice with butt edges

## 5. CONCLUSION

The transmission factors of orifice plates commonly used in high-head power plants indicate that the plates do not significantly hinder the transmission of pressure waves. Distribution of pressure waves in the tunnel or the surge tank deviates only marginally, depending on the pressure wave gradient, from the theoretical branch system approach. The calculations also show that the generally accepted steady throttle headloss for throttle ratios 1:5 to 1:16 are very suitable and that additional approaches to describe the unsteady resistance behaviour on the orifice plate are necessary only for orifice area/pipe cross-section area ratios of 1:50 and smaller.

## REFERENCES

[1] WYLIE, E.B. and STREETER, V.L.: „Fluid Transients in Systems".Prentice-Hall, New Jersey, 1993.

[2] JAEGER, CH.: „Théorie générale du coup bélier". Dunod, Paris, 1933.

[3] ZIENKIEWICZ, O.C. and HAWKINS, P.: „Transmission of Water-Hammer Pressures Through Surge Tanks". Proceedings of the Institution of Mechanical Engineers, 1954, Vol. 168, Nr. 23. pp. 629-642.

[4] SETH, H.B.S.: „Pressure wave transmission at an orifice surge tank". Water Power, August 1973, pp. 305-308.

[5] BERNHART, H.H.: „Einfluß der Schließzeit auf die Druckstoßtransmission durch Wasserschlösser". Versuchsanstalt für Wasserbau und Kulturtechnik, Theodor - Rehbock - Flußbaulaboratorium, Universität Fridericiana Karlsruhe, Mitteilung Heft Nr. 164, 1977.

[6] ALLIEVI, L.: „Teoria generale del moto perturbato dell'acqua in pressione". Annali della Societa degli Ingegneri ed Architetti Italiana, Milano, 1903.

[7] CONTRACTOR, D.N.: „The reflection of Waterhammer Pressure Waves from Minor Losses". Transactions of ASME, Journal of Basic Engineering, June 1965, pp. 445-452.

[8] TRENGROUSE, G.H., IMRIE, B.W. and MALE D.H.: „Comparison of Unsteady Flow Discharge Coefficients for Sharp-Edged Orifices with Steady Flow Values". Journal Mechanical Engineering Science, Vol.8 No.3 1966, pp. 322-329.

[9] FUNK, J.E., WOOD, D.J. and CHAO, S.P.: „The Transient Response of Orifices and very Short Lines". Transactions of the ASME, Journal of Basic Engineering, June 1972, pp. 483-491.

[10] BERGER, H.: „Druckstossverhalten an Blenden bei Berücksichtigung der Anlaufströmung". Institut für Wasserbau und Wasserwirtschaft, Technische Universität Berlin, Mitteilung Nr. 89, 1978.

# Surging Phenomena of Twin Surge-Tanks of the Okukiyotsu No.2 Power Station

Koichi FUJINO, Kaoru HAGA
Electric Power Development Co.,Ltd. (EPDC) Tokyo, JAPAN

This paper presents the knowledge gained from measurements during the construction, model tests, and design related to the surging phenomena of two restricted-orifice surge tanks at the Okukiyotsu No.2 Power Station in Niigata Prefecture, Japan. The coefficient of restriction for the orifice, or port, was found to be more desirable than the value used in design as the scale of the tests and actual port grew in size. In the load rejection test, and during the automatic frequency control operations test, it was confirmed that the two surge tanks performed better than expected.

## 1. INTRODUCTION

In recent years, surge tanks of pumped storage power stations built in Japan are commonly equipped with a restriction orifice that can absorb the water level variations over a wide area in order to respond to transient phenomena during accidents as well as operations of automatic frequency control. For designing a ported surge tank, a coefficient of restriction for the port must be determined so that restrictions due to both the surging phenomenon and increased water hammer pressure are satisfied. However, there is no established theory for the relationship between the coefficient of restriction and the geometric shape of the port. Normally, each designer individually establishes the relationship through experiments based on a hydraulic model.

## 2. OUTLINE OF OKUKIYOTSU No.2 WATER CONDUIT

The Okukiyotsu No.2 Power Station utilizes the upper and lower reservoirs of the existing Okukiyotsu Power Station, which was completed in 1978 with a maximum installed capacity of 1,000MW. The additional water conduit and power station resulted in an addition of capacity of 600MW as shown in **Table 1**.

Table 1  Outline of the Project

| Power Station | Okukiyotsu(Existing) | Okukiyotsu No.2 |
|---|---|---|
| Maximum Discharge | 260 m³/s | 154 m³/s |
| Effective Head | 470m | 470m |
| Number of Units | 4 | 2 |
| Installed Capacity | 1,000MW | 600MW |

Construction of the No.2 power station was started in October 1992 and completed in June 1996, when it went into operation. Total maximum installed capacity of 1,600MW is currently Japan's largest hydropower station. The profile and typical cross sections of conduit are shown in **Fig.1**. The pressurized water conduit is single, but the upstream and downstream portions of the two pump turbines fork into two separated streams. The shapes of the two surge tanks set up for both the headrace and trailrace are shown in **Fig.2** and **Fig.3**.

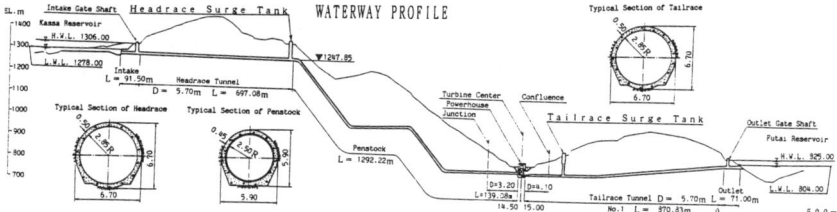

**Fig.1** Profile and Cross Sections of Conduit

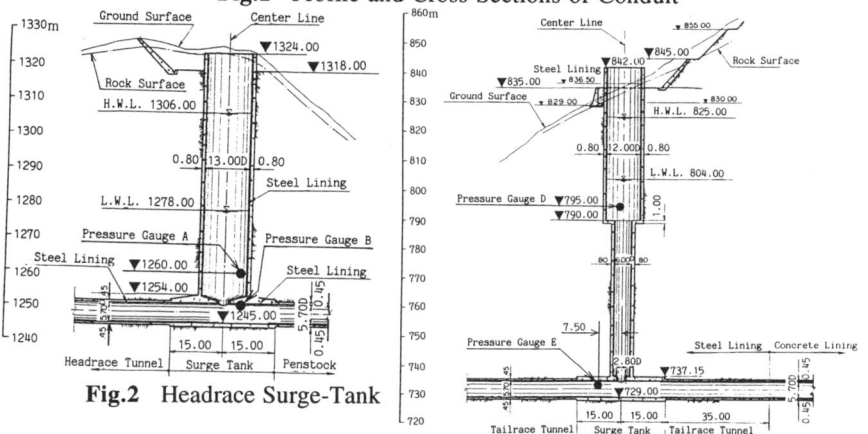

**Fig.2** Headrace Surge-Tank

**Fig.3** Tailrace Surge-Tank

## 3. COEFFICIENT OF RESTRICTION FOR PORT

### 3.1 Definition the Coefficient of Restriction

The coefficient of restriction, Cd, for the restriction orifice is a coefficient which quantifies the extent of loss of the pressure head due to the restriction port. A definition of the coefficient of restriction is formula (1) and **Fig.4**:

$$Cd = Qp/Fp/\sqrt{(2gk)} = Vp/\sqrt{(2gk)} \qquad (1)$$

where
  Qp: Flow passing through Orifice (m³/s)
  Fp: Cross Sectional Area of Orifice (m²)

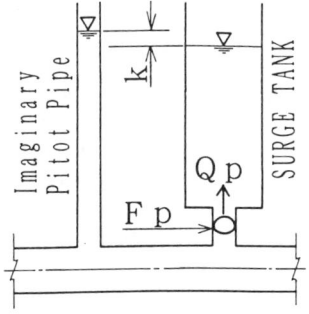

**Fig.4** Coefficient of Port

Vp: Velocity passing through Orifice (m/s)
k: Loss of Pressure Head by Orifice (m)
g: Acceleration of Gravity (m/s$^2$)

3.2 Coefficient of Restriction used for Design
In Japan, the Gardel and Fujimoto methods are normally used to determine the coefficient of restriction. The former is to estimate the coefficient by means of applying the result of experiments on head loss in a straight pipe with corn-shaped orifice and alteration of cross section to the actual branch between conduit and base of surge tank. The latter is based on the application of the experimental results on head loss between two perpendicular pipes jointed through a restricted orifice to actual conduit and tank crossing at right angle.

3.3 Experiments Based on the Hydraulic Model
As the detailed design of the turbine generator progressed after starting construction, there found the possibility of pressure fall in the draft tube during load rejection occurred simultaneously for two turbine pumps with different characteristics leading a water column separation. It was requested to know more accurate value of the coefficient of tailrace surge tank in order to cope with this problem by means of changing the shape of the port, if necessary. In response, a large-scale (1:30) model test was carried out. However, the desirable results shown in section 3.5 showed that there was no need to change the shape of the port and that the gradient of the bottom surface of the surge-tank adjacent to the port gave only slight influence to the results.

3.4 Measurements
Since December 1995 various tests on the turbine generators were conducted by passing water through the completed conduit. Measurements of water level changes and water pressure were taken for each set of test conditions. From the results, we could detect the "Vp" in formula (1) from the change of water level inside the surge tank. We could also detect "k" from the pressure measurements above and bellow the restriction port. Then we could estimate "Cd" as described in section 3.5.

3.5 Comparison of the Coefficient of Restriction
Results from above-mentioned actual measurements and calculations are shown in **Table 2**. As made clear in formula (1), the higher the value of Cd is, the less influence has the port on the passing water. Compared with the Gardel values used in the design, both the model and actual restriction port are more effective during periods of inflow or "up-surging" and less effective during outflow or "down-surging". In other words, looking at the surge tank for the trailrace, when the reversible pump turbine's draft tube water pressure drops, the supply of water coming out of the surge tank exits more smoothly than expected and a drop in pressure can be avoided. In turn, when water flowing in to the surge tank, the water level does not increase as much as expected.

**Table 2** Comparison of the Coefficient of Restriction

| Surge-Tank | Direction | Gardel | Fujimoto | Model Test | Actual Port |
|---|---|---|---|---|---|
| Headrace | Inflow | 0.84 | 0.86 | 0.86 | 0.70 |
|  | Outflow | 0.85 | 0.85 | 0.98 | 1.03 |
| Tailrace | Inflow | 0.93 | 1.00 | 1.02 | 0.91 |
|  | Outflow | 0.78 | 0.98 | 1.09 | 1.11 |

## 4. MEASUREMENTS DURING LOAD REJECTION TESTS

The results of the measurements for the two surge tanks at the time of load rejection are shown in **Fig. 5**. A and B show water pressure variations in the headrace surge tank above and below the restriction port. C shows the water pressure variation at the entrance of turbine #1. In the trailrace, water pressure variations were recorded above and below the port as shown in D and E. For the draft tube for turbine #1, water pressure measurements were taken as shown in F.

At the early stage, A and B had different movements because there is difference of water pressure derived from the flow passing through the port. There were slight water pressure vibration even in A because of the water hammer pressure propagated to the surge tank's shaft. Large water hammer pressure more than 150 meters at the entrance of the turbine had almost completely dissipated in the surge tank.

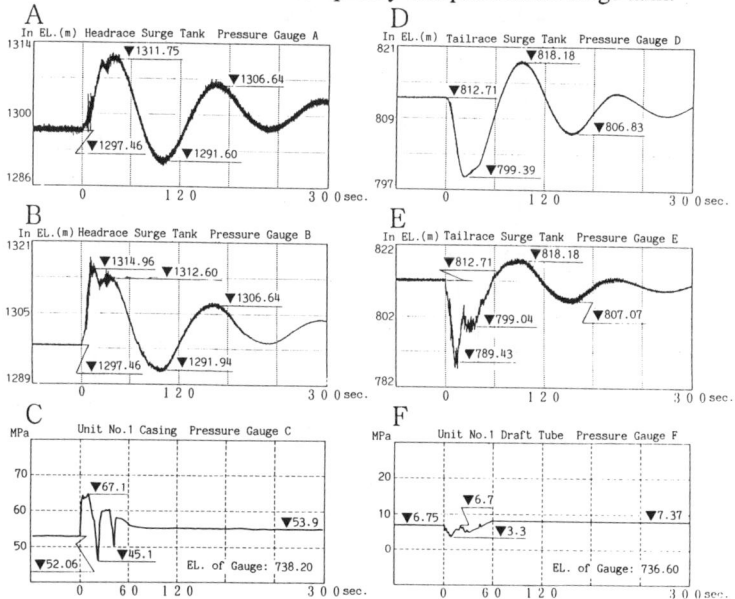

**Fig. 5** Measurement during Load Rejection Test

Measured surging period of the headrace tank was around 130 seconds, almost equals to the characteristic period of 128 seconds calculated by the formula (2).

$$T = 2\pi \sqrt{\frac{LF}{gf}} \qquad (2)$$

where
- T: Characteristic Period (sec)
- L: Length of Water Conduit (m)
- F: Surge Tank Cross Sectional Area (m$^2$)
- f: Water Conduit Cross Sectional Area (m$^2$)

## 5. AUTOMATIC FREQUENCY CONTROL

AFC operation is to stabilize the frequency variations in the total power system by synchronizing the power plant's output with detected random variations. On the other hand, there is a characteristic period shown in formula 2 for the water conduit. When the period of output variation agrees with the characteristic period, it is feared that the surge tank's water level variation would cause resonance. To solve this problem, we conducted a test to force the power plant's output to have the same characteristic period as that for the water conduit. We then measured water level changes.

As shown in **Fig.6**, the magnitude of water level changes was gradually increased. After the 4th wave, water level variation stabilized and it was confirmed that there was no divergence. From this, the restriction orifice evidently absorbed the energy from the changes in water level and the design for the surge tanks looked quite reasonable.

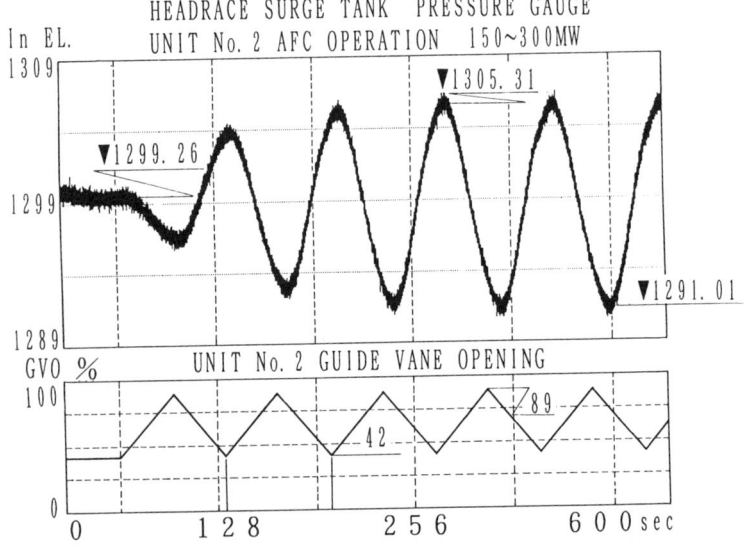

**Fig.6** Automatic Frequency Control Test

Up to the present time, there has been no interference between the two surge tanks, nor any indication of mutual acceleration.

## 6. NUMERICAL SIMULATION

**Fig.7** shows a comparison between measurements of surging in the headrace surge tank in the same case as in section 4 and the values given in the numerical simulation. Two results agree well with each other. There is, however, room for improvement of the expression for boundary conditions of the pump turbine. We are planning to sophisticate the numerical model and apply the above-mentioned coefficient of restriction to carry out numerical simulation and verify that dangerous conditions will not be created in any transient phenomena.

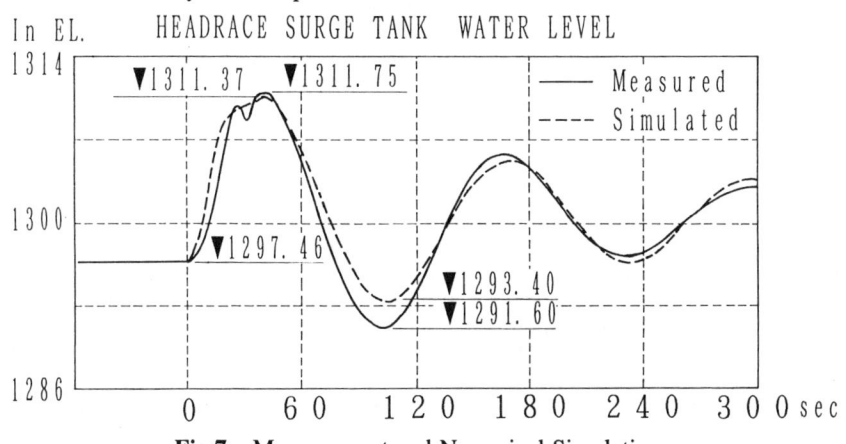

**Fig.7** Measurement and Numerical Simulation

## 7. REFERENCES

Haga,K. & Fujino,K. 1996 Construction of Okukiyotsu No.2 pumped storage power plant. Proceedings of The International Conference Hydropower '96 pp.866-874

Gardel, A. 1962 Perte de charge dans un etranglement conique. Bulletin Technique de la Suisse Romande. Vol.88, No.21,22

Fujimoto,T. 1981 Smoothed Edge of Orifice and Coefficient of Restriction. Research document of Central Research Institute of Electric Power Industry (in Japanese)

# Optimal Unit Commitment Scheduling in Hydropower Systems

J. YI, J. W. LABADIE
Department of Civil Engineering
Colorado State University
Ft. Collins, CO 80523-1372 USA
S. STITT
Hydroelectric Research and Technical Services Group
U.S. Bureau of Reclamation
Denver, CO 80225-0007 USA

## ABSTRACT

The hydropower unit commitment problem is a complex decision-making process involving the integrated hourly scheduling of generators in a multi-project hydropower system. The purpose is to satisfy power load demands, water demands, reliability constraints, operational restrictions and security requirements. A dynamic programming (DP) optimization model is developed which maximizes basin-wide operating efficiency subject to: (1) plant constraints on total generation requirements, generation shape requirements, and spinning reserve; and (2) operational restrictions on rough zone avoidance, minimum up and down time requirements and unit outage modes. Application of the DP model to the hydropower system of the Lower Colorado River Basin demonstrates the capabilities of real-time optimal unit scheduling in satisfying complex operational constraints while improving basin-wide generation efficiency.

## INTRODUCTION

Optimal commitment of electrical generation units is an important task in the short-term scheduling of power system operations. According to Grimes, et al. (1989), a 1% reduction in operating costs can result in savings of 10 to 30 million US $ per year for an electrical utility with 10,000 MW of installed capacity. Demand for electricity exhibits large variations between weekdays and weekends, and between peak and off-peak hours, making it uneconomical to continuously maintain all available generating units on-line. The problem is to determine which, if any, generating units should be out of service to obtain maximum benefits. The unit commitment procedure is designed to search for the most economical and feasible schedules of start and stop times for all units in a system with specified load schedules. A feasible schedule is one for which the forecast load, spinning reserve requirements and various operating and system constraints are satisfied at all times within the given operational horizon.

A wide range of optimization methods have been applied to solution of the unit commitment problem, including mixed integer programming (Dillon, et al., 1978); dynamic programming (Ouyang, et al., 1991); hierarchical optimization (Van den Bosch, 1985); and Lagrangian relaxation (Ruzic et al. (1991); Peterson et al. (1995); Baldick (1995). Mixed integer programming methods require extensive computational resources and can only be implemented for scheduling limited numbers of power plants and turbine units in real-time. Nonlinear programming methods based on Lagrangian relaxation and its variants are difficult to apply to unit scheduling problems with nonconvex constraint sets and *if-then* conditional logic in operating rules. Dynamic programming can be an effective technique, but suffers from the curse of dimensionality which has restricted its application to large-scale systems.

Ouyang et al. (1992) applied neural networks to short term unit commitment, proposing a two stage process consisting of an artificial neural network (ANN) and a modified dynamic programming model. Although a promising approach, considerable adjustment is required on the trained neural network for real-time implementation. Kothari et al. (1995) presented a hybrid expert system dynamic programming approach to the unit commitment problem, with scheduling output from the usual dynamic programming procedure supplemented with a rule based expert system. Kazarlis et al. (1996) presented a genetic algorithm (GA) solution to the unit commitment problem. The disadvantages of applying GA to the unit commitment problem include lack of guarantees in finding optimal solutions and excessive computer execution times.

## DYNAMIC PROGRAMMING FORMULATION

### BASIN OPERATING EFFICIENCY
Daily basin-wide operating efficiency for all hydroelectric powerplants is defined as:

$$E_{total} = 100 \cdot \frac{\sum_{p \in P} \sum_{t=1}^{24} G_{pt}}{\sum_{p \in P} \sum_{t=1}^{24} \kappa \cdot Q_{pt} \cdot H_p} \tag{1}$$

where $P$ is the set of all powerplants in the basin, $G_{pt}$ is power generation from hydroelectric powerplant $p$ during hour $t$ in MW, $Q_{pt}$ is flow through plant $p$ during hour $t$ in units of 1,000 cfs, $H_p$ is current day gross head on the plant (ft), defined as the difference between forebay and tailbay elevations, and units conversion $\kappa = 8.46 \times 10^{-5}$. This equation suggests that increasing overall basin operating efficiency is achieved by maximizing generation from water-driven plants and minimizing releases from generation-driven plants. Since water-driven plants supply required downstream water demands, while maintaining power generation shapes within a specified deadband range, maximizing generation for fixed plant releases results in increased basin operating efficiency. Generation-driven plants, on the other hand, are designed to supply a required generation load, so minimizing plant releases for a fixed generation load also increases total basin operating efficiency.

### SYSTEM STATE EQUATIONS
In this study, the unit commitment problem is formulated as a dynamic programming (DP) problem with utilization of a successive approximations procedure for overcoming the dimensionality problem. In contrast with usual DP applications, the decision stages represent the sequence of turbine units in the basin, with the system state dimension defined over the operational time horizon. Assuming a total of $N$ units associated with all hydroelectric powerplants in the basin, state variable $x1_{i\tau}$ is defined as the accumulated power generation from unit 1 through unit $i-1$ at the $\tau$ th schedule change (MW). The corresponding decision

variable $g_{i\tau}$ is defined as generation from unit $i$ at the $\tau$ th schedule change (MW). State variable $x2_i$ is accumulated daily release from unit 1 through unit $i-1$ (cfs), with the corresponding decision variable $Q_i$ defined as daily releases from unit $i$ (cfs). The final state variable $x3_{i\tau}$ is the accumulated assigned power capacity from unit 1 through unit $i-1$ at the $\tau$ th schedule change (MW), with the corresponding decision variable $c_{i\tau}$ defined as the power capacity assigned to unit $i$ at the $\tau$ th schedule change (MW). The resulting system state equations are expressed as follows:

$$x1_{i+1,\tau} = x1_{i\tau} + g_{i\tau} \tag{2}$$
$$x2_{i+1} = x2_i + Q_i \tag{3}$$
$$x3_{i+1,\tau} = x3_{i\tau} + c_{i\tau} \tag{4}$$
(for $i = 1,...,N$ ; for all $\tau \in T$ )

where T is the set of discrete times where generation shape or capacity requirements change, which generally occur when:
- required changes in release from water-driven plants exceed a given deadband
- required changes in energy generation from generation-driven plants exceed a given deadband
- required changes in plant capacity from either water or generation-driven plants exceed a given deadband
- a unit running in the previous hour becomes unavailable at the current hour
- a unit down in the previous hour becomes a *must run* unit at the current hour

Other than these cases, the unit schedule is assumed to be the same as in the previous hour. Calculating a unit schedule for only those hours where changes occur reduces the number of state variables at each decision stage (i.e., turbine unit) to 2 × (no. of change hours) + 1.

## OBJECTIVE FUNCTION

The objective is to maximize total generation from water-driven hydroelectric powerplants, while minimizing releases from generation-driven plants, which achieves the result of maximizing overall basinwide efficiency as seen in equation (1):

$$\max \sum_{i \in W}[\sum_{\tau \in T} g_{i\tau}] - \sum_{i \in P} E_i [\sum_{\tau=T} \kappa\, Q_{i\tau} H_i] \tag{5}$$

where $W$ is the set of all turbines in water-driven plants; $P$ is the set of all turbines in the generation-driven plants; $E_i$ is the average efficiency for unit $i$ in a generation-driven plant; $H_i$ is gross head for unit $i$ in a generation-driven plant; and units conversion $\kappa = 8.46 \times 10^{-5}$.

The difficulty with this formulation is that there are three different state variables (i.e., hourly generation, daily release, and hourly assigned unit capacity) calculated using separate state equations, but which are actually closely related. That is, the sum of hourly releases over 24 hours for unit $i$ producing $g_{i\tau}$ should equal total daily releases for unit $i$, $Q_i$. To encourage equality of these variables, the following quadratic penalty term is subtracted from the terms in equation (5):

$$P_1(\hat{Q}_i(g_i) - Q_i)^2 \qquad \text{for all } i \in W \text{ and } i \in P \tag{6}$$

where $P_1$ is a large penalty coefficient (e.g., $1 \times 10^6$) and $\hat{Q}_i(g_i)$ is daily unit flow calculated from generation vector $g_i$ over each hour for unit $i$ (MW). The latter calculations are based on complex tables of turbine characteristics including hourly speed no load flow, hourly break point flow, and generating status (i.e., below rough zone, above rough zone, or motoring). The tabular turbine characteristics required for calculating $\hat{Q}_i(g_i)$ are easily accomplished within the dynamic programming formulation since all variables are discretized in the solution process. An additional penalty term is subtracted from the terms in equation (5) to equate assigned capacity $c_{i\tau}$ with the capacity $\hat{c}_i(g_i)$ calculated from generation $g_i$ in the generation-driven plants:

$$P_2 \sum_{\tau \in T} (\hat{c}_{i\tau}(g_i) - c_{i\tau})^2 \qquad \text{for all } i \in P \qquad (7)$$

## CONSTRAINTS

- Unit generation $g_{i\tau}$ should be greater than the minimum motoring range and less than the unit capacity during its continuation.
- Daily releases from each unit $Q_i$ should lie between specified lower and upper bounds.
- Maintain assigned capacity $c_{i\tau}$ within required minimum and maximum ranges.
- Total generation levels for water-driven and generation-driven plants must be maintained within the specified deadband of a rough generation schedule which is developed using the generation shape; this effectively assigns bounds on state variable $x1_{N+1}$.
- Total release requirements from water driven plants must be satisfied.
- Capacity requirements should be satisfied for all plants.

## SOLUTION PROCEDURE

Details on the dynamic programming successive approximations (DPSA) solution procedure for this problem can be found in Yi (1996). A generalized dynamic programming package CSUDP (Labadie, 1990) is utilized which greatly reduces the programming effort in implementing the DPSA algorithm. The DPSA method is described in Larson (1968), and employs a *one-at-a-time* solution procedure over the state-space. Although convergence to global optima is not assured with DPSA, it is a highly efficient procedure which consistently converges to excellent solutions. The efficiency of the algorithm allows implementation for on-line unit scheduling.

## CASE STUDY: LOWER COLORADO RIVER HYDROPOWER SYSTEM

### BACKGROUND

The Lower Colorado River Basin extends from the Grand Canyon in Arizona (below Glen Canyon Dam) to the U.S-Mexico border. The Lower Colorado River hydroelectric powerplants generate about five billion kilowatt-hours of energy annually (Fig. 1). The Hoover powerplant has 19 turbines with a total generation capacity of approximately 2,000 MW; Davis Dam powerplant five turbines and a total generating capacity of approximately 240 MW; the powerplant at Parker Dam is comprised of four turbines and a total generating capacity of approximately 120 MW. Since the main purpose of Hoover Dam is power generation, it is classified as generation-driven plant, with regulation of the Lower Colorado River as one of its important objectives. The flow regulation schedule is changed often during the day, thereby requiring frequent changes in unit commitment scheduling at Hoover. The main purpose of Davis and Parker Dams is satisfying various water demands downstream, and they are therefore classified as water-driven plants.

Historical data used in the case study were obtained from the Bureau and consists of a 24 hour generation and capacity schedule and total 24 hour releases for Davis and Parker Powerplants from April 17, 1988. The generation and capacity schedules are acquired from actual operation of generators on this day, as well as required water releases from Davis and Parker. The DP formulation of the unit commitment problem for April 17, 1988 results in 27 state variables and 26 stages with each stage representing individual units in the basin. A total of 8 minutes and 49 seconds of CPU time are required on a DECAlpha, resulting in a basin-wide efficiency of 84.78%, which makes the algorithm feasible for real-time implementation.

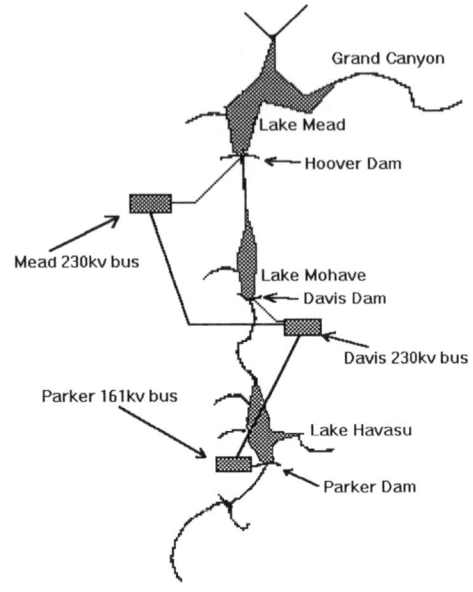

**Fig. 1.** Lower Colorado River Hydropower System

Two mixed integer programming (MIP) models were also formulated for comparison with the DP solution: a suboptimal MIP model which optimizes the water-driven and generation-driven plants separately; and a Combined MIP model which integrates the scheduling of both types of plants. The Combined MIP model has

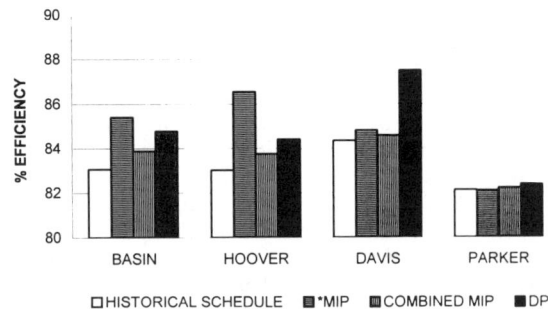

**Fig. 2.** Comparison of Basinwide and Powerplant efficiencies

12,366 constraints and 10,045 variables, with 8,274 of these constrained as binary variables. Although excessive computational times preclude use of the MIP models for real-time scheduling, they are useful for comparison with the DP solution. Fig. 2 provides a comparison of average basin efficiency values obtained from the three models for April 17, 1988 actual data. Although schedules from the MIP model appear to be superior, they are actually infeasible since they require a deadband range of generation schedule that violates the desired range. This is due primarily to the disjointed, nonintegrated procedure embodied in the MIP model. The Combined MIP model overcomes these disadvantages and produces feasible solutions, but at the price of an enormous computation load. Performance of the DP model is consistently high, giving increases in basin efficiency of 1.8% over actual schedules, based on calculations from equation (1). Assuming electricity is sold at 2.5 cents per KWh, annual extrapolation of these higher efficiencies represents annual revenue increases of 2.52 million U.S.$ for the Lower Colorado River Hydropower System.

## REFERENCES

Baldick, R. (1995). "The Generalized Unit Commitment Problem", *IEEE Transactions on Power Systems*, Vol. 10, No. 1.

Dillon, T., K. Edwin, H. Kochs and R. Taud (1978). "Integer Programming Approach to the Problem of Optimal Unit Commitment with Probabilistic Reserve Determination", *IEEE Transactions on Power Apparatus and Systems*, Vol. PAS-97, No. 6.

Grimes, R. and S. Jabbour (1989). "The DYNAMICS Model for Measuring Dynamic Operating Benefits", Technical Report, Decision Focus Inc., Los Altos, CA.

Kazarlis, S., A. Bakirtzis and V. Petridis (1996). "A Genetic Algorithm Solution to the Unit Commitment Problem", *IEEE Transactions on Power Systems*, Vol. 11, No. 1.

Kothari, D. and A. Ahmad (1995). "An Expert System Approach to the Unit Commitment Problem", *Energy Conversion Management*, Vol. 36, No.4, pp. 257-261.

Labadie, J. (1990). "Dynamic Programming with the Microcomputer," *Encyclopedia of Microcomputers*, Vol. 5, Marcel Dekker, Inc., New York.

Larson, R. E. (1968). *Incremental Dynamic Programming*, American Elsevier Publishing Co., Inc., New York.

Ouyang, Z. and S. Shahidehpour (1991). "An Intelligent Dynamic Programming for Unit Commitment Application", *IEEE Transactions on Power Systems*, Vol. 6, No. 3.

Ouyang, Z. and S. Shahidehpour (1992). "A Hybrid Artificial Neural Network-Dynamic Programming Approach to Unit Commitment", *IEEE Transactions on Power Systems*, Vol. 7, No.1.

Peterson, W. and S. Brammer (1995). "A Capacity Based Lagrangian Relaxation Unit Commitment with Ramp Rate Constraints", *IEEE Transactions on Power Systems*, Vol. 10, No. 2.

Ruzic, S. and N. Rajakovic (1991). "A New Approach for Solving Extended Unit Commitment Problem", *IEEE Transactions on Power Systems*, Vol. 6, No.1.

Van den Bosch, P. and G. Honderd (1985). "A Solution of the Unit Commitment Problem via Decomposition and Dynamic Programming", *IEEE Transactions on Power Apparatus and Systems*, Vol. PAS-104, No. 7.

Yi, J. (1996). "Decision Support System for Optimal Basin-Wide Scheduling of Hydropower Units," Ph.D Dissertation, Dept. of Civil Engineering, Colorado State University, Ft. Collins, CO.

# REAL TIME CONTROL OF RIVER WATER QUALITY FOR HYDROELECTRIC OPERATION

by:

F. WELT
HMS Énergie Inc.
1801, Mc. Gill College, Suite 1000, Montréal, Québec
Canada, H3A 2N4

and

R. KAHAWITA
École Polytechnique de Montréal
P.O. Box 6079, City Center Station, Montréal, Québec
Canada, H3C 3A7

## ABSTRACT

A fully 2-dimensional hydrodynamic model based on a finite volume formulation using a Riemann solver resolution scheme is used for real time applications of hydroelectric plant operation. The impact of hourly changes in power generation on water quality for a set of hydroplants arranged in cascade along a complete river system is studied. The diffusion of heat from hot water sources such as fossil plants and industries, as well as heat exchanges due to radiation, convection and evaporation with the environment, are modeled. Of primary interest is the effect of peaking on water temperatures and dissolved oxygen levels. Provisions for representing the various power plants and spillways as internal boundary conditions are made, while initial conditions of flow, water levels and temperatures are provided by real-time measurements. Results show that minimum flow requirements should be imposed at all times, and that their values must be continuously adjusted for proper control of the river conditions. The approach is implemented within the context of real-time river management for power plant operation in North America and internationally.

## 1. INTRODUCTION

The amount of flow released by river operators at various hydroplants along a river system, can have a very significant impact on water quality. The parameters of water temperature and dissolved oxygen concentration are particularly susceptible to the operation when processing industries or fossil power plants are present and use the water for cooling purposes. With the growing concern for the environment at the industrial and government level, combined with the objectives of the upcoming energy market deregulation which can lead to high peak and very low off-peak periods, there is a stronger need for the implementation of more powerful water quality forecasting tools in the context of real-time operation.

The selection of a model for such tools should be based on a proper balance between precision, robustness and performance, in order to obtain a reasonable response time and degree of accuracy. More importantly, the input data should be sufficiently accurate to reflect the current river operation. Although this has been a major stumbling block in the past, the emergence of integrated river management systems using real time data acquisition, validation of the water levels and estimation of the tributary inflow, such as the one presently developed at Hydro-Quebec, a power utility in Eastern Canada, provide the proper conditions for successful implementation (1).

A number of solutions to river hydrodynamic problems have been proposed over the years, mostly for the design of new construction or impact study purposes. This includes the simpler one-dimensional models such as HEC-1 or Dambreak, to the more sophisticated 2 and 3-D models based on either the finite difference or finite element approach (2). The finite volume approach, originally researched for solving the compressible flow equations in high speed aerodynamics, has recently been applied to the river problem, and resulted in the development of a commercial package called TASE (3). The well proven robust properties of its numerical solver, which can handle sub as well as supercritical flows, make it well suited for applications where transient conditions are found, as with the presence of peaking hydroplants, rivers with rapids, etc. It has been used to perform studies on a number of rivers in Quebec and North America.

In this study, the finite volume hydrodynamic model, combined with a finite element scheme to solve the transport equation, is applied to the real-time operation and short-term scheduling of power plants. The steps required for such an implementation are first presented. Adequate information regarding the flow and temperature conditions must be provided to obtain useful results. Particular attention is devoted to river modeling, in terms of boundary and initial conditions, to permit a realistic representation of the problem.

## 2. RIVER MODELING

### 2.1 Governing principles

The river hydrodynamics are represented by the shallow water 2-D Saint-Venant's equations, coupled with the mass conservation principle, as presented in the following vector relation,

$$\frac{\partial f}{\partial t} + \nabla.(F - G) = -R - B$$

where f represents the diffusion terms, F the inertia, G the water friction forces, R the gravity field and B the external friction forces at the boundary, including wind and bed roughness. Bed friction is given by the Manning equation. A wind shear stress empirical relation is used and turbulence modeled by the eddy viscosity approach, according to Falconer's formulation (3).

The standard transport equation is then used to solve for the diffusion of heat, or possibly other source terms, into the flow, as

$$h\left(\frac{\partial \phi}{\partial t} + u.\nabla\phi\right) - \nabla.D.\nabla\phi = q$$

where h is the river depth, u is the water velocity, $\phi$ is the average concentration over the depth, D represents a turbulent dispersion coefficient, and q is a heat source or sink term which includes losses due to evaporation and convection, and heat addition caused by solar radiation.

### 2.2 Boundary Conditions

Boundary conditions for the two equations above are of the static (i.e., river configuration) or dynamic types (flow, temperatures or water levels at the boundaries). The static boundary conditions are defined by the dry wall conditions at the river shores, as well as by the so called internal boundary conditions at the power plants and spillways located along the river (Figure 1).

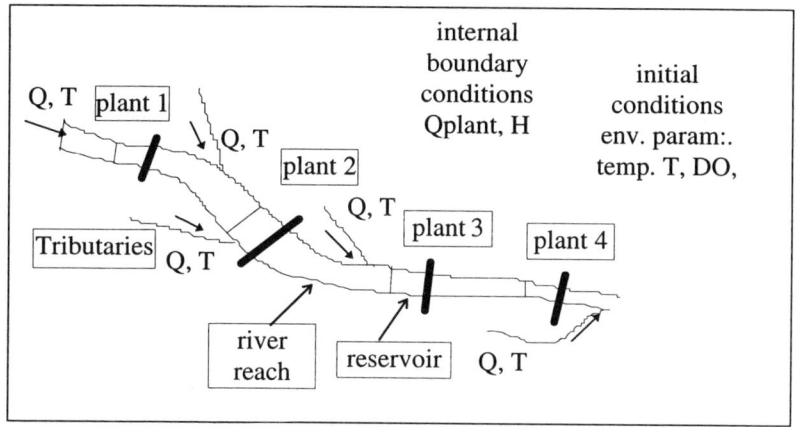

Figure 1: River configuration with cascading hydroplants and spillways

In the context of real-time operation and short-term planning of the operation, the dynamic boundary conditions should be based on a good forecast of the river flow parameters for the scheduled period. Flows and water levels at the power plants and spillways can vary significantly, especially during peak and off-peak hours and are all expressed as functions of time. Tributary inflow and temperatures are also time functions, with natural streams providing the source of cold water, and fossil plants or other industries warm water inflow. Atmospheric conditions can be specified. Ambient air temperature, relative humidity and solar radiation are obtained from short term weather forecasts or climatic data.

## 2.3 Initial Conditions

Initial conditions should be based on real time measurements to reflect the current operation as closely as possible and help provide a good water quality forecast, at least in the next few hours. Information about the current flow, water levels and temperatures are usually available at the power plants. A complete initial condition based on these measurements may thus be generated.

## INTEGRATION INTO A REAL-TIME RMS

Real-time operation of power plants will increasingly be conducted by river management integrated systems to perform the following operations:

- continuous analysis of the real time data, as provided by existing supervisory and data acquisition systems (SCADA), including data validation and estimation
- calculations of the real-time and future inflows based on detailed precipitation forecasts
- optimal scheduling of the hydroplants according to precise plant efficiency curves and spillway characteristics
- water quality forecast and control

The flow of information required by the water quality model, and subsequently returned to the river management system, is shown in Figure 2. Basic data such as present and forecasted flow and water levels at the various hydroplants and tributaries are transferred as inputs, while outputs of water temperatures, or other environmental parameters such as dissolved oxygen (DO) are returned. In the current version of the river management system, the water quality model has its own working PC based environment.

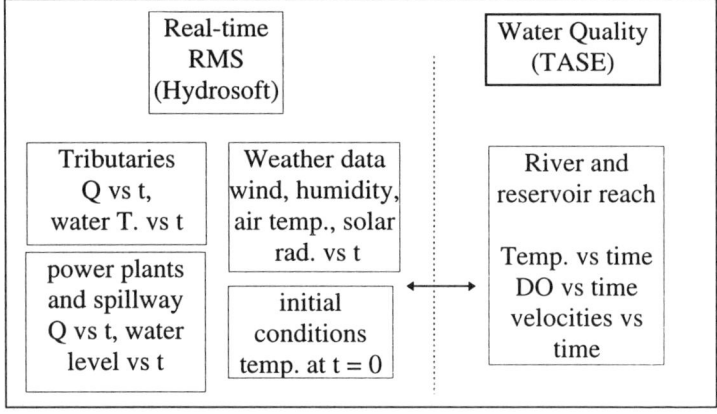

Figure 2: Integration of water quality model into the real-time DSS

## NUMERICAL RESULTS

Simulation were carried out on typical river configurations, such the Jacques Cartier river in Quebec (Figure 3), Canada, bounded by two control sections. The dissipation of hot water on the upper section was simulated for typical scenarios of peaking conditions, as illustrated in Figure 4, and compared against steady flow conditions with the same mean value. For a given mean flow, a significant variation in the temperature profile can be experienced, depending on the peaking and current hydraulic conditions. Imposition of pre-set minimal flow requirements must be therefore quite conservative to cover all the cases. In the example shown in Figure 4, a mean flow of 80 m$^3$/s is required to keep the temperature below 8.6°C under peaking conditions, as opposed to 62.6 m$^3$/s under steady flow, at a distance of 10 m from the heat source.

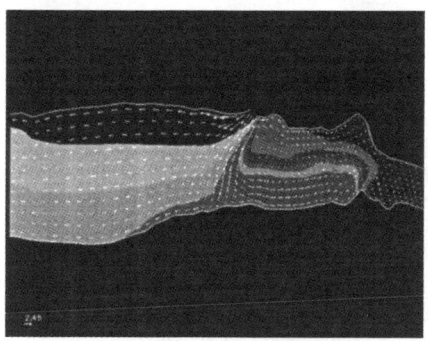

Figure 3:Typical temperature simulation

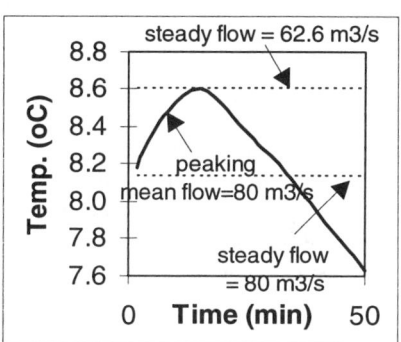

Figure 4: Variation of water temperature versus time

## CONCLUSIONS

By imposing continuous variations of the boundary and initial conditions, numerical simulations show that proper control of the water quality parameters require repeated adjustments to the power plant operation. This is in contrast to the traditional imposition of a minimum flow requirements calibrated prior to implementation. The use of a hydrodynamic model, based on the finite volume scheme, proves to be a reliable tool for real-time applications.

## REFERENCES

1. Robitaille, A., Welt, F., and Lafond, L., Real Time River Management, Water Power and Dam Construction, July 1996, pp. 56.
2. Fread, D.L., Channel Routing, Hydrological Forecasting, John Wiley & Sons, Chapter 14, pp. 437-503.
3. Zhang, H., Kahawita, R., Shen, M., and Li, Z., A PC Based Numerical Package for the Simulation of Two-Dimensional Free Surface Flows, Proceedings of the Second Canadian Conference on Computing in Civil Engineering, Ottawa, Canada, pp. 794-803.

# The Problem of Small-Scale Hydropower Stations on the Rivers of Central Russia

S.LACHTCHENOV, V.SEMENKOV
RAO "EES ROSSII", Moscow, Russia

## INTRODUCTION

The analysis of small-scale HPS operation experience, constructed within 40-50$^{th}$ years on mild slope rivers of Central Russia has been carried out. There have been executed a scope of technical examination work of a number of small-scale HPSs. Cardinal principles of designing and construction, as well as some conceptual positions, aimed on increasing of small-scale HPS competitiveness have been worked out. The results of small-scale HPS designing studies with advanced foreign technics and technologies usage are exemplified.

## HISTORY

Extensive mastering of small-rivers hydroresources of Central Russia, along with the Volga river hydropower plants cascade erection and simaltaneously with familiarization of other potentially powefull types of energy production sources started after the World War II, to provide the economy rehabilitation of European part of the country with electric power supply.

In the Central region of Russia most of the main tributaries in upper part of the Volga, Oka, Don and other rivers have been completely mastered and used for hydropower generation. More than 6000 SSHS were erected having the installed capacity within the range of 250-4000 kW. Only in the drainage basin of the Oka river, which is one of the tributaries of the Volga river, more than 60 small-scale hydropower plants have been constructed.

Most of small-scale plants used to have a riverbed layout, with 3-6 meters static head. Having such a layout the reservoirs were practically completely located in the natural riverbed without permanent flooding of the lower floodplains normally used for agricultural purposes and were characterised by a high water exchange ratio. Apron type spillways sectional construction with rotating girders provided minimum increase of the water levels during the flood periods in comparison with normal supply level and regular wash-out of deposited sediments (see Fig.1).

Local infrastructure was formed around the small-scale hydro power plants, providing operation of the station itself, power transmission lines, utilisation of reservoirs for domestic and industrial water supply, irrigation, fishery and river navigation. The role of small rivers as local transport lines have been considerably increased.

## PRESENT STATUS

Commissioning in 60-70-s of large hydropower plants on the Volga and Kama rivers, putting into operation of thermal power plants with relatively low energy production

*Fig.1 Lykovskaya SSHS on the Zusha river.*

costs as well as large-scale construction of high-voltage transmission lines and distribution facilities which provide centralised power supply to local consumers led to gradual stopping of SSHS operation. Operation of several stations was stopped after destruction of wooden spillway elements or damage of foreign-made hydropower equipment ( Most of SSHS, constructed in 50-s, had been furnished with hydroturbines of the "Voith" Company and "Siemens" Company, which produced electric equipment).

The stopping of SSHS operation including ship-lock facilities, drawdown of reservoirs, destruction of spillways led to desolation of infrastructure, formed while decades.

A stable interest towards the construction of new, reconstruction and rehabilitation of formerly existing SSHS is arising today. The mentioned fact should be considered as a consequence of constantly increasing prices on organic fuel and its transportation, that in its turn generates the raise of electric power unit rates in the local and federal energy markets and practically a state-wide cut-off of local power consumers because of electric power deficit. In this situation the SSHS which do not require fuel purchasing and delivery and are able to produce electric power with minimum maintenace and operation costs turn to be economically efficient as main or reserve power sources.

## MAIN DESIGN AND CONSTRUCTION PRINCIPLES OF SMALL-SCALE HYDROPOWER STATIONS IN CENTRAL RUSSIA

Basing on requirements of economic development in Central Russia and taking into account the improvement of the market economic relations in the state the small-scale hydropower industry must have been developed at a higher rate than today. There are several reasons for that. The main one - is a stagnation in design approaches to small-scale hydro. The practical experience of such design acquired in 40-50-s have been lost by the present time. The existing SSHS projects often tend to be a scale models reproducing big HPS of the Volga-Kama or Dnieper cascades, Hydropower Stations on the rivers of Siberia and Central Asia.

In our opinion in the process of designing and constructioning of SSHS the following main principles should be taken for basis, under conditions of application of modern hydroturbine and flow control equipment, which will provide these principles to be implemented in practice. This approach will allow to solve problems of economic efficiency and environmental safety of small-scale hydro schemes.

- Construction of SSHS must be completed within 1-2 flood-free seasons.
- The structures layout must be mostly of riverbed type without extension of main retaining structures to the flood plains. Power house, spillway structure and bank abutments must be located between steep river banks. The earth-fill structures on the floodplains should be mainly used for the switchyard and civil facilities as well as to create the stream diversion dikes forming the stream around the power house during the passing of extreme floods along the floodplains.
- The spillway structure must be designed to pass the discharges corresponding to the floods with 5-10 per cent of annual probability. The control equipment of the spillway structucs placed on a lower elevation sill must provide practically complete opening of the cross section of the river bed to allow the passing of flood discharges.
- For extreme flood events of annual probability lower than designed the spillways constructions, bank abutments, ship-lock chambers and mechanical equipment should admit a complete submergence. Thus it should be possible to pass the flood discharges not only through spillway structures and power house but also around the main structures along the floodplains. In this case a hydraulic head on structures should be minimal and provoke no threat for structures of being scoured or damaged. The flood pass mode on the river should be very close to the natural one.
- Power house design must be simplified in construction relation and admit the submergence during the disastrous flood events.
- The diversion canals for passing the construction period river flow are expedient of being preserved during the scheme operation and can be used for structures repairing.

## TECHNICAL CONCEPT FOR DEVELOPMENT Of SMALL-SCALE HYDROPOWER IN CENTRAL RUSSIA

Engineering and technical aspects of new construction, rehabilitation and reconstruction of SSHS in Central Russia - densely populated region with mild slope rivers may be summarised as follows:
- Development of the concepts main idea as a new approach to designing, construction and operation of small-scale hydropower plants, that will meet requirements of a market economic development in the regions with a complex integration of power generating, energy consuming, manufacturing and transport sectors into entire investment projects which could produce the maximum of both financial and economic effect.

- Designing, construction and operation of small-scale hydropower plants with utilization of advanced world experience in the field of creation of new technologies and equipment, which can meet the contemporary economical and enviromental reqiurements including those claimed to small-scale hydro schemes on mild slope rivers, and considering the main designing and construction principles mentioned above.
- Concept implementation through rehabilitation of several hydropower stations in Central and North-West regions of Russia as well as construction of a pilot experimental project.
- Development of industrial base for small hydro rehabilitation in Russia, training of experts in the field of implementation of new technical and technological approaches, development of designing principles adopted in the concept and corresponding to modern market relations in Russia.

Engineering servey of small-scale hydropower plants constructed in 50-s on the rivers of Oka drainage basin showed that there is an essential number of stations with well preserved concrete structures, the rehabilitation of which can be economically justified. The number of power stations still retain hydrostatic pressure front, so it essentially reduces the costs for creation of reservoirs. The existance of ship-locks and excavated diversion canals will allow to minimise the costs connected with temporary river closure for spillway reconstruction or reparing.

## EXPERIMENTAL SMALL-SCALE HYDRO POWER PLANT

According to the main stages of the concept development "RAO EES ROSSII" has worked out the project of experimental hydropower station on the Nerl' river (Khorobrovskaya HPS) in the region of Yaroslavl. The station is being constructed at the site of an old dam destroyed in 70-s.

At this experimental SSHS RAO "EES ROSSII" plans to test the efficiency of flood control system with application of automatic fusegate - "Hydroplus System" (France) and single-unit submersible hydropower generators developed by "ITT FLYGT AB" (Sweden), in climate conditions of Central Russia. It is anticipated to test other technologies capable to increase the efficiency in construction and rehabilitation of small hydropower station.

Application of "Hydroplus System" fusegates at the spillway structures of the mild slope rivers (see Fig.2) completely reflects the design principles formulated above.

Automatically operated fusegates installed on a low reinforced concrete sill retain the water to create the potential head at power station. During the flood period the fusegates automatically tumble to the downstream and open the spillway spans up to the entire river cross section area.

The fusegates can be easily manufactured, need no gear, what minimises the construction and operation costs. The fusegates can be installed on the existing spillway sills without their radical reconstruction.

A scope of researches, dealt with adaptation of "Hydroplus" fusegates for operation in the number of low pressure spillway dams on mild slope rivers with hard ice conditions have been carried out before designing the pilot SSHS. This type of gates

*Fig. 2 Hydroplus System fusegates installed on the concrete sill at McClure Dam, New Mexico, USA.*

had not been applied in such conditions before.

On the experimental SSHS the gates are mounted above the low sill and while the high-water is passed the flood sometimes partially, or completley inundates the gate. In this mode od operation the gate needs a more precise consideration of power factors, influencing its stability on the sill.

The researches of "Hydroplus"gates in SSHS or analogous conditions have been conducted in AO "Scientific-research Institute of Power Structures" of RAO "EES ROSSII".

- Using a laboratory hydrolic duct the passing of flood with ice slabs on water surface through "Hydroplus" gates, mounted on sill of spillway sector, was explored. There was achived a range of "Hydroplus" various gates applicability on the rivers, when there is a possibility of ice slabs passing through the spillway during high-water period.
- The water discharge characteristics were studied on the experimental SSHS with reference to different extent of gates submersion. Besides, the gates stability on the sill have been studied for the same conditions.
- For conditions of an experimental SSHS the gates freezing, frost penetraition in its drainage system and also freezing of tumbling control hydraulic system has been studied by means of culculations.
- The studies of ice characteristics as well as its travel during a spring flood on the spillway dam, located on the Nerl river in the upperstream of proposed experimental SSHS dam site have been carried out. The reservoir of the mentioned dam has chareteristics close to a designed one.
- On the experimental SSHS spillway, designed to pass a maximum disharge of 420 q.m/c there will be mounted 12 gates of two types: "Classic" type of gates, "Hydroplus"company designed and utilized at a number of projects (see. fig. 2) and those atapted to Russian conditions and designed jointly by russian and french specialists. The spillway structure makes it possible to mount or to remove the gate quite rapidly, or even to use the gate of a revised construction at any convinient time of operation with no negative influence on water reservoir.

Besides an approbation of different types of gates during flood pass at the experimental SSHS the researches of their behavior under influence of ice load and freezing are also planned.

Submersible single-unit hydroturbine generators "FLYGT" are one of the most advanced design solutions for SSHS. Being developed as submersible equipment they do not require construction of a power house and complicated inlet and outlet turbine ducts (see Fig.3), what essentially reduces the power house construction costs. Besides the hydroturbine generators design makes safety a partial or complete submergence of the power house during extreme flood events.

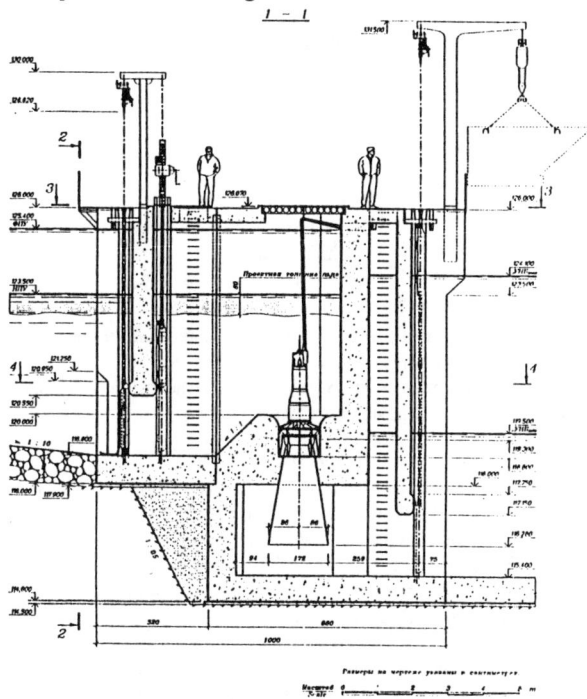

Fig. 3 Experimental small-scale power house cross-section on the Nerl' river with submersible hydroturbine generators.

## CONCLUSION

Analysis of situation with SSHS built at 40-50$^{th}$ on the rivers of European part of Russia, carried out by RAO "EES ROSSII", has shown the necessity to recover most of them as one of conditions for economically successfull solution of the problem of Russian province development.

At present time, basing on complex of research, designing and engineering works a new technical conception of development of small-scale hydro power in Central Russia is being worked out.

# THE DEVELOPMENT OF THE SIMULATED STUDY ON STRUCTURAL VIBRATION INDUCED BY DISCHARGE FLOW ENERGY

Prepared by:

Li Guifen    Yuan Ximin    Liu Shukun
China Institute of Water Conservancy and Hydroelectric Scientific Research
Fu Xing Road    Beijing City
China    100038

Cui    Guangtao
Tianjin    University
Tianjin    City
China    300072

ABSTRACT    The phenomenon of structural vibration induced by high velocity discharge flow always occurs from time to time. This is    an important question concerned by many engineers and scientists. Hydroelastic model, inverse analysis method and neural network method are introduced in this paper. I think that these are the nearest studies by now.

## INTRODUCTION

Some projects and buildings are usually destroyed due to structural vibration induced by high velocity water flow, for example, destroyed diversion wall, flood gates, piers and overflow dams, etc. Nobody makes it clear till now, because the vibration system is a kind of complex interaction system of water body-foundation-structure-water fluctuating pressure. And then, the consequence is very important and public interest and economy will be met with great disaster if the project is destroyed by vibration induced by discharge flow.

We all know that the dynamic responses can be certain easily if the loads acting on the structure are clear, but it is so difficult that it is unable for us to do. Following the study of National Eighth Five Key Project of China    " The high arch dam's vibration induced by discharge flow of XIAO WAN arch dam " , hydroelastic model was setup and inverse analysis method was adopted to study the water-structure coupled vibration system. Based on the prototype measured results of WU JIANG DU diversion wall of China, neural network model was made to analyze and forecast the vibration of the right diversion wall.

## RESEARCH ONE

The height of XIAO WAN arch dam, which is being designed now, is H=292 meters and the total crest length is L=900 meters. From structural joint of view the dam body is shaped as a double-curved thick shell structure. The crest thickness is 8 meters and the bottom thickness is about 67 meters. The discharge flow power is 34000 MW. This arch dam is the highest arch dam and concrete dam in the world by now, including all being built and designed dams.

## OBJECTIVE OF THIS RESEARCH

The energy that induces the dam's vibration includes three parts, the fluctuating pressure of outlets, wave pressure on dam body in plunge pool and the acting of nappe impacting to plunge pool. So this dynamic system is a multi-input and multi-output structure vibration system. The main objectives of these simulate studies are the evaluation of prototype arch dam's dynamic properties accurately, through setting up hydroelastic experimental model, computing model and dynamic inverse method. Using a set of strong motion instruments, dam's natural frequencies, corresponding mode-shapes, damping characteristic and responses of discharge vibration are measured from the hydroelastic model. These data allow a correlation of experimental analytically derived data and provide important information for accurate mathematical models' formulation. Dynamic inverse method may predict the responses of the parts that cannot be tested and is able to obtain prototype arch dam's dynamic properties accurately correcting the model's error.

## SETTING UP EXPERIMENT MODEL

High arch dam's vibration induced by discharge flow energy is a reservoir water-dam body-foundation coupled dynamic system. It is necessary to be content with hydro-dynamics and structural-dynamics similarity in model, the similitude ratio of model size $\lambda_l=1 : 150$, the similitude ratio of material unit weight $\lambda_\gamma=1$, the ratio similitude elastic modulus $\lambda_E=\lambda_l$, the similitude ratio of damping coefficient $\lambda\eta=1$, and the similitude ratio of Poisson coefficient $\lambda\mu=1$. According to that of above, heavy rubber material is developed and used to build model in state of concrete material in this study. Damping coefficient of this material are not similar and their influence is corrected by dynamic inverse analysis calculation. And, it is important to certain the scope and dynamic properties of arch dam's foundation. Through three dimension finite element computation and comparison, results are obtained: a) The foundation is suitable when it's depth is 0.75H(H is dam's height), upstream foundation is 0.5H, downstream foundation equals to H, and the shoulder of dam's crest is 1 / 3 H. b) The influence of unit material weight of foundation is evident to the dam's vibration. Fig.1. c) The dam's dynamic response reduces with the increase of elastic modulus, Fig.2.

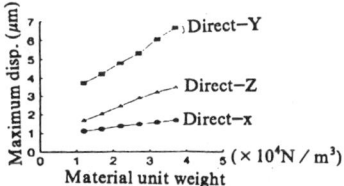

Fig.1 The curves of the maximum vibration displacement mean square root value varying with material unit weight

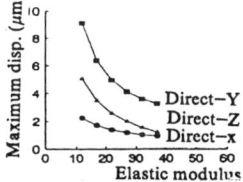

Fig,2 The curves of the maximum vibration displacement mean square root value varying with elastic modulus

## MEASURING OF VIBRATION RESPONSE

Measuring points of Xiao Wan arch dam's Vibration induced by discharge flow are located at the upper and middle parts on the downstream face, where dynamic displacement is bigger than the other. All testing points are divided into A, B, C, D four water levels, 1245.0 meters, 1210.0 meters, 1170.0 meters, 1130.0 meters. Eighteen testing points are located at layer A and A10 is the central point of the dam's crest. The responses of the dam's vibration are measured in three cases, shallow outlets discharge flow, middle outlets discharge flow and shallow and middle outlets discharge flow simultaneous. Point A9 vibration displacement history and power spectrum see Fig.3. The points of the bigger displacement are in the center of the dam and the vibration displacements present same phase position. The equal value curves of mean square root value of vibration displacement see Fig.4. The biggest vibration displacement is at point A10 and mean square root value is 1.23μm, matching 184.5μm in prototype.

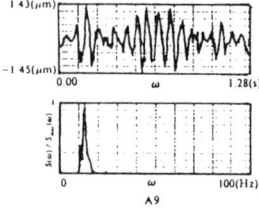
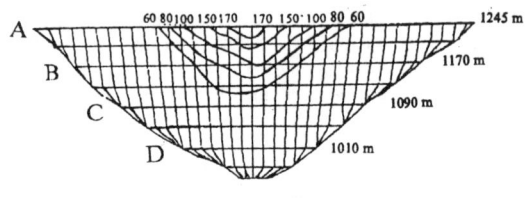

Fig.3 The curves of time history and spectrum of vibration displacement

Fig.4 Equal value curves of vibration displacement mean square root value on the dam's downstream surface

All three cases, the vibration displacement of shallow outlets discharge flow is the biggest, that of shallow and middle outlets discharge flow is the next and that of middle outlets discharge flow is the smallest. That is because when discharge flow is big, the pulse pressure acting to the dam is big too. When the discharge flow comes from only the shallow, the loads act at the center of dam's crest, the vibration sensitive position, and the distance of water pounding pool is the nearest to dam. The water random wave and fluctuating pressure will lead to the dam's vibration easily. But if the discharge flow comes from only the middle outlets, the pulse pressure, water pounding load and the acting position not do good to induce dam's vibration than before. When shallow and middle outlets discharge flow at the same time, the correlation of them is important to the dam's vibration. If the phase positions of them are same, the dam's vibration responses will become bigger than that of single one, or the responses will be smaller. The vibration displacement mean square root values of shallow and middle outlets discharge flow is smaller that of the shallow discharge flow, because the vibration displacement of them appears negative correlation.

## DYNAMIC INVERSE ANALYSIS

The structural vibration induced by discharge flow is decided by two parts, structural system and load system. The relationship is:

$$G_y(f) = |H(f)|^2 G_f(f) \tag{1}$$

$G_y(f)$ is the output, $G_f(F)$ is the input, $H(f)$ is the frequency response function.
When fluctuating load induced by discharge flow is a kind of stationary random process, structure bodies are linear elastic system and the feed back of flow fluctuating pressure acting to dam body is neglected, the differential equation f dam's vibration may be express:

$$[M]\{\ddot{V}\} + [C]\{\dot{V}\} + [K]\{V\} = \{P\} \tag{2}$$

Where [M] mass matrix, [C] damping matrix, [K] stiffness matrix, [V]、[$\dot{V}$]、[$\ddot{V}$] displacement vector, velocity vector, acceleration vector, {P} random load vector.
Adopting mode superposition, eigenequation is:

$$|[K] - \omega^2[M]| = 0 \tag{3}$$

Resolving this equation, eigenvalue $\omega_1, \omega_2, \cdots, \omega_n$ and eigenvector $\phi_1, \phi_2, \cdots, \phi_n$ are certain. $\omega_n$ and $\phi_n$ are modal frequencies and mode shapes.

Simple freedom degree body's vibration equation is written:

$$\ddot{\delta}_j + 2\xi_j \omega_j \dot{\delta}_j + \omega_j \delta_j = F_j(t) / M_j \qquad (j=1,2,\cdots,q) \tag{4}$$

Here, $\delta_j$ is generalized coordinate, q is stage number of modal.

From (4), transient is response of mode shape can be gained

$$\delta_j(t) = \int h_j(t - \eta) F_j(t) \, d\eta \tag{5}$$

In (5), $h_j(t)$ is unit impulse response function:

$$h_j(t) = 1/(m_j \omega_{dn}) * e^{-\zeta \omega t} \sin \omega_{dn} t \quad (t>0), \qquad h_j(t) = 0 \quad (t<0)$$

$$\omega_{dn} = \omega_j (1 - \zeta_j^2)^{-0.5}$$

Displacement response history of point k can be written:

$$V_k(t) = \Sigma \phi_{ki} \Sigma \phi_{ij} \int h_j(t - \eta) p_i(t) \, d\eta \tag{6}$$

In this formula, $\phi_i$ is mode vibration function. If every modal is independence each other, formula (6) can be made Fourier transfer into displacement power spectrum:

$$S_k(\omega) = \Sigma \Sigma \phi^2_{kj} \phi^2_{ij} S_p(\omega) |H_j(\omega)|^2 \tag{7}$$

According to equation (7), reverse question of fluctuating pressure may be considered to resolve load power spectrum $\{S_{pi}\}$:

$$\{S_{pi}\}^T = \{S_{p1}, S_{p2}, \cdots, S_{pn}\} \tag{8}$$

It is necessary to make the sum of variance of weighting averages minimum:

$$S = \Sigma \omega_k (S^2_{\gamma k} - S^{*2}_{\gamma k}) \to \varepsilon \tag{9}$$

Where, $S_{\gamma k}$ is computed spectrum value, $S_{\gamma k}^*$ is measured spectrum value.

Based on (7) and (8), using least mean square error, the effective load spectrums may be certain. If load spectrums are used as input loads, the dam's dynamic response distribution may be obtained. Dynamic inverse analysis computing program is based on finite element method. A corresponding three-dimensional block element model with 8 nodes and each with three degree-of-freedom has been applied, Fig.5.

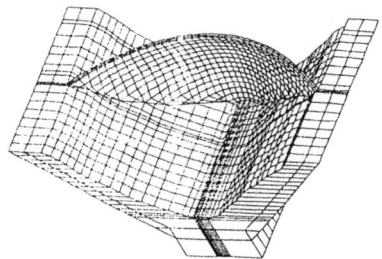

Fig.5 Finite element model

From the results of mode shape computing, the data of finite element joint coordinates, element parameters, material parameters and other information is analytically obtained, $M=0.5 \rho_0 \sqrt{hl}$, $\rho_0$ is water's density, $h$ is water's depth, $l$ is the distance from the point to water surface. In order to evaluate dam body-reservoir water-foundation-fluctuating water load this multi-input and multi-output system, effective loads are adopted to replace the real loads. This is important, because the real loads acting on dam body are complex and impossible to be determined accurately and these effective loads include the effective parts of total dynamic loads, fluctuating pressure of outlets, nappes pounding loads and wave's action. The calculated effective load power spectrums see Fig.6. Inverse analysis results present that most of the effective loads come from the loads of shallow outlets discharge flow and nappes pounding. That is to say that shallow outlets discharge flow and nappes pounding are the main foci inducing dam's vibration. According to effective loads, vibration displacement distribution of dam may be computed. Comparison of the computed values and measured results of the vibration displacement mean square root values of the dam's crest see Fig.7. Therefore, the influence of some parameters to the dam's vibration may be modified.

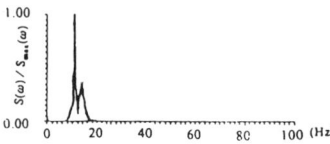

Fig.6 Effective load spectrum in the case of shallow outlets discharge flow

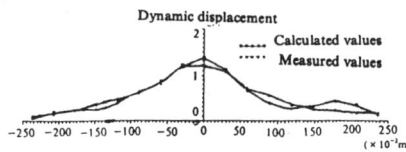

Fig.7 The displacement comparison between the testing results and computed values

## RESEARCH TWO

WU JIANG DU diversion wall is a kind of high and thin structure, which height is 10 meters and thickness is 2 meters. The wall's vibration induced by high velocity discharge flow is obvious. How to certain the response is a key problem to us. Neural network method is adopted to research the diversion wall's vibration in the case of different gate opening controlling water head. The main objective of this study is to express the action of neural network in structure and water dynamic energy.

### SETTING UP NEURAL NETWORK MODEL

Artificial neural network is consisted of many neural elements and its fundamental functions are learning and thinking. The neural element model may be expressed as Fig.8. There are two kinds of neural element, network neural element and memorial neural element. Memorial neural network is constructed by network neural elements and memorial neural elements. Based on the relationship of input and output, there are single input and single output network and multi-input and multi-output network. Considering the complexity and characteristics of interaction of fluid and structure, the single input and single output memorial neural network model of diversion wall is set up, Fig.9. Back-propagation learning method is used to train neural network. So, intelligent analysis system of diversion wall's vibration is made up.

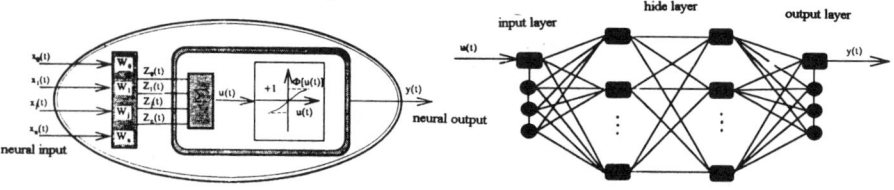

Fig.8 Neural element model

Fig.9 Neural network model

### FORECASTING THE DIVERSION WALL'S VIBRATION

The neural network modal of WU JIANG DU diversion wall includes one input layer, two hide layers and one output layer. The diversion wall's vibrations on point $1^{\#}$ of the gate's opening 2 meters,4 meters,6 meters, 8 meters, 10 meters are used as training samples. Through 2000 cycles, the total error of this neural network is 0.003. When the gate opens completely, the vibration on the point is forecasted by neural network's thinking. The comparing curves of forecasted results and measured values see Fig.10.

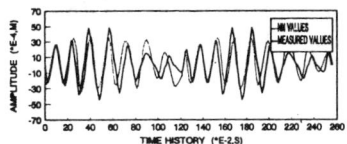
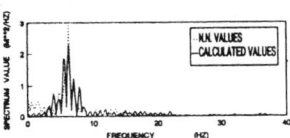

Fig.10 The comparison of vibration's time history and spectrum

## CONCLUSIONS

By the research of hydroelastic model, dynamic inverse analysis and artificial neural network, simulate models of structural vibration induced by dynamic energy of discharge flow are set up. Used to experiment results, prototype measured dada, theory analysis and numerical calculation, new methods are developed about the influence of flow dynamic energy to structures, which are more accurate to emulate the cases of prototype than the before. These also present the important value of dynamic inverse analysis and artificial neural network in the field of civil engineering.

KEYWORDS:
the structural vibration induced by flow energy, hydroelastic model, dynamic inverse analysis, artificial neural network.

## MAIN REFERENCES

1. Paklov T.A. Petrovski J.T. and Jurukovski D.V. Full-scale forced vibration studies and mathematical model formulation of arch concrete dams. Dams and earthquake. Proceeding of a conference held at the Institution of civil Engineers, London on 1-2 October 1980 275-285.
2. Cui Guangtiao. The hydroelastic model study of Er Tan arch dam's vibration induced by discharge flow. Journal of Tianjin University, China 1990(1), 1-10.
3. Cui Guangtao, Peng Xinmin and Yuan Ximin, The hydro elastic model study of Xiao Wan arch dam's vibration induced by discharge flow. Journal of Hydraulic Eng. CHES, 1996(4).
4. Zhu Befang. Inverse analysis of rock initial earth stress, Journal of Hydraulic Eng., China, 1994(10) 30-35.
5. Yuan Ximin. The simulated study on flow induced vibration of high water head and large discharge. Dissertation for Doctor's Degree, Tianjin University, China, 1996.9.

# Structural Modifications at Hydro Dams: An Opportunity for Fish Enhancement

DILIP MATHUR AND PAUL G. HEISEY
Normandeau Associates; Drumore, Pennsylvania USA

JOHN R. SKALSKI
University of Washington; Seattle, Washington USA

STEVEN G. HAYS
Mid Columbia Consulting, Inc.; East Wenatchee, Washington USA

and

MARK R. SMITH
U.S. Army Corps of Engineers; Portland, Oregon USA

## ABSTRACT

Spillways and sluiceways at hydroelectric dams were constructed strictly as conduits for transporting excess river flow or debris with little focus on potential for safe fish passage. However, the declining salmonid populations have helped emphasize a critical re-examination of spillways and sluiceways as effective fish passage routes. Consequently, spillways at some hydro dams have been modified either to take advantage of surface oriented behavioral patterns to bypass fish via installation of prototype overflow weirs or installed with flow deflectors to reduce total dissolved gas levels in the river. These spillway modifications have opened up a new set of fish passage survival issues. Controlled tag-recapture experiments at hydroelectric dams on the Columbia River show that not all spillway structural modifications are 100% fish friendly; differences in survival at unmodified spillbays between sites also occurred. Estimated survival of juvenile chinook salmon in spillway passage ranged from 95.5 to 100%. It appears that depth of the tainter gate opening, amount of gate opening, discharge volume, obstructions in flow path (e.g., dentates, end walls), excessive turbulence, presence of boulders, etc. may affect fish survival. Hydraulic modeling and detailed physical examination of spillways, in combination with fish survival information, may open new opportunities and economic impetus to incorporate appropriate structural modifications that afford safer fish passage at spillways.

## INTRODUCTION

Historically, spillways and sluiceways at hydroelectric dams were constructed strictly as conduits for transporting excess river flow or debris with little focus on their potential for safe fish passage routes. In recent times, however, these conveyances are increasingly viewed as viable fish passage routes, particularly in the overall recovery plan for the declining salmonid populations. Consequently, at hydro dams on the Columbia River Basin, spill is used as an alternative to turbine passage because of reported higher survival rates for juvenile salmonid emigrants. However, spill is expensive in terms of lost power generation and with some spillway configurations can result in potentially lethal levels of total dissolved gas in the river. Additionally, many spillways are equipped with bottom opening tainter gates and the surface oriented salmonid emigrants may not be effectively guided by these spillways. The emigrants would be required to sound approximately 20 to 60 ft to exit at such a conventional spillway.

To alleviate the dissolved gas saturation problem and to take advantage of the surface oriented behavioral patterns of salmonid emigrants, two major structural modifications of spillways have occurred at some hydro dams: installation of flow deflectors for total dissolved gas abatement, installation of top flow structures such as overflow weirs or vertical slots to improve spill effectiveness for attracting surface oriented salmonid emigrants. Flow deflectors are concrete sills installed on the downstream face of a spillway that direct the spilled flow towards the surface of the tailwater rather than allowing it to plunge to the bottom of the stilling basin. This action reduces the pressure gradient that forces atmosphere gases into the water at supersaturation levels.

To take advantage of the surface oriented behavioral pattern of salmonid emigrants, prototype structural modifications designed to create a surface attraction flow have been installed as a means to improve passage effectiveness. One prototype design uses a weir (bulkhead) upstream of the bottom spilling tainter gates to draw water from the surface rather than from the depths of the gate openings (typically $\geq 40$ ft). However, the subsequent fate of fish exiting such a route is unknown. From a practical viewpoint, improving spillway designs to enhance survival may require estimates of the direct effects embodied in spillway configurations.

The U.S. Army Corps of Engineers, Portland District, Oregon sponsored a research investigation to estimate survival and condition of hatchery-reared chinook salmon (average length 124 mm) immediately after passage through spillbays equipped with flow deflectors, without flow deflectors, with overflow weir, and "I"-slot bulkhead. HI-Z Turb'N tag-recapture (balloon tag) technique was used to quantify fish survival and injury rates/types (Heisey *et al.* 1992) at Bonneville and The Dalles dams on the mainstem Columbia River; 270 to 280 chinook salmon smolts were released through each of the tested spillbays along with accompanied control releases. The use of this technique has allowed researchers to physically examine a high proportion ($\geq 0.90$) of fish after passage and thus permitting quantification of injury and/or mortality (Heisey *et al.* 1992). The statistical treatment of data as well as procedures for tag-recapture, handling, etc. are detailed in Normandeau Associates *et al.* (1996). Survival estimates presented in this paper represent the immediate effects of passage, *i.e.*, fish which survived for 48 h regardless of its condition (Mathur *et al.* 1996).

## STUDY SITES

Bonneville Dam, built in 1938, is the first dam upriver (river mile 145) on the Columbia River. The Dalles Dam, built in 1957, is the second dam upriver (river mile 191.5). The spillway at Bonneville Dam consists of 18 bottom spilling leaf gates, each 50 ft wide. The bottom of these gates is approximately 40 ft below normal head pond elevation. A double row of concrete dentates extend across the end of the stilling basin to aid in energy dissipation. These energy dissipaters are 14 ft wide at the base and slope up to a 2 ft width at the top. They are 6 ft above the bottom sill of the spillbay and their tops were approximately 26 ft below the water surface when the study was conducted. Flow deflectors were installed in 1975 on the downstream face of 13 of the 18 spillbays for gas abatement. Water discharged through a bottom spilling leaf gate drops about 6 ft onto the flow deflector.

The configuration of The Dalles Dam is such that the spillway is perpendicular to the river flow while the powerhouse is parallel. The spillway consists of 23 bottom opening tainter gates, each 50 ft wide. The bottom of these gates is approximately 40 ft below normal forebay elevation. Water discharged through the bottom opening tainter gates plunges over an ogee approximately 50 ft to the stilling basin. A single row of 9 ft high by 10 ft wide concrete dentates are located in the stilling basin to dissipate energy. Three baffles lie downstream of each spillbay and they remain approximately 9 ft below the normal tailrace elevation. Another energy dissipation structure (end sill) lies about 45 ft downstream of the baffles. This structure is a 13 ft high continuous vertical wall and lies approximately 5 ft below the normal tailrace elevation. Two spillbays were modified by placing bulkheads upstream of tainter gates in the bulkhead slot to more effectively entrain surface oriented salmon smolts in spill. One spillbay was modified such that water is entrained through an "I"-shaped opening while another spillbay was modified so that the bottom half of the spillbay was blocked and water then passed through the top 20 ft (overflow weir).

## RESULTS

Although survival probabilities exceeded 0.95 under all tested conditions, some variations occurred (Table 1). At Bonneville Dam, estimated survival was identical (1.0) at the two spillbays. However, at The Dalles Dam survival was at least 0.038 lower at the unmodified spillbay than at the modified spillbays. These differences were not significant ($P>0.05$).

The potential confounding effects of dentates and the vertical end sill downstream of spillbays at The Dalles masked the isolation of differences in survival due to modifications at the spillbay and spill volume. Although the estimated survival probability was about 0.038 higher at "I"-slot spillbay configuration than at the unmodified spillbay (10,500 cfs spill) it was virtually identical to that at the overflow weir with a lower spill volume (4,500 cfs). However, it was suspected that a portion of the differences in survival at The Dalles could have arisen due to entrapment of some fish in the downstream dentated area or collision with energy dissipaters.

Table 1  Number of fish released (N), discharge rate tested, and estimated survival probabilities of chinook salmon in passage through modified and unmodified spillbays at Bonneville and The Dalles dams. 90% confidence intervals shown in parentheses.

|  | N | Discharge (cfs) | Survival Probability |
|---|---|---|---|
| *Bonneville Dam* | | | |
| Unmodified spillbay | 280 | 12,000 | 1.0 (0.970-1.0) |
| Spillbay with flow deflectors | 280 | 12,000 | 1.0 (0.991-1.0) |
| *The Dalles Dam* | | | |
| Unmodified spillbay | 270 | 10,500 | 0.955 (0.927-0.983) |
| "I"-slot | 271 | 10,500 | 0.993 (0.972-1.0) |
| Overflow weir | 270 | 4,500 | 0.990 (0.951-1.0) |

Although the estimated survival probabilities at Bonneville Dam were 1.0, a small proportion of fish exhibited injuries which were not lethal in all cases; the injury rate was estimated at 1.1% for the modified spillbay and 1.4% for the unmodified spillbay. At The Dalles Dam, the lowest injury rate (0.5%) was observed at the unmodified spillbay that also had lower survival (0.955); the injury rate was the highest (2.3%) at the overflow weir and intermediate (1.5%) at the "I"-slot spillbay. Primary injuries observed were hemorrhaging, protruding eyes, and bruises. None of the observed injuries proved fatal over the 48 h observation period.

## DISCUSSION

Although flow deflectors at spillbays reduce gas supersaturation, the magnitude of their effects on fish survival relative to passage through standard spillbays can vary with site configuration and hydraulics, species, and experimental protocols. No differences were noted at Bonneville Dam, survival was identical (1.0) at the two tested spillbays. In contrast, survival probability of chinook salmon smolts was 0.957 in passage through a spillbay equipped with flow deflectors and 0.996 at a standard spillbay at Wanapum Dam on the Columbia River (Normandeau Associates et al. 1996); this difference was significant ($P<0.05$). A portion of this difference could have been due to some fish contacting these structures. Long et al. (1975) reported a substantial difference in survival probability of steelhead smolts in passage between spillbays with flow deflectors (0.974) and standard spillbays (0.745) at Lower Monumental Dam on the Snake River. However, a recent study by Muir et al. (1995) at the same site reported a smaller, but opposite survival difference for chinook salmon in passage over a spillbay with flow deflectors. Survival was estimated at 0.927 over the flow deflector and 0.984 in passage over standard spillbays; the difference was not significant ($P>0.05$). Fish at this project must sound at least 50 ft to exit the bottom spilling tainter gates. It should be noted that all the studies cited above, except that by Normandeau Associates et al. (1996), utilized standard tagging methodologies (e.g., freeze brands, PIT tags) that required recapture of fish at dams many miles downstream and over several days or weeks. Both the treatment and control groups may have been exposed to other indirect sources of mortality downriver and thus the resulting estimates most likely contained direct as well as indirect passage effects. Our estimates represent only the direct effects of passage and do not account for indirect passage effects which are generally difficult to isolate (Mathur et al. 1996).

Because of the relatively recent development of the overflow or "I"-slot weir concept to attract surface oriented salmonid emigrants, survival data are limited to provide a broad perspective on the results from The Dalles Dam. A recent study at Wanapum Dam on the Columbia River reported survival of chinook salmon smolts in passage through a spillbay modified with an overflow weir configuration (Normandeau Associates et al. 1996). The overflow weir consisted of stacked bulkheads positioned upstream of the tainter gate. The lower 40 ft of the spillbay was blocked and only surface flow was allowed to pass through up to four rectangular bulkhead doors (6 ft x 20 ft deep) positioned near the surface. The estimated survival probability was 0.920 for juvenile chinook at a discharge of 2,000 cfs and 0.969 at 4,000 cfs. The difference was significant ($P<0.05$). Both estimates are lower than recorded (0.99) for The Dalles overflow weir. Differences in the degree of turbulence in the area between each overflow weir and the tainter gate suggest one reason for the disparity in survival rates at The Dalles and Wanapum dams. Although at The Dalles backroll turbulence occurred at both types of top flow openings, the water appeared substantially less "violent" than downstream of the "doors" at the Wanapum overflow weir. Overflow weirs entrain surface dwelling emigrants, but the fish must still sound to greater depth to pass under a tainter gate. Whether the area under the tainter gates poses additional hazards is unknown. However, fish transported through the two leaf gate spillbays tested at Bonneville Dam also had to sound to exit under the gates, but the resulting survival estimates were identical (1.0) with minimal injury. This suggests other factors, such as depth of gate opening, amount of gate opening, stilling basin structures (e.g., dentates, end walls), and approach angle of fish may affect fish survival.

Injury rates and types may vary with site-specific exit routes. At The Dalles Dam, 2.3% of chinook salmon smolts exhibited injuries in passage through the overflow weir. Most were primarily hemorrhaging eyes and body injuries. At the "I" slot configured spillbay, the injury rate was 1.5%; at the unmodified spillbay the injury rate was 0.5%. At Bonneville Dam, the injury rates of chinook salmon smolts at the spillbays were similar (1.1% at flow deflector spillbay and 1.5% at spillbay without flow deflector). In contrast, injury rates at Wanapum Dam ranged from 3.0% to 5.8% depending upon the passage route and spill volume; the overflow weir exposed fish exhibited higher (5.8%) injury rates at 4,000 cfs than at 2,000 cfs (4.4%), but the difference was not statistically significant ($P>0.10$; Normandeau Associates et al. 1996). The difference in injury rates between the standard spillbay (5.5%) and spillbay equipped with flow deflector (3.0%) was about 2.5%, also non-significant ($P>0.10$). The spill volume tested was about 4,300 cfs.

The principal causal mechanisms for injury/mortality to fishes transported via spillways have been attributed to shear forces, turbulence, rapid deceleration after attaining a high terminal velocity, impact against the base of the spillbay, scraping against the concrete face of the spillbay, and rapid pressure change (Bell and DeLacy 1972). Although experiments have not been conducted to quantify the relative importance of these factors in affecting fish condition/mortality at most spillways, reported injury types sustained have included eye damage, embolism, hemorrhaging, and abrasions. The scrape, cut, and bruise wounds could have been caused by the fish physically contacting structural components at the spillbay including the frame of the overflow weir, tainter gate, and flow deflectors. Fish attracted by surface currents to the overflow weir must sound up to 60 ft to exit; thus, these fish may be exposed to pressure changes. However, the potential effects of pressure changes may be site-specific. None of the injured fish showed symptoms of pressure-related effects at The Dalles and Bonneville dams, but some fish in the Wanapum Dam study did exhibit such symptoms (Normandeau Associates *et al.* 1996). An insufficient number of spillbay and overflow weir investigations have been conducted to point out specific features and configurations of fish exit routes that may be benign. In their review of spillway fish passage, Bell and DeLacy (1972) noted that water velocities exceeding 50-60 ft/s may inflict injury or mortality. This critical velocity threshold is undoubtedly exceeded at some projects. However, it appears that other factors such as excessive turbulence, high exit velocities, obstructions in the flow path, end sill, plunge depth, or boulders in the plunge pool may also affect survival and injury rates.

## LITERATURE CITED

Bell, M. C. and A. C. DeLacy. 1972. A compendium on the survival of fish passing through spillways and conduits. Report prepared for U.S. Army Corps of Engineers, Portland, OR.

Heisey, P. G., D. Mathur, and T. Rineer. 1992. A reliable tag-recapture technique for estimating turbine passage survival: application to young-of-the-year American shad (*Alosa sapidissima*). Can. Jour. Fish. Aquat. Sci. 49:1826-1834.

Long, C. W., F. J. Ossianer, T. R. Ruehle, and G. M. Matthews. 1975. Survival of coho salmon fingerlings passing through operating turbines with and without perforated bulkheads and of steelhead trout fingerlings passing spillways with and without a flow deflector. Final Progress Report to the U.S. Army Corps of Engineers., National Marine Fisheries Service, Seattle, WA.

Mathur, D., P. G. Heisey, E. T. Euston, J. R. Skalski, and S. Hays. 1996. Turbine passage survival of yearling fall chinook salmon (*Oncorhynchus tshawytscha*) at a Columbia River dam. Can. Jour. Fish. Aquat. Sci. 53:542-549.

Normandeau Associates, J. R. Skalski, and Mid Columbia Consulting. 1996. Fish survival in passage through the spillway and sluiceway at Wanapum Dam on the Columbia River, Washington. Report prepared for Grant County Public Utility District No. 2, Ephrata, WA.

Muir, W. D., et al. (five co-authors). 1995. Relative survival of juvenile chinook salmon after passage through spillbays and the tailrace at Lower Monumental Dam, 1994. Report to U.S. Army Corps of Engineers. Contract E86940101, Walla Walla, WA.

# METHODOLOGY OF ECOLOGICAL FISHWAY DESIGN

Tetsuro TSUJIMOTO[1] and Noriko HORIKAWA[1]
[1] Dept. of Civil Eng., Kanazawa Univ., Kanazawa, Japan

ABSTRACT

Previously the fishways were designed to assist the migration of only commercial fish that were interrupted by transverse structures. However, recently ecological system preservation has become one of functions of rivers to be considered in river management, and targets of fishways are no longer commercial fish but various species of fish. In this paper, the objective and evaluation criteria of fishway based on a new concept of river management are discussed, and the methodology of "ecological fishway" design is proposed. The "Instream Flow Incremental Methodology", which has derived for habitat evaluation, is refined and applied to the fishway which is not a comfortable habitat but an emergency path.

INTRODUCTION

The new concept of river management is to keep or improve the multiple functions of rivers, which are not only flood protection, water resources utilization (water supply and hydropower), navigation, but also amenity for human life and eco-system preservation. Particularly the new concept for the latter is "sustainable habitat for diversity of species." Fishways were previously designed only for commercial or fishery purposes, but now based on the new concept of rivers, it must play a role in keeping a biopath or an ecological corridor for fish. However, the target species to save must be many. Then, the methodology of fishway design must be changed.

In this paper, "ecological fishway" is defined in a close-to-nature river improvement, and target fish and designed discharge are discussed. Then, IFIM (Instream Flow Incremental Methodology) proposed for habitat evaluation is applied to the fishway design. Fish habitat implies the space for fish to continue their sustainable lives, which must support various stages of life and generation changes. Fishway is a part of habitat, and it provides only emergency passage. The procedure of applying IFIM to fishway evaluation is proposed, where IFIM is refined by adding a quantitative evaluation of time series. Finally, a project of a test fishway is introduced with sample calculation of the proposed method.

POTENTIAL NATURE OF RIVERS AND FISH ECOLOGY

To precede the river management based on the new concept, close-to-nature river improvement projects are attempted. Then, what is a good model for close-to-nature rivers. Most parts of rivers have been already artificially degenerated. One has to postulate "the potential nature" of a river, as follow: One has to divide a river into several "segments" by the bed slope, bed material, the morphological units to characterize each segment, vegetation bordering along a stream, and maybe by organisms. For example, the segment of fluvial-fan reach is characterized by the slope 1/100~1/400, gravel-beds, willow bush, alternative bars, and so on. In other words, many elements are potentially interrelated for each segment. Photo 1 demonstrates a difference between two segments: one is a gravel-bed stream with rapid flow, bordered by willow (*Salix gilgia*); and the other is a sand-bed stream with mild slope, bordered by reeds (*Phragmites japonica*).

Photo 1 Difference of landscape due to segment

Fish ecology is also related to the segment. Fig.1 shows the observed species along some river (the Asano River in Ishikawa Pref.; Sano & Yamamoto, 1996). Habitat of some species is limited in some segment. Others use all the segments along a river. Some of them migrate upward for spawning, and others migrate downward for spawning. While, juvenile fish migrate to opposite direction.

The construction of many structures must affect such an interrelating system, and they often degenerate the potential interrelating system. The pattern to degenerate the system depends on the location of the structure as well as its size and type. Some structures divide one segment into two, and some are located just on the boundary of two segments (see Fig.1). The most important point in river restoration is to gate such degeneration to keep a potential interrelation for each segment.

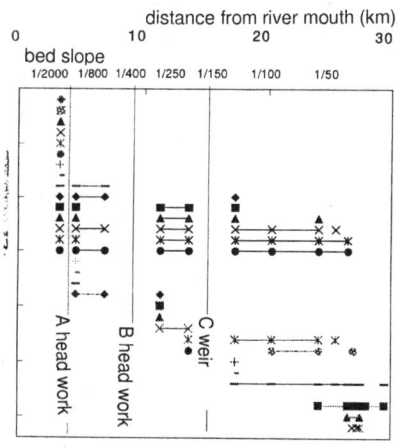

Fig.1 Change of observed species of fish

## TARGET FISH AND DESIGN DISCHARGE

As shown in Fig.1, there are in general a few dozens of species in rivers. From the view point of potential nature of rivers, the target fish for designing a fishway attached to the structure must be considered as follows: the species of which habitat is interrupted by that structure. They are (1) the species of which migration to the keep generation sustainability is interrupted by the structure; (2) the species that may have a chance to be flushed to the downstream segment during flood and cannot return upstream to their inherent segment; and (3) the species of which inherent segment is divided into limited parts by the structure.

There are still many species. They might be classified into several categories. At first, they are classified by their habitat segment as follows: (A) the species of which habitat is limited to some segment (A-n; $n$=segment number); (B) the species to migrate upstream (B-1) or downstream for spawning (B-2); and (C) the species to live anywhere. They are further classified by the migration season. Next, they are classified by their vertical space to stay: (i) near the bottom (benthotic fish); and (ii) swimming far from the bottom. Moreover, the shape and the size may be the factors for classification: several typical categories of silhouette and several ranks of size. The combinations of these criteria for classification may provide several categories of fish. Then, one has to choose species to represent these categories. They are target fish. In this stage, commercially important fish, endangered species, species popular to the local citizens may be chosen as target fish.

In hydraulically designing fishways, the target fish are often characterized by proper velocity and depth respectively. For example, the flow in a fishway is designed slower than the speed of fish on emergency and deeper (or more spacious) enough than the size of fish. If the species are limited to a spindle-type silhouette, the speed of fish $U_e$ in emergency is proportional to the size of fish as well as the normal speed $U_N$ of fish (on ordinary swimming). From the hydraulic view point, the fishway design must satisfy the condition of the proper velocity and depth. In general, the slope of fishway is very steep and the discharge to release the fishway is limited. Thus, the key of design is how to reduce the velocity and/or how to increase the depth under the given limited discharge. Previously, the fishways were designed empirically. But now, the target species are only a few, and the design guidelines provide to make a fishway suitable to those species for designed discharge (the discharge for lower stage is often adopted). We have to focus now various categories of fish. It means that a fishway has to be designed to have variety of velocity and depth in its cross section. Furthermore, the design discharge must be considered with the season important for target fish that are different from each other. The time series of discharge has better to be taken into account.

## OUTLINE OF IFIM

IFIM (Insrteam Flow Incremental Methodology) was a methodology for habitat evaluation (Platts *et al.*, 1983; Nestler *et al.*, 1989). The outline of the main part of this methodology is explained as follows (somehow modified from the original): The reach for investigation is divided in to mesh of the area $\Delta A_k$ ($k$=1~$K$) and the flow properties of each mesh are estimated, for example by using two-dimensional flow analysis. The flow properties in this scheme are the depth $h_k$ and the depth-averaged velocity, $U_k$. If necessary, more detailed structure can be taken into account (Tsujimoto, 1997). In addition to these hydraulic parameters, the the information of substratum (cover) and so on are investigated for each mesh. The value of each parameter for each mesh is to be represented here by $\xi_{jk}$ ($j$=1~$J$).

On the other hand, one has to know the relation between the values of HSI (Habitat Suitability Index) $\eta$ and $\xi$ for each parameter ($j$). HSI-values ranges (0, 1) for each $j$. The integrated evaluation for each mesh is often expressed as $\Pi(\eta_j)$ (one among some proposals). And the evaluation for the reach investigated is expressed as follows, and it is represented by *WUA* and termed "weighted usable area" for habitat.

$$WUA = \sum_{k=1}^{K} \left[ \left( \prod_{j=1}^{J} \eta_{jk} \right) \Delta A_k \right] \qquad (1)$$

$WUA$ changes with the discharge $Q$. According to the time series of $Q(t)$, the time series of $WUA$ is calculated, and the result is used for habitat evaluation.

## APPLICATION OF IFIM TO FISHWAY EVALUATION

Fishway is a part of habitat, and IFIM is modified to apply it to fishway evaluation. In case of fishway, the cross-sectionally distribution of hydraulic parameters must be parameter to evaluate the fishway. The flow can be treated as longitudinally averaged one in case of stream-type fishway; while the control section such as a slot of vertical slot-type fishway is focused. When only one species is a target, the flow condition in a cross section should be designed for the appropriate condition. However, now target species are various, and the variety of hydraulic parameters must be required in a cross section. In the fishway, the parameters to subject to the suitability as a fishway must be the depth and the velocity. For the evaluation of the suitability as fishway, the suitability curves for fishway versus the velocity and the depth, $\eta_U(U)$ and $\eta_h(h)$, must be known in advance. By consulting the knowledge on fish behavior (experience and/or observation), they should be formulated.

When the spatial distributions of hydraulic parameters such as the depth-averaged velocity and the depth, $U$ and $h$, are estimated and suitability-index curves for respective parameters ($\eta_U(U)$ and $\eta_h(h)$) are known, the suitability index value for fishway, $\Xi_n$, can be defined as follows:

$$\Xi_i = \sum_{k=1}^{K} \left[ \left( \prod_{j=1}^{J} \eta_{ijk} \right) \Delta B_k \right] / \left\{ \sum_{k=1}^{K} \Delta B_k \right\} \tag{2}$$

where $\Delta B_k$=width of a series of transversely cut strip in a cross section; and the subscript $i$ means the species of fish. $\Xi_i$ is here normalized, and ranges in (0,1). Since $U$ and $h$ vary with the discharge, $\Xi_i$ value is a function of the discharge in the fishway, $\Xi_i(Q)$.

Applying the relation between $\Xi_i$ and $Q$ to the time series of daily discharge $Q(t)$, the time series of the suitability index value, $\Xi_i(t)$, is obtained. By the way, the significance of the fishway changes seasonally for each species. This significance factor ranging (0, 1) is also defined and represented by $\gamma_i(t)$, which is different according to the species. For example, the season of migration for spawning must be ranked highly. The suitability of the fishway through a year ($T$) is evaluated as follows:

$$\Phi_i = \int_0^T \left[ \Xi_i(t) \cdot \gamma_i(t) \right] dt / \int_0^T \gamma_i(t) dt \tag{3}$$

The integrated suitability index might be calculated by considering the weight for each species $\lambda_i$, as follows:

$$\Psi \equiv \sum_{k=1}^{K} \left[ \lambda_i \Phi_i \right] \tag{4}$$

$\lambda_i$ is more highly ranked for the higher necessity of the fishway for sustainable life of each species, but it may be modified by the importance of the species from the ecological vie point (endangered species) or the fishery purpose. The higher value of Y, the fishway is highly evaluated. However, from the view point of species, the fishway that minimizes the differences of $\Phi_i$ due to the species, from the vie point of "diversity of species." Though the final decision making needs more discussion, the quantitative comparisons of several alternatives for fishway designs become possible by the present methodology.

## PROCEDURES AND EXAMPLE

As the simplest case of fishway that has variety of depth and velocity transversely, a fishway with a triangular cross section is taken as an example, as shown in Fig.2. If one employs some hydraulic model, the relation among $H$ (maximum depth in the cross section), $Q$ and $U$ (depth-averaged), is obtained as shown in Fig.3(A) and (B). For simplicity, the subscript $i$ to distinguish the species is abbreviated. Applying the suitability index curves (SIC) $\eta_U$ and $\eta_h$ as shown in Fig.3(C) and (D), $\Xi(Q)$ is calculated as shown in Fig.4(E). Next, one has to collect data or designed time series of discharge $Q(t)$ as shown in Fig.4(F), and the time series of the fishway suitability $\Xi(t)$ is obtained which is depicted by solid curve in Fig.4(G)). The weighting function for the seasonal necessity of the fish-way is illustrated in Fig.4 (H), and the value of $\bar{\Xi}$ is shown by a dotted curve in Fig.4(G).

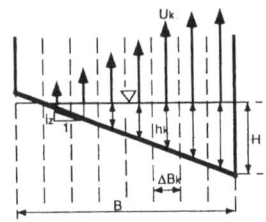

Fig.2 Simplified fishway with triangular cross section

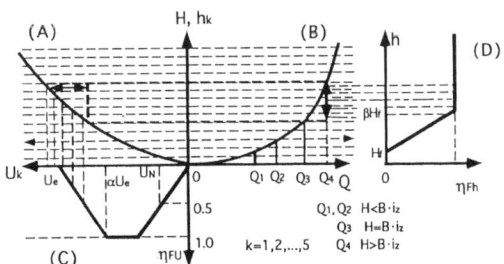

Fig.3 Relations among Q, h U and suitability index

Photo 2 shows the experiment in a test fishway with a triangular cross section (3.0m wide,

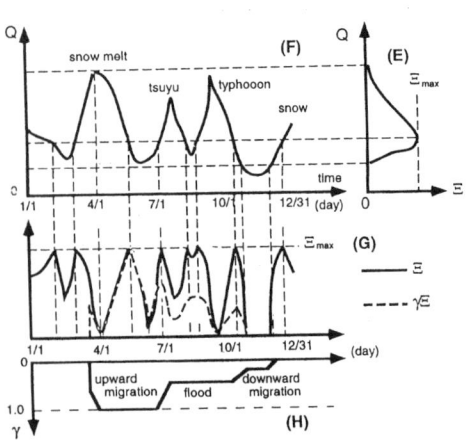

Fig.4 Time series of Q and $\Xi$

Photo 2 Test fishway in Uomi River

15m long, and the slopes of the bed are 1/15 in the longitudinal and transverse direction in the Uomi river, a tributary of the Kuzuryu river (Fukui Pref.). In order to keep the sufficient depth, boulders (the diameter is around 25cm) are fixed on the slope. Fig.5 is a calculated result of the transverse distributions of depth and velocity for each lane of $\Delta B_k$ ($\Delta B$=0.125m), where the flow was calculated based on the balance of the longitudinal component of the gravity of the water and the drag by boulders considering the volume and projected area of the submerged parts of boulders within the strip of the width of $2\Delta B$ including each lane of $\Delta B_k$. In this case, there are two lanes without arranging the boulders to provide the variety of flow condition. Then, the fishway suitability index for some species (spindle-type, size is 20cm), $\Xi$, was calculated. $\Xi(Q)$ is shown in Fig.6, and $\Xi(t)$ is overlaid in Fig.6 by a dotted curve. The experiment in this site has started. The spatial distributions of the depth and the depth-averaged velocity are measured by a point gauge and an electromagnetic current meter. On the other hand, fish behavior is recorded on a map of the fishway.

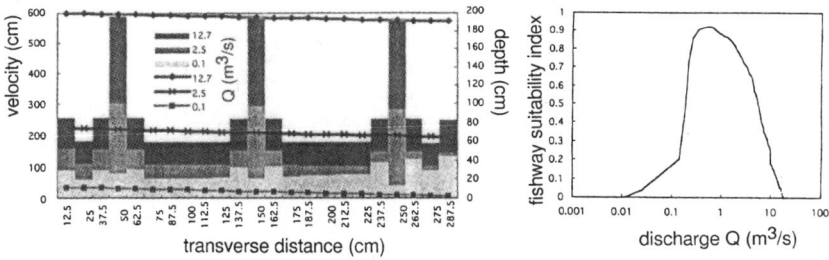

Fig.5 Transverse distributions of velocity and depth

Fig.6 Calculated example of $\Xi(Q)$ fortest fihway

## CONCLUSION

In this paper, the fishway design under the new concept of river management is discussed, where the view point of ecological fishway or biopath is necessary, and fishway has a cross section to induce variety of hydraulic conditions for many species, though the target species are carefully chosen. IFIM is applied to evaluation of fishway, and its procedure is explained.

## REFERENCE

1. Nestler, J.M., R.T. Milhaus and J.B. Layzer : Instream habitat modeling techniques, *Alternative in Regulated River Management*, edited by J.A. Gore & G.E. Petts, CDC Press, 1989.
2. Platts, W.S., W.F. Megahan, G.W. Minshall : *Methods for Evaluating Stream, Riparian, and Biotic Conditions*. General Technical Report, INT-138, USDA, 1983.
3. Sano, O. and K. Yamamoto : *Freshwater Fish in Ishikawa Prefecture*, Ishikawa Pref., 1997.
4. Tsujimoto, T. : Secondary flow and fish habitat, *Proc. Conf. Managements of Landscapes Distributed by Channel Incision*, Oxford, Mississippi, USA, 1997 (in printing).
5. Tsujimoto, T. and T. Kitamura : Numerical calculation of flow in vertical slot and its application, *Proc. Int. Sym. Fishways '95*, Gifu, Japan, 1995.
6. Tsujimoto, T. and Y. Shimizu : Flow structure of stream type fishway, *Proc. 2nd Int. Sym. Habitat Hydraulics*, Quebec, Canada, pp.843-854, 1996.

# The Virtual Fish Concept: Numerical Prediction Of Fish Passage Through Hydraulic Powerplants

FOTIS SOTIROPOULOS AND YIANNIS VENTIKOS
School of Civil and Environmental Engineering
Georgia Institute of Technology
Atlanta, GA 30332-0355, USA

## ABSTRACT

This paper presents a brief overview of the latest developments in the area of numerical simulation of fish passage through hydropower installations. An advanced Computational Fluid Dynamics (CFD) method has been developed for simulating complex hydroturbine flows. The output of this method provides the virtual flow environment through which the motion of "virtual" fish is investigated. This is accomplished by using a novel fish passage model for tracking the trajectories of three-dimensional fish-like bodies through the precomputed flow field. The model yields detailed quantitative information about forces acting locally on the fish body, potential for scraping and descaling, dizziness effects, and other sources which could cause injury and/or mortality. The potential of the proposed approach is demonstrated by applying it to track fish trajectories through TVA's Norris Dam draft tube.

## INTRODUCTION

Increasing environmental awareness regarding the impact of hydropower installations on aquatic ecosystems has recently stimulated intense research, among the entire hydropower community, aimed at: i) identifying fish-unfriendly elements in existing turbine designs; and ii) developing new design concepts for the next generation of environmentally improved high-efficiency hydroturbines. The work described herein has been carried out in the context of these efforts.

The major difficulty for evaluating the fish-friendliness of a given hydroturbine design is the lack of knowledge regarding the fate of fish as they swim through the powerplant. Most available work in the area of fish mortality (Normandeau Associates, 1995) has focused on documenting the frequency and type of injuries suffered by various fish species but little, if anything, is known today as to the actual causes of these injuries--in fact most such knowledge is based on after-the-fact-speculation rather than in depth understanding of the mechanisms that govern the interaction of fish with the local flow. This state of affairs should be attributed to insufficient understanding of the physics of hydroturbine flows along with lack of fundamental knowledge of fish behavior in such highly complex flow environments. Furthermore, the site-specific nature of the problem in conjunction with the complexity of the geometries involved make laboratory studies expensive, difficult to set up and carry out, and of limited use for drawing conclusions that are applicable beyond a specific hydropower project. Advanced computational fluid dynamics methods, which have been recently applied with a great deal of success to several real-life hydroturbine flows, offer the only viable alternative today for developing a general numerical model for predicting fish passage through hydropower installations.

Existing fish passage models, are based on oversimplifying assumptions and can only give gross estimates about the probability for fish to impact the turbine blades. These models have built into them little or no detailed flow physics and can not be used to predict mortality which is the combination of numerous poorly understood phenomena. The objective of this work is to present a general framework for developing an advanced fish passage model which takes into account all flow complexities encountered in real-life hydroturbine geometries. The input for such a model consists of the geometrical and physical characteristics of the fish species under consideration and a complete three-dimensional solution of the flowfield through the subsystem in which fish trajectories need to be predicted. Assuming that the presence of fish does not affect the local flow characteristics, the model can predict three-dimensional fish trajectories, thus directly yielding information about mechanical strike and potential for abrasion as well as the distributions of the various forces and moments exerted by the flow on the fish body. Such information may be used to estimate fish mortality and evaluate the fish-survivability potential of a given design. Obviously, crucial prerequisite for the success of the proposed approach is the development of CFD flow solvers capable of accurate quantitative predictions of real-life flow phenomena in hydroturbines. The accuracy of the present fish-passage model--no matter how refined, sophisticated and well-calibrated it may be--would ultimately depend on the accuracy of its input: the calculated three-dimensional flowfield through various subsystems of a hydropower installation.

In what follows, we first discuss recent advancements in the area of numerical modeling of hydroturbine flows with emphasis on the critical role of high-accuracy numerics and advanced turbulence modeling for quantitatively accurate predictions. Subsequently, we present a general outline of the proposed fish-passage model and underscore areas that require further research. The model is applied to track the trajectories of virtual fish through the TVA Norris Dam draft-tube. The results underscore the potential of the proposed approach and the need for in parallel advancements in both flow and virtual fish modeling should a reliable fish-passage model is to be developed.

## ADVANCEMENTS IN FLOW MODELING

The numerical method employed herein is based on the work of Lin and Sotiropoulos, 1997) who developed an efficient time-marching procedure for solving the three-dimensional Reynolds-averaged Navier-Stokes (RANS) equations, in conjunction with two-equation, near-wall, turbulence closures, in generalized curvilinear coordinates. Pressure-velocity coupling is achieved using the artificial compressibility approach. The governing equations are discretized on a non-staggered computational mesh using high resolution finite-volume discretization schemes (Lin and Sotiropoulos, 1997). The discrete mean flow and turbulence closure equations are integrated in time using a four-stage explicit Runge-Kutta algorithm enhanced with local time-stepping, implicit residual smoothing, and multigrid acceleration (Lin and Sotiropoulos, 1997).

Several improvements were required in order to make the method of Lin and Sotiropoulos (1997) applicable to complex hydroturbine geometries. These include, among others, the ability to handle multiple connected domains, use of variable residual smoothing coefficients, and implementation of Total Variation Diminishing schemes (see Sotiropoulos and Ventikos (1997) for more details). This latter development was found critical for accurate high Reynolds number turbulent flow simulations for geometries involving stagnation points (such as the piers of a draft tube). None of the high-resolution non-monotone schemes tested by Lin and Sotiropoulos (1997) were robust enough to handle draft-tube flows. In fact stable simulations could be carried out only when first-order accurate flux-difference splitting upwind was implemented for discretizing the convective terms. The resulting solutions, however, are contaminated due to excessive numerical viscosity. The dramatic effect of spatial accuracy in predictions of draft-tube flows is demonstrated in Fig. 1 which compares solutions obtained using the first-order upwind scheme with those obtained using the up-to-second order accurate symmetric TVD scheme of Yee and Harten (1987). Both calculations were carried out on the same mesh using the two-layer k-ε model. The particle traces in both figures have been released from exactly the same points located just upstream of the right pier.

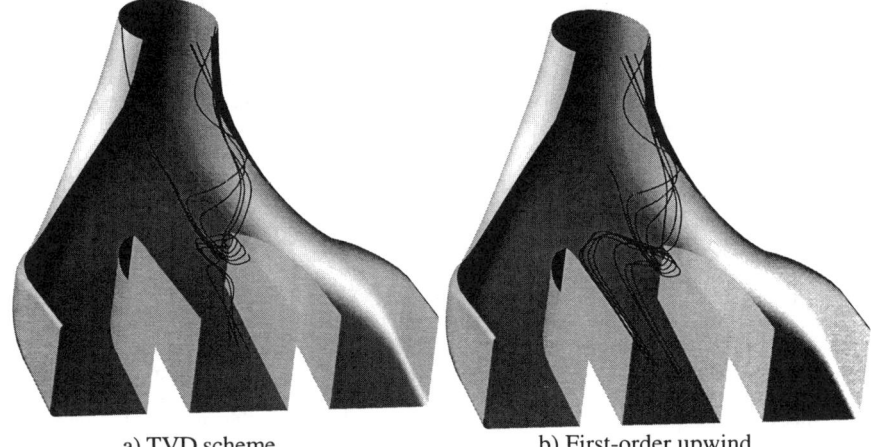

a) TVD scheme     b) First-order upwind

Figure 1. Calculated particle paths for the TVA Norris Draft Tube

The overall accuracy of the present method is demonstrated in Fig. 2 which compares computed axial mean velocity profiles with measurements (Hopping, 1992) at several locations inside the three bays of the Norris draft-tube. These results were obtained on a mesh consisting of $1.2 \times 10^6$ grid nodes--the finest mesh to be used so far in hydroturbine flow simulations--using the low-Reynolds number k-ω model of Wilcox. This model was selected after careful evaluation of various near-wall turbulence modeling alternative (Sotiropoulos and Ventikos, 1996) which demonstrated the superiority of k-ω based models in predictions of three-dimensional shear flows. As seen in Fig. 2, the calculations capture most experimental trends with reasonable accuracy. To the best of our knowledge this is the first time that CFD predictions are successfully compared with mean velocity measurements for a complex draft-tube geometry (see Ventikos et al (1996) for details).

## THE VIRTUAL FISH MODEL

The proposed fish-passage model is based on the assumption that a fish swimming through a complex, three-dimensional, turbulent flow field--which has been obtained via a separate calculation--can be approximated as a body of a simplified, yet fish-like, geometry (see Fig. 3) moving through, but not affecting by its motion, the aforementioned pre-computed flow field. The motion of such a fish-like body can be described in terms of a sequence of translations along and rotations around of each one of the three Cartesian axes. Thus, a total of six differential equations (for the Cartesian components of the linear and angular acceleration vectors) are necessary for describing the entire spectrum of possible motions. The proposed model assumes that the following forces, comprising the source terms of the differential equations to be integrated numerically, are acting on the fish-body as it moves through the flow: 1) inertial forces, which account for the fish weight and for added-mass-type corrections for the acceleration; 2) lift and drag forces due to viscous flow; 3) pressure forces caused by ambient pressure gradients in the fluid; 4) empirical corrections for deviation from steady flow; 5) fish "free-will" forces which account for swimming and overall response of fish to a given flow environment.

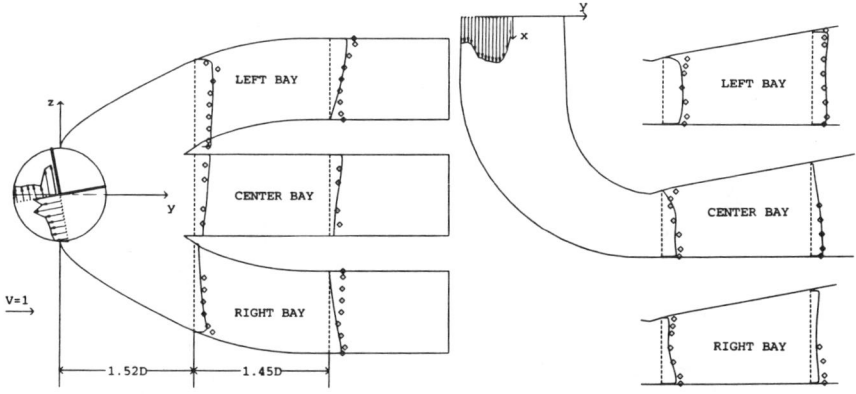

Figure 2. Comparisons of measured and computed velocity profiles

Due to space considerations, the specifics regarding the implementation of the various forces in the equations of motion are omitted from this paper--a comprehensive description will be given in a future publication (Ventikos and Sotiropoulos, in preparation). Here it suffices to say that items 1 to 4, from the above list, have already been implemented--albeit to varying degree of completeness. Perhaps, the most significant modeling challenge is to account for the forces due to fish "free-will" (item 5 in the above list). The magnitude and direction of these forces depend on the manner in which various fish species react to stimuli from the surrounding flow environment.

Figure 3. Fish and fish-like body

Feedback from fish biologists is necessary here and detailed controlled environment experiments with real fish are due. It should be emphasized that no attempt has been made so far to implement such forces. That is, the current version of the model accounts only for the motion of "dead" fish-like bodies. Another important aspect of the model which is currently under development is its ability to allow for fish to strike a wall and bounce back into the flow. This is a critical capability as it would facilitate modeling of scrape and potential of descaling.

The potential of the model is demonstrated in Fig. 4 which depicts calculated locations of a fish-like body, released at the entrance of the TVA Norris Dam draft tube, at four instants in time--the fish-body surface has been discretized using 11x11 grid nodes. As seen, this particular fish passes very near the right pier of the draft-tube and exits from the right bay. The enlarged plots of the fish-body geometry, shown in Figs. 4a to d, are included in order to clarify the local orientation of the fish body as well as demonstrate the ability of the proposed model to yield the distribution of various flow-induced forces on the fish body. For demonstration purposes, we have included in Fig. 4 pressure and shear magnitude (defined approximately as the difference of velocity magnitude between two points on the fish body) distributions--these have been computed by interpolating the known flow quantities on the surface of the fish body.

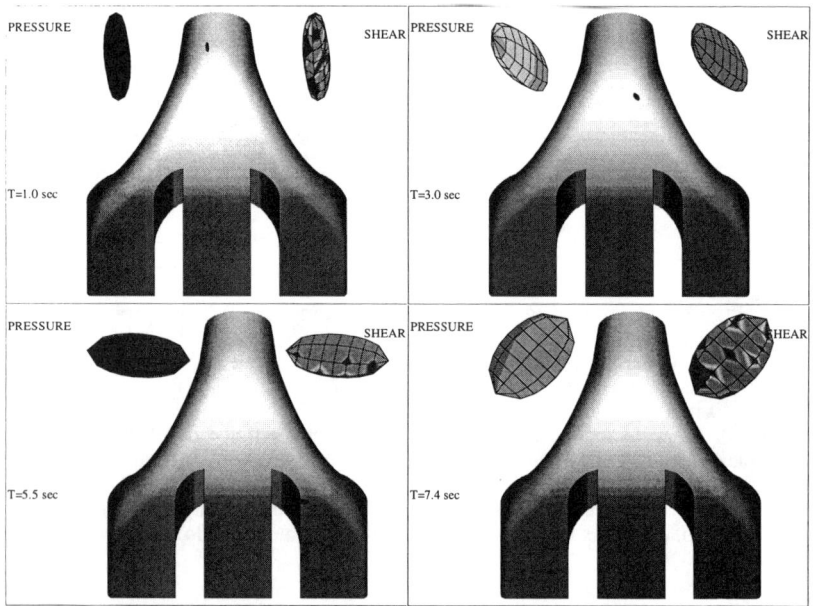

Figure 4. Locations of fish-like body along a computed trajectory

These and other interpolated flow quantities may be used to estimate, among others, pressure variations on crucial fish body parts (head, bladder, etc.), direction and magnitude of shear stresses on fish skin, moments that tend to bend and/or rotate the fish body, dizziness effects due to intense spinning, potential for cavitation-related damage, intensity of turbulence fluctuations in the vicinity of the fish, etc. With input from fish biologists this information could be translated to estimates of injury and/or mortality rates. Finally it should be noted that although only draft-tube geometries were considered herein, the proposed model is formulated in a general manner so that it can be readily applied to any hydroturbine component for which a three-dimensional CFD solution is available. In fact, the model can be employed to predict fish trajectories through the entire hydropower installation: from the forebay to the tailrace.

## SUMMARY AND CONCLUSIONS

A general framework was presented for developing a numerical model capable of predicting the trajectories of virtual fish through arbitrarily complex hydroturbine geometries. Although at its preliminary stage of development, the proposed approach is a promising alternative for developing a practical numerical tool for assessing the fish friendliness of existing and proposed turbine designs. Future work will focus on: i) refining the accuracy of the existing flow solvers; ii) developing a fish bounce-back model in order to predict abrasion and potential for descaling; and iii) implementing fish free-will related forces in the existing version of the model. The two latter objectives will require close collaboration with fish biologists and input from carefully designed controlled environment experiments with real fish. Such experiments are also needed in order to validate the proposed model.

## ACKNOWLEDGMENTS

The initial phase of this work was funded by the U.S. Department of Energy, under the Advanced Hydroturbine Project. The development of the virtual fish model is supported by Voith Hydro Inc.. The authors are most grateful to Richard K. Fisher, Jr., of Voith Hydro, Inc., and Patrick March and Paul Hopping, of TVA, for their advice and support. Calculations were carried out on the Cray C90 supercomputer of the San Diego Supercomputer Center.

## REFERENCES

Hopping P. (1992), "Draft Tube Measurements Of Water Velocity And Air Concentration In The 1:11.71 Scale Model Of The Hydroturbines For Norris Dam", Report No. WR28-2-2-116, TVA, Engineering Laboratory

Lin, F., and Sotiropoulos, F. (1997), "Strongly-Coupled Multigrid Method For 3-D Incompressible Flows Using Near-Wall Turbulence Closures," to appear in the *ASME Journal of Fluids Engineering*.

Normandeau Associates (1995), "Fish Survivability in Passage Through Propeller Type Turbines," Report prepared for Voith Hydro, Inc. (Normandeau Associates, 1921 River Rd, P.O. Box 10, Drumore, PA 17518).

Sotiropoulos, F., and Ventikos, Y. (1997), "A High Resolution Numerical Method for Predicting Complex Draft-Tube Flows Using Near-Wall Turbulence Models," to be submitted.

Sotiropoulos, F., and Ventikos, Y. (1996) "Assessment Of Some Non-Linear Two-Equation Turbulence Models For Flows Through Curved Ducts And Pipes," Flow Modeling and Turbulence Measurements VI, pp. 331-338, Balkema, Rotterdam.

Ventikos, Y., and Sotiropoulos, F. (1997) "Three-Dimensional Numerical Model for Simulating Fish Motion in Hydraulic Turbines," to be submitted.

Ventikos, Y., Sotiropoulos, F., and Patel, V. C. "Prediction of Turbulent Flow through a Hydroturbine Draft-Tubes Using a Near-Wall Turbulence Closure," Proc. of XVII IAHR Symp. on Hydraulic Machinery and Cavitation (Cabrera, Espert, and Martinez, Eds.), vol. I, pp. 140-149.

Yee, H. C., and Harten, A. (1987), "Implicit TVD-Schemes for Hyperbolic Conservation Laws in Curvilinear Coordinates," *AIAA J*. 25, 266.

# THE REGULATED RIVER AND STURGEON SPAWNING MIGRATION

E.N. DOLGOPOLOVA
Water Problems Institute, Russian Academy of Sciences, Moscow, Russia

The effect of the regulated flow of the River Volga on living conditions of the sturgeon stock are considered. We study cross-sectional and vertical prelarvae distributions in the river. A method to estimate the prelarvae density along the river is suggested. This method together with mathematical modelling of the operating of Volga cascade reservoires enables one to find the most appropriate volumes and hydrographs of spring flushes.

INTRODUCTION

Fish stock of Caspian basin plays a considerable role in economy of Russia because more then 80% of the world catch of sturgeon originates here. Preservation of fish stock depends on efficiency of successful natural breeding and feeding migration, this efficiency being dependent on the quantity and quality of the River Volga runoff. Poor water quality in the Lower Volga has an averse effect on the fish stock. Dangerous condition of ecosystem of the Caspian Sea can be illustrated by the following: in spite of abundance of several years (in 1986-1990 the average annual runoff was 275 km$^3$, whereas normally it is 250 km$^3$) biological condition of fish, primarily sturgeon, is considered by specialists to be risk-prone [7]. The presence of chlororganic pesticides, heavy metal salts and other chemical substances, whose content exceeds many times the maximum permissible concentrations, promotes the development of chronic toxicosis in fish [7].

The River Volga runoff has been altered by regulation of the flow. Cascade of hydroelectric power stations radically changed hydrological regime of the river, which negatively influenced sturgeon stock [5,7]. The negative effects of large dams on migrating salmon were considered by J.H. Mundie [2]. They mostly are the same for the anadromous sturgeon activity in the Lower Volga (such as obstruction of upstream passage of adult fish and changes in discharge below dams). However, the greatest Volgograd dam on the Lower Volga is not equipped with fishways. As a consequence, the reach of the river, where spawning grounds were located, was twice shortened. Before the construction of the Volgograd dam in 1958 sturgeon migrated up to Saratov (900 km from Astrakhan) and could spawn along all this reach. Construction of Volgograd hydroelectric power station had bounded the spawning reach to the part of the river from Astrakhan to Volgograd (about 400 km). In this reach there are 16 natural spawning grounds with total area of 425 ha and 7 artificial spawning grounds with total area of 65.9 ha (Table 1) [5]. The young of sturgeon is well adapted for survival in sea water but fry (earlier stage) perishes in the sea. This problem becomes more acute in low water years, when the salinity of the Northern Caspian increases (Fig.1).

Table 1.
The area of spawning grounds in the Lower Volga

| Distance from the dam, km | Number of spawning grounds | Area of spawning grounds | |
|---|---|---|---|
| | | River-bed, ha | Spring-flooded, ha |
| 2 - 47 | 8 | 114.2 | 38.8 |
| 50 - 135 | 4 | 163.7 | 45.0 |
| 150 - 400 | 11 | 89.2 | 40.0 |

The regulation of the River Volga flow led to some reduction of fresh water runoff and re-distribution of it within a year. Prior to runoff regulation of the River Volga, the volume of spring flood in the Lower Volga was about 60% of the total runoff volume and the average value was equal to 150 km$^3$, the duration of flood beeing approximately equal to 80 days. Such spring floods created sufficient and long enough inundation of spawning grounds. At present, accumulation of spring flood water in reservoirs (up to 60 km$^3$) for futher utilization during summer fall and winter dry periods, practically excludes natural volumes, heights and durations of spring flood in the Lower Volga. The average volume of spring flood decreased to 90 km$^3$, and its duration became about 50 days [7].

After construction of the Volgograd dam (1960) special water flushes, simulating spring flood, were made into the tailrace canal of the dam to water the spawning grounds in spring. The volumes of these flushes, their hydrograph and times are defined on the basis of forecast of spring flood flow and concrete economic situation of a certain year. Moreover, the limits on fishing were introduced [5] to create ecological safety for sturgeon stock of Volga-Caspian basin. At the beginning of 60-th fishing of sturgeon in the sea was prohibited. The periods of fishing in river were also limited: the sturgeon fishing is forbidden in spring from ice melt to May 15, and between 1-31 of August. These periods provide the passage of sturgeon for spawning in spring and migration of sturgeon to overwinter below the dam. The portion of sturgeon stock, which overwinters below the dam is also stressed by sharply altered water level and discharge, which has a negative influence on their spawning ability. Therefore, one of the possibilities to improve the ecological conditions at the Lower Volga is to change the regime of reservoirs operation and to make it nearer to the natural cycle.

This problem is discussed in [7], where a mathematical model for calculation of the optimum volumes and hydrographs of spring flushes was suggested. This model includes the data on the runoff regulation of all 11 big reservoirs on the

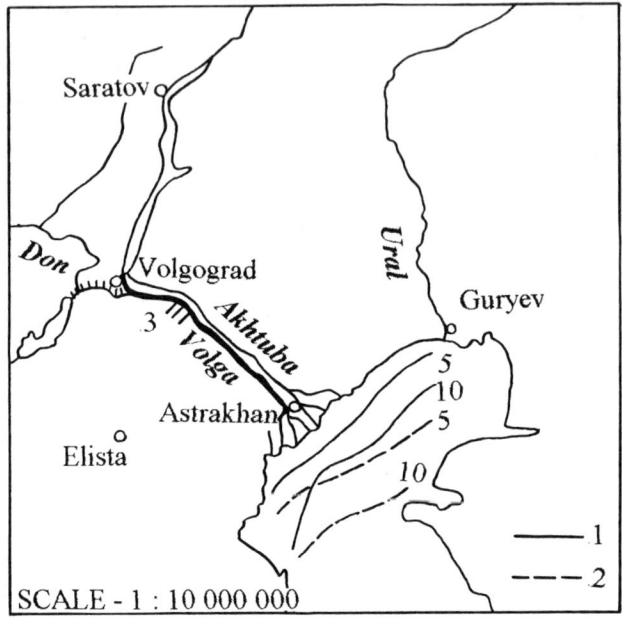

Fig.1 Map of the Lower Volga: 1, 2 -location of isohalines ($^\circ/_{\circ\circ}$) in the low flow year 1977 (year runoff - 187 km$^3$), and in the high water year 1985 (year runoff - 289 km$^3$); 3 - measurements cross-sections.

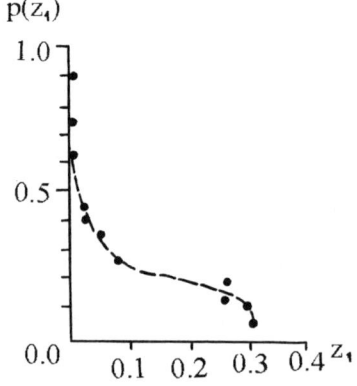

Fig.2 Average vertical distribution of prelarvae on the axis of the river.

River Volga. To obtain the connection between the regime of spring flushes and the amount of sturgeon prelarvae in different cross-sections, the method of calculation of prelarvae distribution in the river was developed.

## CONCENTRATION OF PRELARVAE OF STURGEON IN RIVER

To choose the most suitable regime of flushes for fish reproduction it is necessary to analyze the volumes and times of flushes together with the intensity of spawning of sturgeon. Here we develop a method of evaluation of this intensity with the help of estimation of downstream migration of prelarvae. The downstream migration was examined using the data for concentration of fish-eggs in the Lower Volga. The concentration of prelarvae in the river can be calculated from the results of the control catches annually performed by Central Scientific-Research Institute of Lake and River Fisheries. A conical net with a 0.8-m-diameter opening was used. If the field of the concentration of prelarvae with x, y, z coordinates is considered stationary (in statistical respect), then the amount n(x,y,z) of prelarvae in the net is equal to

$$n(x,y,z) = S \, T \, v(x,y,z) \, c(x,y,z) \tag{1}$$

where v(x,y,z) is the time-averaged velocity of the prelarvae at a given point of the flow, S is the area of the opening of the net, T is the time of its exposure, c is the concentration of prelarvae, x, y, z are streamwise, lateral and vertical coordinates respectively (z=0 is at the bottom). Prelarvae can be regarded as dynamically passive particles, and, according to the results of investigations, for the density $\rho = 1.020\text{-}1.035 \text{g/cm}^3$ they have a settling velocity of about 4 cm/s and can rise with velocity of about 6.5 cm/s. The height and frequency of rises of prelarvae can be considered as random variables. The vertical distribution of the concentration of prelarvae is determined by the height and frequency of rises and also by turbulent diffusion [1]. Because of turbulent mixing of the layers the prelarvae will lag behind the flow:

$$\mathbf{v} = \mathbf{u} - \frac{|u'w'|}{\sigma_w^2} \varpi \tag{2}$$

Here $\varpi$ is the settling velocity of a prelarva, u is local velocity of the flow, u', w', $\sigma_u, \sigma_w$ are fluctuations and standard deviations of the longitudinal and vertical components of flow velocity. Using semiempirical relationships for $\sigma_u(z_1), \sigma_w(z_1)$ [3] and $\omega$ = 4cm/s one finds that the difference between u and v is only about 2.4 cm/s. This enables one to consider these velocities as equal. The vertical distribution of the current velocity of the river was assumed to obey the power law

$$u = (1 + n)<u> z_1^n, \tag{3}$$

where $<u>$ is the depth-averaged velocity of the flow, $z_1 = z/h(y)$, $h(y)$ is the depth of the river. According to measurements in the Lower Volga the dimensionless parameter n is within the interval from 1/9 to 1/6 [3] and with accuracy sufficient for practical purposes can be assumed to be equal to 1/7 [4].

Using the data for concentration of prelarvae in some cross-section we calculated the distribution of the probability density $p(z_1)$ of their distance from the bottom which is given in Fig.2. As can be seen, the bulk of the prelarvae moves in the bottom layer with the thickness of about 0.5 h, and the probability of their rise to the surface of the flow is small. Using the vertical distribution of pralarvae and Eq.(3), one can calculate vertically averaged velocity of prelarvae:

$$<v> = \int_0^1 p(z_1)u(z_1)dz_1 \qquad (4)$$

The result of calculation is $<v> = 0.82 <u>$.

The distributions of prelarvae over the width of the river were analyzed for three cross-sections [1]. The analysis of statistical characteristics of these distributions showed that they are almost similar for different cross-sections and can be approximated by the normal law.

## DISTRIBUTION OF STURGEON PRELARVAE IN THE RIVER

At comparatevely small distances from the source in a quasi-uniform two-dimensional flow, the probability density of the distribution of prelarvae can be described by normal two-dimensional law. Taking the average distance from the spawning ground $\bar{x} = <u>t$, and coefficients of longitudinal and transverse diffusion $\varepsilon_x$ and $\varepsilon_y$ equal to $\varepsilon_x = 2.5\varepsilon_y$, $\varepsilon_y = c_y<u>h$ ($c_y = 0.024 \pm 0.005$ [3]), one obtains:

$$p(x,y) = \frac{1}{\sqrt{20\pi c_y h \bar{x}}} \exp\left[-\frac{(x-\bar{x})^2}{10 c_y h \bar{x}} - \frac{y^2}{4 c_y h \bar{x}}\right] \qquad (5)$$

Here y is the distance from the axis of the stream. In the case of large distances from the source x>>B, where B is the river width, it is natural to consider the one-dimensional problem of the distribution of the pollutant and to determine directly the average concentration of prelarvae c(x) over the river cross-section [6]:

$$c = \frac{M}{A\sqrt{2\pi D_L t}} \exp\left[-\frac{(x-Ut)^2}{2D_L t}\right] \qquad (6)$$

Here A is the cross-sectional area, U is the velocity averaged over the cross-section, $D_L$ is the constant dispersion coefficient [6], M - is the total amount of prelarvae in the initial cross-section. There are several empirical estimates of $D_L$, for example, in [6]:

$$D_L = 11.7\,h\,u\,, \qquad (7)$$

Using this relation and Eq.(6), one can obtain an estimate of the change of the average cross-sectional concentration of prelarvae along the river as a function of the amount of prelarvae migrating downstream from the site of spawning ground.

## CONCLUSIONS

Analysis of the operation of the River Volga's cascade of hydroelectric power stations together with sturgeon stock activity has shown that fish reproduction mostly depends on the regime of artificial flushes. This regime must be carefully adapted for fish necessity. The most sensible variant is to take into account this problem before constructing the dam. Analyzing the movement of sturgeon prelarvae, we have found that the bulk of larvae moves in the bottom layer of thickness 0.5 h, the velocity of prelarvae being less than that of the flow. The distribution of prelarvae over the width of the river can be approximately considered as the normal one. The maximum of the distribution lies in the middle of the river. Characteristics of this distribution from site to site change insignificantly, which enables us to consider the prelarvae movement as stable. In our approach, using the data for initial concentration of prelarvae in the cross-section at spawning ground, we calculated the amount of prelarvae downstream from the spawning ground. The analysis of sturgeon prelarvae migration enables one to choose correctly the places for water intake with the least damage for fish reproduction.

## REFERENCES

1. Debolsky V.K., E.N. Dolgopolova et.al.(1986) Calculation of the distribution of sturgeon prelarvae in a turbulent flow. J. Water Resources, vol.13, No 1.
2. Mundie J.H.(1979) The regulated stream and salmon managment. Proc. of the First Inter. Symp. on Regulated Streams, Erie, Rensylvania, April 18-20.
3. Orlov A.S. E.N. Dolgopolova & V.K. Debolsky, (1985) Some empirical relationships of river turbulence. J. Water Resources, No 6.
4. Schlichting H.(1968) Boundary layer theory. McGraw Hill Book Co. New York.
5. Shekhodanov K.L.(1989) The influence of regulation of fishery on the natural reproduction of Russian sturgeon in the River Volga. Doctoral Thesis of Central Scientific-Research Institute of Sturgeon Economy, Astrakhan.
6. Thackston E.L., P.A. Krenkel, (1967) Longitudinal mixing in natural streams. J. Sanitary Eng. Div. Proc.of ASCE, v.93, No SA5.
7. Voropaev G.V., et.al.(1994) Investigation of a possibility of ecologically favorable spring flushes in the Volga Lower Course. J. Water Resources, vol.21, No 1.

# USE OF VOLGA RIVER FLOW FOR POWER GENERATION AND FISHERY PURPOSES. MITIGATION OF CONFLICTS OF INTEREST

### ALEXANDER ASSARIN

Share-holding company "Institute Hydroproject"
Moscow, Russia

ABSTRACT.
The paper describes the natural water flow regime of the Volga River and its changes after construction of a chain of hydrodevelopments. Information is given on the trends in catches of sturgeon and fine-mesh net fish from 1913 until 1990. The paper reviews the requirements set by ichthyologists for the regime of flows and water stages to be maintained upstream and downstream of the dams and in the lower unregulated reaches of the Volga. It outlines approaches and corrective measures to mitigate conflicts between water users.

The Volga River basin is the most important economic region of Russia. Here about of 1/3 of Russian population lives and it produces 1/3 of agricultural and industrial products. The Volga-Caspian basin accounts for 60% of inland fish and 90% of sturgeon catches. Hydropower plants (HPP) of the coordinated Volga-Kama hydroelectric system, main parameters of which are given in Table 1, generate 1/5 of the electric energy produced by the Russian HPPs.

Table 1

| Dam | Natural average annual run off, cu.km | Full supply level, m | Minimum operating winter time level, m | Active storage, cu.km | Installed capacity of HPP, MW |
|---|---|---|---|---|---|
| Rybinsk | 34 | 102.0 | 97.1 | 16.7 | 360 |
| Gorkey | 60 | 84.0 | 81.0 | 3.9 | 520 |
| Cheboksary | 113 | 63.0 | 63.0 | 0 | 1404 |
| Kama | 60 | 108.5 | 100.0 | 9.8 | 504 |
| Votkinsk | 55 | 89.0 | 85.0 | 3.7 | 1000 |
| Lower Kama | 91 | 62.0 | 62.0 | 0 | 1248 |
| Kuibyshev | 240 | 53.0 | 45.5 | 33.9 | 2300 |
| Saratov | 246 | 28.0 | 27.0 | 1.8 | 1350 |
| Volgograd | 250 | 15.0 | 12.0 | 8.3 | 2530 |
| Total | | | | 78.1 | 11216 |

The navigation of this river basin accounts for about 70% of total carriage on the inland waterways and for the bulk of timber rafting.
Regulation of the Volga flow by the storage reservoirs, primarily in the interest of power generation and navigation (reduction of the spring flood flow by 1/3 and increasing of discharges in the winter time as much as two times) as well as water withdrawals and evaporative losses, have lead to dramatic changes in the water flow regime of the Volga River in its irregulated lower reaches and delta. Apart from these changes, operation of the hydroelectric plants involved in the daily load follow-up functions leads to considerable tail water level fluctuations. The daily range of level fluctuations downstream of some HPP reaches 2.5-3 m.
Long since, the Volga-Caspian region has been known for high fish yields. In 1913, over 28,000 t of sturgeon, over 40,000 t of fine-mesh net fish and 140,000 t of Caspian roach were caught. After the revolution of 1917, sturgeon catches dropped down to 13,000 t in 1930 and to 1,000 t during the period of 1940-1960. After completion of the Volga chain of dam projects and construction of the hatcheries, average sturgeon catches varied from 10-15,000 t . Within the last year, the downward trend has continued in sturgeon catches by the state enterprises, while fish catches by small private fishing co-operatives and by poachers are difficult to assess.
The annual fish catches in the Volga and Kama reservoirs by state fishing establishments and private fishing co-operatives exceed 1,500 t. It is assumed that a similar quantity of fish is caught by anglers. This quantity is as much as several times that which used to be caught in the Volga and its tributaries before constructions of the dams. As a result of implementation of the hatcheries augmenting fish population, fishing yield of the reservoirs may exeed 5,000 t per year. At the same time, creation of the chain of dams has blocked access for high value migratory fish to a considerable part of spawning grounds and has lead to degradation of the environment for spawning and feeding of anadromous and diadromous fish species in the lower reaches and delta of the Volga River which accounts for 10% of the sturgeon spawning area.
The most effective pattern of water resourcers management in the interest of each water user is as a rule controversial. The most vivid example of such conflicts is the conflict between the optimum water management in the interest of power generation and fisheries. The interconnected power system of Russia is interested in using the water resources accumulated in the storage reservoirs primarily in the winter time, i.e. during the maximum energy demands. While fisheries interest insist on preserving the volume and duration of spring flood flows by imposing the maximum possible restrictions winter drawdowns which lead to the fish entrapping by ice cover subsiding at water drawdowns and dail fluctuation in tailwater level. Large drawdowns lead to the

entrapment of fish under the ice cover. The last requirement of ichthyologists has been implemented by prohibiting the normal weekly load-following functions and partially restricting the daily load following function during spring floods (spawning period of many high value fish species) at the Volgograd hydro power plant.

Natural fish reproduction in the Volga-Caspian region and crop raising in the Volga-Akhtube flood plain and Volga delta require their spring flooding. For this purpose, so-called fishery management water releases are made annually from the Volgograd storage reservoir. The maximum discharges of these releases (25-27,000 cu.m/sec) far exceed discharge capacity of the turbines of the three HPP on the Volga lower reaches - Kuibyshev, Saratov, Volgograd, totalling 13-16,000 cu.m/sec.

In some years, the waste flow releases result in the loss of $2\text{-}3 * 10^9$ Kwh in generation at the above mentioned HPP.

To meet the required dates and specified flow volumes of the spring time releases, either filling of the Kuibyshev and Volgograd reservoirs has to be retarded or these reservoirs should be partially drawn down during the flood period. At low inflow or with mistakes in the forecast of spring inflows (which is based on the snow accumulated in the Volga River watershed by March) it may happen that the reservoir would not be filled to the full retention level. The aggregate active storage capacity of the reservoirs in the Volga-Kama cascade of dams totals at present 80 cu.km The sum of storage volumes to be filled up annually amounts to 55-65 cu.km. Failure to fill the reservoirs to the full retention level leads to limitations on navigation releases in autumn time and to reduction in energy production by HPP in the most critical winter time.

Operation of all Russian large storage reservoirs (including those on the Volga River) is regulated by a special document called the "Basic Rules for Water Resources Management". These Basic Rules are worked out by Design Agencies (normally Designers of the respective dam project) and shall be subject to clearing by the National Economy Sectors-water users and these rules are approved by the Ministry of Water Management for 10 years.

Inspite of formal approval of the Basic Rules, water users express from time to time their dissatisfaction with operation of the reservoirs. Fish breeders suggest that (1) the reservoir water level be raised gradually during the spring floods to the elevations ensuring flooding of the spawning grounds to a height of 0.5-1.5 m, (2) receding of the water levels be prevented throughout the entire period of egg incubation, (3) water stages in the reservoirs be surcharged by 0.3-0.4 m above the full retention level during spring time every two years and (4) the reservoirs be drawn down to a minimum depth of 1.0-1.5 m after the flood flows in late July to early August to keep the zones of drawndown from repeated flooding.

It should be stressed that regular surcharging of the reservoir levels above the full retention level at the flood recession is not questioned by the power generation industry. Such a surcharge has positive impact also on reservoir fish husbandry. After water recession, the previously submerged shore zone grows up with grass, which during flooding of the next year serves as substrate for egg spawning.

Surcharging of the headwater level and subsequent releases of the stored water for power generation may in some measure meet the fisheries requirement to draw down the reservoir level after spawning. The spring surcharge of the reservoirs is also beneficial for navigation because such a surcharge increases the dependability of water releases from the reservoir in the interest of navigation and **guarantees** adequate navigation depths in the upstream pool. But surcharging of the upstream water levels in combination with low releases from the reservoirs, and consequently low tailwater levels, may negatively affect the stability of the dams. Partial raising of the full retention levels provided for by the design to accommodate extremely high flood flows may inflict damage to the property located on the shore of reservoirs as well as to arable land by its temporary flooding and ground water table rise and intensify erosion of the shore line.

Some requirements of the fish management organizations are controversial from the view point of interest of fisheries in the entire Volga-Caspian basin. It is practically impossible to simultaneous maintain constant water levels in lower Volga reservoirs during floods, when lake and river fishs spawn, and the specified volume and pattern of water releases in the lower Volga reaches to create the required depths in the spawning areas of migrating and anadromous fish. To evaluate the impact of flow regulation by the storage reservoirs on the natural water levels in the spawning area of the Volga flood plain and delta, we have made computations for routing of the regulated flows and plotted hydrographs of flood flows which would have taken place near Volgograd within the last decades with the absence of dams and reservoirs.

The computations were based on numerical integration of Saint-Venan, one-dimensional equations which solution consists in joint consideration of the flow regime at the begining and end of the design time interval equal to 24 hours. They were conducted for 145 stretches of the Volga River and its major tributaries for which cross-sections of the river channel and flood plain were available, as well as stage-discharge relation curves $Q = f(Z)$ obtained from hydrological records.

The computations have revealed that without dams on the Volga and Kama Rivers in 1961-1990 the maximum flood discharges at Volgograd would have been by 5-15,000 cu.m/sec higher than with a chain of dams, while the duration of flood discharges exceeding 25,000 cu.m/sec would have been sufficient for spawning of the anadromous fish and sturgeon in most years. Based on these data and biohydrological model, fish yield losses (decrease in catches of sevruga, beluga, sturgeon, perch, bream, sazan and Caspian roach) were determined for each year of the above-mentioned period and recommendations were given for spring flow releases in the interest of fisheries.

According to these recomendations, the hydrograph of the spring releases depending on the available water resourses (the forcasted spring flood volume and water accumulated in the Volga cascade reservoirs as on the 1st of april) has to possess the following characteristics.

Table 2

| Characteristic | Value of the flood volume, $km^3$ | | | |
|---|---|---|---|---|
| | 150 | 120 | 100 | 70 |
| Maximum water discharge, cu.m/sec | 30 | 28 | 27 | 25 |
| Duration of maximum discharge, days | 7 | 5 | 5 | 5 |
| Duration of the period with Q>24,000 cu.m/sec | 39 | 12 | 8 | 5 |
| Total spring releases duration, days | 78 | 65 | 60 | 42 |

This regime of releases proposed by the ichtyologists provokes the objections of the power engineering and water transport specialists.

The search for the concent of different water users is still in progress.

# Evaluating the Hydraulic Performance of Fish Passage Structures

## BRENT W. MEFFORD AND JOE P. KUBITSCHEK
U.S. Bureau of Reclamation, Denver Colorado, USA

### ABSTRACT
Many existing fish passage structures are less effective than originally hoped. Often the hydraulics of the fish passes deviate from that desired when all the complexities of operation are brought to bare. As part of Reclamation's fish passage research program, field evaluations, laboratory models, and computational fluid dynamics (CFD) numerical methods are being used to identify and improve the hydraulics of fish passes. A recent case study of a non-salmonid fish passage structure is presented that typifies many of the lessons learned and the value of the various hydraulic evaluation techniques.

### INTRODUCTION
There are hundreds of fish passage structures in the United States. Most are found on coastal rivers that contain anadromous fish, being mainly salmonid species. There are far fewer fish passage structures designed to provide passage for non-salmonids, although the number is growing. Good fish passage performance is derived from a well integrated marriage of engineering and fishery science. A common problem is often the realization long after construction of less than expected fish passage efficiency. To confront this problem, U.S. Bureau of Reclamation engineers and fishery biologists working under its fish passage research program are actively evaluating the hydraulic performance of many existing ladders. This paper summarizes recent work on a non-salmonid fish passage facility with a long history of poor performance.

### MARBLE BLUFF DAM FISHWAY
Marble Bluff Dam and fish passage facilities are located on the Truckee River, approximately 3 miles upstream of Pyramid Lake, near Reno, Nevada. The dam and fish passage facilities were built by Reclamation and turned over to the U.S. Fish and Wildlife Service (FWS) to operate in the early 1970's (6). The 40 ft high dam was built to stabilize an incising river and provide fish passage. The dam provides water to the Pyramid Lake fishway, which starts at Marble Bluff Dam and extends northerly about 3 miles to Pyramid Lake. The lake supports several fish species including endangered cui-ui lake suckers and threatened Lahontan cutthroat trout. Both species migrate up the Truckee River to spawn during high spring flows.

In addition to the fishway channel, a mechanical hoist-type fish lift was designed to provide passage for fish migrating up the river to the toe of the dam. Neither of the fish passage facilities has functioned fully as intended. Cui-ui were found to be incapable of passing the fishway ladders as designed. The ladder design was based on then typical salmonid style ladders and available biological studies (Koch 1972, 1973, 1976; Ringo and Sonnevil 1977) of the cui-ui physical and behavioral attributes.

Adult cui-ui congregate in March and April near the mouth of the river prior to migration. Spawning runs begin in April or May, depending upon timing of runoff, river access, and water temperature. Most spawners migrate less than 6 miles upstream spending only a few days in the river. Spawning runs may continue for 4 to 8 weeks, but most fish migrate during a 1- to 2-week period. During the peak of the run tens of thousands of cui-ui enter the passage facilities in a 24 hour period.

## FISHWAY LADDER HYDRAULICS

The fishway channel as constructed by Reclamation in 1976 had four half-Ice Harbor style ladders along its length, figure 1. The ladders were designed based on "then" standard ladder criteria derived mostly from experience with adult salmon. The ladders are on a 10 percent slope with combination weir/orifice baffles spaced every 10 ft of run. The head drop over each baffle is 1 ft as originally designed. The baffle weirs act as short crested weirs and as such have a higher discharge coefficient than a broad crested weir. However, an estimate of the peak weir velocity can be determined by the brink depth velocity for a broad crested weir. Brink depth is the depth of flow at the downstream brink of a broad crested weir. For a non-contracted rectangular weir, the brink depth equals about 0.71 of critical depth (Boss, 1989). For 1 ft of head the brink depth velocity over the weir is about 5.8 ft/s.

**Figure 1** - A Marble Bluff Dam fishway ladder. As shown, the ladders are modified with intermediate temporary baffles.

The estimated velocity in the orifice is about 5.6 ft/s assuming zero approach velocity head and a loss coefficient of 0.7 (suppressed orifice). However, the as-built baffle design has the bottom passage orifices aligned from baffle to baffle down the ladders. For this baffle geometry, approach velocity head must be considered as the flow moves down the ladder. The bottom orifice openings are 1.5 ft square which is equivalent to a circular orifice of 1.69 ft diameter. Decay of the centerline velocity downstream of each orifice can be estimated as given by Albertson (1);

$$\frac{V_{max}}{v_o} \frac{x}{D_o} = 6.2 \qquad (1)$$

where, $V_{max}$ = core velocity a distance x downstream of orifice, ft/s
$v_o$ = initial orifice core velocity, ft/s
$D_o$ = orifice diameter, ft

Equation 1 applies to $x/D_o$ ratios greater than 6.2. For $x/D_o$ values less than 6.2, the core velocity of the jet at the jet centerline has not started to decay. The ratio of $x/D_o$ between baffles for the original design is about 6. Therefore, the centerline velocity exiting the upstream baffle orifice is not decreased prior to encountering the next downstream baffle orifice. Accounting for velocity head increases the estimated orifice velocity to about 7.0 ft/s.

## AS-BUILT LADDER PERFORMANCE

Operation of the ladder quickly revealed many problems that yielded very poor cui-ui passage. Flow velocity and turbulence were too strong for the majority of cui-ui despite biological swimming test data to the contrary. Field data revealed many older age classes and egg laden female cui-ui were unable to negotiate the ladder. This resulted in gender selective passage and fish stacking at the entrance to the ladder. Large fish kills occurred as weaker swimmers packed together crushing fish and depleting dissolved oxygen levels.

Study of the fishway performance in conjunction with modifications to the baffle design has, over many years, identified important design features for this non-salmonid ladder. Fishway ladders must be designed to pass the weaker swimmers efficiently. For cui-ui a maximum passage velocity of 3.0 - 4.0 ft/s is now targeted. The ladder baffle designs which help fish maintain orientation to the primary flow direction, are needed. In a water column, fish usually orient themselves and move into the dominant current. This behavior is known as rheoreaction and includes both orientational and locomotory components. Fish orientation into the current is primarily accomplished through the use of optical and tactile sensory systems. Cui-ui like many warm water species migrate during high river flows when river turbidity is normally high. High turbidity impedes the fish's use of optical orientation, thus leaving them to rely primarily on tactile senses. Threshold velocity and spacial distribution of flow are important for sustaining upstream orientation and movement of fish. Pavlov, (1989) in a study of different species found that fish required a threshold velocity of between 1-30 cm/s to maintain a orientation reaction to the current. Ideally, the flow field within a ladder pool should maintain the orientation of the fish to the thalweg flow with velocities ranging between the orientation threshold and the fish's cruising speed. Fish orient to large scale eddy flow in similar manner. Thus, fishway designs that support large areas of recirculating flows bestow greater opportunity for disorientation and passage delay. Spacially, this implies the area within a pool supporting strong eddies should be minimized in the fishway design.

## FIELD MODIFICATIONS TO THE FISHWAY LADDERS

Since initial operation the FWS has experimented to improve passage performance of the existing ladders. Modifications have been made in response to a better understanding of fish physical and behavioral needs and facility limitations. Intermediate wood baffle weirs were installed in an alternating pattern to reduce the head drop per baffle to 0.5 ft. Adding intermediate baffles reduced pool lengths to 5 ft. In addition, bottom orifices were closed to reduce pool turbulence levels and flow through the ladder was reduce to maintain weir flow depth of about 0.5 ft. Excess flow was diverted to an existing bypass system and reintroduced at the bottom of each ladder. These modifications reduced peak passage velocity over the weirs to about 4.1 ft/s. Blocking the orifices meant fish had to pass over the weirs which also raised concern as cui-ui typically move upstream along the channel bottom. Following the modifications cui-ui passage up the ladders significantly improved. However, the ladders remained barriers to many fish. Fish surveys again revealed gender and age class selective fish passage was occurring.

As a further test, in one ladder, side to side alternating bottom orifices 1.5 ft high by 0.8 ft wide were installed. The narrower opening and alternating pattern was designed to reduce flow and increase dissipation of jet velocity within the small pool size available. As a rule of thumb, fish pass orifices and slots are commonly held to 1.5 ft or larger to minimize the chance of fish contact with the structure. For some species small orifices or slots have been shown to cause significant descaling. However, for cui-ui and other heavily scaled species descaling is not a common form of injury. For these fish, narrower slots or orifices can offer hydraulic advantages, especially when working with an existing ladder where pool size is fixed. Reducing the orifice width in the Marble Bluff fishway baffles greatly reduced pool turbulence. A field hydraulic survey of the modified baffles indicated maximum orifice velocities were reduced to between 4.1 to 4.8 ft/s. In the two years since the modification no delay of cui-ui within the ladder has been noted, however no biological evaluation of cui-ui use of the bottom orifices has been conducted. The ladder modifications have improved the passage efficiency to where many thousands of cui-ui pass up the fishway each year.

These modifications are only considered temporary solutions. Cui-ui still experience substantial delays during the spawning run and many fail to negotiate the fishway ladders. Reclamation, FWS and Pyramid Lake Fisheries are jointly working to replace the fish passage structures with better facilities. One component of this effort is developing ladder designs better suited to the cui-ui.

## NEW FISH LADDER BAFFLE DESIGNS

In 1997 Reclamation is replacing one of the fishway ladders. The new ladder will be 8 ft wide, 6 ft deep, with baffles placed every 8 ft of length. To improve flow conditions, the ladder gradient is being reduced to 3 percent and new baffles are being designed. The ladder gradient was selected to target a maximum ladder passage velocity of less than 4 ft/s. Baffles were designed to enhance cui-ui passage by maximizing the downstream flow field area within the pools; providing passage opportunity at any elevation within the water column; maximizing flow within the ladder while meeting the passage velocity criteria; and avoiding strong vertical turbulence. The baffle design was developed using both a physical model and a computational fluid dynamics model (CFD). A dual-vertical-slot chevron shaped baffle was chosen. Figure 2 shows a typical CFD generated velocity field for the ladder for a chevron shaped baffle with a nose offset angle of 15 degrees. Both physical and numerical models were used to investigate the flow field as a function of slot width, chevron angle, and slot wing walls. In the prototype, slot widths will be adjustable between 0.75 ft and 1.0 ft. Short wing walls were required on the upstream inside edge of each slot to achieve the desired flow pattern. The wing walls turn the flow from each slot to the center of the downstream pool creating a large centered downstream flow within each pool. Without wing walls flow through the slot attaches to the downstream pool walls creating a large eddy in the center of the pool. The new baffle design will be field tested in spring of 1998.

## SUMMARY

The design of fish passes for non-salmonids often require different behavioral and hydraulic conditions be considered than typical for salmonid ladders. For many species there is little swimming endurance or behavioral data available to help designers. Our field experience with sucker species has also revealed that fish swimming strength and behavior can vary widely as a function of age, gender, and spawning condition. In many cases the best guidance for passage design comes from a close look at the a streams natural conditions.

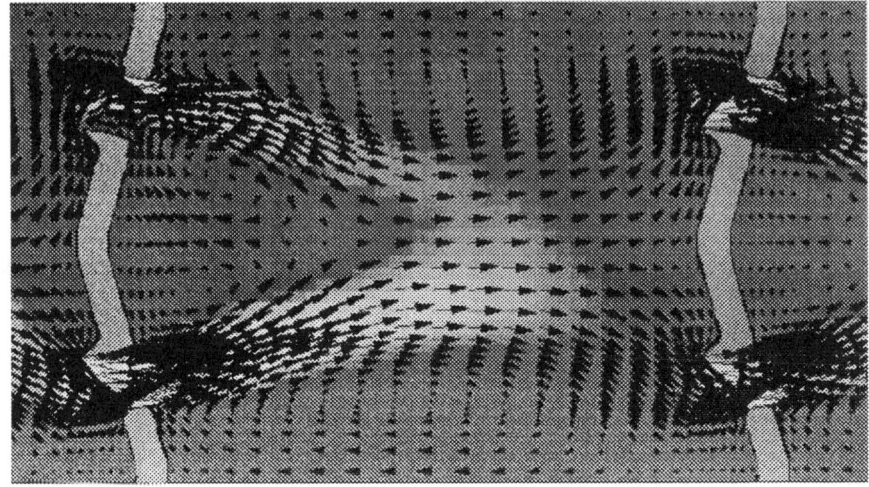

**Figure 2-** CFD velocity vector plot showing the flow pattern within a ladder with dual-vertical slot style ladder baffles.

REFERENCES
1. Albertson, M.L., Dai, Y.B., Jensen, R.A., and Rouse, H.M., "Diffusion of submerged jets", Paper No. 2409, Selected Writings, Hunter Rouse, 1971.
2. Boss, M.G., "Discharge Measurement Structures",ILRI publication 20, 1989.
3. Koch, D. L., "Life History Information on the Cut-up Lakesucker (Chasmites cujus Cope 1883) Endemic to Pyramid Lake, Washoe County, Nevada," PhD. Dissertation, University of Nevada, Reno, pp 343, 1972.
4. Koch, D. L., "Reproductive Characteristics of the Cut-up Lakesucker (Chasmites cujus Cope) and its Spawning Behavior in Pyramid Lake, Nevada," Trans. Am. Fish. Soc., 102(1):145,149, 1973.
5. Koch, D. L., "Life History Information on the Cut-up Lakesucker (Chasmites cujus Cope 1883) in Pyramid Lake, Nevada," Biol. Soc of Nev. Occasional Papers 40:1-12, 1976.
6. Marble Bluff Dam and Fishway Design Specification No. DC-7030, U. S. Bureau of Reclamation.
7. Pavlov, D., "Structures Assisting the Migrations of Non-salmonid Fish: USSR," Food and Agriculture Organization of the United Nations, Rome, FAO Fisheries Technical Paper No. 308, 1989.
8. Ringo, R. D., and Sonnevil, G. M., "Evaluation of Pyramid Lake Fishway: Operation and Fish Passage 1976-1977," U. S. Fish and Wildlife Service special report, Unpublished, Reno, Nevada, 21 pp, 1977.

# LINKING KINEMATIC AND PHYSICAL HYDRAULIC MODELS

## DUNCAN HAY[1], JAMES STRONACH[2], LARRY WEBER[3]
## CHARLES SWEENEY[4]

[1,2] Hay & Company Consultants Inc., Vancouver, B.C., Canada

[3] Iowa Institute of Hydraulic Research, Iowa City, Iowa, USA

[4] ENSR Consultants, Redmond, Washington, USA

## ABSTRACT

This paper presents the use of a computational kinematic model of flow in conjunction with physical hydraulic models where the hydrodynamics of the flow fields were considered to be important relative to fish behavior.

## INTRODUCTION

The abundance of salmon and steelhead in the Columbia River and its tributaries has been declining since the 1800's due to a combination of many factors, including overfishing and the construction of 13 mainstem dams for hydropower. Juvenile salmonids that move downstream during spring and summer outmigrations are at risk since the cumulative mortality that can result from these juveniles passing through several dams may be substantial. Efforts in recent years have been directed to designing means and facilities for the outmigrants to bypass the turbines.

Extensive use has been made of physical hydraulic models for the design and development of bypass entrances, screens, collection and fish transport facilities. Live fish (often small trout) have been used in some hydraulic models, relying upon the innate ability of these fish to quickly identify flow separations or shear zones. Modeling with live fish, however, is very limited in scope, leaving the hydraulic engineer with the challenge of developing and demonstrating hydraulic characteristics within a design that are considered by fisheries biologists to be conducive to fish passage without injury or undue delay.

Although the hydraulic conditions to which fish respond by either avoidance or attraction are not well understood, many classical hydraulic features such as velocity gradients, accelerations, and turbulence appear to play a part in fish behavior. However, to date, there has been only a limited ability to define hydrodynamic features such as accelerations, or subsurface flow paths over a long distance, solely on the basis of observing a physical hydraulic model. This shortcoming has been overcome by linking kinematic and physical models.

## KINEMATIC MODEL

The purpose of the kinematic model is to develop a three-dimensional flow field, based upon limited three-dimensional velocity input data from a physical model. The kinematic model then serves as a platform for the analysis of flow conditions or a graphical presentation of the physical model results. The computational structure of the kinematic model used was similar to a finite difference three-dimensional hydrodynamic model. However the kinematic model interpolates the three-dimensional velocity data, using a $1/r^2$ weighting, to the rest of the flow field through conservation of mass, whereas a hydrodynamic model would interpolate boundary conditions through both conservation of mass and momentum. The flow field characterized by the kinematic model is, or course, limited to the flows and boundary conditions imposed on the physical model at the time of velocity measurements.

## EXAMPLES

ROCK ISLAND DAM, COLUMBIA RIVER - TOP SPILL
Observations of the downstream passage of juvenile fish through sluiceways which draw relatively small amounts of water from the surface of a reservoir, as compared to spillway or turbine intakes drawing flow at depth, have indicated a relatively high efficiency of fish passage as measured by the ratio of the number of fish passed to the flow rate.

A preliminary fish passage evaluation in 1996 using hydroacoustics at the Rock Island Dam suggested an 11 foot "notched" gate was more effective at passing fish than a 30 foot wide "overflow" gate at the same flow, raising the question as to whether any difference in the hydraulic zone of influence between the notched and overflow gates could be quantified, or whether other gate configurations would be more effective for fish passage.

Use was made of a 10-foot wide by 7.5-foot deep flume at the Iowa Institute of Hydraulic Research to obtain 1:20 scale three-dimensional velocity data upstream of the gate configurations shown in Figure 1, at grid points as shown in Figure 2. Velocity measurements, obtained using a 3-D ADV, were input to a kinematic model which filled in the flow field and calculated flow accelerations within the grid.

The zone of influence of a particular gate was defined as the volume of water upstream of the gate within which the acceleration exceeded a particular value. These volumes were extracted from the data files of the kinematic model through a subroutine that summed the grid volumes exceeding a given value of acceleration. These volumes were then weighted based on the vertical distribution of fish measured at the dam in 1996. The results, plotted on Figure 3, indicated that narrow deeper gates had

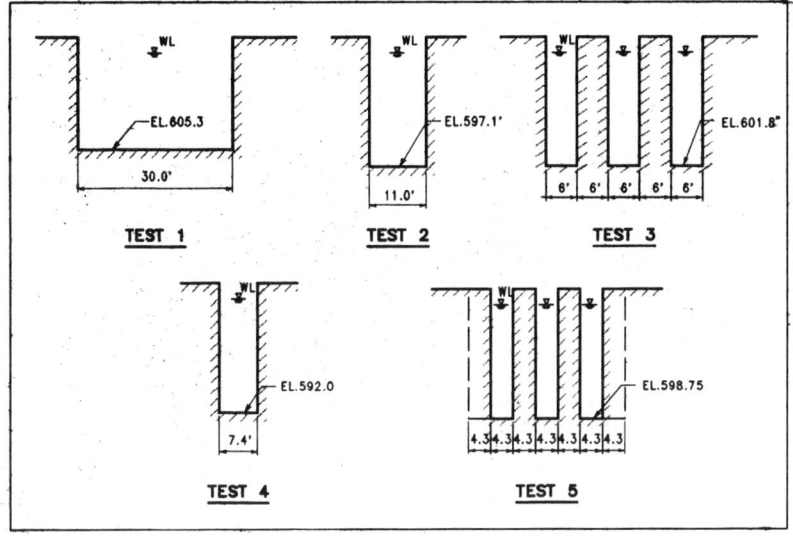

Figure 1 - Top Spill Gates Tested

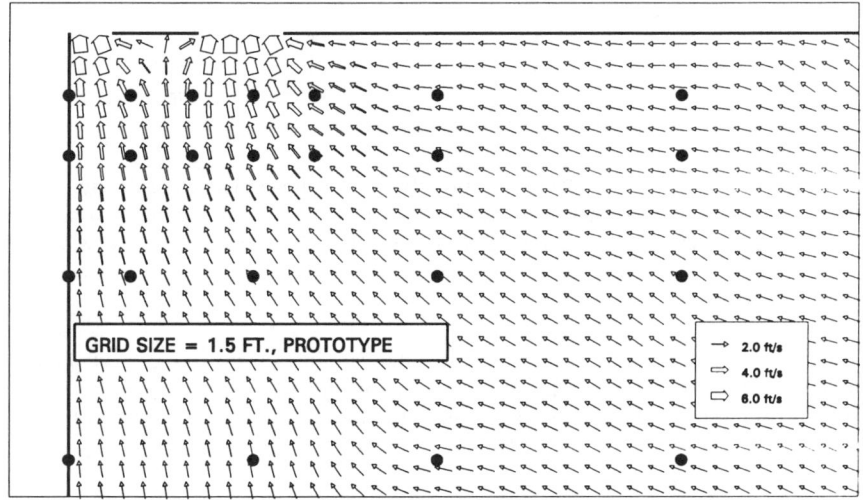

Figure 2 - Typical Interpolated Velocity Field for Top Spill Gates

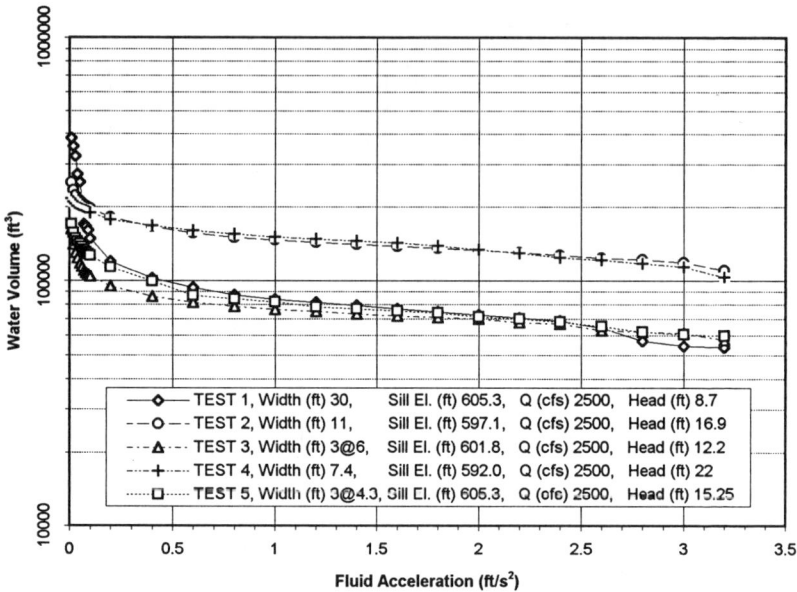

Figure 3 - Weighted Zone of Influence Upstream of Gate

a larger "zone of influence" at Rock Island Dam than slotted, or wider, shallow, gates, confirming conclusions drawn earlier for the hydroacoustic data.

## ROCKY REACH DAM, COLUMBIA RIVER - SURFACE COLLECTOR

A single 15-foot wide by 55-foot deep opening that draws about 2200 cfs is being tested as a juvenile fish bypass facility at the Rocky Reach Dam, Figure 4, with the possibility that additional openings may be added later.

A 1:30 scale model of the forebay fronting the 11-unit powerhouse is actively being used at the hydraulic laboratory of ENSR Consulting and Engineering to assist in design development and evaluation. The model was used to develop the design of the first prototype tested in 1995 and further used to develop modifications implemented for testing in 1996.

Although not quantified in detail, it was the opinion of the project's fisheries biologists that the bypass efficiency was lower in 1996, leading to a detailed examination of the differences in the forebay hydrodynamics between 1995 and 1996 brought about by modifications to the bypass structure and differences in powerhouse loadings.

A kinematic model was used in the analysis. Three-dimensional velocity data was collected from the 1:30 scale model at 5 depths on a 150 ft grid spacing. The model interpolated the input data to a 7.5 foot grid, with conservation of mass at the upstream model boundary, the fish bypass, and each operating powerhouse unit. The model was then used to generate plots of velocity and acceleration fields (both horizontal and vertical) and streamlines of the source of flow entering the fish bypass at various depths to define the differences in flow conditions between the two years of prototype testing. The kinematic model was also used in conjunction with the physical model as a tool to assess what modifications for field testing in 1997. Typical outputs from the kinematic model are shown in Figures 4, 5, and 6 to illustrate the benefits of linking a kinematic model to a physical model.

## CONCLUSIONS

Linking a kinematic model to a physical model expands the utility of physical models permitting better analysis and visualization of physical model data. The kinematic model bridges the gap between physical models and computational fluid dynamic models, giving, to some degree, the best of both modeling techniques.

## ACKNOWLEDGMENTS

Work undertaken on these projects was funded by the Chelan Public Utility District whose staff, Bill Christman, Brett Bickford, Steve Hayes, Chuck Pevan, Robert McDonald, along with others, have participated actively in the design and evaluation of fish bypass facilities.

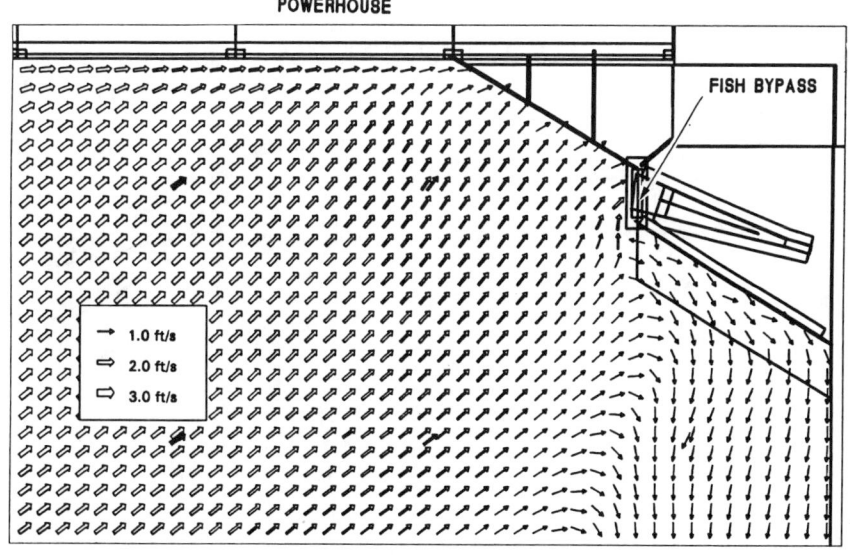

Figure 4 - Interpolated Velocity Field at 50 foot depth, Rocky Reach Fish Bypass

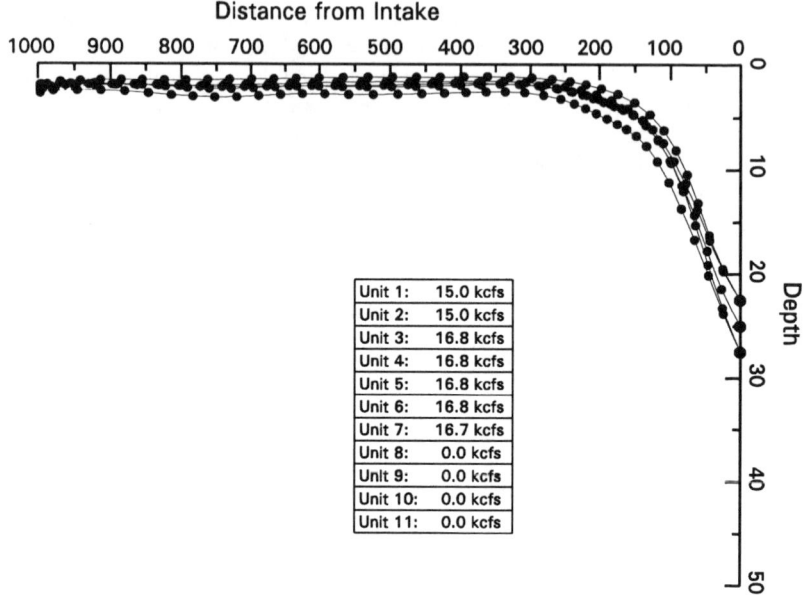

Figure 5 - Source of Flow Entering Bypass at 20 - 30 foot depth, Rocky Reach Fish Bypass

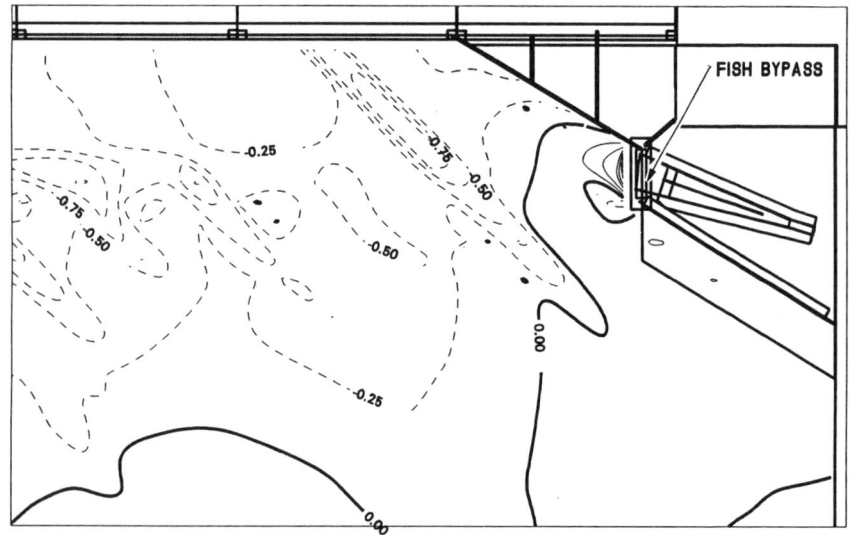

Figure 6 - Vertical Accelerations at 30 foot depth, Rocky Reach Fish Bypass

# JUVENILE FISH PASSAGE SYSTEM AT McNARY DAM

CINDY PHILBROOK, P.E.
U.S. Army Corps of Engineers, Walla Walla District
Walla Walla, Washington, USA

ABSTRACT

Juvenile fish passage systems are widely used at Corps of Engineer (Corps) hydropower dams along the lower Snake and Columbia Rivers. They are an important aspect of anadromous fish survival in the Pacific Northwest. This paper outlines the basic elements of the juvenile fish passage system used at McNary Dam on the Columbia River.

BACKGROUND

Numerous locks and dams constructed within the Columbia River basin provide flood control, power generation, navigation, irrigation, and recreation benefits to the Pacific Northwest. However, these multi-purpose projects also form partial or complete barriers to migrations of highly valued anadromous salmonid species.[1] Steelhead and Pacific salmon spawn in fresh water rivers. Juvenile fish travel (some over 900 miles) past several dams on their downstream migration to the ocean. One to four years later, adult fish return from the ocean to these same rivers.

In order to reduce adverse impacts to migrating fish, most mainstem projects provide both juvenile and adult fish passage systems. The eight Corps of Engineer (Corps) dams on the lower Snake and Columbia Rivers were all completed (between 1938 and 1975) with adult fish passage systems. Many were built without juvenile fish passage (leaving fish to pass over the spillway or through the turbines). Diverting juvenile fish away from turbines has been shown to improve their survival at these projects. Currently, all eight dams provide juvenile fish passage, having been modified or retrofitted where necessary.

The Corps' juvenile fish passage systems have several components. These systems may include turbine intake screening, a collection system, primary dewatering, river bypass, and collection and transportation facilities (where fish can be collected and transported past all dams between the collection site and the ocean). Details of these components are described below and shown schematically in Figure 1. The discussion to follow will focus on the system at McNary Dam, Columbia River at the Washington

---

[1] Turner, et al.,(1993)

and Oregon border. It is the Corps' newest and largest juvenile fish bypass, collection, and transportation system.

## OVERVIEW

The McNary juvenile fish facility replaced an existing facility. Construction costs were $16.2 million (not including the intake screening system). Replacing the existing facility improved juvenile fish survival (reducing facility mortality from 1.5 to 4 percent over the fish season to near 0.5 percent in 1995). This is the third bypass, collection, and transportation facility designed and built by the Corps' Walla Walla District since 1989. It includes improvements over previous designs that reduce fish injury and delay through the system.

## TURBINE INTAKE SCREENING

### GENERAL

Juvenile fish are attracted by the flow passing through the turbines. Turbine intake screening is accomplished by using two screens. These screens are designed to divert juvenile fish away from the turbines, guide them to the collection orifices, and remove excess flows. Due to high unit discharges, heavy debris loads, and the design of the turbine intake, intakes are only partial screened.[2]

### EXTENDED-LENGTH SUBMERGED BAR SCREEN (ESBS)

These 40 foot long screens are located in the turbine intake, and angled to guide flow and fish into the bulkhead slots. Turbine units at McNary Dam are currently being retrofitted with ESBS's. They replace 20 foot long screens previously used and guide substantially more fish into the bulkhead slots. They have bar screen on the face with perforated plate behind to help control velocities through the screen. A traveling brush system runs along the screen face to keep it free from debris. Debris blocking the screen could increase velocities through the screen raising the risk of fish injury.

### VERTICAL BARRIER SCREEN (VBS)

VBS's are located vertically between the (upstream) bulkhead slot and the (downstream) intake gate slot. VBS's have a mesh face with perforated plate behind to help spread velocities evenly along the face of the screen. Screens are designed to guide fish towards the collection orifices (leading out of the bulkhead slot) and remove the majority of water entering the bulkhead slot. Between 400 and 600 cubic feet per second (cfs) of flow enters the bulkhead slots at McNary, but only about 9 to 17 cfs exit through an orifice. The rest passes through the VBS, down the intake gate slot, and through the turbine.

## COLLECTION SYSTEM

The collection system is made up of collection orifices, and a collection channel. Orifices are located near the top of the bulkhead slot (3.5 to 10 feet below the water

---

[2] Turner, *et al.*,(1993)

# JUVENILE FISH PASSAGE SYSTEM 401

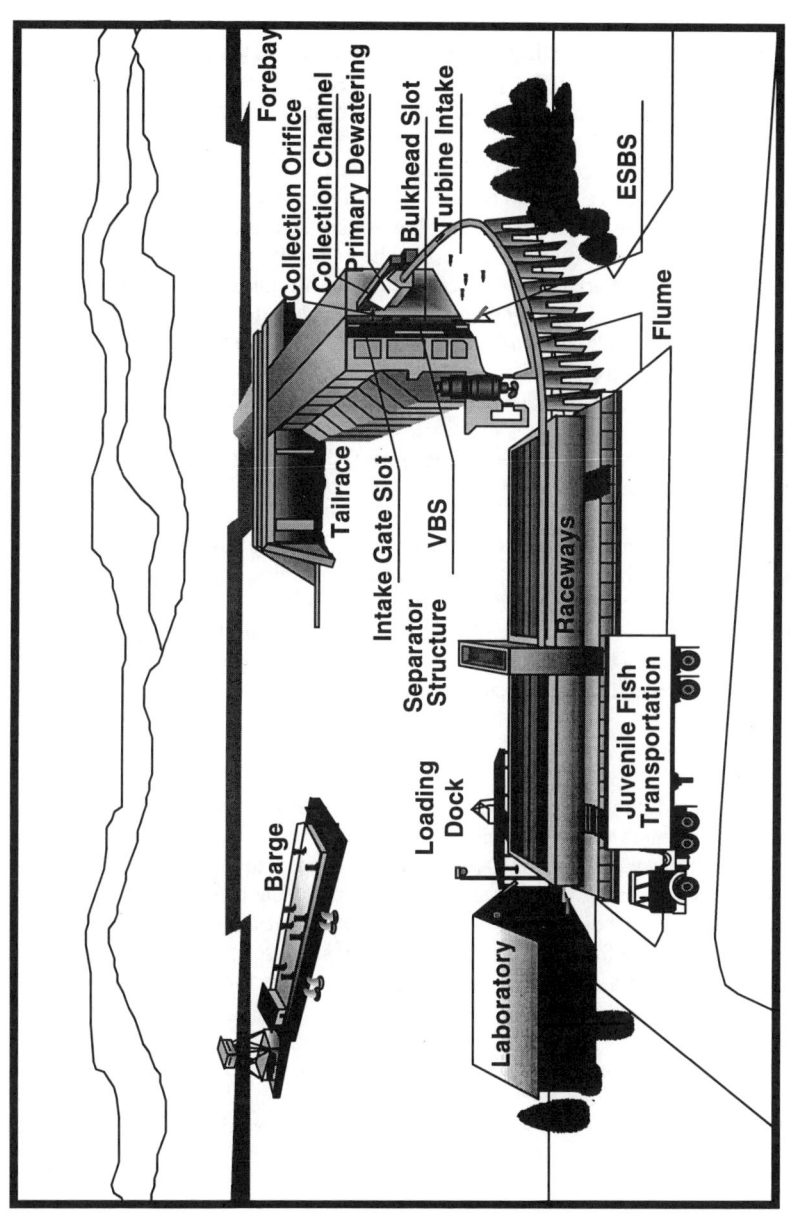

Figure 1. Juvenile Fish Passage System

surface). Knife gate valves control the flow through the orifices and are either full open or full closed.

Orifice flows plunge into a collection channel 9 feet wide. Depths range from about 4 to 13 feet. With 14 turbine units, flows in the collection channel range from 400 to 700 cfs. Fish and flows travel from the collection channel to primary dewatering.

## PRIMARY DEWATERING

Primary dewatering is designed to remove excess flows not needed for transportation of juvenile fish. Approximately 30 cfs of flow is used to transport the juvenile fish from primary dewatering to a river bypass or the collection and transportation facilities. Excess flows are removed through the dewatering screen at low velocities (0.4 fps or less). McNary uses a combination of side dewatering and floor dewatering. Both systems are carefully designed to distribute the flow evenly across the screen area. Water removed from the floor dewatering is controlled by valves that are set to remove a constant flow. Side dewatering is controlled by automatically adjusting slide gates linked to water surface sensors. Since collection orifice elevations are fixed, side dewatering gates adjust for flow fluctuations caused by changes in the forebay elevation of up to 5 feet.

A screen cleaning system is used to minimize plugging of dewatering screens with debris. Water passing through the screens is used for water supply to the collection and transportation facilities, routed to the adult fish passage system (for use as attraction flow), or drained to the river. If there is a problem with the primary dewatering system, an emergency bypass system (located upstream of primary dewatering) can be used to bypass fish and flows directly from the collection channel to the river downstream of the dam.

## RIVER BYPASS

Fish can be routed directly from primary dewatering back to the river (downstream of the dam) to continue their migration. This is normally done early and late in the fish passage season when fish numbers are low, or if there is a problem at the holding and loading facilities requiring them to shut down. Bypass pipe or flume slopes are designed to provide predictable flow conditions for fish passage. Typical velocities in the river bypass range from 9 fps to 11 fps, though some specially designed pipes have been designed to operate at higher velocities. Impact velocities (of fish and flow to the river) are less than 30 fps.

Considerable time and effort is put into selecting a bypass outfall site. River conditions are frequently hydraulically modeled and field verified to insure minimum velocities of 3 fps, good depths, and an absence of eddies and flow disturbances downstream of the site. This is done to reduce predation on the juvenile fish as they are released back to the river.

# COLLECTION AND TRANSPORTATION FACILITIES

## GENERAL
Collection and transportation facilities are used not only to hold fish until they are loaded onto trucks or barges, but also for research on fish passage and survival. Features of collection and transportation facilities include a separator, flume system, laboratory, raceways, river release pipes, and truck and barge loading areas.

## SEPARATOR
The separator's purpose is twofold: One, it separates juvenile fish from adult fish and trash. Two, it separates juvenile fish by size. Separating juvenile fish by size reduces stress on the smaller fish, and allows for research on two groups of fish. The "wet" separator design (widely used on the lower Snake and Columbia Rivers) has bars covered with shallow water (supplied from below the bars). Two bar spacings are used, both too narrow for adult fish to pass through.

This design takes advantage of the natural tendency of the fish to dive. Fish that don't fit through the first set of bars travel downstream to the next set of bars. Fish that don't fit through either set are returned to the river along with any large debris that accumulates on the separator. Once below the separator bars, the juvenile fish enter a flume system and are routed to various areas of the facility.

## FLUME SYSTEM
Flume systems at McNary are generally one foot wide rectangular systems (though some pipes are also used). Slopes are designed to provide velocities between 7 fps and 15 fps at depths of 3 inches or more for the flows used. Dewatering units remove excess flow, where required, at through screen velocities of 0.4 fps or less. Water add-in units add water from below the main flume section. Matching water add-in velocities to flume velocities minimizes flow disturbances. Bends over 30 degrees are spiraled and banked to minimize flow disturbances. Specially designed switch gates and drop gates provide flexibility in flume routings.

## LABORATORY
Periodically (for example, 10 minutes every hour), juvenile fish are routed to the laboratory. Fish are anesthetized, counted, and examined. This information is used to estimate numbers of fish being collected (important for loading raceways, barges, and trucks) and to evaluate the condition of fish passing through the facility. Special tags inserted in some fish allow scientists to track their passage through the river system. After leaving the laboratory, fish are returned to holding tanks to be transported downstream, or released back to the river.

## RACEWAYS
Raceways are, basically, large holding ponds. Fish are held here before being loaded onto trucks or barges. Fresh water is supplied to the raceways at all times. The

amount of flow necessary is based upon the weight of fish in the raceways (one gallon per minute for every 5 pounds of fish).

## TRUCK AND BARGE LOADING

Fish are loaded onto trucks or barges for transportation downstream past the last dam on the Columbia River. The majority of Corps research has shown transportation to produce more adult returns than in-river passage. There is, however, some regional debate over this issue. Over 20 years of research have indicated no significant impact on homing or survival of transported fish.[3] Trucks and barges have controlled environments designed to provide the best possible fish conditions during transportation.

Trucks are used early and late in the fish passage season when fish numbers are low. Driving times from collection to release are about 3.5 hours from McNary and up to 8 hours from Snake River projects. Each tank is specially fabricated and permanently mounted on a semi-trailer. Water is continually circulated through the fish compartments. Temperatures are kept within 3 degrees Fahrenheit of the in-river temperature at the release site. Tankers are equipped with means to aerate and remove gas from transportation water. Fish are released to the river through a flexible hose.[4]

Fish loaded into barges are generally separated by size into different compartments (chambers). Barges are double-hulled construction. River water is pumped through aeration/degassification systems and into the fish holding chambers during loading and transportation. Water flows through these chambers and back to the river. Fish are release through (normally closed) openings in the bottom of the fish chambers.[5]

## RIVER RELEASE

River release routes are provided at the facility for several reasons. One, this allows fish to be returned to the river in an emergency (such as a loss of water supply at the collection and transportation facility). Two, it allows for continuing research on transportation versus in-river survival studies. Three, due to the debate over transportation versus in-river survival, it provides the option to release fish back to the river under normal operating conditions.

## REFERENCES

Turner, A. Rudder, Jr., Ferguson, John W., Barila, Theresa Y., and Lindgren, Mark F., "Development and Refinement of Turbine Intake Screen Technology on the Columbia River", *Proceedings, Bioengineering Section Symposium. 123rd Annual Meeting, American Fisheries Society*, Portland, OR Aug. 29-Sept. 2, 1993.

Ross, Scott S., "Overview of the Fish Program at NPW", *1995 USACE Electrical and Mechanical Engineering Training Conference*, St. Louis, MO June 5-9, 1995.

---

[3] Ross,(1995)
[4] Ross,(1995)
[5] Ross,(1995)

# Hydraulic Investigations Associated with Fish Passage at Red Bluff Diversion Dam, California

JOSEPH P. KUBITSCHEK AND BRENT W. MEFFORD
U.S. Bureau of Reclamation, Denver, Colorado, U.S.A.

ABSTRACT

Two studies associated with fish passage at Red Bluff Diversion Dam (RBDD) have recently been completed by the U.S. Bureau of Reclamation (Reclamation), Water Resources Research Laboratory. These studies include field and laboratory investigations which have been conducted to identify hydraulic performance features associated with the existing fish passage facilities and to evaluate the potential for improving fish passage at RBDD through enlarged ladders. A detailed field evaluation was completed in August 1996 which identifies hydraulic performance deficiencies associated with the existing right abutment fish ladder. In addition, laboratory physical model investigations have been completed to address the potential for improving fish passage through enlarged fish ladders at RBDD. Both studies provide critical information and will be used by the Interagency Fisheries Work Group (IFWG) in the decision making process regarding short and long term solutions to the current fish passage problems at RBDD.

## INTRODUCTION

Red Bluff Diversion Dam (RBDD) is located on the Sacramento River in north central California. The diversion dam is a concrete gated weir structure which is 54-ft high and 740-ft wide. The project was completed in 1964 as part of the California Central Valley Project (CVP). The purpose of the project is to divert Sacramento River water to the west side of the Sacramento River valley for irrigation.

The fish passage facilities at RBDD consist of two primary fish ladders located on the right and left abutments of the diversion dam, and a temporary ladder located at the center of the diversion dam in place of the fixed wheel gate. The right and left abutment ladders were originally designed for 85-$ft^3$/s pool to pool flows with supplemental flows on the order of 265-$ft^3$/s. The supplemental flows are introduced into the downstream portion of each ladder via diffusers located along the right wall. These flows are required to provide attraction to the ladder entrances, in the downstream diversion dam tailrace. The center ladder is somewhat smaller (approximately 100-$ft^3$/s). Limited design and operational details are available.

Ineffective fish passage at RBDD has been identified as a contributing factor in the decline of the anadramous fisheries resource (i.e. primarily Salmon and Steelhead). In response to this problem, the Red Bluff Fish Passage Program was initiated in 1991 to identify and develop viable long term solutions for improving fish passage at RBDD. This program is a combined effort between the U.S. Bureau of Reclamation (Reclamation), the U.S. Fish and Wildlife Service (F&WS), the National Marine Fisheries Service (NMFS), and the California Department of Fish and Game (CDF&G). Recent studies have been completed by Reclamation's Water Resources Research Laboratory (WRRL) in Denver, Colorado which consist of field and laboratory investigations intended to better define the fish passage deficiencies at RBDD and to develop potential long term solutions for those deficiencies.

## FIELD INVESTIGATIONS

A detailed hydraulic field evaluation of the right abutment fish ladder was completed by the WRRL in August 1996. The primary objective of this study was to determine whether a recent modification to the ladder entrance had improved or degraded hydraulic performance.

In 1993, the right abutment ladder entrance was modified in an attempt to improve downstream attraction flow conditions. The idea being that concentrated flows could be sustained for a greater distance downstream of the ladder entrance, in the RBDD tailrace, by changing the gated entrance configuration to a submerged orifice. In doing so, a more compact exit jet geometry could be created and diffusion of such jet minimized, resulting in a greater zone of influence by the ladders, from an attraction flow standpoint. The submerged orifice was cut into the left side of the ladder end sill immediately below the left regulating gate. Operation of this entrance configuration consisted of closing the right side regulating gate while maintaining operation of the left side regulating gate in conjunction with the submerged orifice entrance. Upon completion of this modification video equipment was placed in the downstream most ladder pool, just upstream of the ladder entrance. The results of this monitoring indicated that as fish entered the ladder they tended to turn immediately into the downstream most ladder pool diffuser. Upon encountering the diffuser, they tended to fall back rather than continue up the ladder. This observation alone created concerns. Although attraction flow conditions may have been improved, it was apparent that this improvement may have been at the cost of degraded ladder performance.

The hydraulic field evaluation consisted of determining performance characteristics of the right abutment ladder via ladder pool and attraction flow velocity field mapping under pre and post entrance modification conditions. The results indicate that ladder pool velocity fields in the downstream portion of the fish ladder are strongly influenced by supplemental or diffuser flows. Figure 1 is a typical diffuser velocity contour plot. Supplemental flows are introduced into the downstream portion of the right abutment fish ladder via four diffusers located along the right wall of the downstream most four ladder pools, respectively. The diffusers are unbaffled and exhibit poor velocity distribution characteristics. Furthermore, the 1.0-ft/s diffuser velocity criteria is exceeded over much of the diffuser area. Strong upstream to downstream skewness in the velocity distributions for all diffusers exists under both pre and post modification conditions. This characteristic combined with nonsymmetric diffuser flow introduction into the fish ladder produces significant crossing flow conditions within the ladder itself. This alone has the potential for creating disorientation and delay on the part of the fishery.

Figure 2 represents ladder pool velocity field characteristics of the downstream most ladder pool for pre and post modification conditions. These results indicate that the entrance modification creates crossing flow conditions which are more severe than those exhibited under pre modification conditions. In addition to velocity measurements, air concentration measurements were taken along the diffuser wall in the fish ladder. The results indicate that volumetric air concentrations on the order of 20 percent exist over a portion of the ladder diffusers. This data combined with visual observations indicate further potential for disorientation and delay.

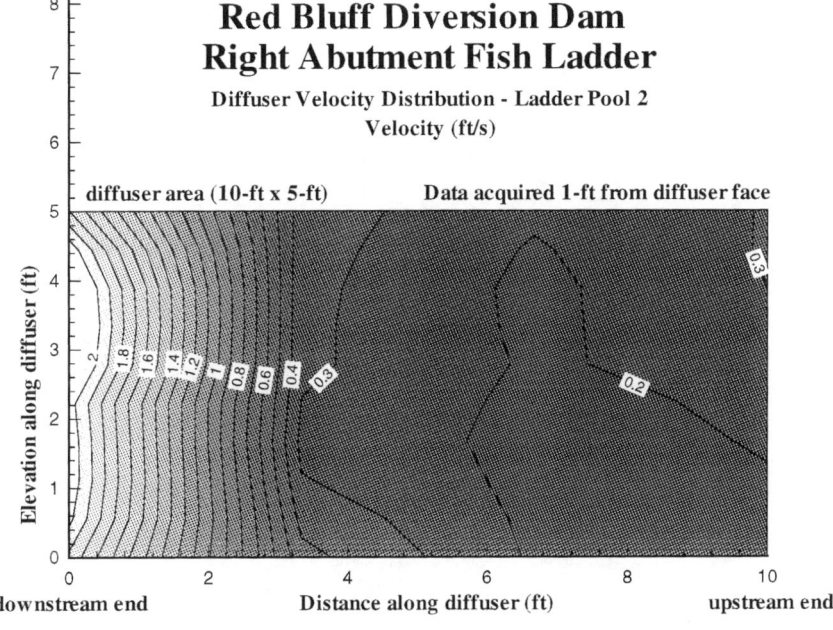

**Figure 1. Typical diffuser velocity distribution contour plot.** *Skewness in the velocity distribution exiting this diffuser is exhibited.*

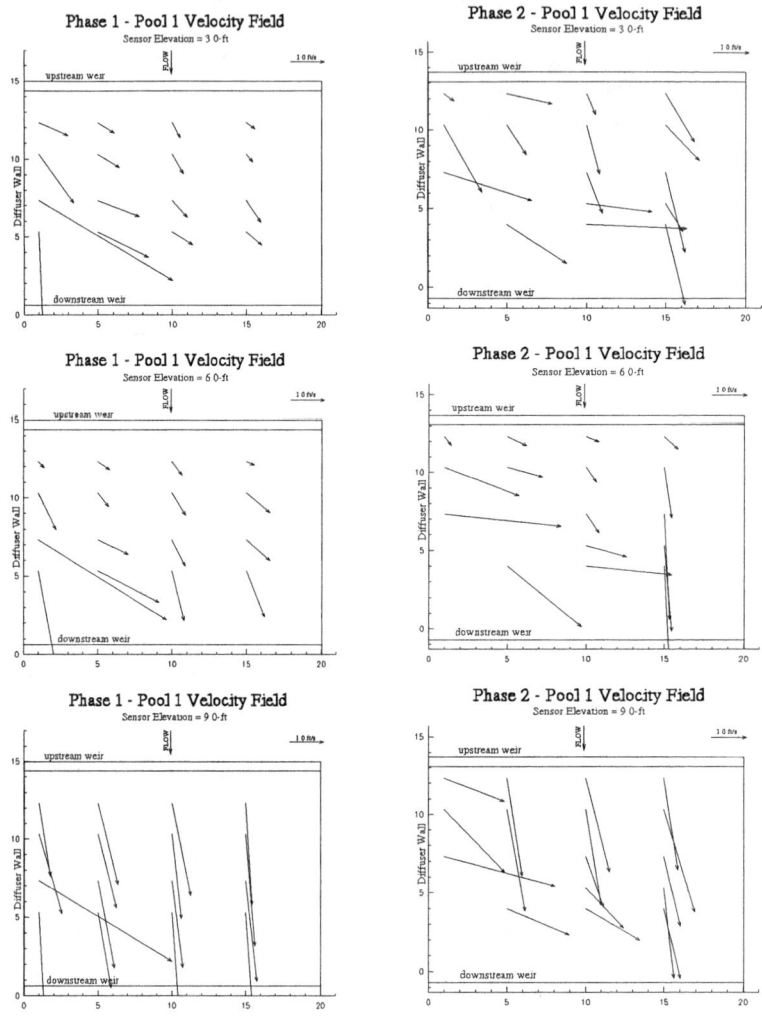

**Figure 2. Ladder pool velocity field results.** *Downstream most ladder pool. Pre and post entrance modification conditions.*

Near and far field velocity results acquired downstream of the ladder entrance in the diversion dam tailrace indicate limited improvement in attraction flow performance of the ladder entrance modification. The zone of influence defined by high velocity regions in the flow field is shown to be approximately 50-ft downstream of the ladder entrance for the entrance modification conditions. This compared with the pre modification zone of influence, found to be approximately 30-ft downstream, demonstrates limited improvement. Thus, in all cases, ladder release penetration into the diversion dam tailrace is less than optimum and likely a result of masking by diversion dam gate releases.

## LABORATORY INVESTIGATIONS

A 1:36 scale physical model including a 900-ft reach of the tailrace has been used to evaluate potential long term solutions for improving fish passage at RBDD. Figure 3 is a photograph of the physical model as constructed in the laboratory. These investigations included evaluation of proposed enlarged ladders (i.e. an 800-$ft^3$/s right abutment ladder, a 1,000-$ft^3$/s left abutment ladder, and a 1,000-$ft^3$/s center ladder) to replace the existing 320-$ft^3$/s right and left abutment ladders. The primary considerations associated with these investigations deal with attraction flow conditions and fish staging locations for various Lake Red Bluff release conditions. Thus, the primary objective was to evaluate the potential for enlarged ladders to produce improved attraction flow conditions. This has been achieved by determining the ladder release interactions with the RBDD tailrace velocity field generated by gate releases for the full range of typical irrigation season release conditions. Analysis of historical and current hydrologic data indicate that typical Lake Red Bluff releases during the gates down period from May 15 to September 15, range from 5000-$ft^3$/s to 20,000-$ft^3$/s. Test points were selected based on this analysis and

Figure 3. **Photograph of RBDD physical model as constructed in the laboratory.** *Scale = 1:36, Froude number similitude.*

included 5,000-ft$^3$/s, 10,000-ft$^3$/s, 15,000-ft$^3$/s, and 20,000-ft$^3$/s. Thus, ladder attraction flow conditions were evaluated for various gate setting configurations consistent with this release range and field operating procedures. Using the physical model results, zones of influence for the enlarged ladder configurations may be compared with those of existing ladder configurations to establish the potential for improved attraction flow performance.

## SUMMARY

These two investigations together represent significant advancement toward implementation of both short and long term solutions for improving fish passage at RBDD. In addition, further understanding of the sometimes complex hydraulic characteristics associated with fish passage structures of this type has been realized. With this experience and the results of these investigations, Reclamation managers, resource agencies, and water users have the necessary information for further developing and implementing solutions which improve fish passage at RBDD. Furthermore, generalized results may be applied elsewhere in the future to further advance the application of fish passage technology.

## REFERENCES

Dodge, R.A. Jr. 1963. "Hydraulic Downpull Studies of the Fixed Wheel Gates for Red Bluff Diversion Dam, Central Valley Project, California." U.S. Bureau of Reclamation, Hydraulics Branch Report No. Hyd-511.

U.S. Bureau of Reclamation. 1991. "Appraisal Report - Red Bluff Diversion Dam Fish Passage Program."

Kubitschek, Joseph P. 1995. "Red Bluff Fish Passage Hydraulic Model Study - Preliminary Report." U.S. Bureau of Reclamation, Water Resources Research Laboratory Report No. PAP-686.

Vogel, David A., Keith R. Marine, and James G. Smith. 1988. "Fish Passage Action Program for Red Bluff Diversion Dam - Final Report on Fisheries Investigations." U.S. Fish and Wildlife Service Report No. FR1/FAO-88-19.

# DISCHARGE EQUATION OF ICE HARBOR TYPE FISHWAY

SHIRO MAENO* , G. SAMPATH KUMAR**, HIROSHI NAGO* and
HIROMICHI SUETSUGU***
*Dept. of Environmental and Civil Engineering, Okayama Univ., Okayama, Japan
**Dept.of Civil Engineering, University of Oulu, Oulu, Finland
***Civil Engineering Dept., The Chugoku Electric Power Co. Inc., Hiroshima, Japan

## ABSTRACT

To get a practical discharge equation of the Ice Harbor type fishway, the main flow features of this type of fishway based on a hydraulic model were studied experimentally. In the Ice Harbor fishway, water flows through the submerged orifice and over the weir simultaneously. The authors measured the detailed velocity profiles of the weir and the orifice flows. Through the experiments, the characteristics of discharges for both the orifice and the weir flows were clarified. An expression for the combined discharge of a Ice Harbor type fishway is also provided.

## INTRODUCTION

Fishways are hydraulic structures that enable fish to overcome the obstructions to their spawning. Of the many fishway types, the three most commonly used are the pool and weir type, the vertical slot, and the Denil. There are probably about 10,000 fishways installed in Japanese rivers [1]. Almost 95 percents of them are the conventional pool and weir type, the others are vertical slot type and Denil type etc. The combination of weir and orifice type fishway has been developed over the years to its present level of sophistication in the Ice Harbor type fishway. These types of fishways are becoming increasingly popular in Japan and elsewhere in the world. However, the available knowledge on understanding the hydraulics of the Ice Harbor fishways is limited.

From this point of view, we performed an experimental study on the Ice Harbor type fishway and proposed the combined discharge equation for the orifice and weir flow to get the total flow rate [2]. To get this combined discharge equation, experiments were carried out separately for the orifice and the weir flows, respectively (so-called uncoupled flow state) as indicated by Rajaratnam et al. [3]. In this paper, to get a more improved discharge equation, we measured detailed velocity profiles for both the orifice and the weir flows under the coupled flow state. The flow characteristics of the Ice Harbor fishway are presented in the present paper. Further, the discharge equation based on the experimental results is also given.

## EXPERIMENTS

Photo 1 shows the 1 : 5 scale model of the Ice Harbor fishway planned at halfway up the Ohta river in Hiroshima. Experimental work on the Ice Harbor fishway was performed in a 0.40 m wide flume with a smooth acrylic bed. A metal sheet painted black was used for the rear wall and the front wall was made of an acrylic sheet to allow the lateral visualization.

Photo 1 View of the Ice Harbor fishway model

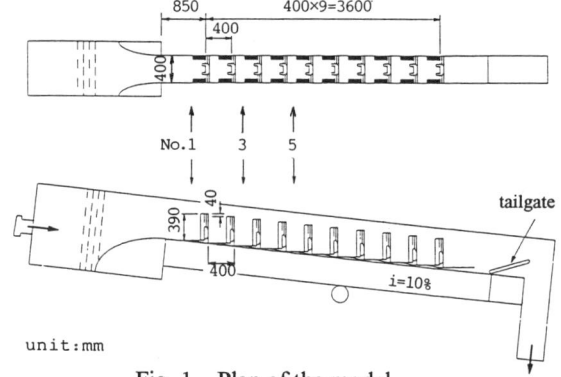

Fig. 1 Plan of the model

The walls were of 0.60 m height and 5.00 m length. The fishway floor had a slope of 1 (vertical) to 10 (horizontal) and was connected to a head tank through a well-designed inlet (Fig. 1). Two weirs of size $0.12 \times 0.19$ m were cut on both sides of the wooden block as shown in Fig. 2. A square orifice of sides 0.06 m were provided in the bottom corners of each weir block. These wooden blocks were fixed in the flume to construct 10 pools with the length of each pool being 0.34 m (Photo 1). A slope of 1 to 5 was given to the entrance of each orifice for increasing the energy dissipation and thereby improving the performance of the fishway.

Water was supplied from the upstream end of the flume by means of a pump, and the flow rates were measured with a standard sharp-crested rectangular notch in the tailwater channel. A tailgate was also provided at the downstream end of the flume for flow control. After the flow settled down to an approximately uniform flow state, the flow depths were measured with a point gauge for a measured flow rate 'Q'. The time-averaged velocity was observed in pool No. 1, 3 and 5 (Fig. 1) with a small pitot tube and a mini-current meter with a propeller of 20 mm diameter. The velocities were measured at the cross-sections shown in Fig. 3. Experiments were performed with three different flow rates ; 7.46 l/s (Case 1), 10.20 l/s (Case 2) and 12.41 l/s (Case 3). Tail water conditions were maintained with the help of a tailgate. The gate was raised to such a level that the flow passes over the last block of the pool.

## RESULTS AND DISCUSSIONS

### Velocity Distribution

Fig. 4 shows the velocity distribution of pools No. 1, No. 3 and No. 5 for Case 2. From this figure, it is shown that the patterns of velocity distribution for each pool are approximately the same except

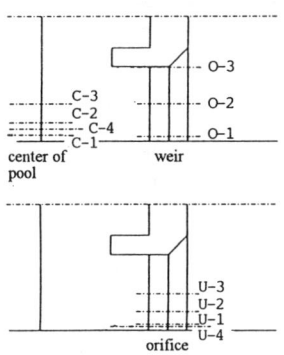

Fig. 3  Measured cross section

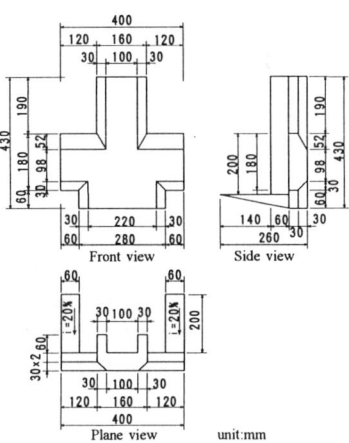

Fig. 2  Details of block

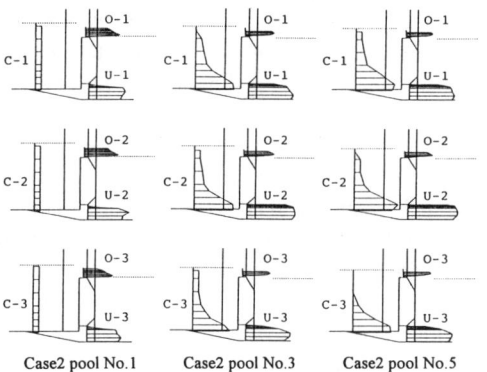

Fig. 4  Velocity profiles

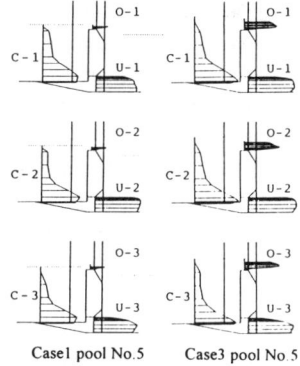

Fig. 5 Velocity profiles (pool No.5)

for that of the upstream end pool No. 1. From the velocity distribution of cross-section C (see Fig. 3), the velocity for pool No. 1 is vertically uniform. But the velocity of the deeper part of pools in No. 3 and No. 5 becomes large. It is considered that the increase of velocity of these parts was caused by the orifice flow of the upper pools. From the velocity distribution over the weir, it is shown that both the water depth over the weir and the magnitude of velocity of pool No. 1 are larger than those of pools No. 3 and No. 5. From the velocity distribution at the orifice, it is noticed that the flow contraction occurs at pool No. 1. But this is not clearly recognized at pools No. 3 and No. 5.

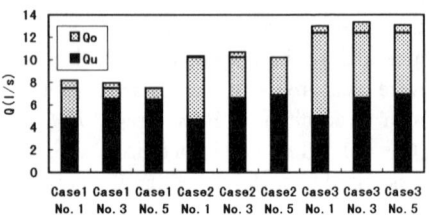

Fig. 6  Discharges of weir and orifice

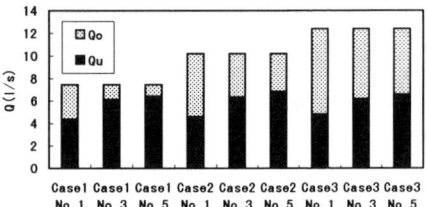

Fig. 7  Corrected discharges

Fig. 5 shows the velocity distribution of pool No. 5 for different discharges. From the velocity distribution of cross-section C, it is shown that the velocity distribution has no relation with the variation of discharge. Even in the case of the largest discharge, the increase of velocity is very small. On the contrary, the velocity distribution over the weir reveals that both the water depth and the velocity increase remarkably with the increase of discharge. The velocity distribution of the orifice flow shows that the effect of the variation of discharge is very small.

Fig. 6 shows the discharges of the weir and the orifice flows which are obtained independently by integrating the measured velocities for these two parts, respectively. We adopted the trapezoidal equation as a integration method. $Q_o$ and $Q_u$ are the discharges for the weir and the orifice flows, respectively. Total flow discharges measured with a standard sharp-crested rectangular notch in the tailwater channel are also shown by solid lines. Although the amount of integrated discharges are slightly larger than the measured discharges, it is considered that the discharges for the weir and the orifice flows were well-described by the integrated discharges. The discharge over the weir increases with an increase in the total discharge. Contrary to these results, the variation of total discharge has a very little effect on the orifice flow. For the upstream end pool No. 1, the ratio of the weir flow rate to the total flow rate is relatively large compared with those for pools No. 3 and No. 5. Fig. 7 shows the corrected discharges. The difference between measured and integrated discharges were distributed to the weir and the orifice flow rates to fit the measured discharge. In the following section, we derive the discharge equation by using these corrected discharges.

## Discharge Equation

First, let us consider the weir flow. We derived the discharge equation based on Itaya & Tejima's formula [4]:

$$Q_o = \frac{2}{3}\sqrt{2g}C_o Bh_0^{3/2}$$
$$C_o = K \cdot C \qquad (1)$$
$$C = 0.605 + \frac{1}{1000 h_0} + 0.08 \frac{h_0}{D} - 0.145\sqrt{\frac{(b-B)h_0}{bD}} + 0.0115\sqrt{\frac{b}{D}}$$

where $C_o$ : discharge coefficient for the overflow. $b$ : width of the flume, $B$ : width of the weir, $D$ : height of the weir, $g$ : acceleration due to gravity, and $h_0$ : the head over the weir (see Fig. 8), $C$ is the discharge coefficient for sharp crest weir by Itaya & Tejima. $K$ is a modification factor which is equal to 1.00 for pool No. 1 and 1.10 for pools No. 3 and No. 5. Fig. 9 shows the calculated overflow rates compared with measured flow discharges. The calculated flow rates agree fairly well with the experimental values.

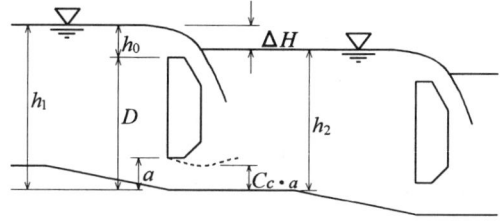

Fig. 8    Schematic sketch

Next, the discharge equation for the orifice flow is represented by

$$Q_u = C_u A\sqrt{2g\Delta H} \qquad (2)$$

where $C_u$ : discharge coefficient for the underflow, $A$ : area of the orifice, and $\Delta H = h_1 - h_2$ : difference of up- and downstream water level (See Fig. 8 ). For the orifice flow, it is well known that $C_u$ for both free and submerged flow is uniquely related with the parameter $(a/h_1)$. However, our experimental results indicate that $C_u$ can be treated as approximately a unique value for each of the pools. This may be due to the range of $(a/h_1)$ which is limited from 0.20 to 0.24 for this type of the fishway. $C_u$ is given by :

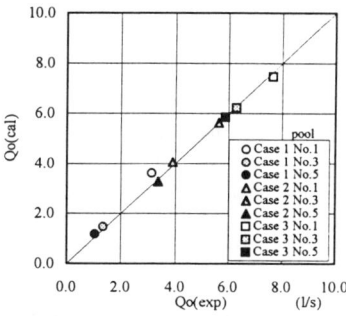

Fig. 9    Discharge for weir

$$\left.\begin{array}{l} C_u = 0.186N + 0.435 \quad (N \leq 3) \\ C_u = 0.993 \quad (N \geq 3) \end{array}\right\} \qquad (3)$$

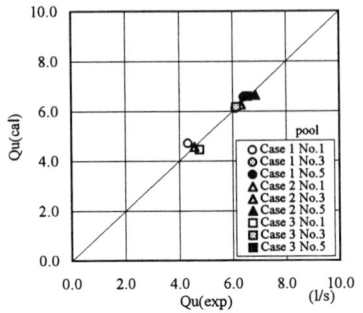
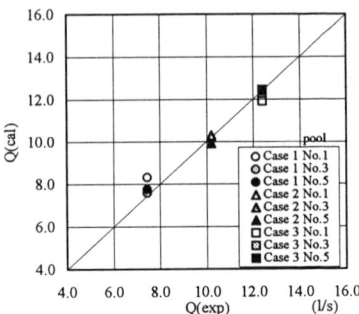

Fig. 10   Discharge for orifice     Fig. 11   Comparison with discharges

where $N$ is the pool number. Fig. 10 shows the calculated orifice flow rates compared with the measured flow discharges. The calculated flow rates agree fairly well with the experimental values.

As described above, the weir and the orifice flow rates were obtained independently. The total flow rate of the Ice Harbour type fishway can be written as :

$$Q = Q_u + Q_o \qquad (4)$$

In Fig. 11 the total flow rates computed from Eq.(4) are compared with the total measured discharges. In general, the deviations are always less than 10%.

## CONCLUSIONS

This paper presents the results of an experimental study on the hydraulics of the Ice Harbor type fishway. The results reveal the characteristics of flow profiles. The suggested total flow rate equation based on the experimental observations estimates fairly well the discharge of the Ice Harbor fishway.

## REFERENCES

[1] S. Nakamura & N. Yotsukura : On the design of fish ladder for juvenile fish in Japan, Proc. of the International Symposium on Design of Hydraulic Structures, Colorado State University, Fort Collins, pp. 499-507, 1987.

[2] G. S. Kumar, H. Nago, S. Maeno and T. Hoshina : Hydraulics of Ice Harbor type fishway, Proc. of the International Symposium on Fishways '95 in Gifu Japan, pp. 79-86, 1995.

[3] N. Rajaratnam, C. Katopodis and A. Mainali : Pool-orifice and Pool-orifice-weir fishways, Canadian Jl.of Civil Engineering, Vol.16, No.5, pp. 774-777, 1989.

[4] S. Itaya & T. Tejima : Discharge coefficient of suppressed weir based on Rehboek's formula, Trans. of the Japan society of Mechanical Engineers, Vol.17. No.56, pp. 5-7, 1951.

# Mechanism of Energy Dissipation and Hydraulic Design for Plunge Pools Downstream of Large Dams

LIU PEIQING   GAO JIZHANG   LI ZHONGYI   LI YONGMEI
China Institute of Water Resources and Hydropower Research, 100044 Beijing, China

## ABSTRACT

The mechanism of energy dissipation for submerged impinging jets downstream of large dams is very complex. This paper is concerned with the main characteristics of submerged impinging jets and their impinging potential on floor slabs in a plunge pool downstream of large dams. Based on model tests and theoretical approaches, the mechanism of fluctuating pressure propagation along cracks of floor slabs, flow behaviors of submerged impinging jets and the stability condition of floor slabs in a plunge pool are investigated in detail. Some hydraulic predicting formulas concerned are derived, which provides scientific basis for hydraulic design of plunge pools downstream of large dams.

## INTRODUCTION

One of the main problems caused by fall jets downstream of hydraulic structures is energy dissipation and scour. Particularly, the problem of energy dissipation downstream of large dams in narrow valleys, where the fall head is high and the specific discharge is large, is one of main difficulties with which we meet in the large dam construction. Therefore, the reasonable selection and design of discharge structures and energy dissipaters are very important to ensure the safety and stability of the large dam. In recent years, plunge pool dissipaters have been adopted in many hydropower projects in narrow valleys, such as Ertan project, Xiaowan project, Goupitan project and Xiluodu project in China etc.. Practice shows that this energy dissipation scheme is quite safe and economic for large dams in narrow valleys. As a result, characteristics of submerged impinging jets in a plunge pool, the mechanism of energy dissipation, and the design of plunge pool structure and floor slabs etc. have been studied by many authors during the last decade. However, due to the complexity of flow structures of submerged impinging jets in a plunge pool the mechanism of energy dissipation of plunge pools is not fully clarified now. This paper, based on the results of model tests and theoretical approaches , studies these hydraulic problems caused by the use of fall jet spillways in hydraulic structures in detail, particularly the mechanism of fluctuating pressure propagation along cracks of floor slabs in a plunge pool and the stability condition of floor slabs, and presents predicting formulas of hydraulic variables concerned.

## CHARACTERISTICS OF SUBMERGED IMPINGING JETS WITHIN A PLUNGE POOL

Flow structures of submerged impinging jets in a plunge pool are quite complex. The jet plunging into water cushion downstream not only produces intense turbulent diffusion along the direction of its central trajectory but also induces two large vortex regions located at upstream and downstream sides of the jet. Fig. 1 is a two dimensional sketch of a submerged impinging jet in a plunge pool. There are three reasonably distinct regions of flow along the main flow direction[1,2]. In region I, the flow characteristics are, for all practical purposes, identical to those of the submerged jet. This region is ,therefore, referred to as the submerged jet region. In region II, the jet undergoes considerable deflection and at end of this region the direction of the jet becomes almost parallel to the wall. This region produces large impinging pressure on the floor in a plunge pool. It is ,therefore, appropriate to refer to region II as the impingement region or impacting zone. In region III, the flow behaviors are similar to those of the wall jet. This region is often referred to as the wall jet region. The regions directly impinged by the jet are the impacting zone and the wall jet region. And the most severe hydrodynamic action of the jet on the floor slabs in a plunge pool occurs in the impacting zone where the impinging pressure and pressure gradients are significant. This region generates intense fluctuating pressure propagation along the cracks of the floor slabs. It is, therefore, the main region which produces the fluctuating uplift on floor slabs and cause a failure of floor slabs. As a result, the impacting zone is the control region of the stability condition for floor slabs in a plunge pool. In the wall jet region, the high velocity main flow is located in the neighborhood of the floor slabs .It has the characteristics of a submerged hydraulic jump. This region mainly controls the length of a plunge pool.

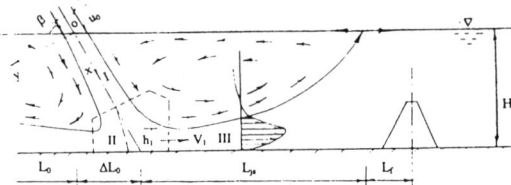

Fig. 1  Conceptual sketch of a submerged impinging jet in a plunge pool

According to the theory of turbulent impinging jets and the analytical results obtained by use of a semi - empirical method in Ref.[3,4,5],as shown in Fig. 1, the maximum velocity of the submerged jet in the neighborhood of the surface of the floor slab could be given by

$$\frac{u_m}{u_0} \propto \left(\frac{h_0}{L_s}\right)^n \quad (1)$$

where $u_m$ is the maximum velocity of the jet near the floor slab, $u_0$ is the jet velocity at the water surface downstream, $L_s$ is the length of the central trajectory line of the jet ( $=H_1/\sin\beta$, where $H_1$ is the depth of water cushion downstream and $\beta$ is the jet angle at the water surface downstream) and n is an exponent. For the deep water cushion $H_1>8.3h_0\sin\beta$, n equals 1.0; for the shallow water cushion $H_1<5.5h_0\sin\beta$, n equals 0.5; for $5.5h_0\sin\beta<H_1<8.3h_0\sin\beta$, n could be estimated by

$$n = 0.5\left(1 + \frac{H_1 - 5.5h_0\sin\beta}{8.3h_0\sin\beta - 5.5h_0\sin\beta}\right) \quad (2)$$

Yu[6] experimentally studied the diffusive characteristics of a submerged impinging jet and presented an empirical formula for $u_m$.

$$\frac{u_m}{u_0} = \left(\frac{h_0}{L_s}\right)^{0.53} \quad (3)$$

For the mean impinging pressure in the impacting zone, many authors, from dimensional consideration, often use the following equation to process experimental data.

$$\frac{\Delta p_m}{\gamma} = f\left(\frac{H_1}{h_0}\right)\frac{u_0^2}{2g} \qquad (4)$$

where $\Delta p_m$ is the maximum pressure differential at the stagnation point (=$p_S$ - $p_{min}$);$p_S$ is the mean pressure at the stagnation point; $p_{min}$ is the minimum mean pressure on the floor slab;$\gamma$ is the specific weight of water; g is the gravitational acceleration. For example,
for $\beta = 40° - 50°$, the empirical equation obtained by Xu and Yu[7] is

$$\frac{\Delta p_m}{\gamma} = 0.475 \exp\left[-0.088\left(\frac{H_1}{h_0}\right)^2\right]\frac{u_0^2}{2g} \qquad (5)$$

for $\beta = 60° - 65°$, the empirical equation obtained by Cui[8] is

$$\frac{\Delta p_m}{\gamma} = 0.74 \exp\left[-0.0013\left(\frac{H_1}{h_0}\right)^2\right]\frac{u_0^2}{2g} \qquad (6)$$

It is shown from Equ. (5-6) that the influence of the jet angle $\beta$ upon the maximum impinging pressure differential $\Delta p_m$ is not negligible. In engineering practice, the following equation is often used to estimate $\Delta p_m$.

$$\frac{\Delta p_{ms}}{\gamma} = \frac{p_m}{\gamma} - H_1 \qquad (7)$$

In China,$\Delta p_{ms}/\gamma \leq 15.0$m is used as the design criterion of the depth of a plunge pool. Liu[9] obtained an empirical formula through processing experimental results of model tests for Ertan project etc..

$$\frac{\Delta p_{ms}}{\gamma} = (0.95\sin^2\beta + 0.175)\frac{q^{\frac{2}{3}}H^{\frac{1}{3}}}{g^{\frac{1}{3}}H_1^{\frac{1}{3}}} \qquad (8)$$

where H is the fall head and q is the specific discharge at the water surface downstream.
Model tests and theoretical analysis show that pressure fluctuations on the floor slab in the impacting zone are quite intense and their probability density functions follow closely the Gaussian distributions. For the mean square root of pressure fluctuations on the floor slab $\sigma_p$, it has been discovered by experiments that the following dimensionalness relationship may be used.

$$\frac{\sigma_p}{\gamma} = f\left(\frac{H_1}{h_0}, \beta\right)\frac{u_0^2}{2g} \qquad (9)$$

Liu et al [4] further analyzed the experimental results of Ref.[7,8] and proposed

$$\frac{\sigma_p}{\gamma} = (0.48\sin^2\beta + 0.025)\exp\left[-0.03\left(\frac{H_1}{h_0}\right)^2\right]\frac{u_0^2}{2g} \qquad (10)$$

Because the wall jet region has the characteristics of a submerged hydraulic jump, we may use the following relationship to estimate the length of a plunge pool, as shown in Fig. 1.

$$L_p = L_0 + \Delta L_0 + L_{js} + L_r \qquad (11)$$

where $L_0$ is the horizontal distance of the jet trajectory,$\Delta L_0$ is the horizontal distance between the impinging point at the water surface downstream and the stagnation point at the floor slab in a plunge pool ($\approx H_1/tg\beta$), $L_{js}$ is the length of submerged hydraulic jump and $L_f$ is the safe length. For estimation of $L_{js}$, it is necessary to calculate the averaged velocity $V_1$ and the effective depth $h_1$ at the section before the submerged hydraulic jump. The velocity $V_1$ may be obtained by use of Equ. (3). The depth $h_1$ is determined by the continuity equation and the value of $V_1(h_1=q/V_1)$. The length of free hydraulic jump $L_j$ is given by[10]

$$L_j = 6.55 h_1 F_{r1}^{-1} \sqrt{\eta - 1}$$
$$\eta = (\sqrt{1 + 8F_{r1}^2} - 1)/2 \tag{12}$$

where $F_{r1}$ is the Froude number at the section before hydraulic jump(=$V_1/\sqrt{gh_1}$). The length of submerged hydraulic jump $L_{js}$ may be expressed as

$$L_{js} = (0.7 - 0.8)L_j \tag{13}$$

## MECHANISM OF FLUCTUATING PRESSURE PROPAGATION ALONG CRACKS UNDER FLOOR SLABS

The possibility that fluctuating pressures are conveyed to the cracks under floor slabs in a plunge pool through the drain system has been observed in the past(Bowers and Tsai[11],Bowers and Toso[12]).Some cases of damage experienced on floor slabs in a plunge pool under flood operation conditions have shown that understanding the hydrodynamic forces involved in the hydraulic design of the lining of plunge pools is very important. It has been concluded from model tests and observations of some prototypes that fluctuating pressures and their propagation along the cracks under floor slabs are two primary factors which cause a failure of floor slabs in a plunge pool. The mechanism of fluctuating uplift generation is quite complex. The pressure fluctuation on floor slabs in a plunge pool possibly causes damage of joint seals of floor slabs and pressure propagation along the rock-slab contact cracks. Although the fluctuating pressures are damped in the propagation through the cracks under floor slabs, the net pressure difference acting on a slab may exceed the sum of the weight of the slab and the drag forces and causes a failure of the slab.

Because the fluctuating pressure propagation along cracks under floor slabs is violent and transient, we can consider the process as a waterhammer or pressure wave phenomenon. Fiorotto and Rinaldo[13] at first studied the process of fluctuating pressure propagation along the cracks under floor slabs and the characteristics of fluctuating uplift on lining slabs by using one-dimensional transient flow equations. Recently, Liu et al [14,15,16] further used one- and two -dimensional transient models to analyze the mechanism of fluctuating pressure propagation within the cracks under floor slabs and the cause of generating fluctuating uplift on floor slabs in a plunge pool in details and presented the theoretical and experimental results for fluctuating uplift on floor slabs in a plunge pool. For example, the two-dimensional model for fluctuating pressure propagation within the cracks under floor slabs derived by Liu[15] is given by Equ. (14). In this equation, $u_x$ and $u_y$ are the velocity components within the crack film, h is the head for fluctuating pressure (h=p'/$\gamma$) and C the speed of a fluctuating pressure wave through the crack film. Fig. 3 shows the process of fluctuating pressures within the crack film under the slab (14 × 10m) calculated by use of this model. In Fig.(3),$p'_s$ is the fluctuating pressure at the entrance of the crack film , $p'_e$ is the fluctuating pressure at the exit of the crack film and $p'_m$ is the fluctuating pressure at the midpoint of the underside of the slab. It is seen from Fig. 3 that because of the larger difference of wave speed ( no phase )between the pressure fluctuations acting on the upper surface and those on the lower surface of the slab, the differential pressures acting on the slab will build up the fluctuating uplift. For the maximum fluctuating uplift on any slab in a plunge pool, Liu et al[14,16] obtained the predicting formula(Equ.(15)).

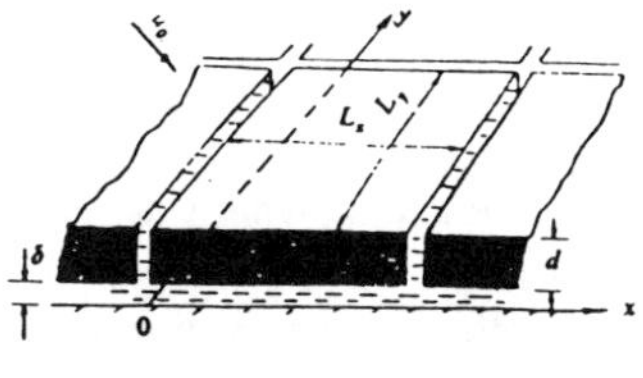

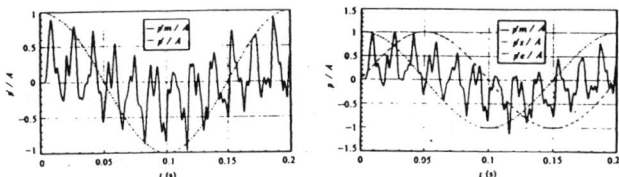

$$\frac{\partial u_x}{\partial t} + u_x \frac{\partial u_x}{\partial x} + u_y \frac{\partial u_x}{\partial y} + g \frac{\partial h}{\partial x} = 0$$

$$\frac{\partial u_y}{\partial t} + u_x \frac{\partial u_y}{\partial x} + u_y \frac{\partial u_y}{\partial y} + g \frac{\partial h}{\partial y} = 0 \qquad (14)$$

$$\frac{\partial h}{\partial t} + u_x \frac{\partial h}{\partial x} + u_y \frac{\partial h}{\partial y} + \frac{C^2}{g}\left(\frac{\partial u_x}{\partial x} + \frac{\partial u_y}{\partial y}\right) = 0$$

Fig. 2  Model for transient flow

$$\frac{A_{max}}{\gamma} = K_p(\beta)\varphi^2 H \frac{q^2}{gH_1^3}\sqrt{1+\alpha_p^2(L_s/L)^2} \qquad (15)$$

$$K_p(\beta) = 4.01\sin^2\beta + 8.85; \alpha_p \approx 1.08$$

where $A_{max}$ is the maximum fluctuating uplift per unit area on the slab, $\varphi$ is the velocity coefficient, L is the horizontal size of the slab and $L_s$ is the turbulent length scale for fluctuating pressures on the upper surface of the slab.

Fig. 3  Process of fluctuating pressure within a rock crack

## STABILITY ANALYSIS OF FLOOR SLABS IN A PLUNGE POOL

Experiments show that The process of vibration for a slab in its matrix is a random vibration. Liu[9], by use of the theory of random vibration, analyzed the vibrating process and obtained the stability condition of floor slabs in a plunge pool.

$$\frac{\sigma_p}{\gamma} \le \frac{\gamma_c - \gamma}{3\gamma} d\sqrt{1 + 0.474 \frac{f_p^2 d}{g}} \qquad (16)$$

where d is the thickness of the floor slab, $\gamma_c$ is the specific weight of the slab, $f_p$ is the principal frequency of the spectrum of fluctuating uplift pressures and $f_p^2 d/g$ stands for the Strouhal number of the vibrating slab. If we only consider the amplitude of fluctuating uplift forces on the slab, the static stability criterion can be inferred by the equilibrium balance for the vertical forces.

$$\frac{\sigma_p}{\gamma} \le \frac{\gamma_c - \gamma}{3\gamma} d \qquad (17)$$

Experiments indicated that at times stability could be achieved by thinner slabs obtained by the static condition Equ.(17).Liu[9], through analyzing the floor slabs of some prototypes, proposed an empirical formula.

$$\frac{\sigma_p}{\gamma} \le 0.2\left(\frac{H_1}{H}\right)^{-1.86} d \qquad (18)$$

where $\sigma_p \approx A_{max}/3$. For $H_1/H<0.3$ and $>0.6$, we approximately use 0.3 and 0.6 respectively as a substitution for $H_1/H$ in Equ.(18).

## CONCLUSIONS

This paper analyzed the main characteristics of submerged impinging jets in a plunge pool, the mechanism of energy dissipation, the behaviors of fluctuating uplift and the stability condition of a floor slab. Based on theoretical and experimental results, hydraulic predicting formulas concerned were derived.

## REFERENCES

[1] Rajaratnam, N., Turbulent jets, Elsevier Sc. Publ. Co., Amsterdam, 1976.
[2] Beltaos, S. and Rajaratnam, N., Plane turbulent impinging jets, J. Hydr. Res., IAHR, Vol. 11, No. 1, 1973, pp. 29-59.
[3] Beltaos, S., Oblique impingement of plane turbulent jets, Proc. of ASCE, Vol. 102, Hy9, 1976.
[4] Liu, P. Q. et al., Characteristics of impinging jets in a plunge pool and scour on the bedrock, J. of Hydraulic Engineering(in Chinese),No. 1, 1995.
[5] Liu, P. Q. et al., Investigation on rock bed scour by ski-jump water jets, J. of Yangtze River Scientific Research Institute(in Chinese),Vol. 11, No. 4, 1994.
[6] Yu, C. Z., Study on the scour by water jets and effect of its dispersion and aeration, J. of Hydraulic Engineering(in Chinese),No. 2, 1962.
[7] Xu, D. M. and Yu, C. Z., Impinging pressures on river bed by planar turbulent jets and its fluctuating characteristics, J. of Hydraulic Engineering(in Chinese),No. 5, 1983.
[8] Cui, G. T. ,et al, Study of forces acting on a river bed and effects by overfall jets downstream of an arch dam, J. of Hydraulic Engineering(in Chinese),No. 8, 1985.
[9] Liu, P. Q ., Study of the mechanism of scour by free jets on rocky river beds, Ph.D. thesis, Tsinghua University, China, 1994.
[10]Liu, P. Q., Discussion of the length of free hydraulic jump on the Horizontal bed, J. of Hydraulic Engineering(in Chinese),No. 1, 1993.
[11]Bowers, C. E., and Tsai, F. Y., Fluctuating pressures in spillway stilling basins, J. of Hydr. Div., ASCE, Vol. 95, HY6, 1969, pp. 2071-2079.
[12]BowerS, C. E., and Toso, J., Karnafuli project, model studies of spillway damage, J. of Hydr. Engrg., ASCE, Vol. 114, HY5, 1988, pp. 469-483.
[13]Fiorotto, V. and Rinaldo, A., Fluctuating uplift and lining design in spillway stilling basins, J. of Hydr. Engrg., ASCE, Vol. 118, HY4, 1992.
[14]Liu, P. Q. et al, Mechanism of the fluctuating pressure transmission in the cracks of bed rock. J. of Hydraulic Engineering(in Chinese),No. 12, 1994.
[15]Liu, P. Q. et al, On the mechanism of fluctuating pressure transmission in cracks of the bed Rock. J. of Hydraulic Engineering(in Chinese),No. 4, 1996.
[16]Liu, P. Q. et al, Experimental study of fluctuating uplift at the bottom of a scour pool, J. of Hydraulic Engineering(in Chinese),No. 12, 1995.

# LOCAL SCOURING DOWNSTREAM OF POSITIVE STEP STILLING BASINS [1]

Corrado GISONNI[*] and Giacomo RASULO[**]
[*] Res. Eng. - Facoltà di Ingegneria - II Università di Napoli - 81031 Aversa (CE) - ITALY
[**] Full Prof. - Facoltà di Ingegneria - Università di Napoli Federico II - 80125 Napoli - ITALY

## ABSTRACT

Local scouring downstream of hydraulic jumps and stilling basins is strongly affected by the geometry of the hydraulic structures and the turbulence character. This phenomenon has been studied at the Department of Hydraulic and Environmental Engineering of Naples since 1985 (Gisonni 1994). LDA measurements of turbulence downstream of the classical hydraulic jump and of the stilling basin with abrupt bottom rise are presented. It was possible to compare turbulence measurements obtained by means of different experimental techniques. Furthermore some scouring tests have been performed. A simple approach is presented to estimate the maximal scour depth as a function of a turbulence parameter and the critical velocity $V_a$ for the sediment movement.

## SEDIMENT TRANSPORT FOR HIGHLY TURBULENT FLOW

The threshold of sediment transport in a highly turbulent flow is strongly connected to turbulence. The critical bed shear stress $\tau_c$ is lowered to $\tau'_c$ (Levi 1978) and the probability of initiation of particles movement is increased. Accordingly, we can refer to the critical velocity instead of the critical bed shear stress concept. Razvan (1971) pointed out an experimental relationship between $V_a$ (critical velocity for initiation of sediment transport at normal levels of turbulence) and $V'_a$ (critical velocity for initiation of sediment transport at high levels of turbulence):

$$V'_a = V_a / (1 + 2.5 K_v) \qquad (1)$$

where the turbulence intensity coefficient $K_v$, defined as:

$$K_v = \frac{1}{U} \lim_{T \to \infty} \sqrt{\frac{1}{T} \int_0^T [u(t) - U]^2 dt} = \frac{1}{U} \lim_{T \to \infty} \sqrt{\frac{1}{T} \int_0^T [u']^2 dt} \qquad (2)$$

is evaluated by means of turbulence measurements (see Tab. I).

---

[1] This research was founded by M.U.R.S.T. 60% (1991)

Thus, the critical flow depth $h'_a$ can be defined as:

$$h'_a = \frac{q}{V'_a} \quad (3)$$

Both relationships (1) and (3) apply to streams with $K_v$ larger than 0.10 which is upper limit of the turbulence intensity coefficient at normal levels of turbulence (Ortiz 1982).

Tests on mobile bed models led to the following conclusions (Nola 1989, Gisonni 1994):
- it is possible to evaluate the maximum scour depth downstream of the classical hydraulic jump or stilling basins (Fig. 1) if we assume:

$$h_2 + e = h'_a \quad (4)$$

- the critical flow depth $h_a$ for average turbulence levels was always measured on the top of the dune downstream of the scour zone and downstream of this end section the sediment transport capacity of the turbulent flow is dampened.

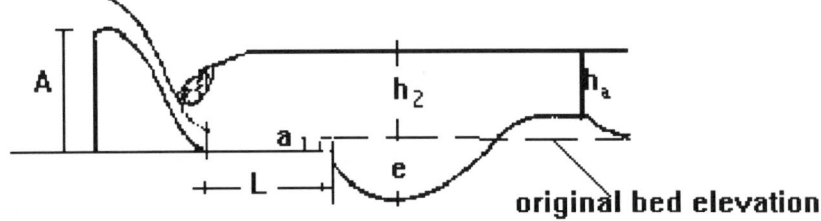

*Fig. 1*    *Typical scour profile downstream of a stilling basin.*

## TURBULENCE MEASUREMENTS IN CLASSICAL HYDRAULIC JUMP

Attention has been focused on stilling basins with a positive step at the end of the concrete floor, as frequently used in Italian practice. Preliminary experiments considered the main character of the flow field downstream of the classical hydraulic jump, to compare other data available from the literature.

Measurements have been performed with the Laser Doppler Anemometer (LDA) installed on a rectangular cross-section channel 0.77 $m$ wide and 18 $m$ long. The hydraulic jump was created by means of two aprons; with a discharge equal to 70 l/s. The sequent depths and the Froude number at the toe of the jump were $h_1=3.3$ $cm$, $h_2=21$ $cm$, $Fr_1 = 4.8$.

The observations included:
- vertical distribution of the instantaneous velocity $u(t)$ in the flow direction at different sections downstream of the main roller,
- water depths $h$.

Thus, the distribution of the average velocity $U$ and the parameter $K_v$ were determined at different sections.

The experimental results were compared with other proposals (Tab. I). All these relationships are derived by experimental data with the velocity measured by micropropeller at one fifth of the flow depth $h$, except for Ortiz (1982) who performed measurements also at elevation $0.4h$.

*Tab. I* -    *Experimental relationships for classical hydraulic jump.*

| Author | Formula | Experimental range |
|---|---|---|
| Kalis (1961) | $\dfrac{U_2}{\sqrt{\overline{u'^2}}} = 0.357 \dfrac{x'}{(h_2 - h_1)} + 10.77 \dfrac{h_2}{h_1}$ | $2.2 < Fr_1 < 6.3$ |
| Koumine (1961) | $\dfrac{\sqrt{\overline{u'^2}}}{U} = \dfrac{1.52}{\beta_x - 1.69\sqrt{(\beta_1 - 4)} - 0.195(\beta_1 - 4)}$ | $6 < \beta_x < 15$ $\beta_1 > 4$ |
| Razvan (1967) | $\dfrac{\sqrt{\overline{u'^2}}}{U} = \dfrac{1.52}{\beta_x - 1.61\sqrt{(\beta_1 - 4)} - 0.145(\beta_1 - 4)}$ | $6 < \beta_x < 15$ $\beta_1 > 4$ |
| Ortiz (1982) | $K_v = 0.1 + \left(\dfrac{31.766}{Fr_1} - 1.43\right) e^{\left[\left(0.217 - \dfrac{3.306}{Fr_1}\right)\dfrac{x}{h_2}\right]}$ | $Fr_1 = 7.6, 9.8,$ $12.4, 14.9$ |

The data from Kalis refer to the distance $x'$ instead of $x$ (see list of symbols) and it was necessary to evaluate the length of roller $L_r$ (Hager 1995) to compare the results from different relationships.

The data analysis indicates:

- $K_v$ decreases with the distance $x$ from the toe of the jump,
- $K_v$ values measured by LDA are always larger than those obtained by formulas of Tab. I, probably because of the low frequency response of the micropropeller that is able to measure only velocity fluctuations caused by large eddies and low wave number (or low frequency).

The LDA data were filtered by applying a frequency cut for frequencies higher than 1 Hz on the power spectrum of the acquired signal; Figure 2 shows (for $Fr_1 = 4.8$) agreement between the different authors and our experimental data.

Turbulence measurements have been performed also downstream of a positive step stilling basin with A=31 cm, L=165 cm, a=11 cm (Fig. 1). The model had a downstream fixed bed; the design discharge was 120 l/s per unit width; the main parameters were $h_1 = 4$ cm, $Fr_1 = 4.5$, $h_2 = 11.5$ cm.

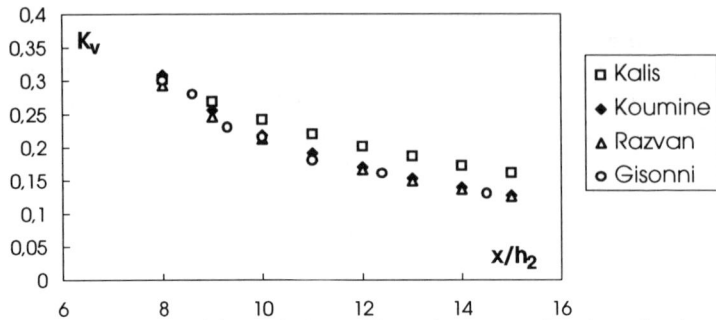

**Fig. 2** $K_v$ values measured by LDA (after filtering), compared to formulas from Tab.1 ($Fr_1=4.8$).

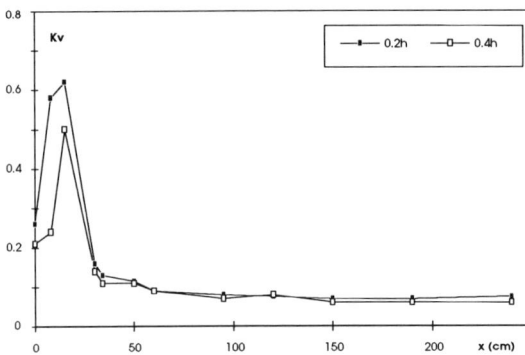

**Fig. 3** $K_v$ values downstream of the stilling basin measured at $0.2h$ and $0.4h$.

In Fig. 3 the values of $K_v$ are plotted against the distance from the end step. The elevation of measuring points is $0.2h$ and $0.4h$ from the bottom, accordingly to the practice of previous authors. The maximum value of $K_v$ was always recorded at a short distance from the step. $K_v$ rapidly decreases downstream because the end step causes a strong dissipation of the turbulent kinetic energy of the flow, and dampens the turbulent fluctuation to average levels of intensity.

As for the measurements relating to the classical hydraulic jump, the data have been filtered by the same frequency cut (1 $Hz$). After the filtering the maximum value of $K_v$ (at $0.2h$) is reduced from 0.62 to 0.36.

## SCOUR TESTS ON STILLING BASIN

In order to validate the proposed approach for the computation of the maximum scour depth, some scouring tests have been performed on the same physical model with a downstream mobile bed.

The sediment diameter was between 1 and 1.5 $cm$, with $d_{50}=1.3$ $cm$; the specific weight was about 27 $KN/m^3$. For such a sediment mixture the threshold value of the flow depth is $h_a=13$ $cm$, corresponding to the threshold value of the flow depth for uniform sediment with $d=1.2$ $cm$. The duration of the tests was at least ten hours to reach the ultimate scour depth. The tests confirmed the conclusions of Hoffmans (1993) and Nola (1989).

The first experimental run was performed with the mobile bed starting just after the end step of the stilling basin. The results are presented in Fig.4; the maximum measured scour depth was $e_m=13$ cm. No dune was observed downstream of the scour hole, because the particles with a diameter smaller than 1.2 cm were moved away for $h_a=13$ cm, according to our prediction.

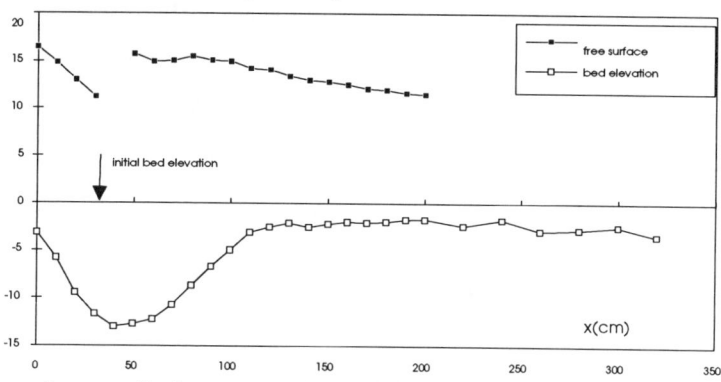

***Fig. 4***   *Scour profile downstream of the stilling basin.*

In the second experiment the mobile bed started 55 cm downstream of the end step; this length was fixed according to Fig. 3, from which it can be seen that the turbulence fluctuations are too small for scour to occur.

## MAXIMUM SCOUR DEPTH

From equations (1) and (4) it is possible to derive the following relationship:

$$e_c = (1+2.5K_v)h_a - h_2 \qquad (5)$$

where $e_c$ is the computed maximum scour depth and $h_2$ is the design flow depth downstream of the stilling basin. Recalling the parameters $h_a=13$ cm, $h_2=11.5$ cm, $K_v=0.36$, the scour depth computed is:

$$e_c = 13.2 \text{ cm}$$

in agreement with the measured maximum scour depth $e_m = 13$ cm.

## CONCLUSIONS

1. $K_v$ values obtained from formulas of Tab. I are in agreement with those obtained with LDA if a frequency cut at 1 $Hz$ is applied;
2. relationship (5) is able to predict the maximum scour depth if we consider turbulence fluctuations with a frequency lower than 1 $Hz$ is considered;
3. The positive step downstream of the stilling basin has a positive effect on scour because it is able to reduce over a short distance the turbulent velocity fluctuations to typical values for average levels of turbulence ($K_v \cong 0.10$).

## REFERENCES

Gisonni C. - "Fenomeni di escavazione localizzata a valle di opere di dissipazione in condizioni di moto piano", Ph. D. Thesis, Università di Napoli 1994.

Hager W.H. - "Hudraulic jump", from 'Energy Dissipators' (D.L.Vischer and W.H.Hager eds.) - I.A.H.R. hydraulic streucture design manual n°9 - Rotterdam, The Netherlands, 1995.

Hoffmans G.J.C.M., Booij R. - "Two dimensional mathematical modelling of local-scour holes", Journal of Hydraulic Research, vol. 31, n° 5, pp. 615-633, 1993.

Kalis J. - "Diminution de la turbulence derriere le ressaut", IX IAHR Congress, Dubrovnik, pp. 43-49, 1961.

Levi E. - "Turbulent processes in natural streams", Centro Internazionale di Idrologia 'Dino Tonini', Istituto di Idraulica (Internal Report), Padova, 1978.

Levy J.J. - "Effet dynamique d'un courant a haute turbulence sur des ouvrages hydrauliques et sur le lit des rivieres", IX IAHR Congress, Dubrovnik, pp. 133-140, 1961.

Nola F., Rasulo G. - "Escavazioni a valle di dissipatori a risalto", Idrotecnica vol.16, n° 2, pp.63-74, 1989.

Ortiz J.P. - "Macroturbulencia de escoamento a jusante de estruturas de dissipaçao por ressalto - Estudo teorico- experimental", Buletin Tècnico DAEE, Sao Paulo, n° 3, 1982.

Razvan E. - "Resultats de l'etude du mouvement macroturbulent en aval du ressaut hydraulique", XII IAHR Congress, Fort Collins, pp. B4/1-B4/10,1967.

Razvan E. - "L'influence de la haute turbulence sur l'entrainement des alluvions", XIV IAHR Congress, Paris, pp. C3/1-C3/7, 1971.

## LIST OF SYMBOLS

|  |  |
|---|---|
| $e_m$ | maximum measured scour depth |
| $e_c$ | maximum computed scour depth |
| $Fr_1$ | Froude number at the toe of the jump |
| $h_1$ e $h_2$ | sequent depths upstream and downstream of the jump |
| $K_v$ | turbulence intensity coefficient |
| $q$ | discharge per unit width |
| $\sqrt{\overline{u^2}}$ | root mean square of the instantaneous velocity |
| $U_2$ | average cross-sectional velocity |
| $U$ | average velocity in the measuring point |
| $V_a, h_a$ | critical velocity and flow depth for initiation of sediment transport at normal levels of turbulence |
| $V'_a, h'_a$ | critical velocity and flow depth for initiation of sediment transport at high levels of turbulence |
| $x$ | distance from the toe of the *classical hydraulic jump* or from the end step of the stilling basin |
| $x'$ | distance from end section of the hydraulic jump |
| $\beta_x$ | relative distance from the toe of the jump ($x/h_2$) |
| $\beta_1$ | sequent depths ratio ($h_2/h_1$) |

## DAM FOUNDATION EROSION: NUMERICAL MODELING

George W. Annandale[1], Todd Lewis[2], Rod Wittler[3], Steve Abt[4], Jim Ruff[5]

## ABSTRACT

This paper summarizes the numerical model for simulating dam foundation erosion downstream of overtopping dams. The model will have the ability to simulate scour of resistant earth materials such as rock. The relative ability of rock to resist erosion is quantified by making use of the Erodibility Index Method (Annandale, 1995), and the relative magnitude of the erosive power of water is quantified by the rate of energy dissipation. The conceptual approach of the model is described, followed by a summary of the Erodibility Index and the method used to calculate the rate of energy dissipation of plunging jets.

## INTRODUCTION

The Dam Foundation Erosion Project is managed by the Bureau of Reclamation, and funded by the Bureau of Reclamation, Pacific Gas and Electric, the Energy and Power Research Institute (EPRI), the Western Area Power Authority (WAPA), the Federal Energy Regulatory Commission (FERC) and the National Weather Service. The objective of the project is to develop engineering methods that can be used to simulate scour of rock and other resistant earth materials downstream of overtopping dams. The research project entails small and near-prototype model studies that are conducted at Colorado State University, Fort Collins. Golder Associates Inc. is responsible for developing software that can be used to simulate scour downstream of overtopping dams. The conceptual approach used by the numerical model is presented by this paper. This is followed by a summary of the Erodibility Index Method (Annandale, 1995) and the approach followed to simulate the erosive power of plunging, aerated jets.

## CONCEPTUAL APPROACH

The conceptual approach used by the numerical model to simulate erosion of rock and other resistant earth materials is presented in Figure 1. The figure shows relationships between elevation below the river bed and stream power. The stream power that is available for scour is calculated by estimating the rate of energy dissipation of a plunging jet that is entrained by the tailwater, and plotted as a function of elevation below the river bed. The stream power that is required to cause scour is determined by the Erodibility Index Method, summarized in the next section. The latter is also plotted as a function of elevation below the river bed. Bed material strength normally increases as a function of elevation below the surface, whereas the erosive power of a jet decreases in the same direction. The maximum depth of scour is determined as the location where the stream power that is available is less than the stream power that is required to cause scour.

## EARTH MATERIAL RESISTANCE

Earth material resistance is determined by making use of the Erodibility Index Method (Annandale, 1995). This method relates the relative ability of earth materials to resist erosion to the erosive power of water.

---

[1] Director: Water Resources Engineering, Golder Associates Inc., 200 Union Blvd, Suite 500, Lakewood, Colorado 80228

[2] Staff Engineer, Golder Associates Inc., 200 Union Blvd, Suite 500, Lakewood, Colorado 80228

[3] Hydraulic Research Engineer, Bureau of Reclamation, Denver, Colorado
[4] Professor, Engineering Research Center, Colorado State University, Fort Collins, Colorado
[5] Professor, Engineering Research Center, Colorado State University, Fort Collins, Colorado

The magnitude of the erosive power of water is determined by its rate of energy dissipation. The relative ability of earth material to resist erosion is determined by the Erodibility Index, defined as:

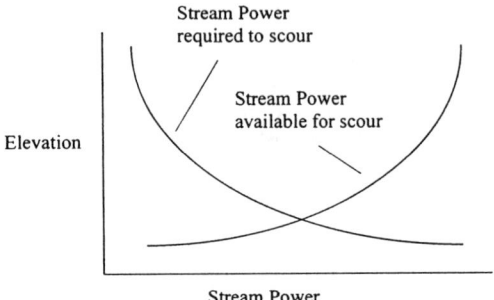

Figure 1. Conceptual approach used to simulate scour of rock and other resistant earth material.

$$K_h = M_s \cdot K_b \cdot K_d \cdot J_s \tag{1}$$

where $K_h$ = Erodibility Index, $M_s$ = Mass Strength Factor, $K_b$ = Block Size Factor, $K_d$ = Shear Strength Factor, $J_s$ = Relative Orientation and Shape Factor. These factors are determined by making use of standardized tables that relate material properties to factor values. The tables can be found in (Annandale, 1995).

The relationship between the rate of energy dissipation (also known as stream power) and the Erodibility Index, defining the erodibility threshold, was determined by analysis of 150 field observations (Annandale, 1995). The graph in Figure 2 displays two types of observations, one set representing events where erosion occurred and the other representing events where erosion did not occur. The boundary between the two sets of points represent the erodibility threshold.

Scour analysis entails determination of the Erodibility Index of the earth material under consideration and the rate of energy dissipation of water discharging over or onto the material. By plotting these two points on Figure 2 it is possible to determine whether scour is likely to occur. Events plotting above the threshold line are likely to scour, whereas events plotting below the threshold line are unlikely to scour.

Figure 2 is also used to compose the curve representing the stream power that is required to cause scour. This is done by indexing the earth material as a function of elevation below the bed surface, and then using the figure to determine the stream power that is required to cause scour.

The software that is developed assists the user in indexing the earth material, and plotting the Erodibility Index as a function of elevation (Figure 3). This is a useful tool to develop understanding of material properties as it pertains to erosion and scour.

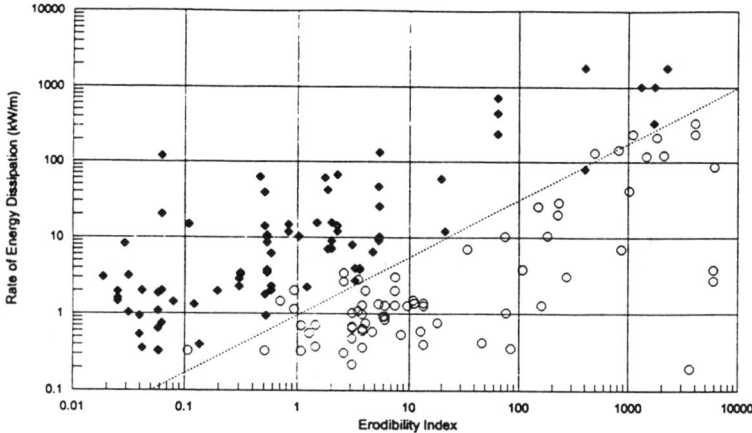

Figure 2. Erodibility Index graphs showing erodibility threshold.

## EROSIVE POWER OF JET

The trajectory of a jet formed by overtopping a dam consists of two primary regions; a free-fall portion and a plunge pool portion, as illustrated in Figure 4.

Prediction of the ultimate depth of scour by the Erodibility Index Method requires that the rate of energy dissipation be tracked as the jet falls through the atmosphere, impacts the plunge pool surface, and travels towards the bottom of the scour hole. Estimation of the rate of energy dissipation at points within the trajectory of the jet is complicated by the degree of jet turbulence and the amount of entrained air present within the jet.

The turbulence intensity of the jet at issuance impacts the rate of energy dissipation in two ways. First, as issuance turbulence intensity grows, the break-up length of the jet lessens, increasing the probability that the jet will consist solely of discrete water droplets at impact with the plunge pool surface. Jet impact velocities are dramatically reduced within broken-up jets, lowering the total energy available (and thus the energy dissipation rate) within the jet to scour bed materials. Second, as issuance turbulence intensity increases, the rate at which air is entrained through free-fall increases. Higher concentrations of air within the jet, which must be treated as a two-phase (air / water) pseudo-fluid, decrease its density, thereby reducing the total energy available within the jet, (Bohrer, 1996; Lewis, 1996).

The complex turbulent nature of two-phase jet flow prevents a direct analytical calculation of the energy dissipation rate of overtopping jets. However, a computer application under development at Golder Associates, Inc., which incorporates historical research and the results of current investigations into overtopping jet characteristics, will enable numerical estimation of the energy dissipation rate within overtopping jets. Once completed, this computer application, combined with the calculation of the Erodibility Index, will permit prediction of the ultimate scour depth. Figure 5 illustrates the density variation within a turbulent jet that was calculated with a preliminary version of the software that is used to calculate the energy dissipation rate.

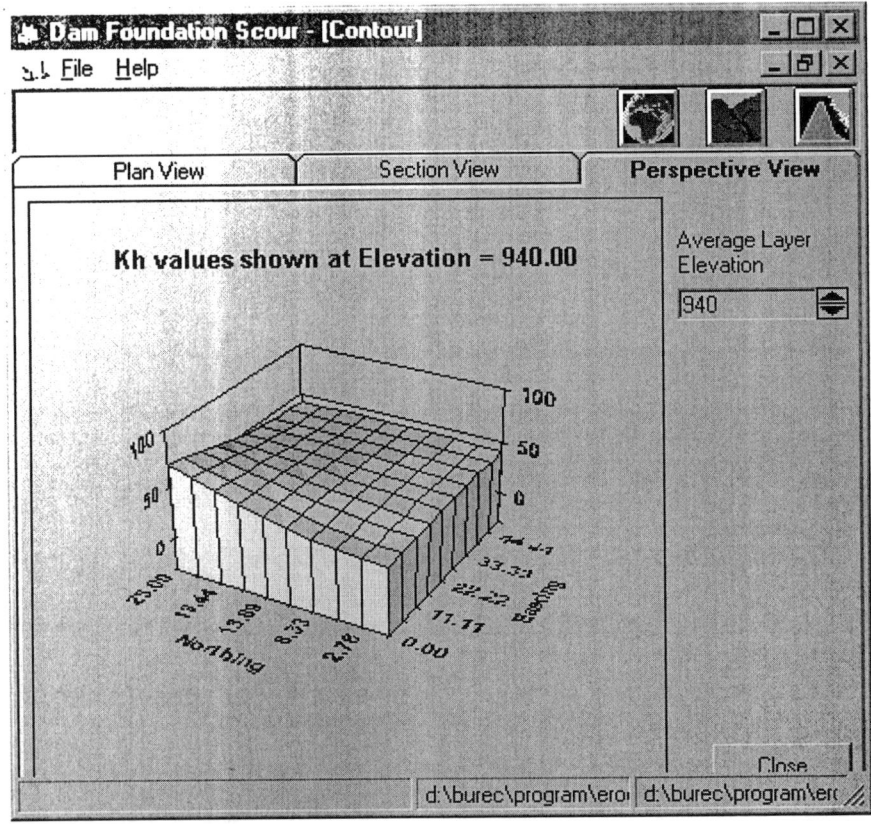

Figure 3. Example of output that can be used to display changes in the Erodibility Index in space.

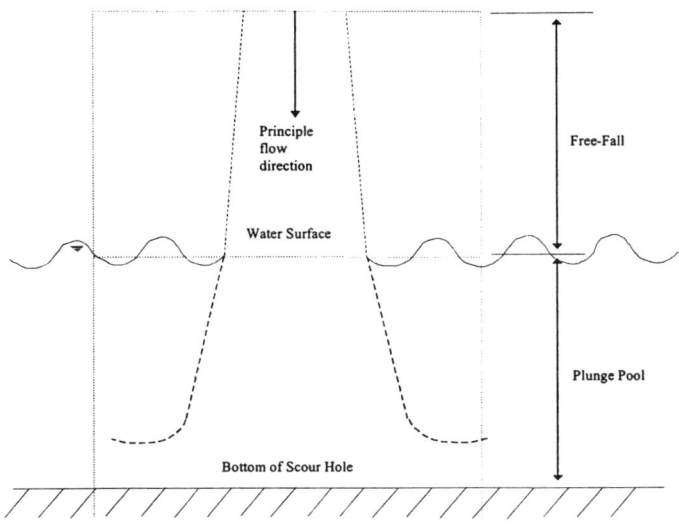

**Figure 4.** Delineation of regions within an overtopping jet (frontal view).

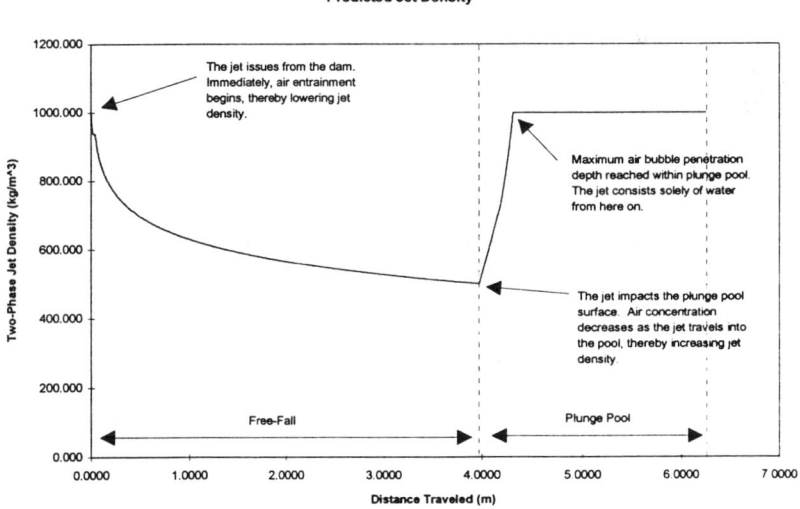

**Figure 5.** Density variation calculated by software to predict rate of energy dissipation of a plunging and entrained aerated jet.

## SUMMARY

The paper summarizes current research and development to create software that can be used to simulate dam foundation erosion downstream of overtopping dams. The software development uses the results of current research into the characteristics of aerated jets together with the Erodibility Index Method to predict extent of scour. The research into the nature of aerated jets provides a basis for calculating the erosive power of plunging and entrained jets. The predicted erosive power of the jet is then used by the Erodibility Index Method to predict whether resistant earth materials, such as rock, would scour. The extent of scour is determined by a procedure that balances the stream power that is required to scour resistant earth material with the available stream power.

## REFERENCES

Annandale, G.W. (1995), Erodibility, Journal of Hydraulic Research, Vol. 33, No. 4, pp. 471-494.

Bohrer, J.B. (1996), Plunge Pool Velocity Prediction of Jets Formed By Overtopping Steep Dams, Master's Thesis, Colorado State University.

Lewis, T.M. (1996), Prediction of Velocities Within Jets Formed By Overtopping Steep Dams, Master's Thesis, Colorado State University.

# Dam Foundation Erosion: Behavior of a Free-Trajectory Jet in a Plunge Basin

JEFFREY G. BOHRER & STEVEN R. ABT
Colorado State University, Ft. Collins, USA

## ABSTRACT

Plunging water jets from overtopping dams have the potential to erode and scour dam foundations. The erosive force of a plunging jet is related to jet velocity. A method was sought to predict the velocity decay of a free falling, highly turbulent, rectangular jet in a plunge pool. The jet velocity decay was determined to be a function of jet impact velocity with the plunge pool surface, jet density at impact with the plunge pool surface, the density of water, gravity, and plunge depth. Three types of jets were classified: a highly turbulent, fully air entrained, highly dispersed (*Fully Developed*) jet, a highly turbulent, fully air entrained, refined (*Developed*) jet and a highly turbulent, non-aerated (*Undeveloped*) jet. A model of an overtopped dam was constructed to simulate a free falling, highly turbulent, rectangular jet. Data were collected from fifty-two tests conducted at the model from which empirical equations were derived for the three jet classifications to predict jet velocity decay through a plunge pool.

## INTRODUCTION

When a dam overtops, the resulting plunging jet dislodges materials by imparting an impact force on the dam foundation and abutment materials. The removal of foundation material may jeopardize the stability of the dam. Impact force is related to velocity. Hence, it is possible to determine the impact force on a dam foundation knowing the velocity of a jet impacting on the foundation. A plunging jet may travel through a plunge pool before impacting the dam foundation, so knowing how the velocity of a plunging jet decays through a plunge pool is essential in predicting the jet impact force on the foundation and abutment areas.

A Dam Foundation Erosion (DFE) study was performed to predict jet velocity through a plunge pool. The DFE study considers three types of rectangular jets: a highly turbulent, fully air entrained, highly dispersed jet (*Fully Developed* jet); a highly turbulent, fully air entrained, refined jet (*Developed* jet); and a highly turbulent, non-air entrained jet (*Undeveloped* jet).

## DEVELOPMENT OF VELOCITY DECAY PREDICTOR

Plunging jets are highly turbulent, making theoretical predictions of velocity decay difficult. An empirical approach was taken to devise a method for predicting the velocity decay of plunging jets through a plunge pool. A dimensional analysis was conducted to provide a means of predicting velocity decay of a plunging jet through a plunge pool. The parameters affecting the velocity decay of a free falling, rectangular jet as it disperses into a plunge pool are $V_i$, the average velocity of a jet at impact with the water surface (L/T), $\rho_i$, the average density of an air entrained jet at impact with the water's surface (M/L$^3$), $\rho_w$, the density of water (M/L$^3$), $L$, the depth a jet has penetrated beneath the water surface (L), and $g$, the gravitational constant (L/T$^2$). Hence, the general expression for the velocity decay of a plunging free jet is:

$$V = f(V_i, \rho_i, \rho_w, L, g) \tag{1}$$

where $V$ is the average velocity of a jet at depth $L$ (L/T). Three Pi quantities result from Equation 1 using the Buckingham Pi Theorem. The Pi quantities are:

$$\Pi_1 = \frac{V}{V_i} \tag{2}$$

$$\Pi_2 = \frac{\rho_i}{\rho_w} \tag{3}$$

$$\Pi_3 = \frac{gL}{V_i^2}. \tag{4}$$

The combination of $\Pi_2$ and $\Pi_3$ as:

$$\frac{\Pi_2}{\Pi_3} = \left(\frac{\rho_i}{\rho_w}\right)\left(\frac{V_i^2}{gL}\right) \tag{5}$$

results in a densiometric Froude number.

### VELOCITY DECAY OF A *FULLY DEVELOPED* JET

A test facility simulating an overtopping dam was constructed at the Hydraulics Laboratory at Colorado State University to conduct a comprehensive testing program. Figure 1 depicts the test facility. Water is pumped to the model through a 20.3 cm delivery pipe. The delivery pipe supplies an orifice assembly, or diffuser, concentric with the delivery pipe. A rectangular orifice emits a highly turbulent, free falling, rectangular jet from the circular diffuser. A rectangular test basin, or plunge pool, contains the plunging jet. Tailwater control is provided using stoplogs at the test basin outlet. Once water passes over the stoplogs, a sediment trap dissipates the plunging flow energy before the flow travels through a wasteway and into a sump.

Data were collected from sixteen tests conducted at the test facility on a *Fully Developed* jet condition. The data included jet issuance velocity, impact velocity, impact air concentration, drop height, depth into a plunge pool, and velocity through a plunge pool. Jet fall height and flow rate were altered for each test. The data were used in Equations 2 and 5 and the results were plotted with Equation 2 on the ordinate and Equation 5 on the abscissa. The plot indicated a logarithmic trend. Equations 2 and 5 were modified to linearize the logarithmic data trend as:

$$\Pi_1 = -\ln\left(\frac{V}{V_i}\right) \tag{6}$$

$$\frac{\Pi_2}{\Pi_3} = \ln\left[\left(\frac{\rho_i}{\rho_w}\right)\left(\frac{V_i^2}{gL}\right)\right] \tag{7}$$

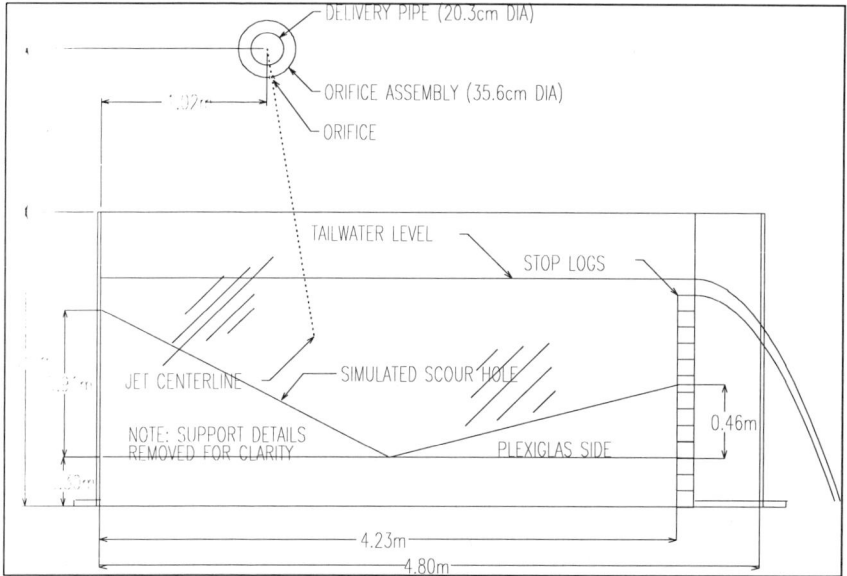

**Figure 1** Side elevation of test facility with orifice

The data from the *Fully Developed* tests were then used in Equations 6 and 7 and the results were plotted in Figure 2. The data trend in Figure 2 suggests a linear relationship.

A standardized residual analysis was conducted on the *Fully Developed* jet data to provide a statistical means of indicating possible outliers. Data with standardized residuals greater than $|\pm 2|$ were considered outliers. The outliers were circled in Figure 2 and were discounted in any further analysis. A linear regression was performed on the remaining data points in Figure 2. The resulting regression equation of the velocity decay through a plunge pool of a *Fully Developed*, rectangular, free falling jet is:

$$-\ln\left(\frac{V}{V_i}\right) = -0.5812 \ln\left[\left(\frac{\rho_i}{\rho_w}\right)\left(\frac{V_i^2}{gL}\right)\right] + 2.107. \tag{8}$$

438  ENERGY AND WATER

The corresponding $R^2$ is 0.724, indicating a relatively good correlation of data to the regression line. Because Equation 8 is an empirical expression, it is verified for

$$-0.29 < \ln\left[\left(\frac{\rho_i}{\rho_w}\right)\left(\frac{V_i^2}{gL}\right)\right] < 2.6.$$

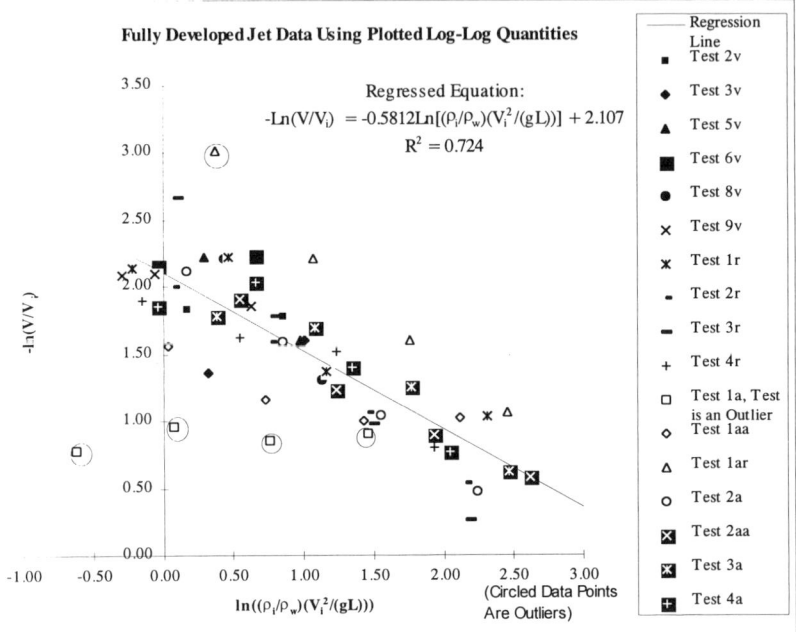

**Figure 2** *Fully Developed* **jet data plotted using logarithmic quantities**

VELOCITY DECAY OF A *DEVELOPED* JET

Data were collected from ten tests conducted on a *Developed* jet condition. The *Developed* jet data were analyzed in a similar manner to the *Fully Developed* jet data, and the results are shown in Figure 3. Four points were determined outliers as indicated in Figure 3. The resulting regression equation of the velocity decay through a plunge pool of a *Developed*, rectangular, free falling jet is:

$$-\ln\left(\frac{V}{V_i}\right) = -0.6381\ln\left[\left(\frac{\rho_i}{\rho_w}\right)\left(\frac{V_i^2}{gL}\right)\right] + 1.848 \qquad (9)$$

The corresponding $R^2$ is 0.875, suggesting a good correlation of data to the regression line. Similar to Equation 8, Equation 9 is empirical and has a limited application range of $-0.42 < \ln\left[\left(\frac{\rho_i}{\rho_w}\right)\left(\frac{V_i^2}{gL}\right)\right] < 2.05.$

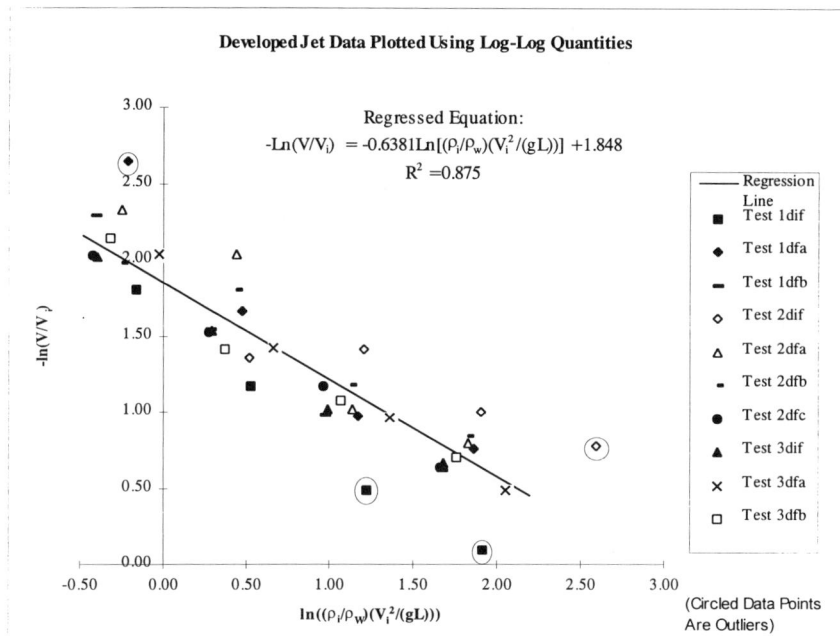

**Figure 3** *Developed* **jet data plotted using logarithmic quantities**

## VELOCITY DECAY OF AN *UNDEVELOPED* JET

Data were collected from four tests performed on an *Undeveloped* jet condition. The data were used in Equations 2 and 5 and the results were plotted as presented in Figure 4. The data from the *Undeveloped* jet test results indicated a positive, linear correlation not a logarithmic trend as exhibited in the *Fully Developed* and *Developed* jet results.

A standardized residual analysis indicated one data point in Test 3lf was an outlier. The outlier was circled in Figure 4 and discounted in further analysis. A linear regression was performed on the remaining data points in Figure 4. The resulting regression equation for the velocity decay through a plunge pool of an *Undeveloped*, rectangular, free falling jet is:

$$\frac{V}{V_i} = 0.0675\left[\left(\frac{\rho_i}{\rho_w}\right)\left(\frac{V_i^2}{gL}\right)\right] + 0.1903. \tag{10}$$

The corresponding $R^2$ is 0.864, indicating a good correlation of data to the regression line. The empirical nature of Equation 10 produces an application range of $0.51 < \left[\left(\frac{\rho_i}{\rho_w}\right)\left(\frac{V_i^2}{gL}\right)\right] < 5.76$.

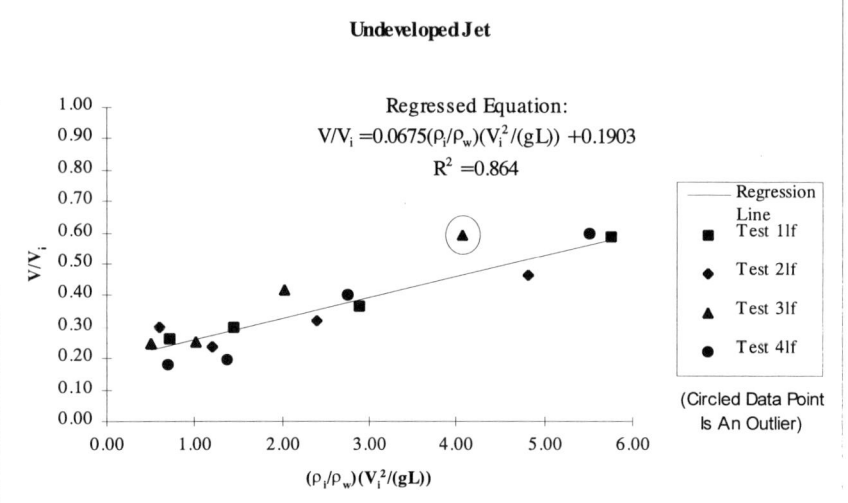

**Figure 4** Dimensionless plot of *Undeveloped* jet data

## CONCLUSIONS

An experimental study was performed to enhance scour prediction methods at the foundation and abutment areas of overtopping dams. Based upon the results and findings of the DFE study, the velocity decay of a *Fully Developed*, *Developed*, and *Undeveloped* jet is a function of impact velocity, gravity, depth into a plunge pool, the density of an impacting jet, and the density of water and can be expressed as presented in Equations 8 through 10. Validation of these findings on prototype structure is encouraged prior to application.

## REFERENCES

Albertson, M. L., Dai, Y.B., Johnson, R. A., and Rouse, H. (1950) Diffusion of Submerged Jets, *Transactions of the ASCE,* Vol 115.

Cola, R., (1965) Energy Dissipation of a High Velocity Vertical Jet Entering a Basin, *Inter. Assc. Hydr. Res. - Proceedings.*, 11th Congress, Vol. 1, pp. 1.52.

Ervine, D. A., and Falvey, H. T., March (1987) Behaviour of Turbulent Water Jets in the Atmosphere and in Plunge Pools, *Proceedings of the Institution of Civil Engineers, Part2,* **83**.

# Dam Foundation Erosion:
# Pit 4 Dam Scale and Prototype Model Test Results and Comparison

R.J. Wittler[1], S.R. Abt, & J.F. Ruff[2], G.W. Annandale[3]

## ABSTRACT

This paper compares the erosion in a 1:3 scale model and a 1:1 prototype scale model with the Mason and Veronese equations. The models simulate flow overtopping a steep dam, impacting at the base of the dam. The purpose of the comparison is to validate either the Mason or Veronese equations for a particular application.

## INTRODUCTION

The Dam Foundation Erosion Study (DFE) is an ongoing cooperative dam safety research study between the US Bureau of Reclamation, Pacific Gas & Electric (PG&E), Western Area Power Administration, the US Federal Energy Regulatory Commission, Colorado State University, and Golder Associates. There are three facets to the study: a scale model, a prototype model, and a numerical model. The primary objective of the study is to create a numerical model that simulates the erosion in the foundation areas of a dam under overtopping conditions. A sub task of the study is a set of experiments simulating specific conditions at a dam in northern California, USA. The model configurations simulate the material near the base of the dam. The report *Pit 4 Dam: Slab and Buttress Foundation Scale Model Simulation* [4] gives details on the first experiments in the scale model facility for PG&E.

## FACILITIES

The Prototype and Scale Model Facilities are located at the Hydraulics Laboratory, Engineering Research Center, Colorado State University, Fort Collins, Colorado, USA. The prototype model is three times larger than the 1:3 scale model, with the exception of basin length. The prototype length is forty-one percent greater than the scaled length of the scale model. Figure 1 shows the plan of the prototype facility. Figure 2 shows the plan of the 1:3 scale model facility. Both facilities include a water supply and flow measurement system, a nozzle assembly, a test basin, wasteway, walkways, and viewing platforms. There is a material recovery system in the wasteway of the scale model preventing large amounts of eroded material from entering the laboratory recirculating water supply. Figure 3 shows relative position of the dam crest and the scale model for the calculated trajectory of a 1.85 ft$^3$/s/ft jet.

---

[1] Hydraulic Engineer, US Bureau of Reclamation, Denver, Colorado, USA.
[2] Professors, Dept. of Civil Engineering, Colorado State University, Fort Collins, Colorado, USA
[3] Director, Water Resources Engineering, Golder Associates, Lakewood, Colorado, USA

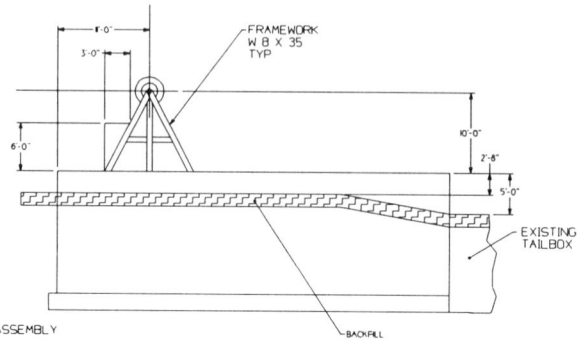

Figure 1. Section of prototype facility.

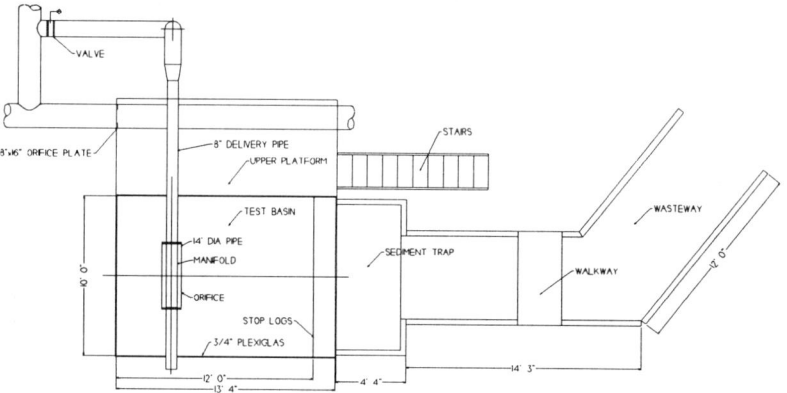

Figure 2. Plan of 1:3 scale model facility.

## MATERIAL SIMULATION

The dam is located in the northeastern portion of Shasta County, California, within extensive volcanic terrain of the Cascade Range geomorphic province. The most widespread and continuous rock unit of the Cascade Range province in Shasta County is the Tuscan Formation. This formation is comprised of andesitic, basaltic, and dacitic breccia, tuff breccia, and interlayered flow rock of various compositions. Bedrock in the dam area mapped by the California Division of Mines are described as andesitic flows and pyroclastic rocks. Mapping shows basalt and tuff at the site as well as surficial units of talus and overburden. Relatively sound, blocky basaltic flow rock is exposed at the right abutment area along with underlying tuff. The contact between the flow rock and tuff is well defined, although mixed as the underlying tuff appears to fill columnar joint spacing at the base of the flow. There are no bedrock exposures in the dam's left abutment area. The left abutment of the dam is occupied by an ancient landslide deposit comprised of basaltic boulders and cobble-sized fragments set in a silty and sandy matrix. Surface relief of the deposit exhibits hummocky topography, closed depressions, and an abrupt change in slope at its upper end. There are no indications of recent landslide activity.

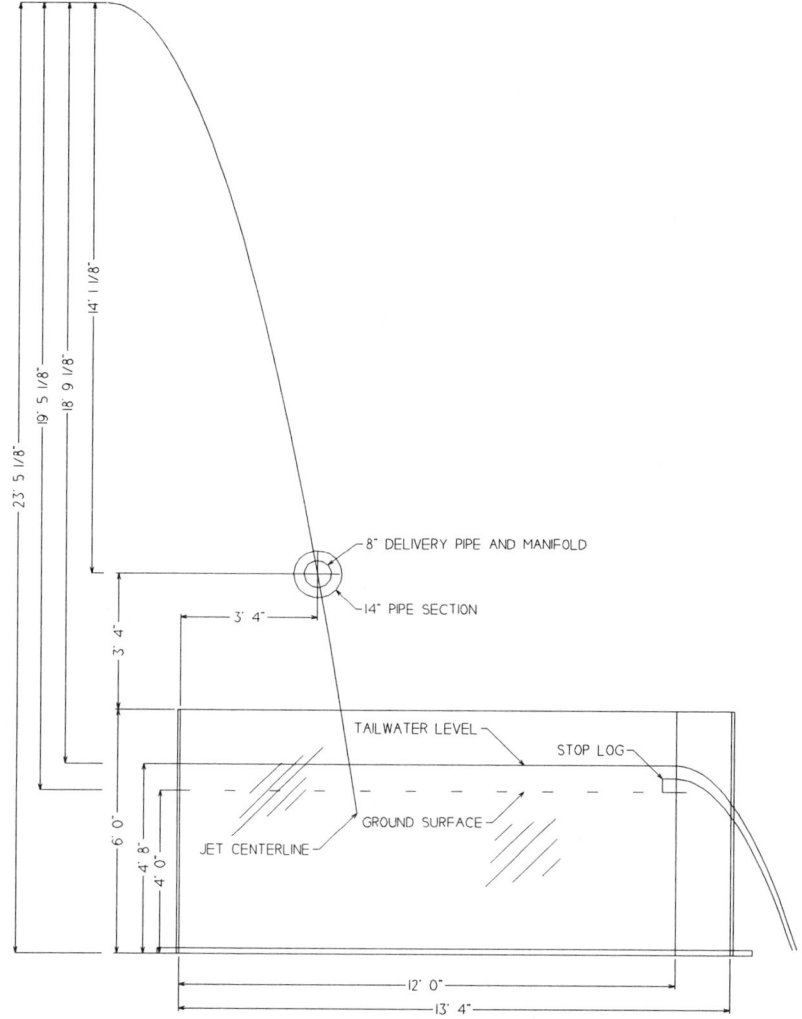

**Figure 3. Section of 1:3 scale model showing jet trajectory.**

## ANALYSIS OF SAMPLE MATERIAL

Table 1 shows the results of the gradation analysis of sample material collected at the dam site. The sample material consists of basaltic boulders and cobble-sized fragments set in a silty and sandy matrix. This is characteristic of the material below the buttress section of the dam, adjacent to the left abutment. The material in the sample has an aggregate specific gravity of roughly 2.65. The basis for the scale model gradation is a 1:3 Froude scaling of the fall velocity.

Table 1. Material and scale model size fractions.

|  | Pit 4 Size (mm) | Pit 4 Fall Vel. (cm/s) | Scaled Fall Vel. (cm/s) | Scale Model Size (mm) |
|---|---|---|---|---|
| $D_0$ | 0.03 | 0.07 | 0.04 | 0.02 |
| $D_{20}$ | 0.19 | 2.10 | 1.21 | 0.12 |
| $D_{40}$ | 0.54 | 6.00 | 3.46 | 0.30 |
| $D_{60}$ | 1.30 | 12.54 | 7.24 | 0.65 |
| $D_{80}$ | 5.00 | 24.60 | 14.20 | 1.65 |
| $D_{100}$ | 100.00 | 110.00 | 63.51 | 33.07 |

$$C_u = \frac{D_{60}}{D_{10}} = 15.000$$

$$C_c = \frac{D_{30}^2}{D_{10}D_{60}} = 0.798$$

## EROSION PREDICTION

Equation 1 is the Veronese [2] equation. The equation yields an estimate of erosion measured from the tailwater surface to the bottom of the scour hole.

$$Y_s = 1.90 H^{0.225} q^{0.54} \tag{1}$$

$Y_s$ = depth of erosion below tailwater (meters)
$H$ = elevation difference between reservoir and tailwater (meters)
$q$ = unit discharge (m³/s/m)

The Veronese equation neglects material properties. Yildiz [3] presents a modified version of the Veronese equation, including the angle of incidence, $\theta$, from the horizontal of the jet.

$$Y_s = 1.90 H^{0.225} q^{0.54} \sin\theta \tag{2}$$

Equation 3 is the Mason [1] equation, based on model data. Note that the Mason model equation is dimensionally homogeneous and includes the median grain size of the foundation material.

$$Y_s = \frac{3.27 q^{0.60} H^{0.05} h^{0.15}}{g^{0.30} d^{0.10}} \tag{3}$$

$h$ = tailwater depth above original ground surface (meters)
$d$ = median grain size of foundation material, $d_{50}$ (meters)
$g$ = acceleration of gravity (m/s²)

The Mason [1] equation for prototype data is, assuming $d$ is constant for prototypes, and using metric units for the coefficients:

$$D = K \frac{q^x H^y h^w}{g^v d^z} \tag{4}$$

$K = 6.42 - 3.10 H^{0.10}$, $x = 0.6 - \frac{H}{300}$, $y = 0.15 + \frac{H}{200}$
$v = 0.30$, $w = 0.15$, $z = 0.10$, $d = 0.25 m$

The Mason equations do include a material factor, $d$. However, it is unlikely that this factor adequately represents the wide variety of materials and material properties found in typical dam foundation. The reasons for endorsing the Mason formulas are the thoroughness of the research, the comprehensive nature of the data, including scale model studies and prototype case studies, and the dimensional analysis and dimensional homogeneity of the formulas.

## RESULTS

Table 2 shows the value of the properties for the Veronese and Mason equations. The difference between the head, $H$, for the model scaled to prototype and the prototype itself is due to the capacity limitation of the 1:3 scale model facility.

Table 2. Variables for erosion formula comparison.

|  | 1:3 Scale Model | Model Scaled to Prototype | Prototype Model | Dam |
|---|---|---|---|---|
| $H$ (m) | 3.55 | 10.65 | 11.90 | 18.20 |
| $q$ m³/s/m | 0.173 | 0.897 | 0.897 | 0.93 |
| $h$ (m) | 0.20 | 0.61 | 0.61 | 3.41 |
| $d_{50}$ (m) | 4.5E-4 | 4.3E-4 | 2.8E-3 | 8.8E-4 |
| $g$ m/s² | 9.81 | 9.81 | 9.81 | 9.81 |

Figure 4 shows an orthographic view of the scour hole contours in the prototype model. The flow in the figure is right to left. Note the scour in the upstream, left corner. This scour, observed during the test, is due to strong recirculation currents.

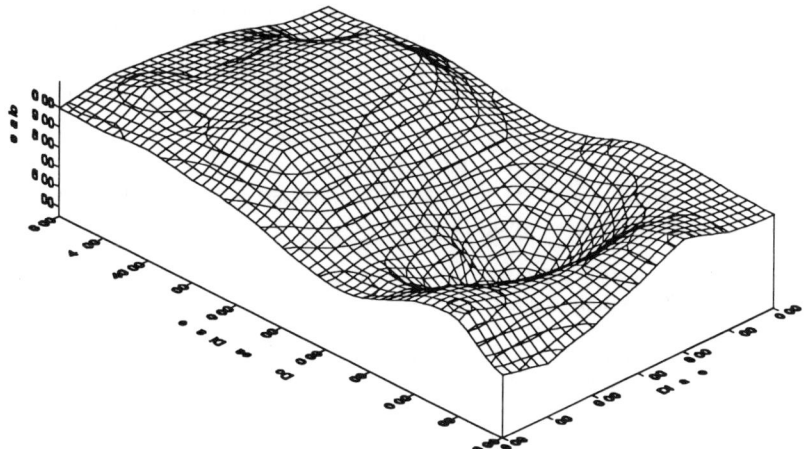

Figure 4. Orthographic contour of a scour hole in the prototype facility. Flow is right to left.

Table 3 shows the results of applying the Veronese and Mason formulae to the Pit 4 simulation and to Pit 4 Dam. Where a formula is not applicable, the symbol N/A appears.

**Table 3. Formula estimates of erosion below tailwater and model results.**

|  | Model Eqtn. (m) | Model Results (m) | Prototype Eqtn. (m) | Prototype Results (m) |
|---|---|---|---|---|
| Veronese | 1.00 m | 0.71 m | 3.13 | 2.57 m |
| Modified Veronese | 0.99 m | 0.71 m | 3.02 | 2.57 m |
| Mason Model | 1.07 m | 0.71 m | N/A (2.92) | N/A |
| Mason Prototype | N/A | N/A | 1.55 | 2.57 m |

The model comparison shows that the Veronese and Modified Veronese equations over predict the scour in the model by roughly 42%. The Mason Model equation over predicts scour by roughly 51%. The prototype comparison shows that the Veronese and Modified Veronese equations over predict scour by roughly 20%. But unlike the model comparison, the prototype results are under predicted by the Mason Prototype equation by roughly 39%. The Mason Model equation, applied to the prototype conditions, over predicts scour by roughly 14%. The results indicate a greater degree of consistency for the Veronese equations between the scale model and prototype model. The Mason prototype equation does not appear to be a good predictor for this application.

## CONCLUSIONS

The Mason model equation compares most favorably with the prototype model results. However, both Veronese equations produce good estimates of erosion, both in the scale and prototype models. Of the two Veronese equations, the Modified Veronese is preferred. The Mason prototype equation under predicts the depth of scour, an unfortunate event. The scale of the prototype model may not be large enough for proper application of the Mason prototype equation.

## REFERENCES

[1] Mason, P.J., Arumugam, K., "Free Jet Scour Below Dams and Flip Buckets," *Journal of Hydraulic Engineering*, vol. 111, No. 2, ASCE, February 1985.

[2] Veronese, A., "Erosioni de Fondo a Valle di uno Scarico," *Annali dei Lavori Publicci*, vol. 75, No. 9, pp. 717-726, Italy, September 1937.

[3] Yildiz, D., Üzücek, E., "Prediction of Scour Depth From Free Falling Flip Bucket Jets", *Intl. Water Power and Dam Construction*, November, 1994.

[4] Wittler, R.J., et. al., "*Pit 4 Dam: Slab and Buttress Foundation Scale Model Simulation.*" PAP 681, US Bureau of Reclamation Water Resources Research Laboratory, April 1995.

# Overview and Planning Considerations for Improving Reservoir Releases

### W. GARY BROCK AND J. STEPHENS ADAMS
Tennessee Valley Authority, Norris, Tennessee USA

## INTRODUCTION

For over 20 years, the Tennessee Valley Authority (TVA) has been making improvements in the water quality of reservoir releases, primarily by providing minimum flows and increases in dissolved oxygen (DO) levels. In 1979, TVA formally initiated the Reservoir Releases Improvements (RRI) program to plan and implement solutions to water quality problems at specific projects. Under this program, several minimum flow improvements were made, and advancements in aeration technologies were researched, developed, and implemented.

In 1991, the TVA Board of Directors approved a 5-year Lake Improvement Plan (LIP) that included a component to improve the quality of the releases at 16 dams by increasing DO levels and sustaining minimum flows. Minimum flows were implemented the day the recommendations were approved, and enhancements have been implemented to extend river reaches with instantaneous flows. Implementation of aeration systems at all 16 projects were completed in 1996. Since provisions for water quality were not included in the original project design for most projects, retrofits and modifications to operations were required to achieve the desired improvements.

The aeration and minimum flow systems implemented under the LIP program have resulted in an overall improvement in the quality of the tailwaters downstream of TVA dams. DO levels have been improved in over 300 miles (482 km) of rivers, and minimum flows are already helping in the recovery of over 180 miles (290 km) of aquatic habitat lost from intermittent drying of the river bed. Surveys have confirmed that a biological response is occurring in tailwaters where water quality improvements have been implemented, with an increase in diversity and numbers of invertebrate and small fish species. Visits at many tailwaters have increased significantly, and some of the cold-water fisheries are now considered to be of national significance and are having positive local economic impacts. Additionally, some local waste treatment facilities have been able to renegotiate discharge requirements because of the higher minimum flows and DO improvements.

The success of the LIP program can be attributed to several factors including the support of the management team, the dedication of sufficient staff and resources to the program, research and development efforts, development and implementation of a planning process, development and utilization of analysis tools, communication, coordination, and teamwork.

## PLANNING PROCESS

Based on experience, the planning process has proved to be key to the program's success. During the program, a process was developed to facilitate planning efforts to achieve broad-based support for the recommendation and implementation of the most appropriate improvement options for each LIP project. The importance of the planning process was demonstrated time and time again due to the unique challenges presented by each project, the complexity and variety of potential aeration and minimum flow options considered, and the need for cooperation and coordination with resource agencies, interested parties, and other stakeholders. Other factors incorporated into the planning process included costs, research and development of technological solutions, quantification of expected tailwater effects for various improvement options, and intangibles or effects difficult or impossible to put a dollar value on. The remainder of this paper provides a description and discussion of the planning process used by TVA for LIP projects.

## GROUP APPROACH

Planning studies for LIP projects were conducted using a group approach. At the beginning of a specific study for most projects, a study team was formed. The TVA team was composed of TVA staff with expertise in all areas relevant to the specific project. Typical composition of this team would include staff with expertise in project management, hydroplant management, power system operation, research and testing, water quality, fisheries, design and construction, cultural resources, environmental review, and recreation. The purpose of the study team is to identify and evaluate potential aeration and minimum flow options, and to select the most appropriate options for recommendation for implementation.

Many advantages have been realized by utilizing the group approach. The diverse composition of the study team provides a look at the project from the perspective of all program participants. This helps in assuring that all viable options are identified, leaving a minimal chance that a good option has been overlooked. In addition, all relevant issues associated with the project can be identified and addressed early in the planning process, hopefully eliminating surprises and unnecessary delays later on. Appropriate environmental review is accomplished during the study process. All staff with input to the review are involved in the project from the beginning, allowing formal environmental review documentation to occur smoothly at the end of the process. The group approach also results in widespread support for the final study team recommendations. This is attributed to participants being a part of the process from the beginning, having the opportunity to provide their input, and hearing all the arguments associated with all options. Utilizing the group approach also helps to maintain good communication among staff and organizations.

The study team also actively seeks to obtain input during the planning process from interested outside sources. In preparing the Environmental Impact Statement (EIS) for the LIP, TVA sought to ensure that the environmental effects of reservoir operating alternatives were thoroughly investigated, and ample opportunities for public review and comment were provided. Broad public participation was essential to ensure that the needs and desires of the regional communities were fulfilled. A similar approach was continued into the implementation phase of the program. Appropriate State resource agency staff were consulted to keep them abreast of LIP plans and activities, to obtain information on their objectives at specific projects, and to determine their preferences for implementation options. This allowed the States to be a partner and helped ensure that TVA considered the fisheries management plans of the States. In addition, local officials were contacted to obtain their input.

## PLANNING PROCESS STEPS

The following list presents the basic steps used for conducting the LIP planning studies.

- Identify project objectives
- Identify potential options
- Narrow the number of options using a screening process
- Prepare cost estimates
- Evaluate intangibles
- Select and recommend most appropriate options for implementation

## IDENTIFY PROJECT OBJECTIVES

TVA prepared an EIS for the LIP, and objectives regarding minimum flow and DO levels were specified. Options considered for each project had to meet these objectives. It is important that project objectives are clearly stated and agreed upon up-front by all parties involved to avoid misunderstandings and disappointments later in the project.

## IDENTIFY POTENTIAL OPTIONS

One of the first tasks in the planning process is to identify all potential minimum flow and aeration options to be considered. This is done by compiling a listing of all possible options, regardless of perceived feasibility, and noting advantages and disadvantages. This provides a basis for a review of each option. A full range of options were considered by TVA for providing minimum flows and improving DO. The following is a listing of typical options considered.

Minimum Flows Options:

- reregulating weir
- small generating unit
- turbine pulsing
- sluicing

Aeration Options:

- aerating weir
- oxygen injection
- compressed air
- turbine venting
- surface water pumps
- autoventing turbines

One of the challenges of planning LIP projects was that significant research and development efforts took place during implementation of solutions. This resulted in the creation of state-of-the-art systems that represent both the development of new technologies and the improvement of existing technologies. Significant improvements to surface water pump and oxygen injection systems were made during the 5-year implementation period. In addition, a variety of aerating weir innovations were conceived and introduced as alternatives. These included an extended-crest labyrinth weir installed at the South Holston project and a completely new design, the infuser weir, installed at the Chatuge project. At the Norris project, new turbine runners were installed that were the first in the world designed to both aerate turbine releases and increase the plant's power generation performance.

## NARROW THE NUMBER OF OPTIONS

The list of potential options is usually much too long for a detailed evaluation of each option to be conducted. It is preferable to reduce the number of options to about four or five if possible using a screening process, and study these further.

In the initial study team meeting, each option is briefly discussed to determine whether further investigations are warranted. Sometimes more than four or five options are selected during this meeting, and further study is required to narrow the options to the desired number.

An important objective in the initial study team meeting is to discuss existing conditions at the project and in the tailwater downstream. These discussions are intended to provide all participants with a good understanding of project features, operating policy, purposes, water quality, tailwater uses, and other considerations so that all relevant issues can be identified and addressed early in the process. This information is also used to avoid consideration of options that would adversely impact existing project purposes and uses.

## PREPARE COST ESTIMATES

Detailed cost estimates were prepared for each of the remaining options. Estimated costs included capital costs, operation and maintenance expenses, and power system impacts. Total capital costs included expenses for design, procurement, and installation of each option including any land purchase requirements. Interest and amortization costs were the annualized value of total capital costs assuming specific evaluation periods and discount rates. Operation and maintenance (O&M) costs were the annualized value of O&M expenses expected to be incurred over the evaluation period. Power system costs were the annualized value of all power system impacts expected to be incurred by operation of each option. These impacts included capacity losses, lost energy generation, and shift of energy from peak to offpeak periods.

## EVALUATE INTANGIBLES

Even though costs are important, intangibles, or items difficult or impossible to put a dollar value on, are also important considerations in option selection. The following is a typical list of intangibles considered in the selection evaluation.

- Safety
- Water quality
- Reliability
- Public interest
- Private land impacts
- Cultural resources impacts
- Miles of tailwater improved

Proper evaluation of intangibles can involve significant effort. To assist in the evaluation of water quality effects and impacts to aquatic life, TVA developed analysis tools during the course of the LIP and RRI programs. These included multi-dimensional hydrodynamic, thermal, and water quality models of reservoirs and tailwaters that helped quantify the tailwater effects due to various improvement options. In addition, a suite of linked models for tailwater analysis was developed to explore tradeoffs among various release improvement options in terms of their affect on downstream river physical habitat and biological response. These evaluations included minimizing adverse impacts from water quality variables such as temperature that are outside the scope of the project mandate.

Testing in a hydraulic laboratory was conducted as necessary to evaluate the safety of potential solutions, and to provide designs for minimizing safety impacts. Assessing private land impacts involved dealing with land owners and providing judgments to the study team regarding specific land related impacts. Evaluating cultural resources impacts often involved field and office investigations as well as close coordination with appropriate State agencies.

## SELECT AND RECOMMEND MOST APPROPRIATE OPTIONS

The last study team meeting was convened when all information needed for selecting the most appropriate minimum flow and aeration options had been obtained including an economic analysis. The goal of the selection process is to select the most appropriate minimum flow and aeration option for implementation. Even though cost is a very important issue, options for some projects may be selected that were not the least cost due to other important intangible considerations. Each option is thoroughly discussed and options for recommendation for implementation are selected.

## CONCLUSIONS

The planning process described in this paper has been successfully utilized by TVA for several of its LIP projects. Probably the most important factor contributing to this success is use of the group concept during the entire planning process. Involving the appropriate staff and other interested parties from the beginning of the process allows a more complete analysis to be made, promotes cooperation, improves relations both inside and outside TVA, results in the most appropriate options to be recommended, and provides the best chance for concluding with the support of all participants for the recommended options.

## REFERENCES

Adams, J. Stephens and W. Gary Brock, "A Process for Selecting Options to Improve Water Quality Below TVA Hydro Projects," Waterpower '95, San Francisco, CA.

# EDF EXPERIENCE IN IMPROVING RESERVOIR RELEASES FOR ECOLOGICAL PURPOSES

Ph. GOSSE, C. SABATON, F. TRAVADE, J.EON*
Electricité de France, 6 quai Watier 78401 CHATOU Cedex FRANCE
*La Défense 6, cedex48 92068 PARIS FRANCE

**ABSTRACT:**

This paper presents some recent research projects on optimizing releases from dammed reservoirs, in terms of water quantity and quality alike, with a view to improving living conditions for fish populations in river environments.

## 1. INTRODUCTION

Some three-quarters of the water stored in dammed reservoirs in continental France - nearly 7 billion m3 - is managed by EDF. Intended at the outset essentially for hydroelectric purposes, these reservoirs are increasingly solicited for other uses; notable examples include regional and national agreements on releases recently signed, with the regional Water Agencies to provide complementary river discharge in low-water periods (1990-1992), with the French government to ensure sufficient water supply in drought periods (May 1990), or with the French federation for kayaking, to enable engaging in this sport, especially for competitions.
The need to improve water releases from reservoirs so as to guarantee good quality in aquatic ecosystems was underscored in an official agreement concluded between EDF and the Ministry of the Environment in 1982, in the fishing law of 1984 and in the water law of 1992.
This article presents some major recent research projects on optimizing releases from EDF reservoirs, in terms of water quantity and quality alike, with a view to improving the state of fluvial ecosystems without overly limiting the potential of hydroelectric plants which represent close to 15% (60 TWh) of all electricity generated in France.

## 2. IMPROVING HABITAT QUALITY FOR NON-MIGRATING FISH

The management mode adopted at a hydroelectric reservoir modifies the flow regime downstream in the river, creating conditions which are more or less favorable to equilibrium in the aquatic ecosystem: low discharge ( base or guaranteed flow) in the by-pass section (between the dam and the point of restitution of turbined water), discharge fluctuations downstream of the restitution point due to hydropeaking.
Following studies to highlight the ecological impact of this management, research has focused on the means of diminishing it. Below are the results of some recent studies, conducted in collaboration with various French research laboratories, on salmonids, a species that populates the majority of rivers on which EDF has hydroelectric dams.

### 2.1. By-pass section
How can one determine the minimum discharge in a by-pass section, or the guaranteed flow at the dam, which can satisfy both salmonids and the plant operator, who understandably can consider this discharge as lost(Gras *et al*, 1994)?
To define the best value, the "microhabitat" method (Bovee,1982) was first validated and then adapted to French rivers. This method determines the habitat quality of a river for salmonids - calculation of the WUA, or weighted usable area - on the basis of three hydraulic parameters: velocity, depth and granulometry of the river bed (the first two

parameters being directly linked to the discharge). As Bovee had done previously for American rivers, experiments were conducted on French rivers ranked in fishing category 1; for each of the hydraulic parameters studied, habitat preference curves were traced for brown trout at three stages in the life cycle (fry, juveniles and adults) on the basis of both electrofishing surveys (Belaud et al,1989; Fragnoud,1987) and expert opinions (Chaveroche et al,1989).
The results obtained enabled verifying the benefits of the method; the consistency between the curves obtained and those proposed by Bovee for brown trout confirm the concept of habitat preference (Sabaton et al,1993). The experiments also enabled proposing a set of preference curves well suited to French rivers, both for brown trout and for salmon (in this latter case, fry, juveniles and reproduction stage), and drawing up a precise protocol so as to provide teams wishing to apply the method with a pertinent, reliable analytical tool (Sabaton et al,1995).
The microhabitat method has been applied on a number of occasions, particularly in attempts on several rivers to preserve a habitat favorable to recolonization by salmon. Examples include the proposal of guaranteed flow values to be adopted downstream of Rabodanges reservoir on the Orne river in Normandy (Courot,1988) and downstream of Queuille reservoir on the Sioule, in the Massif Central region (Sabaton et al,1996a). On a larger scale, the method was used to determine a satisfactory guaranteed flow for trout downstream of a number of dams, in the framework of renewal of their operating licenses. An inter-institution French working group was set up to monitor eight of these sites over the longer term; the aim is to see whether the evolution in habitat quality subsequent to implementation of the new discharge values corresponds to a real improvement in fish biomass (Merle et al,1996). For the purposes of illustration, below are the results of the application of the microhabitat method to Beyrède dam on the Neste d'Aure in the Pyrenees (Baran et al,1996). Present guaranteed flow is 500 l/s; the future value must represent a good trade-off between the increase in WUA for adult trout at the chosen discharge and the reduction in WUA for young life stages (Figure 1). Discharge of 1.5 m3/s should enable ensuring both virtually optimum conditions for fry and juveniles and very satisfactory conditions for adults (with WUA values significantly higher than at present). If it is economically acceptable, a guaranteed flow around this value should no doubt be adopted when the dam license is renewed, and the planned hydrobiological monitoring will enable verifying the resulting improvement.
These 8 sites are sufficiently high in altitude to preclude any limiting factors for the fish habitat other than the hydraulic parameters considered in the microhabitat method. In the case of Rabodanges, at a low altitude (125 metres), hydrobiological monitoring between 1990 and 1995 demonstrated concretely (Salignat et al,1996) that the increase in the guaranteed flow in 1990 to provide for a more favorable habitat was not sufficient to ensure growth of salmon fry, as the poor water quality in the Orne river has extremely negative effects.

## 2.2. The hydropeaking zone

A number of studies have also been conducted to determine the impact of hydropeaking on the aquatic ecosystem (Sabaton et al,1996b). They focused on various environmental compartments: water quality and thermal regime, invertebrate population and fish populations. The diversity in results showed the complexity of the phenomena at work. However, the importance of river bed morphology was clearly pointed up: in particular, the more abundant the hydraulic shelters, the lesser the impact of hydropeaking on fish populations. The role of the base "non-hydropeaking" discharge is also significant: the better the habitat conditions at base discharge, the lesser the impact. As a result, it may be necessary to optimize the guaranteed flow, not only in the by-pass section but downstream of the turbined water restitution as well.
This is the case downstream of Queuille reservoir on the Sioule, where efforts are being made to promote recolonization by salmon. The presence of pebbles and large rocks in all the riffle zones favorable to growth of salmon parrs should limit the impact of high discharge. The choice was therefore made to concentrate on the value of the base discharge. The WUA as determined by the discharge released at the plant depends on the season, given the major variations in inflow along this segment during the year. In fact, it would be preferable to give priority to the reproduction stage from November to April, while maintaining good conditions for the development of young fish; from May to October, it is development of the young which should take priority. Adopting a discharge of around 5 m3/s to 5.5 m3/s in winter and 4 m3/s in summer would appear to be acceptable.

## 3. IMPROVING PASSAGE OF MIGRATING FISH

The 1984 law mandates "free circulation" for migratory fish on a certain number of French rivers classified as migration zones. This requires that specific facilities be installed at hydroelectric dams to ensure that fish migrating up rivers can overcome obstacles (fishpasses) and to prevent passage of migrants through the turbines as they descend toward the sea (downstream migration devices). Above and beyond the design of such installations, which will not be dealt with here, one problem is to define the reservoir discharge needed to ensure good efficiency.

### 3.1. Upstream migration devices
#### 3.1.1. Localization
Fishpasses to facilitate upstream migration can be located either by the plant or at the dam. Their localization depends on the discharge respectively turbined at the plant and spilled at the dam (which is to say the guaranteed flow plus overspilling in flood periods) during the periods of migration of each of the migratory species considered, as well as on the technical feasibility of their implantation. In some cases, it is not feasible to position them by the plant, particularly at plants supplied by a long penstock. If the values for mean discharge spilled at the dam and that turbined at the plant are comparable, it may be necessary to build two passes, one by the plant and the other at the dam. In France, poor placing of a number of fish passes has resulted in their failure; this is particularly true on relatively large rivers (Garonne, Dordogne, Rhône, etc.) where, when the hydroelectric plants were built (late 19th and early 20th century), the passes were principally located at the dam rather than by the plant. Fishpasses reinstalled in the past decade on the same sites have been successfully positioned by the plants.
#### 3.1.2. Discharge in the pass
Fish are attracted into the pass only by suitable flow rates. The important factors are the velocity at the entrance to the pass and the relative discharge in the pass in comparison with those exerting a competitive attraction (turbined flow or dam spillage). It is considered in France (Larinier et al,1984) that the discharge in the pass should be at least 1% to 5% of the turbined discharge (passes located by the plant) or dam spillage (passes at the dam); this proportion depends on the site, the migrants concerned and the period in the year. On the large French rivers Garonne and Dordogne, whose annual mean flow is between 250 and 500 m3/s, discharge in the recently-built passes is between 3 and 10 m3/s (Travade et al,1996).
In addition to the need to attract fish into the pass when they are at the foot of the obstacle, there is the question of the discharge needed to encourage them to swim up the by-pass section when the pass is located by the dam. This discharge must fulfill two requirements: attract the fish into the by-pass section while minimizing the length of time they remain below the turbines, and allow for their progress through the by-pass section (water depth and velocity compatible with their swimming capacity). The discharge values that meet both criteria must be determined on a case-by-case basis depending on the site and the species concerned. Radiotracking surveys of salmon at Poutès on the Allier (annual mean flow of 16 m3/s, maximum turbined discharge of 28 m3/s and a 10-kilometer by-pass section) have shown that discharge of 0.5 m3/s in the by-pass section is insufficient to encourage upstream passage of fish, due to physical blockage (riffles less than 15 cm deep) and for behavioral reasons (fish backtrack once they have entered the by-pass section). At the present time, a guaranteed flow of 2.5 m3/s is being tested at Poutès dam.
If it is not acceptable for economical reasons to increase the guaranteed flow to a great extent, it could be envisaged to try occasional spillage at the dam to attract fish and prevent them from being blocked below the turbines. The studies at Poutès have shown that such spillage is effective only on three conditions: discharge of several m3/s at the dam (6 m3/s minimum), simultaneous suspension of turbining, duration of spillage for several hours (12 hours minimum).

### 3.2. Downstream migration devices (DMD)
In general, these are gates (with by-pass conduits) located in the near upstream of the turbine intake, which help prevent fish from passing through the turbines, a potential cause of major mortality. The technology of such devices is as yet poorly understood. Recent studies in France (Larinier et al,1996) on juvenile salmonids (salmon, sea trout)

have enabled fine-tuning surface DMDs placed in the immediate vicinity of water intake screens. One of the fundamental parameters in DMD efficiency is the discharge, which must be defined in relation to the turbined discharge. The minimum efficient range found to date for small plants (with turbined discharge between 20 and 30 m3/s) is from 3 to 5% of the turbined discharge, or from 0.6 m3/s to 1.5 m3/s at the plants studied. We do not yet know whether this discharge range is suitable at installations with higher turbined discharge.

## 4. IMPROVING WATER QUALITY

One large-scale research project has been under way since 1982 to identify and minimize the impact of total reservoir draining, required by French regulations(Cardinal, 1988; Galland et al,1996; Guide vidange,1994).
As concerns normal hydroplant operating conditions,the most notable recent project on improving downstream water quality for fish is probably that of 1995 at Petit Saut reservoir in French Guiana (3.5 billion m3 with a depth of 35 meters at the dam), where the survival of fish populations over a 50-kilometer stretch of the Sinnamary (250 m3/s mean annual discharge) from the dam to the estuary, was at stake.
Here, degradation of the tropical forest (300 km2), flooded as the reservoir was gradually filled (from January 1994 to mid-1995), led to the disappearance of dissolved oxygen (DO) in the reservoir except in a thin surface layer (sometimes less than 20 cm thick in 1994, and now close to 3 meters on average in 1996). It was nonetheless possible to maintain a constant level of 4 mg/l of DO in the Sinnamary downstream, as long as the guaranteed flow of 100 m3/s was obtained by opening the lower gates at the foot of the dam on the left bank. The jet phenomenon and the hydraulic jump in this area created significant mixing of air and water which immediately raised the DO content to the level of equilibrium with the atmosphere (close to 8 mg/l for the temperature observed, around 25° C).
Serious difficulties arose in June 1994 with the first turbining tests, which projected an underflow (with no entrainment of atmospheric air) at the foot of the dam on the right bank (in the plant outlet canal which joins the river 300 metres downstream). It was then thought that dilution of the turbined discharge - containing no oxygen but withdrawn a little less deep - by a more or less equivalent discharge through the bottom gates would suffice to maintain the chosen base level of 2 mg/l of DO in the river. It was found, however, that far more additional discharge was needed from the bottom gates. Which means that the mixing between the atmosphere and the water from the bottom gates was eliminating dissolved gases formed in anaerobic conditions at the bottom of the reservoir, with these gases consuming several mg/l of DO in less than two days in the aerated aquatic medium (Gosse, 1994). It was demonstrated that the gas primarily responsible was methane (CH4) - whose levels exceeded 10 mg/l at the bottom of the reservoir beginning in October 1994, creating a potential DO demand greater than 30 mg/l - and that its degradation in the oxygenated medium was of bacterial origin (Galy-Lacaux et al,1995; Dumestre et al,1996).
To solve the problem, it was decided in July 1994 to set up an overflow weir immediately downstream of the dam in the plant outlet canal, in order to bring the DO and CH4 levels in the turbined water as close as possible to the atmospheric equilibrium. The device chosen uses the entire technically possible head and has two successive nappes (Figure 2). In the absence of a satisfactory general model enabling precise prediction of the effect of such an installation on DO (and even less on methane, for which the targeted elimination by means of a weir was probably a first), we drew on the experience of similar devices throughout the world - three small installations by the Tennessee Valley Authority (TVA) in the United States (Hauser et al,1993)- and on factors identified as being essential to good entrainment and dissolution of air bubbles per unit of water volume: effective fall height of nappe or difference in altitude between the upstream and downstream water levels (with good efficiency between 2 and 3 meters according to Hanisch and Tron who analysed navigation weirs), depth and volume of the downstream plunge pool, optimization of the thickness and width of the nappe, presence of obstacles (such as piles) promoting a break in the falling nappe and the entry of air around the fall (talks with G.Hauser/V.Alavian from TVA; literature).
This device designed in collaboration with EDF/LNH and EDF/CNEH was installed in March 1995 and has shown (measurements of Petit Saut Environmental lab) good efficiency, with close to 80% of the methane eliminated from the turbined water and DO immediately downstream of the weir close or greater than 90% of the saturation

level, for turbined discharge up to 200 m3/s, which is compatible with the present demand for electricity in French Guiana. On October 22, 1996, for example (water temperature at 26° C, all discharge from the dam being turbined), with turbined discharge at 230 m3/s and CH4 and DO levels respectively measured at 5 mg/l and 0.8 mg/l upstream of the weir, the levels measured in the near downstream of the weir were 1.3 mg/l of CH4 and 6.8 mg/l of DO (with 2.1 mg/l of CH4 and 5.2 mg/l of DO in the hexagons before passage of the second fall). With turbined discharge at 80 m3/s and CH4 and DO levels respectively measured at 5.5 mg/l and 0.7 mg/l upstream of the weir, the corresponding levels measured downstream were 1.0 mg/l and 7.1 mg/l (with 2.3 mg/l of CH4 and 6.1 mg/l of DO before passage of the second fall).

## REFERENCES

Baran P., Lauters F., Sabaton C., (1996). Site de Beyrède sur la Neste d'Aure - Application de la méthode des micro-habitats. *Rapport EDF - ENSAT* N° D4160/DTG-REENV/96-017-A
Belaud A., Chaveroche P., Lim P., Sabaton C. (1989). Probability-of-use curves applied to brown trout. *Regulated Rivers : Research & Management*, Vol.3, 321-336.
Bovee K.D. (1982). A guide to instream habitat analysis using the instream flow incremental methodology. *West. Energy and Land Use Team*, US Fish and Wildlife Service, Fort Collins.
Cardinal H. (1988): Bilan/recommandations en vidanges de retenues. EDF/DER/HE31/88/020.
Chaveroche P., Sabaton C. (1989). An Analysis of brown trout (*Salmo trutta fario L.*) habitat : the role of qualitative data from expert advice in formulating probability-of-use curves. *Regulated Rivers : Research & Management*, Vol.3, 305-319.
Courot A. (1988). Débit réservé de l'Orne a Rabodanges - Evaluation de la qualité d'habitat piscicole en fonction du débit. *Rapport EDF/DER* , N° HE-31/88-07.
Dumestre J.F ,Labroue L.(1996). Réservoir de Petit Saut. Bilan des respirations aérobies et anaérobies.*CESAC Rapport scientifique n°2*.Convention GP7573 EDF-CNEH.
Fragnoud E. (1987). Préférences d'habitat de la truite fario (*Salmo trutta fario L..*, 1958) en rivière (quelques cours d'eau du Sud Est de la France).*Thèse.Université Claude Bernard Lyon.*
Galland J. C., Bouchard J. P., Maurel F. (1996). Limiting sedimentologic impact of reservoir emptying, a case study. *Int. Symp. on reservoir sedimentation, Fort Collins.*
Galy-Lacaux C, Delmas R., Jambert C, Kouadio G., Lacaux J.P(1995) Etude de la méthanogenèse et de son influence sur la consommation d'oxygène dans le réservoir de Petit Saut. *Univ.Toulouse. Lab.Aérologie Rap. sci. n°3.* Convention GP7540 EDF-CNEH
Gosse,Ph.(1994). : Hypothèse de l'influence des gaz formés dans le réservoir de Petit Saut sur les baisses d'oxygène dissous dans le Sinnamary aval. *EDF report.* EDF/DER/HE31/94/38
Gras R., Sabaton C., Gosse Ph. (1994) : Le comportement des poissons mis en équation. Aménagements hydrauliques et débits réservés. *Epure* . 43, 3-19.
Guide vidange, 1994 :. Document d'aide à la préparation et à la réalisation des vidanges d'ouvrages hydrauliques. *EDFInternal Report*
Hanisch H. (1980) : Naturmessungen uber den sauersstoffeintrag an verschiedenen wehrformen. *Natur und Modellmessungen Zum Kustlichen Sauerstoffeinstrag in Flussen*
Hauser G. E., Proctor W. (1993) : Performance of prototype aerating weirs downstream from TVA Hydropower dams. *Proc. Natl. Conf. Hydraul. Eng. San Francisco. ASCE*, p. 99.
Larinier M., Porcher J.P., Travade F., Gosset C.(1984) Passes à poissons / *Collection mise au point* - Conseil Supérieur de la Pêche
Larinier M., Travade F.(1996) Smolt behavior and downstream fish bypass efficiency at small hydroelectric    plants in France *Ecohydraulics 2000 - Québec, June1996, Proceedings*
Merle G., Eon J. (1996). Full-scale test to validate the contributionof the IFIM procedure in the choice of guaranteed flows downstream hydrostations. *Ecohydraulics 2000.Québec, June1996*
Sabaton C., Miquel J. (1993). La méthode des micro-habitats : un outil d'aide au choix d'un débit réservé à l'aval des ouvrages hydroélectriques - Expérience d'Electricité de France. *Revue Hydroécologie Appliquée*, Tome 5, Vol 1, pp 127-163.
Sabaton C., Valentin S., Souchon Y. (1995). La méthode des micro-habitats - Protocoles d'application. *EDF/DER - CEMAGREF - Université de Provence - Université Paul sabatier - INP/ENSAT* - Rapport EDF/DER N° HE-31/95-10.
Sabaton C., Demars J., Lauters F. (1996a). Recolonisation de la Sioule par le saumon à l'aval du barrage de Queuille: Etude par la méthode des microhabitats de la qualité d'habitat du milieu pour reproduction et grossissement des tacons *EDF/CSP* Rap. EDF/DER/HE31/96-02.
Sabaton C., Lauters F., Valentin S. et groupe éclusées(1996b). Impact sur le milieu aquatique de la gestion par éclusées des usines hydroélectriques - Synthèse des résultats issus des travaux du groupe de 1990 à 1995. Recommendations pour l'étude d'un site.*EDF/CEMAGREF/INP ENSAT/Universités Provence et Sabatier* RapportEDF/DER/HE-31/95-19.
Salignat O., Gosse Ph., Richard M., Sabaton C. (1996) : Synthèse sur les études de qualité d'eau et de réimplantation de saumons sur l'Orne amont. *EDF* . Rapport EDF/DER/HE31/96/8

Travade F., Larinier M., Boyer-Bernard S., Dartiguelongue J.(1996) Feedback on four fishpass installations recently built on two rivers in southwest France. *International Conference on Fish Migration and Fish Bypass Channels - Vienna 24-26 september 1996*. To be published

Tron H. D., Hollweg U., Hanisch H. (1979 : Untersuchungen uber die technischen Möglichkeiten der Gewässerbelüftug. Bundesanstalt für Gewässerkunde. Koblenz N 10204013.

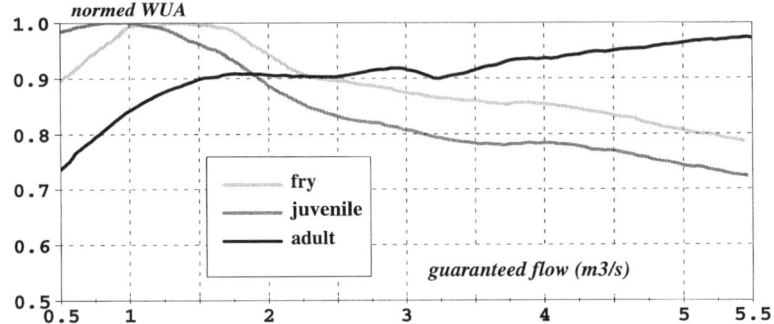

*figure 1 : Beyrède site - WUA for the brown trout according to the guaranteed flow*

*figure 2 : View of the -85 metres wide - aerating weir built downstream of Petit-Saut dam, in the outlet canal of the plant. The metalllic structure consists of 5 hexagones (side = 7 mètres) delimited by upstream and downstream walls which are 5.2 metres and 2.8 metres high (i. e 5.8m and 3.4 m above sea level). At 110 m3/s (resp. 220 m3/s), the water surface altitude is around 6.5 m (resp. 6.9) upstream of the first nappe, 4.0 m (resp. 4.4) in the hexagons and around 1.1 m (resp. 2.4 m) downstream of the second nappe.*

# Status and Vision of Turbine Aeration

Patrick A. March and Paul N. Hopping
Tennessee Valley Authority, Engineering Laboratory, Norris, Tennessee USA
Richard K Fisher, Jr., Vice President, Technology,
Voith Hydro, Inc., York, Pennsylvania USA

## INTRODUCTION

Although of minor concern prior to about 1970, environmental problems stemming from turbine discharges of low dissolved oxygen (DO) continue to receive important attention as a significant issue at many hydroelectric facilities. The U.S. Department of Energy recently found improving low DO, along with providing fish passage and minimum flow, to be one of the most important environmental mitigation issues for the hydropower industry (USDOE, 1991). In the United States, attention to low DO is driven primarily by the Electric Consumers Protection Act (ECPA) of 1986. For a given project, ECPA provides a process by which the development of hydropower shall be balanced with concerns for the protection of wildlife, fish, recreation, and other environmental-related site characteristics. As a result of ECPA, and based on criteria by the U.S. Environmental Protection Agency (USEPA), specifications for monitoring and maintaining DO levels are becoming a regular part of license agreements for affected hydroplants. In these situations, utilities must provide a method to increase the DO in the turbine releases.

To provide dissolved oxygen, a variety of methods are available (e.g., see Bohac and Ruane, 1990). However, where operable, turbine aeration is often the method of choice. Such aeration can be provided by natural aspiration or forced injection. With natural aspiration, air is supplied to openings in the turbine where the pressure is subatmospheric. In some cases, subatmospheric conditions are created by adding small deflectors or baffles on flow boundaries in the turbine. Outside, the openings are vented to the atmosphere, thus providing the pressure difference to draw air into the water. A naturally aspirating turbine is also called an auto-venting turbine (AVT). With forced injection, the pressure in the turbine is above atmospheric, requiring compressors or blowers to push air into the water. In both the AVT and forced injection arrangements, the DO is increased by the transfer of oxygen from the entrained air to the water. Due to the minimal requirements for extra mechanical equipment, as well as reduced expenses for operation and maintenance, the AVT usually is the least-cost alternative for increasing DO in hydro releases.

This paper gives a brief overview of the status and vision for turbine aeration. Included is a short presentation of the DO problem, current practice for providing aeration in existing and new hydroturbines, and comments regarding the testing and analysis of aerating units. The vision focuses on aspects of turbine aeration that are presently under development, or planned for development in the near future. Examples cited in these discussions are based largely on the experience of the Tennessee Valley Authority (TVA) and Voith Hydro, Inc., in providing aerating hydroturbines for projects under TVA's comprehensive Lake Improvement Plan.

## PROBLEM BACKGROUND

The root cause of low DO is well documented in the literature (e.g., see Ruane and Hauser, 1991). Briefly, in many reservoirs, solar heating creates thermal stratification in the summer, yielding a water column with a warm surface layer and a cool bottom layer. This arrangement is hydrodynamically stable, which inhibits mixing of the warm and cool layers and isolates the bottom water from atmospheric oxygen. Concurrently, the respiration of biological organisms and decomposition of organic substances, both in the water and sediments, deplete DO in the bottom layer of the reservoir. For projects with intakes located in the bottom layer, this low DO water is released during hydropower operations, often creating poor water quality in the river downstream from the reservoir.

Summaries of the potential extent of low DO in different parts of the U.S. are given by the USEPA (1989). The summaries are based on statistical analyses of DO data (e.g., STORET) and site questionnaires of dams managed by the U.S. Army Corps of Engineers (USACE), TVA, and U.S. Bureau of Reclamation (USBR). In general, projects with low DO (i.e., DO $\leq$ 5 mg/L) have been reported throughout the country, indicating that no areas are immune from the problem. Reservoirs of depth greater than 50 feet and retention time greater than 10 days, especially those in the warm, humid climates, are more likely to encounter problems with low DO. This trend is supported by data from TVA and USACE, which in some areas report as many as 50 percent of the hydro sites with concerns for low DO. Data collected by the USBR show that low DO is not as problematic in arid climates, where at most only 4 percent of reporting sites express concern. It should be noted, however, that these summaries are based on sparse data. Worldwide, hydro projects with low DO should be identifiable by the same root causes and trends recognized for projects in the United States.

## STATUS

In existing units, the evolution of turbine aeration has led to a number of retrofit designs. These designs typically provide air at one of two locations, either the vacuum breaker outlet or the entrance to the draft tube. Within TVA, the former location is preferred and is currently used in 21 Francis units at 9 projects. All these arrangements provide naturally aspirating, or auto-venting, capabilities. Also, in all but one of these turbines, hub baffles are used to enhance the airflow. The baffles include two basic designs, streamlined and flat plate, and in most cases include a separate air pipe to bypass the vacuum breaker inlet valve (Carter, 1995). Although air is not needed year round, TVA experience has found that the energy loss attributable to baffles is smaller than the cost to install these devices temporarily for the low DO season, if the baffle design is optimized. Cavitation damage has not been significantly increased by the baffles, and in some cases enhanced airflow has reduced adverse surging, reduced load swings, and reduced turbine vibrations.

Examples of hydro facilities where air is provided in existing units at the entrance to the draft tube, including propeller-type runners, are given by Miller and Sheppard (1983). To offset lower velocities and higher pressures typically found at this location, such arrangements usually require a deflector of size larger than that of hub baffles. For propeller runners, furnishing air in the draft tube is at this time the only feasible option for providing AVT capabilities in existing units.

For new and replacement units, an ongoing joint development effort by TVA and Voith Hydro has made substantial improvements in the design of aerating hydroturbines. In this work, requirements for aeration are included as an integral part of the turbine design. To eliminate the need for hub baffles, locations to supply air are selected at sites that will aspirate air as a natural outcome of the turbine geometry. The first units containing this technology are the new replacement Francis turbines at TVA's Norris Dam (Figure 1). These AVT units each contain several methods to oxygenate the flow, including central, distributed, and peripheral aeration outlets at the exit of the turbines. Each method has been tested in single and combined operation for environmental (DO uptake) and hydraulic performance over a wide range of conditions. Results show that up to 5.5 mg/L of DO uptake can be obtained, with efficiency losses ranging from 0 to 4 percent, depending on the operating condition and aeration method. Compared to the original turbines, these units provide overall efficiency and capacity improvements of 3.5 and 10 percent, respectively (March and Fisher, 1996).

As part of the work for the replacement units at Norris, new procedures have been developed for the analysis and testing of aerating turbines. In general, analyses focus on the prediction of three factors (Greenplate and Cybularz, 1993): airflow; DO uptake; and performance effects. Prediction of airflow requires a balance between the computed pressure loss through the air supply passageways and the pressure at the aeration outlets found by computing the flow through the turbine. The latter computation involves the effects of a two-phase air/water mixture. Due to the complexity of these flows, the prediction of airflow at this time relies heavily on dimensionless parameters derived from model and prototype measurements of pressures at the aeration outlets.

Current procedures for predicting DO uptake are based on a scaling relationship presented by Thompson and Gulliver (1993). In the relationship, the oxygen transfer efficiency is given as a function of the concentration of entrained air, runner speed, and runner diameter. Also found is an empirical "transfer" coefficient, which, based on TVA/Voith experience, depends on a variety of factors, including the location of air entrainment and other yet to be determined site-specific turbine characteristics.

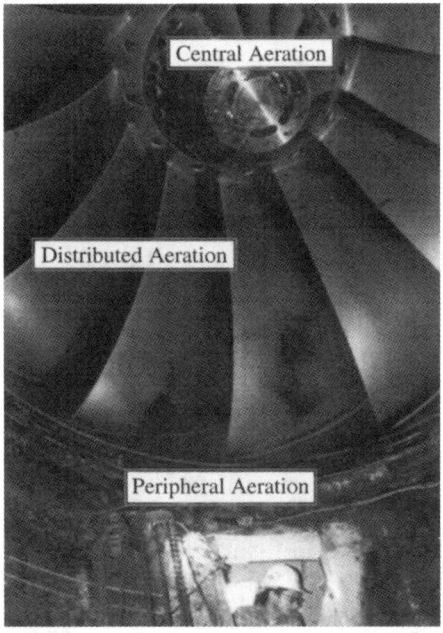

Figure 1. New Aerating Turbine for Norris Unit 2

The prediction of performance effects has been limited to an examination of the relative change in turbine efficiency derived from model tests conducted over a range of operating conditions, with and without air admitted at the various aeration locations (Greenplate and Cybularz, 1993). Results of prototype tests at Norris show that this procedure tends to overpredict the amount of efficiency loss.

Although this leads to a "conservative" estimate of the performance effect, an improved procedure is needed. This is true not only for assessing the quality of the turbine design, but also for effectively evaluating the cost of turbine aeration compared with other methods for enhancing DO.

New procedures for testing aerating turbines include the measurement of the additional parameters, in both model and prototype situations, needed to assess the environmental performance of the units (e.g., Hopping et al., 1996). These include airflow, DO upstream of the turbine, and DO and total dissolved gas (TDG) downstream of the turbine. TDG is needed to verify that aeration will not cause gas bubble disease in tailwater fish populations. In model situations the procedures specify the runner speed and NPSH required to scale DO uptake and performance effects. Procedures also include an improved method for turbine testing that uses a PC-controlled, multitasking operating system to automate and integrate data acquisition, analysis, and reporting (Wolff et al., 1995).

## VISION OF FUTURE NEEDS

Based on the current status of aerating turbine technology and the anticipated trends in the hydropower industry, several aspects of turbine aeration are presently under investigation:

**Problem Identification** - The amount of DO data has increased substantially since the late 80's. Studies to assess the extent of low DO need to be updated using this new information and include a more worldwide view. In these studies, the data needs to be supplemented with detailed information for the physical design, environmental, watershed, and operational factors of the projects in order to develop a more reliable method to evaluate the potential for low DO at sites where water quality data is not available.

**Options for Turbine Aeration** - The new technology developed for Norris is being expanded to include improved options for turbine aeration. These developments are needed to enlarge the range of site conditions applicable to AVT technology and to reduce costs. New designs are to be tested for TVA's Douglas Dam, which includes Francis turbines containing a much higher tailwater submergence than those at Norris. To allow aeration at projects where discharge TDG/nitrogen levels may be too high, experiments are also being conducted with oxygen injection through the turbine aeration outlets. For propeller-type units, entrainment of the amounts of air required to obtain the desired DO uptake can substantially reduce draft tube efficiency. Studies to examine the unique aeration problems for these types of units are in progress.

**Testing of Aerating Turbines** - To address issues regarding the combined hydraulic and environmental performance of aerating turbines, efforts are underway to establish appropriate methods of testing through the test code committee, PTC-18 of the American Society of Mechanical Engineers. These issues should be addressed by TC-41 of the International Electrotechnical Commission, as well.

**Technical Evaluation** - Work currently is underway to improve methods to predict airflow quantities, DO uptake, and performance effects. In part, these improvements will involve a more detailed examination of the exact mechanisms for the transport and distribution of air in the turbine draft tube and tailwater using state-of-the-art numerical algorithms for turbulent, two-phase flow and gas transfer.

**Operational Support** - TVA and Voith Hydro have developed and implemented a machine condition monitoring system to help balance the energy, economics, and environmental requirements for hydro facilities, including DO uptake. For units with multiple aeration options, an optimization module determines which option (or options) should be operated to provide the desired DO level with the least impact on operating efficiency.

## REFERENCES

Bohac, C.E., and R.J. Ruane, "Solving the Dissolved Oxygen Problem," *Hydro Review*, February 1990.

Carter J.C., Jr., "Recent Experience With Turbine Venting at TVA," *Proceedings, Waterpower 95*, ASCE, 1995.

Greenplate, B.S., and J.M. Cybularz, "Hydro Turbine Aeration," *Proceedings, Waterpower 93*, ASCE, 1993.

Hopping, P.N., P.A. March, T.A. Brice, and J.M. Cybularz, "Plans for Testing the New Autoventing Turbines at TVA's Norris Hydro Project," *Proceedings, North American Water and Environment Congress*, ASCE, 1996.

March, P.A., and R.K. Fisher, "TVA's Auto-Venting Turbines Increase Downstream Aeration," *Hydro Review*, September 1996.

Miller, D.E., and A.R. Sheppard, "Current Status of Turbine Aeration Activities at Alabama Power Co.," *Proceedings, Waterpower 83*, ASCE, 1983.

Ruane, J.R., and G.E. Hauser, "Factors Affecting Dissolved Oxygen in Hydropower Reservoirs," *Proceedings, Waterpower 93*, ASCE, 1993.

Thompson, E.J., and J.S. Gulliver, "Oxygen Transfer Similitude for the Auto-Venting Turbine," *Proceedings, Waterpower 93*, ASCE, 1993.

USDOE, "Environmental Mitigation at Hydroelectric Projects, Volume 1. Current Practices for Instream Flow Needs, Dissolved Oxygen, and Fish Passage," USDOE Report DOE/ID-10360, 1991.

USEPA, "Dam Water Quality Study," USEPA Report 506/2-89/002, 1989.

Wolff, P.J., D.B. Hansen, C.W. Almquist, and P.A. March, "An Improved System for Hydroturbine Index Testing," *Proceedings, Waterpower 95*, 1995.

# EFFECTS OF THE SPILLWAYS OPERATION ON THE FISHES HABITAT : STUDY OF SOLUTIONS

C.M. ANGELACCIO, J.D. BACCHIEGA, C.A. FATTOR, H.D. BARRIONUEVO
Laboratorio de Hidráulica Aplicada - INCyTH, Ezeiza, ARGENTINA

## ABSTRACT
The massive mortality of fishes that took place downstream of Yacyretá Dam, resulting from supersaturation of total dissolved gases, generated diagnosis studies of the causes and of possible corrective solutions to avoid re-occurrence. The causes of the environmental conflict are detailed, and changes in the original geometry of the spillway are analyzed through physical modelling, in order to reduce the supersaturation.

## INTRODUCTION
The evaluation of the environmental impacts of large dams must accompany the development of the project within a tending process to identify, evaluate, solve and minimize the negative consequences of the actions of the hydraulic work.

Yacyretá hydroelectric project shows the potencial for an environmental conflict of this nature, actions were adopted for its resolution. Yacyretá is located on the Paraná river, on the boundary between Argentina and Paraguay (Figure 1). The spillway on right margin, on the Añá-Cuá branch (VBAC), is composed of 16 spans of 15 m regulated by gates. The other spillway, the one which is mainly addressed by this, is located on the Principal branch (VBP) and has 18 spans of 15 m. This structure includes a stilling basin of 100 m of length, with a floor 15 m under the natural level of the bed of the river. At the site of the location work, the river is split into a series of branchs with averages depths of 4,50 m for Principal branch, 2,25 m for Añá-Cuá branch and 3,50 m for San José Mí branch.

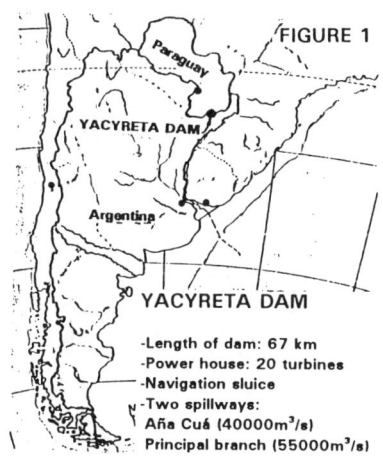

FIGURE 1

YACYRETÁ DAM
-Length of dam: 67 km
-Power house: 20 turbines
-Navigation sluice
-Two spillways:
 Añá Cuá (40000m³/s)
 Principal branch (55000m³/s)

## DESCRIPTION OF THE ENVIRONMENTAL INCIDENTS

In August of 1994, during the filling of the reservoir and as a consequence of operation of the sluices of the VBP, a massive fish mortality downstream of the dam was detected, prevailing mortality ocurred with bottom dwelling species. The Instituto de Ictiología de la Universidad Nacional del Nordeste established that the death of these fishes was provoked by the designated disease "Gas Bubble Disease", characterized by the gaseous bubbles training in the blood vessels and in the conjunctive tissue. The disease is produced in the fish exposed to conditions of supersaturation of gases in water, being critical if the concentration of dissolved total gases (TDG) reaches values of 105 to 140%, and being acute (death in few hours) when these percentage is greater than 140%.

The supersaturation is produced when the gases of the air entrained in the water, are exposed to higher pressure. Under these conditions the solubility of gas in water is greater than the solubility at atmospheric pressur++e. Among the generating causes of this phenomenon is the transport of entrained bubbles to large depths in stilling basin. The Yacyretá spillways generated these phenomena when operating.

If the fish can reach a "compensation depth" that generates pressure conditions that stabilize the dissolved gas in the blood, mortality by this disease would not result. This situation explains the survival of these fishes downstream of large water falls (Iguazú Falls) or dams with ski jumps (Itaipú Dam).

Even though the supersaturation tends to be neutralized by the gases release to the atmosphere, some field measurements conducted by the Entidad Binacional Yacyretá (EBY) determined a notable permanency of the supersaturation downstream of the work. The air dissolved has a behaviour that can be relationed with a pollutant of very decrease transverse and longitudinal dispersion. Since the TDG concentration evaluations were accomplished on 700 Km of the river, it has been determined a problem of regional scale, whose magnitude exceeds the proximity of the project. EBY observed concentrations of TDG of the 151%, 10 Km downstream of Iguazú Falls, while in the reservoir levels of 102 % were observed. Immediately downstream of the VBP, TDG levels reached values of 140%, while 100 Km downstream of the VBP the values of 121 % were observed, showing the persistence of the supersaturation (Figure 2).

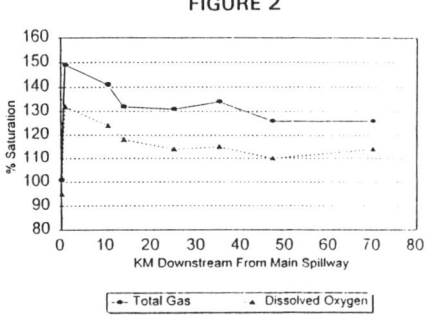

FIGURE 2

Consequently, the problem of supersaturation presents a regional dimension which must be analyzed and valued in this context, particularly for the works planning that are located in a same river.

## CHARACTERIZATION OF THE CURRENT OPERATION

The spillway and the energy dissipator were designed to assure a low residual energy levels, maximizing energy dissipation in the interior of the stilling basin through the formation of a hydraulic jump. In addition, an aireation device was incorporated, in order to prevent cavitation risks.

The energy dissipation in the stilling basin produces the air entrainment through two mechanisms (Figure 3). In first place, air is entrained at the aireation device; the air is driven downstream, toward a mixture zone, remaining concentrated in the zone of the stilling basin in a small band near the boundary. In addition, larger volumes of air are entrained by the interaction of the flow from the spillway with the front of the hydraulic jump; the air entrained by the jump enters to the stilling basin, producing a turbulent mixture, that does not reach the basin floor.

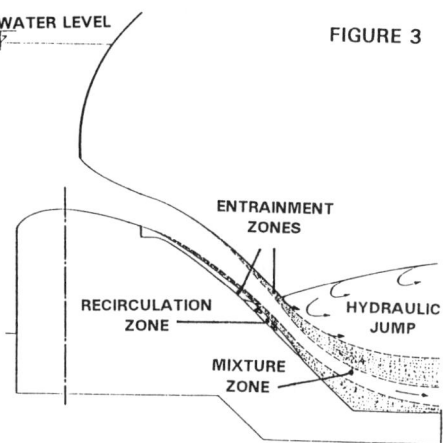

FIGURE 3

Taking into account the uniform operation of the spillway maximizes the air entrainment by both mechanisms and the observed mortality problem, operation of the VBP was modified to reduce the values of supersaturation downstream of the spillway, were opearated only 4 or 5 gates totally open. Nevertheless, the not uniform operation of the spillway originates certain risks in the integrity of the stilling basin, a situation that would be aggravated with the higher reservoir levels. Consequently, the need was identified to analyze feasible devices for reducing the TDG supersaturations to admissible values for the fishes of this species.

## ALTERNATIVES OF SOLUTION

The international experience is scarce, which documents similar phenomena of massive mortality of fish (some works on Columbia and Snake rivers) indicates that the construction of deflectors on the profile of the spillway result in effective reduction of the TDG supersaturation.

In order to minimize the transport of air toward the bottom of the stilling basin, a modification of the original profile of the spillway, eliminating the aireation device and incorporating on the profile a deflector that modifies the stilling basin flow patterns from a classic hydraulic jump to a flow with greater surface velocity. Alternatives were studied on a 1:50 two-dimensional physical model of the VBP, in the facilities of the Laboratorio de

Hidráulica Aplicada of the INCYTH, Ezeiza, Argentina. Four deflector alternatives were analyzed. DH is the difference in elevation between the crest and deflector.

| ALTERNATIVES | DH (m) | Level (m) | Lenght (m) | High (m) |
|---|---|---|---|---|
| ALTERNATIVE I | 4,50 | 59,00 | 3,50 | 2,73 |
| ALTERNATIVE II | 6,50 | 57,00 | 3,50 | 3,17 |
| ALTERNATIVE III | 7,50 | 56,00 | 3,50 | 3,36 |
| ALTERNATIVE IV | 6,50 | 57,00 | 4,00 | 3,67 |

The proposed modification would generate a sustancial change in the downstream flow, eliminanting totally the artificial air entrainment from the aeration device.

The tests on the four deflectors demonstrated that Alternatives II and IV yielded effective performance. Alternative IV, with the longer deflector, established better conditions downstream, with a largely horizontal discharge of the flow and better supersaturation reduction. Consequently, from analysis of the alternatives, the performance of the deflector at level 57 m with a length of 4 m yielded the best performance over the operation range of the spillway. The Figure 4 shows a section through the selected deflector, and Figure.5 shows the flow characteristics of the modified spillway.

With respect to the air entrainment, the two-dimensional 1:50 model adequately reproduces the hydrodinamics conditions of the flow and of the dissipation in the stilling basin, but does not correctly represent air entrainment. Therefore, the model can not be used to measure the quantity of air entrained by the flow. The model can be employed to establish a qualitative relationship between the original geometry and the different alternatives.

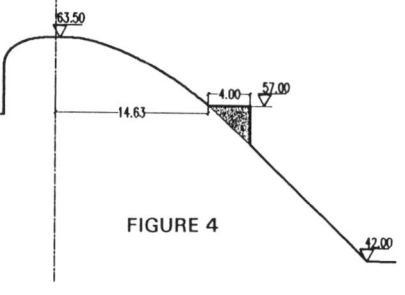

FIGURE 4

According to the characteristics of the flow, for operation with a hydraulic head of 11,3 m on the crest of the spillway and gate openings of up to 6 m, no significant quantities of air were observed compared with the original geometry. With total gate opening, the quantity of air increased, maintaining less air flow than the structure without deflector.

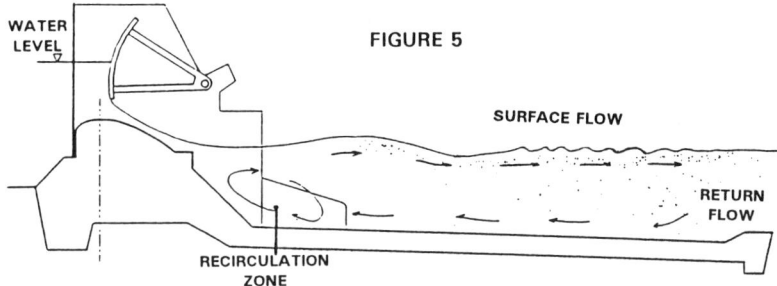

FIGURE 5

Even though results feasible to find a greater quantity of air in prototype, the effected study permits to deduce that the same would result smaller than the resulting from the original geometry, mainly near the bottom.

With a hydraulic head of 18,5 m and with small openings, a mean air entrainment reduction of the order of the 30% was observed for all the alternatives with reference to the original geometry. This situation also occurred with larger gate openings, although the reduction were more meaningful for the alternative IV deflector in the 6 m on the bottom of the stilling basin.

## HYDRODYNAMIC CONDITIONS

With the elimination of the aireation device and the incorporation of the selected deflector, the fluctuating pressure measurements in the energy dissipator do not indicate, a risk of cavitation for pressure pulses, at any point on the surface of the slab of the stilling basin. The most critical zone by the pressure fluctuations was registered on the downstream face of the pier, with nearby values to the limit of beginning of the cavitation.

The analysis of turbulence intensity ($C`p$) compared with the original geometry, shows an increase of $C`p$ in the beginning of the energy dissipator. In the stilling basin, the new configuration presents smaller values of $C`p$ for reduced gate openings and similar values to the original geometry for larger openings. For openings of up to 6 m, energy dissipation is generated only by the macroturbulent effect in the bottom of the stilling basin, yielding little dissipation of the flow discharged from the deflector. For greater openings, the hydraulic jump is the principal dissipation mechanism, with similar values between both geometries and with a displacement of the jump downstream. Even though these results indicate that there is no risk of cavitation, the configuration of the flow shows the existence of a recirculation current that could cause material incorporation in the stilling basing with consequent erosion risk by abrasion.

## CONCLUSIONS

* The original geometry of the spillway, from the hydrodynamic point of view, generated the optimum condition for operation, including the presence of the aireation device as a safety element to prevent the erosion risk by cavitation.

* Given the massive fish mortality of August of 1994 on the Principal branch, alternatives modification of the geometry of the spillway profile were studied which changed the original hydraulic safety criteria in order to reduce the risk of enviromental problems. These modifications consisted of the elimination of the aeration device and construction of a deflector on the profile, that modifies stilling basin performance, generating a surface flow with a return zone near the bottom of the stilling basin, reducing the transport of air toward the bottom of the stilling basin.

* After a series of tests in a 1:50 scale physical model was selected a deflector of 4 m of length located to 6,5 m under the crest of the spillway. This alternative defelctor generated the most adequate behavior in dissipation conditions downstream and air distribution in the stilling basin. Furthermore, it minimized the surface waves training to the exit of the deflector, generating a relatively horizontal flow downstream for gate openings up to 6 m.

* The hydrodynamic behavior of the modified spillway with deflector, indicates that an erosion risk by cavitation does not exist for pressure pulses over all the structure.

* Given the configuration of the resulting flow generated by the deflector and by the creation of a recirculation zone at the bottom of the spillway, is with material present in the stilling basin, autocleaning would not result in the zone of the toe of the spillway, for any analyzed operating conditions. Consequently, the material presence in such zone could generate a erosion process, resulting in abrasion of the concrete surfaces.

## REFERENCES

- JACOBO, W.et: al, Informe Técnico de Auditoría sobre la Mortandad de peces en el río Paraná, Universidad Nacional del Nordeste, agosto 1994.
- EBEL et al., Effect of Atmospheric Supersaturation Caused by Dams on Salmon and Steelhead Trout of the Snake and Columbia Rivers (1974), January 1975.
- EBY-CIDY, Mediciones de Condiciones de Sobresaturación sobre el Río Paraná, 1994.
- LHA-INCyTH, Estudios adicionales para la nueva configuración del vertedero del Brazo Principal, Informes LHA-144-01-95, LHA-144-02-95 y LHA-144-03-96, Ezeiza, Argentina.

# MODELING APPROACHES FOR TAILWATER ENHANCEMENT

GARY E. HAUSER
MING C. SHIAO
JAMES A. PARSLY
BRUCE L. YEAGER
TVA Resource Group, Norris, Tennessee   USA

Enhancement and restoration of hydropower tailwaters requires balancing environmental and economic objectives. Renewed attention to this balance is a result of 1) new hydropower relicensing requirements; 2) power economics favoring modernization and new capacity; 3) increasing tailwater uses and users; and 4) impact on aquatic resources from cumulative development. There is also new awareness among hydro owners and regulators that significant biological recovery is achievable from cost-effective solutions.

Sound solutions require an ability to quantify relative cost-effectiveness of an ever wider array of alternatives. Mathematical models are being used in flexible new ways to identify and evaluate site-specific solutions, including 1) capital or operational changes to meet objectives for minimum flow, temperature, and dissolved oxygen; 2) measures to minimize reduced substances in dam releases, such as iron, manganese, and hydrogen sulfide; 3) in-channel structures and physical habitat; and 4) measures to reduce pollution from upstream sources.

This paper provides an overview of some of the modeling approaches that the Tennessee Valley Authority (TVA) has developed to evaluate alternative solutions to tailwater management issues.

## RESERVOIR MODELING

Seasonal water budgets and reservoir operating policy determine a hydroproject's ability to sustain a target minimum flow for its tailwater. Water budgets, operating policy, reservoir geometry, meteorology, and watershed pollutant loadings all combine to determine seasonal patterns of reservoir and release water quality. TVA employs hydropower scheduling models (PRSYM, WSM) and two-dimensional reservoir water quality models (CE-QUAL-W2, BETTER) to explore interactions that lead to poor reservoir and release quality.

## HYDROPOWER SCHEDULING

In an integrated reservoir system, operational changes to enhance the tailwater at one dam can affect the ability to enhance other tailwaters in the system. Models are used to explore such interactions. For example, scheduling models were used to evaluate the effects of a range of minimum flow targets on generation patterns and pool levels. Minimum operating guides (MOGs) and economy guides that incorporate minimum flow targets were developed for each reservoir.

MOGs mark the bottom of the "zone of flexibility" in operating a reservoir. For pool levels above the MOG, the operator has flexibility, subject to other operating constraints, to generate power in the most economical fashion. Once the pool level is drawn down to the MOG, the operator must restrict generation to the minimum level used in the computation of the MOG, or face an unacceptable risk of being forced to curtail minimum flow or draw the reservoir below normal minimum level.

Economy guides provide the expected future value of hydrogeneration as a function of time and pool elevation. Economy guides account for 1) the seasonal cost of power, 2) the chance that water saved now might be wasted as spill in the future, and 3) the effect of pool levels on generating capacity. When the reservoir is in the zone of flexibility above the MOG, the operator compares the future value of hydrogeneration (economy guide) with the current value of hydrogeneration to guide scheduling decisions.

## WATER QUALITY

Dam release quality is dependent on many reservoir factors, including inflow hydrology and quality, meteorology, residence time, watershed loadings, release magnitude, thermal stratification, and outlet depth. Two-dimensional reservoir water quality models are calibrated to actual field data. They are then used with hydropower scheduling models to understand why specific water quality conditions develop, and how often they can be expected to occur, and to simulate effects of alternative management options for the reservoir.

Multi-year water quality simulations were once avoided due to cost and computational intensity, but they are now becoming more widespread with modern computing capabilities. Long-term simulations allow deterministic models to be used to evaluate the effects of stochastic inputs over a long period of record. In some climates (e.g., southern Africa), long-term climate cycles can dominate the hydrology and water quality. Long-term simulations are essential in such cases to capture the effect of the cycles. Multi-year simulations can demonstrate in what percentage of the year's poor water quality conditions are likely to develop, and the variety of conditions under which they can occur.

At Nottely Reservoir (deep tributary project), a two-dimensional model was used to assess potential causes of a forebay fish kill. Simulations revealed that inflows and inflow loadings created persistent patterns of metalimnetic dissolved oxygen (DO) depletion combined with near-bottom DO depletion in the late summer. Cool, oxygenated water between these two depleted layers provided refuge for forebay fisheries until fall, when large turbine discharges were used to draw down the reservoir in anticipation of high winter and spring inflows. The fall drawdown was more rapid than in previous years due to policy changes that required summer pools to be held later into the summer. The model showed how these heavy withdrawals rapidly removed the refuge water, leaving forebay fisheries without sufficient oxygen for survival.

Other applications for reservoir models include evaluations of 1) the effects of forebay aeration systems on reservoir and release quality; 2) the relative merits of reservoir flow versus in-reservoir aeration for improved release DO; 3) the effects of dam releases to provide cooling water for downstream thermal plants; and 4) the relative effects of watershed loadings and merits of treatment.

## TAILWATER MODELING

During generation at major hydropower dams, water quality and habitat for long distances downstream are controlled by dam releases. Between periods of generation, water quality and habitat are controlled by diurnal instream processes and local inflows. Dynamic flow and water quality patterns in these disturbed systems must be considered in solutions to protect or restore uses of a tailwater.

## HYDRODYNAMICS

TVA's one-dimensional longitudinal river model (ADYN) is used to quantify fluctuations in flow, water surface, travel time, depth, velocity, and wetted area throughout a tailwater resulting from dam releases and local inflows. This model is highly adapted to shallow depths and transient flows of hydropower tailwaters.

At Douglas Dam tailwater, flow and wetted area improvements for various minimum flow options were explored. Model results were used to demonstrate to conservation agencies that turbine pulsing with existing units (half hour generation every four hours) could provide stable flows within short distances of the dam, and that greater minimum flows could be provided at much less cost than sluicing the flows (bypassing the turbines).

The river hydrodynamic model was also used to evaluate the effects of reregulation and aeration weirs. The model helped quantify weir pool drain times, overflow hydraulics, and increased tailwater depths (reduced net head) at the upstream hydroturbine. Hydraulics of other structures such as boulders, sills, rock jetties, and artificial spawning channels have been evaluated.

Model results can be used to identify the incremental environmental benefit associated with incremental minimum flow increases from a hydroplant, leading to establishment of target minimum flows. Environmental benefits (e.g., channel wetted area) increase with increasing target minimum flows, but at a decreasing rate for higher flows after the entire channel bottom is wetted. Cost of providing the target minimum flow increases exponentially with higher flows, due to the increased need for off-peak generation to sustain the minimum flow. The benefit of higher flow has a diminishing return while the cost increases exponentially, so the benefit to cost ratio achieves a peak value at some target minimum flow. This corresponds to the flow where maximum benefit per investment is achieved, and this flow can often produce significant biological improvements in balance with economic objectives.

## WATER QUALITY

A one-dimensional longitudinal river water quality model (RQUAL, linked to ADYN) is used to evaluate fluctuations in temperature, DO, and carbonaceous and nitrogenous biochemical oxygen demand throughout the tailwater. Operational alternatives were developed with this model for tailwater temperature control in important fisheries downstream of Tims Ford Dam (Proctor et al., 1994) and Apalachia powerhouse (Hadjerioua et al., 1996) in the Tennessee River valley. Daily turbine release guidance was developed to maintain desirable river temperatures at index locations in response to a wide variety of meteorological and hydrological conditions.

At Tenkiller Dam in the mid-western U.S., the river model was recently used to explore the effects of an aerating weir designed for a location just downstream from the dam (Proctor et al., 1996). The model quantified expected DO improvements downstream of the weir due to aeration at the weir under various flow regimes. The model was also used to identify changes in weir pool temperature and quality resulting from weekend versus weekday dam operations.

## PHYSICAL HABITAT

The dynamic river model includes a component (RHAB) to quantify weighted usable area for a particular fish species and life stage based on physical habitat variables (depth, velocity, substrate, cover), as determined by the hydrodynamic model and measurement. Suitability curves define relative habitat values over the range of each variable. A two-dimensional array of cell variables is prepared, then weighted with the suitability curves, then summed to provide overall weighted usable area. Because high velocities can control habitat suitability in tailwaters of large hydropower dams, recent improvements to the model included calculation and use of bottom velocity rather than average velocity for bottom-dwelling species.

## FISH GROWTH BIOENERGETICS

Habitat models explore changes in physical habitat but they cannot quantify biological response. Biological response to alternative reservoir release improvement options is evaluated using TVA's bioenergetics fish growth model (Shiao et al., 1993). The model allows prediction of fish growth patterns for various important species as a function of temperature, DO, and food availability. The model requires calibration to fish growth patterns in a particular tailwater. Once calibrated, the model can be used to evaluate fishery responses to reservoir release improvement options, including those that affect temperature or DO.

The model was used to quantify the relative growth that would occur with different DO targets and with different aeration systems. For example, TVA used the model to predict growth response from two different aeration systems, one "throttled" and one "binary". A throttled system (e.g., forebay oxygenation) attempts to provide only the aeration necessary to meet a flat target, and oxygen supply is throttled as necessary to cover the oxygen deficit. A binary system (e.g., turbine venting) is either on or off, and aeration occurs as a function of local operating conditions. The binary system was less expensive at several TVA locations, but it produced DO in excess of targets at times and DO slightly less than targets at other times. When the bioenergetics model was used to compare fish growth responses for these systems, growth resulting from the more economic binary system typically equaled or exceeded that of the more expensive throttled systems.

## CONCLUSIONS

Mathematical models can be used in flexible new ways to identify and evaluate effectiveness of certain alternatives. Use of reservoir water quality models in conjunction with hydropower scheduling models can ensure that system tradeoffs are recognized in balancing environmental and economic objectives. With tailwater models, flexible, site-specific solutions to environmental issues can be developed to maximize the cost-effectiveness of solutions.

## REFERENCES

Hadjerioua, B., A. Sutton, W.D. Proctor, and G. Hauser (1996); "Improving Minimum Flow and Water Temperature for a Trophy Trout Fishery Below Apalachia Powerhouse;" TVA Norris Engineering Lab; WR28-1-15-103; Norris, TN, USA; September.

Proctor, W.D., J. Hoover, and G. Hauser (1996); "Tenkiller Ferry Aerating Weir;" TVA Norris Engineering Lab; WR28-1-590-166; Norris, TN, USA; April.

Proctor, W.D., B. Yeager, G. Lowe, and W.G. Brock (1994); "Using Computer Modeling for Solving Water Management Problems on the Elk River;" TVA Norris Engineering Lab; Norris, TN, USA; draft; May.

Shiao, M.C., and G. Hauser (1995); "Two-Dimensional Water Quality Modeling of Nottely Reservoir;" TVA Norris Engineering Lab; WR28-1-18-103; Norris, TN, USA; December.

## TERMINOLOGY

**anoxic byproducts** undesirable releases from sediments in the presence of zero oxygen (iron, manganese, ammonia, methane, dissolved organics)
**bioenergetics** energy flow kinetics for fish species (effects of temperature, dissolved oxygen, food availability on assimilation, respiration, growth)
**forebay** reservoir zone nearest the dam
**loadings** mass rate of organics and nutrients in the inflow to a water body
**metalimnetic** of the metalimnion (zone of high vertical density gradient in a thermally-stratified reservoir)
**minimum flow** minimum flow rate established for a hydroproject to protect downstream aquatic habitat
**nonpoint source** pollution from distributed sources (e.g., agricultural runoff)
**point source** pollution from a point location, such as sewage outfall
**tailwater** river reach downstream of a dam or weir

# MONITORING DISSOLVED OXYGEN AND TOTAL DISSOLVED GAS IN TAILRACES

## H. RUCKER, D. HARSHBARGER
Tennessee Valley Authority, Muscle Shoals, Alabama, USA

### INTRODUCTION

The accurate measurement of dissolved oxygen (DO) and total dissolved gas (TDG) in reservoirs and in the water released from dams has become increasingly necessary as water quality concerns have grown in recent years. The Tennessee Valley Authority (TVA), recently completed a Lake Improvement Program (LIP) designed to meet established DO and minimum flow goals downstream from 16 major TVA dams. Several types of aeration systems were installed at TVA hydroplants under this program. Studies involving DO and TDG measurements were conducted to help determine the aeration technique to be applied, to evaluate aeration efficiency after the selected systems were installed, and to help determine operating procedures.

In another application of TDG and DO monitoring, measurements have been taken at TVA's Kentucky Dam to aid in mitigating fish kills suspected to be caused by gas bubble disease under certain spill operations.

Accurate, reliable, and economical measurement of DO and TDG requires careful attention to instrument selection, field procedures, and equipment maintenance. Experience has shown that training of field personnel, timely maintenance, and proper calibration and handling procedures help ensure reliable DO and TDG data. In one instance, a relationship between TDG and DO was also established which greatly reduced the time and effort necessary to obtain TDG data.

### DISSOLVED OXYGEN MONITORING

Galvanic and polargraphic membrane electrodes have been used for continuous DO monitoring. "Oxygen sensitive membrane electrodes of the polargraphic or galvanic type are composed of two solid electrodes in contact with supporting electrolyte separated from the testing solution by a selective membrane. The basic difference between the galvanic and polargraphic systems is that the galvanic electrode reaction is spontaneous, while in the later, an external source of applied voltage is needed to polarize the indicator electrode" (Standard Methods, 1995).

Galvanic DO probes were used during the initial testing of the aeration equipment and provided sufficient data for system design. These probes were initially chosen because they were relatively inexpensive (around $800 ) and interfaced readily with existing data collection computers. During the subsequent shift to long-term operation of the aeration systems these probes could not be managed to meet TVA quality standards. Possibly the problems encountered resulted from the specific probe design and construction and from user inexperience with that particular design. Probe malfunctions, excessive time on-site for maintenance and standardization, inconsistent stabilization time after field cleaning, significant amounts of fouling, and erratic readings resulted in data that was unreliable and inconsistent. Field maintenance as simple as wiping the membrane required minimum stabilization times from 2 to 6 hours. Frequently a minimum of 24 hours was needed before field standardizations could be completed. For some projects bio-fouling required probe cleaning as many as three times a week and on some cases the probe would require cleaning again even before standardization could be completed.

The first step taken to reduce DO monitoring problems was to assess field standardization procedures. To ensure that proper procedures were being followed, training was provided on-site by a factory representative. No discrepancies were discovered in the field procedures that were being used and the training resulted in no significant improvement in data acquisition.

Next, a comparison study was conducted with a water quality instrument of a type used on other TVA monitoring projects. These instruments were multi-parameter devices using a polargraphic DO probe. They could also be used to measure pH, conductivity, temperature, depth, and oxidation-reduction potential. Field personnel had considerable experience using this type of instrument at other locations. The multi-parameter probe cost on the order of $4500 and did not readily interface with existing electronics at the site. The multi-parameter instrument was temporarily installed with an external power source and data was retrieved during weekly site visits. As demonstrated in Figure 1, comparison with Winkler data and readings obtained prior to maintenance indicated that the multi-parameter probe provided greater accuracy, reliability, and consistency than the galvanic device.

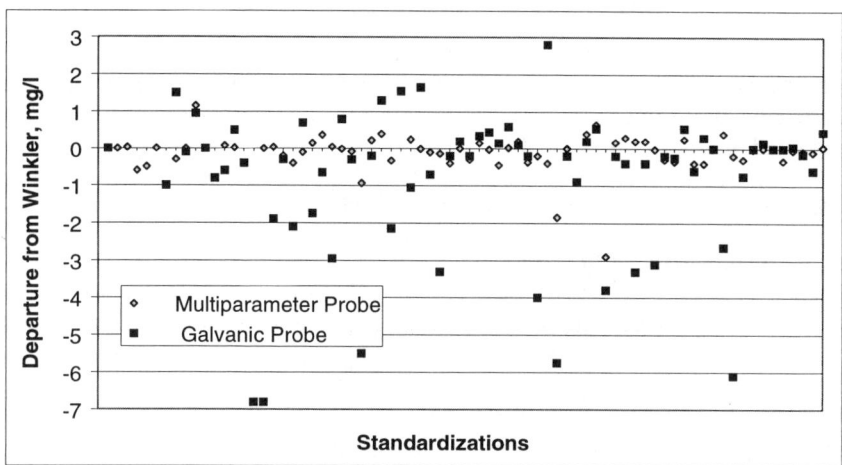

Figure 1: Galvanic and Multi-parameter Probe Comparison

The galvanic probes were replaced with the multi-parameter probes and maintenance costs were lowered by nearly 33 percent by reducing the number of trips to the site and the amount of time spent at the site. The multi-parameter probes could be cleaned and standardized in less than one hour. In addition, personnel were properly trained and had considerable experience with these probes. TVA already had field quality assurance procedures adapted from Standard Methods and factory operating manuals in place, and the probes had proven accurate at other projects.

After the multi-parameter probe was chosen for use, process improvements were also adapted to improve operation and reliability: (1) During cleaning and standardization, an instrument of the same manufacturer was temporarily installed to prevent invalid data from readings taken while the permanent probe was out of the water. (2) The instruments were installed with a permanent power supply which eliminated the problem of missing data from dead batteries. (3) The instruments were replaced every three weeks and thoroughly cleaned by removing the membrane, cleaning inside the cell, and allowing sufficient time for the membrane to relax before standardization. (4) Each instrument was checked electronically, once per year in a laboratory facility, for probe drift. These process improvements while specific to the multi-parameter probe, could be modified to adapt to other situations.

## TOTAL DISSOLVED GAS MONITORING

Dissolved gas supersaturation has long been recognized as a potential threat to fish because of gas bubble disease (Marking, 1987). In 1972, the Environmental Protection Agency established a guideline of 110 percent dissolved gas saturation as the maximum acceptable value (Water Quality Standards Handbook, 1992). The following formula can be used to calculate TDG:

$$\% \text{ TDG} = [BP + \Delta P] \div BP \times 100$$

where    $BP$ = local barometric pressure in millimeters of mercury and
$\Delta P$ = the sum total of all dissolved gas partial pressures measured by a saturometer or tensiometer in millimeters of mercury.

Initially, TDG measurements at aeration projects were made during weekly site visits using hand-held saturometers, which were incapable of automated data logging. However, continuous monitoring was desirable because of the wide range of variability of aeration system operating conditions and the variable TDG results from these conditions. However, proven quality TDG continuous monitors that were compatible with the data collection and transmission systems were unavailable. In 1995, an automated TDG tensiometer was purchased, incorporated with the multi-parameter probe via a commercial data logger, and added to the monitoring system at Tims Ford Dam. This application proved successful, and the availability of on-line TDG levels enabled forced air injection to be safely increased and more expensive oxygen use to be reduced.

The second application for TDG monitoring was a special investigation at TVA's Kentucky Dam during spill operations. Fish kills had occasionally occurred during events when relatively small amounts of water were released over the spillway for flood control purposes. During recently monitored spill operations, TDG concentrations have reached levels in excess of 160 percent (unpublished TVA data).

Recent studies have indicated that modification of spill procedures could mitigate TDG levels, thus several alternatives to modifying spill procedures at Kentucky Dam have been developed and are being investigated. In support of this ongoing investigation, TDG monitoring consists of a series of transect measurements. Each transect consists of a grab measurement at a 0.3-meter depth taken at 25 percent, 50 percent, and 75 percent of the horizontal distance across the stream. Transects are made at 0.3 miles and 1.2 miles below the dam and at several other locations throughout the 22 miles downstream of the dam to the confluence of the Tennessee River with the Ohio River. TDG pressure readings are measured using a saturometer, which require 10 to 30 minutes for the sensor to stabilize for each reading. Data collection by this method is time consuming and flow and TDG conditions may change before data collection from each transect is completed.

The multi-parameter probe has factory-installed programming that calculates and displays percent oxygen based on oxygen solubility. Oxygen percent saturation can also be calculated using the following formula (ECO Enterprises):

$$\% O_2 = [DO/BO_2(0.5318)] - [0.20946 (BP - PH_2O)]$$

where 
- DO = Dissolved Oxygen in milligrams per liter
- $BO_2$ = bunsen coefficient of oxygen
- BP = local barometric pressure in millimeters of mercury
- $PH_2O$ = water vapor pressure in millimeters of mercury

A linear correlation, as shown in Figure 2, has been observed between the percent oxygen values obtained by these two methods using data collected at Kentucky Dam..

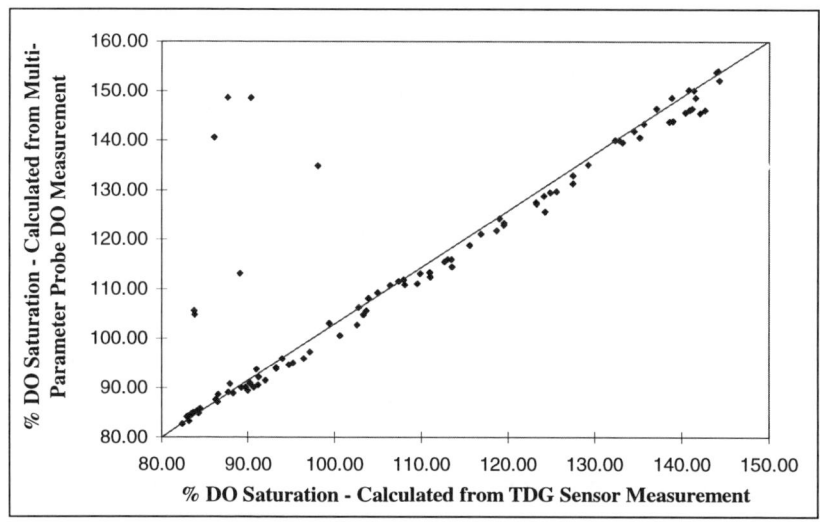

Figure 2: Comparison of Calculated Percent Oxygen Values

Using this correlation, the percent oxygen readings from the multi-parameter probe can be used in the above equation to estimate TDG values. The importance of this relationship is that whereas the TDG sensor method takes 10 to 30 minutes to stabilize for an accurate reading, the multi-parameter probe takes accurate readings in less than one minute. Thus, readings can be taken in shorter periods of time to reduce monitoring costs.

A small percentage of the data did not conform to this correlation as evidenced in Figure 2. These data, verified to be accurate, were obtained during different release conditions (tailwater elevation and gate arrangements). More monitoring and further data analysis are needed to explain this discrepancy. Further investigations may indicate a different linear relationship for each set of release conditions.

## CONCLUSION

DO and TDG monitoring at TVA projects has evolved through a series of process improvements. Real-time data for operating aeration systems at remote sites requires continuous monitors with a high degree of accuracy to minimize operating and maintenance costs. The accuracy and reliability of DO measurements has been improved by standardizing field procedures, using multi-parameter probes, training, and incorporating preventative maintenance measures. These general techniques can be universally applied and adapted to site specific conditions.

The use of aeration systems has resulted in new concerns about TDG and spurred the development of reliable continuous TDG monitors. Experiences at Kentucky Dam indicate that a relationship can be established to estimate TDG by using the percent saturation of oxygen provided by the multi-parameter probe. This technique could reduce future monitoring time for TDG by as much as 75 percent.

## REFERENCES

ECO Enterprises, WEISS Saturometer Field Manual for the ES-2 & ES-3 Instruments, Seattle, Washington.

Marking, Leif L., Gas Supersaturation in Fisheries: Causes, Concerns, and Cures. Fish and Wildlife Leaflet 9, United States Department of Interior, Fish and Wildlife Service. 1987.

Standard Methods for the Examination of Water and Wastewater. 19th Edition. 1995.

Water Quality Standards Handbook: Appendix A, Water Quality Standards Regulation. Second Edition. 1992.

# Modifying Reservoir Release Temperatures Using Temperature Control Curtains

TRACY B. VERMEYEN
US Bureau of Reclamation, Denver, CO USA

## ABSTRACT

Reclamation (Bureau of Reclamation) has constructed four temperature control curtains to reduce release water temperature at structures in the Sacramento and Trinity River drainages in northern California. These curtains provide selective withdrawal at intake structures, control topography induced mixing, and control interfacial shear mixing associated with plunging density currents entering reservoirs. Comprehensive field monitoring has been conducted to measure curtain performance characteristics. Monitoring included continuous temperature profiling, and velocity profiling using an ADCP (acoustic Doppler current profiler). This paper presents and summarizes performance data collected near curtains in Lewiston and Whiskeytown Reservoirs.

## BACKGROUND

During the late 1980s, extended drought in northern California created potentially life threatening conditions for endangered salmon species inhabiting the Sacramento River. Summer and early fall river water temperatures threatened to exceed critical levels for sustaining juvenile salmon populations. High release water temperatures from the reservoirs, coupled with natural in-stream warming, threatened to make downstream waters too warm for egg incubation and juvenile fish survival. As a result, California's water resources agencies and the National Marine Fisheries Service imposed a maximum temperature of 56 °F in the Sacramento River below Keswick Dam. To comply with temperature requirements, Reclamation began an aggressive program to construct selective withdrawal features that would yield cold water releases.

## DESCRIPTION OF RESERVOIR SYSTEM

Water from the Trinity River Basin is diverted to the Sacramento River Basin through two tunnels and three reservoirs. Trinity River water is diverted from Lewiston Reservoir through Clear Creek Tunnel to the Judge Francis Carr Powerplant and discharged into Whiskeytown Reservoir. From there, water flows through the reservoir and into the Spring Creek Tunnel and through Spring Creek Powerplant. Spring Creek Powerplant releases water into Keswick Reservoir, where it combines with water released from Shasta Lake. Water released from Keswick Dam enters the upper Sacramento River. Over the course of this diversion and prior to curtain installation, Trinity River water temperatures commonly increased 10 to 13 °F. To provide cold water releases, Reclamation engineers chose to install temperature control curtains in Lewiston and Whiskeytown Reservoirs; a total of four curtains were constructed. Curtains allow project operators to manage hydropower operations while controlling the temperature of water releases.

## CURTAIN DESIGN

When a reservoir is thermally stratified, water can be selectively withdrawn from distinct horizontal layers. The vertical position and thickness of the withdrawal layer depends on several factors:

- Placement, size, and orientation of the intake
- Degree of thermal stratification
- Frequency of the withdrawal discharges
- Boundary interference from the reservoir water surface, the dam, or topography surrounding the intake

To develop a strategy for providing selective withdrawal, Reclamation engineers conducted a VE (value engineering) study to develop cost-effective selective withdrawal options. During the VE study, temperature control curtains were found to offer potential cost savings compared to structural modifications to existing intake structures. Physical model studies were conducted by Reclamation's Water Resources Research Laboratory to determine both reservoir and river responses to curtain installations in Lewiston and Whiskeytown Reservoirs (Johnson and Vermeyen, 1993).

The VE team recommended three sites for potential curtain installations. In Lewiston Reservoir, a small, shallow and weakly stratified impoundment, a curtain was recommended to provide selective withdrawal for a near-surface intake to the Clear Creek Tunnel. In Whiskeytown Reservoir, two curtains were recommended: 1) the Carr Powerplant tailrace curtain would minimize interfacial shear mixing which occurs when cold water entering the reservoir plunges below the epilimnion, and 2) a second curtain would provide selective withdrawal for the Spring Creek Tunnel intake.

Model and prototype performance of the Lewiston Reservoir curtain and the Carr Powerplant tailrace curtain will be summarized in this paper.

## LEWISTON RESERVOIR CURTAIN MODELING

A physical model study indicated that a curtain surrounding the Clear Creek Tunnel intake was ineffective because substantial mixing with the epilimnetic water occurred upstream of the curtain when cold water passed through a restricted cross section in the reservoir. Mixing occurred in a shear zone, which developed between the withdrawal layer and the deeper epilimnion that formed upstream of the curtain. Locating the temperature control curtain upstream from the "narrows" reduced interfacial shear mixing by lowering curtain approach velocities. In addition, warming caused by mixing in the narrows was reduced by isolating this mixing zone from the main reservoir body, which limited the supply of warm water to the mixing zone. The recommended curtain design called for an 830-ft-long, 35-ft-deep, surface-suspended curtain.

## CARR TAILRACE CURTAIN MODELING

A physical model study was conducted to develop a curtain design for Whiskeytown Reservoir that minimized mixing associated with plunging inflows. The model study indicated that a curtain located 2 miles downstream from the Carr Powerplant would effectively limit interfacial shear mixing. The recommended curtain design called for a 600-ft-long, 40-ft-deep, surface suspended curtain.

## INSTALLING THE LEWISTON AND WHISKEYTOWN CURTAINS

In the summer of 1992, crews installed a temperature control curtain in Lewiston Reservoir based on the model study research. Total time for engineering, procurement, and construction of the Lewiston Reservoir curtain was five months. Reclamation's Northern California Area Office was responsible for the design and construction of the curtain. Costs for this curtain totaled $650,000.

Two temperature control curtains were installed in Whiskeytown Reservoir during the summer of 1993. The Carr tailrace curtain was fabricated and installed in one month at a cost of $500,000. A 100-ft-deep, 2,400-ft-long surface suspended curtain which enclosing the Spring Creek Tunnel intake was installed over a four-month period at a cost of $1.8 million.

## LEWISTON RESERVOIR CURTAIN PERFORMANCE

Lewiston curtain performance was evaluated by analyzing temperature data collected in the Clear Creek Tunnel intake. In 1992, data collected before and after curtain installation indicated that for similar operational conditions the average temperature of water entering the Clear Creek Tunnel was reduced by about 2.5 °F after the curtain installation.

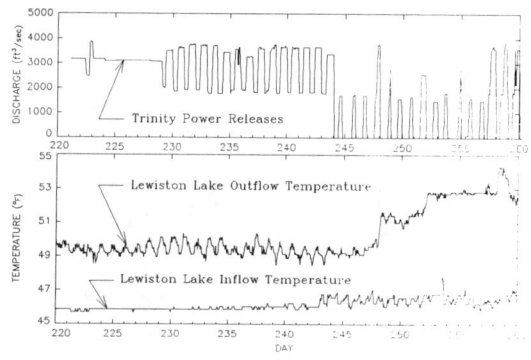

Figure 1 - Lewiston Reservoir operations along with inflow and outflow temperatures for August 8 - September 17, 1994. These data illustrate the temperature gain of Trinity River water diverted through the reservoir.

During August and September 1994, curtain performance was evaluated for three types of power operations at Trinity and Carr Powerplants (see figure 2a): 1) during days 220 through 228, baseload power releases were held constant at 3,200 ft³/s; 2) days 229 through 243 had partial peaking power operations, during which flows fluctuated between 1,800 and 3,500 ft³/s; 3) days 244 through 250 were strictly peaking power operations, with one and occasionally two turbines operated for 10 to 12 hours. A comparison of Trinity Dam outflow and Clear Creek Tunnel temperatures showed a consistent 3.5 °F temperature gain through the reservoir for days 220 through 243 regardless of the operations (figure 1). However, when operations were changed on day 244, a steady increase in outflow temperature was observed. After day 254, temperature gain through Lewiston Reservoir had stabilized at 6.5 °F. This additional 3 °F temperature gain occurred because warm water accumulated upstream and downstream from the curtain during no-flow periods. Then, during peaking operations, warm water was entrained and released. Warm water accumulation is shown in figure 2c, which is a plot of the hourly temperature profiles collected 200 ft upstream of the curtain.

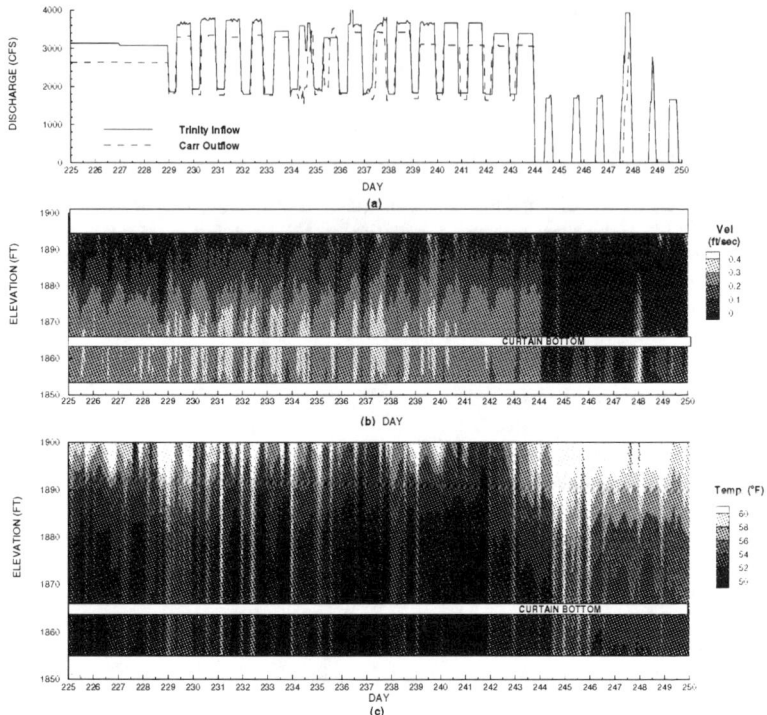

**Figure 2.** (a) Lewiston Reservoir operations, (b) isovel plot of ADCP data, and (c) temperature profile data collected 200 ft upstream of the curtain. All three plots cover the period from August 13 to September 7, 1994.

Figure 2b is an isovel plot of ADCP (acoustic Doppler current profiler) data collected 200 ft upstream from the curtain for the same time period that temperature profiles were collected. Figure 2b illustrates several interesting selective withdrawal characteristics. For days 225 through 228, variations in the upper limit of withdrawal are caused by diurnal fluctuations in the epilimnion thickness. During the day, the epilimnion expands and forces the withdrawal zone to narrow. Conversely, after sunset, the epilimnion contracts and the withdrawal zone expands upward. For days 228 to 244, the expansion and contraction of the withdrawal zone was magnified by peaking flows during daylight hours. For days 245 to 250, the withdrawal zone was slow to develop and was restricted by the thickened epilimnion.

Temperature profiles shown in figure 2c illustrate how the partial and full peaking power operations generated strong mixing upstream from the curtain. Mixing was strong enough to break down the thermal stratification. Two forms of mixing were responsible for the break down: 1) when peaking was initiated, the withdrawal layer extended through the thermocline and entrained epilimnetic water, and 2) plunging inflows generated interfacial shear mixing where the cold, denser water plunged beneath the epilimnion. When flow

When flow rates were reduced, reservoir stratification was quickly reestablished. These data and increased temperatures in the Clear Creek Tunnel intake lead to the recommendation that full peaking operations be avoided during periods when release temperature restrictions are in effect.

## CARR TAILRACE CURTAIN PERFORMANCE

Curtain performance was evaluated by analyzing temperature data collected in the tailrace below Carr Powerplant and temperature profiles collected downstream from the curtain. The main objective was to determine the reduction of inflow warming attributed to the curtain. In May 1994, temperature profiles collected before and after curtain installation showed dramatic modifications to the reservoir stratification. After curtain installation, the temperature of water flowing into the hypolimnion was reduced from 56 to 53 °F. The upstream epilimnion was reduced to a depth of 10 to 15 ft, and the downstream epilimnion expanded to a depth of 20 to 25 ft. In August 1994, the two curtains reduced the overall temperature gain of water routed through Whiskeytown Reservoir by 4 °F compared to pre-curtain temperatures collected in August 1988. The majority of the temperature reduction was attributed to the Carr tailrace curtain.

Plots in figure 3 illustrate the hydraulic performance of the Carr tailrace curtain. The data were collected over an 11-day period when the power operations were baseload for 4 days and partial peaking for 7 days (figure 3a). Figure 3b shows ADCP data collected 150 ft downstream from the curtain. For baseload operations, velocities varied with diurnal fluctuations in underflow temperatures. For partial peaking operations, velocity variations were caused by displacement of the thermocline (figure 3c). When diversion flow rates through Carr and Spring Creek powerplants were increased a surface seiche formed. The seiche displaced the thermocline upward near the tailrace curtain and downward at the Spring Creek curtain. Conversely, when flow rates were reduced, the seiche quickly subsided. These variations to the thermal stratification did not adversely affect the performance of the Carr tailrace or the Spring Creek intake curtains. It is notable that during reduced flow periods the curtain did not appear to hydraulically control the underflow; water was flowing into the hypolimnion as a density current.

## CONCLUSIONS

- Temperature control curtains have been successfully used to provide selective withdrawal and control reservoir mixing at Lewiston and Whiskeytown Reservoirs.

- Curtains have reduced temperature gains of Trinity River water diversions by 4 to 5 °F during late summer and early fall, which allows some flexibility in how Shasta Dam powerplants are operated.

- Hydro-power operations have a strong influence on curtain performance. Peaking power operations increased temperature gains in Lewiston Reservoir by 3.5 °F.

- Thermally stratified physical models were valuable tools for developing effective temperature control curtain designs.

Continuous temperature monitoring and ADCP data were used to document curtain performance and hydraulic characteristics for a wide range of reservoir operations.

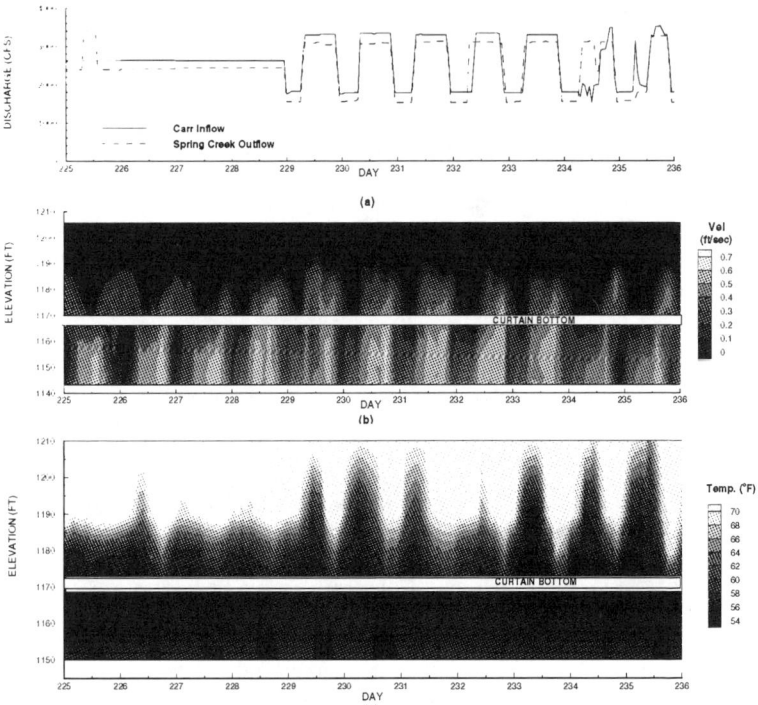

**Figure 3**. (a) Whiskeytown Reservoir operations, (b) isovel plot of ADCP data, and (c) temperature profile data collected 800 ft downstream from the curtain. All three plots cover the period from August 13 to August 24, 1994.

## REFERENCES

Vermeyen, T.B. and Johnson, P.L. (1993). "Hydraulic Performance of a Flexible Curtain Used for Selective Withdrawal: A Physical Model and Prototype Comparison." Proceedings of the Hydraulics Division ASCE National Conference, San Francisco, CA, July 25-30, 1993.

Johnson, P.L. and Vermeyen, T.B. (1993). "Flexible Curtain Structure for Control of Vertical Reservoir Mixing Generated by Plunging Inflows," Proceedings of the Hydraulics Division ASCE National Conference, San Francisco, CA, July 25-30, 1993.

# OXYGEN UPTAKE AT BARRAGES OF THE ELBE CASCADE

P Novak* - P Gabriel**
*Em. Prof., Department of Civil Engineering,
University of Newcastle upon Tyne, UK
**Prof., Faculty of Civil Engineering,
Czech Technical University, Prague, Czech Republic

## ABSTRACT

Measurements at three barrages of the Elbe cascade were utilised to test predictive equations for oxygen uptake at overfalls. The Avery-Novak equation gave good results and has now been incorporated into the operating rules of the whole 170 km cascade. The boundary conditions and other aspects of the use of the equation are discussed in some detail.

## INTRODUCTION

During recent decades anthropogenic influences have caused serious pollution of the river Elbe (Labe) as well as destruction of its riverine zones. A series of investigations was carried out during 1991/96 with the aim of suggesting measures for substantially improving the river water quality and contributing to the revitalization of the adjacent region. Although the main thrust of the work was the identification of the main sources of pollution, aeration at barrages of the cascade was also investigated with the aim of formulating procedures for operating rules. The paper discusses measurements at three localities used to test predictive equations and the analysis of the effect of various hydraulic parameters.

## THE ELBE CASCADE

The Elbe is the largest river of the Czech Republic and is intensively used for water supply, navigation and hydropower. During the last hundred years the river was canalized by constructing twenty seven low-head barrages forming a continuous cascade stretching over 170 km. This together with the adjacent 40 km canalized stretch of the river Vltava (Moldau) forms the Elbe-Vltava cascade (Fig. 1). By slowing down the flow velocities the cascade can contribute to the reduction of self-purification which together with the main culprit - discharges of effluents without treatment - has lately caused a frequently unacceptable fall in oxygen levels in the river.

The head on the Elbe barrages varies from 2 to 9m. At each barrage there is a navigation lock and in most cases a power station. After modernisation of the barrages there is the possibility of passing the discharge by overflow over the gates. The whole cascade is operated from a central control room. Fig. 2 shows a cross-section of a typical barrage (Brandýs) with three vertical lift gates and attached flaps, 23.5m span each. The navigation lock is located in a short canal and the power station has two Kaplan turbines with capacities of $45 m^3/s$ each.

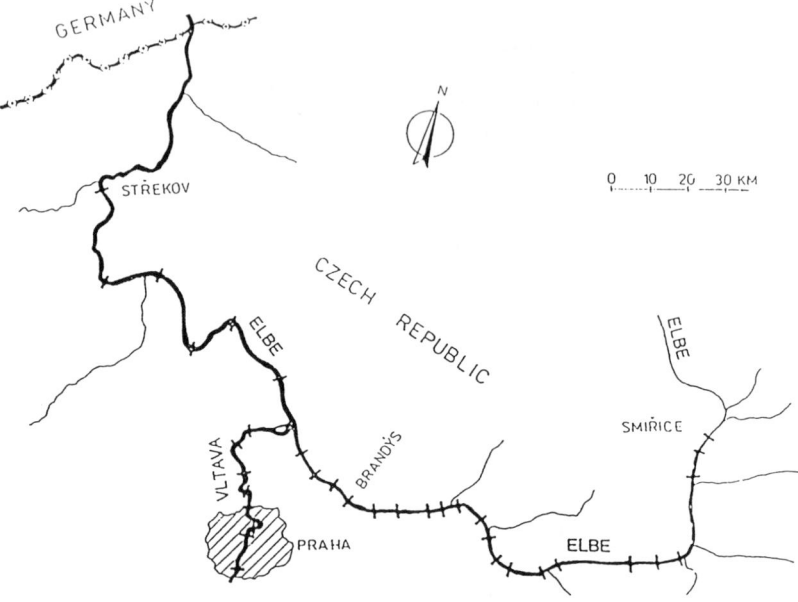

Fig.1 The Elbe cascade

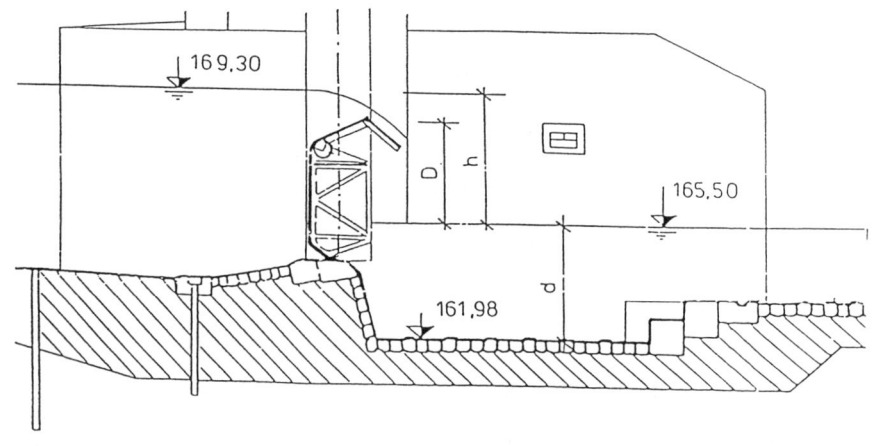

Fig. 2 Brandýs barrage

## MEASUREMENT OF OXYGEN CONCENTRATION

Several oxygen measurements (Gabriel et al. 1995) were carried out along the whole Elbe cascade using an OXI325 meter. Measurements were taken at 0.6m below the water surface upstream and downstream of each barrage, above and below each outflow/confluence, in the reservoirs at intervals of 500m and also in the vicinity of moving barges. Overall the results indicated an increasing (improving) oxygen concentration in the downstream direction with significant improvement at barrages without power stations; when the whole flow passed through the turbines there was no significant local improvement in the oxygen content. In the case of part of the flow passing over the gates and part through the power station, the higher the proportion of the overflow, the bigger the improvement in the oxygen content which was evident throughout the whole upstream reach of the next barrage on the cascade. The economic appraisal showed that aerating by overfall is more effective than aeration of the draft tubes of the turbines in spite of partial losses in the power production. At three locations (Smiřice, Brandýs, Střekov) - see Fig. 1 - more detailed measurements of oxygen concentration upstream and downstream of barrages were carried out with the main aim of testing predictive equations so that these can be utilised as a basis for formulating procedures into the operating rules of the cascade. The measurements (Table 1) were carried out during low river flows with above normal temperatures and steady hydrological conditions.

## HYDRAULICS OF OXYGEN TRANSFER AT OVERFALLS

As the measurements and their intended use in operation covered only cases of water flowing over barrage gates, we are not considering other flow configurations such as flow under the gates, spillways of dams and weirs, hydraulic jumps, etc. The results were, therefore, compared with two predictive equations: 1) the equation proposed by Avery and Novak (1978) and 2) the equation(s) by Nakasone (1987).

In the aeration of water at overfalls many parameters influence the result: the height of fall of the water jet (nappe) h, the unit discharge (discharge per unit width) q, the downstream depth (d), salinity (or presence of sodium nitrite in the water), the shape of the downstream water pool, water temperature T, access of air to name just the more obvious ones. Of these parameters the first three are the most important ones.

Avery and Novak (1978) proposed for the oxygen deficit ratio r at 15°C the equation:

$$r_{15} - 1 = k \ Fr_j^{1.78} \ Re_j^{0.53} \tag{1}$$

where $k = 0.627$ E-4 (for water without salinity), $Fr_j = (gh^3/(2q_j^2))^{0.25}$ (jet Froude number at impact into the downstream pool), $Re_j = q_j/\nu$, where $q_j = R \ (2gh)^{1/2}$ with R the jet hydraulic radius at impact and $r = (C_s - C_u)/(C_s - C_d)$, where $C_s$, $C_u$ and $C_d$ are the saturated, upstream and downstream concentrations of dissolved oxygen (at T°C).

Avery and Novak used a temperature correction based on the experiments of Gameson et al. (1958):

$$(r_T - 1)/(r_{15} - 1) = (1 + 0.046T)/1.69 \tag{2}$$

or
$$r_T - 1 = (r_{15} - 1)\left[1 + 0.0271(T - 15)\right] \qquad (2a)$$

The boundary conditions for use of eq. 1 were established as (Avery and Novak (1978)):

$$h \leq 6q^{0.33} \text{ (m) and } d \geq 0.0041 \, Re_j^{0.39} \, Fr_j^{0.24} \text{ (m)}$$

For wide overfalls, width b, with air access on both sides of the nappe the wetted perimeter at impact becomes 2b and thus $q_j = Q/2b = q/2$ (q being the unit discharge at the crest).

Nakasone (1987) proposed a set of four equations for the oxygen deficit ratio at 20°C according to the value of $(D + 1.5H_c)$ $(= h)$ and q (m$^3$/hm); in our case the following equation applies:

$$\ln r = 5.92 \, (D + 1.5H_c)^{0.816} \, q^{-0.363} \, d^{0.31} \qquad (3)$$

Nakasone recommends for the temperature corrections:

$$\ln r_t = \ln r_{20} \left[1 + 0.0168(T - 20)\right] \qquad (4)$$

Table 1 summarises the results of measurement and computations using eq. 1-4. Measurements 1-6 were carried out at Smiřice (km 171,728), 7-9 at Brandýs (km 27,878) and 10 at Střekov (km 40,400) (Fig. 1).

## Table 1

| Meas | T (C°) | h (m) | Q (m$^3$/s) | q (m$^2$/s) | d (m) | $C_U$ (mg/l) | $C_S$ (mg/l) | $C_d$ (mg/l) | r | Ave - Nov | | Nakasone | |
|---|---|---|---|---|---|---|---|---|---|---|---|---|---|
| | | | | | | | | | | r | $C_d$ (mg/l) | r | $C_d$ (mg/l) |
| 1 | 19.9 | 9.02 | 9.5 | 0.864 | 5.07 | 5.0 | 9.12 | 8.5 | 6.61 | 6.18 | 8.44 | 26.28 | 8.96 |
| 2 | 19.9 | 9.02 | 9.5 | 0.864 | 5.07 | 5.0 | 9.12 | 8.6 | 7.88 | 6.18 | 8.44 | 26.28 | 8.96 |
| 3 | 20.3 | 9.02 | 9.5 | 0.432 | 5.07 | 5.9 | 9.02 | 8.4 | 4.21 | 7.73 | 8.62 | 64.31 | 9.06 |
| 4 | 20.4 | 9.02 | 9.5 | 0.432 | 5.07 | 5.9 | 9.02 | 8.4 | 4.32 | 7.73 | 8.62 | 64.31 | 9.06 |
| 5 | 20.5 | 9.02 | 28.8 | 0.400 | 5.07 | 4.0 | 8.99 | 7.7 | 3.09 | 7.94 | 8.36 | 72.08 | 9.23 |
| 6 | 20.7 | 9.02 | 28.8 | 0.400 | 5.07 | 4.6 | 8.96 | 8.1 | 4.87 | 7.97 | 8.41 | 72.08 | 8.93 |
| 7 | 14.7 | 3.82 | 21.0 | 0.298 | 3.52 | 7.2 | 10.14 | 8.8 | 2.15 | 3.11 | 9.19 | 6.44 | 9.71 |
| 8 | 14.6 | 3.83 | 23.2 | 0.494 | 3.58 | 7.2 | 10.16 | 9.0 | 2.49 | 2.76 | 9.09 | 4.81 | 9.57 |
| 9 | 14.5 | 3.84 | 37.2 | 0.528 | 3.58 | 7.2 | 10.17 | 9.1 | 2.67 | 2.32 | 8.89 | 4.38 | 9.53 |
| 10 | 22.4 | 8.20 | 115.0 | 1.600 | 2.70 | 9.3 | 8.67 | 8.8 | 6.00 | 4.88 | 8.80 | 7.79 | 8.77 |

## SOME COMMENTS ON THE RESULTS IN TABLE 1

a)   The results using the Avery - Novak procedure (eq. 1) show fairly good agreement with the measured values of $C_d$ (and hence r); eq. 1 certainly performs better than Nakasone's equation 3. Measurement 10 showed supersaturation.

b) Large discrepancies in r values (e.g. measurement 3: 4.21 - 7.73 - 64.31) give sometimes a slightly misleading impression as the percentage "errors" in $C_d$ values are much smaller (8.4 - 8.62 - 9.06); thus r is a correct but extremely sensitive parameter for comparing results of computations.

c) Checking the boundary conditions of eq.1 it can be seen that the downstream depth d was in all cases substantially bigger than the required minimum (optimum) for the Avery-Novak equation and thus should have little effect on the results. In the case of Brandýs (7 - 9) the condition of $h < 6q^{0.33}$ was also satisfied; for Střekov (10) the height of fall was close to the limit (8.2 m > 7 m) but in the case of Smiřice h = 9.02 m was almost double of the limiting value (4.5 to 4.7 m). This limit indicates the height of fall where the jet disintegrates; beyond it the intake of air into the jet itself increases but the air drawn at impact into the pool - the decisive factor in the Avery - Novak analysis - begins to decrease. Thus the boundary is not a point of sharp overall change in the value of r which can even increase slightly before decreasing; this is borne out by the results as well as by analysis of some experiments for flow over a notch with heights of fall $2 < h (m) < 4$ and the limiting value $h = 6.q^{0.33}$ between 1.04 and 1.63 m (Vasel 1995).

d) The correct $q_J$ used in eq. 1 in this case was q/2 whereas in eq. 3 q = Q/b was used. Should $q_J$ be used in eq. 3 the computed results would have increased substantially; using q in eq. 1 would substantially reduce the computed value of r. This demonstrates the need for correct usage of unit discharge in the equations.

e) Although the changes in measured values of r at each site are relatively small, at Smiřice (1-6) (and also at Brandýs (7-9)) there seems to be a tendency of increasing r with increasing q - the opposite of the trend in the computed values according to eq 1 (and 3). The physics supports the contention of increasing r with a decreasing q (as the wetted perimeter at impact increases for the same discharge Q). Analysing in more detail particularly cases 1-4 it will be noticed that the change in q was achieved by passing the same discharge over 1 or 2 gates of the barrage resulting in substantial circulation of aerated flow in the stilling basin in cases 1 and 2 and thus increased contact time on the fluid/air interface (both at the water surface and by increased circulation of air bubbles) and hence increased values of measured r. It is also possible, that overestimation of q in the computation of nos 1 and 2 (the thickness of the nappe at impact is neglected) plays a certain role.

f) The two temperature correction equations 2 and 4 are in fact identical as both are based on experiments by Gameson et al. (1958) showing a 4.6% rise in r per °C. Although these tests covered a wide temperature range [$0 < T (°C) < 40$] and heights of fall $0.6 < h (m) < 3.3$ the downstream pool depth was kept constant at d = 0.09 m - too low according to the Avery - Novak criterion. Using a temperature correction factor

$$f_c = \theta^{(T - T_r)}$$

(5)

where $\theta$ is a temperature coefficient and $T_r$ a reference temperature the range of r in Table 1 for a temperature change from 15 to 20°C results in $1.010 < \theta < 1.018$. These are not unreasonable values (see e.g. Jensen (1991)), but as $\theta$ is a function of both the turbulence of water and diffusivity of air with $\theta$ decreasing with increasing turbulence ($\theta$ approaches 1.0), it is possible that the temperature corrections in measurements 1-6 for equation 1 (and 7-9 in eq. 3) are slightly too high (reducing somewhat the computed values of r for Avery -Novak).

g) Uncertainties remain about scale effects (Novak (1994), Task Committee (1991)). However, considering that eq. 1 was derived from laboratory experiment with notches 0.1-1.3m wide, discharges 0.6-5.8 $l.s^{-1}$ and heights of fall 0.25-2.10m, its performance seems to be quite good.

## CONCLUSIONS

Measurements of oxygen concentration along the 170km long Elbe cascade established the main sources of pollution and the effect of barrages on water quality. Detailed measurements at three localities with different modes of passing discharge by overfall over gates gave data which correlated well with the Avery-Novak predictive equation. This equation was used to formulate procedures (in case of occurrence of large oxygen deficiencies) which were incorporated into the operating rules of the cascade. A more detailed analysis of the hydraulic background of the equation presented in this paper stresses the importance of boundary conditions, air access, geometry of structure and downstream pool and temperature corrections, and concludes that in spite of some uncertainties about scale effects the performance of the predictive equation is quite robust.

## REFERENCES

Avery, S.T. and Novak, P., 1978. Oxygen transfer at hydraulic structures. Journal of the Hydraulic Division, ASCE, **104**, HY 11, pp 1521-1540.

Gabriel, P., Dolecek, J. and Kucerova, J.K. 1995. Vliv labske kaskady na kyslikovou bilanci toku. Vodni cesty a plavba, No. 2, Praha (in Czech).

Gameson, A.L.H., Vandyke, K.G. and Ogden, C.G., 1958. The effect of temperature on aeration at weirs. Water and Water Engineering, **62**, pp 489-492.

Jensen, N.A. 1991. Effect of temperature on gas transfer at low surface renewal rates. In Wilhelms, S.C. and Gulliver, J.S., (editors): Air Water Mass Transfer, ASCE, 2nd Int. Symp. on Gas Transfer at Water Surfaces, Minneapolis, Sept. 1990.

Nakasone, H. 1987. Study of aeration at weirs and cascades. Journal of Environmental Engineering, ASCE, **113**, No. 1 pp 64-81.

Novak, P. 1994. Improvement of water quality in rivers by aeration at hydraulic structures. In Hino, M (ed): Water Quality and its Control. IAHR Hydraulic Structures Design Manual No. 5; ch. 6 pp 147-168, Balkema, Rotterdam.

Task Committee on Gas Transfer at Hydraulic Structures, 1991. Gas transfer at hydraulic structures. In Wilhelms, S.C. and Gulliver, J.S., (editors): Air Water Mass Transfer, ASCE, 2nd Int. Symp. on Gas Transfer at Water Surfaces, Minneapolis, Sept. 1990.

Vasel, Y. L. 1995. Private communication.

# INFLUENCE OF PUMPED-STORAGE POWERPLANT ŻARNOWIEC ON NATURAL LAKE (LOWER RESERVOIR)

WOJCIECH MAJEWSKI
Technical University of Gdańsk, Gdańsk, Poland

## ABSTRACT

Pumped-storage powerplant Żarnowiec of the capacity 716 MW was put into operation in 1983. Artificial upper reservoir was constructed on the top of the nearby hill. Natural lake of postglacial origin of the volume 118 mln $m^3$ was used as the lower reservoir. Several changes were introduced to the lake to fulfill new functions. Comprehensive studies of the lake have been carried out to indicate the changes resulting from the operation of pumped storage powerplant. These studies included: hydrodynamics, thermal regime, hydrochemistry, hydrobiology, and fishery state. It was found out that important influence on lake water quality has the management of its catchment.

## INTRODUCTION

Deficit of electrical energy in Poland in the 1970s stimulated the design of a pumped-storage and nuclear plant. Both powerplants were located about 70 km from Gdańsk in the vicinity of Żarnowiec lake, which was to be used as the lower reservoir of the pumped-storage plant (4 x 170 MW) and as a cooling water reservoir for the nuclear plant (4 x 440 MW). It was recognized that the operation of both projects would result in considerable hydrodynamic, thermal and environmental changes in the lake and therefore comprehensive studies of the lake were undertaken to solve engineering problems. They included water balance, thermal regime and hydrodynamics. The planning and engineering of the Żarnowiec pumped-storage and nuclear projects was initiated in the 1970s. In 1983 the pumped-storage plant was commissioned, but in 1990 the construction of nuclear powerplant was stopped as the result of public protests caused by Chernobyl catastrophe and difficulties in further financing of the project from the state budget.

In 1994, after 10 years of the operation of the pumped-storage plant, it was decided to carry out detailed investigations of the lake to find out what physical, chemical, biological and environmental changes had occurred during this time. The influence of the management of the lake catchment on water quality in the lake was also considered as important factor.

## ŻARNOWIEC LAKE AND ITS CATCHMENT

Żarnowiec lake is in the northern part of Gdańsk province and is one of the largest and most beautiful lakes in this region. The lake is of glacial origin and is situated among moraine hills. In natural conditions at average water elevation 1.34 m above sea level the lake had the surface area of 14 km$_2$ and a total volume of 118 mln m$^2$. The length of the lake is 7.6 km and the maximum width 2.6 km. The maximum depth is 19.4 m and the average depth is 8.4 m. The maximum and minimum recorded water elevations in natural conditions were 1.78 and 0.86 m respectively. Daily water level variations were in the range 1 to 2 cm, and in extreme situations they reached 5 to 7 cm. The bottom of the lake consists of sand and gravel, and in some areas is covered with organic sediments. The northern part of the lake is shallow with flat shores which were inundated during high water levels. This part has a variety of aquatic vegetation, especially in the shore areas (Majewski, 1996). Surface inflow to the lake comes from the direct catchment and 2 small tributaries (Piaśnica Górna and Bychowska Struga). Their catchments amount to 39, 88 and 122 km$^2$ respectively. Thus total catchment of the lake has 249 km$^2$. Surface inflow to the lake amounts to 1.7 m$^3$/s, while groundwater inflow has been estimated at 0.6 m$^3$/s. This inflow is very uniform during the year because of the existence of several small lakes and marches. The natural outflow from the lake (Piaśnica Dolna) discharges directly to the sea. The catchment of Piaśnica Dolna has the surface of 61 km$^2$ and consists primarily of meadows which are under special protection. The catchment of the lake is shown in Fig.1. The catchment of the lake comprises agricultural land, meadows and forests. There is hardly any industry in the area. Lake inflows are used for agricultural water supply and fish farms. In recent years more and more land has been used for recreational purposes, which create danger of pollution. The average annual precipitation amounts to 620 mm and shows considerable variation in space and time.

Predominant wind directions are along the lake axis and their maximum speed has reached 20 m/s causing waves up to 1.5 m high and wind currents of 0.1 m/s. Average annual lake water temperature was in the range of 8 to 9°C. The maximum water surface temperatures during summer have reached 23°C in the shore areas. During severe winters the lake has been completely covered with solid ice of the thickness up to 0.5 m. During hot summers thermal stratification has been frequently observed. The annual average evaporation amounts to 600 mm and is usually balanced by precipitation. The lake waters are of good quality, which has been advantageous for fish breeding. More than 20 species of fish have been detected in the lake. Some of them are very useful, but others are of minor value. Regular fish management was carried out on the lake. During vegetation season water from the lake is used for irrigation of the meadows with maximum discharge 1.8 m$^3$/s. These data are based on comprehensive studies of the lake carried out initially during 1973-75 (Majewski, 1983). They were mainly related to engineering aspects but included also the status of ichthyology and fisheries as well as hydrochemistry and hydrobiology.

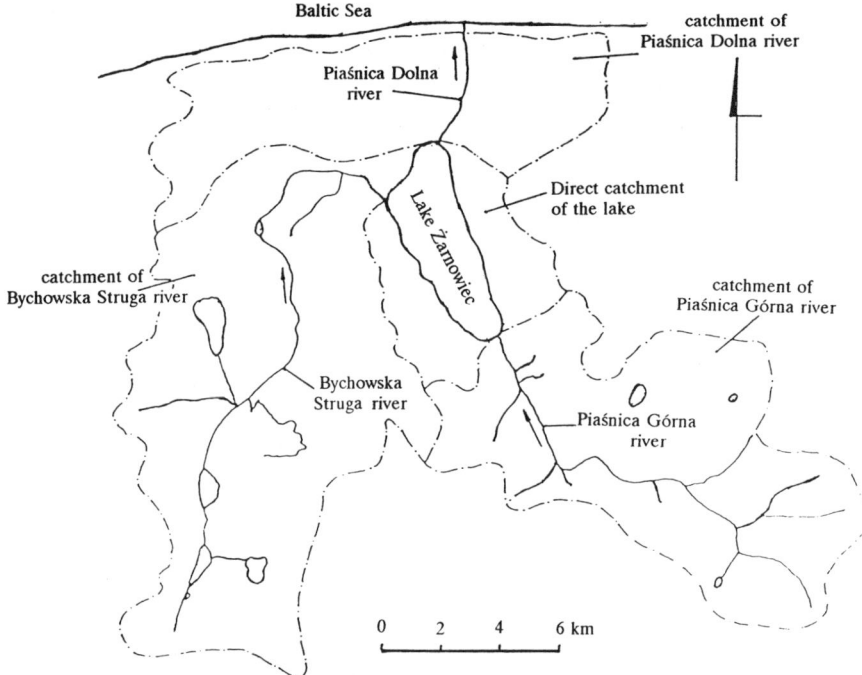

Fig. 1. Catchment of Żarnowiec lake

## CHANGES IN LAKE MANAGEMENT CAUSED BY THE PLANT

Operation of pumped-storage plant results in the exchange of water between upper reservoir and the lake of the volume 13 to 14 mln m$^3$ every day which causes variation of water levels of 1 m. Pumping of the water from the lake is mainly during the night (6 hours), while turbine operation (discharge to the lake) is generally in the morning and evening (5.5 hours). The use of the lake as the lower reservoir required the construction of several new hydraulic structures (Fig.2). The construction of the nuclear powerplant reduced the surface of the lake, and part of the shoreline was protected by sheet piling and other concrete structures. At the same time the lake area increased by about 12 ha as the result of the construction of the channel connecting the lake with the powerplant. Both surface inflows to the lake and the irrigation channel have been reconstructed to take into account varying water levels in the lake. The flat terrain north of the lake is now protected from inundation by dikes. Because of the rapid water level changes in the lake, it was necessary to construct a control weir at the outflow from the lake. The minimum biological discharge downstream from the lake has been determined as 0.4 m³/s. On the west coast of the lake near Nadole, a sewage treatment plant was constructed. This discharges treated waste water to the lake through a bottom pipe extending about 100 m from the shoreline. Daily water level changes due to the operation of pumped-storage plant result in the flooding of and subsequent exposure of a belt 10 to 15 m wide in some parts of the lake. Much attention has focused on environmental aspects of this scheme. Special attention was directed to reduce water pollution and to keep water in the lake of a very good quality.

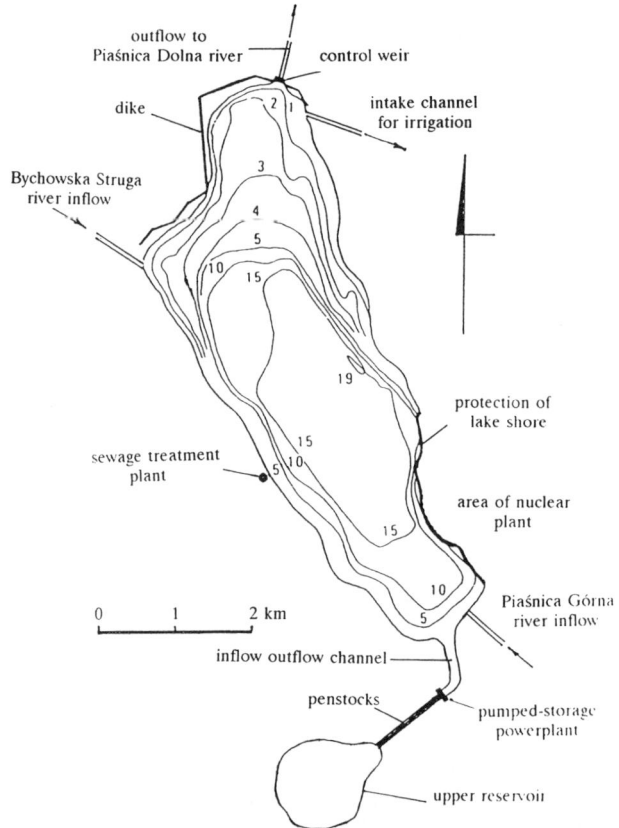

Fig. 2. Management of Żarnowiec lake

## RECENT STUDIES OF LAKE ŻARNOWIEC

The need to analyze ecological problems and to protect the environment in view of the pumped-storage plant led to detailed studies of the lake after 10 years of the operation of the plant. It has been assumed that water quality

and its natural values are not only associated with the operation of the plant, but also depend on the present management of the catchment. The aim of the study was to determine all factors which negatively influence the water quality of the lake and thus its landscape and recreational values. The scope of the study included the following aspects:
* the hydrodynamics and thermal regime of the lake
* physical and chemical analysis of water in various parts of the lake and inflows,
* sanitary state of the lake and surface waters in the catchment,
* hydrobiological state in various parts of the lake,
* condition of the fish in the lake,
* influence of daily water level variations on the shore region of the lake.

The studies were mainly based on direct measurements in the lake, as well as laboratory analysis of samples taken from the lake. During the whole monitoring period (April to September) an automatic meteorological station was in operation. Studies were carried out by a multidisciplinary team from Technical University of Gdańsk (Faculty of Environmental Engineering) and two other Institutes headed by the author of this paper (Majewski, 1996 a)

## RESULTS OF STUDIES

The primary effect of the operation of pumped-storage plant is the change of lake hydrodynamics (magnitude and pattern of lake currents). These in turn result in changes of thermal regime and thus influence water quality, the state of ichthyology, degree of eutrophication, and the changes in the shoreline.

The present pattern of currents in the lake changes depending on the mode of the operation of powerplant. In turbine cycle maximum velocities at the outlet of the channel reach 1 m/s. Discharged water has the form of a long jet of the length of about 1500 m. During pumping cycle, the range of influence of water withdrawal from the lake is about 600 m but occupies much wider area of the lake in the form of a triangle. Maximum currents near the inlet to the channel amount to 0.8 m/s. Lake currents in natural conditions did not exceed 0.1 m/s. The operation of the powerplant results in the lake oscillation which has the period of about 9 min. and the amplitude of 12 cm at the ends of the lake.

Thermal regime of the lake depends predominantly on meteorological conditions but changed considerably because of the intensive mixing due to water currents. Thermal stratification in summer does not occur or lasts for a very short time during May or June. The average water temperatures in shallow near shore areas are now in the range 24-25 °C that is 1.0-2.0 °C higher than temperatures observed before the powerplant operation. In deep parts of the lake differences between surface and bottom water temperatures did not exceed 1.0-1.5 °C.

Water quality of the lake varies considerably over the lake area. Pelagial waters in the deep central part are very clean, and their quality has improved in comparison with the state which existed before powerplant was put into operation. This state depends very much on the intensive currents, mixing and good aeration. Water quality in the littoral zone is much worse and varies considerably in space and time. This situation is created by varying water quality from two surface inflows, discharge from sewage treatment plant as well as decay of plants and weeds in the shore area. The trophy of pelagic waters estimated on the basis of majority of hydrochemical indices, water transparency and chlorophyll content can be defined as transitory between mezo- and eutrophy.

The status of fish in the lake deteriorated mainly as the result of the destruction of spawning grounds by water level variation. The amount of fish species did not change, however, the amount of valuable fish fell down. The amount of less valuable fish increased. Inspection of the upper reservoir revealed that no fish move into the intake channel and are transported with the water to the upper reservoir. The conditions for fish reproduction deteriorated due to the disappearance of plants and destruction of spawning grounds for the majority of fish species.

Water level variations of about 1 m daily caused erosion and destruction of the roots of trees growing within the belt around the lake where water level fluctuates. In the littoral areas dead biomas has been deposited, which accelerates the eutrophication of the lake. Zooplankton composition is similar as before powerplant operation, resembling planctonic fauna of limnetic waterbodies of a mesotrophic type.

## CONCLUSIONS

The operation of pumped-storage plant brought considerable changes in lake hydrodynamics, thermal regime, shoreline and status of fish. Negative impacts concern the deterioration of shoreline and decrease of the amount of valuable fish. Water quality in the central deep part of the lake improved due to mixing and intensive aeration. Increase of lake eutrophication is mainly due to the inflow of pollutants from the catchment. In general it can be stated that the operation of pumped-storage powerplant has not resulted in visible deterioration of the lake.

## REFERENCES

Majewski W., 1983, (Editor), Study of Żarnowiec lake for nuclear and pumped-storage powerplants (in Polish), PWN Warszawa - Poznań

Majewski W., 1996, Żarnowiec: Poland's largest pumped-storage powerplant. Hydropower & Dams, Issue 2

Majewski W., 1996a, (Editor), State of Żarnowiec Lake after 10 years of the operation of pumped-storage plant (in Polish), Monograph of the Committee for Water Resources Management, Warsaw 1996

# TURBULENCE AND STABILITY AT INTERFACES IN STRATIFIED RESERVOIR FOR POWER GENERATION

SOTOAKI ONISHI*, MITIHIRO SUGII*, YUICHI KITAMURA**
*Science University of Tokyo, Noda city, Chiba, Japan 278
**Electric power development Co. 6-15-1, Ginza, Chuo-Ku, Tokyo,

## ABSTRACT

In a stratified reservoir for power generation, a couple of stable interfaces are usually formed at the depth several meters below the water surface and at the level of intake bottom in the period of spring through autumn. Turbulent features of these interfaces may affect water environment of the reservoir and its downstream region. In this paper the authors discuss mixing length, eddy viscosity and instability at the interfaces with data obtained in a practical field.

## INTRODUCTION

Water in a large reservoir is usually stratified in a period of spring through autumn due to heat exchange, and steep gradient of vertical distribution of water density is formed near the water surface. Besides, in the case of reservoirs for electric power generation an other stable interface is often formed at the level of intake bottom as well. At a time of heavy flood, which convey a volume of solid particles into the reservoir and make sediments at the bottom raise up, the stratification is destructed with turbulent mixing.
Since the density stratification is generally recovered quickly after finish of the flood and the stratification resist settlements of the solid particles floating in the reservoir, the condition of high turbidity remain in the reservoir for a prolonged period after the flood. Such prolongation of high turbidity condition may cause deterioration of water quality in the downstream water regions. To

establish a plan to reduce the deviation of water turbidity from natural state, it is important to seize substantially the turbulent features and the instability condition of interfaces in the reservoir. Most of existing studies on these subjects have been done by by means of model tests so far. In this study we discuss these with data obtained through in-field observations at the Kazaya reservoir in Japan.

## STUDIED RESERVOIR AND OBSERVATION METHOD

The Kazaya dam is exclusively for the power generation, and of 101 m height and 330m crest length, and its reservoir has total storage capacity of $130 \times 10^6 m^3$ (effective capacity of $86 \times 10^6$ m$^3$), a flooding area of 553 km$^2$ and an annual inflow of approximately $1150 \times 10^6$ m$^3$. Ratio of annual inflow to the storage capacity is about 5 to 10. An intake is of 3.2m height and 3.2m width and the bottom is opened at the elevation level of 240m, which is nearly 90m depth below the maximum level of water surface. To measure the turbulent features of the reservoir, we have installed observation system at four stations through a year since August of 1992. The measurements of data including water turbidity temperature, three components of current velocity, conductivity, chlorophyllin, transparency and wind are done from the reservoir surface to bottom vertically with an ordered depth intervals. The details of observation methods and some results of obtained data have been reported previously [1].

## MIXING LENGTH AND EDDY VISCOSITY

The Kazaya reservoir is stratified in the period between the mid of May and the beginning of November, and two distinguished interfaces are formed usually at water depth of less than 10m and at a level of the intake bottom. The upper interface is produced by heat exchange at the water surface and the water above it flows towards a spillway at the dam crest. And the water between the upper and lower interfaces is withdrawn into the submerged intake. Then we assume a three layered flow model as shown in figure 1. It has been known that physical parameters relating to hydraulic behavior of the flow are densimetric Froude numbers and Reynolds number defined as follows;

$$F_{iU} = q/(\Delta \rho_U \, g \, d_1^3/\rho_i)^{1/2} \quad : \quad \Delta \rho_U = \rho_i - \rho_s \tag{1}$$
$$F_{il} = q/(\Delta \rho_m \, g \, d_2^3/\rho_i)^{1/2} \quad : \quad \Delta \rho_m = \rho_i - \rho_l \tag{2}$$
$$Re = q/\nu \tag{3}$$

where, Re:Reynolds number, $F_{iU}$ and $F_{il}$:densimetric Froude numbers of the upper and middle layers, q:flow rate per unit width of channel, $\nu$:coefficient of kinematic viscosity, $\rho_s$, $\rho_i$ and $\rho_b$:water densities of the upper, middle and bottom layer, $d_1$ and $d_2$:depths of the upper and middle layers and g:gravity acceleration. Flow in a practical field is of large value of Reynolds number and its turbulent nature is described by mixing length $l$ and shear stress "$\tau$" represented as follows.

$$\tau/\rho = u_*^2 = \overline{u'w'} = l^2 \ |\partial u/\partial z| \ \partial u/\partial z = \varepsilon \ \partial u/\partial z \quad (4)$$

in which, $\varepsilon$:eddy viscosity coefficient, $u_*$:friction velocity, $u'$ and $w'$:horizontal and vertical component of velocity, z:water depth. Then $l$ and $\varepsilon$ are described by

$$\varepsilon = l^2 \ |\partial u/\partial z| = u_*^2/|\partial u/\partial z| \quad (5)$$
$$l = u_*/|\partial u/\partial z| \quad (6)$$

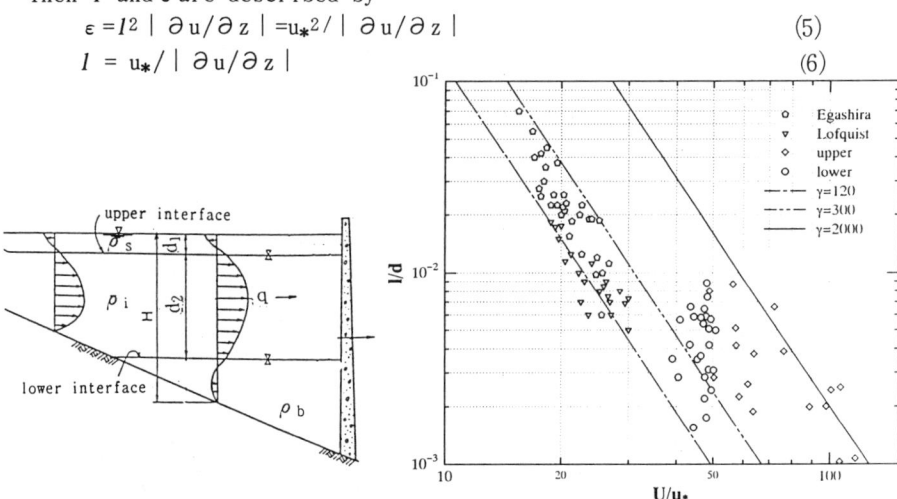

Figure 1 Three layered flow model the intake and spillway

Figure 2 Non-dimensional mixing length $\beta = l/d_2$

Mixing length is normalized with the water depth ($d_2$), as follows;
$$\beta = l/d_2 = \gamma \ (U/u_*)^{-3} \quad (7)$$
where, $\beta$: normalized mixing length, U: mean velocity and $U/u_*$ is velocity factor.
In the case of three layer flow model, the velocity factors of upper and lower interfaces are represented respectively as follows;

$$U_m/u_{*u} = (2/f_{iU})^{1/2} \quad \text{:for the flow of upper interface} \quad (8)$$
$$U_m/u_{*l} = (2/f_{il})^{1/2} \quad \text{:for the flow of lower interface} \quad (9)$$

in which, $U_m$:mean flow velocity of the middle layer, $u_{*u}$, $u_{*l}$:

friction velocity of each layer, $f_{iU}$, $f_{il}$ :coefficients of shear stress at the upper and lower interfaces. Referring to Egashira and Ashida $f_{iU}$ and $f_{il}$ are presented as follows[2];

$$f_{iU} = 15/Re + 0.00207F_{iU}^2 + 0.0015d_2/(d_1+d_2) \quad (10)$$
$$f_{il} = 15/Re + 0.00207F_{il}^2 + 0.0015d_2/(d_2+d_3) \quad (11)$$

Using these equations together with the data obtained in the Kazaya reservoir, mixing lengths and coefficients of eddy viscosity for the upper and lower interfaces are calculated and the results are shown in table 1. One can see that the mixing lengths and the values of $\varepsilon$ at the lower interface are larger than those at the upper one, but the mixing lengths have order of 10cm. At the time of 5am when the data were obtained, the power station is stopped, hence, turbulence in the reservoir are usually small. Then the relations between the normalized mixing lengths and the velocity factor are shown in figure 3, in which the values obtained by model tests by Egashira & Ashida and Lofquist previously are presented as well.

Table 1 Mixing lengths and coefficients of eddy viscosity (1993)

|  | 5am;JUNE 10 | 5am;JULY 1 | 5am;AUG. 5 | 5am:SEPT. 21 |
|---|---|---|---|---|
| $d_1$ (m) | 8.39 | 4.56 | 5.46 | 11.14 |
| $d_2$ (m) | 27.5 | 32.5 | 35.0 | 27.5 |
| $d_3$ (m) | 34.0 | 44.0 | 39.0 | 36.5 |
| $(du/dz)_u$ | 0.0044 | 0.0024 | 0.0063 | 0.0020 |
| $(du/dz)_l$ | 0.0032 | 0.0045 | 0.0029 | 0.0034 |
| $\tau_u$ (N/m$^2$) | 7.44E-05 | 2.43E-05 | 3.51E-04 | 2.28E-04 |
| $\tau_l$ (N/m$^2$) | 1.36E-4 | 1.03E-04 | 1.18E-03 | 3.82E-04 |
| $l_u$ (m) | 6.20E-02 | 6.50E-02 | 9.41E-02 | 2.39E-01 |
| $l_l$ (m) | 1.15E-01 | 7.15E-02 | 3.75E-01 | 3.82E-01 |
| $U_m/u_{*u}$ | 58.65 | 98.01 | 89.67 | 56.46 |
| $U_m/u_{*l}$ | 43.44 | 47.57 | 48.89 | 43.64 |
| $\varepsilon_u$ (m$^2$/s) | 1.69E-05 | 1.00E-05 | 5.58E-05 | 1.14E-04 |
| $\varepsilon_l$ (m$^2$/s) | 4.00E-05 | 2.30E-05 | 4.08E-04 | 4.96E-04 |

The values of $\gamma$ at the upper interface are larger and lay scattered in an extent wider than those at the lower interface. This indicate that the upper interface located near the water surface is under less stable condition than the lower interface is, because of forces such as wind and others added at the reservoir surface. Other noteworthy thing is that the values of $\gamma$ are in a range of 300 to 2000, which are larger than these obtained previously by the model tests [2]. Indeed the practical reservoir contains eddies of larger scale than those measured in the model tests.

## STABILITY OF STRATIFIED INTERFACE

The interfaces are stable in the period between the mid of spring and the end of autumn. But when a heavy flood strikes the reservoir, the interfaces may become unstable. The stability of stratification can be defined the Brunt-Vaisala frequency defined as follows;

$$N=\sqrt{(g/\rho_0)(-d\rho/dz)} \tag{12}$$

In the period between the July and September of 1993, three heavy floods were recorded in the studied area as shown in figure 3.
Time series of N at the lower interface through a year of 1993 are calculated and its results are shown in figure 3. The values of N increase rapidly in March and reach $N \fallingdotseq 0.04$ at the beginning of July. Thereafter these are maintained nearly at this level by the mid of October, though they decrease temporarily to 0.03 at the

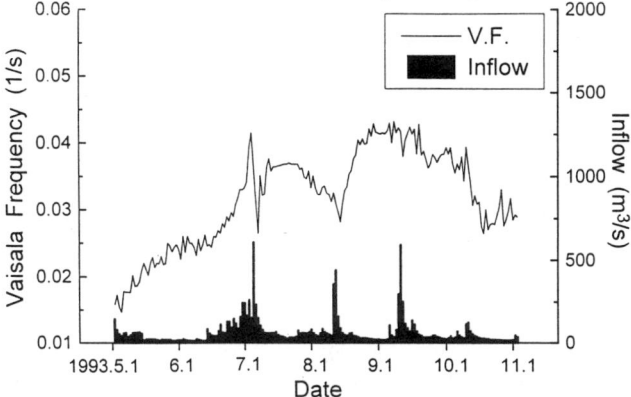

Figure 3  Records of floods and Brunt-Vaisala frequency

flood of the July of 1993. Then figure 4 is the time series of water temperature and turbidity observed in a year of 1993. We can see that the whole of water above the level of intake bottom increases its turbidity and the state of high turbidity remain till the next food visits. However, the vertical distributions of water temperature recover very quickly after the floods passed over, especially the lower interface is not disturbed by the floods. From the results indicated in figures 3 and 4, we consider that the stability of lower interface can be stable at the N-values of order larger than at least 0.03. Then vertical mixing across the lower interface is considerably limited, and the water body below the lower interface may suffer from oxygen starvation. If some harmful wastes, which may be discharged to the reservoir from its drainage areas, were accumulated on the reservoir bottom, the whole of reservoir would be suffered from it through the turning over of the reservoir water which occur at the end of every autumn.

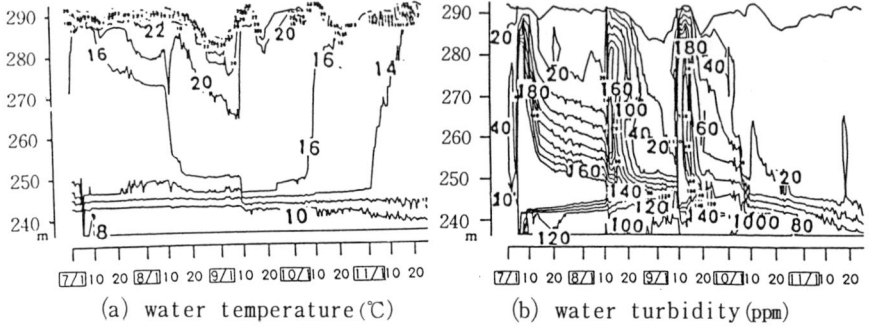

(a) water temperature (℃)    (b) water turbidity (ppm)
Figure 4   Time series of water temperature and turbidity

## CONCLUSION

In the Kazaya dam reservoirs for power generation a couple of stable interfaces are commonly formed at depths near the water surface as well as the level of a submerged intake bottom in a period of the beginning of spring and the end of autumn.   Condition of high turbidity of reservoir water caused by the destruction of the lower interface at the time of flood prolong for a considerable long period after the flood, which may create deterioration of water environment at the downstream region.  Then turbulent features of current in the reservoir and the instability of the lower interface are important subjects to be studied to find proper measures to reduce the deviation of water turbidity from natural state in the water environment. We measured them in the Kazaya dam reservoir and indicated that the mixing length and the eddy viscosity are of 10cm and of $10^{-5}$ m$^2$/sec respectively, and the Brunt-Vaisala frequencies are in a range of 0.03 to 0.04.

## REFERENCES

1) Onishi, S. and Kitamura, Y. : Study on turbid water current penetrating into dam reservoir through field measurements, Proc. of hydraulic Engineering, JSCE, vol. 30, (1986).
2) Egashira, S. and Ashida, K. : Structures of density stratified flow, Disaster prevention research Inst. Annual. No. 22 B-2, Kyoto Univ.

# A NETWORK OPERATION OF RESERVOIRS FOR ENHANCEMENT OF THE ECOLOGICAL FLUSHING DISCHARGE

NOBUYUKI TAMAI, YOSHI EMURA and HIRONORI MATSUZAKI
Department of Civil Engineering, University of Tokyo
7-3-1, Hongo, Bunkyo-ku, Tokyo 113, JAPAN

## ABSTRACT

Water supply of Tokyo Metropolis is heavily dependent on water transfer from the Tonegawa River basin. The Tonegawa River serves as a major source of water for south part of the Kanto region where total population amounts to 32 millions. Main eight multipurpose reservoirs are all located in the upstream catchment of the Tonegawa River basin. In recent years, environmental and ecological aspects of reservoir construction and operation have attracted higher and higher concern. Special attention should be paid to ecosystem in rivers. It is necessary for benthos in rivers to flush silt deposited on cobbles according to experiences of ecologists and inland fishermen. This discharge which keeps desirable condition of substrate is called ecological flushing discharge in this paper. The purpose of this study is to derive a network operation rule among the reservoirs in order to enhance instream flow to provide the ecological flushing discharge in the upper Tonegawa River basin.

## INTRODUCTION

It is so important to conserve the Tonegawa River that has the biggest river basin in Japan and runs through the area in which more than 30-million people and so many facilities are concentrated. Since 1960s, according to the increase of water demand in Tokyo Metropolis, many multipurpose reservoirs have been constructed in the upstream catchment of the Tonegawa River basin. Today, eight reservoirs are operated. Main purposes of these reservoirs are, for example, flood control, water utilization, and generation of electricity, and we enjoy benefits already. On the other hand, it is also true that reservoir construction contains adverse effects especially on environment and ecological systems. Recently, changes in river regime which were caused by reservoir construction and consequently deteriorated fluvial ecosystem have been given higher notice. The purpose of this study is to evaluate impacts on river regime caused by the reservoir construction in Tonegawa River basin from environmental point of view and to suggest better scheme for reservoirs operation. First, we analyzed the river regime of the upstream catchment of the Tonegawa River basin, which formed the foundation of the evaluation. Then a better scheme of reservoir operation is sought to enhance environmental flow.

## CHARACTERISTICS OF RIVER REGIME IN SUB-CATCHMENTS

Generally the upstream catchment of the Tonegawa River basin can be divided into five sub-catchments according to their hydrological or geological features. Usually we classify Okutone catchment, Agatsuma catchment, Karasu-Kanna catchment, Watarase catchment, and residual catchment. Each catchment has diversity of their discharge pattern. Figure 2 shows seasonal variability in natural discharge at three gauging stations. While Okutone catchment which has source area in mountainous region with much snowfall has so much discharge in May by snow melt, discharge in autumn is the largest in Watarase catchment due to rainfall brought by typhoons.

Moreover Yagisawa and Sonohara have seasonal diversity even if both of

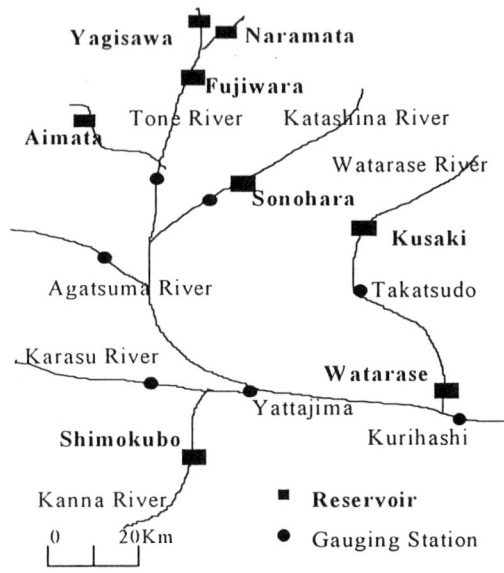

Fig.1 A schematic map of the study area

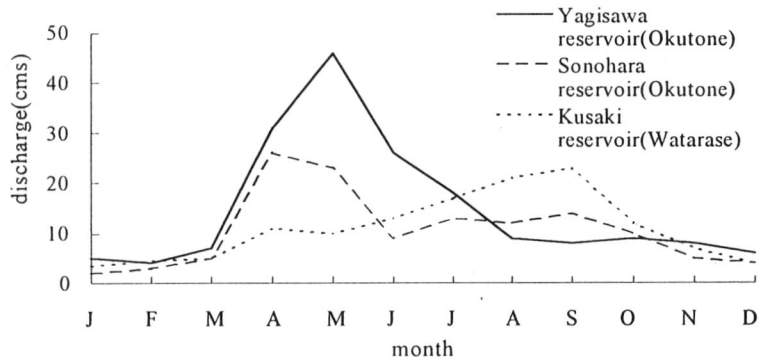

Fig.2 Seasonal variability of discharge in sub-catchments

them belong to Okutone catchment. Figure2 shows possibility that some catchments can provide water when other sub-catchment are confronted with shortage of water.

## CHANGES IN RIVER REGIME DETECTED BY TIME SERIES ANALYSIS

Figure3 shows the comparison of autocorrelation function before and after the Kusaki reservoir construction (1977) in Watarase catchment, based on daily discharge at Takatsudo gauging station, about 20Km downstream of the dam. The peak around 90 days is observed and time lag at which autocorrelation function becomes 0 is longer before the construction. On the other hand, after the construction, the peak around 90 days disappeared, although we can distinguish the peak around 40 days, and autocorrelation function approaches to 0 more quickly. This tendency can be interpreted as the collapse of about three months period in discharge, which is often observed in river regime in Kanto region where river runoff is highly dominated by snow melt (March to April), the rainy season (June), typhoons (September to October), by artificial reservoir operations.

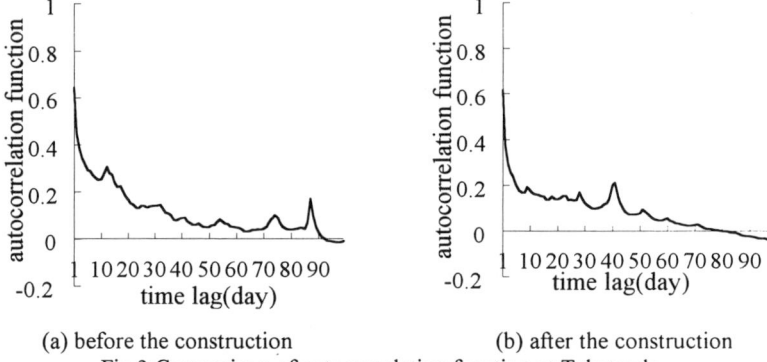

(a) before the construction  (b) after the construction
Fig.3 Comparison of autocorrelation function at Takatsudo
before and after the Kusaki reservoir construction

## ECOLOGICAL FLUSHING DISCHARGE

We discussed changes in river regime using time series analysis in the previous section. In this section, we introduce another measure, "ecological flushing discharge", to describe river regime. One hint of this measure is a biological comment that algae which grow on the surface of cobbles on river bed will be deteriorated without appropriate flushing for one month after reservoir construction. This means it is desirable to have 12 small-scale or medium-scale floods in one year. Let us denote these small-scale or medium-scale floods which are effective for flushing deposited silt on the river bed as an ecological flushing discharge. If we assume that duration of this ecological flushing discharge is 5 days, daily discharge exceeds this threshold value for 60 days in one year. Analyzed results explained in Fig.4. In this paper floods which exceed $Q_{60}$ (the 60$^{th}$ biggest discharge in one year) are called the ecological flushing discharge.

Figure4 shows comparison of annual occurrence of the ecological flushing discharge at Kurihashi on Tonegawa river and Takatsudo on Watarase river. Takatsudo is located just downstream the Kusaki reservoir, while Kurihashi is away from upstream reservoirs and is weakly affected by reservoirs construction. An example of time series of daily discharge is given in Figure5. As far as Kurihashi data are concerned, remarkable change can not be observed before and after the construction, on the other hand, annual occurrence of the ecological flushing discharge at Takatsudo had reduced by 3 or 4 times on average. One of the cause of this reduction is assumed that artificial reservoir operations smoothed out flood discharges and consequently made the duration of each ecological flushing discharge longer.(Refer to Fig.5(a))

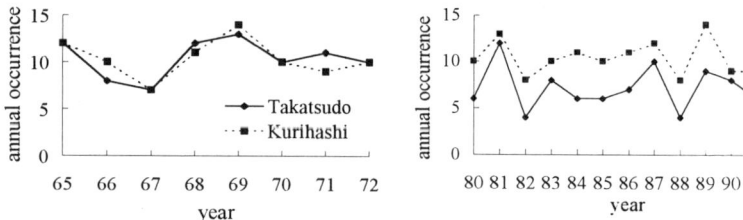

(a)before the construction             (b)after the construction
Fig.4 Comparison of annual occurrence of ecological flushing discharge

## A SUGGEESTION OF ENVIRONMENTALLY BETTER SCHEME OF RESERVOIR OPERATION

While we defined the ecological flushing discharge and paid attention to its annual occurrence in the previous section, the interval of two consecutive flushing discharges is considered in this section. Table 1 shows average and standard deviation of the interval of two consecutive flushing discharges for both summer (July to September), when algae need more ecological flushing discharge than the other seasons, and other seasons (October to June) at Takatsudo.

This table clearly shows that little change is observed for summer events (flood season), but in a dry season (from October to June) conspicuous prolongation of the interval are observed. This may be attributed to exploitation of water resources. In the period between March and June we

Table 1  Average and standard deviation of the interval of ecological flushing discharge

| July to September | before | after |
|---|---|---|
| average(day) | 16.6 | 15.6 |
| standard deviation(day) | 3.6 | 3.6 |

| October to June | before | after |
|---|---|---|
| average(day) | 41.5 | 63.8 |
| standard deviation(day) | 6.6 | 8.8 |

have relatively large amount of precipitation in Kanto region.But demand of irrigation water is large. Therefore, spill from reservoirs is small. In the period between October and February we experience natural drought period. Furthermore, demand of municipal water is constant throughout the year. Therefore, in natural drought period we have difficulty to release water from a reservoir. If we assume that former river regime (before the reservoir construction) is better than present one (after the reservoir construction) environmentally, better features should be restored by reservoir operations. If it is possible to augment discharge from a reservoir, ecological flushing discharges will be restored as in the previous condition. In actual reservoir operation, however, each reservoir may have no enough capacity for ecological use. Then a network operation of reservoirs linking between several reservoirs by pipes will be useful because we are able to use regional variability in discharge advantageously as mentioned in the former section. Figure5 shows natural flow and simulated flow, which will be realized by linking between reservoirs, for daily discharge at Takatsudo gauging station.

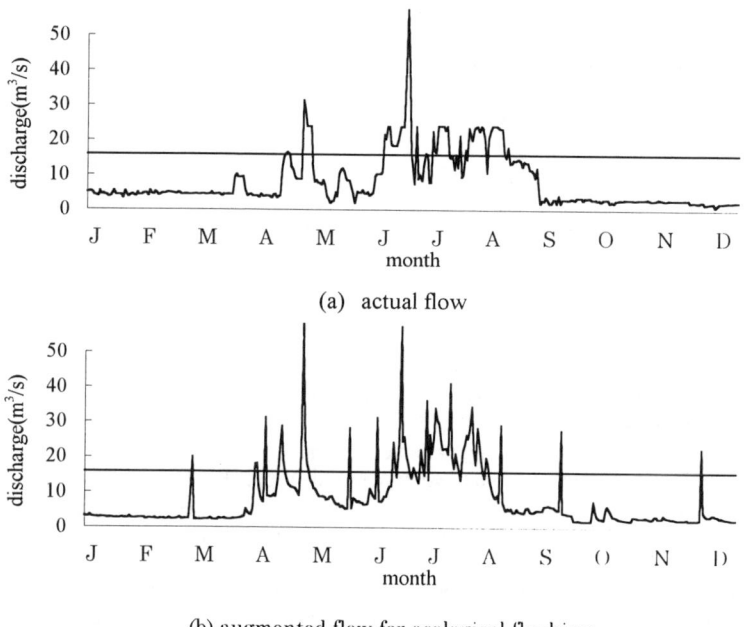

Fig.5  Actual flow and augmented flow at Takatsudo for daily discharge(1984)

In this simulation, Kusaki reservoir, Sonohara reservoir and Shimokubo reservoir were connected by pipes for water transfer, and, system dynamics was used for actual calculation. Figure 5 shows the result of simulation of available release from the reservoir by this water transfer. It is shown that occurrences of small-scale or medium-scale floods are increased by augmentation for ecological flushing and river regime is improved environmentally.

## CONCLUSIONS

We studied changes in river regime before and after the reservoir construction from environmental point of view. Through this study following concluding remarks are extracted.
1. The concept of ecological flushing discharge is formulated. With this concept we are able to judge changes in river regime, and at just downstream of the reservoir its annual occurrence considerably reduced after reservoir construction.
2. To try to restore the river regime of before the reservoir construction, time series treatment of the river regime is important. The characteristics of the interval of the ecological flushing discharge is clarified at Takatsudo hydrometric station
3. Potential advantage of a network operation among several reservoirs is explained for the upstream area of Tonegawa River basin.

## ACKNOWLEDGMENT

This study is partly supported by Grant-in-Aid for Scientific Research of the Ministry of Education, Japan (Principal Researcher, Prof. Hanaki, 07408013).

## REFERENCES

(1) N.Tamai and Y.Emura, Changes in flow regime by reservoir construction, International conference on aspects of conflict in reservoir development and management,pp641-649,1996.
(2) Allan G. H. Locke, Recommending variable flow value for fish,. Second international symposium on habitat hydraulics, Vol.A, pp559-570,1996.
(3) Robert T. Milhous, Modeling of instream flow needs: The link between sediment and aquatic habitat,. Second international symposium on habitat hydraulics, Vol.A, pp319-331,1996.
(4) G. Mathias Kondolf and Peter R. Wilcock, The flushing flow problem: Defining and evaluating objectives,. Water resources research, Vol.32 No.8, pp2589-2599,1996.

# TOTAL DISSOLVED GAS IN THE NEAR-FIELD TAILWATER OF ICE HARBOR DAM

STEVEN C. WILHELMS AND MICHAEL L. SCHNEIDER
US Army Engineer Waterway Experiment Station, Vicksburg, MS USA

## ABSTRACT

Total dissolved gas (TDG) saturation levels were measured and recorded along 3 lateral transects in the immediate area downstream of the Ice Harbor Spillway. The dissolved gas levels were dependent on total spillway discharge and spill pattern, reaching nearly 140 percent at the navigation lock guide wall for the largest discharge. The data showed significant and rapid gas absorption in the stilling basin, with a maximum TDG level of 162 percent at the stilling basin endsill for the largest discharge. A rapid desorption of TDG occurred over the next 200 ft of tailrace, reducing saturation to that measured at the end of the guide wall. These results provide a basis to develop alternative designs to reduce TDG produced at hydraulic structures.

## INTRODUCTION

Structural and operational alternatives are being developed to reduce dissolved gas levels generated during spillway operations on the Snake and Columbia Rivers. Field studies are one of several approaches being used to define gas exchange at these hydraulic structures. The objective of this field study (Schneider 1996) was to determine the amount and locations of dissolved gas exchange in the stilling basin and tailrace area of the Ice Harbor Dam Spillway. Thus, instruments to measure total dissolved gas (TDG) were deployed in the high-velocity, extremely turbulent, highly-aerated bubbly flow of the stilling basin, end sill area, and near-field tailrace area. These data could then be analyzed to determine the relative importance of gas exchange processes within the stilling basin and the downstream tailrace. Study results will help quantify processes contributing to dissolved gas transfer during spillway releases and help identify operational and structural measures that reduce dissolved gas supersaturation.

The transfer of dissolved gas is generally thought to be related to several operational and geometric parameters, but due to the limited duration of this study (30 hours), only spillway discharge and spill pattern were varied. The TDG instruments were located in the tailwater channel as shown in Figure 1. Three instruments were on the bottom along Transect SB, just downstream of the stilling basin; three instruments were on the

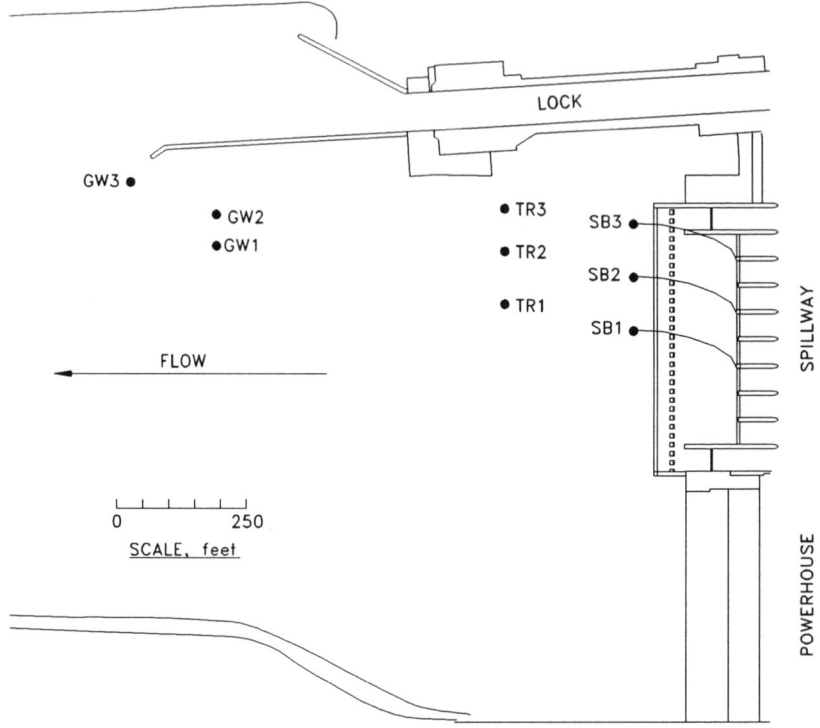

Figure 1. Location of monitors in Ice Harbor spillway tailwater

bottom along Transect TR, about 200 ft downstream of Transect SB; and three instruments were at mid-depth on Transect GW near the end of the lock guide wall.

## PROJECT AND STUDY DESCRIPTION

The Ice Harbor Project consists of a powerhouse (6 turbines; 105,000 cfs capacity), a 10-bay, 590-ft-long gated spillway, and a navigation lock. A schematic of the spillway stilling basin is shown in Figure 2. The spillway nappe plunges into the stilling basin and sweeps along the bottom of the stilling basin, creating a surface roller back toward the dam. The highly-turbulent aerated shear flow along the spillway floor is broken up at the baffle blocks and end sill, which direct flow vertically (Figure 2) into the 25-ft-deep channel downstream of the stilling basin. The very shallow tailrace downstream of the powerhouse causes generation and spillway discharges to converge to the navigation channel.

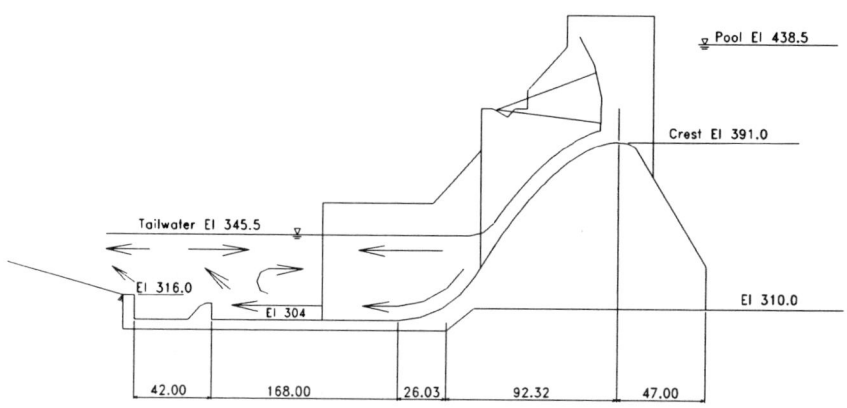

Figure 2. Schematic cross section of Ice Harbor Dam Spillway
(all elevations are in feet)

During the study, several major operational changes were implemented and the resulting TDG levels were recorded at 5-minute intervals. The tailwater at the beginning of the testing period was at el 345.5 ft, the forebay elevation was 438.5 ft and the water temperature was approximately 11 °C.

## RESULTS

The TDG pressures, recorded by the instruments, were converted to TDG saturation by dividing by the local barometric pressure (751 mmHg). The TDG saturation levels from selected monitors are shown in Figure 3 along with spillway operations. Changes in TDG correspond closely with operational changes. Low spillway discharges with a gate settings of 1 stop, actually caused degassing to approximately 116 percent from the forebay concentration of about 122 percent. Higher spillway discharges resulted in higher levels of TDG. The TDG levels at the stations farthest downstream ranged from the 116 percent with the low spillway discharge to nearly 140 percent with the highest spillway discharge.

The TDG at the endsill (Transect SB) was generally greater than that observed downstream on Transects TR or GW for high spillway discharges (greater than 20,000 cfs). For low spillway discharges, TDG levels for all three transects were similar. The maximum TDG at the endsill was recorded at 162 percent for the large spillway flow (40,000 cfs). However, this maximum decreased to about 140 percent on Transects TR and GW, which were similar throughout the testing period.

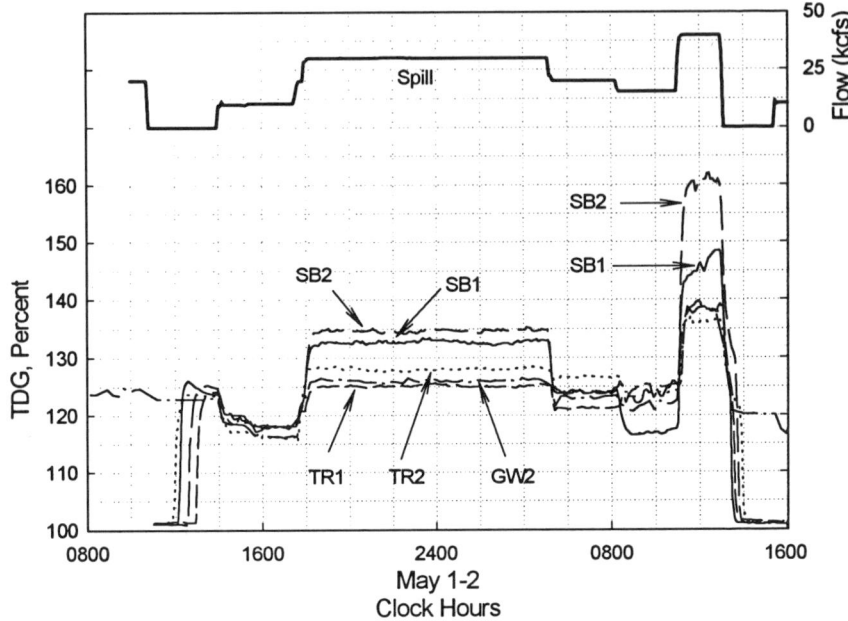

Figure 3. Observed TDG Levels and Total Spillway Flow

For flows up to 40,000 cfs, the similarity of TDG levels on transects TR and GW suggests that there is very little gas transfer beyond Transect TR, which is approximately 240 feet downstream of the stilling basin end sill. The data indicate a rapid and extensive <u>absorption of gas in the stilling basin</u> followed by rapid and extensive <u>desorption of dissolved gas in the tailrace channel</u>. Clearly identifying the absorption process in the stilling basin and desorption in the tailrace provides a basis for several alternatives for gas abatement.

## CONCLUSIONS

This study of dissolved gas levels in the tailwater at Ice Harbor Dam successfully demonstrated the viability of remotely monitoring TDG levels in high velocity flow near hydraulic structures. The results of this study document large spatial variations in TDG levels in the immediate vicinity of the stilling basin, showing that gas absorption occurs in the stilling basin, followed by desorption in the immediate tailrace area. Gas exchange farther downstream appears to be minimal.

An additional study should be conducted in the Ice Harbor for a better understanding of the spatial variation in dissolved gas exchange processes. Instruments should be sited at multiple depths to provide information about vertical TDG gradients. Higher spillway releases (up to 100,000 cfs) should be investigated for base line data on the influence of discharge and channel depth on TDG levels. A similar study should be conducted at a project with a shallow stilling basin for gas production and shallow tailrace for gas loss. This information would help design raised stilling basins/tailrace channels at other projects. A near-field study of a spillway with deflectors should be conducted to determine their gas production/loss characteristics for comparison to the conventional stilling basin. With this information, combinations of alternatives are possible, such as spillway deflectors with a raised tailrace channel, to reduce TDG levels.

The tests described and the resulting data presented herein, unless otherwise noted, were obtained from research conducted under the Civil Works Program of the United Stations Corps of Engineers by the Waterways Experiment Station. Permission was granted by the Chief of Engineers to publish this information.

## REFERENCES

Schneider, M. L. 1996. "Total Dissolved Gas Data Documentation and Preliminary Analysis - Near-Field Study of the Ice Harbor Tailwater," Memorandum for Record dated 13 June 1996, US Army Waterways Experiment Station, Vicksburg, MS.

# A UNIQUE APPROACH FOR PHYSICAL MODEL STUDIES OF NITROGEN GAS SUPERSATURATION

LARRY J. WEBER[1] and CARL MANNHEIM[2]
[1]Assistant Professor and [2]Graduate Research Assistant
Iowa Institute of Hydraulic Research, University of Iowa, Iowa City, IA 52242, USA

## ABSTRACT

High levels of dissolved gas downstream of Wanapum Dam on the Columbia River have required Public Utility District Number 2 of Grant County (the District) to investigate various approaches to reduce gas supersaturation levels. To evaluate the effectiveness of flow deflectors mounted on the spillway face the District had a physical model built of three bays of the Wanapum Spillway. The model was used to both qualitatively and quantitatively assess the performance of the flow deflectors. The qualitative evaluation was performed by visual observations of the flow patterns generated by each flow deflector downstream of the spillway. Whereas, the quantitative evaluation was performed by collecting an extensive data set describing velocity fields and bubble distributions. This data set was then analyzed numerically to predict the downstream concentration of gas supersaturation. The purpose of this paper is to describe the physical model study and the qualitative evaluation of the flow deflectors. A companion paper by Orlins and Gulliver (1996) describes the numerical model developed to analyze the velocity and bubble data.

## INTRODUCTION

During recent years spillway discharges at many Columbia River dams have increased because of supplemental spill flows to enhance juvenile fish passage, and also, spill flows resulting from river discharges exceeding powerhouse capacity. As a result of these increased spillway discharges there has been an increase in dissolved gas concentration downstream of many Columbia River hydropower projects. The cumulative effect on the Columbia and Snake River systems has been total dissolved gas concentrations that are approaching, or exceeding, state water quality standards.

Increased spillway discharges, and consequent, increased levels of gas concentration have been particularly evident at Wanapum Dam on the Columbia River. In an effort to reduce the level of gas concentration the District had a physical model built of three spillway bays of Wanapum Dam. The model was constructed based on Froude scaling laws, and hence, accurately replicated the hydrodynamics of the flow field downstream of the dam. However, it is widely known that Froude-scaled models do not correctly reproduce the entrainment of air bubbles. Therefore, the physical model could not be used to directly determine the change in downstream gas concentration for various flow deflector configurations.

The benefit of flow deflectors could be quantified if the bubble distribution from the resulting flow could be modeled. Plastic particles with the correctly scaled rise velocity were released upstream of the spillway gate and their vertical distribution was recorded at several locations downstream of the spillway apron. This information of bubble distribution was then combined with hydrodynamic data, of mean velocity and turbulence, within the tailrace to form a data set capable of predicting levels of gas concentration downstream of a spillway.

## MODEL DESCRIPTION

Three spillway bays of Wanapum Dam were reproduced in the model using an undistorted scale of 1:21.5. The width and height of the flume in which the model was constructed was 10 feet and 7.5 feet, respectively. All dimensions, except elevations, will be given in model units. The tainter gates, spillway crest, and downstream apron, see figure 1, were fabricated in accordance with the as-built construction drawings. The length of the flume available for flow development upstream of the spillway was about 30 feet and the flume length downstream of the spillway was approximately 40 feet.

The three spillway bays in the model occupied essentially the entire width of the flume. Since the flume is equipped with Plexiglas walls, clear visual observations and photographic documentation were possible in the two outer bays. For this reason, tests with the conceptual flow deflector designs were performed in the right side bay (looking downstream).

## MODELING APPROACH

Initial evaluations of flow deflectors were based solely on a visual comparisons of the flow field downstream of the deflector. It was quickly observed that two obvious flow patterns were possible. The first flow pattern resulted from a deflector with a relatively low depth of submergence (less than approximately 13 feet). This configuration led to an expanding surface oriented jet with a numerous waves on the water surface. Also, a clockwise rotating recirculation zone was present on the apron beneath the expanding jet which likely entrains air and places it at depth. A second possible flow pattern resulted from a deflector with a relatively high depth of submergence (greater than approximately 14 feet). This configuration led to the spillway jet being bulked by the downstream tailwater and deflected upward toward the water surface. The water surface had to prominent waves, one where the jet was being deflected upward and a second downstream wave near the energy dissipater on the apron. Based on these visual observations, the District decided to install a single flow deflector at elevation 473 ft for prototype evaluation in 1996. Figures 1 and 2 show the schematics of the configurations and the downstream flow patterns.

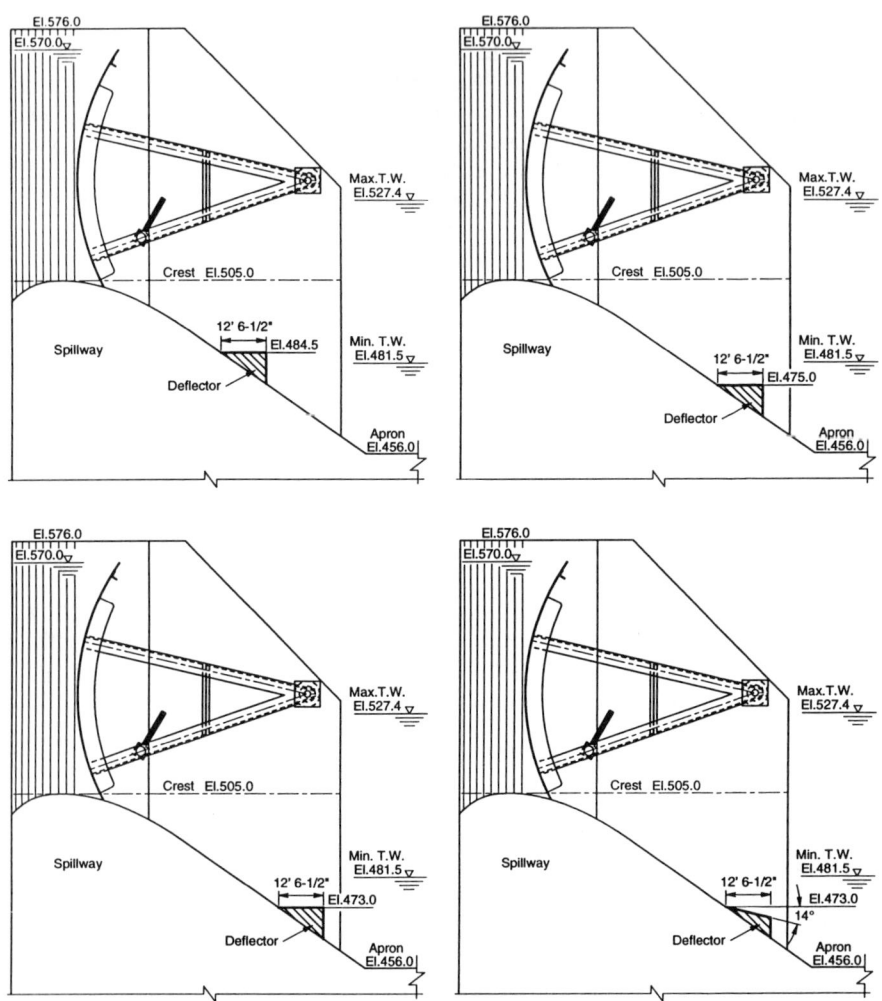

Figure 1. Schematic of Various Deflector Configurations Tested.

After viewing the above flow conditions an intermediate condition was sought, and found, in which the jet flow quickly expanded both upward and downward. In this condition nearly all of the flow on the apron was directed downstream and very little flow was entrapped in two small recirculations zones, one just below the deflector and the second at the plunge point of the entering jet. It was speculated that this condition with maximum energy dissipation may be beneficial in terms of reducing downstream gas levels. To obtain this flow the deflector is located at an elevation of 473 feet with the upper face oriented 14 degrees downward from horizontal. Figure 1 shows the geometric configuration and the flow pattern is shown in figure 2. This configuration may be prototype tested in 1997 by the District, however, questions remain regarding the stability of this flow condition.

To further understand nitrogen supersaturation in the tailrace and to determine if the flow deflector configurations chosen for field testing were the best possible configuration, the District decided to develop a method for quantitative comparison of conceptual designs. The numerical method used for quantitative comparison is described in the companion paper by Orlins and Gulliver (1996). Briefly, given hydrodynamic data of mean velocity and relative turbulence, along with information of the bubble distributions within the tailrace, the model can estimate the downstream gas concentration. The following sections describe the data collection program.

## BUBBLE DISTRIBUTIONS

As stated previously, the Froude-scaled model did not accurately replicate the distribution or dynamics of the prototype bubbles. Therefore, a model equivalent was used to simulate the prototype bubbles in the laboratory. Linear polyethylene beads 1/8 inch in diameter were found to have the same rise velocity in the model as the mean bubbles present in the prototype.

The beads were loaded into 3/16 inch diameter, 100 foot long, tubing and released upstream of the gate by flowing water through the tube. The beads were then photographed at four locations downstream of the apron, the photographs were digitized and the bead distributions were recorded from the images. The cameras were mounted approximately 6, 14, 22 and 30 feet downstream from the toe of the spillway. Because the depth perception on the photographs made determining the elevation of the beads nearly impossible from the photographs a solid white wall was placed six inches from the flume wall to serve as a backdrop for the photographs. Also, to determine the percentage of beads that have left the flow, skimmers were placed just below the water surface to collect the 'model bubbles' that have left the flow. This allowed the total bubble distribution to be determined with distance.

## VELOCITY DATA

A Sontek Acoustic Doppler Velocimeter (ADV) was used to collect velocity data within the six inch channel (both horizontal and vertical components) at four foot intervals from 6 feet to 30 feet downstream of the spillway toe. The ADV was

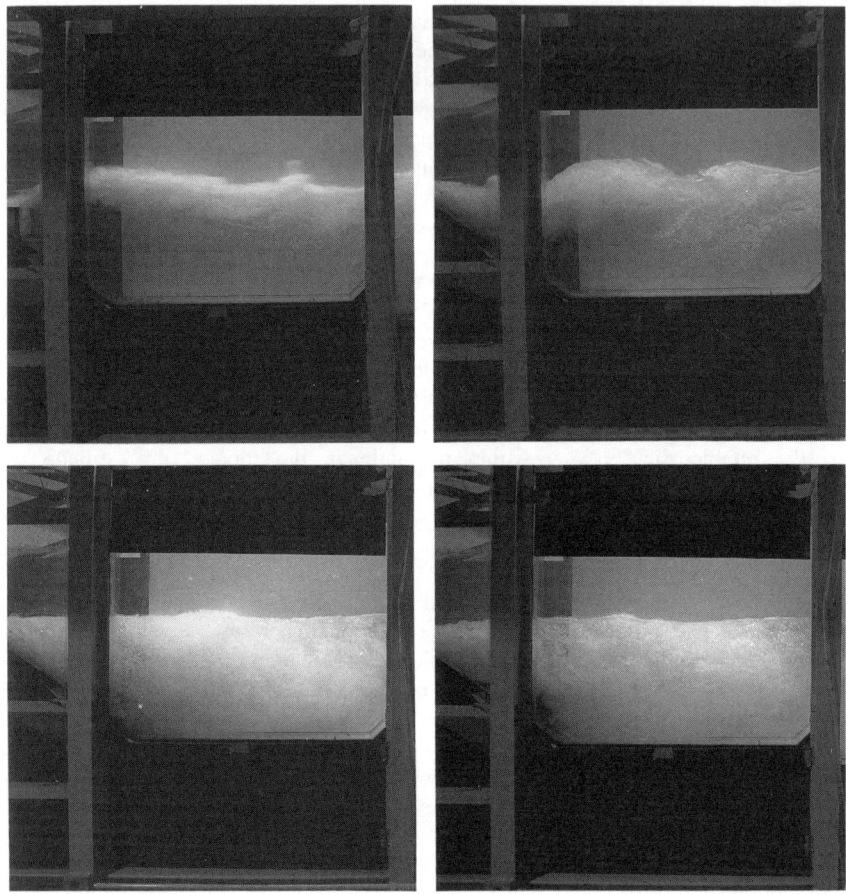

Figure 2. Photographs of Flow Above the Apron for Various Deflector Configurations. Deflector at 484.5, 473.0, 473.0 Sloped 14° and No Deflector (beginning in the upper left and going counterclockwise around the page.

unable to collect velocity data upstream of 6 feet because the flow on the apron had too many entrained bubbles resulting in inaccurate velocity measurements. Therefore, to obtain velocity measurements in this region Particle Image Velocimetry (PIV) was applied to collect the entire two-dimensional flow field, for a description of the technique see Ettema et al. (1996).

## SHORTCOMINGS

Although this approach was very useful for comparative estimates of downstream gas concentrations, improvements will be necessary before a reliable predictive analysis will be available. The following difficulties were noted. The six inch channel, necessary to photograph the beads, destroyed large three-dimensional turbulent structures downstream of the spillway. This resulted in the channel conveying a two-dimension flow field with lower levels of turbulence and beads that rose out of the flow too quickly. The beads, although scaled with the correct rise velocity, were geometrically too large and thus would be affected differently by the turbulent fluctuations. The entrained air, with relatively large rise velocities, may have affected the beads or water velocities. Finally, the PIV technique did not yield accurate turbulence values and must be applied to a two-dimensional flow field.

## CONCLUSIONS

The physical model was used to qualitatively compare various flow deflector designs at Wanapum Dam. A numerical method for prediction of downstream gas concentrations has been developed and used successfully for comparative purposes. Further development of the experimental data collection is necessary prior to finalizing the numerical approach.

## ACKNOWLEDGMENTS

This research is supported by Public Utility District Number 2 of Grant County, represented by Mr. Stephen Brown. The authors are grateful to Mr. Rex Elder, Mr. Duncan Hay and Dr. Jacob Odgaard for their valuable suggestions. The authors would also like to acknowledge Dr. John Gulliver and Mr. Joe Orlins of the St. Anthony Falls Hydraulics Lab, University of Minnesota, for their input throughout the data collection program. The field measurements were collected by Parametrix Inc., Dr. Don Weitkamp, Project Manager.

## REFERENCES

Ettema, R., Fujita, I., Muste, M. and Kruger, A. (1996). *Particle-image velocimetry for whole-field measurement of ice velocities*, submitted to Cold Regions Science and Technology.

Orlins, J.J. and Gulliver, J.S. (1996). *Prediction of dissolved gas concentration downstream of a spillway*, submitted to the XXVIIth IAHR Congress: Water for a Changing Global Community, August, 1997, San Francisco, CA USA.

# PREDICTION OF DISSOLVED GAS CONCENTRATION DOWNSTREAM OF A SPILLWAY

JOSEPH J. ORLINS[1] and JOHN S. GULLIVER[2]
[1] *Graduate Research Assistant and* [2] *Professor*, St. Anthony Falls Laboratory,
Department of Civil Engineering, University of Minnesota, Minneapolis, MN 55414 USA

## ABSTRACT

The increase in dissolved gas concentration downstream of hydraulic structures such as dam spillways can be harmful to juvenile salmonids. Such increases have been noted at Wanapum Dam on the Columbia River in Washington State. Modifications to the spillways at this dam will be installed to help lower the concentration of total dissolved gas downstream of the dam. These modifications were designed and optimized using a combination of physical and numerical models. The physical model provided information about the hydraulics associated with different spillway modifications. The numerical model calculated the concentration of total dissolved gas based upon hydrodynamic data from the physical model and mass transport relations for air-water flows. This article describes the numerical model development and application. A companion article describes the physical modeling efforts and field measurements made to evaluate the performance of the modifications installed at the dam.

## INTRODUCTION

Elevated dissolved gas levels downstream of hydraulic structures such as dam spillways can be harmful to fish. The increase in total dissolved gas concentration (TDG) is due to entrainment of air along the spillway face and at the location where the spillway flow enters the plunge pool or stilling basin. This air is entrained as bubbles, which have a high surface area for air-to-water mass transfer. When these bubbles are pulled to depth, the bubble-water equilibrium increases proportionate to the bubble pressure. Thus, relative to atmospheric conditions, supersaturation of dissolved gas is possible.

Public Utility District No. 2 of Grant County has experienced elevated gas levels downstream of the spillways at Wanapum Dam. The dam is located on the Columbia River in east-central Washington State, with 12 spillway bays and ten hydroelectric generating units.

As part of an effort to alleviate high TDG levels downstream of the spillway, Grant County PUD has installed a prototype deflector in one spillway bay. A second prototype deflector will be installed in 1997. The concept of the deflector is to prevent flow over the spillway from entering the deepest parts of the stilling basin. Deflectors of this style have been tested at a number of dams owned by the U.S. Army Corps of Engineers on the lower Snake and Columbia Rivers. Deflector performance, however, is dependent upon site-specific parameters and design constraints, necessitating an independent study.

The hydraulic design of the deflectors for Wanapum Dam has been aided by the use of physical model studies conducted at the Iowa Institute of Hydraulic Research (IIHR). A 1:21.5 scale sectional model of three spillway bays has been used to evaluate deflectors of differing sizes and placements on the spillway face. The physical model included the tainter gates, spillways, stilling basin apron, and approximately 300 meters of the downstream river reach.

Unfortunately, a reduced-scale physical model can not simulate the entrainment of air into the stilling basin. While the mean flow fields can be well reproduced at a reduced scale, the distribution of air bubbles (including their size and quantity of air entrained) can not be accurately simulated. This implies that the air-water mass transfer (and thus concentration of dissolved gas downstream of the spillway) can not be determined directly from the physical model. Information from the physical model can be used as input to a computational model which represents the turbulent mixing and air-water gas transfer, and can thus be used to predict dissolved gas concentrations for differing flow conditions.

## MODEL DESCRIPTION

### MASS TRANSFER BACKGROUND

The concentration of total dissolved gas downstream of a spillway can be described by the two-dimensional, laterally averaged equation

$$\frac{\partial C}{\partial t} + U\frac{\partial C}{\partial x} + V\frac{\partial C}{\partial y} = D_z(\frac{\partial^2 C}{\partial x^2}+\frac{\partial^2 C}{\partial y^2}) + (K_L a)_{bubble}(C_{eq} - C) + (K_L a)_{surface}(100 - C) \quad (1)$$

where  $C$ = concentration of total dissolved gas in water (% saturation)
  $C_{eq}$ = equilibrium gas concentration at interfaces considering hydrostatic pressure (% saturation)
  $x$ = horizontal distance (L)
  $y$ = vertical elevation (L)
  $U$ = temporal mean velocity component in horizontal direction (LT$^{-1}$)
  $V$ = temporal mean velocity component in vertical direction (LT$^{-1}$)
  $(K_L a)_{bubble}$ = overall transfer coefficient for exchange with bubbles (T$^{-1}$)
  $(K_L a)_{surface}$ = overall transfer coefficient for exchange with atmosphere (T$^{-1}$)
  $K_L$ = mass transfer rate coefficient (LT$^{-1}$)
  $a$ = specific surface area, or surface area per unit volume (L$^{-1}$)
  $D_z$ = eddy diffusivity (L$^2$T$^{-1}$)

Here, it is assumed that the flow is well-mixed laterally (*i.e.*, there are no transverse gradients in velocities or concentration). For this application, we wished to estimate the gas concentration, $C_d$, at some point downstream of the stilling basin, given an upstream concentration, $C_{in}$, and information about the flow field and distribution of air bubbles entrained in the flow.

### MODEL INPUT DATA

The numerical model developed required input of both hydrodynamic data and information about the distribution of air bubbles in the flow. Some of these parameters could be measured directly on the physical model, while others had to be calculated based upon physical model measurements.

***Physical Model Measurements Used as Input***  The temporal mean and turbulent velocity fluctuations were measured in the stilling basin and river reach downstream of the spillways. Measurements were made using an acoustic doppler velocimeter (ADV) downstream of the apron. In the highly turbulent, aerated region in the stilling basin, velocities were measured using particle image velocimetry (PIV). The combination of these techniques allowed characterization of the mean flow field (U, V) and root-mean-square (rms) turbulent velocity fluctuations (u', v'). The distribution of air bubbles was estimated to be uniform with depth, with an exponential decay with distance downstream from the spillway.

***Calculated Input Parameters*** The turbulent diffusivity ($D_z$) was calculated based on a relationship for open channel flow presented by Nezu and Nakagawa (1993): $D_z \approx v'l$, where v' is the rms vertical velocity fluctuation and $l$ is a vertical length scale of the largest eddies. We assumed the mixing length was equal to the Prandtl mixing length, and utilized a ramp function described by Nezu and Nakagawa for calculation purposes:

$$l = \kappa y, \quad y/h < 0.3; \quad l = \beta h, \quad y/h \geq 0.3 \quad (\kappa = 0.4, \quad \beta \approx 0.12) \tag{2}$$

where y is the distance from the river bottom and h is the flow depth. In the turbulent region of the stilling basin, the resulting eddy diffusivity is fairly large (roughly 10-20 times that of an open channel flow) due to the high rms velocities. In the downstream reach, where v' is very small, the eddy diffusivity approaches that expected for fully developed open channel flow.

The mass transfer coefficients $K_L a$ were calculated based upon the literature and experience in air-water flows at hydraulic structures. For transfer from the bubbles, we used a relationship similar to that proposed by Azbel (1981) for rising bubble plumes:

$$K_{L(bubble)} = \alpha \cdot \frac{D}{h} \cdot \frac{(1-\phi)^{0.5}}{(1-\phi^{5/3})^{0.25}} \cdot Re^{\eta} \cdot Sc^{0.5} \tag{3}$$

where  Re = Reynolds number, v'h/v (dimensionless)
 Sc = Schmidt number, v/D (dimensionless)
 v = kinematic viscosity of water ($L^2T^{-1}$)
 D = diffusivity of air in water ($L^2T^{-1}$)
 h = flow depth (L)
 v' = turbulent velocity fluctuations in the vertical direction ($LT^{-1}$)
 $\phi$ = volumetric air concentration (volume air / total volume)
 $\alpha$ = fitted coefficient
 $\eta$ = empirical coefficient of 0.66 (Geldert, 1996)

The specific surface area of the bubbles, (a), is the ratio of the total surface area of the bubbles to the total volume of air: $a = A_{bubble}/V_{bubble}$. The bubble surface area and volume were estimated based on the bead distribution data collected from the physical model and a calculated maximum bubble size. The maximum bubble diameter was estimated using relationships presented by Hinze (1955).

The overall mass transfer coefficient for exchange with the atmosphere were calculated based on the vertical velocity component (V) and the surface area at the free surface. The mass transfer velocity was calculated as $K_{L(surface)} = \beta v^2$, where $\beta$ is a coefficient selected to match results from field investigations.

## MODEL DISCRETIZATION

The computational domain of the numerical model started at the intersection of the spillway face and stilling basin apron and extended to a point approximately 240 m downstream from this point, and included the full river depth. This physical domain was divided into a computational grid of 160 horizontal cells by 10 vertical cells, with a nominal cell size of 1.5 x 1.3 m. The measurements made on the physical model at discrete locations were interpolated to the computational grid nodes using a kriging algorithm.

A two-dimensional power law scheme (Patankar, 1980) was used to solve equation (1). For this method, an assumed initial concentration field in the river reach is used to predict the steady-state concentration based upon an upstream gas concentration, the hydrodynamic parameters and bubble distributions.

## MODEL APPLICATION

The model was used to evaluate the performance of the existing spillway and three different spillway modifications for a range of discharges and water levels. At the time of writing, three of the configurations had been run with the model, and data was being collected from the physical model for the fourth.

For initial evaluation purposes, the model was run for one discharge and two water levels. The specific discharge selected (9.3 m$^2$/s) is at the lower end of the operating range for the spillway. The water levels tested (depths of 12.8 and 15.2 m) are at the middle and high end of the expected operation range.

The spillway modifications considered consisted of installation of flow deflectors installed on the face of the spillway, as shown in Figure 1. Two of the deflectors had a horizontal top surface, with submergence of 5.8 - 8.2 m (a "low" deflector elevation) and 2.3 - 4.7 m (a "high" deflector). The third deflector had a downward sloping surface with submergence of 5.8 - 8.2 m.

The two calibration coefficients, $\alpha$ (for mass transfer from bubbles) and $\beta$ (for atmospheric exchange) were adjusted to match low tail water level field measurements of TDG taken at Wanapum Dam for the no deflector and low deflector arrangements, and adjusted from field measurements at Lower Granite Dam for the high deflector arrangement (Geldert, 1996). For all cases, the flow field information used was based on data from the physical model. A representative flow field in the stilling basin region is shown in Figure 2.

The depth-averaged total dissolved gas concentration calculated for six test cases is shown as a function of distance downstream from the spillway toe in Figure 3. At the low tail water level, the low deflector performance is quite similar to that of the high deflector with regard to TDG. Both deflectors had TDG levels significantly below that which occurred without the deflector. In addition, for the original spillway configuration (*i.e.* no deflector) and the "low" horizontal deflector, there is a decrease in TDG level with water level. In contrast, for the "high" horizontal deflector, there is an increase in TDG with water level.

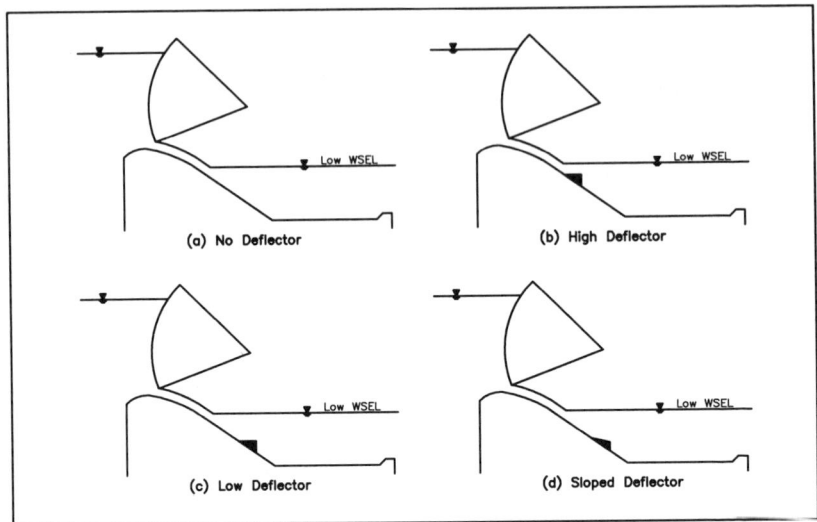

**Figure 1** Spillway Configurations Tested

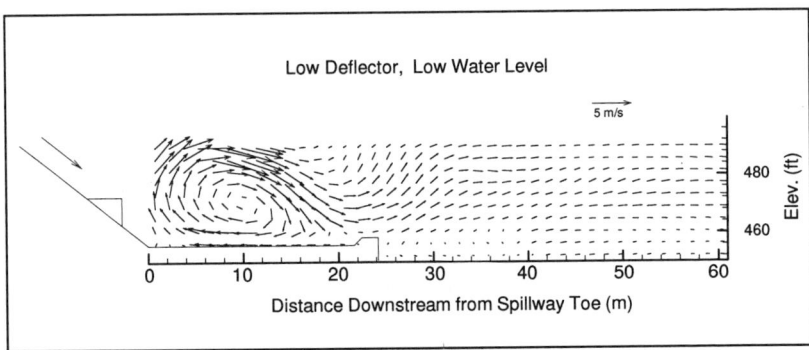

**Figure 2** Typical Flow Field in Stilling Basin

## CONCLUSIONS

The numerical model described above has been used to evaluate the effectiveness of different spillway design modifications at Wanapum Dam. Model input consists of hydrodynamic parameters and the distribution of air entrained in the flow, based upon measurements and observations on a physical model. While the physical model provides information regarding the acceptability of the hydraulics associated with the modifications, it cannot be used directly to assess the impacts on dissolved gas concentrations in the river. The combination of the physical and numerical models in the design process has allowed the selection of the design modifications with optimum hydraulics and gas concentrations. The low deflector preforms as well, and possibly better than the high deflector currently in use at the Lower Granite and Little Goose dams on the Snake River. The sloped deflector will be field tested in 1997.

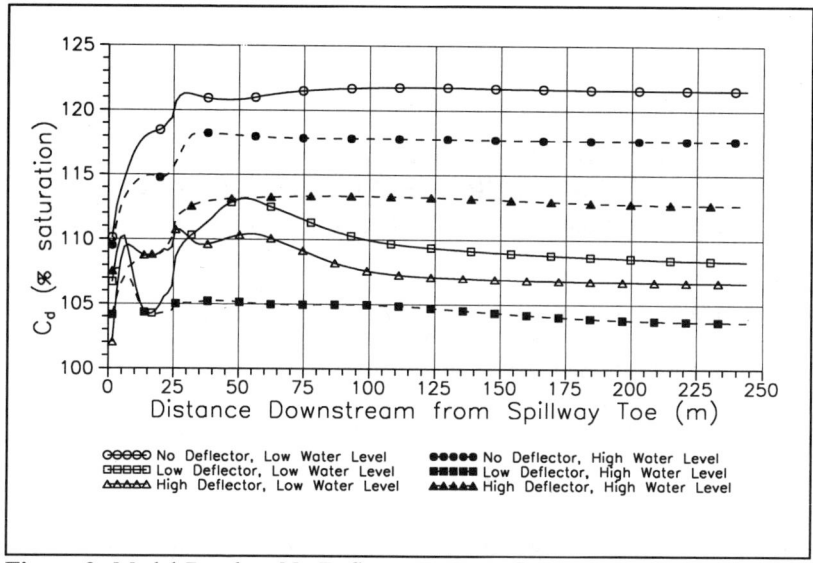

**Figure 3** Model Results: No Deflector, Low Deflector, High Deflector

## REFERENCES

Azbel, D. (1981). *Two-phase flows in chemical engineering.* Cambridge University Press, New York.

Geldert, D.A. (1996). *Parametric relations to predict dissolved gas supersaturation below spillways.* M.S. Thesis, University of Minnesota, Minneapolis, MN.

Hinze, J.O. (1955). "Fundamentals of the hydrodynamic mechanism of splitting dispersion processes." *AIChE J.*, 1(3), 289-295.

Nezu, I., & Nakagawa, H. (1993). *Turbulence in open channel flows.* A.A. Balkema, Rotterdam, Netherlands.

Patankar, S.V. (1980). *Numerical heat transfer and fluid flow.* McGraw Hill, New York.

## ACKNOWLEDGEMENTS

The development and application of the studies described have been a collaborative effort involving many individuals and organizations. The authors would like to acknowledge the guidance and direction of Steve Brown of Grant County PUD, Rex A. Elder, consulting hydraulic engineer and Duncan Hay of Hay and Company. The physical model studies at the Iowa Institute of Hydraulic Research were conducted under the supervision of Drs. Larry Weber and Jacob Odgaard. Field measurements of TDG at Wanapum Dam used for calibrating the numerical model were collected under the direction of Don Weitkamp, Ph.D. of Parametrix, Inc.

# PHYSICAL MODEL STUDY OF JET-BREAKER BAFFLE CONCEPTS IN PUMP INTAKE STRUCTURES WITH PIPE INFLOWS

HARTMUT ROSENBERGER[1], DIETER-HEINZ HELLMANN[1], MAHMOOD NAGHASH[2]

[1] Institute of Turbo Machinery, Department of Mechanical Engineering, University of Kaiserslautern, Kaiserslautern, Germany.
[2] Bechtel Corporation, Gaithersburg, Maryland, USA.

## ABSTRACT

The physical model study of several jet-breaker concepts was performed in a two-pump intake model with pipe inflow to find the most suitable configuration for providing a uniform flow condition at the pump. The concepts investigated were: weir-curtain wall baffle arrangement, curtain wall-weir baffle arrangement, rows of small-size baffle columns, half rows of large-size baffle columns, and horizontal baffle wall. Several arrangements of each concept were tested for six pump operating conditions and water levels. For each recommended final configuration, the baffle geometry, surface and sub-surface vortex classification, pump bell throat swirl angle, and pump bell throat axial velocity measurements are reported.

## BACKGROUND

Production of electricity has become a very competitive market globally. Frequently, new power plants are awarded through a lump-sum bidding process and early and reliable cost estimates are vital. In thermal power plants with once-through cooling systems, the intake structure is a major cost component requiring the design to be nearly complete even during the bidding process. Efforts toward standardization of water intakes are the key elements in providing a reliable design during this process.

For sea water intakes with vertical pumps, the intake structure consists of two major sections: the pump sump section, which hosts the pumps and screens; and the transition section, within which the flow is transformed from a non-uniform condition to a uniform approach flow condition to the pump sumps. Although the design of pump sump section has nearly reached maturity [Prosser, 1977] [Padmanabhan, 1987] [Hydraulic Institute, 1991] [Mellville et al., 1994], the flow transition section is handled on a case by case basis depending on the geometry and flow. The majority of hydraulic problems at pumps can be attributed to this section where eccentric flow conditions are generated and are conveyed to the pump sumps.

One type of intake for coastal waters requires offshore buried pipes to discharge to a transition section of an onshore pump intake structure. The inflow from the pipe imposes an eccentric flow distribution in the pump bays that is not desirable. The eccentric flow distribution must be transformed to a more uniform flow distribution through the transition section before reaching the pumps. The objective of this investigation was to investigate the suitability of several jet-breaker baffle concepts for such intakes. Emphasis was placed on using baffles that are easy to construct and are economical. Also, the optimization (reduction) of the size of transition area was an objective.

## TEST FACILITY AND PROCEDURE

Physical modeling of several jet-breaker baffle concepts was performed in a two-pump intake model as illustrated in Figure 1, to find the most suitable configuration for providing a uniform flow condition free from free- and sub-surface vortices at the pump with smallest swirl angle. The concepts investigated were: no-baffle as base case, weir-curtain wall baffle arrangement, curtain wall-weir baffle arrangement, rows of small size baffle columns, half rows of large size baffle columns, and horizontal and vertical baffle wall (see Figures 2-6).

The model consisted of two bays; one bay was completely modeled while the other bay was added to model pump flow only. In the model bay, floor and backwall splitters, floor and backwall fillets, and a curtain wall were installed as vortex suppressers [Sweeny, et al., 1982]. Traveling screens were modeled by fine-mesh screens using Reynolds number similarity [Padmanabhan, 1978]. The scale model was 1:10, and prototype pump rated and runout flows were 6 and 7.7 $m^3/s$, respectively.

Each jet-breaker baffle concept was tested in the transition section of the intake model. The baffle was placed between the inlet pipe and the pump sump, and several arrangements were tested for six pump operating conditions. The pump operating conditions used were: one- and two-pumps operation with three different sump water levels. The pump bell throat swirl angle measurement combined with a dye study (for vortex type classification) were used to identify the most satisfactory arrangement of the concept.

The intake dimensions in Figure 1 are normalized to D, where D is nominal pump bell diameter [JSME, 1984] [Mellville et al., 1994]. An optimized transition length of 3.8D was obtained by trial and error and was used for all tests. As the base concept, no baffles were installed within the transition section (Figure 2). Figures 3 and 4 illustrate the weir-curtain wall and curtain wall-weir baffle arrangements with notation for parameters varied for these two concepts. The geometry of vertical baffle columns and their arrangement is shown on Figure 5. The baffle columns are commonly used as energy dissipaters in hydraulic structures [Peterka, 1978]. Figure 6 illustrates the size and location of the horizontal baffle wall spanning across the entire width.

## MEASUREMENT TECHNIQUES

The intensities of free-surface vortices were visually assessed from the dye study and classification of Hecker [1987]. Type 1 free-surface represents a weak swirl, Type 2 a surface dimple, Type 3 is a dye core towards the intake, Type 4 pulls trash but not air, Type 5 pulls air bubbles, and Type 6 pulls air continuously towards the intake.

Inside the pump column at the elevation of the pump impeller, a vortimeter was installed indicating the strength of the rotational component of flow entering the pump bell. The swirl angle $\theta$ was obtained from $\theta = \arctan(V_t/V_a)$ where $V_t$ represents the vortimeter's blade tip velocity and $V_a$ is the axial mean velocity inside the pump throat. The blade tip velocity is determined by counting the revolutions per unit time. The axial mean velocity is obtained from the discharge rate measured by inductive flow meter and related to the cross sectional area of the pump throat. Generally, swirl angles should be lower than 5 degrees [Hecker, 1987].

Pump throat velocity traverse measurements were taken by means of a Prandtl-tube. The commonly accepted deviation of the local velocity along the circumference of each ring of equal radius is ±5% about mean velocity [Knauss, 1978].

## RESULTS AND DISCUSSION

Table 1 contains the results of a selected number of tests. Column 1 describes each concept, Column 2 contains water levels in terms of D, pumps modes of operation are in Columns 3 and 4, Column 5 is swirl angle, Column 6 is observed free-surface vortex, and Column 7 is maximum deviation for velocity traverse measurements. Free-surface vortices were only detected upstream of the curtain wall. Free-surface vortex type other than 1 was considered unacceptable. No sub-surface attached vortices were detected within this investigation.

In the base test, non-uniform approach flow conditions were detected especially for one pump operation. Strong free-surface vortices were formed due to lateral flow along the curtain wall imposed by circulation around the divider wall (Figure 7).

The weir-curtain wall baffle arrangement provides acceptable flow conditions at the pump (Table 1). However, this concept provided a non-uniform velocity distribution upstream of the pump curtain wall which may not be acceptable to the traveling screens performance (Figure 8). Furthermore, this concept provided a high head loss across the baffle (6 cm in the model).

The curtain wall-weir combination did not provide acceptable flow conditions at the pump (Table 1). It provided, however, a better flow condition at the traveling screens compared to the weir-curtain wall concept (Figure 9).

The rows of small square columns baffle concept with 50% clear opening (Figure 5) did not provide any acceptable flow condition at the pump for all configurations studied. No attempts were made to change the clear opening or to change the width to length ratio of the columns. The results for a typical test is shown in Table 1.

The horizontal baffle wall concept provided acceptable flow conditions at the pump with no vortex formation of any kind and for all cases tested (Table 1). The water surface in the transition section was wavy (Figure 10) but had no effect on the pump. The maximum intake model head loss for this concept was 4 cm.

Two other baffle concepts tested include i) a vertical beam and ii) two one-half rows of large square columns located near the pipe exit. The flow at the pump for both concepts were not acceptable and the results are not reported here.

In summary, for the baffle configurations tested, the weir-curtain wall baffle and horizontal wall baffle concepts provided acceptable flow conditions at the pump. The horizontal wall baffle is preferred since it also provides satisfactory flow conditions at the traveling screens and is easy to construct.

## REFERENCES

Hecker, G.E, 1987, *Fundamentals of Vortex Intake Flow*, Chapter 2 in *Swirling Flow Problems at Intakes*, IAHR Hydraulic Structures Design Manual 1, J. Knauss, editor.

Hydraulic Institute, 1991, *Hydraulic Institute Standards,* 14$^{th}$ Edition, Cleveland, Ohio.

JSME, 1984, *Standard Method for Model Testing the Performance of a Pump Sump*, The Japan Society of Mechanical Engineers, JSME S 004-1984.

Knauss, J., (Editor), 1987, *Swirling Flow Problems at Intakes*, IAHR, Hydraulic Structures Design Manual 1, A.A. Balkema Publisher, Rotterdam.

Melville, B.W.; Ettema, R., and Nakato, T., 1994. *Review of Flow Problems at Water Intake Pump Sumps*, Electrical Power Research Institute Report TR-103474, June.

Padmanabhan, M., 1978, *Pressure Drop Due to Flow Through Fine Mesh Screens,* Journal of Hydraulics Division, Vol. 104, No. HY8, ASCE.

Padmanabhan, M., 1987, *Pump Sumps*, Chapter 6.2 in *Swirling Flow Problems at Intakes*, IAHR Hydraulic Structures Design Manual 1, J. Knauss, editor.

Peterka, A.J., 1978. *Hydraulic Design of Stilling Basins and Energy Dissipaters*. Bureau of Reclamation, Engineering Monograph No. 25, Washington, D.C..

Prosser, M.J., 1977. *The Hydraulic Design of Pump Sumps and Intakes,* BHRA, July.

Sweeney, C.E., Elder, R.A., and Hay, D., 1982. *Pump Sump Design Experience: Summary.* Journal of Hydraulics Division, Vol. 108, No. HY3, ASCE, pp 361-377.

## Table 1 Experimental Results of Model Study

| 1 | 2 | 3 | 4 | | 5 | 6 | 7 |
|---|---|---|---|---|---|---|---|
| Jet-Breaker Baffle Concept | Baffle Arrangement | Water Level [times D] | Operating Mode | | Swirl Angle [degree] | Free Vortices Upstream Curt. Wall | Throat Velocity Variation [%] |
| | | | Pump 1 | Pump 2 | | | |
| 1<br>Base Test<br>No Baffle | Transition Length<br>= 3.8D | 2.27<br>2.27<br>2.76<br>2.76<br>3.24<br>3.24 | rated<br>run-out<br>run-out<br>rated<br>rated<br>run-out | rated<br>off<br>off<br>rated<br>rated<br>off | 1.82<br>2.42<br>0.97<br>0.57<br>1.47<br>2.39 | 1<br>2.5<br>3<br>1<br>0<br>4 | |
| 2<br>Weir - Curtain Wall | a=0.89D<br>b=0.53D<br>c=1.69D<br>d=0.31D | 2.27<br>2.49<br>2.76<br>2.76<br>3.24<br>3.24 | run-out<br>rated<br>rated<br>run-out<br>run-out<br>rated | off<br>rated<br>rated<br>off<br>off<br>rated | 1.29<br>1.66<br>3.51<br>4.98<br>4.45<br>4.68 | 1<br>0<br>0<br>1.5<br>1<br>1.5 | |
| 3<br>Weir - Curtain Wall | a=0.89D<br>b=0.53D<br>c=1.16D<br>d=0.84D | 2.27<br>2.27<br>2.76<br>2.76<br>3.24<br>3.24 | rated<br>run-out<br>run-out<br>rated<br>rated<br>run-out | rated<br>off<br>off<br>rated<br>rated<br>off | -<br>1.23<br>0.61<br>0.89<br>1.62<br>0.80 | 0<br>1<br>1<br>0<br>0<br>0 | 3(9)<br>3(11)<br>2(11)<br>3(11)<br>2(11)<br>2(10) |
| 4<br>Weir - Curtain Wall | a=0.89D<br>b=0.53D<br>c=2D<br>d=0 | 2.27<br>2.49<br>2.76<br>2.76<br>3.24<br>3.24 | run-out<br>rated<br>rated<br>run-out<br>run-out<br>rated | off<br>rated<br>rated<br>off<br>off<br>rated | 0.47<br>1.33<br>1.21<br>0.93<br>1.82<br>1.66 | 1<br>0<br>0<br>1.5<br>1<br>0 | 3(12)<br>3(11)<br>4(8)<br>4(12)<br>3(11)<br>7(10) |
| 5<br>Curtain Wall - Weir | a=0.89D<br>b=0.53D<br>c=0.52D | 2.27<br>2.27<br>2.76<br>2.76<br>3.24<br>3.24 | run-out<br>rated<br>run-out<br>rated<br>run-out<br>rated | off<br>rated<br>off<br>rated<br>off<br>rated | 0.24<br>0.57<br>0.90<br>0.55<br>0.17<br>0.91 | 1<br>1<br>1.5<br>1.5<br>2<br>0 | |
| 6<br>Curtain Wall - Weir | a=0.89D<br>b=0.53D<br>c=0.52D<br>d=0.44D | 2.27<br>2.27<br>2.76<br>2.76<br>3.24<br>3.24 | run-out<br>rated<br>rated<br>run-out<br>run-out<br>rated | off<br>rated<br>rated<br>off<br>off<br>rated | 0.33<br>1.05<br>1.09<br>0.41<br>0.35<br>0.48 | 1.5<br>0<br>0<br>1.5<br>2.5<br>0 | |
| 7<br>Baffle Columns | a=1.11D<br>b=1.78D<br>Column Spacing<br>=0.11D | 2.27<br>2.27<br>2.76<br>2.76<br>3.24<br>3.24 | run-out<br>rated<br>run-out<br>rated<br>run-out<br>rated | off<br>rated<br>off<br>rated<br>off<br>rated | 0.94<br>3.51<br>0.66<br>2.30<br>1.21<br>2.14 | 1<br>1<br>1.5<br>3<br>2<br>3 | |
| 8<br>Horizontal Wall | | 2.27<br>2.27<br>2.76<br>2.76<br>3.24<br>3.24 | run-out<br>rated<br>run-out<br>rated<br>run-out<br>rated | off<br>rated<br>off<br>rated<br>off<br>rated | 1.13<br>0.83<br>0.25<br>1.05<br>0.72<br>0.97 | 0<br>0<br>0<br>0<br>0<br>0 | 2(11)<br>3(9)<br>2(11)<br>2(12)<br>3(12)<br>4(9) |

Note: 3(9) in Column 7 means: 3% max. variation in ring of equal radius, 9% max. overall variation

# JET-BREAKER BAFFLE CONCEPTS

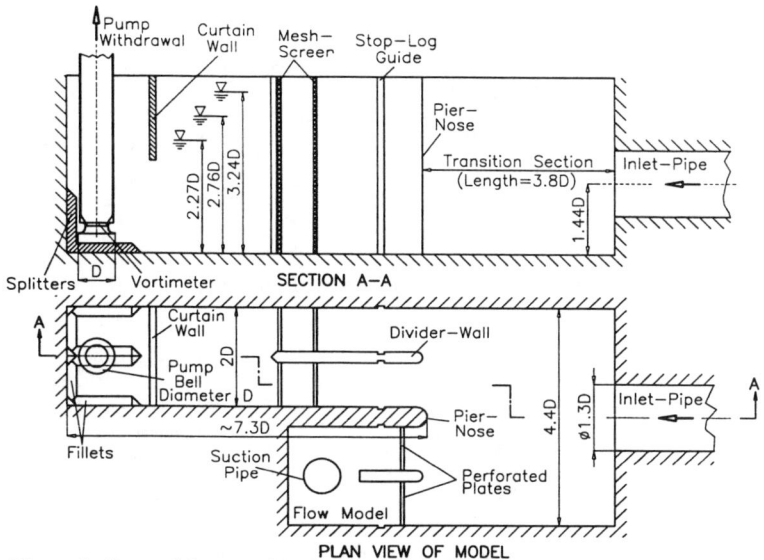

*Figure 1: General Layout of Jet-Breaker Model*

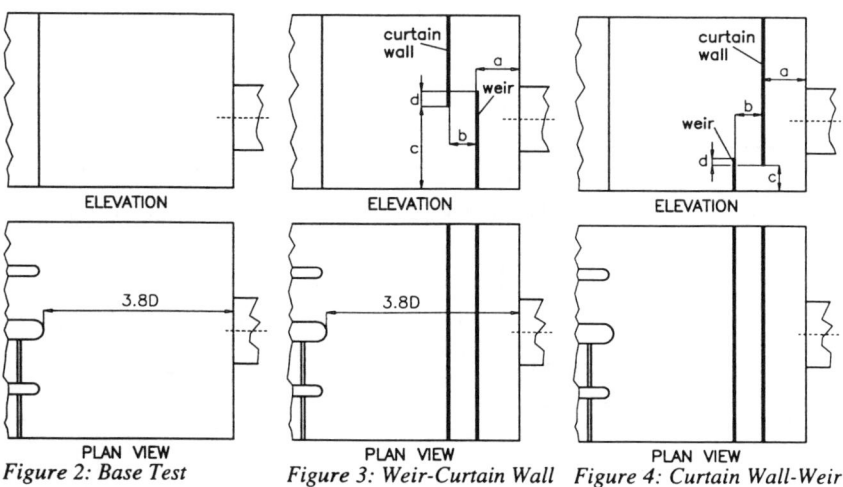

*Figure 2: Base Test*

*Figure 3: Weir-Curtain Wall Concept*

*Figure 4: Curtain Wall-Weir Concept*

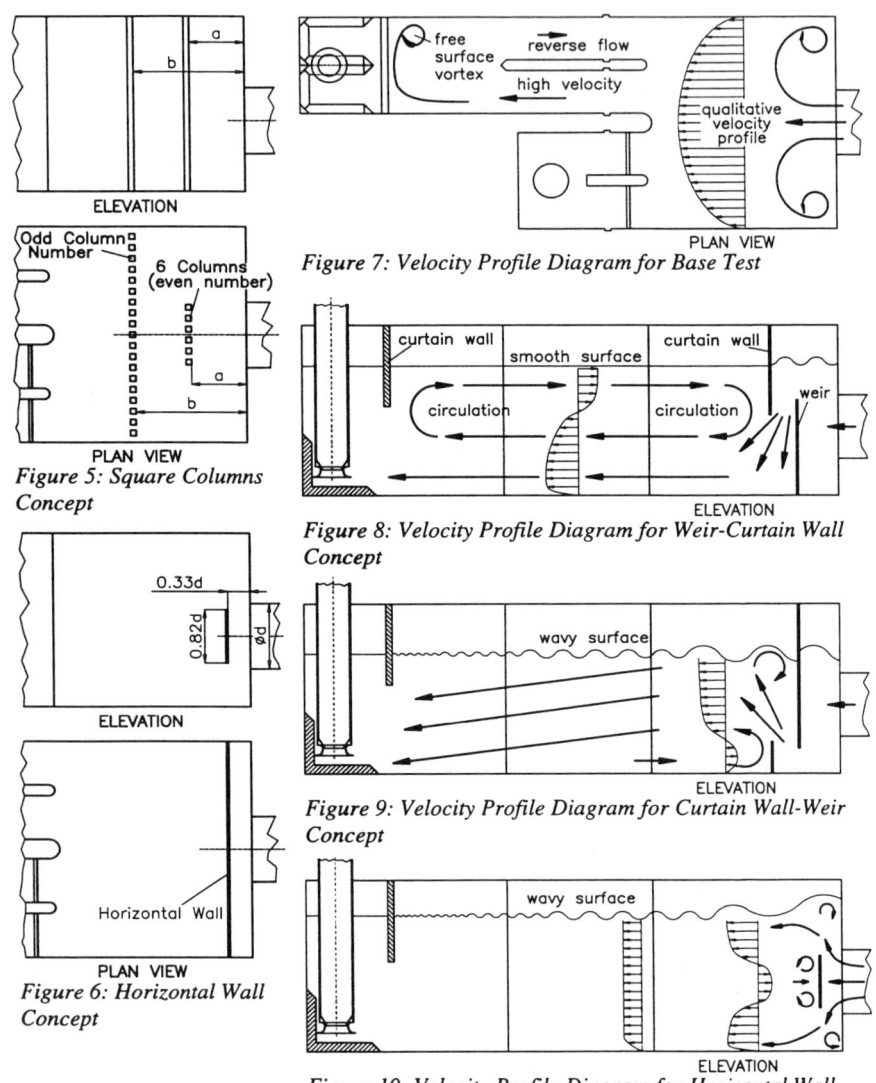

Figure 5: Square Columns Concept

Figure 6: Horizontal Wall Concept

Figure 7: Velocity Profile Diagram for Base Test

Figure 8: Velocity Profile Diagram for Weir-Curtain Wall Concept

Figure 9: Velocity Profile Diagram for Curtain Wall-Weir Concept

Figure 10: Velocity Profile Diagram for Horizontal Wall Concept

# INFLUENCE OF APPROACH FLOW NON UNIFORMITY ON VORTICES AT A PUMP INTAKE

by G. CONSTANTINESCU and V.C. PATEL
Iowa Institute of Hydraulic Research, University of Iowa, Iowa City, IA 52242, USA

## ABSTRACT

A computational fluid dynamics (CFD) model is applied to study the effect of non uniformity in the approach flow on vortices and swirl at a pump intake. The model is based on the solution of the Reynolds-averaged Navier-Stokes equations and a near-wall turbulence closure, and seeks steady-state solutions with different approach flow conditions, including non-symmetrical positioning of the pipe relative to the intake channel and imposed mean shear at the channel inlet. It is found that the number, location, size and intensity of free-surface vortices and swirl in the intake pipe are greatly affected by intake flow non uniformity.

## INTRODUCTION

There is no theoretical model for prediction of the different types of vortices that arise as a result of wall proximity and non uniformity of the approach flow in a hydraulic pump sump. Melville et al. (1994) give a comprehensive review of the flow problems encountered in pump sumps, including vortices and swirl, and list the parameters that affect the flow. They also describe guidelines proposed by various agencies for designing intakes for vortex-free operation and the significant variations among them. The reliability of these guidelines is severely limited, as evidenced by the continuing need for laboratory model studies to verify designs or remedy post-construction problems.

Computational fluid dynamics (CFD) offers a potentially powerful approach to simulate pump-intake flows because, in principle, a properly validated CFD model could be used for any given layout and operating conditions to obtain a detailed description of the flow field. The development and validation of CFD models for complex three-dimensional flows, such as that in a pump intake, is a subject of intense research in a number of fields. A CFD model for pump-intake flows is described in Constantinescu and Patel (1996) along with some example solutions for a simple symmetric rectangular intake and fully-developed flow in the approach channel. In this paper, the model is applied to study the effect of the approach flow non uniformity on intake vortices and swirl.

## PREVIOUS WORK

The principal sources of vorticity in the pump sump are non uniformity in the approach flow, wakes of obstructions or 'flow conditioners' placed in the approach channel, and the boundary layers on the walls. Asymmetric location of the pump bell relative to the side walls tends to amplify existing vorticity in different ways and leads to changes in the vortex structure.

Most previous studies of pump-intake flows allude to the importance of non uniformity of the approach flow. Based on studies of air-entraining vortices at horizontal pipe intakes, Quick (1970) concluded that formation of such vortices was determined by the vorticity and unsteadiness in the approach flow as well as the intake geometry. Chang (1980) made a similar study with vertical intakes and found that the main source of vorticity was the boundary layer on the exterior of the suction pipe. Knauss (1987) studied the influence of circulation in the approach flow on the critical submergence depth, and found a significant increase in the strength of vortices when the swirl in the incoming flow (roughly defined as the ratio of the transverse to the longitudinal velocity component) was increased. He proposed an empirical relation between the critical submergence depth, circulation in the approach flow, and pipe diameter. Similar relations were obtained by Daggett and Keulegan (1972) and Jain et al. (1978) for free-surface vortices in a cylindrical tank with a bottom orifice. Quite different methods were adopted for inducing swirl. For example, Denny (1956) allowed the flow to enter through part of the channel width, while Tagamori (1980) used guide vanes to create the swirl. The experiments of Tagamori (1980) and Amphlett (1978) were used to determine the constant in Knauss' correlation.

Hecker (1987) found that subsurface vortices can be just as damaging to pumps as free-surface vortices, but little attention has been given to their study. He noted that subsurface vortices are more likely to form, or grow in intensity, when the ratio of the intake-pipe velocity to the approach flow velocity is increased, or the circulation in the approach flow is increased, or the distance to the lateral walls is decreased.

The principal conclusion from these previous studies is that vortices in pump intakes depend on swirl in the approach flow, but the term itself is vaguely defined and there are no quantitative data that relate the vortices to non uniformity of the approach flow. This paper demonstrates the use of a CFD model to obtain quantitative information on pump-intake vortices and swirl.

## CFD MODEL AND FLOW GEOMETRY

The CFD model described in Constantinescu and Patel (1996) uses the numerical method of Sotiropoulos and Patel (1992), which solves the Reynolds-averaged Navier-Stokes equations with the two-layer k-$\varepsilon$ turbulence closure of Chen and Patel (1988). In the momentum equations, the viscous and pressure terms are discretized using second-order, three-point central differences, while the convective terms are discretized with second-order accurate upwind differences. A Poisson equation for pressure is derived by substitution of the velocities into the continuity equation. The momentum and pressure equations are solved using an ADI method. The turbulence model equations are treated in a similar way to the momentum equations. The accuracy and efficiency of the numerical method was established in previous studies of diverse flows. The two-layer turbulence model which resolves the near-wall flow was found to be critical in solutions of flows involving separations and vortices.

Constantinescu and Patel (1996) presented calculations for the pump-intake configuration shown in Figure 1. A single pump, with an infinitely thin circular pipe in a long rectangular intake channel was chosen as the model problem. There are no flow-training devices to further complicate the geometry. In spite of the simple geometry, however, the model problem includes the essential fluid dynamics features and enables parametric studies to be performed by varying the geometry and flow variables.

The flow is assumed steady and the free-surface is treated as a plane of symmetry (i.e., with no tangential stresses and surface tension effects). As a first approximation, the calculated pressure distribution on this plane can be converted into a free-surface deformation assuming hydrostatic variation in the vertical direction.

The effect of non uniformity of the approach flow is investigated here by comparing the results of three new calculations with that presented in Constantinescu and Patel (1996) in which the pipe was placed symmetrically in the channel. All calculations were carried out on a grid of 550,000 points (Figure 2) for turbulent flow at Re=60,000. The relative dimensions of the intake (Figure 1) are as follow: back wall clearance, X/D = 0.4, submergence, S/D = 1.25, and distance from lateral walls to the pipe center, L1/D = L2/D = 1.3 in the symmetric case and L1/D = 0.9, L2/D = 1.7 in the asymmetric case. The channel inlet is situated 6D upstream of the pipe, the length of the intake pipe modeled is 12D, and the water depth, (S+C)/D = 2.0.

In the symmetric case (S), fully-developed turbulent flow was prescribed at the inlet, simulating a long straight fetch of upstream channel. The new calculations were made with the asymmetrically located pipe and three different upstream flow conditions, one with fully-developed flow (NS), and two with superimposed shear (NS1, NS2) of opposite sign, the streamwise velocity distribution at the inlet being specified by

$$u = u_D \pm C_s \cdot y / 0.5(L1 + L2) \cdot u_D$$

where $u_D$ is the velocity distribution in fully-developed flow, y is distance across the channel from the center, and $C_s$ =0.7 is a measure of the imposed shear. The latter two cases simulate the effect of a transverse flow, such as might be present in a channel connected to a river, with different flow direction in the two cases.

## RESULTS

Limitations of space preclude a detailed analysis and discussion of all of the flow features that are revealed by the CFD model (see Constantinescu and Patel (1996) for a discussion of the reference case, S). The "streamlines" constructed from the horizontal components of the velocity vector in the plane of the free surface and in a plane near the pipe inlet are shown in Figures 3 through 6 for the four sets of calculations. They reveal a variety of vortical flow structures outside and inside the pipe and significant changes in these among the four cases. The numerical solutions can be analyzed to determine the location, strength and other properties of these structures at various depths, as well as their origins at the channel walls, and determine if they can be classified as vortices. For the present purposes it suffices to discuss the flow on the basis of Figures 3 through 6. The vortical features in these figures are identified either as vortices or swirls, depending on whether the streamlines spiral into or out of the foci. In topological terms, these are stable and unstable foci, respectively, and longitudinal vorticity is associated with both.

Figure 3a shows two pairs of symmetrically located vortices at the free surface outside the pipe (labeled V1-V4 for later convenience) and a pair of swirls (S1 and S2) inside the pipe. At the level of the entry to the pipe, the swirl inside the pipe is not well developed and vortices V2 and V3 disappear into the boundary layer on the interior wall due to the strong radial flow at this location. The corner vortices V1 and V4 are clearly seen. Below this level, these vortices change their orientation and feed or connect with corner vortices near the floor and parallel to the back and side walls, and finally feed vortices attached to the side walls which are not visible in these figures.

Comparison of Figure 3 (S) with 4 (NS) shows that moving the pipe closer to a side wall leads, at the free surface, to V2 being pulled around the pipe, V3 attaching to the boundary layer on the exterior of the pipe, and V4 moving closer to the back-wall, becoming similar to V2 in the symmetric case. The main difference between the two cases is at the level of the entry to the pipe where there is swirl in only one direction (S1) along the entire pipe.

Comparing Figures 4, 5 and 6 gives information on the effects of non uniformity of the upstream flow. When the circulation is increased by superimposing a positive shear (NS1), the vortex V4 moves away from the wall, while V2 and V3 decrease in size and are pushed into the boundary layer on the exterior of the pipe. The corner vortex V1 remains essentially unchanged in location, but appears to entrain more fluid at the free surface. Interestingly, the effect of negative shear (NS2) leads to a flow pattern that is similar to the symmetric case, canceling the effect of the asymmetrical position of the pipe. The location of the pipe relative to the side walls and approach flow non uniformity also influences the intensity of swirl inside the pipe. Table 1 shows the variation of the axial component of vorticity along the foci of the swirls S1 and S2, non dimensionalized by the pipe diameter and bulk velocity, over a distance of more than 6.5D starting just above the entry at z/D=0.75. It is seen that the intensity of swirl, and its persistence along the pipe, vary greatly among the four cases. This particular feature of the flow is, of course, of paramount importance in intake design.

Much further analysis of the model results is obviously needed to provide quantitative information on the vortices at the free surface. As already mentioned, the present model does not account for free-surface deformation and surface tension. However, it is of interest to note that local pressure minima are associated with the cores of vortices V2 and V3 in case S, V2 in NS, V1 and V4 in NS1, and V3 in NS2. These vortices will intensify with appropriate changes in geometry and flow variables (a decrease in submergence, for example) and may become strong enough to break the free surface and eventually entrain air.

## CONCLUDING REMARKS

This paper has demonstrated the use of a CFD model to calculate the flow in a pump sump with different types of flow non uniformity in the approach channel. The solutions clearly reveal the complexity of the flow, including a variety of vortices and swirls which have hitherto fore defied quantitative description. It is found that non uniformity of the approach flow influences the level of swirl in the intake pipe. Although the present CFD model can be used to make parametric studies to better understand vortex formation in pump sumps, and perhaps identify the most effective means for vortex-free conditions, considerable further development will be required to numerically model the flow in practical intake geometries.

## ACKNOWLEDGMENTS

This research was partially supported by a grant from EPRI, monitored by Mr. J.L. Tsou. The calculations were performed on the SGI Power Challenge Supercomputer of IIHR.

## REFERENCES

Amphlett, M.B. (1978). "Air-entraining vortices at a vertically inverted intake," Hydraulic Research Station, Wallingford, Report No. Od/17.

Chang, E. (1980). "An investigation into vortex formation in rectangular channels," *Proceedings of the Winter Annual Meeting of ASME*, Chicago, IL, 77-86.

Chen, H.C. and Patel, V.C. (1988). "Near-wall turbulence models for complex flows including separation," *AIAA Journal*, 26(6) 641-648.

Constantinescu, G. and Patel, V.C. (1996). "A numerical model for simulation of pump-intake flow and vortices," submitted to *J. Hydr. Eng.*, ASCE.

Daggett, L. and Keulegan, G.H. (1972). "Similitude conditions in free surface vortex formation," *J. Hydr. Div.*, ASCE, 100(11), 1565-1580.

Denny, D.F. (1956). "An experimental study of air-entraining vortices in pump sumps," *Proceedings I. Mech. E.*, U.K., 170( 2), 106-116.

Hecker, G.E. (1987). "Fundamentals of vortex intake flow," IAHR Hydraulic Structures Design Manual on *Swirling Flow Problems at Intakes,* J. Knauss, Balkema / Rotterdam, 12-38.

Jain, A.K., Raju, K.G.R. and Garde, R.J. (1978). "Vortex formation at vertical pipe intakes," *J. Hydr. Div.*, ASCE, 104(HY10), 1429-1445.

Knauss, S.C. (1987). "Prediction of critical submergence," IAHR Hydraulic Structures Design Manual on *Swirling Flow Problems at Intakes,* J. Knauss, Balkema / Rotterdam, 67-89.

Melville, B.W., Ettema, R. and Nakato, T. (1994). "Review of flow problems at water intake pump sumps," *EPRI Research Project RP3456-01 Final Report,* Iowa Institute of Hydraulic Research, Univ. Iowa, Iowa City, IA.

Quick, M.C. (1970). "Efficiency of air-entraining vortex at water intakes," *J. Hydr. Div.*, ASCE, 96(HY7), 1403-1415.

Sotiropoulos, F. and Patel, V.C. (1992). "Flow in curved ducts of varying cross section," *IIHR Report No. 358,* Iowa Institute of Hydraulic Research, Univ. Iowa, Iowa City, IA.

Tagomori, M. (1980). "Flow pattern and air-intake vortex in an intake sump, Part 2 - Influence of inflow conditions," Ebara Seisakusho Central Laboratory, Japan, *Turbomachinery,* Vol 8, 45-77.

Table 1. Axial vorticity at the swirl foci inside the pipe

| $z/D$ | S | | NS | NS1 | NS2 | |
|---|---|---|---|---|---|---|
| | S1 | S2 | S1 | S1 | S1 | S2 |
| 0.90 | 1.94 | 1.94 | 2.80 | 3.28 | 2.21 | 3.43 |
| 1.20 | 1.30 | 1.30 | 2.08 | 2.65 | 1.64 | 1.79 |
| 1.60 | 0.80 | 0.80 | 1.51 | 1.90 | 1.12 | 1.10 |
| 2.00 | 0.50 | 0.50 | 1.10 | 1.36 | 0.73 | 0.73 |
| 2.70 | 0.24 | 0.24 | 0.80 | 1.34 | 0.46 | 0.55 |
| 4.10 | 0.17 | 0.17 | 0.63 | 0.95 | 0.38 | 0.21 |
| 6.50 | --- | --- | 0.52 | 0.80 | 0.30 | --- |

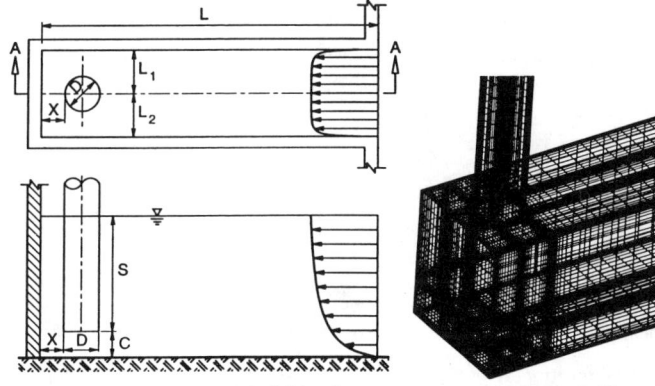

Figure 1 Model intake

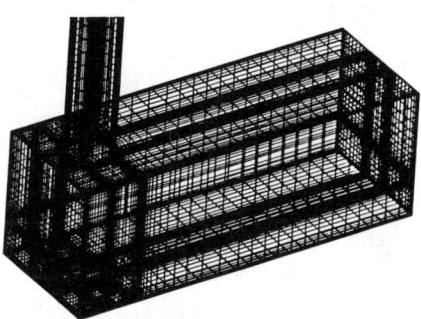

Figure 2 Computational mesh

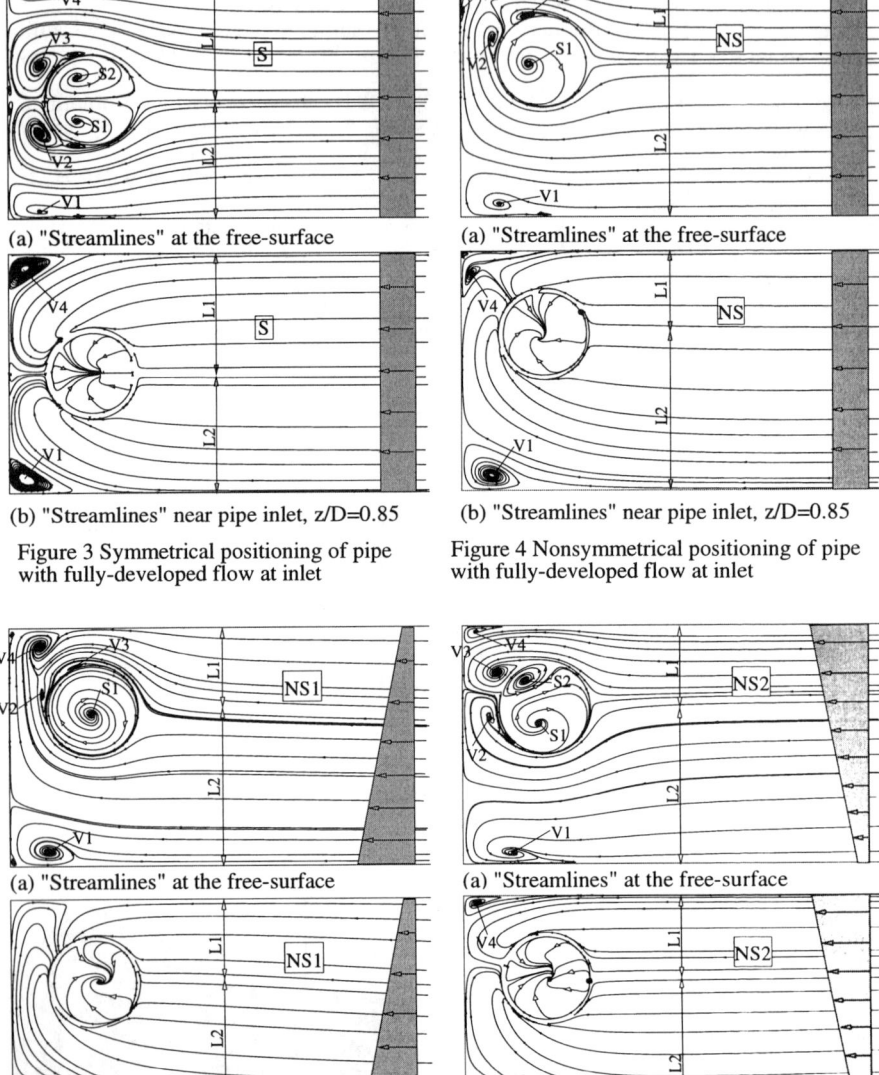

(a) "Streamlines" at the free-surface

(b) "Streamlines" near pipe inlet, z/D=0.85

Figure 3 Symmetrical positioning of pipe with fully-developed flow at inlet

(a) "Streamlines" at the free-surface

(b) "Streamlines" near pipe inlet, z/D=0.85

Figure 4 Nonsymmetrical positioning of pipe with fully-developed flow at inlet

(a) "Streamlines" at the free-surface

(b) "Streamlines" near pipe inlet, z/D=0.85

Figure 5 Nonsymmetrical positioning of pipe with fully-developed flow and positive imposed shear

(a) "Streamlines" at the free-surface

(b) "Streamlines" near pipe inlet, z/D=0.85

Figure 6 Nonsymmetrical positioning of pipe with fully-developed flow and negative imposed shear

# EFFECT OF DUAL FLOW SCREENS ON THE HYDRAULIC PERFORMANCE OF PUMP INTAKES

## M. PADMANABHAN, D.K. WHITE AND J. LARSEN
Alden Research Laboratory, Inc., Holden, MA 01520, USA

### ABSTRACT

Dual flow screens are used in cooling water and service water intakes of power plants to replace the once-through travelling screens. Dual flow screens may result in a skewed approach flow to the pumps with concentrated flow regions and backflow eddy regions. When the once-through travelling screens are replaced with dual flow screens, objectionable vortexing and swirl are likely to be induced. Hence, modifications to the intake involving introduction of vortex suppressing and flow straightening devices will be required. Based on several hydraulic model studies conducted at the Alden Research Laboratory, adverse flow patterns due to dual flow screens of the in-flow type and possible remedial measures are discussed.

### INTRODUCTION

Dual flow screens are designed to have two screen surfaces placed parallel to the intake bay wall, dividing the bay into three sections. As shown in Figure 1, the dual flow screens may be designed as an in-flow type or out-flow type. In the design of the in-flow type, the flow enters the two side sections, passes through the respective screens, merges and exits in the middle section. In the out-flow type, the flow enters the middle section, passes through the respective screens and exits through the side sections.

Compared to once-through travelling screens, dual flow screens may be advantageous in that they do not produce "carry-over" of debris as both screen surfaces always face the flow. Further, it is possible to provide a greater screen area with dual flow screens for a given intake bay width.

While a once-through travelling screen may help straighten the flow approaching the pumps, a dual flow screen generates a skewed flow with a concentrated flow region in the middle and eddies with backflow at the sides or vice versa, depending on whether the dual flow screen is the in-flow or out-flow type. The instability of the concentrated flow region is likely to result in the flow favoring one side of the bay and may result in a strong circulation at the pump.

For pump intakes, which at present have once-through flow screens, and perform satisfactorily, the hydraulic performance of the intake has to be reevaluated, when the once-through travelling screens are replaced with dual flow screens. Objectionable vortexing and swirl are likely to be induced by the dual flow screens, and modifications to the intake such as introduction of vortex suppressing and flow straightening devices will be required.

Several model studies of pump intakes with in-flow type dual flow screens have been conducted at the Alden Research Laboratory, Inc.(ARL). The purpose of this paper is to discuss the flow patterns due to in-flow type dual flow screens and possible remedial measures derived from model studies.

## FLOW PATTERNS DUE TO DUAL FLOW SCREENS

Flow patterns, both upstream and downstream of an in-flow type dual flow screen are sketched in Figure 1, as obtained from observations in a hydraulic model. The flow at the screens is not uniform, as the flow has to turn sharply to enter and exit the screens. Flow separations occur, which could result in regions of higher velocities at the downstream end of the screenface and regions of low or even negative velocities at the upstream end, limiting the effective area of the dual flow screens. Figure 2 shows typical velocity distribution along the screenface and at the exit from the screen for an intake with an average approach velocity in the bay of 1 ft/sec.

The flow patterns as the flow exits the dual flow screen structure are shown in the photograph of Figure 3, as obtained from flow visualization using confetti in a hydraulic model. The flow exits the screen structure in the middle of the bay as a high velocity jet, resulting in eddies and back flow on either side of the middle jet. The jet is generally unstable, and as the flow expands, the flow is directed to one side of the bay, resulting in a strong circulation in the vicinity of the pump, contributing to strong vortices and swirl.

In general, geometric parameters of pump intakes (such as bay width, floor and wall clearances to bell) are selected using available standards, e.g., Hydraulic Institute Standards, which assume uniform flow approaching the pumps. With dual flow screens, the approach flow to the pumps is skewed, which may result in unacceptable vortices and swirl at the pump entrance. Hence, remedial modifications will be needed.

Recent hydraulic model studies of single bay intakes with dual flow screens, conducted at ARL, showed presence of strong sub-surface floor vortices and a very high swirl at the pump entrance, as much as 15 to 25 degrees. Any coherent-core free-surface vortices present were weak and not well organized due to the high turbulence and surface waves in the flow exiting the dual flow screen. Extensive modifications were needed to suppress the sub-surface vortices and reduce the swirl to acceptable levels of 5 degrees or less.

## REMEDIAL DESIGNS BASED ON MODEL STUDIES

Based on ARL's experience, the modifications to the intake are likely to involve installation of one or more of the following devices, depending upon the extent of vortexing and swirl:

1) A flow distributor, at least one bay width downstream of the dual flow screens to reduce the skewness of approach flow to pumps. Flow distributor (head loss device) formed of vertical columns providing about 30 to 40% open area is suggested.

2) A gradual expansion (transition) from the middle opening exiting the flow from dual flow screens to the full bay width, to reduce flow separation and allow gradual expansion of flow.

3) A curtain wall across the bay at some distance upstream of the pump, projecting into the water by more than half the depth, to reduce free surface vortexing.

4) Floor and back wall splitters, and side and back wall fillets to guide the flow and to eliminate or reduce sub-surface vortexing and resulting swirl.

As obtained from single bay hydraulic models tested at ARL, Figures 4 and 5 show conceptually two typical modification schemes, found effective in producing acceptable hydraulic performance. These designs eliminated free and sub-surface vortices and reduced swirl from as high as 15 to 25 degrees to less than 5 degrees.

To reduce flow separations and resulting high velocity regions of the flow entering the dual flow screens, vanes may be provided upstream of the screens to guide the turning flow smoothly to the screens. In addition, to minimize flow separations at the blocked area upstream of the screens, a "nose" may be provided.

Even though this paper discusses the in-flow type dual flow screens, similar hydraulic problems and remedial devices are also expected for the out-flow type dual flow screens.

## CONCLUSIONS

The following conclusions are drawn pertaining to the effect of in-flow type dual flow screens on the hydraulic performance of pump intakes:

- Dual flow screens cause skewed approach flows to the pumps with a high velocity region in the middle and eddies with backflow at the sides, which may contribute to objectionable vortices and swirl.

- The flow at the screens themselves is skewed, as the flow has to turn sharply to enter and exit the screen. Flow separations occur, which could result in backflow regions at the screen, limiting the effective area of the screens.

- Due to skewed approach flows generated by the dual flow screens, even pump intakes with geometric parameters (such as bay width, floor and wall clearances to bell) selected using available standards, have been observed to give objectionable vortexing and swirl problems.

- Modifications should be derived using a model study, and may involve installation of flow distributors or vortex suppressing and swirl reducing devices (such as curtain walls, splitters, and fillets), depending upon the extent of vortexing and swirl.

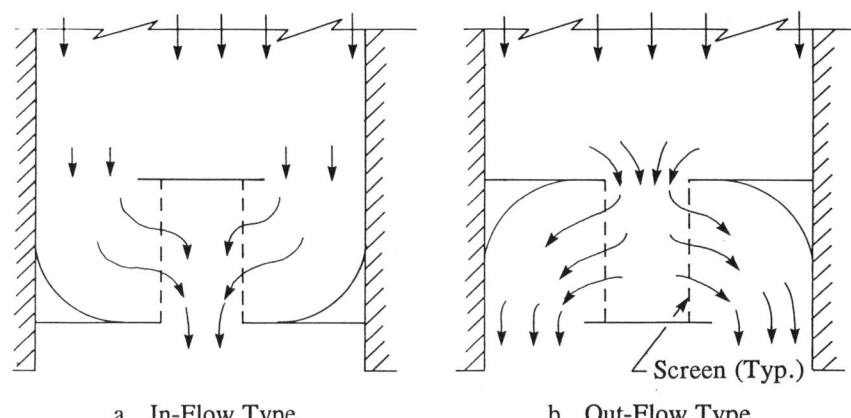

a. In-Flow Type    b. Out-Flow Type

FIGURE 1  TYPES OF DUAL FLOW SCREENS

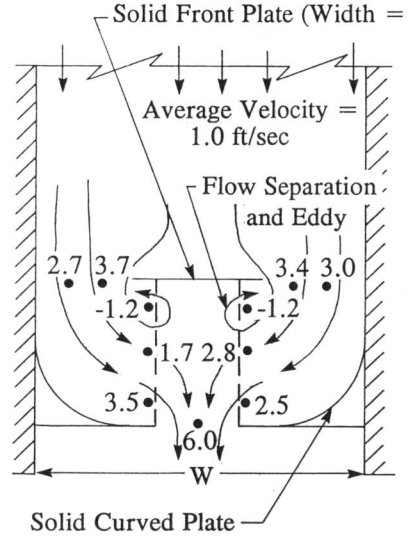

FIGURE 2  VELOCITY DISTRIBUTION AT THE SCREEN

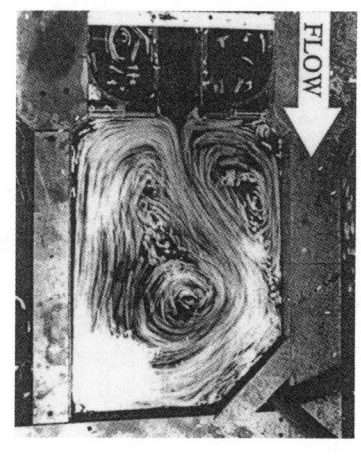

FIGURE 3  DOWNSTREAM FLOW PATTERNS

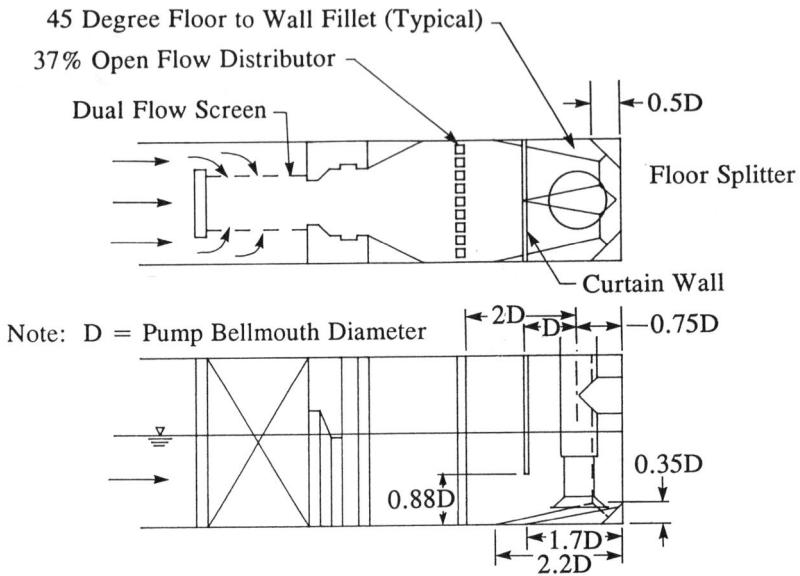

FIGURE 4  INTAKE MODIFICATION SCHEME 1

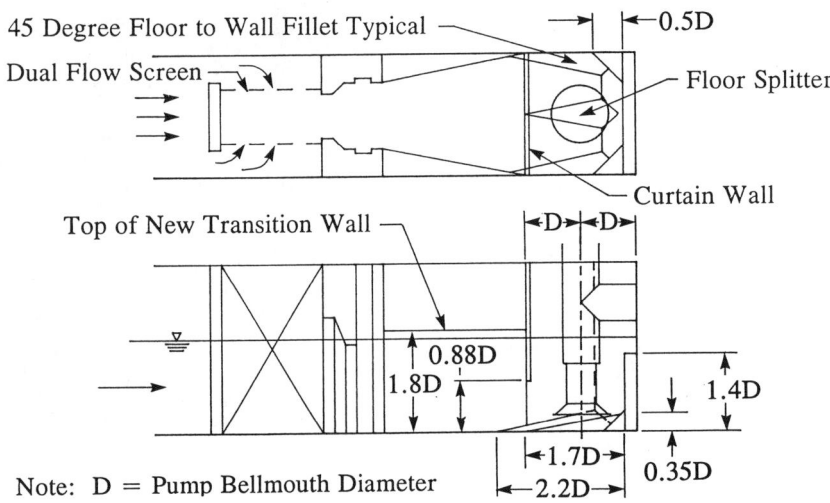

FIGURE 5  INTAKE MODIFICATION SCHEME 2

# VORTEX SUPPRESSION IN MULTIPLE-PUMP SUMPS

DEBORAH I. BAUER, TATSUAKI NAKATO, MATAHEL ANSAR
Iowa Institute of Hydraulic Research, The University of Iowa, Iowa City, IA 52242
USA

## ABSTRACT

Two laboratory investigations on rectangular pump sumps with multiple vertical pumps were conducted, one with a specific practical application in mind and another with more generic applications in mind. In the first study hydraulic modeling of an as-designed rectangular intake which consisted of seven pumps was carried out. The purpose of this model study was to identify any undesirable features of the intake design and to modify the layout, if necessary, in order to provide hydraulically acceptable pump-approach flows to individual pumps. The second study, motivated by the first one, was a generic study of flow problems associated with a rectangular pump sump with multiple pumps.

## INTRODUCTION

Although some design guidelines are available for general-purpose pump intakes, achievement of the proper design of even a single-pump intake for satisfactory pump operation is not an easy task because of various site-specific, geometrical, and hydraulic constraints. Unfortunately, engineers, who have little experience in pump-intake design or are not familiar with various pitfalls in pump-intake design, cannot visualize the undesirable hydraulic consequences of their intake design. Most common mistakes include improper sump layouts and their dimensions, sharp corners which produce stagnant flow areas, various structural obstacles within the pump sump which distort pump-approach flow distributions, inadequate pump-submergence depth, etc. There are two principal hydraulic goals which must be considered in pump-sump design. The first goal relates to a far-field pump-approach-flow distribution. The lateral distribution of the approach flow upstream from the pump must be as uniform as possible in order to avoid formation of swirling flows. The second goal relates to suppression of both free-surface and sub-surface vortices surrounding the pump bell, which is a near-field goal. These two goals appear to be quite simple. However, even experienced design engineers cannot predict exact flow distributions within the pump sump during the design phase. Therefore, it is a common practice to resort to hydraulic model testing before final pump-intake design is completed.

Despite the fact that a great deal of progress has been made recently in the numerical simulation of pump-approach flow distribution at water intakes, laboratory model studies still remain the best way to solve these problems. A comprehensive review of swirling flow problems at water intakes by Melville et al. (1994) shows that there is a considerable amount of work that has been done on the prevention and suppression of vortices at water intakes. However, most of these studies focused on vortices that form in a single-pump intake. Little is known about the prevention and suppression of vortices in multiple-pump intakes in a single sump. This paper describes two independent laboratory studies in which solutions to swirling flow and vortex problems in a single sump with multiple pumps were developed.

## LABORATORY STUDY 1 – FIX OF AS-DESIGNED INTAKE

The first study was a 1:4-scale undistorted hydraulic model of a rectangular pump intake for the East Kentucky Power Cooperative Spurlock Station, which is located in Maysville, Kentucky, and houses seven vertical pumps (Nakato 1991). These seven pumps include two papermill-makeup (P) pumps, two power-station makeup (M) pumps, and three future (F) pumps, whose column diameters are 10 in. (25.4 cm), 1 ft 4 in. (40.6 cm), and 1 ft 8 in. (50.8 cm), respectively. Four pumps, one of each P pump and M pump, and two F pumps, would be operated at the same time. The rated discharge of each P, M, and F pump was 2,600 gpm (0.16 $m^3$/s), 5,000 gpm (0.32 $m^3$/s), and 11,000 gpm (0.70 $m^3$/s), respectively. The purpose of this model study was to modify as-designed pump-sump layout to provide hydraulically acceptable pump-approach flows to individual pumps. The model was operated successively under twelve possible pump-operating combinations, and in accordance with the Froude-similarity law. Because river water from the Ohio River to the 41-ft 6-in. (12.65-m) long and 34-ft-3 in. wide (10.44-m) rectangular pump sump was designed to be withdrawn through two 54-in. (1.37-m) pipes, the initial design was not able to provide much needed, uniform pump-approach flow distributions within the sump. A number of swirling-flow and vortex-formation problems were detected during the preliminary tests which were conducted under the as-designed condition, which included large-scale swirls within the rectangular intake bay, nonuniform pump-approach-flow distributions, a high rate of prerotation in the F pumps, strong floor-attached subsurface vortices under each pump bell, strong backwall-attached subsurface vortices for each pump bell, weak sidewall-attached subsurface vortices, and weak air-entraining free surface vortices around all pump columns.

The far-field nonuniform flow distribution problem was resolved by means of three rows of staggered, 8-ft (2.4 m) tall, 6-in. (15.2 cm) by 14-in.(35.6 cm) rectangular baffle bars, which were installed 20 ft (6.1 m) upstream from the backwall of the sump, as shown in figure 1. Solutions to the near-pump field problems, including flow swirling around each pump column, interactions between pumps, formation of free-surface and subsurface vortices, were also achieved by the following fixes:

1. Flow circulation around the pumps, and interactions between adjacent pumps, were suppressed by means of six 8-ft (2.4-m) tall vertical perforated-plate dividing walls (2-in. <5.1-cm> perforations staggered on 2.5-in. <6.4-cm> centers: 58% opening) between the seven pumps, as shown in figure 1. Use of solid dividing walls caused flow separation at the leading edges of the dividing walls and increased the strengths of the free-surface vortices.
2. Floor-attached subsurface vortices were suppressed using circular grating (2-in. tall, 2-in. by 2-in. <5-cm x 5-cm x 5-cm> grating) at the inside bottom of each pump strainer, as shown in figure 2. This device was chosen because there was no space available between the strainer and the sump floor to install a floor-vortex splitter.
3. Backwall-attached vortices for M1 and M2 pumps were suppressed using two 4-ft (1.2-m) tall backwall splitters. The base of this triangular splitter was 12 in. (30.4 cm), and its depth was 3-3/4 in. (9.5 cm), as shown in figure 2.
4. The dead zone at the right corner of the sump (looking downstream), and formation of sidewall or backwall-attached vortices in the vicinity of this region, were suppressed by a 4-ft (1.2-m) tall, 8-in. (20.3-cm) by 8-in. (20.3-cm) triangular vertical corner fillet.

# VORTEX SUPPRESSION IN MULTIPLE-PUMP SUMPS

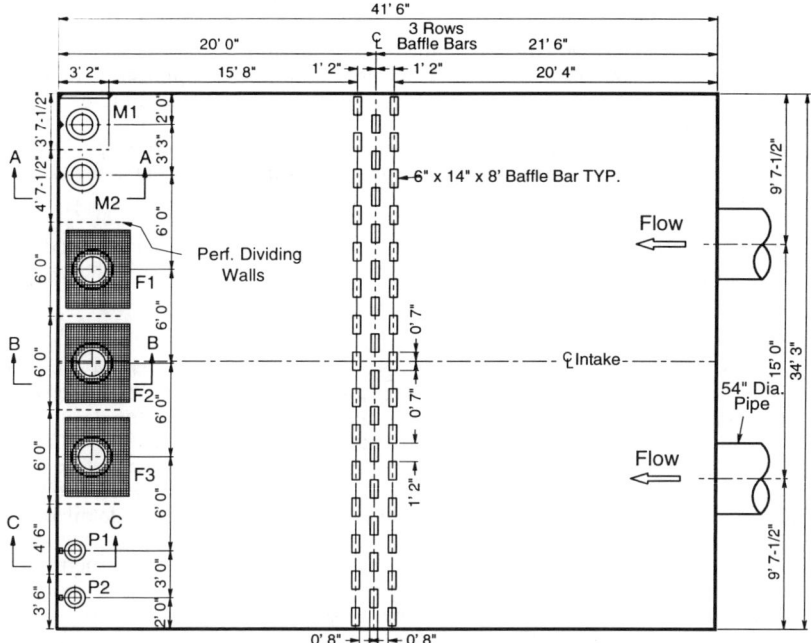

**Figure 1** A plan layout of the Spurlock Station pump-intake model and various fixes developed in Study 1

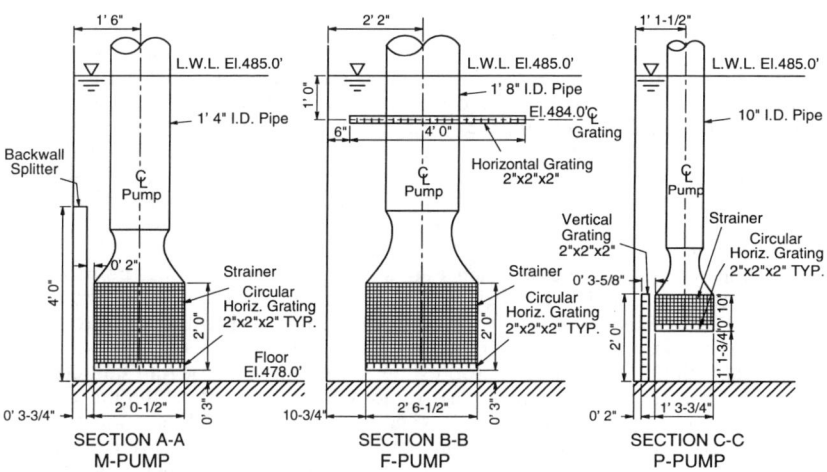

**Figure 2** Section views of M, F, and P pumps with various vortex suppressors

5. Sidewall-attached vortices on the right sidewall of the sump were suppressed by using a 4-ft 4-in. (1.3-m) long, 8-in. by 8-in. (20.3-cm by 20.3-cm) triangular horizontal floor corner fillet along the right sump floor.
6. Backwall-attached vortices near P1 and P2 pumps were suppressed by using a 16-in. (40.6-cm) wide and 2-ft (60.9-cm) tall vertical grating behind each pump bell, as shown in figure 2.
7. Air-entraining free-surface vortices around each of F1, F2, and F3 pumps were suppressed using a horizontal grating, which was installed about 1 ft (30.5 cm) below the low water level, as shown in figure 2.

The prototype pumping station built on the basis of the laboratory recommendations has been operating superbly without any maintenance problems during the past five years.

## LABORATORY STUDY 2 – GENERIC SUMP

The first study illustrates typical swirling-flow and vortex-formation problems associated with a nonuniform pump-sump inflow condition and different operating combinations of pumps in different sizes that would be encountered in a real multiple-pump sump. Unfortunately, the solutions developed for this project are not universal, and different solutions might be needed for different cases. This led to undertaking a more basic laboratory study of pump-approach-flow distributions at a rectangular water intake with four pumps (Bauer 1996). In this study, detailed time-averaged two-dimensional, planar, velocity measurements were taken within the pump sump using an electromagnetic flow meter; swirl angle in each suction pipe was measured using a vortimeter; and flow visualization surrounding individual pumps as well as the entire sump was achieved using food dye. The experimental results were analyzed and used as a basis for obtaining relatively simple modifications to the sump layout for suppressing formation of subsurface vortices.

### MODEL DESCRIPTION
A schematic of the model layout, which was built at an undistorted geometric scale of 1:16, is shown in figure 3(a). The 4-ft (1.22-m) wide and 5-ft (1.52-m) long rectangular sump consisted of four identical, 4-in. (10.2-cm) diameter, vertical pump columns equally spaced at 12-in. (30.5-cm). The pump-bell diameter was 6-1/8" (15.6 cm). The multiple-pump sump was connected to a large model basin such that the flow at the entrance of the intake was fairly uniform with no cross flow. The model was operated in accordance with the Froude-similarity law for a rated discharge of 107,000 gpm (6.8 m$^3$/s) each. Although nine different pump combinations were tested, the emphasis herein is on only two cases, i.e., Case 1 with all four pumps running and Case 2 with three pumps (Pumps 1, 2, and 3) running.

### EXPERIMENTAL RESULTS
The locations of velocity measurements are shown in figure 3. Measurements were taken at nine, seven, and five equally-spaced points, in the longitudinal, transverse, and vertical directions, respectively. Profiles of the lateral distribution of the depth-averaged streamwise velocity, U, for Cases 1 and 2 are shown on figures 4 and 5, respectively. Note that only five profiles are shown therein to avoid cluttered data points. In these figures, the streamwise velocity is normalized by the cross-section averaged velocity, $U_{av}$ (= Q/A, where Q is the total discharge withdrawn and A is the cross-sectional area). The streamwise and lateral distances x, and y, were normalized by the length, L, and the width, b, of the sump, respectively. These figures show nearly-parabolic velocity profiles developing in the first few sections of the intake. The depth-averaged streamwise velocities at the section immediately upstream of the

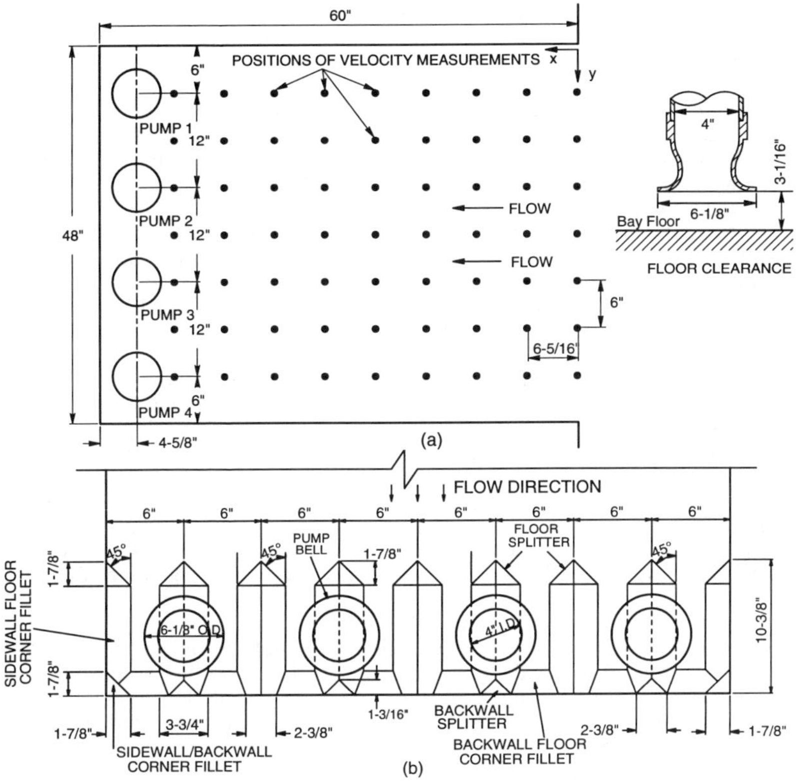

# VORTEX SUPPRESSION IN MULTIPLE-PUMP SUMPS

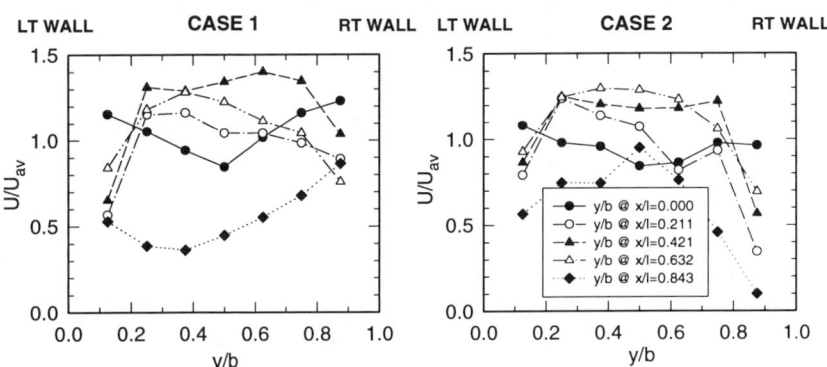

**Figure 3** A sump layout and various fixes developed in Study 2

**Figure 4** Lateral distributions of depth-averaged streamwise velocities for Case 1

**Figure 5** Lateral distributions of depth-averaged streamwise velocities for Case 2

suction pipes were noticeably smaller than the streamwise velocities at the other sections. This is due to a significant increase in the vertical components of velocity in the vicinity of the suction pipes.

Dye injection tests revealed formation of several vortices in the vicinity of the suction pipes for the sump layout tested. For Case 1, there was a strong floor-attached subsurface vortex below each pump bell. Strong backwall-attached subsurface vortices were present behind Pumps 2 and 3. Weak backwall and sidewall-attached vortices were observed around Pumps 1 and 4. For Case 2, there was a strong floor-attached vortex below each pump, as with Case 1. Around each of these pumps there was a dye core free-surface vortex and a backwall-attached subsurface vortex. The swirl angle measured in each pump column varied between 0.0 and 6.4 degrees. In both cases, pumps that were located in the center showed smaller swirl angles.

## SUPPRESSION OF VORTICES

The above results were used as a basis for developing relatively simple vortex-suppressing devices. Figure 3 (b) shows a plan view of the modified sump layout. Triangular-shaped horizontal floor splitters were placed beneath each pump bell along the axis of longitudinal symmetry, and between neighboring suction pipes. A vertical triangular-shaped backwall splitter was installed on the backwall behind each pump column. Corner fillets were placed in all corners near the pump columns, including the sidewall-backwall corners, the backwall-floor corner, and sidewall-floor corners. The details on the dimensions of these devices are given by Bauer (1996). With these modifications all the subsurface vortices were eliminated in all operating conditions.

## CONCLUSIONS

Two experimental studies, in which vortices at multiple-pump water intakes were suppressed, are described. In the first prototype-application study, sump swirling and near-pump vortices were eliminated using a combination of flow-straightening and vortex-suppressing devices. In the second study, a more basic understanding of pump-approach flow distribution was first obtained through detailed measurements of velocities in the approach flow and measurements of swirl in the suction pipes. These measurements were used as a basis to develop relatively simpler vortex-suppressing devices in a multiple-pump sump. Through more basic studies like Study 2, more generalized solutions to swirling-flow and vortex problems at water intakes could be attainable.

## ACKNOWLEDGMENTS

The first study was sponsored by East Kentucky Power Cooperative, Inc., and the second study was sponsored by the Electric Power Research Institute (EPRI) and Iowa investor-owned power utilities, EPRIa.

## REFERENCES

Bauer, D. I. (1996). "Subsurface vortex suppression in water intakes with multiple-pump sumps." *Master's Thesis*, Iowa Institute of Hydraulic Research, The University of Iowa, Iowa City, Iowa

Nakato, T. (1991). "Hydraulic model study of the water intake structure, Spurlock Station Papermill Steam Supply, East Kentucky Power Cooperative, Inc.," *Final Report*, Iowa Institute of Hydraulic Research, The University of Iowa.

Melville, B.W., Ettema, R., and Nakato, T. (1994). "Review of flow problems at water intake pump sump." *EPRI Project RP3456-01 Final Report*, Iowa Institute of Hydraulic Research, The University of Iowa, Iowa City, Iowa.

# Measurement of subsurface vortices in a model pump sump

V.P. RAJENDRAN and V.C. PATEL
Iowa Institute of Hydraulic Research, and
Department of Mechanical Engineering
The University of Iowa, Iowa City, IA 52242-1585

ABSTRACT
Experiments were conducted in a simple model pump sump with a vertical intake pipe in a rectangular channel. The geometry and flow conditions were selected to avoid free-surface vortices but generate a system of subsurface vortices attached to the floor, the back wall, and the side walls. Vortices of varying strength, core shape and size were observed. A counter-rotating pair of vortices attached to the back wall, and single vortices attached to the floor and the two side walls were found. Quantitative measurements were made using particle image velocimetry (PIV). The floor-attached vortex was found to be the strongest for the conditions studied.

## 1. INTRODUCTION

The comprehensive review by Melville, Ettema and Nakato (1994) of flow problems at water intake pump sumps shows that numerous studies have been conducted to investigate the conditions in which free-surface vortices form in pump sumps and that guidelines have been suggested by various authors and organizations for clearances between the pipe intake and the sump walls and floor, and the critical submergence of the intake, to avoid air-entraining vortices. Although subsurface vortices attached to the walls of the sump also generate swirling flow in the intake pipe and may be equally detrimental to pump performance, Hecker (1987) has noted that there are very few studies related to such vortices. Only a classification based on visual observations is made, with vortices ranging from a weak swirling mass of water made visible with dye to those with cores strong enough to entrain dissolved air into the intake.

A research program is underway at the Iowa Institute of Hydraulic Research (IIHR) to obtain quantitative information on pump sump vortices. Experiments are conducted in a simple rectangular pump sump model over a range of conditions, varying the submergence and the location of the pipe relative to the channel walls. This paper provides a glimpse of the results obtained with a single set of conditions in which there are simultaneously occurring vortices attached to the back and side walls, and the floor, in the absence of free-surface vortices. Particle image velocimetry (PIV) is used to obtain instantaneous maps of the velocity field and the data are processed to provide quantitative information about the average structure, size, shape, strength and location of the vortices.

## 2. EXPERIMENTAL ARRANGEMENT AND PROCEDURES

An outline of the experimental arrangement is shown in Figure 1. The model pump sump comprises a 0.3 m wide, 0.3 m deep, and 1.2 m long rectangular channel. The walls and the floor of the channel are made of Plexiglas to facilitate flow visualization and imaging. Water is fed to the channel from a head box, followed by a contraction and screens to provide a nearly uniform flow at the channel entrance. Water is withdrawn through a vertical intake pipe of inside diameter, d = 76 mm (3 in), with a bell mouth of diameter, D = 114 mm (4.5 in), by pump driven by a 1 hp motor, and returned to the head box. The water is thus circulated in a closed loop. The flow rate is measured with a calibrated orifice.

The particle-image velocimeter (PIV) developed by TSI Inc. is employed to obtain velocity vectors in selected planes of the flow. This system uses a SureLite™ Model Y200-1 double-pulsed 200 mJ Nd:YAG laser as the light source, a Kodak MEGAPLUS™ 1.4 camera with a resolution of 1317 x 1035 pixels and a maximum speed of 7 frames/sec, and an image and data processing software that runs on an IBM RISC 6000 workstation. Optimage™ 30-microns seed particles are used to mark the fluid. The camera, and laser are synchronized to capture double-exposed particle images. The pulse rate (separation time between the two images) could be varied from 0.5 sec to 1 micro sec. The system is capable of storing 32 consecutive images.

Each camera image is divided into a number of interrogation spots and the software uses an auto-correlation technique to find the displacements of particles. While processing double-exposed images, directional ambiguity arises in regions of flow recirculation (as in vortices) due to the symmetric nature of the auto-correlation function. This directional ambiguity is resolved using an image-shifting mirror oscillating at a prescribed frequency. The effect of this oscillating mirror is to give an artificial displacement between the first and second images of a particle. The amplitude of oscillation of the mirror is selected such that the most negative displacements are offset and all displacements are shifted to one direction. A vector is drawn between the shifted second image and the first image of a particle, and the shift or bias is vectorially subtracted before displaying the vector field. The PIV software provides a plot of velocity vectors on a specified regular grid (30 x 30, 50 x 50, etc.) within the field of view which is determined by the optical arrangement and camera location. An uncertainty of 5 percent of maximum velocity was estimated from the uncertainties in the measurement of pulse separation and particle movement.

## 3. EXPERIMENTAL CONDITIONS

As mentioned earlier, the geometry and flow conditions were selected so as to avoid free-surface vortices and generate a number of subsurface wall-attached vortices. The results presented and discussed below were obtained with the intake centered in the channel, with the center of the bell mouth located at 0.8D from the side walls, 0.4D from the back wall, 0.6D from the floor, and 10D from the channel entrance, with a submergence of 3.2D. The discharge was held constant at 0.0087 m$^3$/s, corresponding to an average velocity of 0.065 m/s in the approach channel and a pipe Reynolds number of $1.45 \times 10^5$. These conditions are also such that viscous and surface tension effects are negligible.

## 4. RESULTS AND DISCUSSION

For each vortex, 32 double-exposed PIV images were recorded at intervals of 0.5 sec. for a total duration of 16 sec.. From the instantaneous velocity vectors in the plane of view, instantaneous streamlines could be plotted and instantaneous vorticity component normal to the plane of view could be determined by differentiation. The circulation was found by identifying the grid point closest to the vortex center, and performing the usual contour integration of the velocity around a square or a rectangular closed curve around the vortex. The closed curve was chosen to include the vorticity associated with a particular vortex but to exclude contributions from neighboring vortices, if they were present. The circulation is defined as positive if the sense of rotation is counterclockwise in the field of view.

As the instantaneous measurements clearly indicated varying levels of unsteady behavior of all of the vortices, the velocities and vorticity were averaged over the 32 time steps to obtain the averages. However, these results should be regarded as preliminary to the extent that studies to determine the proper number of samples required to obtain meaningful statistics of the motion of the vortices are still in progress. As an example of the unsteady motion, Figure 2 shows the meandering of the center of the vortex attached to the floor, and its mean position denoted by ($\bar{x}$, $\bar{y}$). In the following, we shall be concerned only with the averaged results, however.

Figure 3 shows the average velocity-vector field and streamlines constructed from it across the floor-attached vortex looking up from under the channel, with the laser light sheet placed 20 mm (0.17D) above the floor. The average vorticity field (determined from the instantaneous vorticity) and the field of view are shown on the figure. The dimensions and location of the field of view relative to the channel were chosen so as to capture the floor-attached vortex. A single vortex, with a well defined core, is found. The strength of this vortex is listed in Table 1, along with that of other vortices.

Figure 4 shows the average structure of the vortices attached to the back wall as viewed from the end of the channel. The laser light sheet was placed 20 mm (0.17D) from the wall. The figure shows two counter-rotating vortices. Their strengths, listed in Table 1, are almost equal, as might be expected from the symmetry of the channel and intake. This structure was found to persistent at almost all times.

Figure 5 shows the average structure of the vortices attached to the two side walls. The laser light sheet was placed 20 mm (0.17D) from side wall 1 (Figure 5a). To visualize the vortex attached to side wall 2, the laser light sheet was placed 5 mm (0.04D) from side wall 2 (Figure 5b). It is clear that the flow is not symmetric in spite of the symmetry of the oncoming flow and intake geometry. While the discrepancy in the average strengths may be due to some asymmetry in the placement of the intake pipe or in the upstream velocity profile, there is a significant difference in the structure of the vortices on the two walls, with side wall 1 showing a single well formed vortex while side wall 2 indicating a pair of co-rotating vortices. However, the latter turns out to be an illusion created by the averaging of a single vortex that switched in time from one location in the mean to another. This last result illustrates the need for more careful examination of the temporal evolution of the vortex structure.

Table 1. Vortex Strengths

| Vortex | Circulation ($m^2/s$) x $10^3$ | |
|---|---|---|
| Floor | -9.13 | |
| Side wall 1 | -4.75 | |
| Side wall 2 | 2.01 | |
| Back wall | 2.41 | -2.3 |

## 5. CONCLUSIONS

The experiments reported here reveal a complex pattern of non-stationary vortices attached to the walls of a model pump sump. PIV was employed to reveal the structure of these vortices. Although the data have been averaged over a period of time to obtain information on the shape, strength, size, location, and movement of the vortices, further experiments and analysis of the data are required to determine the number of samples required to obtain statistically meaningful results. Such data will be useful in the guidance and validation of computational models of these complex flows.

## 6. ACKNOWLEDGMENT

This research was sponsored by the Electric Power Research Institute under the supervision of Mr. John Tsou.

## 7. REFERENCES

Hecker, G.E. (1987). "Swirling flow problems at intakes," Chapter 8, Fundamentals of vortex intake flow, *Hydraulic Structures Design Manual*, IAHR.

Melville, B.W., Ettema, R., and Nakato, T. (1994). "Review of flow problems at water intake pump sumps," *EPRI Report TR-103474 (Research Project RP3456-01 Final Report)*, Iowa Institute of Hydraulic Research, The University of Iowa, Iowa City, IA.

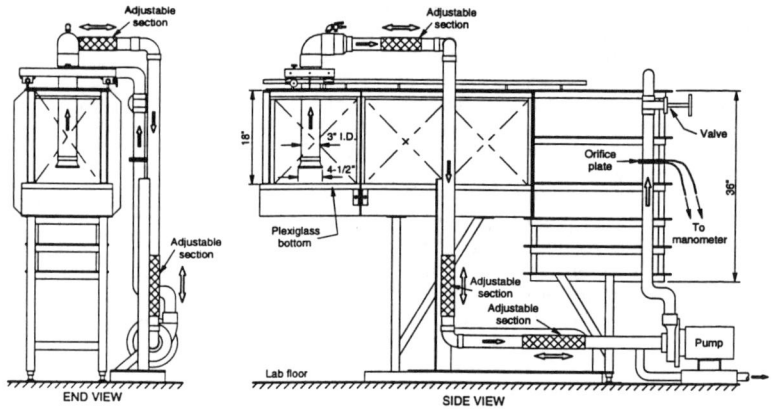

Figure 1. Schematic of the model pump sump.

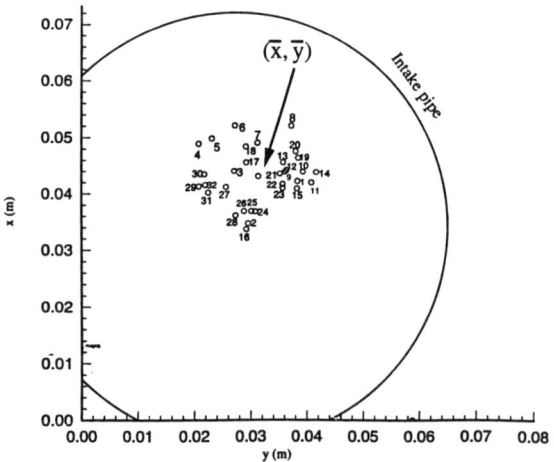

Figure 2. Meandering of floor attached vortex.

SUBSURFACE VORTICES MEASUREMENT 559

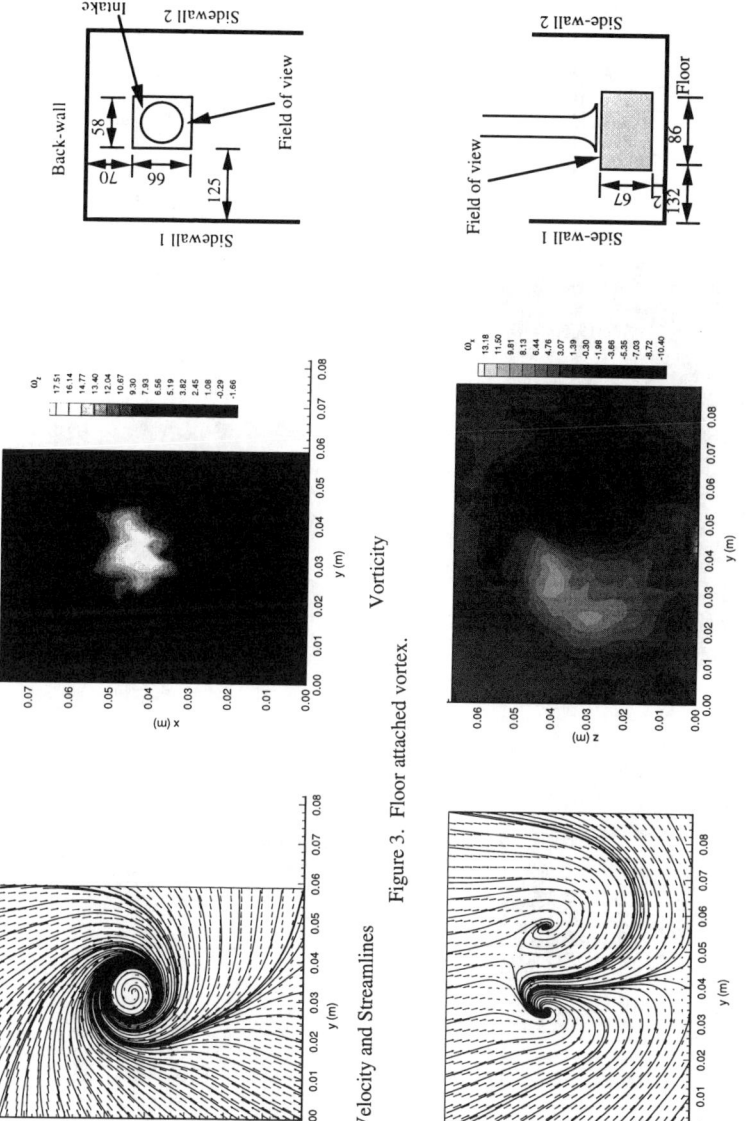

Figure 3. Floor attached vortex.

Figure 4. Vortices attached to the back-wall.

560 ENERGY AND WATER

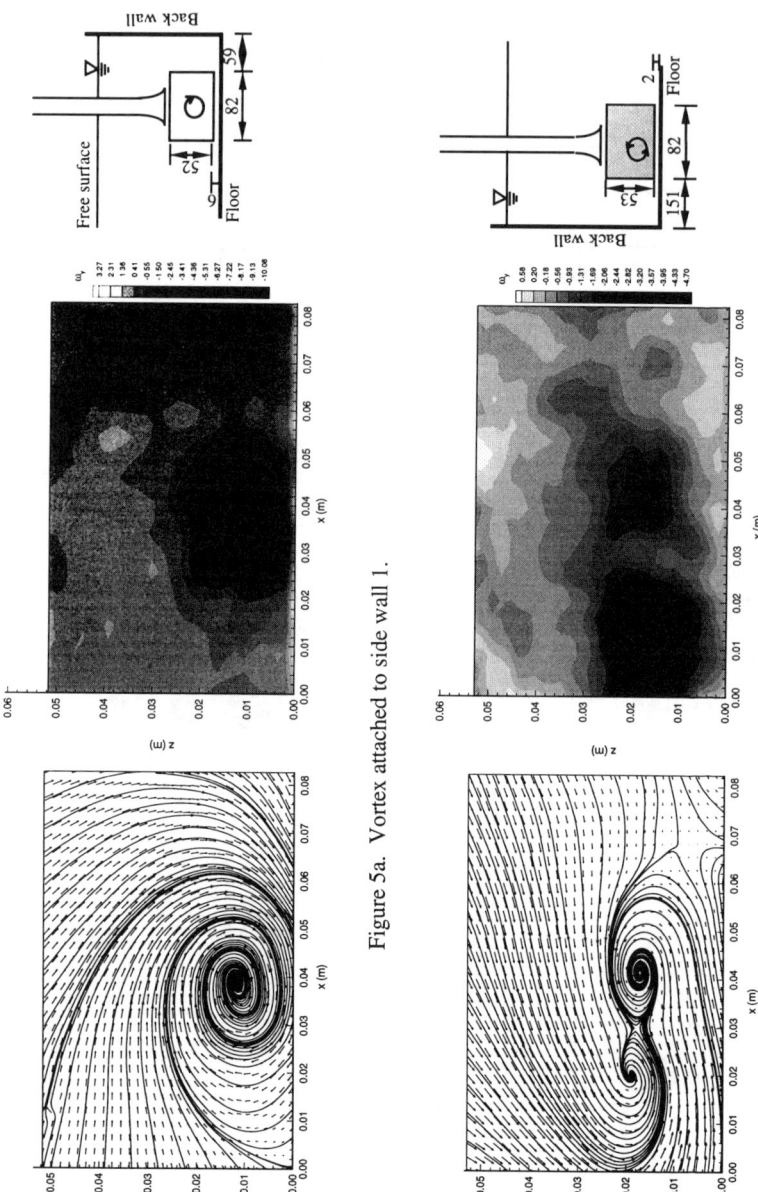

Figure 5a. Vortex attached to side wall 1.

Figure 5b. Vortex attached to side wall 2.

# A HYBRID SYSTEM OF WAVE POWER EXTRACTION AND SHORE PROTECTION

## HIDEO KONDO

Department of Civil Engineering & Architecture, Muroran Institute of Technology
27-1 Mizumoto-Cho, Muroran, 050 Japan

## ABSTRACT

A system is proposed which has capability of wave power extraction and coastal protection at erosive coasts. It consists of a high efficiency power extractor of pendulum type atop pile supporting frame structure and a submerged backwall which will act like a detached breakwater.

## INTRODUCTION

A large number of devices have been developed for wave power extraction since the 1973 oil crisis. Fields experiments to several of them have revealed few of them to be prospective. Nevertheless wave power utilization has not essentially been commercialized yet except small size navigation buoys. The reason is simply that the cost of electricity from waves is still much higher compared with those from fossil fuels and of nuclear. However, it is required to remove unfavorable gases like $CO_2$ etc. exhausted in the process of burning fuels from the standpoint of environmental protection. This process needs an additional cost which we may call the environmental cost. The cost is known presently to be no less than that of power production itself. It means when we compare the total cost which contains the environmental cost and the social cost besides the production one, the cost of wave power is close to that of fossil-fired.

On the other hand, continuous retreat of coasts by wave and meteorological tidal actions has become serious at most coastal countries in the world. Coastal engineers have toiled to it by constructing coastal structures or by feeding sand in eroding beaches. Extraction of wave power has an effect to weaken beach erosion similar to the above two methods.

In the present study the author proposes a hybrid system of wave power extraction and shore protection applying a successful power extracting device.

## PENDULOR WAVE POWER EXTRACTOR

We have studied the pendulor power extractor of bottom standing type which can be installed in upright caisson breakwaters since 1980 ( Kondo et.al, 1984 and Yano et.al, 1985 ).

A basic arrangement of it is sketched in Fig.1. It consists of pendulum plate in a concrete caisson with seaside open and a hydraulic power transmission system. Two sidewalls and the bottom plate of caisson together with the pendulum plate forms a water chamber where a standing wave system is generated by oscillation of the plate.

Laboratory and field experiments has proved it to have high efficiency and to lower the cost compared with the other types ( Watabe, 1993 ).

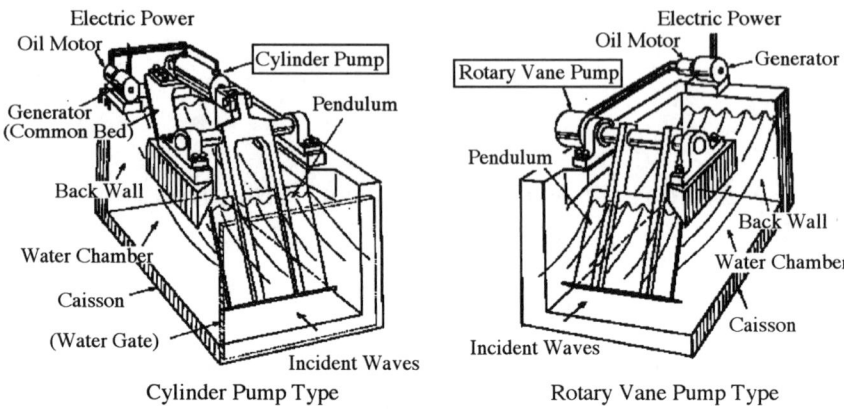

Fig.1 Pendulor Wave Power Extractor in Caisson

## RECENT TREND OF BEACH EROSION AND PROTECTION FACILITIES

Beach erosion has come serious to all coastal countries for the following reasons ;
(1) Shortage of sediment supply due to stock of river sediments by dams.
(2) Decrease of longshore transport due to long breakwaters, jetties, etc.
(3) Effect of the mean sea level rise
The first two of above mentioned have emerged severe for most of the developed coastal countries which had constructed many civil and industrial works. Meanwhile the last one does so for the developing countries with large low lands.
Several countermeasures are sought against erosion problems due to these reasons. They are;
(1) Supplying sand by filling or bypassing to eroded beaches
(2) Detached breakwaters for preventing sediments runoff offshore
(3) Groins to trap longshore sediments
Since these are expensive, we must find other effective ways to solve the problem.

## HYBRID SYSTEM AND ITS FUNCTIONS

A hybrid system proposed here can extract wave power and protect shore at erosive coasts. The system consists of Pendulor power extractor, a supporting pile frame structure for it and a submerged solid backwall to reflect shoreward waves from the oscillating pendulum plate or directly incoming oblique waves ( see Fig. 2 ).

(1) WAVE TRANSFORMATION
When incident waves act on the pendulum plate hung from the top of supporting structure, a part of it is reflected while the major part is absorbed by the oscillating motion of it with a standing wave motion generated within the water area between it and the solid backwall.
A part of the incident power is transmitted shoreward of system by allowing overtopping and

penetration through the backwall. The ratio of transmission becomes larger for higher waves. Approximate ratios of reflected, absorbed and transmitted will be designed to 20, 40 and 10 %, respectively.

(2) STRUCTURES

In the present system the main structure consists of vertical piles and horizontal beams made of steel and/or r.c. The backwall is the only solid one. No sidewalls and bottom plate contrary to the original type of which major structure is r.c. caisson.

In order to decrease the heavy wave forces during storms, the backwall may be submerged by setting its crest below sea level at the expense of higher power absorption.

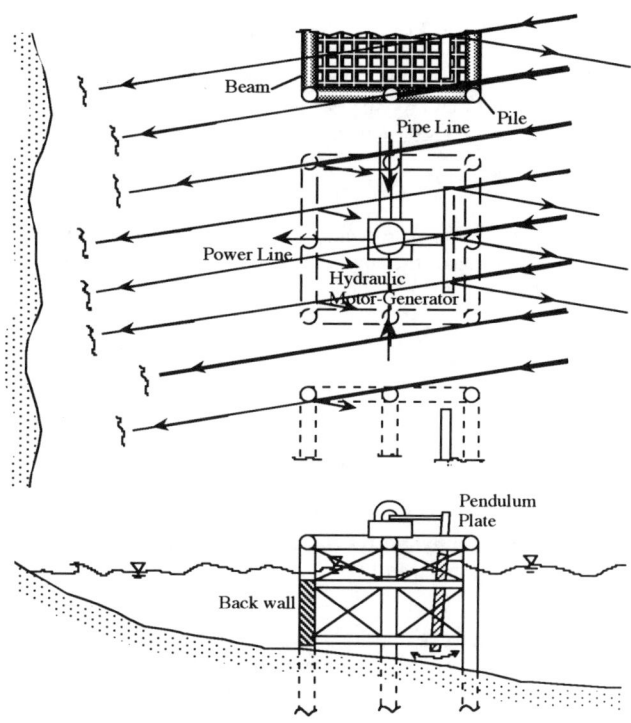

Fig.2  Concept of The Hybrid System

## (3) CONTROL OF SEDIMENT MOVEMENT

Since supply of sediment to shores from landside has been decreased recently, the countermeasures have been focused to prevent sediment runoff to offshore from inshore. The Pendulor system for caisson type can use as a detached breakwater and apparently prevent sediment runoff and protect erosive shores. On-off shore movement can be also shut out with the present system with the solid backwall parallel to shoreline.

The perfect stopping on-off motion of sediment may, however, be not always favorable, especially from the marine biological standpoint. A vertical clearance between the bottom of backwall and the ground surface may be acceptable for those situations.

Littoral drift has usually a predominant direction parallel to a coastline, interception of which will inevitably bring an accretion on updrift side and an erosion on downdrift side of structure at a sandy shores. In order to make this geometrical change small, the structure may have no side walls in principle to sustain major part of longshore sediment motion.

## COST EFFECTIVENESS OF THE SYSTEM

The author studied the cost of elecricity from wave power to Japanese coasts( KONDO,1996). In the study, the socalled total cost or true cost was was estimated besides the production or the market cost. The total cost $Pt$ (¥/ kWh) consists of the market cost $Pm$, the environmental cost $Pm$ and the social cost $Ps$ as ( Kondo ct.al, 1993 )

$$Pt = Pm + Pe + Ps \quad (1)$$

$Pm$ is estimated with conventional procedures. Defining $E_{p,j}$ as the environmental value of $j$th factor, $Pe$, the cost to repair the environmental damages occurred in power production process is estimated as,

$$Pe = \Sigma \Delta P_{i,j} = \Sigma \frac{\partial E_{p,j}}{\partial x_i} \Delta x_i \quad (2)$$

In the above equation $\Delta P_{i,j}$ is the repair cost due to the variation of $i$th environmental factor' $\Delta xi$ by the variation of the environmental factor $j$ in the course of production.

For the cases of renewable energies it can be minus when they improve the environment by power extraction. It was found that $Pm$ was about 4 times of oil-fired on the basis of interest ratio of 0.06 in Japanese coasts . However, $Pt$ was decreased to 1.4 times of that ( KONDO, 1996 ). For the case of present hybrid system $Pm$ is expected to be less than that of caisson type, and $Pt$ will possibly be less than 1.0.

## CONCLUSIONS

A hybrid system for wave power extraction and shore protection has been proposed to get clean renewable energy at the same time to prevent beach erosion. Employing Pendulor power convertor which has the highest mechanical efficiency which will result in lower cost of electricity production.

The structure has been modified from the original one of caisson installed type which act as a solid detached breakwater to pile supported porous structure except solid backwall. It thus will be less resistive to longshore sediment movement and may expect control on-offshore movement too, with a few manufacturing.

The total cost of power extracted with this system is expected to be less than that of oil-fired.

## ACKNOWLEDGMENTS

The author would like to thank Dr. Tomiji Watabe, T-Wave Consulting Volunteer for his information on hydraulic system of Pendulor. Thanks go to Mr.Noriyuki OOTA, technical staff of Muroran Institute of Technology for his assistance to prepare the manuscript.

## REFERENCES

Kondo, H. T.Watabe and K.Yano (1984): Wave power extraction at coastal structure by means of *moving body in the chamber, Proc. of 19th International Conf. on Coastal Eng.,Vol.4,* ASCE, pp. 2875-2891.

Kondo, H., S.Osanai and I. Sugioka (1993): The concept of true cost of energy and its application to ocean energies, *Proc. of International Symposium on Ocean Energy Development,* Muroran Inst. Tech. and Cold Region Port & Harbor Eng. Research Center, pp.101-106.

Kondo, Hideo (1996): Cost effectiveness of wave power extraction at erosive coasts, Proc. 25th *International Conf. on Coastal Engineering,* ASCE ( in press ).

Watabe, Tomiji (1993): Pendulor wave power convertor -15 years study and its future prospect -, *Proc. of International Symposium on Ocean Energy Development,* Muroran Inst. Tech. and Cold Region Port & Harbor Eng. Research Center, pp.101-106.

Yano, K., H.Kondo and T.Watabe (1985): A device for wave power extraction in coastal structures - Field test of a pendular system - , *Coastal Engineering in Japan,* 28. 243- 254.

# STUDY ON WAVE PUMPING-UP POWER STATION

Seiyo Shigemitsu    Akira Hiratsuka
Professor    Associate Professor
Department of Civil Engineering, Osaka Sangyo University
Daito-shi, Osaka 574 Japan

## ABSTRACT

The prospects of supplying a new electric power from the sea waves have been studied hydraulically in the case of wave pumping-up station. According to the fundamental study by simplified hydraulic model on some hydraulic phases, the station is one of the effective methods use to generate electric power indirectly from the sea waves, especially along the prominent wind waves coast.

The overtopping discharge of the station also can be estimated by use of the existing method for the case of the overtopping over an uniform slope multiplied by some modified correction factors.

## INTRODUCTION

The energy consumption by human being has been exceeded one sixtieth of the total energy of wind and waves in nature, and also will increase rapidly with the increase of the world population in near future. This will cause meteorogical variation including abnormal weather in the global environments. There are many alternative energy sources such as solar, geothermy, ocean, wind and differential thermal energy instead of the existing coal, petroleum and nuclear power generation. Between them, the type of ocean energy generation(wave pumping-up electric power generation) is adopted for the study according to our past fundamental examination.[1],[2]

The elemental layout of this pumping-up station is shown in Fig.1. The incident waves will build up and break on the complex levee which consists of uniform slope and convergent walls to increase the run-up height and overtopping discharge into a retention pool. The low head between the retention pool and the background static sea level is used to generate hydropower energy. The first writer named this type of electric generation as " Wave Current Hydro Powerplant". The depth of the construction site is considered to be near 10 meters or a little more because of economical construction cost for such a gravity type structure. The generation efficiency of this type is strongly dependent on the magnitudes of both overtopping discharge and head. The basic scheme of the pumping-up levee for the experimental study are shown in Fig.2. The convergent angle and slope are also determined by our past hydraulic model study. In this study regular waves were adopted into all performance.

# WAVE PUMPING-UP POWER STATION

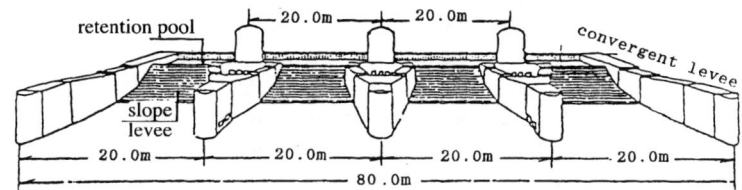

Fig.1 Layout of the wave pumping-up station

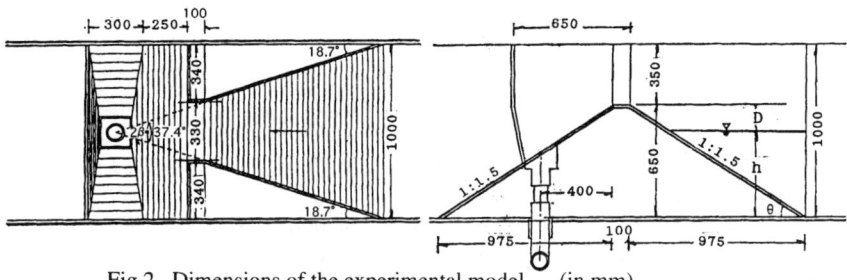

Fig.2 Dimensions of the experimental model (in mm)

By treat some arrangements and assumptions, we derived the following expressions to estimate the overtopping discharge volume per unit width and wave.[3],[4],[5]

$$Q = a \cdot b \cdot A' \cdot \left(\frac{B_c}{B_0}\right) \quad \cdots (1)$$

$$A' = \left\{\frac{1}{2} \frac{(1+\cot^2\theta)}{(\cot\gamma - \cot\theta)}(R-H_c)^2 + 0.15H(R-H_c)\right\} \quad \cdots (2)$$

$$\cot\gamma = 67\left(\frac{H}{L}\right)(\cot\theta)^{1.6}, \quad \text{for } \cot\theta \geq 1 \quad \cdots (3)$$

$$\cot\gamma = \left\{n + \frac{n(n-1)}{2}\cot^2\theta\right\}^{\frac{1}{2}}\cot\theta, \quad \text{for } \cot\theta < 1 \quad \cdots (4)$$

$$n = -3.224 \log_{10}\left\{\frac{1}{1+\left(67\frac{H}{L}\right)^2}\right\} \quad \cdots (5)$$

$$a = 7.6(\cot\theta)^{0.73}\left(\frac{H_0}{L_0}\right)^{0.83} \quad \cdots (6)$$

$H_0/L_0 < (2\theta/\pi)^{\frac{1}{2}} \cdot (\sin^2\theta/\pi)$, (non-breaking)

$$\frac{R}{H_0} = \left\{\left(\frac{\pi}{2\theta}\right)^{\frac{1}{2}} + \delta\right\}\frac{H}{H_0}\left(\frac{B_0}{B}\right)^{\alpha} \quad \cdots (7)$$

$H_0/L_0 > (2\theta/\pi)^{\frac{1}{2}} \cdot (\sin^2\theta/\pi)$, (breaking)

$$\frac{R}{H_0} = \left\{\left(\frac{\pi}{2\theta_c}\right)^{\frac{1}{2}} + \delta\right\}\left(\frac{\tan\theta}{\tan\theta_c}\right)^m\left(\frac{H}{H_0}\right)\left(\frac{B_0}{B}\right)^{\alpha} \quad \cdots (8)$$

$$H_0/L_0 = \left(\frac{2\theta_c}{\pi}\right)^{\frac{1}{2}} \cdot \left(\frac{\sin^2\theta_c}{\pi}\right) \quad \cdots (9)$$

in which, Q=overtopping volume per unit width and wave, a=discharge coefficient, b=correction factor, A'=modified spatial run-up wave profile in Takada's expression[3], Bc=width of overtopping weir, Bo=initial width of convergent walls, B=width at the end of convergent walls, R=run-up height, Hc=weir crest height from static sea level, Ho and Lo= deep water wave height and length respectively, H and L = incident wave height and length, tan $\theta$ =uniform slope of levee, $\delta$ = nonlinear term for finite amplitude wave, m and $\alpha$ =exponent (=2/3).
Then, the average overtopping discharge per unit width and wave "q" is represented by

$$q = Q/T \quad \quad (10)$$

where, T is wave period.

## EXPERIMENTAL EQUIPMENT AND CONDITIONS

The wave flume is 50m in length and 1.0m in width and height. A piston type wave generator is used to generate wave height up to 20cm and period from 0.8 to 2.5 sec. The model is located at the another end of the flume. The reduced scale is one to twenty(1/20) which is selected by our thinking scale of the prototype. The conversion ratio is according to the Froude's similarity. The experimental conditions are shown in Table-1.

Table-1 Experimental conditions

| Item | | Model | Prototype |
|---|---|---|---|
| Incident wave height | H | 0-20cm | 0-4m |
| wave period | T | 0.894, 1.342, 1.789, 2.236, 2.683sec | 4, 6, 8, 10, 10, 12sec |
| Overtoppig weir height | D=Hc | 5, 10,15,17.5cm | 1, 2, 3, 3.5m |
| Depth in front of pumping up levee h | | 60.5, 55.5, 50.5, 48.0cm | 12.1, 11.1, 10.1, 9.6m |
| Slope of pumpping up levee | | tan $\theta$ =1:1.5 | tan $\theta$ =1:1.5 |
| Angle of convergent walls | | $2\beta$=37.4° | $2\beta$=37.4° |
| Slope of sea bottom | | tan$\alpha$=0 | tan$\alpha$=0 |

## EXPERIMENTAL RESULTS AND DISCUSSIONS

The relations between the measured discharge and wave height can be represented by the following relation(11) using period as parameter. Table-2 shows their values of regression coefficients $a_o$ and $b_o$. All the discharges were converted into prototype value. All of their correlation coefficients are larger than 0.9.

$$q = a_o H^2 + b_o \quad \quad (11)$$

From Table-2, one can find out that the largest discharge occured under the wave period T=8sec and next is T=6, 10, 4, 12sec in turn for both D and h.
Along the Japan sea coast, the excellent wave periods at winter season are between 6 to 8 sec, so that the proposed geometrical dimensions of the pumping-up levee is an adequate unit for construction.
From the relationship between the observed and the calculated discharge by the

existing expression before any modification, we found that they have a linear relationship between them. Then the obtained correction factors "b" in the Eq.(1) relates to the relative depth "$h/L_0$" are shown in Fig.3. The comparison between the observed and the estimated discharge from Eq.(1) are shown in Fig.4 for the case of D=2m.

## EXPECTATION OF ELECTRIC POWER PRODUCTION

The expected electric power production per unit width of the station "P(kw/m)" for the head D=2 and 3m are shown in Fig.5. The combined efficiency of turbine and generator is assumued to be 0.64.

Fig.6 shows the variations of the expected power prodution for different head by using wave height and period as parameters.

The predominent significant wave height and period along the Japan sea coast in winter season is 1.5m and 8sec respectively, then the expected power production is about 2.7kw/m for D=2m, then the total power production will be up to 200kw for the total width of the station. The conversion rate of the wave energy to the electric energy will be up to 35%. This means that the proposed pumping-up station is one of the alternative tools for new energy production.

Table-2 Regression coefficients of the overtopping discharege

| $D_{(m)}$ | $h_{(m)}$ | $T_{(sec)}$ | $q = a_0 H^2 + b_0$ (m³/ sec / m) | |
|---|---|---|---|---|
| | | | $a_0$ | $b_0$ |
| 1 | 12.1 | 4 | 0.0872 | 0.0379 |
| | | 6 | 0.1469 | -0.0082 |
| | | 8 | 0.2090 | -0.0355 |
| | | 10 | 0.1370 | -0.0291 |
| | | 12 | 0.0909 | -0.0028 |
| 2 | 11.1 | 4 | 0.0482 | 0.0165 |
| | | 6 | 0.1020 | -0.0481 |
| | | 8 | 0.1410 | -0.0985 |
| | | 10 | 0.0741 | -0.0860 |
| | | 12 | 0.0421 | -0.0420 |
| 3 | 10.1 | 4 | 0.0227 | 0.0072 |
| | | 6 | 0.0781 | -0.0874 |
| | | 8 | 0.0885 | -0.1230 |
| | | 10 | 0.0474 | -0.1160 |
| | | 12 | 0.0255 | -0.0706 |
| 3.5 | 9.6 | 4 | 0.0175 | 0.0024 |
| | | 6 | 0.0667 | -0.1090 |
| | | 8 | 0.0735 | -0.1420 |
| | | 10 | 0.0434 | -0.1320 |
| | | 12 | 0.0201 | -0.0838 |

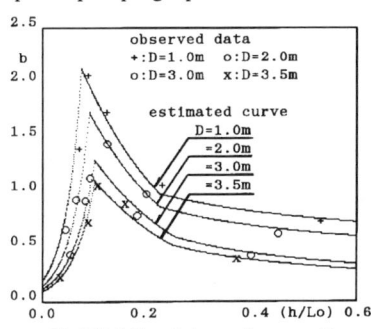

Fig.3 Relations between the correctin factor'b' and the relative depth '$h/L_0$'

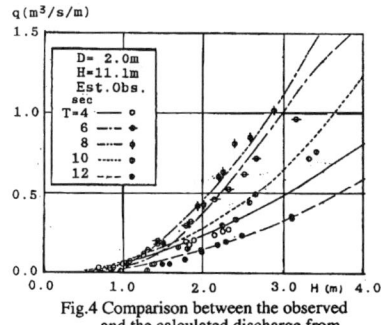

Fig.4 Comparison between the observed and the calculated discharge from the Equation (1).

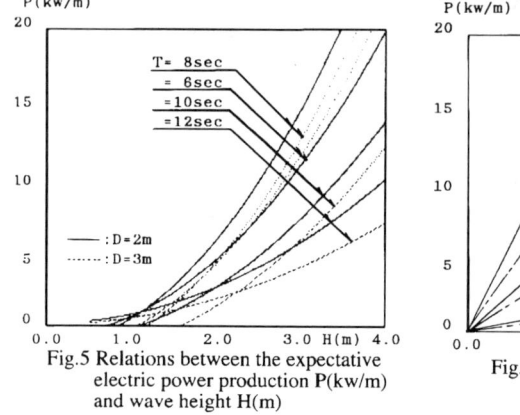
Fig.5 Relations between the expectative electric power production P(kw/m) and wave height H(m)

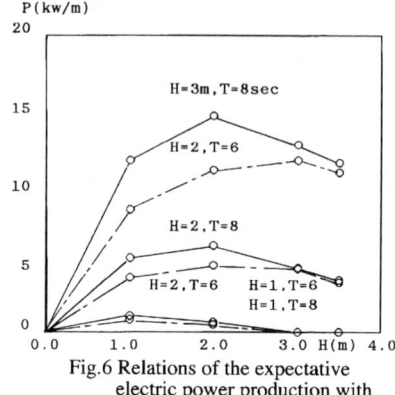
Fig.6 Relations of the expectative electric power production with the head of retention pool

CONCLUSIONS

The overtopping discharge can be estimated by the modified expression from the existing uniform slope relations. The proposed geometrical layout of the wave pumping-up station is one of the optimal figures for the meteorogical condition along the Japan sea coast. The conversion rate of the wave energy to the electric power is very high for a new energy production.

REFERENCES
1) S.Shigemitsu et al : Experimental study on wave pumping-up station, Outline paper for annual technical convention of JSCE, Vol.2, 1980.
2) S.Shigemitsu : Effectiveness of wave pumping-up power station, Outline paper for annual technical convention of JSCE, Vol.2, 1981.
3) A.Takada : Wave Overtopping Quantity Correlated to the Spatial Profile of Run-up Waves, Proceedings, JSCE, No.212, April, 1973.
4) Y. Iwagaki et al : Shoaling and deflection of finite amplitude wave theory, Coastal Engineering in Japan, Vol.28, 1981.
5) Y.Tsuchiya et al : Some considerations on water particle velocitys of finite amplitude wave theories, Costal Engineering in Japan, Vol.15, 1972.

# AN EXPERIMENTAL STUDY ON EFFECTIVE CONVERSION OF WAVE ENERGY INTO POTENTIAL ENERGY

T. KOMATSU[1], T. OKADA[1], N. MATSUNAGA[2], M. HASHIDA[3]
and K. FUJITA[1]
1) Dept. of Civil Engineering, Kyushu University, Fukuoka 812-81, JAPAN.
2) Dept. of Earth System and Technology, Kyushu University, Kasuga 816, JAPAN.
3) Dept. of Civil Engineering, Nippon Bunri University, Oita 870-03, JAPAN.

## ABSTRACT

The effective conversion of wave energy into potential energy has been investigated through laboratory experiments. In the conversion system, the structure which has a slope in front is used to activate wave overtopping and to gain much head difference. Wave overtopping is generated by progressing two-dimensional regular waves to models of the structure. The gradient of the slope, the wave parameters and the crown height have been varied over a wide range. The relations between the dimensionless wave overtopping rate and the slope have been obtained with the aid of a dimensional analysis. They are normalized by using dimensionless representative quantities, which are related to the dimensionless crown height and the wave steepness. The wave overtopping rate can be estimated by using the obtained dimensionless empirical diagrams.

## INTRODUCTION

Man-made structures such as a sea-wall and a breakwater have been constructed to protect a coastal region against big waves, storm surges, Tsunamis and so on. On the other hand, technology developments to utilize wave energy have been performed by many researchers and engineers (McCormick, 1981; Hysing, 1983; Takahashi, 1993; and so on). These techniques and studies were mainly the ones to convert the wave energy into the electric power. Therefore, in practical use, the wave-activated power generation system faces some technical difficulties, that is to say, the stable power supply, the simplification of the generation system and so on.

Our conversion system is composed of two parts. One is a structure to activate the wave overtopping and to gain effectively much water head difference. It has a slope in front. The other is a V-shaped vertical wall to converge widely distributed wave energy. This system enables us to make a unidirectional flow by using the gained head difference. That the system does not have any mechanically moving parts is also one of advantages. The unidirectional flow is useful for the water purification in stagnant waters such as a lake and a closed bay. For example, it becomes possible to discharge fresh water of an outside sea to a closed stagnant sea by connecting each other by a pipe, and also to send water of the surface layer including a large amount of dissolved oxygen (DO) into a bottom layer in which DO is lacking (see **Fig.1**). This technique may give a possibility for the development of an artificial fisher ground by discharging nutrient water in a deep sea near the surface.

Though several techniques analogous to our conversion system have been proposed by Miyae and Tesaguri (1984) and Kawamura *et al.* (1989), their estimations for the gained quantity of wave overtopping and the head difference are not so universal. The purposes of this study are to determine the form of the structure so as to gain

**Fig. 1** System to discharge sea water near a surface layer into a bottom layer.

the overtopping rate as much as possible and to estimate the overtopping rate and the head difference gained by an arbitrary form of the structure.

## EXPERIMENTAL METHOD

**Fig. 2** shows schematically an experimental set-up. The experiments were carried out by using a wave tank. It was 16 m long, 60 cm high and 25 cm wide. A 1/30-sloping bed was placed at the end of the tank. The mean water depth was 36 cm at the horizontal bed section. A model of the structure to activate the wave overtopping and gain the potential energy was fixed on the sloping bed. A wave gage was set at the position where the mean water depth was 10.6 cm. Two-dimensional regular waves made by oscillating a flap progress on the slope of the model. $H$ and $L$ are respectively the wave height and the wave length at the edge of the model, $h_C$ the crown height, $\theta$ the angle between the slope of the model and the horizontal, and $h$ the water depth at the edge of the model. The values of $H$ and $L$ were calculated from the wave height and the wave length measured by using the wave gage. The wave overtopping quantity was measured for 5 waves from 6th to 10th wave after starting to make waves in the case

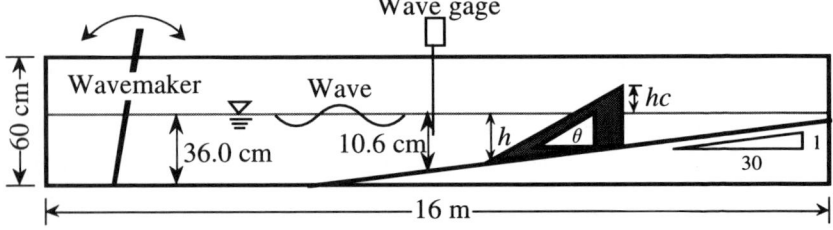

**Fig. 2** Schematic diagram of experimental set-up

# WAVE ENERGY CONVERSION

**Table 1** Experimental parameters.

| RUN | | $T$ (s) | $h$ (cm) | $L$ (cm) | $h/L$ | $H$ (cm) | $H/L$ | $hc$ (cm) | $hc/H$ | $\cot\theta$ | |
|---|---|---|---|---|---|---|---|---|---|---|---|
| 1 | a | 0.95 | 8.6 | 81.6 | 0.11 | 1.5 | 0.019 | 0.4 | 0.27 | 0.00 | 1.00 |
| | | | | | | | | 0.8 | 0.53 | 1.41 | 1.73 |
| | | | | | | | | 1.1 | 0.73 | 2.00 | 2.30 |
| | | | | | | | | 1.5 | 1.00 | 2.60 | 3.00 |
| | | | | | | | | 1.9 | 1.27 | 4.00 | |
| | b | 0.95 | 8.6 | 81.6 | 0.11 | 2.3 | 0.028 | 0.6 | 0.26 | 0.00 | 1.00 |
| | | | | | | | | 1.2 | 0.52 | 1.41 | 1.73 |
| | | | | | | | | 1.7 | 0.74 | 2.00 | 2.30 |
| | | | | | | | | 2.3 | 1.00 | 2.60 | 3.00 |
| | | | | | | | | 2.9 | 1.25 | 4.00 | |
| | c | 1.16 | 8.6 | 102 | 0.08 | 4.5 | 0.044 | 1.1 | 0.24 | 0.00 | 1.00 |
| | | | | | | | | 2.1 | 0.47 | 1.41 | 1.73 |
| | | | | | | | | 3.2 | 0.71 | 2.00 | 2.30 |
| | | | | | | | | 4.2 | 0.94 | 2.60 | 3.00 |
| | | | | | | | | 5.3 | 1.18 | 4.00 | |
| | d | 0.97 | 8.6 | 83.6 | 0.10 | 4.8 | 0.057 | 1.2 | 0.25 | 0.00 | 1.00 |
| | | | | | | | | 2.4 | 0.50 | 1.41 | 1.73 |
| | | | | | | | | 3.6 | 0.75 | 2.00 | 2.30 |
| | | | | | | | | 4.8 | 1.00 | 2.60 | 3.00 |
| | | | | | | | | 6.0 | 1.25 | 4.00 | |
| | e | 0.89 | 8.6 | 75.7 | 0.11 | 5.4 | 0.071 | 1.4 | 0.26 | 0.00 | 1.00 |
| | | | | | | | | 2.8 | 0.51 | 1.41 | 1.73 |
| | | | | | | | | 4.2 | 0.77 | 2.00 | 2.30 |
| | | | | | | | | 5.6 | 1.03 | 2.60 | 3.00 |
| | | | | | | | | 7.0 | 1.29 | 4.00 | |
| | f | 0.86 | 8.6 | 72.8 | 0.12 | 5.9 | 0.080 | 1.5 | 0.26 | 0.00 | 1.00 |
| | | | | | | | | 3.1 | 0.52 | 1.41 | 1.73 |
| | | | | | | | | 4.6 | 0.78 | 2.00 | 2.30 |
| | | | | | | | | 6.2 | 1.04 | 2.60 | 3.00 |
| | | | | | | | | 7.7 | 1.30 | 4.00 | |
| | g | 0.80 | 8.6 | 6.8 | 0.13 | 6.2 | 0.093 | 1.7 | 0.27 | 0.00 | 1.00 |
| | | | | | | | | 3.3 | 0.53 | 1.41 | 1.73 |
| | | | | | | | | 4.9 | 0.79 | 2.00 | 2.30 |
| | | | | | | | | 6.6 | 1.06 | 2.60 | 3.00 |
| | | | | | | | | 8.2 | 1.32 | 4.00 | |
| 2 | | 0.8 | 3.5 | 45.1 | 0.077 | 1.4 | 0.031 | $0.25 < hc/H < 1.4$ | | $\cot\theta = (\cot\theta)_{max}$ | |
| | | | | | | 3.0 | 0.067 | | | | |
| | | | | | | 4.0 | 0.088 | | | | |
| | | | | 5.2 | 54.0 | 0.096 | 1.3 | 0.024 | | | |
| | | | | | | 2.8 | 0.052 | | | | |
| | | | | | | 3.7 | 0.068 | | | | |
| | | | | 6.8 | 60.6 | 0.112 | 1.3 | 0.021 | | | |
| | | | | | | 2.7 | 0.044 | | | | |
| | | | | | | 3.5 | 0.058 | | | | |
| | | | | 8.5 | 66.5 | 0.128 | 1.2 | 0.018 | | | |
| | | | | | | 2.6 | 0.039 | | | | |
| | | | | | | 3.4 | 0.052 | | | | |
| | | | | 10.1 | 77.1 | 0.131 | 1.2 | 0.016 | | | |
| | | | | | | 2.6 | 0.034 | | | | |
| | | | | | | 3.4 | 0.044 | | | | |

when much quantity was obtained, or for 10 waves from 6th to 15th wave in case of a small wave overtopping quantity. The overtopping quantity per a wave and per unit width $Q$ (cm$^2$/wave) was calculated by averaging the values obtained from five measurements.

With the aid of a dimensional analysis, the dimensionless overtopping rate $Q/HL$ is expressed by

$$\frac{Q}{HL} = f\left(\frac{H}{L}, \frac{hc}{H}, \cot\theta, \frac{h}{L}\right) \tag{1}$$

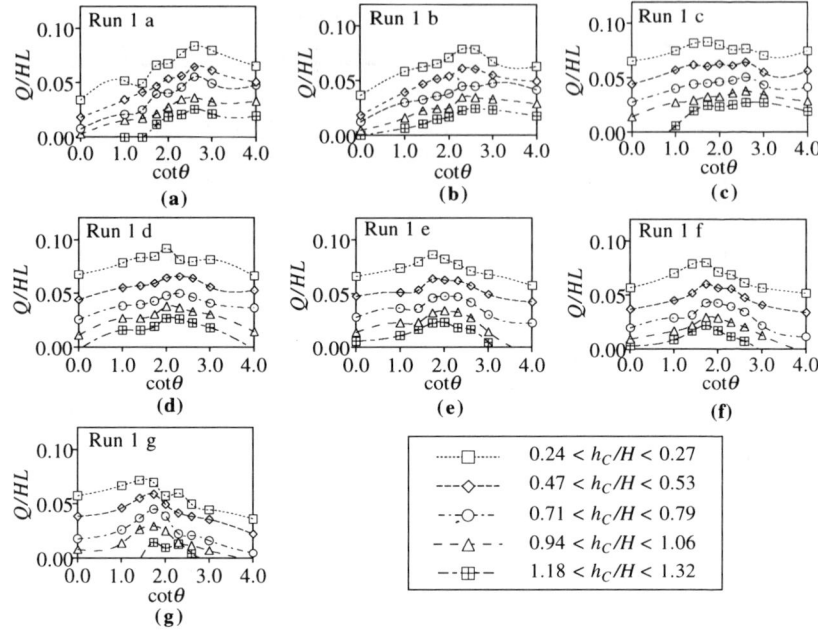

**Fig. 3** Relations between $Q/HL$ and $\cot\theta$. Figures (a) to (g) correspond to experiments of Run 1 a to Run 1 g, respectively. ( **(a)** $H/L$ = 0.019, **(b)** 0.028, **(c)** 0.044, **(d)** 0.057, **(e)** 0.071, **(f)** 0.080, **(g)** 0.093 ). Values of $h_C/H$ in Table 1 are classified by the ranges given in the legend.

The relations between $Q/HL$ and the dimensionless parameters have been investigated by varying systematically the parameters as given in **Table 1**. The experiments to determine the optimum form of the structures were carried out under the conditions of **Run 1**. The dependence of $Q/HL$ on $h/L$ was examined in **Run 2**.

## RESULTS AND DISCUSSION

**Fig. 3** shows the dependence of $Q/HL$ on $\cot\theta$. The values of $Q/HL$ increase with increase of $\cot\theta$, and decrease after taking a maximum value. As a natural result, $Q/HL$ decreases with increase of $h_C/H$.

As shown in **Fig. 4**, let us represent the maximum value of $Q/HL$ by $(Q/HL)_{max}$, $\cot\theta$ at $Q/HL = (Q/HL)_{max}$ by $(\cot\theta)_{max}$, and $Q/HL$ at $1/2(\cot\theta)_{max}$ by $(Q/HL)_*$. **Fig. 5** shows a relation normalized by using $\Phi$ and $\Psi$, where

$$\Phi = \frac{(Q/HL) - (Q/HL)_*}{(Q/HL)_{max} - (Q/HL)_*}, \quad \Psi = \frac{(\cot\theta)}{(\cot\theta)_{max}} \qquad (2).$$

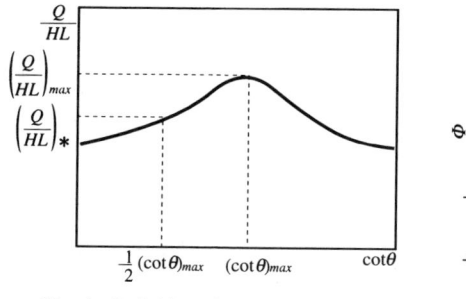

**Fig. 4** Definition of representative quantities of $Q/HL$ - $\cot\theta$ profiles.

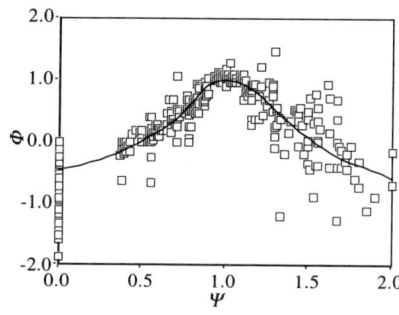

**Fig. 5** Normalized $Q/HL$ - $\cot\theta$ profiles.

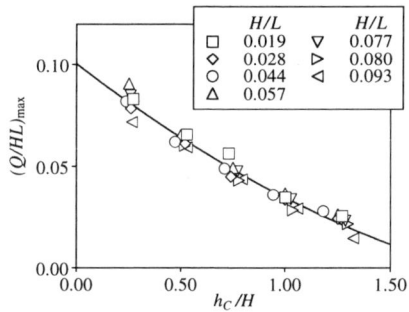

**Fig. 6** Relation between $(Q/HL)_{max}$ and $h_C/H$.

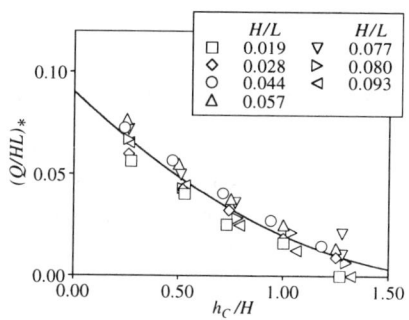

**Fig. 7** Relation between $(Q/HL)_*$ and $h_C/H$.

Though a relatively large scattering is seen in the ranges of $\Psi < 0.5$ and $\Psi > 1.3$, a good normalization is performed near the peak of $\Phi$. This normalization is favorable for us because our purpose is to gain the overtopping rate as much as possible. The solid lines drawn in **Fig. 5** and the following **Figs. 6** to **8** are obtained by the least-square fit method. The representative quantities of $(Q/HL)_{max}$, $(Q/HL)_*$ and $(\cot\theta)_{max}$ should be evaluated in order to predict the wave overtopping rate by using **Fig. 5**.

**Figs. 6** and **7** show the relations between $(Q/HL)_{max}$ and $h_C/H$ and between $(Q/HL)_*$ and $h_C/H$, respectively. Not depending on $H/L$, $(Q/HL)_{max}$ and $(Q/HL)_*$ decrease monotonically with increase of $h_C/H$. The values of $(Q/HL)_{max}$ and $(Q/HL)_*$ are dependent on $h_C/H$ only. On the other hand, the dependence on $H/L$ is seen in the quantification of $(\cot\theta)_{max}$ (see **Fig. 8**). The values of $(\cot\theta)_{max}$ decrease with increase of $H/L$. As shown by the line in the figure, $(\cot\theta)_{max}$ varies from 1.47 to 4.13 ( from 34.2° to 13.6° ) for $0.01 < H/L < 0.1$.

**Fig. 9** shows the dependence of $Q/HL$ - $h_C/H$ profiles on $h/L$. In these experiments, the values of $\cot\theta$ at which the wave overtopping rate becomes maximum for given values of $H/L$, i.e., $(\cot\theta)_{max}$, were calculated from **Fig. 8**, and were used as a gradient of the

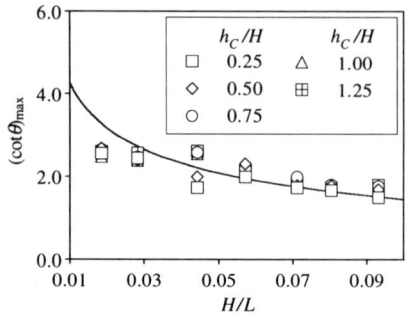
Fig. 8  Relation between $(\cot\theta)_{max}$ and $H/L$.

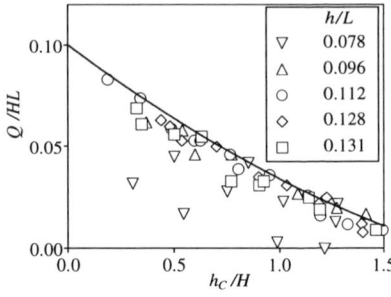
Fig. 9  Dependence of $Q/HL$ on $h/L$.

slope. Therefore, the values of $Q/HL$ in this figure mean $(Q/HL)_{max}$. The solid line is the one drawn in **Fig. 6**. The data collapse well onto the solid line without depending on $h/L$ when the wave shoaling effect is not large, that is, $h/L > 0.096$.

## CONCLUSIONS

In this study, the optimum form of the structure was determined so as to gain the wave overtopping rate as much as possible. The wave overtopping rate and the water head difference were estimated for an arbitrary form of the structure. The main results are summarized as follows.

(1) $(Q/HL)_{max}$ depends only on $h_C/H$.
(2) $(\cot\theta)_{max}$ is determined by $H/L$.
(3) $(Q/HL)_{max}$ is independent of $h/L$ when the effect of the wave shoaling is not large.
(4) The maximum overtopping rate and the optimum form of the structure can be obtained by using **Fig. 6** and **8**.
(5) The overtopping rate for various forms can be estimated by using **Fig. 5, 6, 7** and **8**.

## ACKNOWLEDGMENTS

The authors are grateful to Mr. K. Masuwa for his technical help. A grant from the Research Project of Artificial Upwelling Development and Use is also acknowledged.

## REFERENCES

1) McCormic, M. E. (1981): *Ocean Wave Energy Conversion*, A Wiley - Interscience Publication, 233p.
2) Hysing T. (1983): TAPCHAN Wave Power Plants, *Norway A S*.
3) Takahashi S. (1983): Recent Development of Wave Power Converters, *Lecture Notes of the 29th Summer Seminar on Hydraulic Engineering, JSCE*, B-1, (in Japanese).
4) Miyae S., Tesaguri N. (1993): *1st Symposium on Wave Energy Utilization*, pp.145-154, (in Japanese).
5) Kawamura T., Komatsu H., Yamamoto A., Nakano S., Mitsui H. (1989): On the Acceleration of the Wave Overetopping by V-shaped Wave Concentrator, *Proceedings of Coastal Engineering, JSCE*, pp.623 - 627, (in Japanese).

# OCEAN WAVE-POWERED DESALINATION

MICHAEL E. McCORMICK
Department of Civil Engineering, The Johns Hopkins University, Baltimore, Maryland 21218

YOUNG C. KIM
Department of Civil Engineering, California State University, Los Angeles, California 90032

## ABSTRACT

According to the World Bank, the World's supply of clean, potable water is rapidly decreasing due to many abuses of the resource. This is particularly a problem in the island communities of the World, which number more than one-hundred thousand. To supply remote communities such as these islands, a floating wave-powered reverse-osmosis (RO) desalination system is discussed. It is shown that the system can supply potable water at a cost of less than $1 per 1000 gallons when operating in a sea having an average wave height of 1.5 meter and a period of 7.5 seconds.

## INTRODUCTION

There are a number of ocean wave energy conversion systems that now show promise for the production of both electricity and fresh water for remote locations. For the production of potable water, the topic of this paper, a promising system is called the McCabe Wave Pump (MWP). The McCabe Wave Pump has been designed to produce potable water by exploiting ocean waves to directly (mechanically) produce the high pressures required for reverse osmosis. The design of the wave-powered desalination system has been performed by Dr. Peter McCabe and a team of engineers from Hydam Technologies Limited in the Republic of Ireland. That system, sketched in Figure 1, and shown in Figure 2, was deployed in the Shannon estuary in August, 1996. The MWP is ideal for supplying potable water to remote communities, such as the more than 100,000 inhabited islands in the world's oceans.

For general discussion of wave energy conversion, the reader is referred to the books of McCormick (1981), McCormick and Kim (1987), Seymour (1992) and Elliot and Caratti (1994), while hinged-pontoon wave energy conversion systems, such as the MWP, are discussed in the papers of Newman (1979), Haren and Mei (1982) and McCormick (1987), among others. In this paper, both the feasibility and economics of wave-powered potable water are discussed.

## THE MCCABE WAVE PUMP

Referring to Figure 1, the MWP system is comprised of three-pontoons, high-pressure pumps, a high-pressure delivery hose, a filtration sub-system and the reverse-osmosis (RO) desalination plant. Waves excite rotational (pitching) motions of the forward and after barges which are both hinged to the central inertial barge. The inertial barge has relatively small motions due to the

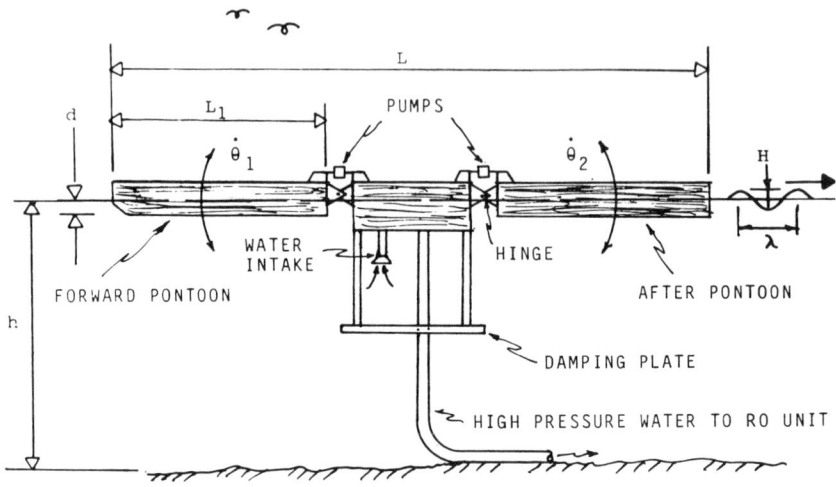

Figure 1. Elevation Sketch of the McCabe Wave Pump

Figure 2. Photograph of the MWP1 in the Shannon Estuary, Kilbaha, Ireland.

inertial-damping plate mounted below the structure. The barge motions, in turn, excite high-pressure pumps on the inertial barge, creating hydraulic pressures that can vary up to 1200 psi (8.29 x 10 N m², or 81 atmospheres). Prior to entering these pumps, the water is pre-filtered and treated on the low-pressure side of the pumps. The MWP deployed in the Shannon estuary is designed to operate most efficiently in waves having an average height of 1.5 meter and a period of 7.5 seconds. The natural period, motions of the system and impedance can be controlled by applying back-pressures on the hydraulic system.

## THE NEED

The destruction of the rain forests has resulted in changes in weather patterns throughout the world, resulting in decreased rain fall in countries that have had adequate rain. In many of the developing countries, excessive farming has resulted in either polluted water supplies or dangerously low water tables. In the contiguous United States, the use of agricultural chemicals in regions having high water tables has caused a large number of cancers to appear. In fact, the School of Public Health of the Johns Hopkins University has identified a contaminated water supply as one of the causes of the high incidence of cancer on the Delmarva Peninsula - the eastern boundary of the Chesapeake Bay. The concern of the supply of water was dramatized by the publication of a special edition of the *National Geographic Magazine* in November, 1993. The title of that special addition is "Water -The Power, Promise and Turmoil of North America's Fresh Water". In March, 1994, at a United Nations sponsored conference -UN World Water Day - in the Netherlands, Elizabeth Dowdeswell, executive director of the United Nations Environmental Program, stated that the "world water demand is now triple what it was in the 1950's, nine times the level of 1900, and is set to double again by the year 2050."

## SYSTEM PERFORMANCE ANALYSIS

The requirements for a wave-powered desalination system are determined by the RO subsystem to be used. There are a number of commercially available RO systems. These systems normally incorporate spiral-wound, semi-permeable membranes. These membrane units require both pretreatment and prefiltering of the resource water. The instantaneous power in the hydraulic system is the product of the pressure and the flow velocity in the supply line to the RO system. Double-acting pumps maintain pressures above the threshold pressure, the minimum pressure required to force the resource water through the membrane unit.

McCormick and Murtagh (1992) determine the time-averaged power available from a random sea. Applied to deep water conditions, where the depth ($h$) is greater than half of the wavelength ($\lambda$), the power available is obtained from the following expression:

$$P_{wave} = 0.051 \frac{\rho g^2}{\pi} H^2 T B_{wave} \qquad (1)$$

where $\rho$ is the mass-density of sea water (1,030 kg/m³), $g$ is the gravitational constant (9.81 m/s²), and $B_{wave}$ is a nominal crest length. The wave height, $H$, and period, $T$, in the equation are the mean values. The conversion by a MWP designed to operate at the average wave conditions should be controlled (by applying back-pressures in the hydraulic system) to satisfy impedance-matching conditions. Ideally, this condition is satisfied if the damping due to power extraction is equals the radiated power from the system. In addition, as discussed by Newman (1979), diffractive focusing will occur. The amount of energy focused can be represented by a "capture length", $\ell$, of crest. The power available can be considered to be coming from a crest length

$$B_{wave} = B + \ell \approx B + \xi L \tag{2}$$

where $B$ is the beam of the MWP, $L$ is the total length and $\xi$ is the capture-width factor. The relationship between the capture width and the length of the floating system is based on the assumption that the MWP motions are in a single even mode, from Newman (1979). This assumption is based on the fact that the center pontoon is approximately inertial. The power extracted by the system is, then,

$$P_{MWP} = \epsilon P_{wave} \approx \epsilon \, 0.051 \, \frac{\rho g^2}{\pi} H^2 \, T \, (B + \xi L) \tag{3}$$

where $\epsilon$ is the efficiency. This power is available to the pumps that both pressurize the resource water and drive the pretreatment and filtering system.

The motions of the pontoons drive high-pressure pumps. The instantaneous volume rate of flow from $n$ pumps on the $i\text{-}th$ pontoon, assuming a sinusoidal piston motion, is

$$n \, q_i(t) = n \, U_i(t) \, A_{pump} = n \omega S_{0i} \, \cos(\omega t) \, A_{pump} \tag{4}$$

where $U_i(t)$ is the time-dependent flow (piston) speed in a pump, $\omega = 2\pi/T$ is the circular frequency of the piston motion. $S_{0i}$ is the amplitude (half-stroke) of the piston motion and $A_{pump}$ is the flow area of the a single pump. The average volume flow rate from the $i\text{-}th$ pontoon is

$$Q_i = \frac{1}{T} \int_0^T q_i(t) \, dt = \frac{n}{T} \int_0^T |U_i(t)| dt \, A_{pump} = \frac{n S_{0i} A_{pump}}{T} \tag{5}$$

The transfer of energy from the waves to the pontoons to the pumps to the RO system can be theoretically analyzed. Since we have some prototype experience, we shall use observed results to determine the efficiency ($\epsilon$) of the MWP system in capturing the available wave power.

## APPLICATION TO THE MWP1

To illustrate, let us analyze the capabilities of the MWP1, deployed in the Shannon estuary in August, 1996, in 20 meters of water ($h$). Referring to the sketch in Figure 1, the floating system has an overall length ($L$) of 41.15 meters and a beam ($B$) of 4 meters. The forward and after power pontoons are 14.65 meters in length ($L_1$ and $L_2$) and draw 1.25 meter ($d$). The design sea for the system has a mean wave height ($H$) of 1.5 meter and a mean period ($T$) of 7.5 seconds. Note: This period is not equal to the natural pitching period of a pontoon. Because of the power takeoff, the pitching motions are heavily damped. From the motions analysis, the undamped natural period is found to be approximately 4 seconds. The damped natural period under the operational conditions with the RO system attached is approximately 7.5 seconds.

In late September, 1996, following a long period of calm water at the location in the Shannon estuary, 500 meter offshore of Kilbaha, Ireland, a 2.0-meter, 7.0-second sea arose. The mean deep-water wavelength value for this sea, assuming a Rayleigh probability distribution of the wavelength, is approximately

$$\overline{\lambda} \approx 1.49 \overline{T}^2 = 73.0 \ meters \tag{6}$$

For the MWP1 geometry, the value of the capture-width factor ($\xi$) for the wavelength-to-system length ratio is approximately 0.215, from Newman (1979). The power available ($P_{wave}$) to the MWP1 in this sea is, then, approximately 579 kW from the results in equation (1).

A measure of the power captured by a component barge is the power represented by the pitching inertia of the pontoon. For the *i-th* pontoon, that power is

$$P_i = \frac{4}{T} \int_0^{T/4} I_{hinge} \frac{d^2\theta}{dt^2} \frac{d\theta}{dt} dt = \frac{4}{T} \int_0^{T/4} I_{hinge} \Theta_{0i}^2 \omega^3 \cos(\omega t) \sin(\omega t) \ dt = 6.70 \ x \ 10^6 \ \Theta_{0i}^2 \tag{7}$$

where $\Theta_{0i}$ is the pitching amplitude of the *i-th* pontoon. For either pontoon, the total mass moment of inertia with respect to the hinges ($I_{hinge}$) is approximately $2.91 \times 10^7$ n-m-s², including the added-mass moment of inertia. Results of the observed pontoon motions of the MWP1 in that sea ($H = 2.0 \ m$, $T = 7.0 \ s$) showed that the vertical amplitude of the free end (bow) of the forward pontoon was approximately 3.5 meters, while that of the free end (stern) of the after pontoon was about 1.5 meter. These respective values correspond to pitching amplitudes of $\Theta_1 = 0.220 \ radian$ and $\Theta_2 = 0.096 \ radian$. From the results of equation (7), the converted power valued of the respective pontoons are $P_1 = 324 \ kW$ and $P_2 = 62.0 \ kW$ and the total power available to the pumps is 386 kW. This represents a conversion efficiency of $\varepsilon = 66.7\%$, based on the ideal wave power available (579 kW).

The efficiency value of 66.7% resulting from observations in a 2.0-meter, 7.0-second sea is used to determine the performance of the MWP1 in a design sea of 1.5 meter and 7.5 seconds. There are six pumps on the MWP1, three on each pontoon, located 1.5 meter above the hinges and 15.15 meters from the free ends of the pontoons. These pumps are part of a closed hydraulic system designed to both control and monitor the pontoon motions. When producing water, these pumps are to be replaced with high-pressure water pumps. The diameter of each of these pumps is 0.3 meter. For the design sea, the average wavelength is 83.8 meters, and the wavelength-to-system length ratio is 2.06. From Newman's (1979) results, the capture-width factor is $\xi \approx 0.320$. Hence, from equation (1), the available wave power is 466 kW. For the 66.7% capture efficiency, then, the MWP1 absorbs 311 kW. For the design sea, the pontoon power expression of equation (7) becomes

$$P_i = 5.45 \ x \ 10^6 \ \Theta_{0i}^2 \tag{8}$$

By comparing the captured power in the design sea (311 kW) with that in the 2.0-meter, 7.0-second sea (386 kW), we see a 19.4% reduction in the design sea. The application of this power reduction percentage to each pump results in a motion power value of 261 kW for the forward pontoon and 50.0 kW for the after pontoon. By using these values in equation (11), we find that the respective pitching amplitudes of the pontoons are 0.219 radian and 0.096 radian. The corresponding piston half-strokes are 0.334 meter and 0.144 meter, respectively, resulting in flow rates in each of the (0.0707 $m^2$ flow area) pumps of 0.0126 $m^3$/s and 0.0054 $m^3$/s. The total flow rate in the supply line to the RO system from the six pumps (three on each pontoon) is 0.054 $m^3$/s, or 14.3 US-gallons per second. Assuming half of this resource water becomes potable water, the daily potable water production is approximately 618,000 gallons. If the total flow rate of 0.054 $m^3$/s is against an average RO system pressure of 4.14 x $10^6$ $N/m^2$ (600 psi), the power required for the water production is 224 kW, leaving 87 kW to produce electricity, if desired.

## ECONOMICS OF WAVE POWERED DESALINATION

The total cost of a MWP desalination system, including the RO unit, is approximately $1,600,000 based on 1996 dollars. The maintenance and monitoring of the system is estimated to be approximately $20,000 per year. Assume that the system is amortized over ten years. The cost per thousand gallons of potable water is approximately $0.80, while the cost of energy production is $0.066 per kW-hr. Both of these costs are quite reasonable in developed, non-remote areas.

## REFERENCE

Elliot, G. and G. Caratti, editors (1994), Proceedings, 1993 European Wave Energy Symposium, National Engineering Laboratory, East Kilbride, Scotland

Haren, P. and C. Mei (1982), "An Array of Hagen-Cockerell Wave Power Absorbers in Head Seas", Applied Ocean Research, Vol. 4, No. 1, pp. 51-56.

McCormick, M.(1981), Ocean Wave Energy Conversion, Wiley-Interscience, New York.

McCormick, M. (1987), "A Normal-Mode Analysis of the Critically Damped Motions of Hinged Barges in Regular Long-Crested Waves", J. Ship Research, SNAME, Vol. 31, No. 2, pp. 91-100.

McCormick, M. and D. Hayden (1995), "Recent Advances in Wave Energy Utilization", Proceedings, 24th American Towing Tank Conf., November, College Station, TX, pp. 125-133.

McCormick, M. and Y. Kim, editors (1987), Utilization of Ocean Waves, ASCE, New York

McCormick, M. and J. Murtagh (1992), "Large-Scale Experimental Study of the McCabe Wave Pump", U. S. Naval Academy Engineering Report EW-03-92, January.

Mei, C. (1983), The Applied Dynamics of Ocean Surface Waves, John Wiley & Sons, New York.

Newman, J. (1979), "Absorption of Wave Energy by Elongated Bodies", Applied Ocean Research, Vol. 1, No. 4, pp. 189-196.

Seymour, R. J., editor (1992), Ocean Energy Recovery, ASCE, New York.

# CHARACTERISTICS OF FLOW CONDITIONS ON STEPPED CHANNELS

IWAO OHTSU and YOUICHI YASUDA
Dept.of Civ.Engrg., Nihon Univ., Coll.of Sci.& Tech.
Kanda Surugadai 1 - 8, Chiyoda-ku, Tokyo, 101, Japan

ABSTRACT   The flows over stepped channels are classified into three types: skimming flow, nappe flow, and transition flow. The hydraulic condition for the formation of each type is determined. Also, each flow type is characterized by the pressure at the step corner. For the skimming flow, characteristics of an eddy near the step corner are clarified from the velocity distribution on the step.

## INTRODUCTION

Stepped chutes have been used for dissipating the energy of supercritical flows. Recently, out of concern for water environments, stepped flows have also been used for improving water quality and in planning river landscapes. The flow conditions of stepped channels have been classified into the skimming flow [Fig. 1(a)] and the nappe flow [Fig. 1 (b)]. The hydraulic conditions required to predict the boundary between skimming and nappe flows have been reported by many researchers [Chanson(1994); Essery and Horner(1978); Montes(1994); and Rajaratnam (1990)]. However, the general agreement for the boundary has not been obtained, because a transition flow might exist between these two flow conditions. Furthermore, the pressure and velocity characteristics of the stepped flow have scarcely been shown. This paper discusses the hydraulic conditions for the formation of skimming and nappe flows. The pressure and velocity characteristics of the stepped flow have been studied on the basis of a systematic experiment.

## FLOW CONDITIONS

The flow condition of a stepped channel depends on the channel slope $\alpha$, the step height s, and the discharge-per-unit width q. In the region of a quasi-uniform flow, the stepped flow is characterized as follows:
The flow with the formation of an eddy as shown by Fig. 1(a) is called a skimming flow, and the flow with the formation of an air pocket shown by Fig.1(b) is called a Nappe flow. Between the skimming and the nappe flows, there exists the transition flow in which the air pocket is formed in some steps, whereas in the others the cavity is totally filled with water. Considering the mechanisms of the turbulent diffusion and the energy dissipation for each flow condition, the validity of the classification might be understood. The hydraulic conditions

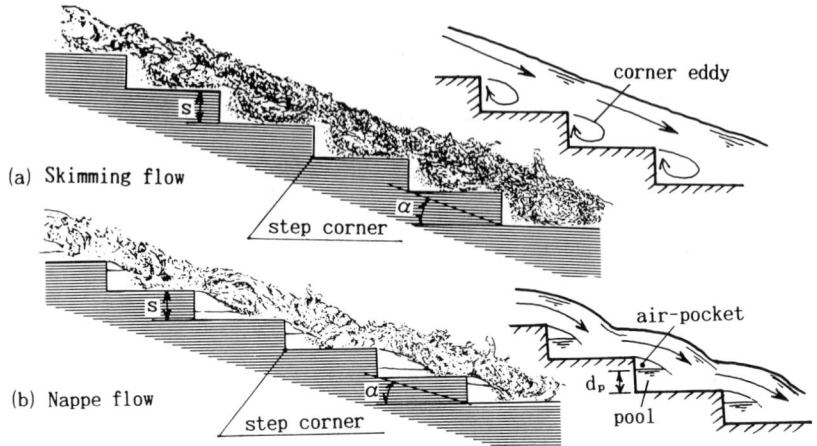

(a) Skimming flow

(b) Nappe flow

Fig. 1 Flow condition of stepped flow

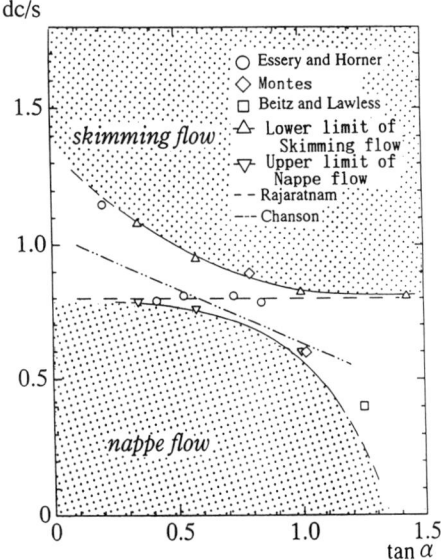

Fig. 2 Classification of flow conditions

for the formation of each flow condition are shown in Fig. 2. In this figure, the solid line indicates the upper or lower limit of the transition-flow region. Considering the existence of the transition flow, the boundary between the skimming and nappe flows proposed by other researchers [Beitz and Lawless (1992); Chanson (1994); Montes (1994);

Essery and Horner (1978); and Rajaratnam (1990)] is located almost in the region of the transition flow, and the reason why the general agreement is not obtained can be understood. In addition, for $\tan\alpha \geq 1.4$ ($\alpha \geq 55°$), the main flow does not always impinge on the step because of the steep slope. Even for small discharge (at least $d_c/s = 0.18$), an air pocket is formed only in some steps. From this, a transition flow is formed for $0.18 \leq d_c/s \leq 0.8$ and $\tan\alpha \geq 1.4$.

## PRESSURE AT STEP CORNERS

Figs. 3, 4, and 5 show the pressure at each step corner for the skimming and nappe flows. In these figures, p= pressure at step corner, n = step number, counting from the first step; $H_{dam}$ = dam height (see Fig. 6); $d_c$ = critical depth ($d_c = \sqrt[3]{q^2/g}$); and N = number of steps.

(1) SKIMMING FLOW

According to the results of the experiments, the main flow touches the step, and the curvature of the stream line passing over the step is not always constant, even if the flow becomes a quasi-uniform flow. As shown by Fig. 3, the pressure p alternately changes for each step corner. In the case of a skimming flow, such an alternate change of pressure occurs, even if the water surface does not seem to change alternately. In addition, for $\tan\alpha \geq 1.4$ ($\alpha \geq 55°$), because the main flow does not always touch the step regularly, the pressure irregularly changes for each step corner.

(2) NAPPE FLOW

The main flow always impinges on the step, and the surface profile depends on the step shape as shown by Fig.1 (b). When the surface profile alternaterly changes for each step, the pool depth below the air pocket changes correspondingly. Simultaneously, the pressure at the step corner changes as shown by Fig.4. Then the pressure at the corner point becomes hydrostatic, and p/w is equal to the pool depth $d_p$ (Fig. 5).

## VELOCITY CHARACTERISTICS OF SKIMMING FLOW

Figs. 7, 8, and 9 show distributions of velocity and turbulent intensity for a skimming flow. In these figures, $\bar{u}$, u' = mean velocity and turbulent intensity for direction x; $\bar{v}$, v' = mean velocity and turbulent intensity for direction y; and $V = \sqrt{\bar{u}^2 + \bar{v}^2}$. The velocity was measured by using a one-dimensional fiber laser-doppler velocity meter (FLV) [15 mW He-Ne; sampling frequency: 25 Hz; sampling time: 164 s; and number of data: 4,100], and the velocity measurement was taken at the upstream position of the start of air entrainment. Fig. 7 shows that a remarkable eddy is formed near the step corner, and that the curvature of the stream line passing over the step is not constant for each step. Also, as shown by Figs. 8 and 9, the turbulent intensities for the x and y -directions are high near the boundary between the main flow and the corner eddy. In addition, for the nappe-flow case, such an eddy is not formed, and the velocity in the pool is negligibly small.

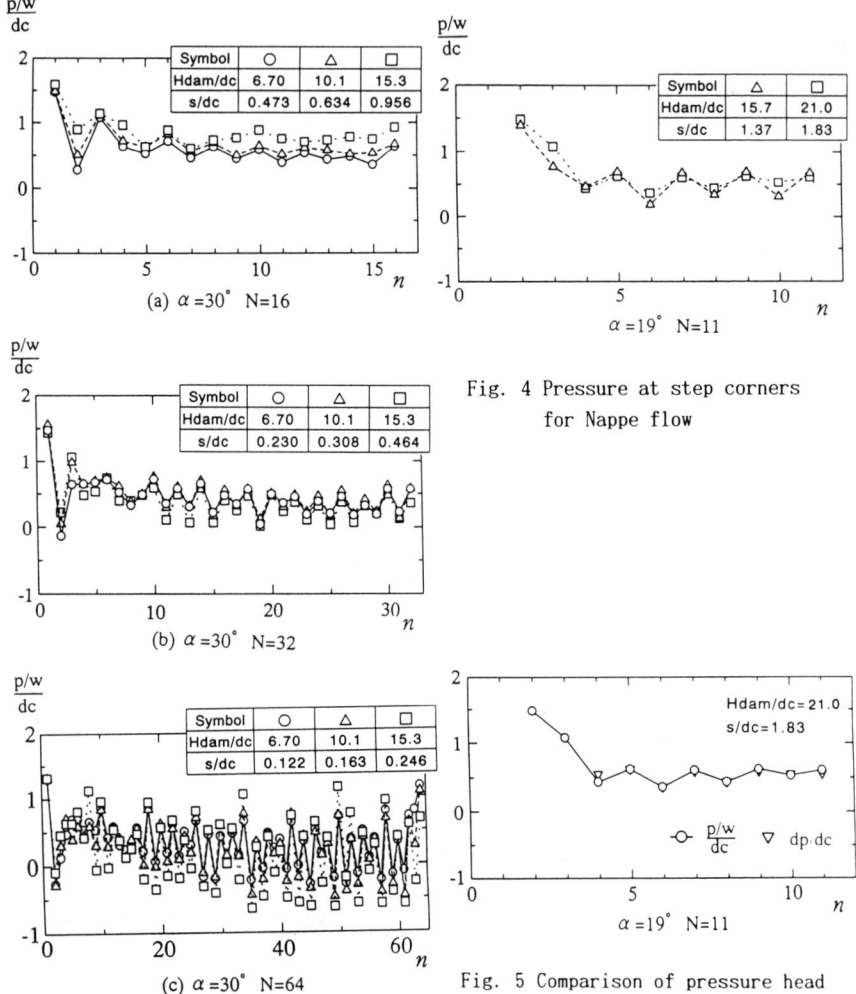

Fig. 3 Pressure at step corners for Skimming flow

Fig. 4 Pressure at step corners for Nappe flow

Fig. 5 Comparison of pressure head at step corner with pool depth (Nappe flow)

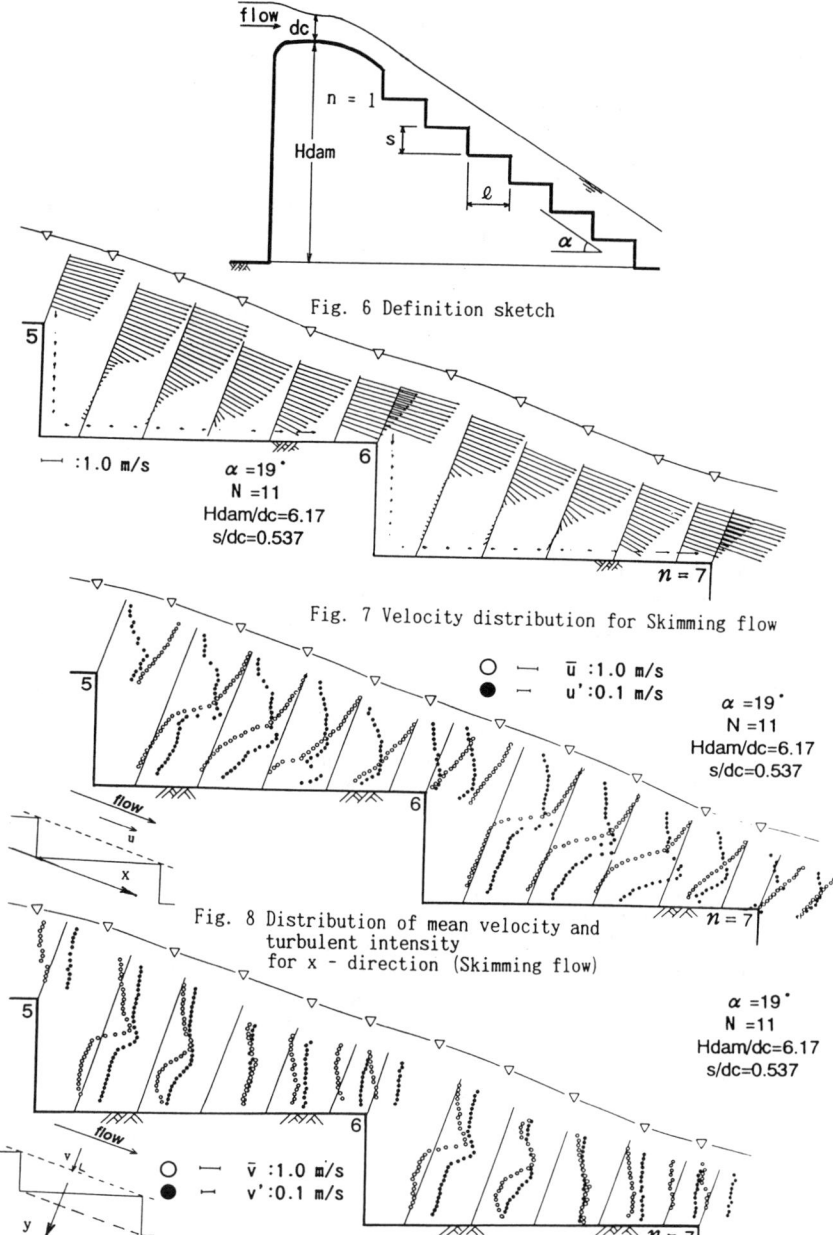

Fig. 6 Definition sketch

Fig. 7 Velocity distribution for Skimming flow

Fig. 8 Distribution of mean velocity and turbulent intensity for x - direction (Skimming flow)

Fig. 9 Distribution of mean velocity and turbulent intensity for y - direction (Skimming flow)

## CONCLUSION

Stepped channel flows were systematically investigated, and the results are summarized as follows:
(1) The flow conditions of stepped flows are classified into 3 flow conditons (i.e., skimming flow; nappe flow; and transition flow). Also, the hydraulic conditions required to form skimming and nappe flows have been clarified.
(2) The pressure at the step corner of the nappe flow is characterized by the formation of an air pocket, and the pressure for the skimming flow is characterized by the formation of an eddy near the step corner.
(3) The velocity distribution of a skimming flow has been shown, and the characteristics of an eddy near the step corner have been made clear.

## ACKNOWLEDGEMENT

This study has been supported by Nihon University President's Grant (University Research Center, Nihon University).

## REFERENCES

Beitz, E. and Lawless, M. (1992) "Hydraulic model study for dam on GHFL 3791 Isaac River at Burton Gorge," Water Resources Commision Report, Ref. No REP/24.1, Sept., Brisbane Australia [see reference of Chanson (1994)].

Chanson, H. (1994) "Comparison of energy dissipation between nappe and skimming flow regimes on stepped chutes," Jour. of Hydr. Res., IAHR, Vol. 32, No. 2, pp. 213-218.

Essery, I. T. S. and Horner, M. W. (1978) "The hydraulic design of stepped spillways," CIRIA Report No. 33, 2nd edition, Jan, London, UK.

Montes, J. S. (1994) see the references of Chanson (1996) [JHR, IAHR, Vol. 34, No. 3, pp. 421-429]; and Moacyr and Andrél (1995) [JHR, Vol. 33, No. 1, pp. 119-122].

Ohtsu, I. and Yasuda, Y. (1995) Discussion of "Comparison of energy dissipation between nappe and skimming flow regimes on stepped chutes," Jour. of Hydr. Res., IAHR, Vol. 33, No. 1, pp. 115-119.

Rajaratnam, N. (1990) "Skimming flow in stepped spillways," Jour. of Hydr. Engrg., ASCE, Vol. 116, No. 4, pp. 587-591.

# INITIATION OF AERATION IN STEPPED SPILLWAYS

Cristobal MATEOS IGUACEL, Director,
and Víctor ELVIRO GARCIA, Director of Programe,
Hydraulics Laboratory,(CEDEX), SPAIN.

## ABSTRACT

Determining the natural aeration inception zone is an important factor in preventing cavitation. This article presents experimental work carried out on the initiation of aeration in stepped spillways, which was performed in the Hydraulics Laboratory (CEDEX) Madrid (Spain).

## 1. INTRODUCTION

High negative pressures that can cause serious cavitation damage to high-speed hydraulic structures often occur in some engineering projets. It is well known to all that one of the most effective ways of preventing cavitation is by introducing air into the water flow, thereby achieving a fluid which is a mixture of water and small air-bubbles in such a way that the proportion of air is greater than 8% close to the spillway face. For this purpose the air can either be inducted into the water mass naturally or artificially.

The development of the boundary layer in a stepped spillway is shorter than that of a conventional spillway, owing to the high flow turbulence, and once the inception point is reached self-aeration takes place. So a high air-content is maintained in the fluid mass.

This paper provides an account of the experimental work conducted by the Hydraulic Laboratory (CEDEX), regarding the initial aeration position in stepped spillways.

## 2. THE CONFIGURATION OF FLOW AERATION IN A STEPPED SPILLWAY

If one observes the flow in a scale model of a stepped spillway from the free surface side, a clear and transparent flow can be seen, suddenly followed by a white, flat "U"-shaped flow, i.e., the aeration first begins close to the side wall, but it develops virtually at the same elevation in the rest of the spillway. If the flow is observed from the side or the bottom, a transparent zone can be discerned that progressively reduces its thickness so that air bubbles appear on a certain step, which are confined by the centrifugal forces of the corresponding eddy. At a distance downstream from this cell, function of the unit discharge,the entire flow fills up with air bubbles, considerably increasing the depth of the water . In the observations carried out, the average elevation of the first step where the bubbles remain confined was taken as the aeration inception point, given that this elevation is the same as the white line marking the initiation of aeration that can be seen in the flow, observed from the free surface side.

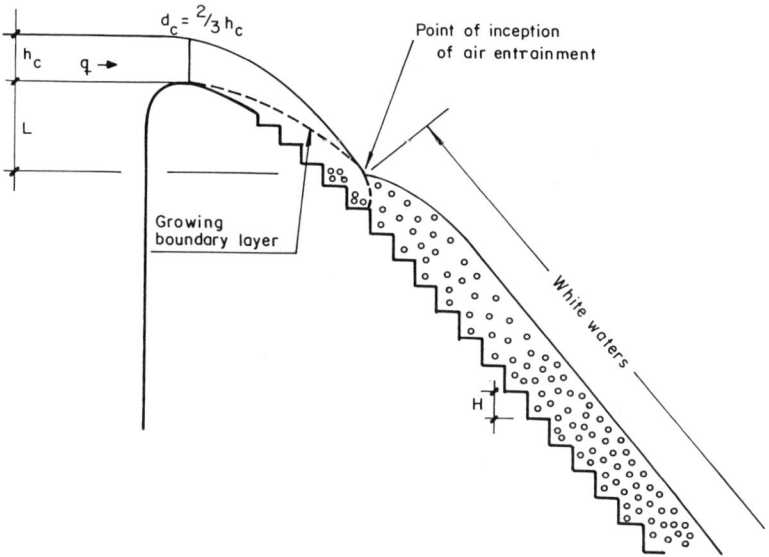

Fig. 1 Air entrainment on stepped Spillways

When quantifying the results, the aeration inception point is located by the difference between its elevation and the crest, instead of using the length throughout the spillway, as is usually done. This is justified in such a case, on the one hand because it is the number of jumps (and not that the tread is a little bit larger or smaller) which is the parameter most closely linked to turbulence development, and on the other hand because it allows the flow speed for the elevation at which aeration commences to be obtained more directly, permitting the cavitation risk to be estimated by determining a correlation in the face of the velocities and the possible existence of negative pressures.

Stepped spillways have a highly turbulent flow, while nappe instability brings about impacts on the step which is conducive to air entraintment. This air remains on the step thanks to the horizontal axis vortices induced by the step shape. At the same time, this considerable agitation makes up sufficiently for the effects of the pressure gradients present in the triangular cell between the vertical and horizontal faces which tend to separate the bubbles from the face.

## 3. MODELS USED

The data presented, were obtained from 10 scale models of 6 Spanish dams, together with 3 models of different scales of one single typified structure, all of which were constructed at the Hydraulics Laboratory (CEDEX).

All the models were of dams with 0.90 m. or 1.20 m. steps, and the scales used ranged from 1/6 to 1/25. With the exception of one arch dam having a 0.333/1 slope, all of them were gravity dams with slopes ranging from 0.75/1 to 0.8/1. Unit discharges of 40 m$^3$/sec. m. were experimentally achieved in the prototypes, although this value is generally well above the maximum design discharges for this type of spillway.

## 4. PRESENTATION OF THE RESULTS. PROPOSED ADJUSTMENT

As has already been stated, the criterion adopted for fixing the elevation at which aeration commences, was to determine the highest step on wich the air bubbles remain stable. "L" is the elevation difference between the spillway crest and the central point on this step, and "H" is the height of the steps in the chute. The unit discharge is "q" and "g" is the gravity acceleration. The data are presented in Fig. 2, using the non-dimensional terms, L/H and $q/(g^{1/2}.H^{3/2})$.

As can be seen an excellent adjustment is obtained with the following equation:

$$\frac{L}{H} = 5.6 \left( \frac{q}{g^{1/2} H^{3/2}} \right)^{0.8} \qquad (1)$$

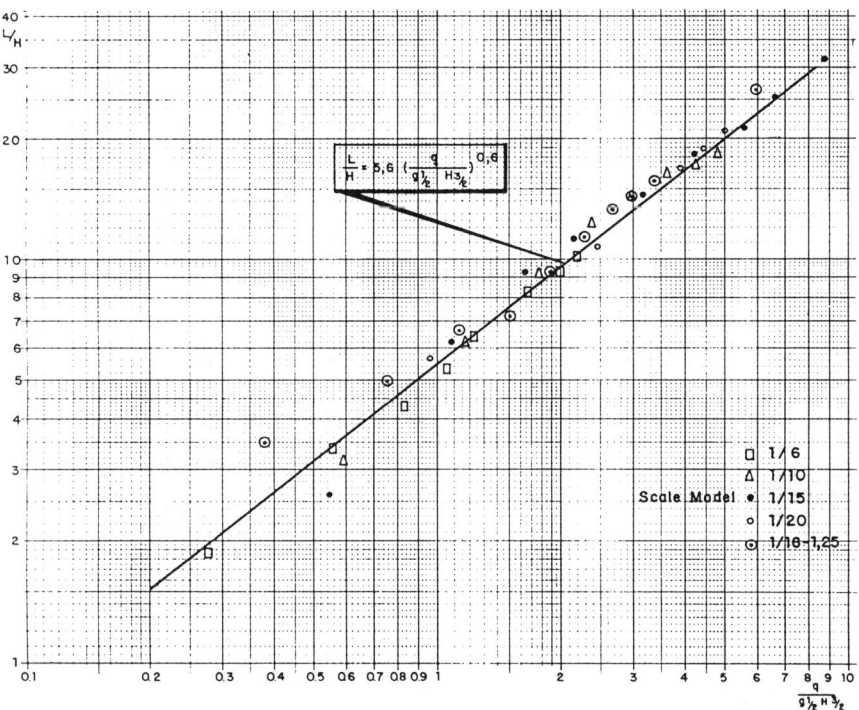

Fig.2. Determination of the inception point. Experimental data.

It was confirmed that the scale effect is negligible for all scales used, including 1/25, although in the latter case the difference with the law (1) is somewhat greater when the discharges are low, and this could be due to the fact that for this situation the nappe is aerated in the Creager profile zone and there the size of the steps is different. Furthermore, no cavitation risk can exist in this zone.

Law (1) can be expressed as follows:

$$L = 5.6 \frac{q^{0.8}}{g^{0.4}} \cdot \frac{1}{H^{0.2}} \quad (2)$$

It can be seen in this equation that the variations of H (within reasonable limits) hardly alters the value of L. However, as is only to be expected, the turbulence increases when H does, and so L is reduced. Thus for example, if the step size is duplicated the aeration distance is shortened by only 15%.

## 5. CONCLUSIONS

1. 5 different scales, 1/6, 1/10, 1/15, 1/20 and 1/25 were tested, to analyse the zone where the air-inflow of a stepped spillway takes place.

2. Regardless of the size of a stepped spillway like the one depicted in Fig. 1, the air inception point can be determined with a law of the following type:

$$\frac{L}{H} = 5.6 \left( \frac{q}{g^{1/2} H^{3/2}} \right)^{0.8}$$

3. For nappe values of the size of the step height or greater, the models are representative even at a scale of 1/25.

4. A 1/6 scale model yields results which are consistent with this general law, even with nappes which are half the step-height.

5. As is only to be expected, it can be deduced from the law, that the aeration distance can be shortened by increasing the size of the steps, but when the step-size is duplicated it is only shortened by 15%.

6. Nappe instabilities cause irregular impacts against the steps which are conducive to intermittent air-inflow, and the eddies on the steps facilitate air retention. The high flow turbulence, gives rise to a very complete and rapid mixing of air. There is then sufficient air to prevent cavitation on the step next to the one where the air enters.

## BIBLIOGRAPHY

1. CHAMANI and RAJARATNAM. Jet flow on stepped spillways. Journal of Hydraulic Engineering. Vol. 120, n° 2, February 1994.
2. CHANSON. Stepped spillways flows and air entrainment. Canadian Journal of Civil Engineering. Vol 20, n° 3, June 1993.
3. ELVIRO and MATEOS. Spanish research into stepped spillways. Hydropower & Dams. Vol 2. September 1995.
4. FRIZELL. Stepped overlays proven for use in protecting overtopped embankment dams. Water Operation and Maintenance. Bulettin n° 174.
5. MATEOS Y ELVIRO. Regularidad del flujo en aliviaderos escalonados. XVI Congreso Latinoamericano de Hidráulica. (Chile). Octubre 1994.
6. MATEOS and ELVIRO. Stepped spillways. Desing for the trasition between the spillway crest and the steps. Hydra 2000. Vol 1. London 1995.

# ENERGY DISSIPATION IN STEPPED WATERWAY

Hubert CHANSON and L. TOOMBES
Dept. of Civil Eng., The University of Queensland, Brisbane QLD 4072, Australia

## 1. INTRODUCTION

Stepped waste-waterways (also called 'byewash') were commonly used to assist with energy dissipation of the flow during the 19th century and early 20th century (CHANSON 1995a). Nowadays stepped spillways are often associated with roller compacted concrete (RCC) dams. The stepped geometry is appropriate to the RCC placement techniques and enhances the rate of energy dissipation compared to a smooth chute design (CHANSON 1994).
A stepped channel geometry is also commonly used in small-slope channels : for river training, in sewers, in storm waterways, at bottom outlets channels. Unfortunately there is little information on the rate of energy dissipation in such flat channels.
The authors present the results of a new series of investigations conducted in a 25-m long channel with a 4-degree slope. The flow characteristics and rate of energy dissipation with a smooth bed and with a stepped bottom are compared.

## 2. EXPERIMENTAL SET-UP

Experiments were performed in a 25-m long channel with a 4.0 degree slope located at the University of Queensland (fig. 1). The flume (0.5-m wide) is made of planed wooden boards ($k_s$ = 1 mm) and the sidewalls are 0.4-m high. Waters are supplied by a pump controlled by a variable-speed electronic controller enabling a fine discharge adjustment in a closed-circuit system. Flow to the flume is fed by a smooth convergent. At the nozzle, the velocity, depth and width are respectively $V_o$, $d_o$ = 0.03 m and W = 0.5 m.
The water discharge is measured with a Dall™ tube flowmeter, calibrated on site. The accuracy on the discharge measurement is about 2%. Clear-water depths and velocities are recorded with pointer gauges and a Pitot tube. Air-water flow properties are measured with conductivity probes. Full details of the instrumentation were reported by CHANSON (1995b) and CHANSON and CUMMINGS (1996).
A first series of experiments was conducted with the smooth-bed geometry (fig. 1(A), table 1) Subsequently 12 identical steps (h = 0.17 m, l = 2.4 m) were installed in the flume. A second series of experiments was then performed with the stepped channel profile (fig. 1(B), table 1)

## 3. EXPERIMENTAL RESULTS - FLOW PATTERNS
*Smooth channel flow*
For the smooth chute experiments, the flow pattern consists of a developing flow region followed by a fully-developed flow region in which the flow is decelerated (fig. 1) At the upstream end of the channel, velocity measurements showed that the boundary layer growth is best correlated by :

Fig. 1 - Sketch of the channel
(A) Smooth bed geometry

(B) Stepped geometry

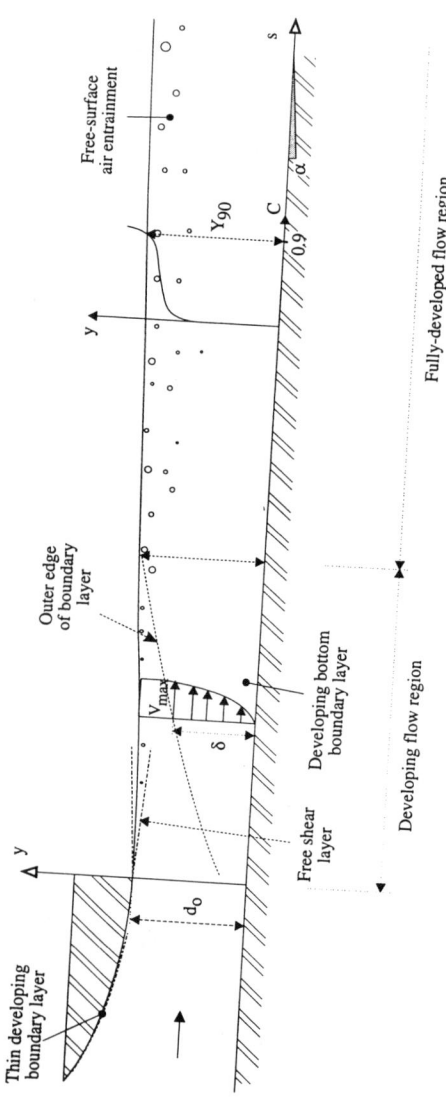
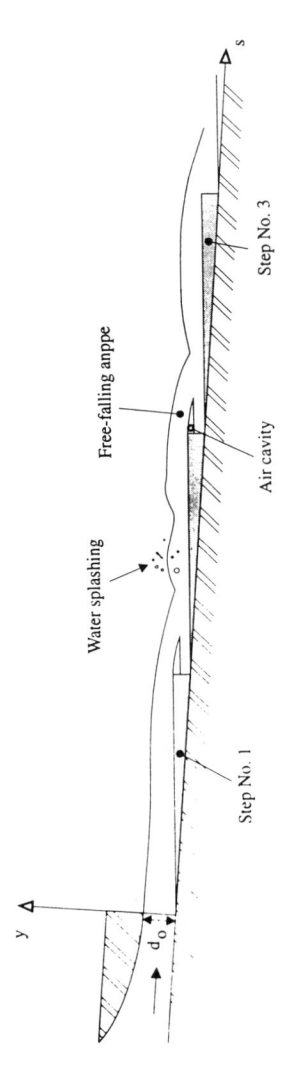

Fig 2 - Free-surface profile for $q_w$ = 0.08 m²/s, $d_o$ = 0.03 m.

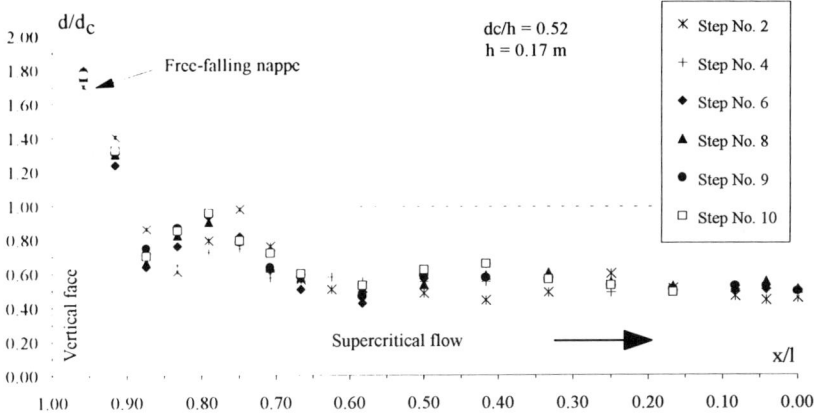

Table 1 - Experimental flow conditions

| Ref. | Slope<br>α (deg.) | $q_w$<br>m²/s | $V_o$<br>m/s | $d_o$<br>m | Comments |
|---|---|---|---|---|---|
| (1) | (2) | (3) | (4) | (6) | (7) |
| Smooth chute | 4.0 | | | 0.03 | W = 0.5 m. Painted timber ($k_s$ = 1 mm). |
| | | 0.142 | 4.7 | | Run MC2 [a]. |
| | | 0.150 | 5.0 | | Runs P5 [a] and PDC1 [b]. |
| | | 0.156 | 5.2 | | Run MC3 [a]. |
| | | 0.164 | 5.5 | | Run MC4 [a]. |
| Stepped chute | 4.0 | | | 0.03 | W = 0.5 m. Horizontal timber steps (h = 0.17 m, l = 2.4 m). |
| | | 0.038 | 1.27 | | Nappe flow regime NA3 (without Hydraulic Jump). |
| | | 0.080 | 2.7 | | Idem. |
| | | 0.130 | 4.3 | | Idem. |
| | | 0.150 | 5.0 | | Idem. |
| | | 0.163 | 5.4 | | Idem. Loud noise generated by air cavity at first drop. |

References : [a] : CHANSON (1995b); [b] : CHANSON and CUMMINGS (1996).
Notes : $d_o$ = approach flow depth; h = step height; $k_s$ = equivalent roughness height; l = horizontal step length; $q_w$ = water discharge per unit width; $V_o$ = approach flow velocity.

$$\frac{\delta}{k_s} = 1.020E\text{-}2 * \left(\frac{s}{k_s} + 757\right)^{0.973} \qquad \text{\{Smooth chute flow\}} \quad (1)$$

where δ is the boundary layer thickness, $k_s$ is the roughness height, and s is the distance from the nozzle.
Free-surface aeration was measured along the channel (CHANSON and CUMMINGS 1996).

The flow is characterised by a rapid aeration in the first section of the flume (i.e. s < 4 m). For $q_w = 0.15$ m$^2$/s, the maximum mean air concentration is about 12% (defined in terms of 90% air concentration). For s > 4 m, the flow is very-gradually de-aerated, with still about 8.5% of mean air content at s = 23 m (for $q_w = 0.15$ m$^2$/s).

*Stepped channel flow*
With the stepped chute configuration, experimental observations indicate that the flow is *supercritical* all along the 25-m long flume for all the flow conditions (table 1). I.e., the waters flow as a nappe flow regime without hydraulic jump (regime NA3, CHANSON 1995a). No hydraulic jump is observed. After the first three drops, the free-surface profiles become nearly identical at each downstream step (fig. 2).
Note that the air cavity below the free-falling nappes was not ventilated. For some particular discharges (i.e. $q_w = 0.163$ m$^2$/s), loud noise was generated by the air cavity at the first drop (i.e. between steps No. 1 and 2). The noise could be stopped by introducing a rod in the nappe, acting as a splitter device.
At the first drops, the jet impact induced significant water splashing and jet deflection, followed by the propagation of *shock waves* intersecting further downstream on the channel centreline. *Sidewall standing waves* were observed also at the impact of the nappes. Both flow patterns are sketched on figure 3.
For $q_w = 0.13$ m$^2$/s, the shock waves developing on the horizontal face of step No. 2 intersected at the brink of the step (i.e. edge of second drop). And a "rooster tail" wave was observed on the centreline of the second free-falling nappe, "riding" over the upper nappe free-surface.
For $q_w < 0.13$ m$^2$/s, the shock waves intersected upstream of the step brink on step No. 2 and no "rooster tail" wave was observed. For $q_w > 0.13$ m$^2$/s, the shock waves did not intersect before the brink of the step No. 2. And on step No. 3, the shock waves were observed to intersect at the step edge, inducing a "rooster tail" wave on the centreline of the third free-falling nappe.

## 4. EXPERIMENTAL RESULTS - ENERGY DISSIPATION

The rate of energy dissipation in smooth and stepped channel flows is shown on figure 4. The data are plotted as $\Delta H/H_o$ versus $s/d_c$ where $H_o$ is the upstream total head, s is the longitudinal distance from nozzle (fig. 1) and $d_c$ is the critical flow depth. Note that the results are based on equivalent clear-water flow depth data for the smooth chute (white symbols, $\Delta d/d < 0.5\%$) and on pointer gauge data for the stepped chute ($\Delta d/d < 5\%$), where d is the flow depth measured normal to the bottom.
Figure 4 indicates that the rate of energy dissipation is important in both smooth and stepped chute flows. In the upstream channel section (i.e. $s/d_c < 150$), larger energy dissipation is observed with the stepped geometry, in particular at large flow rates. At the downstream end of the chute, the rate of energy dissipation is nearly comparable between smooth chute data (i.e. $\Delta H/H_o \sim 0.8$) and stepped chute data ($\Delta H/H_o \sim 0.85$ to $0.9$).
With stepped chute flows, it is thought that the absence of hydraulic jump might limit the rate of energy dissipation. At the first steps, significant losses may be associated with jet deflection, shock wave generation and splashing. Further downstream, energy dissipation occurs by nappe impact and skin friction in the supercritical flow.

Fig. 3 - Free-surface flow patterns at the first steps (stepped chute experiments)

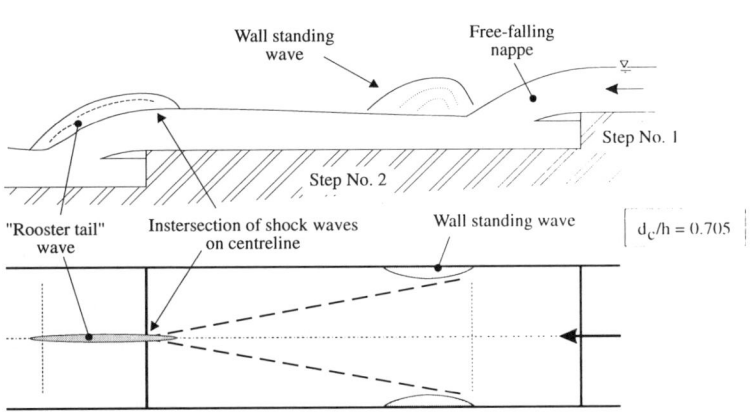

## 6. CONCLUSION

New experiments were performed to compare the flow characteristics between a smooth bed and a stepped bottom in a small-slope channel ($\alpha = 4$ degrees) downstream of a gated intake. Experimental results show that the stepped chute flow is a nappe flow regime without hydraulic jump. Shock waves, jet deflection and standing waves are observed on each step. Energy dissipation takes place rapidly in the upstream part of the flume. Further downstream the rate of energy dissipation is comparable for smooth-bed and stepped-bottom experiments.

The results may be applied to storm waterways and stepped channels downstream of bottom outlets. They suggest that existing theories, derived for steep stepped chutes, cannot be applied to small-slope channels.

## ACKNOWLEDGEMENTS

The authors acknowledge the financial support of Australian Research Council.

## REFERENCES

CHANSON, H. (1994). "Comparison of Energy Dissipation between Nappe and Skimming Flow Regimes on Stepped Chutes." *Jl of Hyd. Res.*, IAHR, Vol. 32, No. 2, pp. 213-218. Errata : Vol. 33, No. 1, p. 113. Discussion : Vol. 33, No. 1, pp. 114-143.

CHANSON, H. (1995a). "Hydraulic Design of Stepped Cascades, Channels, Weirs and Spillways." *Pergamon*, Oxford, UK, Jan., 292 pages.

CHANSON, H. (1995b). "Air Bubble Entrainment in Free-surface Turbulent Flows. Experimental Investigations." *Report CH46/95*, Dept. of Civil Engineering, University of Queensland, Australia, June, 368 pages.

CHANSON, H., and CUMMINGS, P.D. (1996). "Air-Water Interface Area in Supercritical Flows down Small-Slope Chutes." *Research Report No. CE151*, Dept. of Civil Engineering, University of Queensland, Australia, Feb., 67 pages.

Fig. 4 - Rate of energy dissipation : comparison between the smooth and stepped chutes

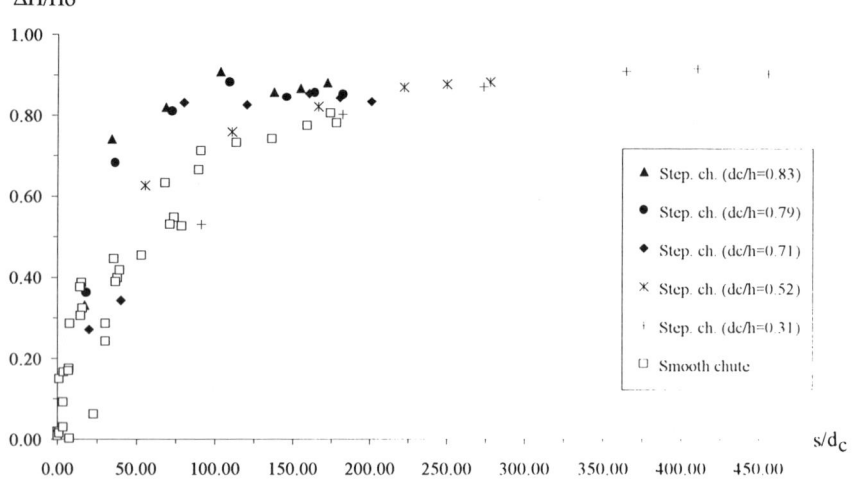

# The Effect of Nappe Impact Angle on Aerator Performance

### JALAL ATTARI
Power & Water Institute of Technology, Tehran, Iran

### AMIR R. ZARRATI
Amir Kabir University of Technology, Tehran, Iran

### ABSTRACT

One of the important parameters affecting the performance of chute spillway aerators is the impact angle of the jet as it reattaches to the surface of the spillway. In this experimental study, the impact angle was increased by installing false beds on the invert of a 1:15 scale model. Results of the tests showed that by increasing the impact angle, the flow rate of air supplied by ducts decreased and the pressure beneath the nappe rose towards atmospheric pressure. A greater recirculation of air in the cavity under the jet was believed to be the main agent of these effects.

### INTRODUCTION

Aerators are increasingly employed for reducing the threat of cavitation damage to spillways of high head dams (Pinto [1]). An aerator usually consists of a ramp followed by a step (Fig. 1). Flow separates at the downstream end of the ramp and forms a nappe. Owing to high turbulence in relation to surface tension forces, air is entrained into the flow along the upper and lower surfaces of the nappe. Normally, air is supplied to the lower nappe by air ducts on each side of the chute (Volkart et al, [2]). In the nappe zone, where the pressure gradients are very low, air bubbles freely diffuse into the nappe. The aerated nappe then impacts on the spillway surface and high dynamic pressures are generated. Downstream of the impact zone, a hydrostatic pressure distribution is re-established.

The characteristics of the air water mixture in the impact zone are influenced by turbulence intensity and the other flow parameters upstream of the aerator as well as the geometry of the ramp and the chute. The undernappe pressure has a marked influence on the behaviour of flow in this zone (Zarrati, [3]). The following factors characterise the flow condition at the impact zone: (i) the impact velocity, (ii) the jet thickness at the impact and (iii) the impact angle. A literature review shows that these factors require more systematic study to describe the complex behaviour of aerated flow. This paper investigates the effect on aerator performance of nappe impact angle relative to the invert of the chute (Attari, [4]).

## AIR CIRCULATION

The high pressures which arise in the impact zone cause a migration of entrained air to zones of lower pressure, i.e. upwards to the free surface and back to the undernappe cavity. There is thus a circulation of air within the cavity as air is entrained in a downstream direction along the lower surface of the nappe and a fraction of that flow is detrained at the impact point and forced back upstream.

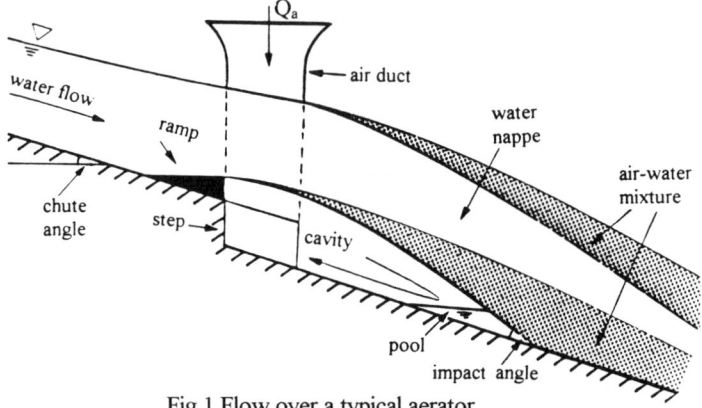

Fig.1 Flow over a typical aerator

Circulation of air under the nappe has been reported before. Measuring the air concentration, Chanson [5] observed that the amount of air entrained by the lower nappe was more than the air supplied through the ducts owing to this circulation. Zarrati [6], measured the concentration of air in the nappe and calculated the amount of air entrained through the jet length ($Q_j$) and compared it with the measured air supplied through the air ducts ($Q_a$). The results showed that $Q_j > Q_a$ and this difference was intensified as the undernappe pressure rose. He concluded that this was due to the effect of jet impact on the spillway invert. Ervine et al [7] confirmed that for jets which strike the boundary at angles greater than about 10 degrees, the net air transport will be less than the amount of air entrained by the lower nappe surface. There will also be a circulation of water in the cavity where a pool at the impact zone forms. Since aerators are designed to entrain as much air as possible, every effort should be made to reduce air detrainment in the impact zone.

## EXPERIMENTS

### EXPERIMENTAL FACILITY

The effect of impact angle on the performance of an aerator was studied in a 1:15 scale sectional model of Karun 1 dam spillway, located in south east Iran. This spillway has an upper and a lower aerator and model experiments were carried out on the former. The distance between the first aerator and the ogee was 9 m in the model. The longitudinal slope of the spillway at the downstream end of the aerator was 35 degrees. The model was 200 mm wide and the detail of the first aerator model is shown in Fig. 2.

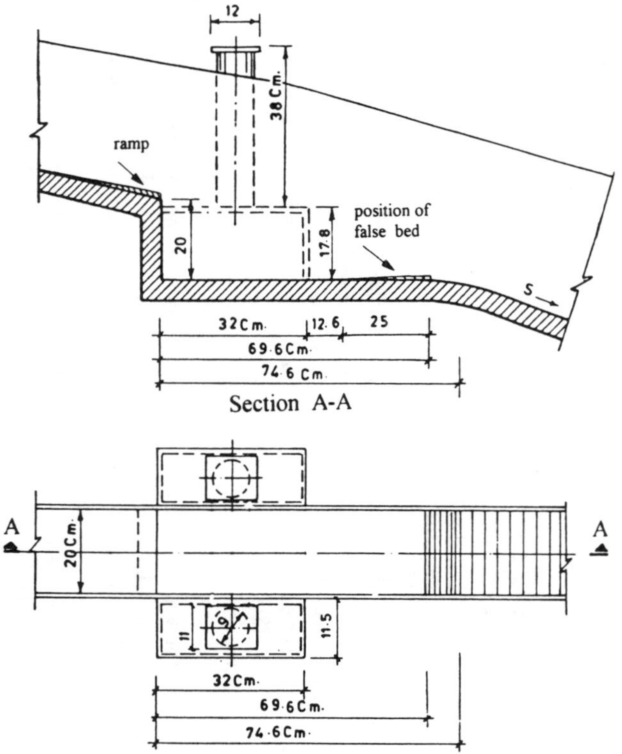

# NAPPE IMPACT ANGLE EFFECT ON AERATOR

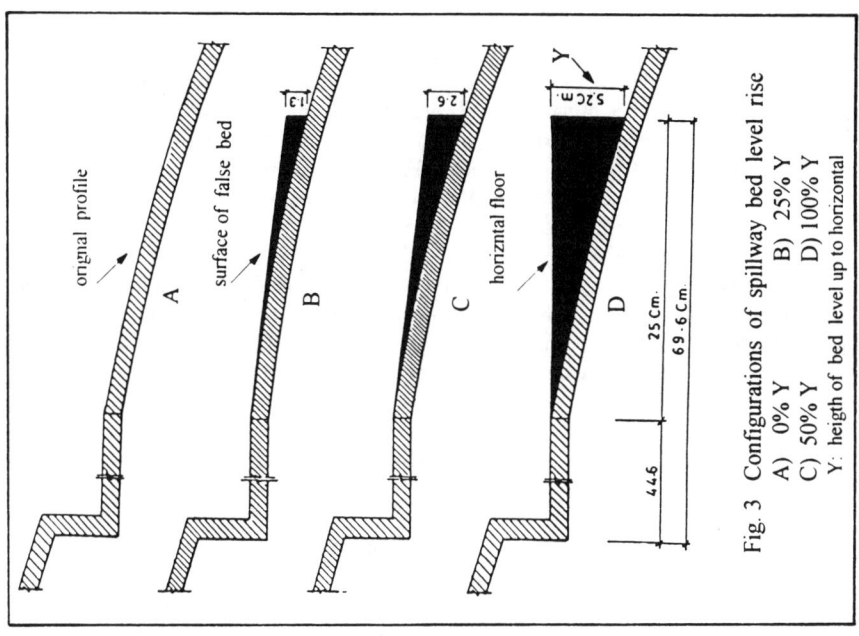

Fig. 3 Configurations of spillway bed level rise
A) 0% Y   B) 25% Y
C) 50% Y  D) 100% Y
Y: heigth of bed level up to horizontal

Fig. 2 Physical Model of the aerator

For water discharges up to 117 l/s, the highest velocity measured on the ramp was 4.3 m/s and Froude numbers were between 3.79 and 4.27. To deduce the air discharge in each experiment, the air velocity in one of the symmetrical supply ducts was measured with a pitot tube on the centreline of the duct and connected to a micro manometer. Air discharge was determined assuming a logarithmic velocity distribution. A representative undernappe pressure was measured with an 1:16 inclined alcohol manometer at a pressure tapping mounted on the vertical face of the step on centreline of the flume.

## EXPERIMENTAL PROCEDURE

To increase the nappe angle of impact on the invert of the chute (without changing other parameters), the slope of the spillway at the impact zone was reduced by mounting a false bed at the downstream end of the aerator. Three such curved beds of identical length but with varying slopes were tested. The highest made the slope horizontal with a maximum size Y (Fig. 3) and the other two had 25% Y and 50% Y. Four cases of bed level rise were investigated:
(A) no bed level rise (original slope); (B) 25% Y; (C) 50% Y; (D) full bed level rise (horizontal slope).

Two air-supply conditions affecting the under nappe pressure were also investigated:
i) only one air duct open;  ii) both air ducts open.

Air velocity in the duct and the undernappe pressure were measured for a combination of the above mentioned cases with various water discharges from 51 l/s to 112 l/s. For the horizontal bed (case D), the air cavity under the jet collapsed after a short period of operation and no further air was drawn through the ducts.

## ANALYSIS OF EXPERIMENTAL RESULTS

Undernappe pressure and air supply discharge were determined for configurations A to C. The false beds increased the impact angle of the flow without significantly altering the length of jet trajectory for a particular water discharge.

Figs 4(a) and 4(b) show the air discharge supplied with 1 and 2 ducts open respectively. The general trend of the results reveal that by increasing the impact angle, the flow rate of air supplied by the ducts decreases accordingly. A circulation of air back to the cavity which reduces the demand of air through the ducts is believed to be the main agent for this effect; this was more pronounced at the higher water discharges where the longer jet trajectories rendered the impact angle even steeper.

The effect of nappe impact angle on undernappe pressure is shown in Fig. 5 for case A (original spillway with no bed level rise) and case B (raised bed level up to 25% of Y). A comparison of results in these two cases shows that the undernappe pressure rises towards atmospheric pressure with the increased impact angle of case B. The trend was more pronounced with only one duct open. The effect of higher undernappe pressure is to reduce the flow of air into the cavity and these results seem to confirm that an increasing impact angle will decrease the effectiveness of the aerator.

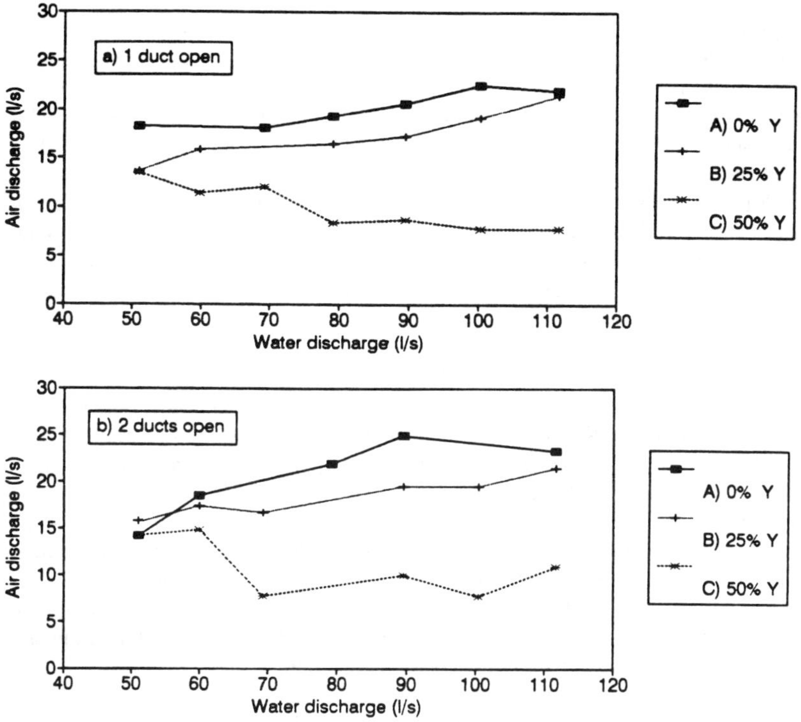

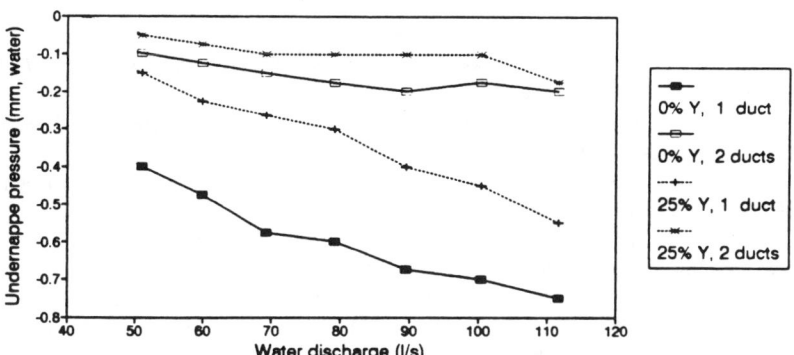

Fig. 4 Effect of impact angle on air discharge

Fig. 5 Effect of impact angle on undernappe pressure
Y: height of bed level up to horizontal

## VISUALISATION OF AIR FLOW

To visualise the flow pattern, smoke was injected from the side wall into the cavity by a kerosene smoke generator. In the vicinity of the ramp, smoke moved upward into the jet, whereas at the impact zone, it flowed backward and to the sides. Flow visualisation for the two limiting cases A (original spillway bed) and D (horizontal floor) confirmed that by increasing the impact angle, the air/water circulation was intensified. In case A, the jet impacted at an acute angle and the greatest entrainment of air into the jet occurred. A small part of the mixture circulated back to the cavity to form a pool of water with a height of about 10 % of the step. In case D, the jet impacted at a steeper angle and created a form of a hydraulic jump. A large portion of the flow recirculated into the cavity; about 80% of the height of the cavity was filled by water and no further air was entrained via the ducts.

## CONCLUSION

The results of the tests showed that by increasing the impact angle, the air flow in the ducts decreased and the undernappe pressure rose towards atmospheric pressure; there was a corresponding increase in the depth of the pool in the cavity. Flow visualisation revealed that the air/water circulation in the cavity intensified with increasing impact angle, further reducing the efficiency of the aerator. It would appear that the most effective aerator is a geometrical compromise. An increasingly steep angle of ramp increases the length of the jet trajectory and hence the air-entraining capacity of the under nappe. At the same time, however, there is a corresponding increase in air detrainment at the impact zone as shown in this paper. Improvements in aerator design will be obtained by a careful balancing of these two influences to achieve an optimum.

## ACKNOWLEDGEMENT

The authors wish to acknowledge the Water Research Centre (affiliated to Ministry of Energy) in Tehran, Iran, for providing the necessary equipment and for their kind co-operation.

## REFERENCES

1) Pinto, N. (1991), "Prototype aerator measurements", *Hydraulic Structures Design Manual No. 4*, IAHR, Ed. I.R. Wood.
2) Volkart, P., Rutschmann, P. (1991), "Aerators on Spillways", *Hydraulic Structures Design Manual No. 4*, IAHR, Ed. I.R. Wood.
3) Zarrati, A.R. (1993), "Mechanism of air entrainment in spillway aerators ", proceedings ASCE *Hydraulic Engineering* .
4) Attari, J. (1995), "Effect of impact angle on aeration in the spillway aeration", *MSc. Thesis*, *Tarbiat Modares University*, Tehran, Aug.
5) Chanson, H. (1994), "Aeration and deaeration at bottom aeration devices on spillways", *Canadian Journal of Civil Engineering*, Vol. 21.
6) Zarrati, A.R. (1991), "Studies of air-water mixtures for spillway aerators", *Thesis submitted to the university of London for the degree of Doctor of Philosophy*.
7) Ervine, D.A., Falvey H., Khan, A. (1995), "Turbulent flow structure and air uptake at aerators", *International Journal on Hydropower & Dams*, Vol. 2, Issue 5, Sep.

# Design of Spillway Deflectors for Ice Harbor Dam to Reduce Supersaturated Dissolved Gas Levels Downstream

### JAMES D. CAIN, P.E.
U.S. Army Corps of Engineers, Walla Walla District
Walla Walla, Washington USA

## ABSTRACT

Dissolved gas supersaturation occurs downstream of the eight Federal dams on the lower Snake and Columbia Rivers when spillway flows are released. This paper describes a new approach to optimize spillway deflector performance to reduce supersaturation levels downstream of Ice Harbor Dam.

## INTRODUCTION

In the last few years, voluntary spills have been used as a method of routing downstream migrating juvenile salmonids away from the powerhouse turbines at the eight Federal dams on the lower Snake and lower Columbia Rivers. This has been done in an effort to increase the overall survival of these juvenile fish. However, dissolved gas supersaturation results when water is passed over the spillways. This is because air entrained as small bubbles in the spillway flow is forced into solution by the high hydrostatic pressures within the stilling basin. Conditions downstream of the stilling basin are such that the high dissolved gases levels continue for many miles down river. These high supersaturation levels can be injurious and even lethal to migrating salmon, both juveniles and returning adults, as well as resident fish and other aquatic organisms in the river.

The best way, identified by the Corps of Engineers, to reduce the dissolved gas supersaturation levels caused by spills is the installation of spillway deflectors on the lower part of the spillway ogee.[1] (See Figure 1c.) The objective of spillway deflectors is to direct (at least for lower rates of spill) the spillway flow across the water surface in the stilling basin instead of allowing it, with the entrained air, to plunge deep into the basin pool. The effectiveness of deflectors in achieving this objective is dependent on the discharge being spilled, the size and shape of the deflector, and the submergence of the deflector (tailwater elevation minus the elevation of the deflector).

---

[1] U.S. Army Corps of Engineers, (April 1996)

## SPILLWAY DESCRIPTION

Ice Harbor Dam is the last dam on the Snake River before the river joins the Columbia River near Pasco, Washington, USA. Its spillway has ten bays which are each 50 feet wide and are separated by 10-foot wide piers. The stilling basin is 170 feet long and has a 12-foot high end sill. The basin also has one row of baffle blocks, located 42 feet from the end sill. The baffle blocks are 10-foot wide and 8-foot tall. The crest of the spillway is at elevation 391 feet mean sea level (fmsl) and the stilling basin floor is at elevation 304 fmsl. The average streambed elevation just downstream of the stilling basin is approximately 327 fmsl. The forebay is operated between 437 and 440 fmsl. The tailwater elevation varies with the total river discharge but is generally between 339.5 and 345.0 fmsl. During the spillway design flood of 850,000 cubic feet per second (cfs), the forebay will surcharge to elevation 446.4 fmsl and the tailwater elevation will be about 374 fmsl.

## HYDRAULIC MODELS USED

As a general rule, direct measurement of gas supersaturation for scaling from a physical model to the full scale structure is not possible. This is because processes that dictate gas transfer performance, including bubble size of entrained air, are not scaleable parameters in physical hydraulic models. However, physical models of hydraulic structures are valuable tools for visualizing flow patterns, evaluating hydraulic performance, and estimating hydraulic conditions that will exist in a full-scale structure.[2]

Two hydraulic models were built at the Corps of Engineers, Waterways Experiment Station (WES) in Vicksburg, Mississippi. The first one was a 1:40 scale, sectional model which reproduced three full spillway bays flanked by a half bay on each side. This model was used to optimize the deflector's size, shape, and elevation. In addition, the model provided insights into the movement of rock debris drawn into the stilling basin by the return flow of the eddy action beneath the deflector induced surface flow. The second model was a 1:55 scale, general model which reproduced the 6-unit powerhouse, 10-bay spillway and stilling basin, adult fishway entrances, navigation lock and guidewall, and riverbed for approximately 2,500 feet downstream of the project. The general model was used to evaluate the potential impacts that spillway deflectors might have on navigation and on adult fish entering into the fish ladders.

## OBSERVED HYDRAULIC CONDITIONS

Observations of sectional model tests with spillway deflectors installed indicated that up to six different hydraulic conditions could occur in the stilling basin for a given spill discharge per bay. The resulting stilling basin condition was dependent on tailwater elevation. Starting with a low tailwater elevation, the spillway flow would

---

[2] U.S. Army Corps of Engineers, (April 1996)

be launched by the deflector and would plunge into the stilling basin, (Figure 1a). Usually, this condition was associated with the aeration of the underside of the nappe at the deflector. As the tailwater elevation was raised, the tailwater became high enough that a mixed skimming and plunging condition occurred, which has been termed unstable, (Figure 1b). With an even higher tailwater elevation, a point was reached where the spillway flow would only skim along the top of the tailwater, (Figure 1c). This skimming flow condition was found only at lower spill discharges. Eventually, the tailwater elevation would become high enough that the skimming flow would start to become an undular surface flow, (Figure 2a). With increases in the tailwater elevation, the initial hump of the undular surface flow would move upstream towards the spillway. The undular surface flow would then become an elevated hydraulic jump above the deflector, (Figure 2b). Finally, when the tailwater was high enough, the hydraulic jump would become a submerged jump on top of the deflector, (Figure 2c).

## DEFLECTOR SELECTION

Future voluntary spills for passing juvenile fish are expected to be between 2,500 and 5,000 cfs per bay. Because the occurrence of these lower levels of spill will be the most frequent, the deflectors at Ice Harbor Dam were designed to optimize their skimming performance at these spill levels. The selected deflector protrudes 12.5 feet horizontally from the spillway ogee at elevation 338 fmsl and has a 15-foot radius fillet between the deflector and the ogee. The fillet provides better skimming action and is likely to be better for juvenile fish. Figure 3 displays the performance of the selected deflector for various spill discharges and tailwater elevations. The shaded area shows the range of spill discharges and tailwater elevations for the most likely juvenile fish passage spill scenarios.

The selected deflector was also tested in the sectional model at discharges of 42,000 and 85,000 cfs per bay, which correspond to the standard project flood and spillway design flood respectively. Observations of these test conditions indicated that energy dissipation within the stilling basin was as good as, and possibly better than, that of the spillway without deflectors.

## REFERENCES

U.S. Army Corps of Engineers, North Pacific Division, Portland District and Walla Walla District, April 1996, *Dissolved Gas Abatement Study, Phase I, Technical Report*.

U.S. Army Corps of Engineers, Walla Walla District, June 1996, *Ice Harbor Lock and Dam, Lake Sacajawea, Washington, Feature Design Memorandum No. 34, Spillway Deflectors*.

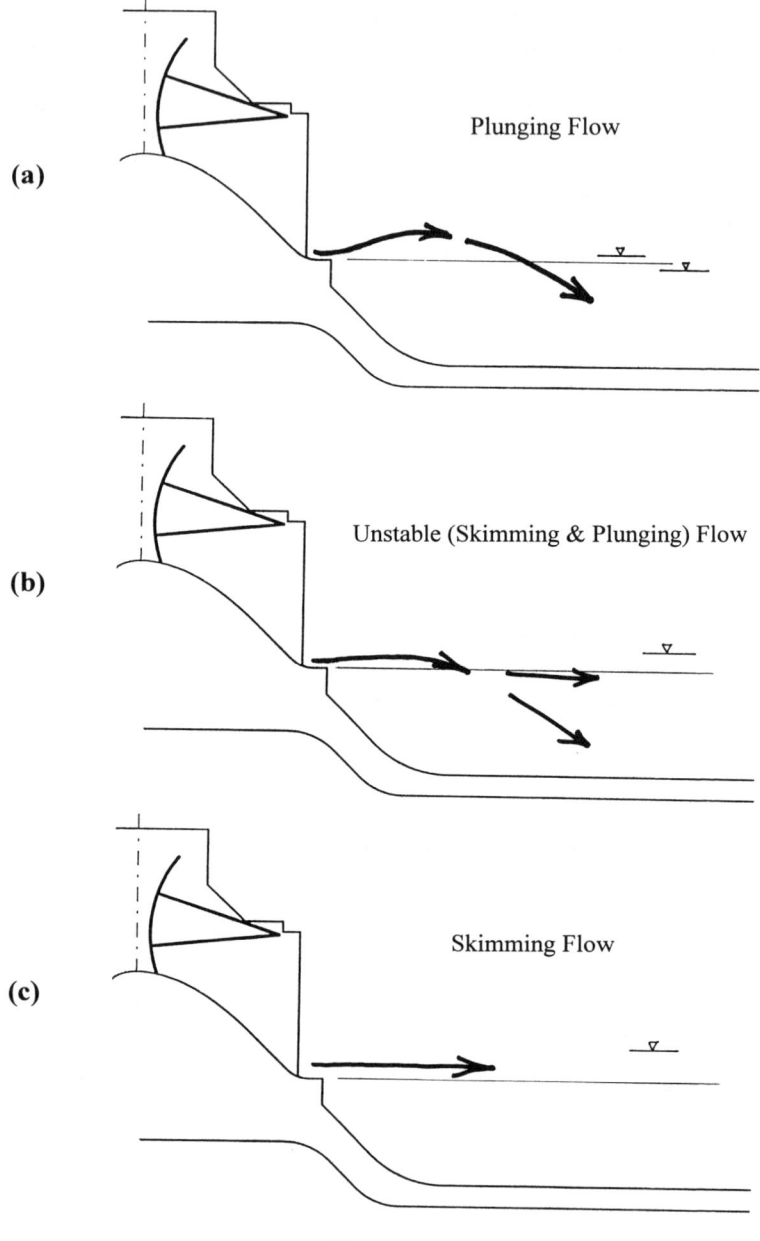

Figure 1

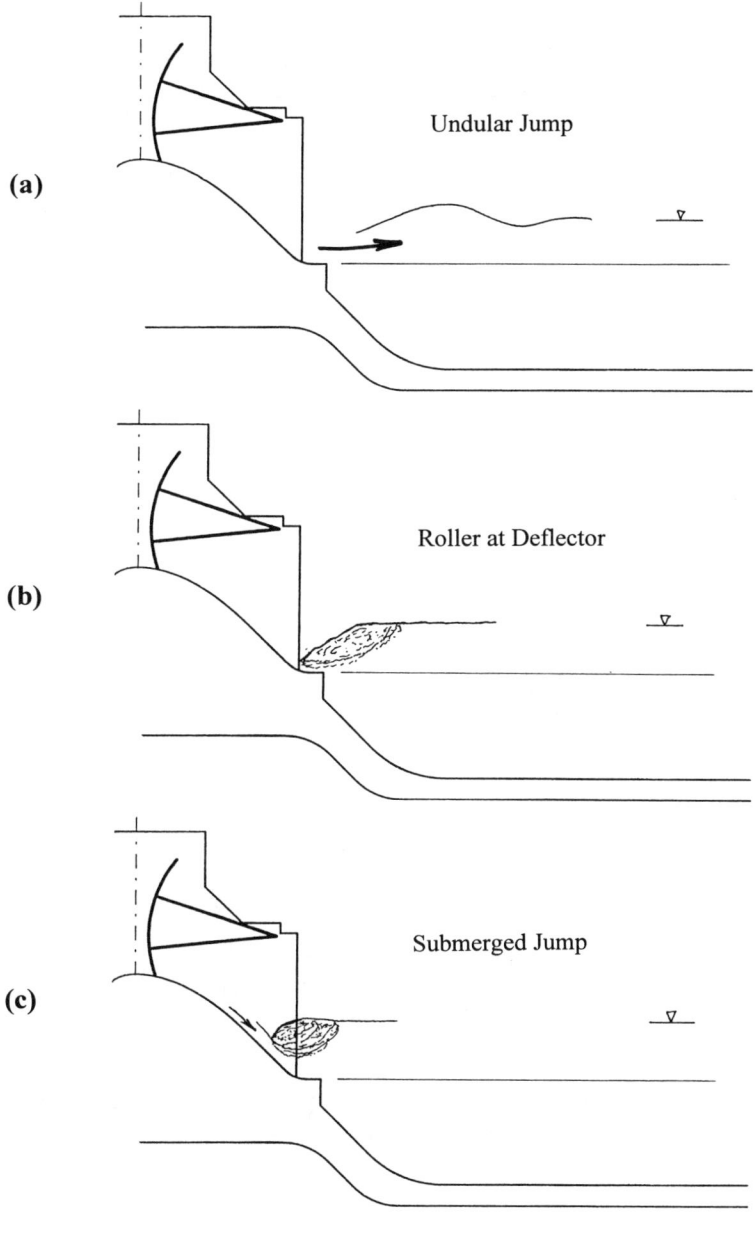

Figure 2

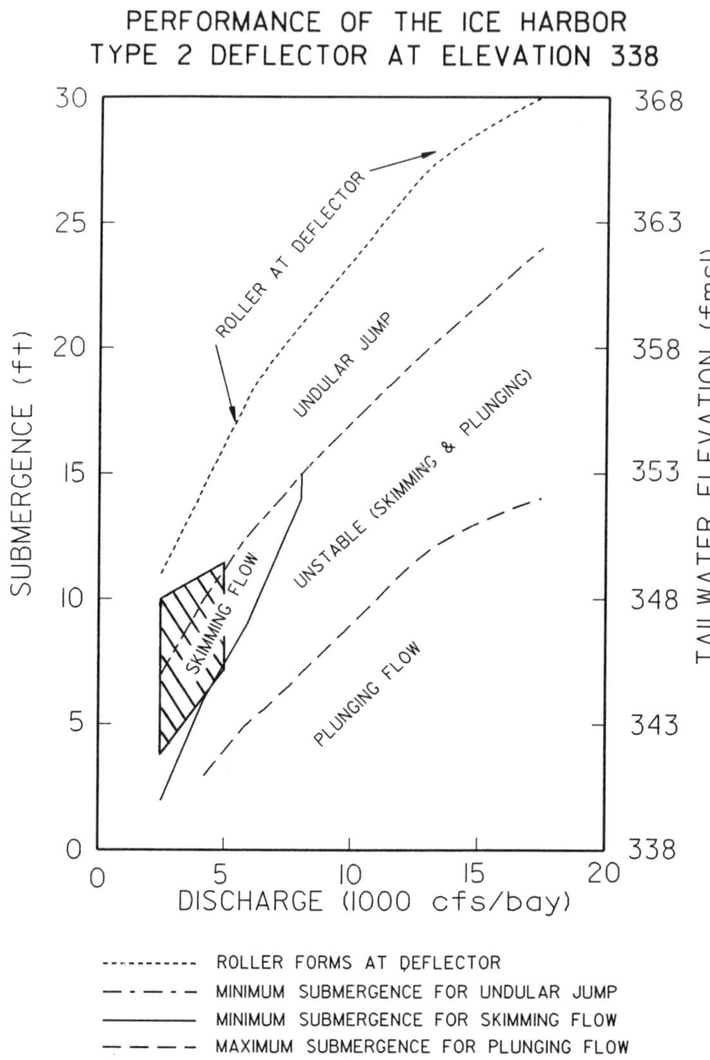

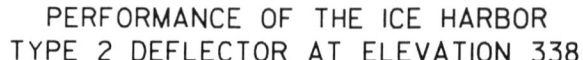

Figure 3

# AIR ENTRAINMENT IN BOTTOM OUTLET TAILRACE TUNNELS

JUERG SPEERLI AND PETER U. VOLKART
Laboratory of Hydraulics, Hydrology and Glaciology (VAW)
at the Swiss Federal Institute of Technology
8092 Zurich, Switzerland

## ABSTRACT

By means of both new prototype measurements and data from existing literature, the main parameters of air entrainment in bottom outlet tailrace tunnels will be discussed. In particular, the emphasis will be on the interaction of the water-air two-phase-flow and the geometry of long bottom outlet tailrace tunnels.

## INTRODUCTION

Downstream of gates at dam heights of 300 m and more, discharge velocities of up to 200 km/h are to be expected in bottom outlet tailrace tunnels, leading to air entrainment into the supercritical and highly turbulent water jet. Unpleasant consequences thereof are, in the first place, an increasing flow depth of the water-air mixture and, in the second place, negative pressures in the air space downstream of the service gate which may favor vibrations of the gate as well as cavitation. The latter encourages cavitation erosion of the tailrace tunnel, thus jeopardizing the safety of a dam. An air supply system normally connects the area downstream of the service gate with the outside atmosphere in order to limit negative air pressures.

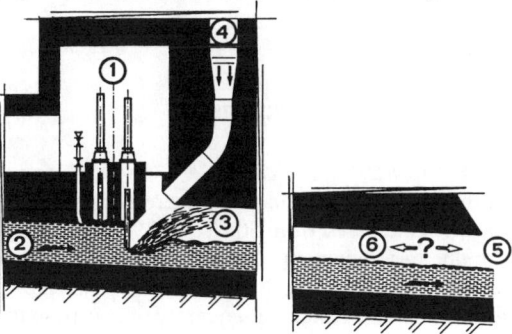

*Fig. 1: Bottom outlet with a long tailrace tunnel: (1) bottom outlet chamber with two slide gates, (2) approaching pressure flow, (3) tailrace tunnel, (4) air supply system, (5) outlet and (6) air flow in or against the direction of the water-air mixture.*

To design an air supply system, both the air entrainment in the water flow as well as the amount of air dragged along the surface of the water-air mixture need to be known. To date, however, formulas which are based on laboratory and field data insufficiently consider or even entirely neglect the influence of the bottom outlet tailrace tunnel (Falvey [1980], Billeter and Speerli [1991]).

The influence of the tailrace tunnel on the air entrainment capacity of the water jet will now be shown by combined presentation of new measurements and data extracted from existing literature.

## PROTOTYPE MEASUREMENTS

### DATA FROM EXISTING LITERATURE
In the nineteen forties and fifties, the U.S. Army Corps of Engineers (USCE) conducted a series of field measurements on bottom outlet tailrace tunnels, e.g. Campbell and Guyton [1953], USCE [1954A] and USCE [1954B].

In the bottom outlets investigated by the USCE, the tailrace tunnel is filled before the gate is completely opened because the cross-sectional area of the gate equals that of the tailrace tunnel.

At the Lumiei Dam in northeastern Italy, Dettmers [1953] carried out field measurements on three outlets with comparable structures for 30, 60 and 90 m of pressure head. The cross-sectional areas of the tailrace tunnels included in these investigations were larger than those of the gates.

### DATA FROM NEW MEASUREMENTS
The Laboratory of Hydraulics, Hydrology and Glaciology (VAW) at the Swiss Federal Institute of Technology in Zurich was given the possibility to conduct extensive investigations of the bottom outlet at the Panix Dam in southeastern Switzerland during the initial filling of the reservoir (Volkart and Speerli [1994] and Volkart, Speerli and Benesch [1995]).

## RESULTS

### INFLUENCE OF THE AIR SUPPLY SYSTEM
Figure 2 shows that the air flow is considerably decreased as the cross-sectional area of the air supply system is increased from 50 % to 75 % blockage. An increase of the negative air pressure due to a reduction of the air supply system only takes place at larger relative gate openings. These findings, therefore, indicate that in lesser gate openings a pressure compensation and, consequently, an air flow through the tailrace tunnel will have to occur.

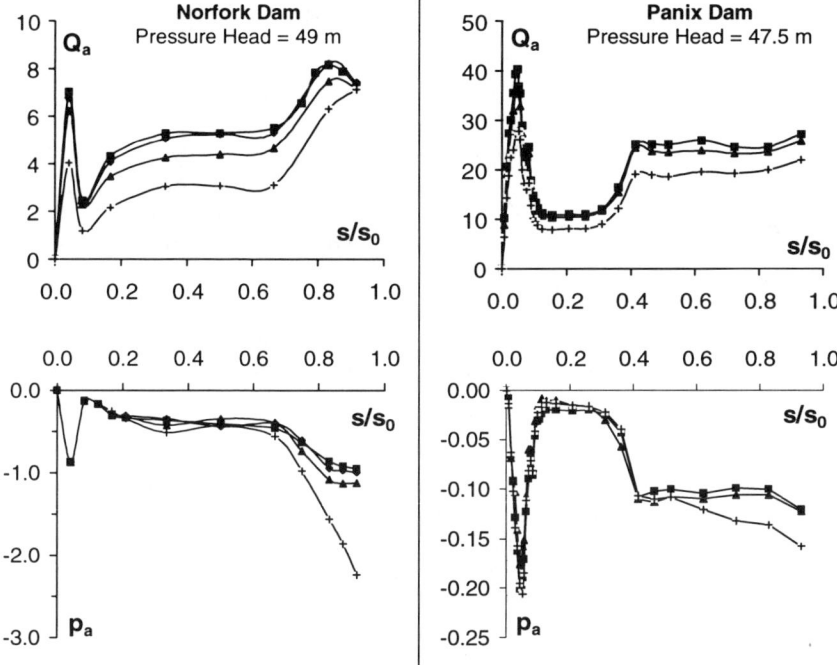

Air supply system: ■ unblocked, ♦ 25 % blocked, ▲ 50 % blocked and + 75 % blocked

*Fig. 2: Air flow $Q_a$ [$m^3/s$] through the air supply system and air pressure $p_a$ [0.1 bar] dependent on the relative gate opening $s/s_0$ [-] at the Norfork Dam in the USA and the Panix Dam in Switzerland.*

## INFLUENCE OF THE GEOMETRY OF THE TAILRACE TUNNEL

After having reached their first maxima, thought to be caused by the so called spray flow (Sharma [1973]), the three curves of the air demand in figure 3 demonstrate a greatly varying courses from each other:

- At the Panix Dam, the air demand remains practically constant for relative gate openings of approximately 0.12 to 0.30. Then it steeply increases again until a relative gate opening of 0.40 is reached to finally settle on an almost constant level thereafter.
- At the Lumiei Dam, the air demand remains almost constant for relative gate openings of approximately 0.25 to 0.70 and then significantly increases until the gate is fully opened.
- At the Pine Flat Dam, the air demand increases continuously for relative gate openings of approximately 0.12 to 0.65 until it reaches a second maximum. Before

the total opening of the gate is accomplished, however, the air demand decreases to zero.

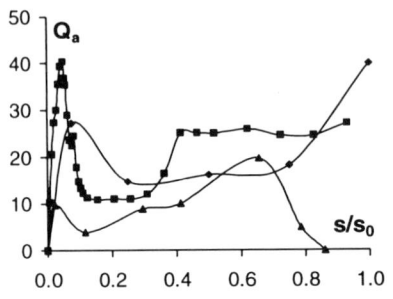

- Panix Dam, Switzerland: Pressure Head = 47.5 m,
- Lumiei Dam, Italy: Pressure Head = 60.0 m,
- Pine Flat Dam, USA: Pressure Head = 77.4 m

$Q_a$ : air flow in the air supply system [m³/s]
$s/s_0$: relative gate opening [-]

*Fig. 3: Influence of the tailrace tunnel on the air flow $Q_a$.*

At the Pine Flat Dam, the tailrace tunnel has the same cross-sectional area as the service gate. The tailrace tunnel fills up before the total opening of the gate is reached, leading to a blockage of the air supply from the outlet of the tailrace tunnel. By increasing the opening of the gate, the entire tailrace tunnel is under pressure, thus causing the blocking of the air supply system so that the water jet can no longer absorb air.

At both the Lumiei and the Panix Dams, the cross-sectional areas of the tailrace tunnels are larger than the ones of the service gates. When the gate opening is large enough, spray flows become free surface flows allowing air to enter from the outlets of the tailrace tunnels. At Lumiei Dam, shortly before reaching 100 percent gate opening, the tailrace tunnel is filled up, thereby cutting the air supply from the outlet, Dettmers [1953]. Contrary to the situation at the Pine Flat Dam, however, the entire tailrace tunnel is not pressurized so that the discharge downstream of the service gate can still entrain air. At the transition of the free surface flow to the pressure flow, a hydraulic jump entrains air into the water flow. This results in a second and higher peak of the air demand than the one previously found in the spray flow. At the Panix Dam, the tailrace tunnel does not fill when the gate is fully opened, so that the air supply from the outlet of the tailrace tunnel is possible.

Thus, the geometry of the bottom outlet tailrace tunnel has a significant influence on the air demand of the water flow in a tailrace tunnel. These findings are of concern for both the cross-sectional area and the length of the tailrace tunnel.

## OPERATING CONDUCT OF THE PANIX AND NORFORK BOTTOM OUTLETS

By reducing the capacity of the air supply system, less air flows through the supply system. As long as the tailrace tunnel is sufficiently large, air will flow from the outlet through the tailrace tunnel to the negative pressure region downstream of the service gate. If the air supply system is reduced, the negative pressures for small relative gate openings will thus increase only slightly. This can be observed at both the Norfork and the Panix Dams (fig. 4).

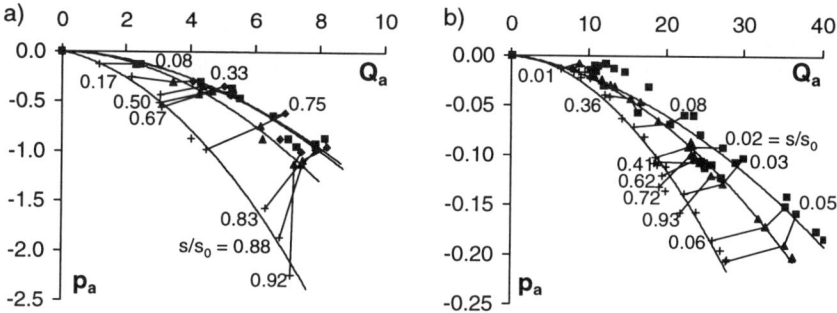

Air supply system: ■ unblocked, ♦ 25 % blocked, ▲ 50 % blocked and + 75 % blocked
$Q_a$ : air flow in the air supply system [m³/s]
$p_a$ : air pressure in the tailrace tunnel [0.1 bar]

*Fig. 4: Pressure flow diagram: The air supply for various blockages is outlined. The results of the identical relative gate openings $s/s_0$ are connected with each other. a) Norfork Dam, USA: pressure head = 49 m and b) Panix Dam, Switzerland: pressure head = 47.5 m.*

Both dams show different operating conducts since at the Norfork Dam the tailrace tunnel fills up before a complete gate opening is reached. Starting at a relative gate opening of approximately 0.3 at the Norfork Dam, the influence of the air supply system on the negative pressures grows steadily stronger. At a relative gate opening of approximately 0.8, the tailrace tunnel fills up, preventing an air flow from the outlet and thus also a pressure compensation. The negative air pressures increase due to a reduction of the air supply system according to the above results, e.g. for a blockage from 50 % to 75 % and a relative gate opening of 0.92 from approximately -0.11 bar to -0.22 bar (fig. 4a). In comparison, the negative pressures at the Panix Dam for medium to large relative gate openings increase only slightly since a pressure compensation through the tailrace tunnel may still take place (fig. 4b).

## CONCLUSIONS

By means of various field measurements, the influence of bottom outlet tailrace tunnels on the air demand of a water flow and the negative air pressures downstream of the service gate were presented. It was shown that there is interaction between the air supply system and the bottom outlet tailrace tunnel. According to the specific layouts, pressure compensation and thus the air flow take mostly place through either the air supply system or the tailrace tunnel.

Previously published formulas do not account for the geometry of the tailrace tunnel and very often also neglect the influence of the air supply system. For high pressures and long tailrace tunnels, these formulas are, therefore, to be interpreted with caution.

The Laboratory of Hydraulics, Hydrology and Glaciology (VAW) in Zurich, Switzerland, will continue to systematically investigate the influence of tailrace tunnels on air demands and pressures in test sites. The extensive results of these investigations will be published in 1997.

## ACKNOWLEDGEMENT
The authors would like to thank Irene Billeter Sauter for the English translation of the manuscript.

## REFERENCES
Billeter, P., Speerli, J.; "Prototype investigation of the air entrainment in a high head bottom outlet"; Proc. 24th Congress, IAHR, Madrid, Vol. D, pp. 321-330, 1991.

Campbell, F.B., Guyton, B.; "Air demand in gated outlet works"; Proc. 5th Congress, IAHR, Minnesota, pp. 529-533, 1953.

Dettmers, D.; "Beitrag zur Frage der Belüftung von Tiefschützen"; Mitteilungen der Hannoverschen Versuchsanstalt für Grundbau und Wasserbau, Franzius-Institut der Technischen Hochschule Hannover, Heft 4, pp. 22-68, 1953.

Falvey, H.T.; "Air-Water Flow in Hydraulic Structures"; Engineering Monograph No. 41, United States Departement of the Interior, Water and Power Resources Service, Denver, Colorado, 1980.

Sharma, H. R.; Air Demand for High Head Gated Conduits; The University of Trondheim, The Norwegian Institue of Technology, 1973.

USCE; "Vibration pressure and air-demand tests in flood-control sluice, Pine Flat Dam, Kings River, California"; Miscellaneous Paper No. 2-75, U.S. Army Corps of Engineers, Waterways Experiment Station, Vicksburg, Mississippi, 1954.

USCE; "Slide gate tests, Norfork Dam North Fork River, Arkansas"; Technical Memorandum No. 2-389, U.S. Army Corps of Engineers, Waterways Experiment Station, Vicksburg, Mississippi, July 1954.

Volkart, P., Speerli, J.; "Prototype Investigation of the High Velocity Flow in the High Head Tunnel Outlet of the Panix Dam"; Proc. 19th Congress, ICOLD, Durban 1994.

Volkart, P., Speerli, J., Benesch, M.; "Grundablass Panix, Strömungsmessung zur Betriebssicherheit", Wasser - Energie - Luft, Heft 10, pp. 251-254, 1995.

# The Submerged Hydraulic Jump Downstream Sluice Weir

Pan Ruiwen[1]    Xu Yiming[2]

(1)    Professor, Yunnan Polytechnic University, Kunming 650051, China
(2)    Associate Professor, Yunnan Polytechnic University, Kunming 650051, China

Abstract: The hydraulic characteristics of submerged jump in the stilling basin with a horizontal bottom in the downstream of a camel-back-type sluice weir has been researched based on experimental model. It can be considered that when the Froude number $F_{r1}$ before submerged jump is from 3.4 to 7.3 and the submergence coefficient is less than 0.7, the flow condition in stilling basin and its downstream area will not worsen even the free jump becomes submerged. The energy dissipation effect of the submerged hydraulic jump is better than that of the free hydraulic jump.

## 1. INTRODUCTION

Energy dissipation by under current is one of the main dissipation ways in the downstream of discharge works of hydroelectric projects and it got a wide application in many projects. But in general, the quality of construction work for this energy dissipation way is great and the project costs are relatively high. For ensuring forming stable hydraulic jump in the stilling basin as well as restricting the size of basin not too large, the common method in the project is to form submerged hydraulic jump in stilling basin. The submergence coefficient S of a hydraulic jump can be defined as follow (Fig.1):

$$S = \frac{h_t - h_2}{h_2} \quad (1.1)$$

where, $h_t$ is the actual tailwater depth, $h_2$ is sequence depth after the critical jump.

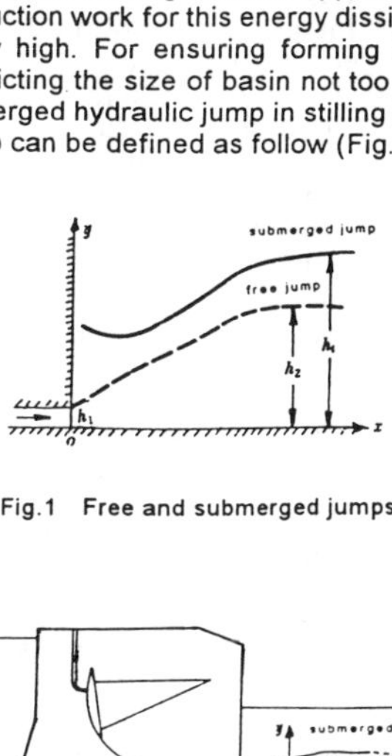

For the submerged hydraulic jump, many achievements have been got by some researchers. Their investigations indicated that the submerged jump would hamper shooting flow to spread out at bottom, thereby increasing the length of jump, causing scour and reducing energy dissipation rate. The velocity near the bottom in the area about the end section of the submerged jump increases with the increasing of Frouds number $F_{r1}$ before the jump and with the increasing of submergce coefficient S.[1][2] Zhang shengming studied the hydraulic characteristics of the submerged hydraulic jump of spillway dam and held opposite opinions.[3]

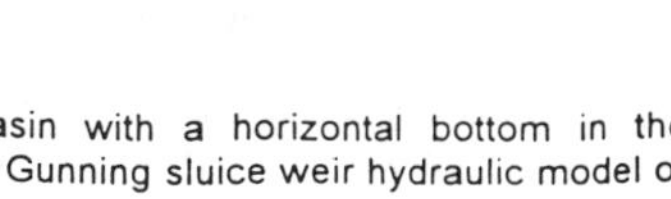

Fig.1 Free and submerged jumps

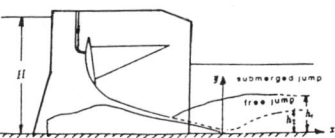

Fig.2 Experimental model arrangement

To get a deeper understanding of the submerged hydraulic jump, we further researched the hydraulic characteristics of submerged hydraulic jump in the stilling basin with a horizontal bottom in the downstream of a sluice weir in conjunction with Gunning sluice weir hydraulic model of Nanpan river.[4]

The experimental model is a camel-back-type sluice weir with three opennings controlled by arched gates (Fig.2). The net width of the openning in the middle is three times of pier's and the width of two side openning is the half of the middle openning's.

They are all made of plexiglass and installed in a glass flume,which is 21-m long,0.5-m wide and 0.64~0.89-m high. Model discharge is controlled by a thin-plate weir in upstream and the tail water level is controlled by a plate gate. Flow depth and water surface wave are measured by a point gage. The mean value and fluctuating features of flow velocity are measured by supersomic velocity meter (CDL-86). In consideration of the characteristics of submerged jump depending on the Floude number $F_{r1}$ before the jump and the submergence coefficient S, 25 cases of hydraulic jump with $F_{r1}$=3.38, 3.94, 4.63, 5.81, 7.34 and S=0, 0.15, 0.3,0.5, 0.7 are conducted on the model (among them, 5 cases of jump with $F_{r1}$=3.38 are from overflow weir, the others are from sluice weir).

## 2. VELOCITY DISTRIBUTION AT THE DOWNSTREAM SECTION OF SUBMERGED JUMP

To contrast out model results with other researcher's, the measuring section in our test is chosen to the section of $5h_2$ downstream $h_1$, ie. $x=5h_2$ ($h_1$ and $h_2$ are sequent depths before and after the aritical free jump). The velocity near the flow bottom and velocity distribution along depth are measured for each running case.

Fig.3 is the relationship between $U_b$ and the Froude number $F_{r1}$ for different submergence coefficient S. $U_b$ is the velocity near the bottom at the section $x=5h_2$, $V_2$ is the mean velocity of this section for free hydraulic jump. From this figure, it can be seen that in the test range ($F_{r1}$=3.38~7.34; S=0~0.7), $U_b$ doesn't increase with the increasing of submergence coefficient S and $F_{r1}$. On the contrary, it has a tendency to decrease with the increasing of S and $F_{r1}$. For example, when $F_{r1}$=3.38 and 7.34, if S=0, 0.3, 0.7 then $U_b/V_2$=1.24 and 1.12, 1.05 and 0.95, 0.82 and 0.72, respectively.

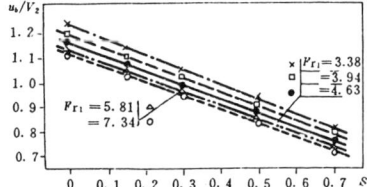

Fig.3 The curve of $U_b/V_2$ and S for diferent $F_{r1}$

Fig.4 is the velocity distribution along depth at the section of $x=5h_2$ for different $F_{r1}$ and S. In Fig.4, $h_t$ and h are mean depth and the depth of measuring point at the section. U is the velocity corresponding to the point with depth h. $V_2$ and $V_t$ are mean velocity of cross section at $x=5h_2$ for critical free jump and for jumps with different S, respectively. Fig.4 shows that both submerged jumps and critical free jumps have the same kind of velocity distribution along the depth. That is, velocity are none uniform with big values near the bottom and small values near the flow surface. The values of none-uniform coefficient β (β=$U_b/V_t$) of velocity distribution for both submerged and critical free jumps are comparatively alike (with the increasing of S, the value of β is from 1.16 to 1.31). Test results indicate that velocity distribution after the submerged jump in a stilling basin with a horizontal bottom downstream sluice weir is not worse even with the increasing of the submergence of coefficient. Further more, with the increasing of submergence coefficient, velocity near the bed at the end of jump decreases. This conclusion is contrary to the results from reference[1] and similar to the results from reference[3]. The first reason is that when the hydraulic jump downstream a sluice weir becomes submerged in the action of higher downstream water level, the head of jump extends upward. This is equivalent to increase the length of stilling basin. Hence, though the length of submerged jump is increased, the jump end position almost remains unchanged. The controlled section $x=5h_2$ is about the end section for each jump with different submergence coefficient S. The second is that the horizontal slope of bottom under the jump enables jet flow to spread intensively in flow direction. The third reason is that there are two piers on sluice weir, the flow over weir has lateral spreading after the piers, so it is three dimensional flow. All these reasons affect flow conditions at the entrance and outlet of stilling basin. The boundary condition in this model is obviously different from reference[1], thereby causing different congnition and conclusions.

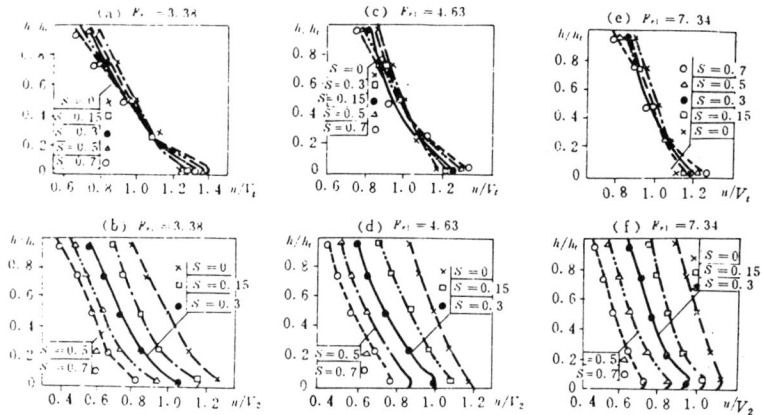

Fig.4 Velocity Distribution in vertical line after jump

## 3. FLOW TURBULENCE CHARACTERISTICS AFTER THE SUBMERGED JUMP

For trying to know a complete picture of submerged hydraulic jump, we have measured some hydraulic parameters relating to flow turbulence of submerged jump on the model.

### 3.1. Turbulence intensity at the end section of submerged jump

The turbulence intensity of flow can be defined as follow:

$$C_v = \frac{1}{\bar{u}}\sqrt{\frac{\sum_{i=1}^{n}(u_i - \bar{u})^2}{n}} = \frac{\sqrt{\overline{u'^2}}}{\bar{u}} \quad (3.1)$$

where, $C_v$ is turbulence intensity, $u_i$ is instantaneous velocity, $u'$ is fluctuation velocity, n is the numbers of $u_i$ in the period of measuring time, $\bar{u}$ is mean velocity in the period of measuring time.

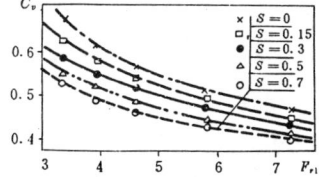

Fig.5 The curve of $C_v$ and $F_{r1}$ for different S

Figure 5 is the relationship between $F_{r1}$ and $C_v$ near the bottom at the section $x=5h_2$ in the stilling basin for deferent S. It is can be seen that $C_v$ is in inverse proportion to $F_{r1}$ and $C_v$ decreases with the increasing of S. For example, when $F_{r1}=3.38$, 4.63 and 7.34, if S=0, then $C_v=0.67$, 0.57 and 0.46; if S=0.3, then $C_v=0.58$, 0.53 and 0.43 and if S=0.7, then $C_v=0.53$, 0.47 and 0.40.

Figure 6 is the distribution of $C_v$ along flow depth. The big value of $C_v$ takes place near the bottom of stilling basin and $C_v$ decreases with the increasing of depth. The distribution curves of $C_v$ along flow depth are the same kind for both submerged and critical free hydraulic jumps. In our test range, turbulence intensity near the free surface is about 70~80% of that near the bottom at the section end of jump nearby.

### 3.2 Turbulence energy at the end of jump

The turbulence energy for per unit weight of liquid is:

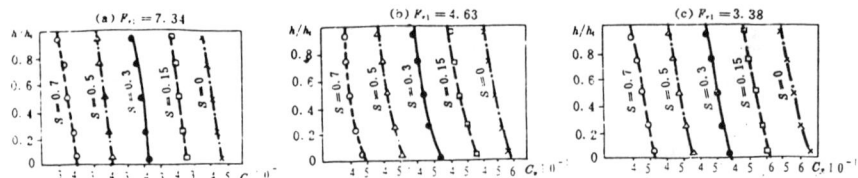

Fig.6 Cv Distribution along depth

$$E_t = \frac{\overline{(u+u')^2}}{2g} - \frac{\overline{u}^2}{2g} = \frac{1}{2g}(\overline{u}^2 + \overline{2uu'} + \overline{u'^2} - \overline{u}^2) \tag{3.2}$$

where $\overline{2uu'} = 0$, then

$$E_t = \frac{\overline{u'^2}}{2g} \tag{3.3}$$

The value of $\overline{u'^2}$ can be obtained from equation (3.1).
Figure 7 is the relationship between $F_{r1}$ and $E_t/E$ for different S. The value of $E_t/E$ is at the section x=5h$_2$, $E = \dfrac{V^2}{2g}$, V

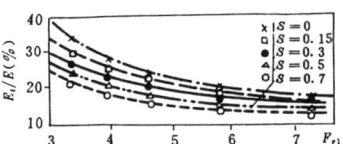

Fig.7 The curve of $E_t/E \sim F_{r1}$ with different S.

is the mean velocity of the same section.
It can be seen that $E_t$ is in inverse proportion to $F_{r1}$ and to S. In other words, $E_t$ decreases with the increasing of $F_{r1}$ and S. Within the range of $F_{r1}$=3.38~7.34, $E_t/E$=33.6~16.9%, 26.3%~15.1%, 21.6%~12.7% for S=0, 0.3, 0.7, respectively.

Above results indicate that flow turbulence after the submerged hydraulic jump is damped with the increasing of $F_{r1}$ and S. The decreasing of turbulence energy in flow means that the energy dissipated by submerged hydraulic jump is increased. That is the flow condition after submerged jump has been improved. Because the turbulence energy (also called the rest energy) involved in the flow after the jump may cause scour in downstream.

4. ENERGY DISSIPATION RATE OF SUBMERGED HYDRAULIC JUMP

The general definition of energy dissipation rate $K_j$ is the ratio of energy loss $E_j$ by hydraulic jump to the specific energy $E_1$ just before the jump, without considering the influence of turbulence after the jump.

$$K_j = \frac{E_j}{E_1} = \frac{E_1 - E_2}{E_1} \tag{4.1}$$

where, $E_2$ is specific energy at the end section of the jump.
For a submerged hydraulic jump, the specific energy before the jump is:

$$E_1' = h_1' + \frac{V_1^2}{2g} \tag{4.2}$$

where, $V_1$ is mean velocity at the cross section before critical jump in the stilling basin, $h_1'$ is the submerged depth at this section.

The specific energy at the section after the submerged jump is:

$$E_2' = h_t' + \frac{V_t^2}{2g} \qquad (4.3)$$

where, $h_t'$ and $V_t$ are depth and mean velocity of the section at $x=5h_2$ corresponding to different S.
Therefore, the energy dissipation rate of the submerged jump is:

$$K_j' = \frac{E_1' - E_2'}{E_1'} \qquad (4.4)$$

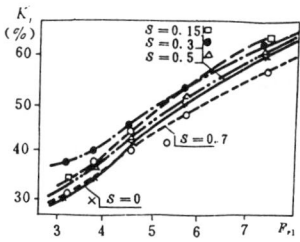

Fig.8 The curve of $K_j' \sim F_{r1} \sim S$

Figure 8 is the relationship between and $F_{r1}$ for different S. within the test range of $F_{r1}=3.38\sim7.34$, the energy dissipation rate of critical free jump is 29.5%~60.4%. When $S \leqslant 0.5$ and when $S<0.7$ for lower Froude number, the energy dissipation rate of submerged jump is greater than that of free jump.

Fig.9 is the relationship between S and relative energy dissipation index $\eta = K_j'/K_j$ (the ratio of energy dissipation rate of submerged jump to free jump). For each $F_{r1}$, $\eta$ increases with the increasing of S at first, and comes to its maximum value, and then decreases. But for most submerged hydraulic jumps ($S \leqslant 0.5$ and for lower $F_{r1}$, $S<0.7$), $\eta \geqslant 1$.

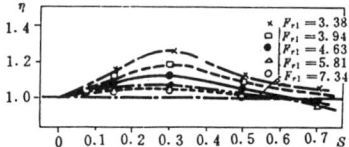

Fig.9 The curve of $\eta \sim S$

Test results indicate that, within the test range, especially for lower Froude number, the energy dissipation effect of the submerged hydraulic jump is better than that of the free hydraulic jump.

## 5. FLUCTUATION OF WATER SURFACE AFTER THE SUBMERGED HYDRAULIC JUMP

The fluctuation degree of water surface in downstream river of stilling basin is an important index to judge energy dissipation effect of hydraulic jump and to reflect the amount of rest energy in flow after the jump. In this model research, fluctuation of water surface at the downstream sections of 1 to 3 times of the jump length $L_j$ away from the entrance of stilling basin are measured for difference $F_{r1}$ and S. Figure 10 indicates that the relative wave height $\Delta h = \bar{h}/H$ (the ratio of mean wave height $\bar{h}$ to total head H before the weir) on water surface after the submerged hydraulic jump decreases with the increasing of $F_{r1}$ and S. The wave on flow surface is damping gradually along flow direction. For example, At the end section

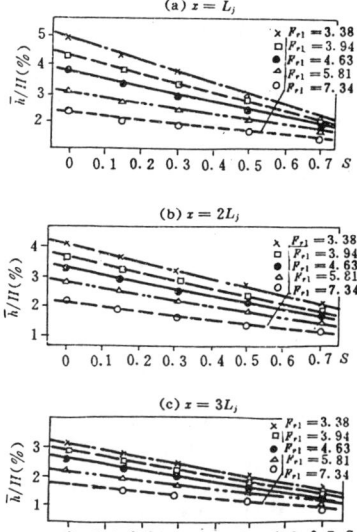

Fig.10 The curve of $\bar{h}/H \sim F_{r1} \sim S$

of the hydraulic jump ($x= L_j$), when $F_{r1}$ increases from 3.38 to 7.34, then $\Delta h$ decreases from 5.0% to 2.4%, 3.8% to 1.9% and 2.3% to 1.5% for S=0, 0.3 and 0.7, respectively. At the Sections of $x=2 L_j$ and $x=3 L_j$, the tendency of $\Delta h$ is as the same to the section of $x= L_1$. The reason is that after the increasing of S, water depth in downstream increases, therefore the turbulence of flow is weakened and mean velocity of cross section as well as surface velocity is reduced. Meanwhile velocity near the bottom dose not increase. It can be seen from above that the energy dissipation effect of submerged hydraulic jump is better.

## 6. CONCLUSIONS

(1) In the stilling basin with a horizontal bottom in the downstream of a camel-back-type sluice weir, the head of hydraulic jump extends upward after the free jump become submerged, therefore flow to stilling basin spreads intensifyly in flow direction and the flow after the piers has lateral spreading. Hydraulic condition at the entrance of stilling basin for the submerged jump has obviously changed compared to free jump. The position of the end section of submerged jump is close to that of free jump. At the section $x=5h_2$, about the end of free jump, velocity near the bottom for the submerged jump does not increase with the increasing of submergence coefficient S, on the contrary, it decreases with the increasing of $F_{r1}$ and S. Meanwhile, with the increasing of water depth downstream, flow turbulence is weakened after the end of submerged jump and mean velocity of cross section as well as surface velocity reduces, with a smaller fluctuating flow surface. All these are beneficial to improv the energy dissipation effect of submerged jump and have been demonstrated by the practice operation of Gunin sluice weir.

(2) When the Froude number $F_{r1}$ at the section before the jump is 3.4~7.3 and submergence coefficient S is less than 0.7, the energy dissipation effect of the submerged hydraulic jump is better than that of the free jump. In the past, some researchers had worried about that the submerged jump would hamper shooting flow to spread at bottom, thereby reducing energy dissipation effect and causing scour downstream. This consideration has its limitation and not accuracy enough.

(3) For reducing the qunatity of construction work of stilling basin, the smaller submergence coefficient S can be chosen in design of stilling basin (such as S=0.05~0.10). But in practice if relatively large water depth downstream are encountered (such as 0.1~0.2<S<0.5~0.6), it is certain to the effectiveness of energy dissipation of the submerged hydraulic jump.


ACKNOWLEDGMENT

This study was supported by Yunnan applied and basic science foundation.


REFERENCES

1. Govinda Rao N S, Rajaratnam N. The Submerged Hydraulic Jump. J.Hyd. Div. proc. ASCE, 1963,89(Hy)
2. The Changjiang River academy of sciences at al, Problems of Energy Dissipation And Erasion Prevention In Downstream Of discharge Works, water conservancy publishing house. Oct. 1980
3. Zheng Shenming, Inquires of characteristics of submerged hydraulic jump in stilling basin with a horizonted bottom in downstream of sluice weir, Technique of water conservancy & water power, No.5, 1989
4. Pan Ruiwen, Hydraulic model Report on spillway dam of Gunin Reservoir in Nan Pan river. Yunnan Institute of Technology. Feb.1990.

# Control of Hydraulic Jump by Abrupt Drop

J. E. Richardson
Flow Science, Inc., Los Alamos, NM, USA

## ABSTRACT

A powerful computational method based on fractional-volume concepts is briefly described and then applied to the simulation of hydraulic jumps and structures which are used to control them.

## BACKGROUND DISCUSSION

The hydraulic jump is a classical, well studied hydraulics phenomenon. Relations exist to predict its height, length, and location in isolation. When control structures are added, the situation moves beyond classical analysis because of splashing, turbulence, and other unsteady flow phenomena. It is the use of a hydraulic jump to increase the flow beneath a sluice that is the subject of this paper.

The flow beneath a sluice may be free or submerged, depending on the depth of the tailwater. When the tailwater is low, the flow is free, and the discharge through the sluice is maximized. When the tailwater is increased, the flow beneath the sluice may become submerged and the discharge through the sluice will be reduced. For many applications it is desirable to maintain a free discharge through a sluice. For these situations a hydraulic jump may be used to hold back the tailwater. In situations where the downstream depth is larger than the sequent depth for a normal jump, a drop in the channel floor may be used to ensure a jump.

To investigate these phenomena mathematically, it is necessary to resort to computational methods. In particular, one must use a computational method capable of treating transient flow phenomena involving overturning, fluid turbulence, and complex geometric regions. The commercial software package FLOW-3D (Flow Science, Inc, 1987) has all these requirements.

In the next section we briefly describe the salient features of FLOW-3D that give it the ability to treat complex, transient, free-surface flows. Then we describe an application of FLOW-3D to the simulation of hydraulics jumps in which we investigate the influence of geometry on the flow rate through a sluice gate.

## OVERVIEW OF NUMERICAL METHOD

Overturning free surfaces, splashing, and other free-boundary distortions are not easily modeled using moving grids (e.g., Lagrangian) techniques. Such problems are better treated using fixed Eulerian grids coupled with volume tracking, not surface tracking, methods such as the Marker-and-Cell method (Harlow and Welch, 1965), or the Volume-of-Fluid (VOF) method (Hirt and Nichols, 1981).

## BASIC VOLUME-FRACTION CONCEPTS

The basic idea of a volume-fraction method is to imagine the region to be modeled as subdivided into two components. For example, it may be composed of a collection of fluid and gas subregions, or a collection of fluid and solid subregions. In either case, we introduce a generalized, Heavyside function (Lighthill, 1958) that assumes the value of zero in one region and unity in the other region. Then the fluid equations of motion, say the Navier-Stokes equations, are multiplied by this generalized function and integrated over a typical control volume (i.e., the cells in an Eulerian grid). Using simple approximations (Hirt and Sicilian, 1985) the resulting discrete equations are similar to those used for flow in porous media, which is not surprising since this is one way to derive those equations.

It is, of course, necessary to supplement the resulting equations with additional terms to account for interfacial effects such as drag and mass or heat transfer. Considerable latitude exists in choosing the form of those interfacial contributions to represent different types of region distributions and material properties. In fact, this latitude is one of the more powerful features of volume-fraction methods because it permits a great variety of complex physical phenomena to be easily modeled.

In the FLOW-3D program two types of volume-fraction techniques are used. One of these corresponds to the original VOF method for the treatment of free surfaces and two-fluid interfaces. Of course, a many improvements to both the VOF method itself and to the application of free-surface boundary conditions have been incorporated into the program over the past ten years.

A second volume-fraction technique is referred to as FAVOR (Fractional Area Volume Obstacle Representation) is used in the FLOW-3D program to model complex geometries (Hirt and Sicilian, 1985). In this case a solid volume faction (or the complement "open" fraction) is supplemented with area fractions in each of the three coordinate directions. The incorporation of these area/volume fractions into discrete control-volume flow equations automatically forces rigid, free-slip boundary conditions at all solid surfaces. This technique is particularly useful in connection with free surfaces because it allows one to more easily satisfy free-surface boundary conditions in control volumes that contain both types of boundaries.

Volume-fraction methods as used in the FLOW-3D program have many advantages. For instance, they require much less memory than moving grid methods, they allow arbitrary boundary displacements without grid distortion, and they permit the creation and destruction of bounding surfaces without the need for complicated numerical logic. Furthermore, fractional volume methods can be generalized for representing other types of physical processes such as time-dependent geometry, unsaturated flow in porous media, and flows through mounds of rubble.

## APPLICATION TO SLUICE GATE

For this study we have chosen to model a sluice gate that is 70 cm high and provides a 7 cm opening for the flow. The water elevation upstream of the sluice is 57 cm, the water elevation downstream of the sluice is 35 cm, and the channel is horizontal.

## THE COMPUTATIONAL MODEL

For our computational model, which is two dimensional, both the upstream and downstream fluid elevations are fixed. These conditions are applied as boundary conditions to the left (upstream) and right (downstream) sides of the computational region. Hydrostatic pressures are also assumed to exist at these boundaries.

For an initial condition the upstream and downstream fluid heights are assumed constant on either side of the sluice. The fluid upstream moves with a uniform velocity of 0.35 m/s, and the fluid downstream moves with a velocity of 0.57 m/s. This initial condition was chosen for simplicity and is not meant to model any particular physical situation.

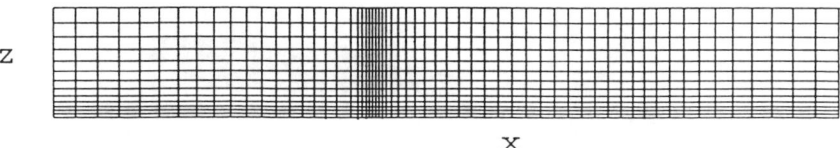

Fig. 1a. Computational Mesh (every 3$^{rd}$ grid line)

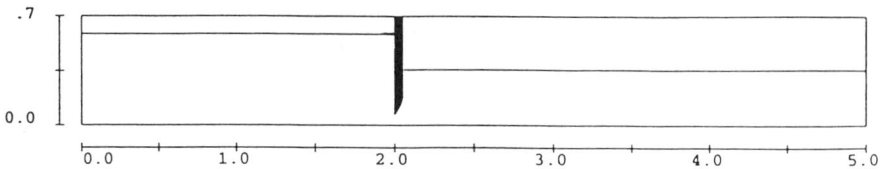

Fig. 1b. Initial Fluid Configuration

Figure 1a shows the computational grid used for the base calculation (every third grid line plotted), and Fig. 1b shows the initial fluid configuration. The grid contains 180 cells in the horizontal direction and 38 cells vertically. Cell sizes are smallest in the vicinity of the sluice opening where the greatest resolution is desired.

All calculations reported include the Renormalization Group (RNG) model of turbulence (Yakhot and Orszag, 1986). This model is effectively an extension of the more common k-ε model except that it has many fewer empirical parameters and is more accurate in regions of strong shear.

## FREE DISCHARGE

Computations were carried out for a period of 20 seconds at which time the computed flow rate is nearly steady (the jump, an unsteady phenomenon, exists in a state of quasi-equilibrium), Fig. 2a-b. The computed discharge rate is 0.15 m$^2$/s, which is in good agreement with a theoretical value of 0.14 m$^2$/s (Chow, 1988). That the computed flow rate is 7% greater than the theoretical value may be explained by the fact that the flow contraction which occurs beneath the sluice is not fully resolved by the grid (additional computations using a finer mesh were performed and the difference between the calculated flow rate and the theoretical value was indeed reduced).

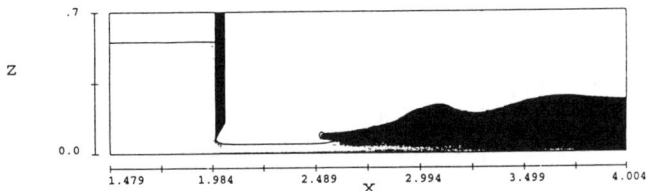

Fig. 2a. Free Discharge Fluid Configuration (colored by turbulent energy)
$F \cong 4.0$

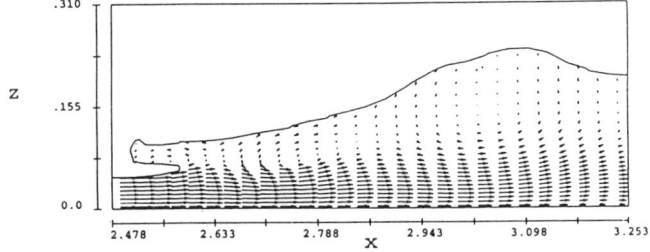

Fig. 2b. Velocity Vectors within Jump (maximum velocity = 3.11m/s)

Inspection of Fig. 2a. reveals the formation of the vena contracta beneath the sluice and the formation of a hydraulic jump downstream. The flow depth within the contraction is reduced to 68% of the sluice opening dimension. In theory, the flow should be reduced to about 62% of the sluice opening height. This difference accounts for the increase in the calculated flow rate as previously noted. The dimensions of the jump as well as its behavior, though not reported, also compare closely to experimental values.

Finally, we note that in the region where the flow under the sluice enters the jump there appears to be large holes of air penetrating the jump, Fig. 2b. These regions, in fact, contain a mixture of air and water. The plotted free surfaces in Fig. 2 are contours of fluid fraction equal to 0.5. Furthermore, fluid velocities have only been plotted in mesh cells that are more than half filled. This fluid-fraction of 0.5 boundary is the proper location of the free surface when the fraction has a zero-to-one discontinuous distribution. Other plots can be made to show, for example, the distribution of fluid fraction using continuous color shading. In any case, the calculation correctly attempts to model the entrainment of air, but it does not do this accurately because the computational model does not include dynamical effects of air.

## SUBMERGED DISCHARGE

Let us now suppose that the operating conditions have changed and that the elevation of the tailwater has risen. In a second calculation the downstream elevation was increased to 38 cm. Because of this change the flow beneath the weir becomes submerged, Fig. 3. The calculated discharge through the weir for this situation is 0.1 m$^2$/s (a reduction of 33% compared to the first calculation).

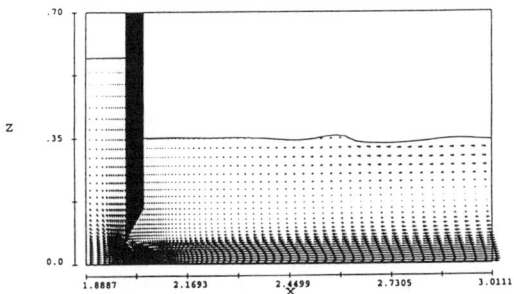

Fig. 3. Computed Flow in Submerged Jump

## CONTROL BY AN ABRUPT DROP

When the downstream depth is larger than the sequent depth for a normal jump, a drop in the channel floor may be used to ensure a jump. In a third calculation a drop in the channel floor has been added. For the operating conditions used in the second calculation, it is seen that the jump is moved downstream away from the sluice opening. The rate of flow beneath the sluice is once returned to its original value of 0.15 m²/s.

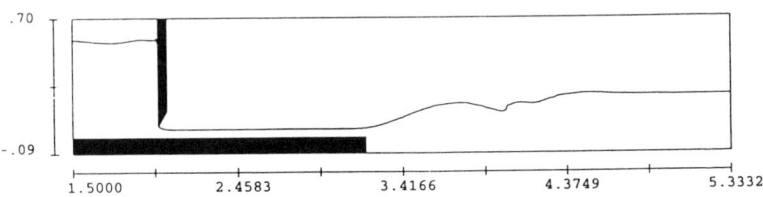

Fig. 4. Fluid Configuration with Abrupt Drop

## SUMMARY COMMENTS

Whether or not a drop in the channel is feasible depends on other factors such as local geology, economics of construction, etc. The point, however, is that computational simulations of the type described in this paper offer a rapid and inexpensive way to investigate the performance of hydraulic structures under varying operating conditions and configurations.

## REFERENCES

Chow, V.T., 1988. *Open-Channel Hydraulics.* New York: McGraw-Hill Publishing Company.

Flow Science, 1987. *FLOW-3D: Computational Modeling Power for Scientists and Engineers.* Flow Science, Inc. Report FSI-87-00-01.

Harlow, F.H. & J.E. Welch, 1965. Numerical Calculation of Time-Dependent Viscous Incompressible Flow. *Physics of Fluids,* 8:2182-2189.

Hirt, C.W. & B.D. Nicols, 1981. Volume of Fluid (VOF) Method for Dynamics of Free Boundaries. *Jour. Comp. Phys.* 39:201-225.

Hirt, C.W. & J.M. Sicilian, 1985. A Porosity Technique for the Definition of Obstacles in Rectangular Cell Meshes. Fourth Inter. Conf. Ship Hydrodynamics, Wash. D.C. Sept. 1985.

Lighthill, M.J., 1958. *Fourier Analysis and Generalized Functions.* Cambridge: Cambridge Uni. Press.

Yakhot, V. & S.A. Orszag, 1986. Renormalization Group Analysis of Turbulence I. Basic Theory. *Jour. Sci. Computing* 1:1-51.

# APPLICATION OF TELEMAC 2D SOFTWARE TO DIMENSIONING HYDRAULIC STRUCTURES: CASE OF THE MESCE DAM

E. LAPERROUSAZ*, C. MOULIN**, J.P. BLAIS*, E. CHIESA*
* Electricité de France, National Hydro Engineering Centre, Bourget-du-Lac, France
** Electricité de France, National Hydraulics Laboratory, Chatou, France

## ABSTRACT

This article describes parallel study of a spillway by physical model and by numerical model using the TELEMAC 2D software. Through comparison of measured vs. computed results, the validity of the numerical model as an aid in design of hydraulic structures is demonstrated.

## INTRODUCTION

The TELEMAC 2D software program, developed by Electricité de France's National Hydraulics Laboratory (LNH), reproduces currents and changes in free surface for 2D flows. It considers a range of physical phenomena including Coriolis force, friction, wind and turbulence (JM. HERVOUET Proceedings of the APCOM 96 : Third asian pacific conference Seoul South Korea 16-18/09/96)). It has been applied in many studies. On the strength of this success, EDF, and in particular its National Hydro Engineering Centre (CNEH), have attempted to enlarge the field of application of TELEMAC 2D to hydraulic structures.

A numerical tool to simulate operation of hydraulic structures such as spillways helps both in design of new schemes and in maintenance or rehabilitation of existing ones. In study of new spillways, numerical simulation can be used to achieve:
- more refined preliminary design than is possible with charts. Project feasibility and cost often depend on charts as more expensive physical models can only be set up in the detailed design phase;
- optimization through study of several alternatives, whereas the difficulty in modifying a physical model limits the number of alternatives that can be studied;
- the greatest possible consideration of inflow conditions to the structures. In fact, because of rules of similarity and limits on size of test facilities, the use of physical models is limited. When upstream conditions influence discharge capacity because of their effect on inflow to the spillway, the model's scale must be restricted. Either the entire area desired is not covered, or the structures are reproduced at too small a scale.

Many hydro structures are aging, and spillway dimensions are based on the practices and hydrological data available at the time of design. However, practices change and hydrological knowledge grows with longer records allowing improved extreme flood estimates. Discharge capacities must therefore be corrected. Past studies on physical models rarely included these new values and the physical models no longer exist. Numerical modeling can therefore be of great assistance in checking discharge capacity of existing spillways and defining corrections, with confirmation by physical model if required.

The CNEH and the LNH decided to test TELEMAC 2D in ongoing study of the spillway at Mesce dam in France. The present article sets out the conditions of the study, compares measurements taken on the physical model to the results of numerical simulation and concludes with an initial validation of TELEMAC in spillway dimensioning or determination of maximum flow on an existing spillway.

## CHARACTERISTICS OF THE WORK

Mesce dam, 77 m high and 145 m long at the crest, was built in 1917 on the Bionia river, in the southeast of France. The Full Supply Level (FSL) and Maximum Water Level (MWL) are both presently 1368.75 m a.s.l. Useful reservoir storage is 1.31 hm$^3$, fed by a 61 km$^2$ catchment area. Floods are discharged by a right bank spillway composed of (i) a channel 86 m long and 11.50 m wide, controlled upstream by two tainter gates with an automatic opening system (floater/counterweight) from FSL, and (ii) a set of 13 manually opened side sluices in the left bank guide wall.
In September 1993, a torrential rain caused exceptional flooding. Since the 13 small gates could not be opened manually, and the water level rose by 6.75 m in 1 ½ hours due to an inflow of 1.6 m$^3$/s.

## THE HYDRAULI MODEL STUDY

A hydraulic model study of the spillway was conducted to:
- improve safety by replacing the small gates with a free overflow sill;
- increase the spillway chute's discharge capacity for the updated design flood (340 m$^3$/s instead of the present 300) and avoid any spilling over the sides;
- limit the risks of erosion downstream by designing an energy stilling bucket.

The hydraulic model study first confirmed the observations made during the 1993 flood (overspilling in the downstream part of the chute when the tainter gates were fully open and the small gates closed); checked the present discharge capacity at MWL with all gates open; and established the level reached by the new design flood, El.1369.20 a.s.l. (i.e. 45 cm above present MWL). Then, the required modifications were defined to include replacement of the manually opened gates with a continuous ungated sill whose crest level will determine the new FSL, required to maintain the existing MWL, enlargement of the chute and heightening of its sidewalls as required to discharge 340 m$^3$/s, and establishment of new operating conditions by study of the various flow states in the chute, laws governing operation of the tainter gates, etc.

## PHYSICAL MODEL AND MEASUREMENTS

The physical model was built to Froude similarity laws at a scale of 1:75, i.e. a velocity and time scale of 1:8.7 and a flow scale of 1:48,700. The model covers the entire dam crest, with a margin of 50 m on either side (i.e. 250 m in all). Topography was represented precisely from 180 m upstream the dam to 150 m downstream. The model's stilling zone extended 1.5 m upstream and 2 m downstream.

Flows were measured upstream using an electromagnetic flow meter and a rotameter. Level measurements were taken from point gauges and pressure cells distributed over the model (reservoir and spillway chute). Current measurements were taken by Particle Image Velocity (PIV) to give the velocity field in any plane at a given instant; photographs were also taken. Measurement accuracy was estimated at ±0.1 l/s on the model (±4.8 l/s life-size) for flows and 0.5 mm on the model (0.03 m life-size) for depth. The tests were done by setting an upstream water level, and then in each test varying flow to achieve a stable reservoir water level.

## PHYSICAL MODEL VS. TELEMAC 2D COMPUTATIONS

Studies on the physical and numerical models were run interactively. The scale model was used to validate the computer model, which was then used to test modifications to the spillway before they were made on the physical model. Only validation of TELEMAC 2D is described here by comparising measured vs. computed results.

## THE TELEMAC 2D NUMERICAL MODEL

The TELEMAC 2D software was developed to solve Saint Venant equations by finite element computations. The model provides the depth and velocity of water columns at any instant, with velocity assumed constant along the vertical.

The area modeled from topographical and bathymetric maps made it possible to take flows into account both in the inlet upstream of the gates and in the spillway. The mesh (3000 nodes) was dense in the spillway channel, with a mesh size of about 1 m.

Steady state conditions were obtained by convergence over time of an unsteady calculation. The model's boundary conditions are (i) sliding (zero friction) on solid vertical walls; (ii) imposed depth and free velocity on the upstream liquid limit (reservoir), and (iii) depth and velocity at the downstream end of the chute.

## MEASUREMENTS VS. COMPUTATIONS WITH GATES OPEN

The first case described here is a configuration with the existing structures, 1 or 2 gates open and sill closed.

The numerical model was calibrated, at a reservoir water level of 1368.75 m, using a Strickler coefficients of 30 for the entire model. Because of the low flow speeds observed in the reservoir, this value has very little influence on the upstream part of the model. For the chute, a value of 30, although low for concrete, is justified because it compensates for zero friction on vertical walls. In the calibration process, the depths of water measured at pressure cells P1 to P7 are practically identical to those given by TELEMAC, the numerical model gives a discharge capacity close to that of the physical model, i.e. 195 m$^3$/s vs. 200 m$^3$/s; and the free surface cross-slope measured at the bend is 0.76 m, vs. a computed cross-slope of 0.59 m.

Using the calibrated model, the measurements and computations were compared with the reservoir at 1368.00 m a.s.l. and both gates open. For both waterlines and discharge capacity, the numerical model gives results close to those of the physical model (154 m$^3$/s vs. 161 m$^3$/s). The free surface cross-slope calculated at the elbow, i.e. 0.49 m vs. a measured cross-slope of 0.70 m, is satisfactory.

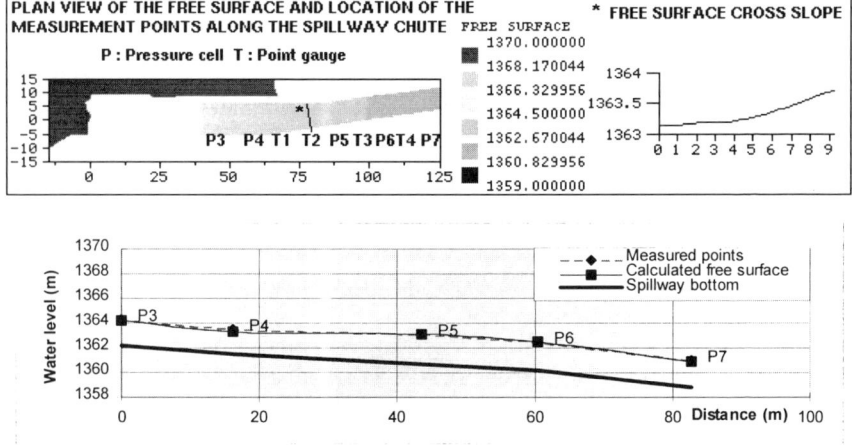

Waterlines measured at pressure cells P1 to P7 and generated by TELEMAC

Pressure cells in the model correspond to a life-size diameter of 5 m, and that the 5 m variation in the free surface is of the same order of magnitude as the maximum deviation noted between the measurement and the computation at the points of adjustment.

The case only one gate open was then studied. For both waterlines and discharge capacity, the numerical model again gives results that are practically identical to those of the physical model (97 m³/s vs. 102 m³/s).

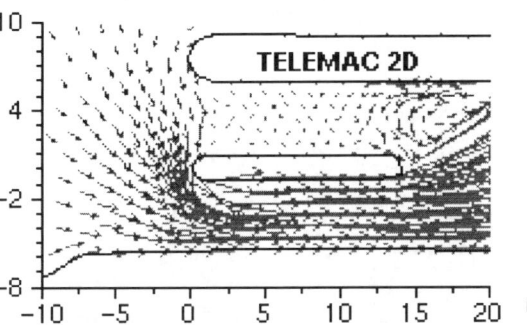

Flow lines photographed on the physical model and generated by TELEMAC

## MEASUREMENTS VS. COMPUTATIONS WITH SILL OPEN

This second case concerns a configuration with the existing structures, water level at 1368.75 m a.s.l., both gates and the sill open. The model was adjusted in terms of sill flow coefficient, using a figure of 0.4. Comparisons were made for discharge capacity, where the numerical model gave 278 m3/s vs. 305 m3/s for the physical model, and water level deduced from pressure cells P1 to P7 and plotted by TELEMAC.

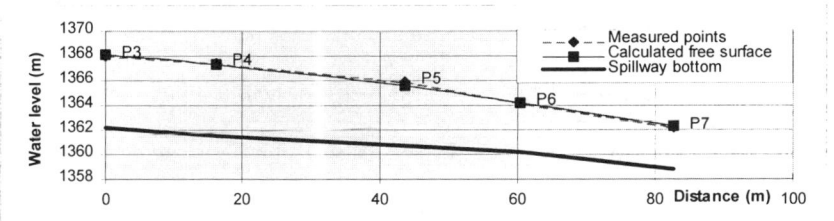

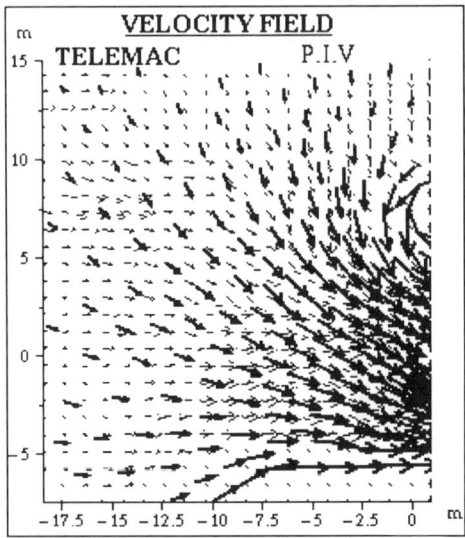

Current measurements taken at the gate inlet, in a horizontal plane located in the first third of the depth under the free surface, were compared with the TELEMAC computations, done with a scatter coefficient of 0.1 m²/s (isotropic scatter that remains constant in time and space).

Since TELEMAC averages the speeds over the vertical, the profile of the logarithmic velocities over the vertical were multiplied by a coefficient of proportionality. In the present case, this coefficient, which is a function of elevation z, is 1.8.

## CONCLUSION

The present study has shown the advantages to using the TELEMAC 2D program as an aid in study of spillways. An initial validation of the software has been achieved through comparison of numerical results with measurements made on a physical model of the Mesce dam. This validation process, and the interactive studies of physical and numerical models, are continuing with study of the new spillway design for Mesce, and will be pursued on other hydraulic structures. However, the considerable benefits that can be expected from numerical simulation are already worth mention. They include better understanding of structures in the preliminary design stage, better optimization through study of alternatives, and, for existing structures, confirmation or modification of discharge capacities at lower cost. The power of the TELEMAC 2D simulation, as demonstrated here, does not exclude the use of a physical model, but rather benefits it in two ways:
- through better guidance of the physical model and modifications to it when the studies are carried out interactively;
- better focusing (and in particular helping in selection of the most appropriate scale) of the physical model on the elements it alone can address, e.g. study of a flip bucket to avoid scouring downstream.

# Numerical Analysis of Unsteady Flow on Rotating Buckets for Optimal Operation of Pelton Turbines

T KUBOTA and Y NAKANISHI
Kanagawa University, Yokohama, Japan

## ABSTRACT

Unsteady free water sheet flow on a rotating bucket of a Pelton turbine is numerically analysed by introducing a new idea of animated cartoon method. The sequential results obtained from frame to frame reveal that the method can be an effective design tool through the study of the distribution of streak patches and thickness of unsteady free sheet flow in rotating buckets.

## 1. INTRODUCTION

Under the atmospheric environment in a housing of a Pelton turbine, the rotating runner buckets successively penetrate into the free water jets shot from the nozzles. Since the penetration of a rotating bucket into a straight-going jet is intermittent in space and time, the location, velocity and incidence angle of inflow to the bucket varies unsteadily with the rotation of bucket. The location and the thickness distribution of the localized water sheet flow having a free surface also varies unsteadily on the three dimensional bucket wall with the bucket rotation.

Normally, the free sheet flow on the bucket wall unsteadily discharges from the brim while shifting its location and velocity along the brim with bucket rotation, whereas it tends to emerge from the cutout of bucket under the specific operating conditions of the reduced head and the high needle stroke. Different from the outflow at the brim, the outflow at the cutout has the large angular momentum due to the high cutout blade angle, and reduces the specific hydraulic energy of runner (differential angular momentum between inlet and outlet of bucket). In high specific speed multi-nozzle Pelton turbines, the outflow from the cutout (so-called CutFlow) may interfere with the coming jet from the next nozzle resulting in the falaise effect[1]. Since the CutFlow tends to increase with increasing Reynolds number from a model to its prototype, we can not predict the jet interference of the prototype units from its model tests. It is very important, therefore, to theoretically clarify the dynamic behaviour of unsteady flow on the rotating buckets to make a correction of scale effect on the serious jet interference in high specific speed Pelton turbines.

So far, the only trial was made by Nakase et al[2] to numerically analyze the free sheet flow on the bucket, however, their sheet flow was impractically steady because they gave the inflow conditions as if the nozzle would rotate synchronously with the runner. This study tries to analyze the actual unsteady free sheet flow on the rotating bucket by introducing an animated cartoon method.

## 2. NUMERICAL ALGORITHM FOR UNSTEADY FLOW ON BUCKET

### 2.1 ANIMATED CARTOON METHOD

In a Pelton turbine, a runner bucket rotates steadily with the constant angular speed, while it intermittently penetrates into a free water jet shot from the stationary nozzle as shown in Fig. 1. Since the penetration of a bucket into a straight-going jet is intermittent in space and time, the inflow of jet into the bucket is unsteady or time dependent, of local and partially filled in. The location, velocity and thickness of free sheet flow on the bucket wall, then, varies unsteadily with the rotation of bucket. If the nozzle rotated synchronously with runner in keeping the relative position between the nozzle and bucket, the inflow of jet into the bucket would become steady or time independent, and the successive free sheet flow on the bucket wall would also be steady.

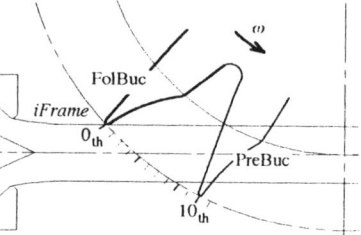

Fig. 1 Definition of frames

To numerically analyze the unsteady free sheet flow in a bucket, we introduce a new idea of an *animated cartoon method*. In the animated movie film, the dynamic movement of an event can be simulated by displaying a sequence of gradually changing stationary *frames* with the specified constant speed. In our case, an animated *frame* represents the infinitesimal time-interval corresponding to the incremental rotating angles $\Delta\theta$ of bucket around the turbine shaft. The number of frames is selected by dividing the spatial angle $\theta_B$ between the adjacent buckets into, say, ten frames. The start of frames (0-th order) is defined as the positioning where the splitter tip of bucket touches the inner-most streamline of the free jet as shown in Fig. 1. Within a frame, the free sheet flow in the bucket is assumed to be steady by introducing the steady inflow supplied by the *rotating nozzle* in keeping its relative position with the bucket. At every beginning instant when the bucket rotates from frame to frame with the constant angular speed of $\omega$, the position of the nozzle is *returned back* to its original place. In the animated cartoon method, the sheet flow is always unsteady at the beginning instant of animated frames, whilst the flow is steady within a frame.

### 2.2. UNSTEADY INFLOW CONDITIONS

Let's find the respective unsteady inflow conditions at the beginning instant of animated frames. When the jet diameter $d_j$ is large with respect to the reference (jet pitch circle) diameter $D_{ref}$, the shape of the velocity triangles is different between the inner- and outer-most streamlines in the free jet as shown in Fig. 2. The relative velocity $W_{ji}$ at inner-most streamline is rather high and less inclined to the splitter due to the lower peripheral speed, whereas the velocity $W_{jo}$ at outer-most line is lower and much inclined to the splitter. Thus the relative velocity distribution in a given frame varies with the peripheral speed at the landing position of jet streamlines on the splitter. The cylindrical solid jet is discretized into the many thin jet-layers as shown in the dashed lines in Fig.2. The discharge for each layer $\Delta Q_j$ can be computed from its segmental cross sectional area $\Delta A_j$ in the circular

Fig. 2 Inflow conditions

jet times the jet velocity $c_j$. The respective jet-layer is represented by the stream plane at the middle of the layer. The number of jet stream planes that can land on the bucket is detected by finding the intersection between the stream plane and the splitter line of bucket at the beginning instant of the respective frames. At the respective inflow points thus obtained for the every frame, the relative velocity $W_1$ is computed and the thickness of inflow sheet $h_1$ at the splitter is determined from the discharge $\Delta Q_j$ divided by the product of the oblique thickness $b_1$ of layer along the splitter and the relative velocity $W_1$ as shown in Fig. 3.

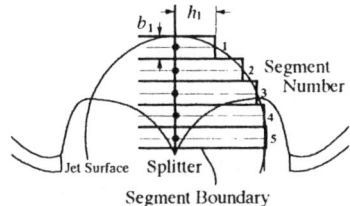

Fig. 3 Jet-layers landing on splitter

## 2.3. GOVERNING EQUATION FOR SHEET FLOW ON BUCKET

Let's consider the steady free sheet flow on a bucket within a frame. The flow is relative to the rotating coordinate system having the constant angular speed of $\omega$. The free sheet flow has its thickness between the solid wall and the free surface as a scalar value, however, is two dimensional constraint on the three dimensional curved surface of the bucket. The flow is inviscid, at the moment, neglecting the shearing stress on the wall. The relative acceleration per unit mass of the flow can be obtained by the following equation;

$$\frac{d^2 s}{dt^2} = \alpha_{cons} + \alpha_{pres} + \alpha_{grav} - \alpha_{cent} - \alpha_{cori} \quad (1)$$

where $s$ is the distance along path line to be obtained on the bucket surface from the point at the beginning instant in an animated frame, $\alpha_{cons}$ and $\alpha_{pres}$ the accelerations due to the constraint on the curved wall and the lateral pressure difference, $\alpha_{grav}$ the acceleration due to gravity, $\alpha_{cent}$ and $\alpha_{cori}$ the apparent accelerations based on the centrifugal force due to the bucket rotation and Coriolis force in the rotating coordinate system, respectively, within the animated frame. The respective component accelerations above can be obtained as follows;

$$\alpha_{cons} = \frac{W^2}{R_c} \quad (2) \qquad \alpha_{pres} = \frac{\Delta p_{late}}{\rho} \quad (3)$$

$$\alpha_{grav} = g \quad (4) \qquad \alpha_{cent} = R\omega^2 \quad (5) \qquad \alpha_{cori} = 2\omega W_{r\text{-}\theta} \quad (6)$$

where $W$ and $R_c$ are the relative velocity of the sheet flow and the principal radius of curvature of the bucket surface along the path line, $\Delta p_{late}$ and $\rho$ the lateral pressure difference on the bucket surface and the density of water, and $W_{r\text{-}\theta}$ the relative velocity component on the r-θ plane, respectively. By numerically integrating Eq. (1), the relative velocity and the position of path line in question can be obtained within a frame.

Starting from the first frame, the number of jet stream planes that have landed on the splitter of bucket increases with increasing order of frames. The movement of the frontal line of free sheet flow during a progress of frames can be simulated by solving Eq. (1) resulting in a streak patch of sheet flow. The thickness distribution along the frontal line of the patch is determined to conserve the original discharge $\Delta Q_j$ of jet layers. The number of streak patches increases with the order of frames until the earliest patch exits the bucket at the brim.

## 3. OPERATING AND BOUNDARY CONDITIONS

### 3.1 PELTON TURBINE AND BUCKET

A Pelton turbine was devoted to verify the availability of the animated cartoon method for the numerical analysis of unsteady free sheet flow in a bucket. The turbine has four nozzles, and the bucket specific speed $B/D_{ref}$ of 0.294 in which $B$ is the inner width of bucket and $D_{ref}$ the reference (jet pitch circle) diameter. The runner has 18 buckets. The shape of half a bucket is shown in Fig. 4 with contour lines. To simulate the three dimensional curved surface of bucket inner wall, the originally designed coordinate points (12 x 12) were increased to 48 x 48 grid points as shown in Fig. 5 with the high-order spline interpolation. To interpolate the coordinates, the angles of splitter, brim and cutout were specified as the gradient at the both end points.

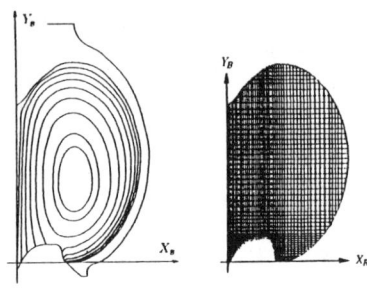

Fig. 4 Contour lines    Fig. 5 Grid points

### 3.2. OPERATING CONDITIONS

The numerical analysis was executed at the best efficiency operating point where the discharge coefficient per a nozzle $\phi_{opt}$ and the energy coefficient $\psi_{opt}$ is 0.0123 and 4.61, respectively, and the needle stroke $S_n/D_{ref}$ of 0.0343. The optimum jet diameter $d_j/B$ amounts 0.261. Since the jet is divided into 9 layers, a streak patch includes 9 path lines.. The rotating angle per a frame is selected as 2 degrees.

## 4. NUMERICAL ANALYSIS OF UNSTEADY FREE SHEET FLOW

### 4.1 FLOW AT 5th FRAME (10 degrees)

At the 5th frame corresponding to the rotating angle of 10 degrees from the 0th frame, the splitter tip of bucket penetrates into the jet as shown in Fig. 6(a). The distribution of streak patches in the bucket is illustrated in Fig. 6(b) in which exist the four patches from 1st to 4th. The lines approximately parallel to the splitter line show the frontal lines of the patches, and the line segments approximately perpendicular to those frontal lines represent the streak lines having the same identity number as the jet layers. The number of streak lines in a patch increases with increasing order of patches. As a typical example, the thickness distributions of free sheet flow at the front lines of 1st and 5th patches are depicted in Fig. 6(c). At the earlier stage of frames, the flow enters into the bucket from the splitter tip. The free sheet flow in the bucket directs toward the inner-most brim.

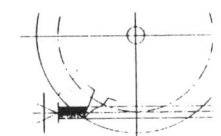

(a) Bucket with approaching jet

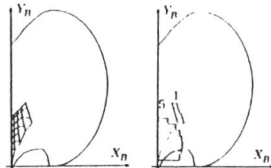

(b) Streak patches  (c) Thickness distribution

Fig. 6 Flow pattern at 5th frame

## 4.2. FLOW AT 10th FRAME (20 degrees)

When the frame proceeds to the 10th (20 degs rotation from the 0th) order, the relation between the buckets and the jet can be confirmed in Fig. 7(a). The streak patches and the thickness distributes in the bucket as shown in Fig. 7(b) and (c). The full jet enters at the splitter, however, the 1st patch does not reach the brim yet.. In spite that the earlier streak patches have less number of streak lines, they spread laterally in the bucket due to the lateral gradient of thickness distribution. On the second front line, there are three streak lines. The first streak has the minimum thickness, whereas the third streak has maximum. Since the fourth streak does not exist on the front line, the lateral gradient of thickness on the third streak is steeper in right hand side than left hand side toward the flow direction.

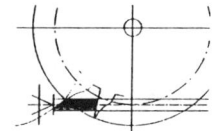
(a) Bucket with approaching jet

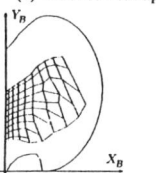 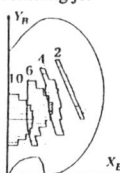
(b) Streak patches (c) Thickness distribution
Fig. 7 Flow pattern at 10th frame

## 4.3. FLOW AT 15th FRAME (30 degrees)

At the 15th frame (30 degs rotation), the relative position of buckets to the jet, and the distributions of patches and the thicknesses are shown in Fig.8. Since the jet reaches the splitter approximately perpendicularly in Fig. 8(a), the streak lines in the 15th patch are also approximately perpendicular to the splitter line in Fig. 8(b), though the patches of earlier order stay at the inside position close to the brim. The streak patches from the 1st to the 3rd have already exited out of the brim. Since the following bucket is apart from the preceeding bucket by 20 degree rotation or 10 frames, the flow in the following bucket corresponds just to the 5th frame.

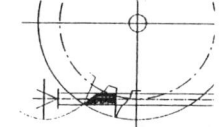
(a) Bucket with approaching jet

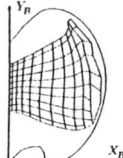 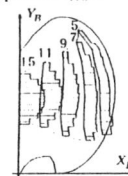
(b) Streak patches (c) Thickness distribution
Fig. 8 Flow pattern at 15th frame

## 4.4. FLOW AT 20th FRAME (40 degrees)

At the 20th frame, the jet lands on the most inner position along the splitter line. Fig. 9(a) shows that this frame is close to the final stage of full jet entry. The 9th streak lines tend to approach to the cutout in Fig. 9(b). The 8th patch is just leaving the brim as shown in Fig. 9(c). The 9th to 20th full streak patches are working effectively in the bucket. At this moment, the flow in the following bucket is in the 10th frame.

## 4.5. FLOW AT 25th FRAME (50 degrees)

At the 25th frame, the only jet stream planes from the 7th to 9th enters into the splitter as shown in Fig. 10, because the 1st to 6th jet planes have already been captured by the following bucket. The relative velocities at splitter are

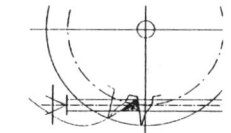
(a) Bucket with approaching jet

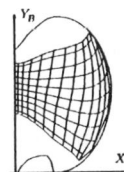 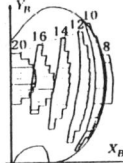
(b) Streak patches (c) Thickness distribution
Fig. 9 Flow pattern at 20th frame

directing toward radially outward, because the bucket is now in the fourth quadrant in Fig. 10(a). The position of the 9th streak lines is very close to the cutout. The 14th to 21st full streak patches are still working effectively in the bucket.

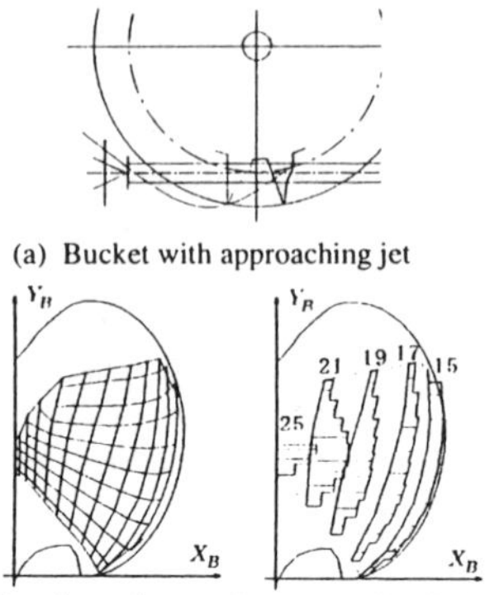

### 4.6. FLOW AT 30th FRAME (60 degrees)

In this frame, there is no jet entry into the bucket as shown in Fig. 11, because the last 9th jet plane has already entered at 26th frame. The patches from 20th to 26th resides on the bucket. The outer-most streak lines of the patches from 23rd to 25th discarge out of the cutout. This is the first indication of CutFlow (spilt flow from cutout) and very important to know the quantity of CutFlow $Q_{cut}$ for the investigation on the scale effect of jet interference in Pelton turbines.

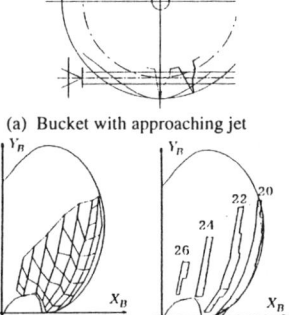

(a) Bucket with approaching jet

(b) Streak patches  (c) Thickness distribution

Fig. 10  Flow pattern at 25th frame

## 5. CONCLUSIONS

So far, there has been no report of the numerical flow analysis for the unsteady free water sheet flow on the three dimensional curved surface of rotating buckets in Pelton turbines. A new idea of animated cartoon method offers how to discretize the real unsteady free sheet flow by a sequence of gradualy time shifting frames. At the beginning instant of each frame, the actual unsteady inflow is given to the splitter, whereas within the frame, the front line of the streak patch of free sheet flow for the respective frame order on the bucket can be obtained by integrating the relative flow acceleration. From the distributions of streak patches and the thickness in the rotating bucket thus obtained, the amount of spilt water at the cutout of bucket can be predicted. The information concerning the flow at the cutout will be usefull to clarify the scale effect of jet interference.

(a) Bucket with approaching jet

(b) Streak patches  (c) Thickness distribution

Fig. 11  Flow pattern at 30th frame

## REFERENCES

1. Kubota, T.; Observation of jet interference in 6-nozzle Pelton turbine, Journal of Hydraulic Research, 27-6 (1989), pp.753-767.
2. Nakase, Y. et al.; An Analysis of the Flow on the Bucket of a Pelton Wheel (in Japanese), 1st Report: Trans. JSME 51-470 (1985-10), pp3355-3359; 2nd Report: Trans. JSME 51-471 (1985-11), pp3680-3684.

# SUBJECT INDEX

Page number refers to the first page of paper

Accuracy, 216
Advection, 216
Aeration, 447, 453, 459, 477, 513, 524, 589, 601
Aerators, 601
Africa, 132
Air entrainment, 613
Algorithms, 204
Analytical techniques, 222
Artesian aquifers, 36
Artificial islands, 180

Baffles, 530
Bank erosion, 162
Barges, 577
Beach erosion, 561
Bed load, 138
Benefit cost analysis, 78
Boundary conditions, 489
Breaking waves, 228
Bureau of Reclamation, 441, 483

California, 84
Cameras, 228
Cascade, 489
Case reports, 387
Catchments, 507
Cavitation, 287, 314, 555, 589, 601
Channel design, 108
Channel flow, 246, 252, 583
China, People's Republic of, 72, 126
Cofferdams, 269
Columbia River, 518
Computation, 114, 204, 210, 216, 240, 302, 370, 387
Computer software, 631
Conflict of interest, 382

Construction, 21
Containers, 637
Contaminants, 216
Contamination, 96
Control, 483
Cost effectiveness, 471
Cost minimization, 168
Currents, 246
Curtain walls, 483
Czech Republic, 489

Dam breaches, 42, 48, 204, 258
Dam design, 78
Dam failure, 48, 54
Dam foundations, 429, 435, 441
Dam safety, 36, 54
Damage, 60, 589
Damping, 293
Dams, 15, 21, 30, 36, 186, 263, 358, 393, 399, 405, 417, 477, 489, 513, 518, 524, 607, 613, 631
Dams, arch, 352
Dams, embankment, 42, 48
Decision support systems, 334, 340
Deflection, 358
Demolition, 21
Desalination, 577
Design, 15, 287, 328, 346, 364, 530, 607, 631
Dewatering, 399
Diffusion, 120
Disasters, 263
Discharge, 42, 72, 352, 411, 507, 613, 619
Discontinuities, 222
Dissolved oxygen, 340, 459, 477, 513

Drawdown, 108, 114, 120
Dredging, 96, 132
Dynamic programming, 334

Ecology, 364, 507
Economic analysis, 78
Economics, 471
Ecosystems, 507
Eddies, 60, 228
Eddy viscosity, 501
Electric power plants, 566
Electrical equipment, 334
Energy conversion, 561, 566, 571
Energy dissipation, 417, 435, 524, 583, 595, 619
Environmental impacts, 168, 465, 495
Environmental quality, 393
Equations of motion, 258, 293, 411
Equilibrium profile, 108
Erosion, 42, 48, 60, 90, 108, 120, 429, 435, 441
Erosion control, 84
Estimation, 376
Evaluation, 314, 387, 405
Experimentation, 595

Failures, 66
Finite differences, 246
Finite elements, 66
Fish habitats, 465
Fish management, 453
Fish protection, 358
Fish reproduction, 376
Fish screens, 399
Fisheries, 382, 399, 495
Fishways, 364, 370, 387, 393, 399, 405, 411, 453, 513, 524
Flood control, 30, 126, 275
Flooding, 382
Floods, 204, 429, 501
Flow, 358

Flow characteristics, 583
Flow control, 60
Flow patterns, 595
Flow profiles, 9
Fluid dynamics, 204, 281, 302, 370, 387, 537, 625
Fluid flow, 314
Flushing, 96, 108, 114, 120, 144, 507
France, 114, 453
Free surfaces, 210, 222, 246
Freezing, 180
Friction coefficient, hydraulic, 269

Gas distribution, 607
Gates, 21, 30, 66, 78, 275, 613, 625

Harbors, 411
Head loss, 308
Heating, 186
Hydraulic design, 234, 314, 364, 417
Hydraulic jump, 9, 222, 228, 423, 465, 619, 625
Hydraulic models, 30, 48, 60, 393, 543, 549, 607
Hydraulic performance, 15, 387, 405
Hydraulic properties, 168, 314
Hydraulic structures, 54, 96, 411, 589, 601, 625, 631
Hydraulics, 192, 240, 320, 530, 566, 577
Hydrodynamics, 340
Hydroelasticity, 352
Hydroelectric power, 1, 150, 156, 174, 234, 281, 358, 364, 370, 393, 453, 465, 471, 501, 619
Hydroelectric power generation, 287, 382
Hydroelectric powerplants, 15, 138, 192, 252, 293, 308, 320, 328, 334, 340, 346, 352, 495, 543
Hydrographic surveys, 102
Hydrographs, 156, 376

# SUBJECT INDEX

Ice control, 168, 180, 186
Ice cover, 180, 192
Ice flow, 174
Ice jams, 156, 162, 168, 174, 180, 186, 192
Ice loads, 180, 198
Imaging techniques, 555
Impingement, 417
Inflow, 144
Instream flow, 364
Intake structures, 530
Intakes, 21, 399
Islands, 577

Japan, 328
Jet diffusion, 530
Jets, 417, 435

Kinematics, 393

Laboratory tests, 549, 571, 589
Lakes, 495
Landslides, 90, 263
Loading, 198
Locks, 186
Losses, 314
Low head, 308, 423

Mathematical models, 376
Meandering streams, 246
Measurement, 114, 198, 328, 489
Methodology, 364
Mixing, 501
Mixtures, 126
Model studies, 518
Model tests, 144, 328, 555
Modeling, 126, 222, 471, 524
Models, 216, 269, 619
Monitoring, 1, 54, 477
Movable bed models, 120

Navier-Stokes equations, 281
Networks, 507
Neural networks, 352
Nitrogen, 518
Numerical analysis, 302, 637
Numerical models, 48, 150, 216, 234, 240, 246, 263, 281, 370, 429, 441, 537, 625, 631

Ocean waves, 577
Offshore platforms, 198
Open channel flow, 204, 222, 240, 595, 625
Open channels, 258
Operation, 346
Optimal design, 308
Optimization, 234, 287, 607, 637
Optimization models, 334
Overflow, 21
Overtopping, 42, 441, 566, 571
Oxygen demand, 489

Parameters, 54
Peaking capacities, 156
Pendulums, 561
Performance, 601
Performance evaluation, 54
Petroleum, 198
Pile structures, 561
Piping systems, 234
Planning, 1, 447
Plunging flow, 417, 435
Poland, 192
Potable water, 577
Power supplies, 577
Predictions, 48, 275, 524
Pressure pipes, 320
Project management, 1
Projects, 132
Pump intakes, 530, 537, 543, 549, 555
Pump turbines, 302

Pumped storage, 328, 495
Pumping, 566

Real-time programming, 340
Regression analysis, 293
Rehabilitation, 21, 30, 287
Research, 1, 453
Reservoir management, 84, 96
Reservoir operation, 72, 447, 483, 507
Reservoir sedimentation, 72, 78, 84, 90, 96, 102, 108, 114, 120, 126, 132, 138, 144
Reservoir system regulation, 156
Reservoirs, 263, 275, 376, 459, 477, 495, 501
Reverse osmosis, 577
Reversing flow, 302
Riprap, 15, 162
River flow, 376, 382
River regulation, 376
River systems, 1
Rivers, 174, 180, 192, 340, 346, 405, 489, 583
Roller compacted concrete, 36
Rotation, 637
Rubber, 30
Russia, 346, 376, 382

Safety, 114
Sand, 138, 144
Sandstone, 102
Saturation, 518
Scale models, 150, 393, 405, 465, 631
Scheduling, 334
Scour, 9, 15, 429
Scouring, 423
Screens, 543
Sea water, 566
Secondary flow, 234
Secondary systems, 9

Sediment control, 72, 84, 90, 96, 138, 144, 150
Sediment deposits, 138, 281
Sediment discharge, 102
Sediment transport, 126, 132, 423
Sediment yield, 84, 90, 102, 126
Sedimentology, 102
Settling basins, 150
Shallow water, 204, 258, 263
Shear, 537
Shear stress, 269
Shock waves, 595
Shore protection, 162, 561
Silts, 120, 132
Simulation, 174, 210, 240, 263, 275, 352, 625
Simulation models, 252
Slopes, 162
Small structures, 346
Socioeconomic data, 78
Spillways, 9, 15, 30, 36, 66, 358, 429, 465, 513, 518, 524, 583, 589, 601, 607, 631
Stability, 15
Stilling basins, 9, 36, 60, 423, 607, 619
Strain gages, 66
Stratification, 501
Structural response, 352
Subcritical flow, 222
Sumps, 530, 543, 549, 555
Supercritical flow, 222, 240
Surge, 258, 320, 328
Suspended sediments, 144, 275

Tailwater, 252, 308, 471, 513, 613
Tanker ships, 198
Tanks, 320, 328
Temperature effects, 483
Three-dimensional flow, 252, 275
Three-dimensional models, 281
Time factors, 210

## SUBJECT INDEX

Trap efficiency, 138, 144
Trashracks, 293
Tunnels, 613
Turbidity, 501
Turbines, 210, 287, 308, 320, 399, 459, 637
Turbulence, 228, 269, 281, 302, 370, 423, 537
Turbulent flow, 252, 269, 583
Two phase flow, 613

Unsteady flow, 156, 210, 258, 637

Valves, 314
Vibration, 66, 352, 555
Vibration control, 293
Videotape, 228
Vortex shedding, 549
Vortices, 537, 543, 549, 555

Water demand, 334
Water discharge, 447, 453, 459, 477, 483
Water flow, 411
Water hammer, 320
Water pollution, 216
Water quality, 447, 495
Water quality control, 340, 459, 477
Water resources, 1
Water storage, 150
Water surface profiles, 126
Water use, 471
Water waves, 258, 561, 566, 571
Watershed management, 84
Watersheds, 90
Waterways, 595
Wave energy, 561, 566, 571
Wave generation, 263
Weirs, 411, 619

# AUTHOR INDEX
Page number refers to the first page of paper

Abt, S. R., 441
Abt, Steve, 429
Abt, Steven R., 435
Adams, J. Stephens, 447
Angelaccio, C. M., 465
Annandale, G. W., 441
Annandale, George W., 429
Ansar, Matahel, 549
Arakawa, C., 302
Asai, Koji, 216
Assarin, Alexander, 382
Attari, Jalal, 601

Bacchiega, J. D., 465
Bareis, Steve, 21
Barrionuevo, H. D., 465
Bauer, Deborah I., 549
Bendahou, Habib, 132
Blais, J. P., 631
Bohrer, Jeffrey G., 435
Bollman, Frank H., 78
Borland, S., 162
Bouchard, J. P., 114
Brekke, Hermod, 287
Brock, W. Gary, 447

Cain, James D., P.E., 607
Carey, Kevin L., 180
Chanson, Hubert, 595
Chen, Bihong, 252, 275
Chiesa, E., 631
Cohen, Elisabeth, 60
Coleman, S. E., 42
Constantinescu, G., 537
Crissman, Randy D., 174
Crossley, A. J., 204

de Oliveira Carvalho, Newton, 102
De Souza, Podalyro A., 228
Delis, A. I., 222
Di Silvio, Giampaolo, 90
Dolgopolova, E. N., 376
Donnelly, C. J., 162
Dorrer, G., 293
Drobir, H., 293, 320
Dum, T., 144, 234

Egashira, S., 246
Eibl, F., 138
Elviro Garcia, Víctor, 589
Emura, Yoshi, 507
Eon, J., 453
Errih, Mohamed, 132

Fattor, C. A., 465
Fisher, Richard K., Jr., 459
Fratino, U., 314
Fujino, Koichi, 328
Fujita, K., 571

Gabriel, P., 489
Galland, J. C., 114
Gao, Jizhang, 417
Gerbig, Lee, 21
Gildenblat, Mikhail Ja., 150
Gisonni, Corrado, 423
Gosse, C. Sabaton, 453
Guan, Jianyong, 252
Gulliver, John S., 524

Hadano, Kesayoshi, 216
Haga, Kaoru, 328
Hager, Willi H., 258
Hanna, Leslie, 60

Harshbarger, D., 477
Hashida, M., 571
Hauser, Gary E., 471
Hay, Duncan, 393
Haynes, F. Donald, 186
Hays, Steven G., 358
Heigerth, G., 144
Heisey, Paul G., 358
Hejl, N., 293
Hellmann, Dieter-Heinz, 530
Hinokidani, Osamu, 263
Hiratsuka, Akira, 566
Hopping, Paul N., 459
Horikawa, Noriko, 364
Hotchkiss, Rollin H., 78
Howard, Charles D. D., 1

Jack, R. C., 42
Janssen, Robert H. A., 108
Jin, H. S., 246

Kahawita, R., 340
Kanestrøm, Ø., 198
Kattelmann, Richard, 84
Kerenyi, K., 293
Kim, Young C., 577
Kitamura, Yuichi, 501
Kjellesvig, Hilde Marie, 281
Klasinc, R., 234
Knapp, W., 308
Knoblauch, H., 144, 234
Komatsu, T., 571
Komatsu, Toshimitsu, 216
Kondo, Hideo, 561
Kubitschek, Joe P., 387
Kubitschek, Joseph P., 405
Kubota, T., 637
Kumar, G. Sampath, 411

Labadie, J. W., 334
Lachtchenov, S., 346
Lai, Jihn-Sung, 120

Laperrousaz, E., 631
Larsen, J., 543
Lauber, Guido, 258
Lever, James H., 168
Lewis, Todd, 429
Li, Guifen, 352
Li, Yongmei, 417
Li, Zhongyi, 417
Liu, B. Y., 246
Liu, Jian, 269
Liu, Peiqing, 417
Liu, Shukun, 352
Løset, S., 198
Lu, Shunan, 174

Maeno, Shiro, 411
Majewski, Wojciech, 192, 495
Mannheim, Carl, 518
March, Patrick A., 459
Marion, Andrea, 90
Marold, W. James, P.E., 15
Mateos Iguacel, Cristobal, 589
Mathur, Dilip, 358
Matsunaga, N., 571
Matsuzaki, Hironori, 507
Maurel, F., 114
McCormick, Michael E., 577
McCorquodale, J. A., 9
Mefford, Brent W., 387, 405
Melville, B. W., 42
Michiue, Masanori, 263
Mochkaai, Y., 308
Molinas, Albert, 126
Molls, Thomas, 240
Moulin, C., 631
Moulton, R., 9
Muller, Bruce C., Jr., 30

Naghash, Mahmood, 530
Nago, Hiroshi, 411
Nakanishi, Y., 637

# AUTHOR INDEX

Nakato, Tatsuaki, 549
Novak, P., 489

Ohgushi, Koichiro, 216
Ohtsu, Iwao, 583
Okada, T., 571
Olsen, Nils Reidar B., 281
Onishi, Sotoaki, 501
Orlins, Joseph J., 524

Padmanabhan, M., 543
Pan, Ruiwen, 619
Parsly, James A., 471
Patel, V. C., 537, 555
Petitjean, A., 114
Philbrook, Cindy, P.E., 399
Piccinni, A. F., 314
Prenner, R., 320
Proudovsky, Alexander M., 150

Qian, Y., 302

Rae, Peter, 9
Rajendran, V. P., 555
Rasulo, Giacomo, 423
Reitbauer, R., 234
Richardson, J. E., 625
Riedel, Norbert, 210
Rodionov, Victor B., 150
Rosenberger, Hartmut, 530
Rucker, H., 477
Ruff, J. F., 441
Ruff, Jim, 429

Saranchev, Vladimir O., 150
Scheuerlein, H., 138
Schilling, R., 308
Schilling, Rudolf, 210
Schneider, Ch., 308
Schneider, Michael L., 513
Semenkov, V., 346
Shen, Hsieh Wen, 108, 120

Shen, Hung Tao, 174
Shiao, Ming C., 471
Shigemitsu, Seiyo, 566
Skalski, John R., 358
Skeels, C. P., 222
Smith, A. F., 9
Smith, Mark R., 358
Sodhi, D. S., 162
Song, Charles C. S., 275
Sotiropoulos, Fotis, 370
Sowlati, Taban, 9
Speerli, Juerg, 613
Stanley, J. M., 162
Stanton, Doug, 36
Stateler, Jay N., 54
Staubli, T., 293
Stitt, S., 334
Stronach, James, 393
Suetsugu, Hiromichi, 411
Sugii, Mitihiro, 501
Sweeney, Charles, 393

Takahashi, Tamotsu, 228
Tamada, Kikuo, 228
Tamai, Nobuyuki, 507
Todd, Robert V., 66
Tominaga, Akihiro, 269
Toombes, L., 595
Travade, F., 453
Tsujimoto, Tetsuro, 364
Tuthill, Andrew M., 180

Ventikos, Yiannis, 370
Vermeyen, Tracy B., 483
Volkart, Peter U., 613

Wahl, Tony L., 48
Weber, Larry, 393
Weber, Larry J., 518
Welt, F., 340
Westrich, Bernhard J., 96
White, D. K., 543

Wilhelms, Steven C., 513
Wittler, R. J., 441
Wittler, Rod, 429
Wright, N. G., 204
Wu, Baosheng, 126

Xiaoqing, Yang, 72
Xu, Yiming, 619

Yamashiki, Yosuke, 228
Yasuda, Youichi, 583

Yeager, Bruce L., 471
Yi, J., 334
Yuan, Weixing, 210
Yuan, Ximin, 352

Zarrati, Amir R., 601
Zhao, Gang, 240
Zhide, Zhou, 72
Zhou, Fayi, 275
Zufelt, Jon E., 156